PLANETS AND SATELLITES

The Solar System

(IN FIVE VOLUMES)

I

THE SUN

1953

II

THE EARTH AS A PLANET

1954

III

PLANETS AND SATELLITES

1961

IV

THE MOON, METEORITES, AND COMETS

V

PLANETS, INTERPLANETARY MEDIUM

PLANETS AND SATELLITES

Edited by

GERARD P. KUIPER

and

BARBARA M. MIDDLEHURST

CONTRIBUTORS

D. Barbier
Dirk Brouwer
B. F. Burke
G. M. Clemence
Audouin Dollfus
W. S. Finsen
Sigmund Fritz
Roger M. Gallet
Yusuke Hagihara
Daniel Harris
Milton L. Humason
Gerard P. Kuiper
Cornell H. Mayer
M. Minnaert
Edison Pettit
William M. Sinton
Clyde W. Tombaugh
H. Wexler
Rupert Wildt

THE UNIVERSITY OF CHICAGO PRESS

CHICAGO & LONDON

This publication has been supported in part by the Geophysics Research Directorate of the Air Force Cambridge Research Center, Air Research and Development Command

ISBN: 0-226-45927-6

Library of Congress Catalog Card Number: 54-7183

THE UNIVERSITY OF CHICAGO PRESS, CHICAGO 60637
The University of Chicago Press, Ltd., London

Published 1961. Third Impression 1971. Printed in the United States of America

Preface

THE first two volumes of this series have dealt with the Sun and the Earth, respectively. The remaining volumes are intended to collect and evaluate the principal empirical data and theoretical concepts concerning the planets and the smaller bodies of the solar system. About two-thirds of the material on planets and satellites to be included is contained in Volume **3**. A critical survey of visual observations made during the past several decades, additional planetary photography, a review of planetary spectroscopy, and several chapters on the Moon will follow later, as will chapters on the asteroids, meteorites, meteor craters, comets, and interplanetary material.

A considerable amount of material presented in Volume **3** has not been published before, and the data of some of the chapters were assembled specifically for the purpose. The material is arranged according to the research methods used, enabling the reader to make a critical evaluation of all steps. Summaries of the physical data for each planet will be found in Volume **5**.

The division of the subject matter by research method has made it possible for each chapter to be written by an active astronomer who would have access to unpublished material and be in a position to evaluate the literature critically. This is important in a field where the literature is widely scattered and contains material of very uneven quality. At a time when planetary data are needed by an increasing number of non-astronomers, such authoritative chapters, giving full particulars rather than mere summaries and conclusions, appeared desirable. They will illustrate the surprising power of some of the methods used.

The growth of planetary astronomy has occurred in three distinct phases. Until the introduction of photography with large telescopes and the birth of astrophysics, both toward the close of the nineteenth century, astronomy was almost wholly concerned with the planets. The system of the "fixed" stars was of profound interest philosophically; but to the practicing astronomer it was largely a convenient frame to which the complex motions of the planets could be referred. The great discoveries in astronomy were made in efforts to interpret the planetary motions.

Copernicus, Kepler, Newton, Euler, Lagrange, Laplace, Gauss, and Poincaré, to name but the greatest, created the concept of modern natural science while studying the planets.

The phenomenal growth of astrophysics and the exciting explorations of the Galaxy and the observable universe led to an almost complete abandonment of planetary studies. The number of astronomers has always been small and, for better or for worse, has not been subject to the recent enormous increases in numbers of physicists and chemists. Celestial mechanics began to look like a nearly finished discipline in which further progress was hopelessly difficult. Physical observations of planetary surfaces, particularly of Mars, led to controversies and speculations that may have been appreciated by the public but hardly by the professionals. More and more this branch of planetary work, including the study of the Moon, became the topic *par excellence* of amateurs—who did remarkably well with it. The *Memoirs* of the British Astronomical Association became the chief record for the development of planetary surface markings. Astronomers with large telescopes were so occupied with the engaging problems of stars, nebulae, clusters, the Galaxy, and the universe that astronomy became almost entirely the science of the stars.

Meanwhile, those astronomers who occasionally took time off for planetary studies (at the risk of mild scorn from their colleagues) found a rich field for investigation. The techniques and ideas developed in stellar studies found wide application with the planets and broadened the research beyond the purely dynamical approach. Spectrographic studies of planetary atmospheres brought the chemistry of the planets within reach. The masses determined dynamically, from planetary and satellite motions, could be combined with accurate measures of diameter, and the resulting densities interpreted with the aid of the physics of compressed matter. The masses and compositions themselves could be understood when it became clear that the process of planetary origin was but a special case of the almost universal process of binary-star formation. The processes of energy generation in the Sun and the stars became known, and the time scales of stellar evolution could be determined. The age of the Earth and that of the meteorites could be measured and agreed with the approximate age of the Sun. Thus the chronology of the planets as part of the solar system and of the solar system as part of the Galaxy was outlined, and the physical problems of the evolution of the planetary system became capable of solution. The physical theory of comet tails found solar corpuscular rays sweeping ionized material out of the solar system —and the process of the mass loss of the protoplanets became clear, as well as the origin of planetary obliquities.

At the same time the development of geophysics stimulated corresponding developments in planetary astronomy. The deepened knowledge of one member of the planetary family illuminated the problems common to all members. Volume **2** bears testimony to the richness of this source. This development and interaction will, no doubt, continue at an increased rate, with lunar problems expected to come into special prominence.

The third phase of planetary astronomy is that of the 1960's, during which entirely new data are expected to become available through the application of research with rockets. The first chapter of this volume shows at once the power and some of the limitations of this new tool.

The first step of this program is likely to consist of close-up photography of the Moon and the planet Mars, with spectroscopic data and eventually composition and structural data to follow. No doubt the planning of these new techniques, which will be several orders of magnitude more expensive than the astronomical observations hitherto made, will continue to depend heavily on ground-based observations, which are the basis of the other chapters in this volume.

The organization of the planetary chapters in this volume will be clear from the Table of Contents. The limits of completeness of present knowledge of the members of the system are examined in chapters 2 and 18. The planetary and satellite masses are derived in chapter 3, while the secular stability of the solar system—the central problem of celestial mechanics (a problem beset with vast difficulties)—is discussed in chapter 4. This chapter elucidates the type of orbits and configurations that can have managed to survive the great age of the solar system, nearly 5 billion years. The theory of the internal constitution of the planets is developed in chapter 5. Chapter 6 deals with the photometry of the Moon, a planet-like object which can be studied in much greater detail than the major planets and whose study therefore is very instructive. Similarly, satellite eclipses are best exemplified by the study of total lunar eclipses, the subject of chapter 7. The general photometry and colorimetry of the planets and satellites is discussed in chapter 8, which includes much hitherto unpublished material. The polarization of the light reflected by the planets at different phase angles is a major source of information, both for planets having atmospheres containing particles and for bodies devoid of atmosphere. Chapter 9 deals with this important topic.

The infrared and radio emissions of the planets—the chief source of information on their atmospheric or surface temperatures—are reviewed in chapters 10–14. The radio observations include the discusson of the various curious non-thermal emissions discovered for the planets Jupiter and Saturn, emissions which appear to be closely related to the planetary magnetic fields and belts of trapped solar particles. The power of visual and photographic studies of planetary surfaces is exemplified by the discussions in chapter 15, which report on the results obtained at one of the most productive planetary observatories. Additional high-quality photographic material is displayed in chapters 16 and 17, with chapter 17 giving some results reproduced in color and chapter 16 containing exquisite photographs obtained with the world's largest telescope.

Differences in nomenclature between the chapters have not been completely removed, since no universally accepted notation exists and since some of the figures included show the symbols used by the authors. It is not believed that the reader will find this troublesome, since each chapter defines its own symbols.

The editors wish to express their indebtedness to the authors for their collaboration. They regret that the lengthy period of preparation of this volume made revision of some chapters necessary. Fortunately, the bulk of this volume is concerned with basic data whose value may be regarded as essentially permanent.

We are indebted to several colleagues for editorial advice; to Dr. Van Biesbroeck for assisting in translations; and to our staff for much typographic work and assistance with the figures. Special thanks are due to the Geophysical Research Directorate of the Air Force for their interest in this publication.

GERARD P. KUIPER
BARBARA M. MIDDLEHURST
Editors

UNIVERSITY OF ARIZONA
December 1960

Table of Contents

CHAPTER 1

Planet Earth as Seen from Space[*]

By S. FRITZ AND H. WEXLER
U.S. Weather Bureau, Washington, D.C.

1. INTRODUCTION

MAN, having scrutinized neighboring planets from his earth-bound platform, has often wondered about the appearance of the Earth as it might be seen from space. Thus H. R. Butler (Russell, Dugan, and Stewart, 1945) portrayed the Earth as it might appear to an observer on the moon. More recently, Wexler (1957) represented the Earth as it might be seen from a satellite located 4000 miles above the Earth's surface. Photographs recovered from rockets have shown limited areas of the Earth both in color and in black and white. Finally, early in 1960, the first of the meteorological cloud-cover satellites, TIROS I, began to supplement man's speculations with pictures of quite large sections of the Earth. This satellite was launched under the direction of the National Aeronautics and Space Administration on April 1, 1960, from Cape Canaveral, Florida.

Pictures of clouds and of the surface can be improved by the use of yellow or red filters, which reduce the contribution to the image of light scattered by the air molecules. This has been done on TIROS I. The combination of a yellow filter and the vidicon on the two television cameras confined the spectrum between about 0.45 and 0.8 μ, with the peak sensitivity between 0.5 and 0.7 μ. Thus the recorded image corresponds to visible light. Within the capability of the system, the cameras took pictures continuously through April and May, 1960, and during its dying phases more infrequently, until the middle of June, 1960. More information on the TIROS system and cameras has been given elsewhere (Sternberg *et al.*, 1960; Stroud, 1960).

* This work has been supported by the National Aeronautics and Space Administration.

Many cloud and terrain features were discernible on the records obtained. When clouds were absent, coastlines (Stroud, 1960), snow-covered mountains, desert sand dunes, and many other features of the Earth's surface were photographed. When clouds were present, as they often were, patterns were revealed associated with processes ranging from the large-scale planetary waves (about 2000 miles or 3200 km across) to individual cumulus clouds (as small as 1 mile in diameter). The reader will be able to identify and allow for defects on the records caused by electronic noise during the reception.

2. LAND

2.1. Coastlines

Among the more prominent features photographed were coastlines of continents. Usually these were discernible because of the contrast between the relatively bright land (albedo usually larger than 10 per cent) and the darker water. Sometimes, however, specular reflection of sunlight from the water highlighted the bright, illuminated water in contrast to the dark land.

Many factors affect the contrast between land and water; these are discussed with the aid of Plates 1 and 2. Plates 1*a*, 1*b*, and 1*c* show three frames of the North African–Mediterranean coast taken during one orbit, while Plate 2 shows the southern part of Africa. The contrast between land and water is quite marked along the Mediterranean, especially east of Tunisia. On the other hand, the South African coast is seen much less clearly because of the smaller brightness contrast between land and water. The approximate locations of the various plates reproduced here are given in Plate 3, which shows 12 land areas and 6 storm areas over oceans. The approximate directions of north (*N*) are indicated on the plates by arrows. The reader will wish to study the coastlines and land masses portrayed here by consulting maps with scale of about 1 to 10 or 20 million.

One factor which may contribute to contrast differences is illumination. Over North Africa the solar elevation was about 53°; over South Africa, about 33°. Therefore, the illumination over the Mediterranean coast was about 1.5 times greater than over South Africa. Other factors are the angle of view of the camera relative to the solar beam, the presence of atmospheric haze, the increasing resolution with distance from the horizon, and differences in photographic processing. But a major factor was undoubtedly the reflectivity of the surface itself. Not far from the North African coast there are desert areas, with high albedos, close to 30 per cent. Near the South African coast no such bright land areas appear.

Another set of photographs which show marked land-ocean contrast are those of the Middle East. Plate 4 shows the Sinai Peninsula flanked by the gulfs of Suez and Aqaba, with the dark Red Sea to the south and the dark Mediterranean to the north. The Nile River Valley is the curving line on the left and represents not only the river itself but the dark vegetation belt along its banks. The white band streaking across Egypt is a band of rather thin clouds; a bright white area represents a cloud mass over the Israel-Lebanon region; still other clouds are visible over the Mediterranean and Turkey. One can distinguish many interesting land features in this picture; and this is also true of many others photographed by TIROS I.

Additional photography of the Red Sea area was accomplished on another orbit of TIROS I. Plate 5, *a–c*, shows a series of pictures stretching from the Mediterranean to the Gulf of Aden. On Plate 6 the Arabian Peninsula stands out in strong contrast to the waters of the Arabian Sea, the Gulf of Oman, and the Persian Gulf. In this area the coasts usually appear very bright by contrast with the waters.

Sometimes, however, a coastal region will appear bright because snow or clouds cover the land, as in Plate 7, which shows northern Japan and the Siberian coastal regions. The island of Hokkaido, Japan, is quite bright, and its north coast is sharply silhouetted against the dark sea. The sharpness of the north-coast boundary suggests that Hokkaido is snow-covered. Elsewhere snow, ice, and cloud apparently combine to outline the islands. It is interesting to note the sharp boundary of the Siberian coast along the sea of Japan. The darker area inland represents the valley which stretches from Vladivostok northward.

Still other coastlines selected from the large number which TIROS has photographed are shown in Plates 8 and 9. Plate 8 contains a photograph of the northwest coast of Australia, with the area near Northwest Cape and Shark Bay evident as darker water indentations into the somewhat brighter land. Plate 9 shows the west coast of South America stretching from about Lima, Peru, to Antofagasta, Chile. In this figure a latitude-longitude grid has been superimposed on the TIROS photographs. This had been done to locate cloud features, and it illustrates, by examination of land detail, the accuracy which can be achieved when recognizable land features are evident in the pictures.

2.2. Snow-covered Mountains

Snow-covered regions also show up well against the darker background. An interesting series of photographs of the Alps was taken by TIROS I, starting April 2, 1960. At that time the Alps were snow-covered, and, although some clouds interfered locally, the mountains themselves could be seen in broad areas, as was obvious from fine, permanent terrain features. Plates 10*a*, 10*b*, 10*c*, 10*d*, and 10*e* show some of the records obtained. Plate 10*a* was taken on April 2; it shows the Alps extending from the French-Italian Mediterranean coast, as a white, bird-shaped array, into Switzerland, Austria, and Yugoslavia, while Italy extends southeastward from the snow-covered area. Imbedded in the snow-covered region over Switzerland, Italy, and France are several terrain features such as the Rhone River Valley, with its sharp bend, and Lake of Geneva, the Aosta Valley in Italy to the south, and the Durance Valley in France on the west side. Lago Maggiore and Lago di Como are also visible. The Alps in Austria and Yugoslavia, although still showing some of the effects of snow, have clouds superimposed on them, the clouds being associated with a weak low-pressure system centered near Turkey. A similar area was photographed the next day, April 3 (Pl. 10*b*), but this time the Austrian and Yugoslavian mountains appeared to be less cloudy, while the Alps over Switzerland and France had a rather dense cloud veil. Nevertheless, one could still see a part of the Rhone River Valley through the western edge of the cloud system and, extending eastward from the Rhone, several other terrain features, imbedded in the snows of Switzerland, Italy, Austria, and Yugoslavia. On April 5 (Pl. 10*c*) the French, Italian, and Swiss Alps were again less cloudy, so that the Rhone River Valley, with its characteristic sharp bend, and some of the additional features are shown. On April 7 (Pl. 10*d*) TIROS photographed the Alps and, indeed, the entire western part of the Mediterranean Basin, from a great distance, and produced the spectacular land array including all of Italy, Sicily, Tunisia, and Lybia. It is interesting to note that where clouds appear over Italy, the contrast with the sea is rather marked; however, on the eastern side of Italy the contrast between land and water is poorer. The African coast, showing the bulge near Tunisia, again appears with marked contrast but not with as much detail as in Plate 1.

It was not long after the picture in Plate 10*d* was taken that TIROS could take pictures only in the Southern Hemisphere; but in May, 1960, TIROS returned to a position where it could photograph the Northern

Hemisphere, and the region over the Alps was again photographed as in Plate 10*e*. The river valleys are not so sharply delineated; the snow seems to have receded somewhat from the French-Italian Mediterranean coast, and there may be some additional clouds over the Alps; nevertheless, the shape of the Alps is still unmistakable. Italy can still be seen with low contrast against the darker sea, while Corsica, apparently cloud-covered, is evident here, as in several other pictures.

Another bright snow-covered area was photographed over the Himalaya Mountains. Plates 11*a* and 11*b* show the Kashmir region, with both the wide-angle and the narrow-angle cameras of TIROS I. The narrow-angle picture (Pl. 11*b*) contains the region included in the rectangle shown in Plate 11*a* and suggests rugged mountain terrain covered by snow, though a few clouds may be present over the darker, lower-lying terrain.

2.3. Sunlit Coastal Areas

Under special circumstances, coastlines may appear dark against the bright sea surface. This occurs when the relation between the angle of view of the cameras and the solar beam is such that the camera "sees" the specular reflection. An example of this is given in Plate 12, which shows the Florida Peninsula, with the northeastern shore highlighted. Also, the coastal regions of Georgia and South Carolina, with their principal irregularities, are evident. The west coast of Florida can be seen with relatively poor contrast against the dark waters of the Gulf of Mexico; various other terrain markings are evident. The bright sinuous line extending from the southern tip of Florida is a cloud only a few miles wide but hundreds of miles long, possibly associated with the Gulf Stream.

An even more striking reflection is seen on Plate 13*a*, which shows the California coast near San Francisco. The relatively calm inland waters appear to provide a more concentrated reflection than do the rougher waters in the open Pacific. The bright band, running parallel to the shore near the right edge, represents the Sierra Nevada with a snow cover still present in May, although a few clouds may be intermixed.

2.4. Miscellaneous Terrain Features

A close comparison of the plates with suitable maps will disclose many additional terrain features. Lakes, both frozen and open, variegated terrain often associated with mountain-valley juxtaposition, and many other land features are evident. But it is interesting to single out one area for further discussion—the sand-dune area in North Africa. In Plate

1 the bright areas away from the coasts are often clearly associated with sand dunes (Bartholomew, 1956). Although the streaky white areas parallel to the coast near Tripoli in Plate 1*b* may be clouds, the bright area southwest of Tunisia is in an area of sand dunes near the Algeria-Tunis border. In Plate 1*c* especially, the sand-dune areas are prominent, as is the dark inland area associated with higher ground. Two of the large bright areas to the west of the dark inland circular area can be clearly identified with sand-dune regions in the Edeyin Ubari and Edeyin Murzuq regions. The dark circular area is in the higher region of El Haruj. Thus the space observer sees the major areas of the Earth delineated by brightness contrasts.

3. CLOUDS

Coastlines and other terrain features can be seen on the records when the atmosphere is clear. Quite often, however, large areas of the Earth's surface are hidden by clouds. Clouds usually result from condensation of water vapor by expansion and cooling as air parcels rise in response to circulations in the atmosphere. Because of the uneven geographic distribution of the absorbed solar radiation, the tropics are warmer than the polar regions. The large-scale circulation of the atmosphere results from the attempt of the atmosphere to transport heat from the equatorial to the polar regions, while at the same time the Earth, with its complicated surface properties, is spinning about its axis. Because of this combination of factors, a predominantly west-to-east wind results in mid-latitudes, where the temperature gradient is largest and the "polar front" is usually located. Eventually these currents, or the "westerlies," as they are called, become dynamically unstable and break down into large eddies—eddies which often start as small waves on the polar front and expand into occluded cyclones of 1000 miles or more in diameter. These are large-scale anticyclonic and cyclonic vortices propelling large masses of cold air toward the tropics and warm air toward the poles. During this process the cold air generally sinks as it moves southward, while the air from the south mainly rises. The rising warm air, being moist, cools and often produces clouds over large areas.

Thus, in viewing the Earth from space, one would expect that large-scale cloudiness, perhaps circling around cyclone centers, would be visible from space. In the cold air, especially over continents where the air is dry, relatively cloud-free areas would appear. However, as cold air flows over the warmer ocean surface, cloud elements smaller than those found in cyclones may appear because of convective patterns produced by thermal instability. Other types of convective clouds may be produced by other processes. For example, heating of land areas during the day often produces cumuliform clouds; in the tropics, convergence of wind currents produces massive cloud systems. Moreover, stratiform clouds may form in air which is thermally stable. Clouds are thus visible manifestations of the exchange of heat, moisture, and momentum between the Earth's surface and the atmosphere in both large- and small-scale processes. The TIROS photographs portray clouds associated with all these scales—from the large-scale spiraling vortices corresponding to planetary cyclones (Fritz, 1960; Fritz and Wexler, 1960; Staff Meteorol. Sat. Lab., 1961; Wexler and Fritz, 1960) to cloud patterns produced by processes associated with much smaller scales of motion (Staff Meteorol. Sat. Lab., 1961).

3.1. Large-Scale Cyclones

Cloud patterns associated with large-scale cyclones appear in many different forms. The state of development of the cyclone has much to do with the appearance of its clouds. In young cyclones, particularly over continents, there would be a large, relatively cloud-free area in the cold air behind the cold front. An example of a young cyclone (Bristor and Ruzecki, 1960) which has just begun its occlusion process is shown in Plate 14*a*. The corresponding weather map on which the cloud cover has been traced over the middle of the United States is shown in Plate 14*b*. The quasi-circular bright area near the top of the picture represents the highly foreshortened cloud associated with the cyclone center and the cloud to the north of it. A bright band of cloud, extending toward the south from the cyclone center, is associated with the frontal system projecting southward from the large cyclone. To the left or west of the frontal cloud, the dark area in Plate 14*a* represents the cold, dry, cloudless air which had entered into the cyclonic circulation from the north. In the photograph this polar air is shown as it spirals into the region occupied by the cloudy air. In the central cloud mass no spiral formation is yet evident. In earlier stages, before the occlusion process had begun, there would perhaps have been a more general cloudy area, with a smaller injection of the cold air into the cyclonic circulation.

As a cyclone develops and the cold air penetrates the cloudy, warm air, small dark areas denoting cloud-free air can sometimes be seen in the center of the cyclone, outlining clearly the complete spiral configuration (Pl. 15). In this figure a broad stratiform cloud in the Gulf of Alaska is seen as a spiral about a vortex center (*A*); the cloud is apparently associated with a broad current of warm, moist air aloft. Entwined with the cloud near the spiral center is a relatively narrow band of cloudless air shown as a dark strip. Various other features, such as a broad frontal cloud further south (*BC*) and a separate stratiform cloud (*DE*) associated with a new cyclonic wave formation on the front, have been analyzed with the aid of the picture (Winston and Tourville, 1961).

In old occluded cyclones the cold air may sometimes occupy a rather large central portion of the cyclone. In that case the spiraling cloud is indicated by a broad stratiform band, more or less around the periphery of the storm; this band may surround a more broken, cumuliform cloud mass. An example of this type of cloud pattern, associated with an old cyclone over the North Atlantic Ocean, is seen in Plate 16, a composite made from several frames photographed by TIROS I during two orbits (cf. Staff Meteorol. Sat. Lab., 1961). A schematic representation, showing the geographic position of the cloud features, is shown in Plate 17. It is evident from these plates that a broad band of stratiform cloud surrounded a more broken cumuliform type of cloud, while a relatively cloudless area separated the two types of cloud areas. This cloud configuration was associated with an old storm centered about 400 miles west of Ireland. In the middle of the storm, very narrow cloud streets suggest an inward-spiraling array into the cumuliform cloud mass.

Another interesting cloud pattern was photographed in the eastern Pacific; a small cutoff cyclone was in the process of occlusion on April 4, when it was photographed for the first time (Staff Meteorol. Sat. Lab., 1961; Winston, 1960); this storm was also photographed about one day later. Plate 18, showing the cloud pattern on April 4 before the storm reached its maximum intensity, indicates several scales of motions. A more or less uniform stratiform cloud near latitude 34° N. denotes a cloud band spiraling into the cyclone center, while a more southerly band near latitude 24° N. corresponds to a broad frontal system. In between are two areas of more broken clouds—clouds which often appear to be arrayed in small circles with cloudless areas in the middle. The leading edge of the first of these secondary cloud bands may represent a secondary front which was difficult to detect on weather maps because of the spareseness of the weather reports in the Pacific Ocean area.

On the second day, when the storm intensity was already decreasing, the smaller-scale cloud systems are much less prominent (Pl. 19), but large stratiform bands are seen swirling in toward the cyclone center.

Some of the large-scale cloud systems observed by TIROS I show remarkable agreement with the classical Norwegian "cyclone family." Such a case is shown in Plate 20*a*, which shows a composite made from pictures taken during two orbits near local noon of May 19, 1960 (Oliver, 1960; Staff Meteorol. Sat. Lab., 1961). In Plate *20b* the cloud features are displayed in their proper geographic locations, together with the appropriate surface synoptic chart. The cloud systems show quite clearly two members of the cyclone family, one far out in the Pacific near the Aleutian Islands and the second near the west coast of North America. Between the two storm centers the cloud patterns show clear areas; and they also show the distinctive rows of cumulus clouds which accompany cold, unstable air and indicate alternating regions of showers and fair weather.

Many other interesting cyclonic cloud systems have been photographed by TIROS I. Pictures of six cyclonic cloud spirals photographed in the Southern Hemisphere are shown in Plate 21. The locations and the approximate times when each of the pictures was photographed are shown in Plate 3. On occasion, in an area which is completely overcast, as in a cyclonic region, clouds will appear above a general overcast area. These individual clouds, which appear at high levels, will cast shadows on the lower cloud. This, too, will indicate the spiral condition of the cyclone. An example of this in the Indian Ocean may be seen in Plate 21 (picture *5*).

TIROS I has also photographed a complete typhoon-cloud pattern for the first time. When the existence of the storm north of New Zealand became known and since solar illumination and the attitude aspect of TIROS I were favorable for photographing the storm, the satellite was programed to televise and store cloud pictures in that area. The result was picture *1*, Plate 21, the first typhoon-cloud system ever photographed in its entirety.

3.2. Meso-Scale Cloud Patterns

The existence of cloud patterns, which suggest small-scale (or meso-scale) processes in the atmosphere, has already been mentioned on a few occasions. Such small-scale cloud patterns are, for example, evident in Plates 18 and 20 and appear many times in TIROS I pictures. Often these patterns consist of numerous circular or semicircular cloud arrays (Krueger and Fritz, 1961; Staff Meteorol. Sat. Lab., 1961); in the center of each circular cloud array the air is essentially cloudless. Examples of these are shown in Plate 22. Especially Plate *22c* looks like, but has important differences from, Bénard cells which have been produced in the laboratory. An important difference concerns the dimensions of the cells. In the laboratory the ratio of the horizontal diameter of the cells to the depth of the cells is about 3 to 1. In the cases of Plate 22 the ratio is more nearly 30 to 1. Other differences can also be cited.

In each of the three cases the vertical distribution of temperature in the atmosphere was similar. The temperature fell more or less rapidly from the surface to a height of about 5,000 feet, and the dew point increased from the surface up to that level, which was near the condensation level. Above that height, the air was very warm and dry, suggesting marked subsidence of the upper air. This indicated that the clouds were imbedded in the region just below the subsiding air and must therefore have been relatively thin, with their tops near a height of 1 mile. This was confirmed from airplane reports (Krueger and Fritz, 1961; Staff Meteorol. Sat. Lab., 1961). The diameters of the cells in Plate 22*c* varied from 20 to 50 miles, giving the ratio mentioned above.

3.3. Summary

Thus we find that, except for Antarctica, TIROS I has been able to photograph all the continents on the Earth's surface, in some cases with considerable detail, showing the coastlines, snow-covered areas, dark outcroppings, rivers, lakes, sand dunes, and many other features. Superimposed on this surface is the continuously changing, kaleidoscopic view of the "cloud-scape"; and TIROS I has photographed cloud patterns over all the non-polar oceans and continents. The clouds represent processes involved in many scales of motion. The large planetary scale, with about three to six circumpolar waves, is represented in the pictures by the cyclonic and anticyclonic cloud systems. Intermeshed with these large-scale clouds are clouds associated with meso-scale phenomena—on a scale up to 100 miles or so; these meso-scale cloud patterns are themselves often composed of individual cumulus clouds a few miles in diameter.

The existence of these high-altitude photographs of both surface features and the overlying clouds opens a new realm in the study of the planet Earth.

REFERENCES

Bartholomew, J. 1956 *The Times Atlas of the World* (Boston: Houghton Mifflin Co.), Pl. 85.

Bristor, C. L., and Ruzecki, M. A. 1960 "Photographs of the Midwest Storm of April 1, 1960," *Monthly Weather Rev.*, **88**, 315–326.

Fritz, S. 1960 "'Cyclone-Prints' from Satellite TIROS I," *Interavia*, **15**, 1384.

Fritz, S., and Wexler, H. 1960 "Cloud Pictures from Satellite TIROS I," *Monthly Weather Rev.*, **88**, 79–87.

Krueger, A. F., and Fritz, S. 1961 "Cellular Cloud Patterns Revealed by TIROS I,"

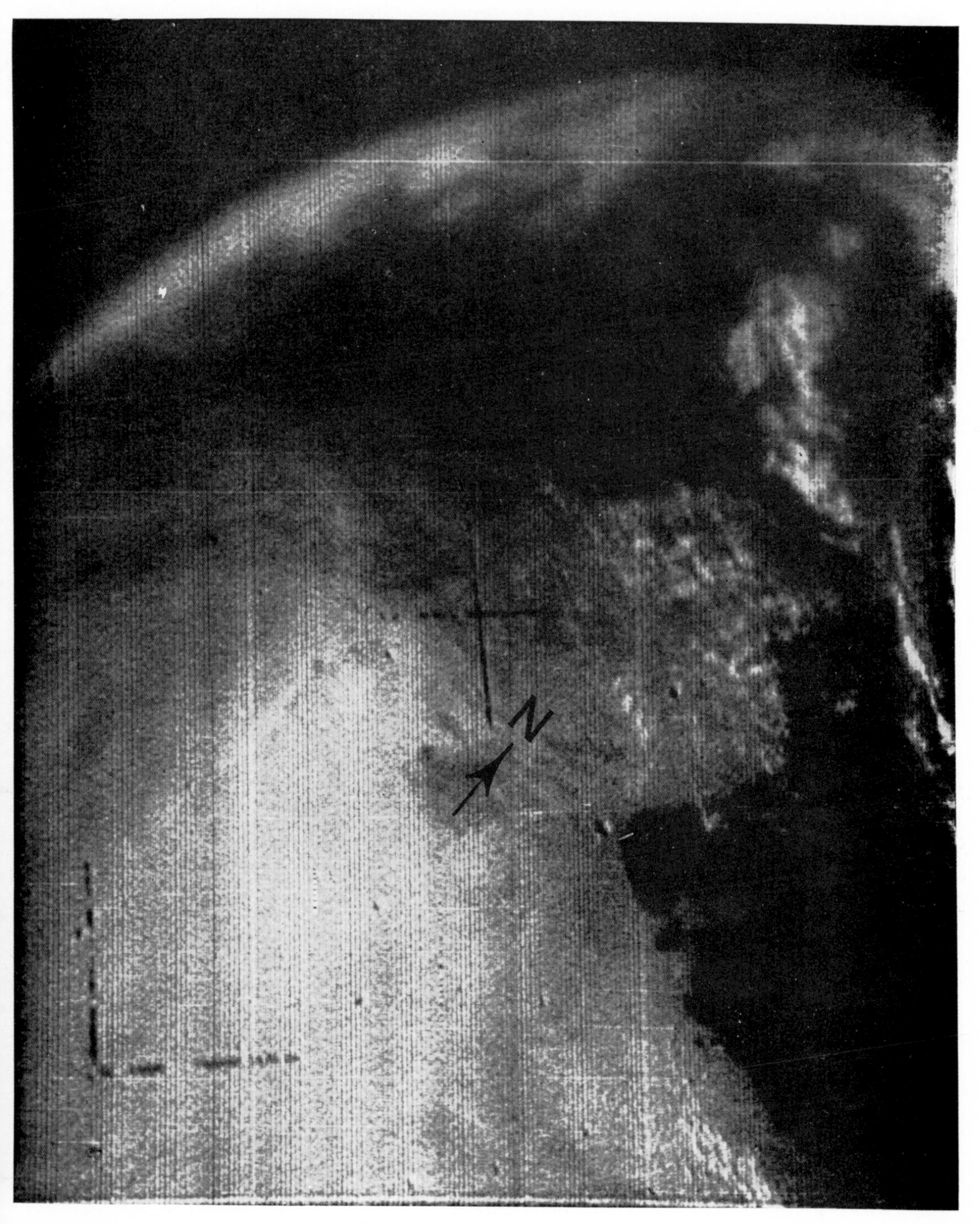

PLATE 1*a*.—Cf. legend for Pl. 1*c*

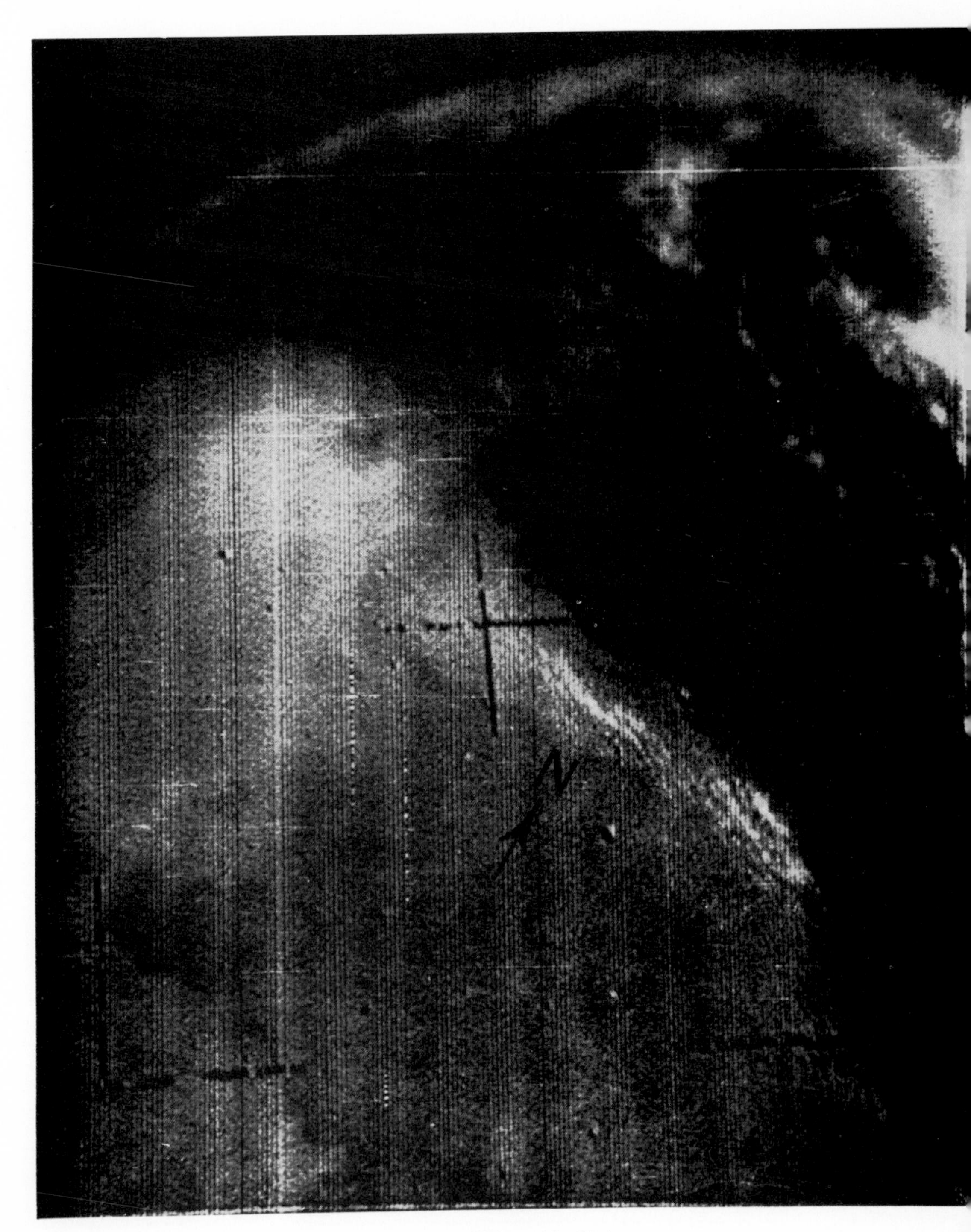

PLATE 1*b*.—Cf. legend for Pl. 1*c*

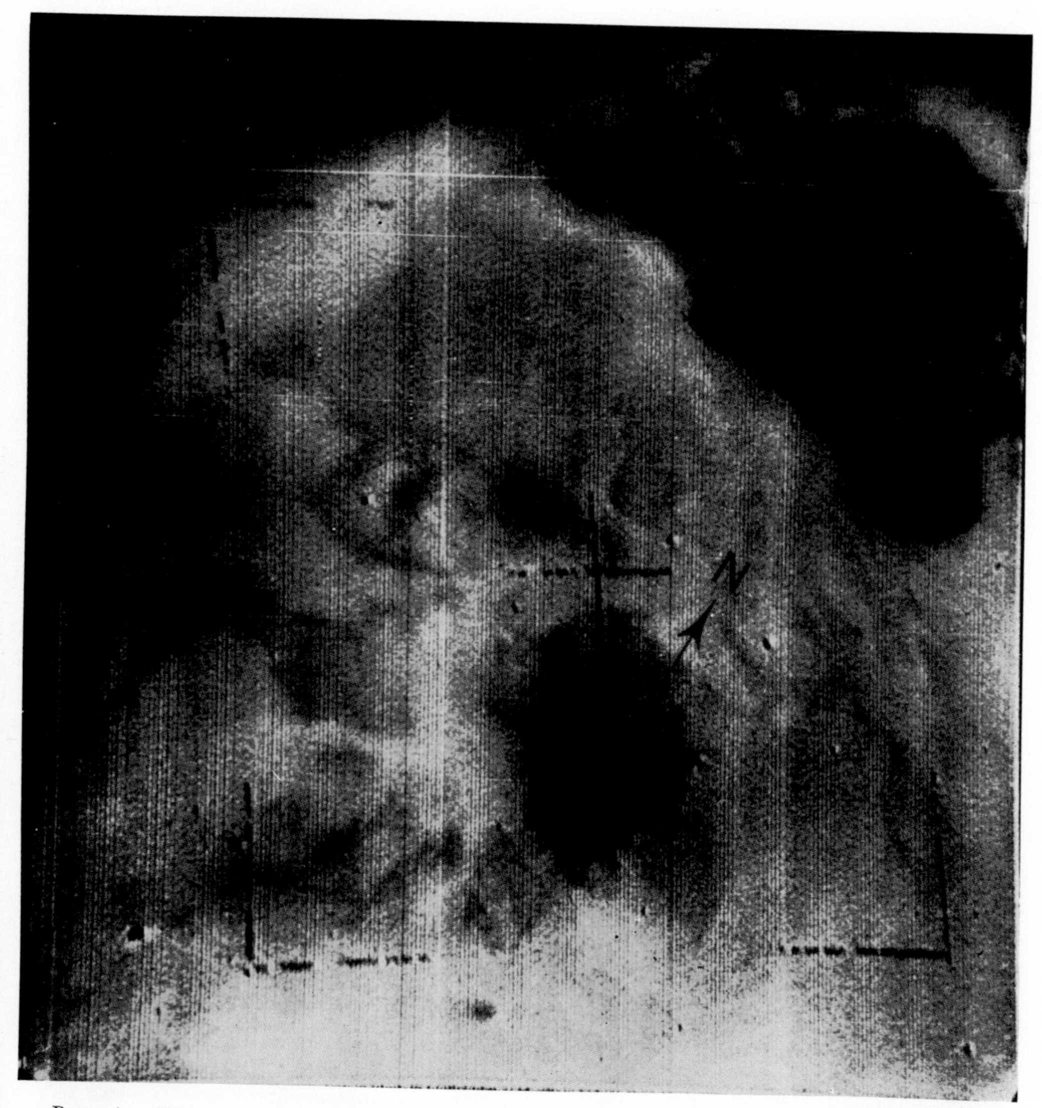

PLATE 1*c*.—Plates 1*a*–1*c* show three non-consecutive frames from one orbit of TIROS I, showing the Mediterranean coastal regions of North Africa (Tunisia and Libya), as well as cloud covers over Sardinia, Corsica, Sicily, and Continental Europe. The dark areas below the center of 1*c* are the mountainous regions on the southern border of Libya. April 2, 1960; 13.10 G.C.T.

PLATE 2.—The south and west coast of southern Africa. An example of strong electronic interference is seen across the picture. April 25, 1960; 08.30 G.C.T.

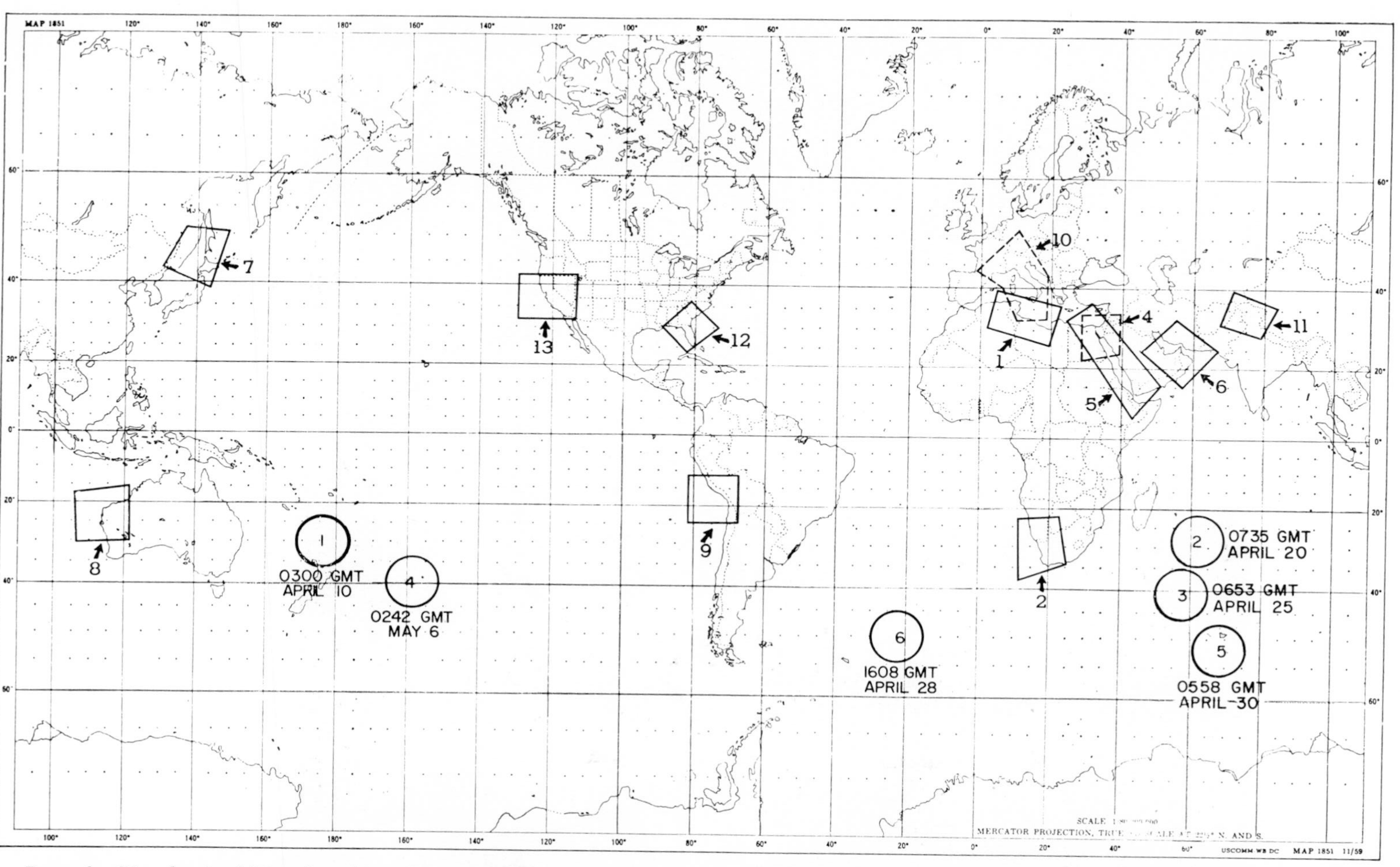

PLATE 3.—Map showing (*a*) locations of land features in rectilinear areas with numbers corresponding to plate numbers of pictures; (*b*) circular areas show locations of storms shown in Pl. 21.

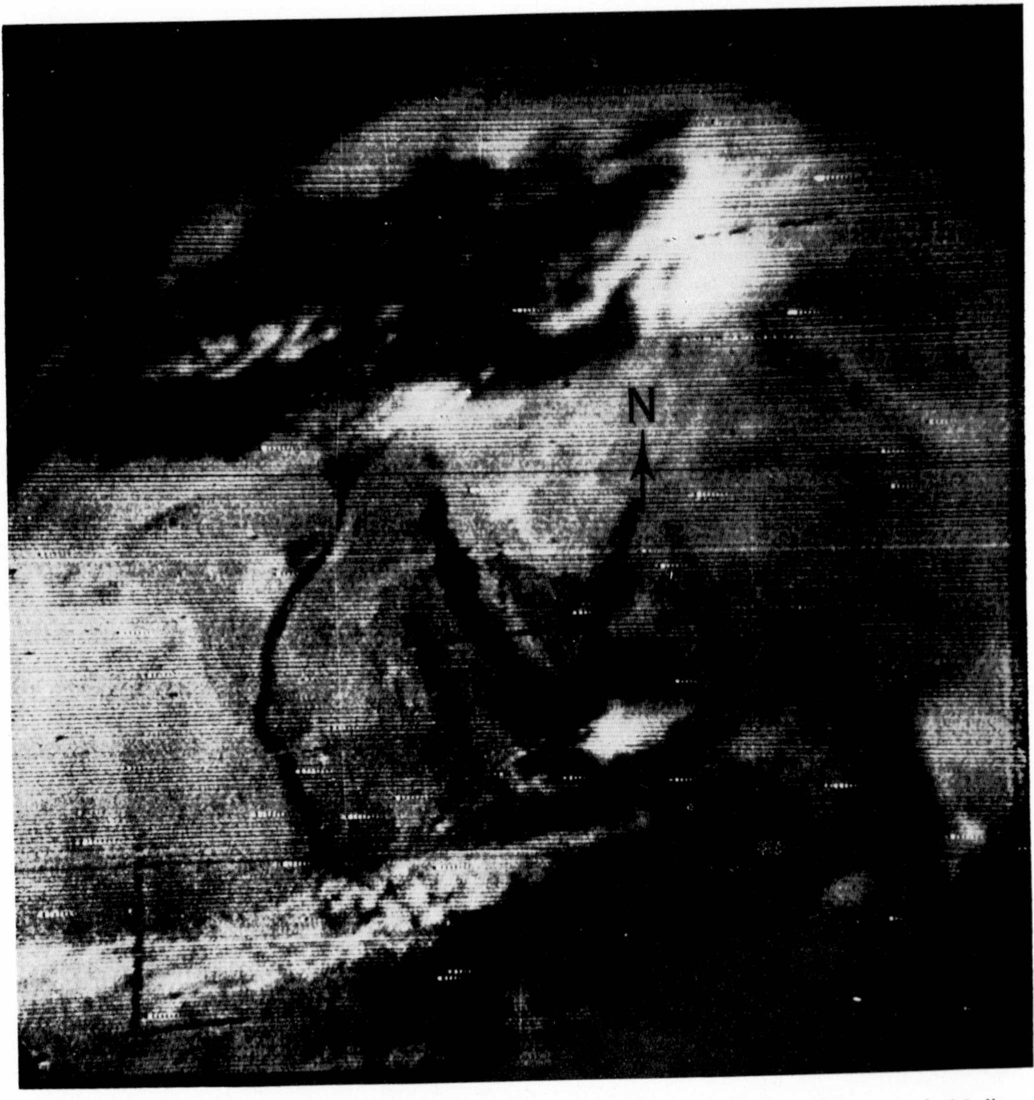

PLATE 4.—The Middle East, showing the Sinai Peninsula, the Red Sea, the Nile River Valley, and the Mediterranean Sea. April 4, 1960; 11.10 G.C.T.

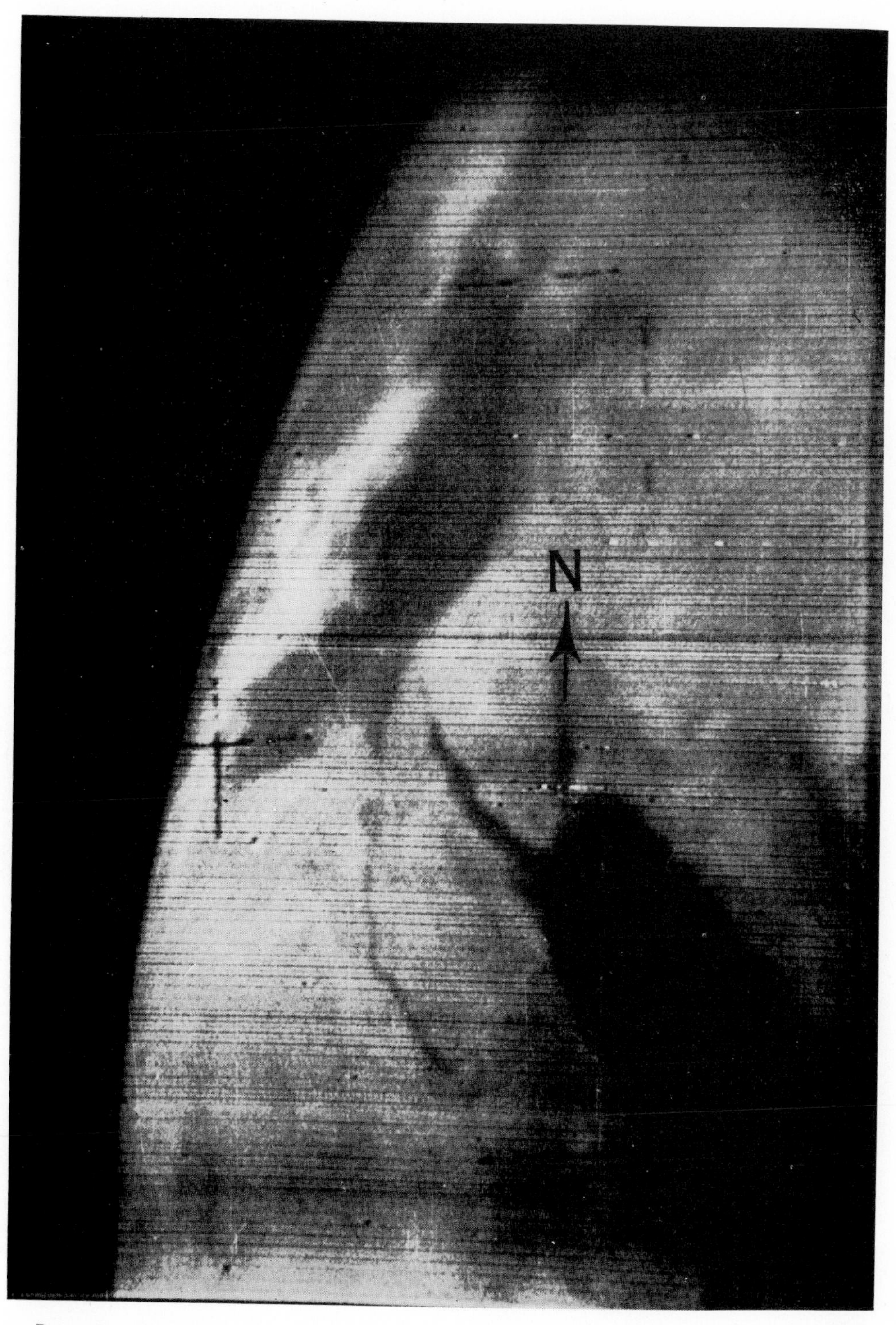

PLATE 5*a*.—Plates 5*a*–5*c* show three non-consecutive frames from one orbit as TIROS pictures moved from the Sinai Peninsula southward over the Red Sea and into the Gulf of Aden. April 11, 1960; 08.30 G.C.T.

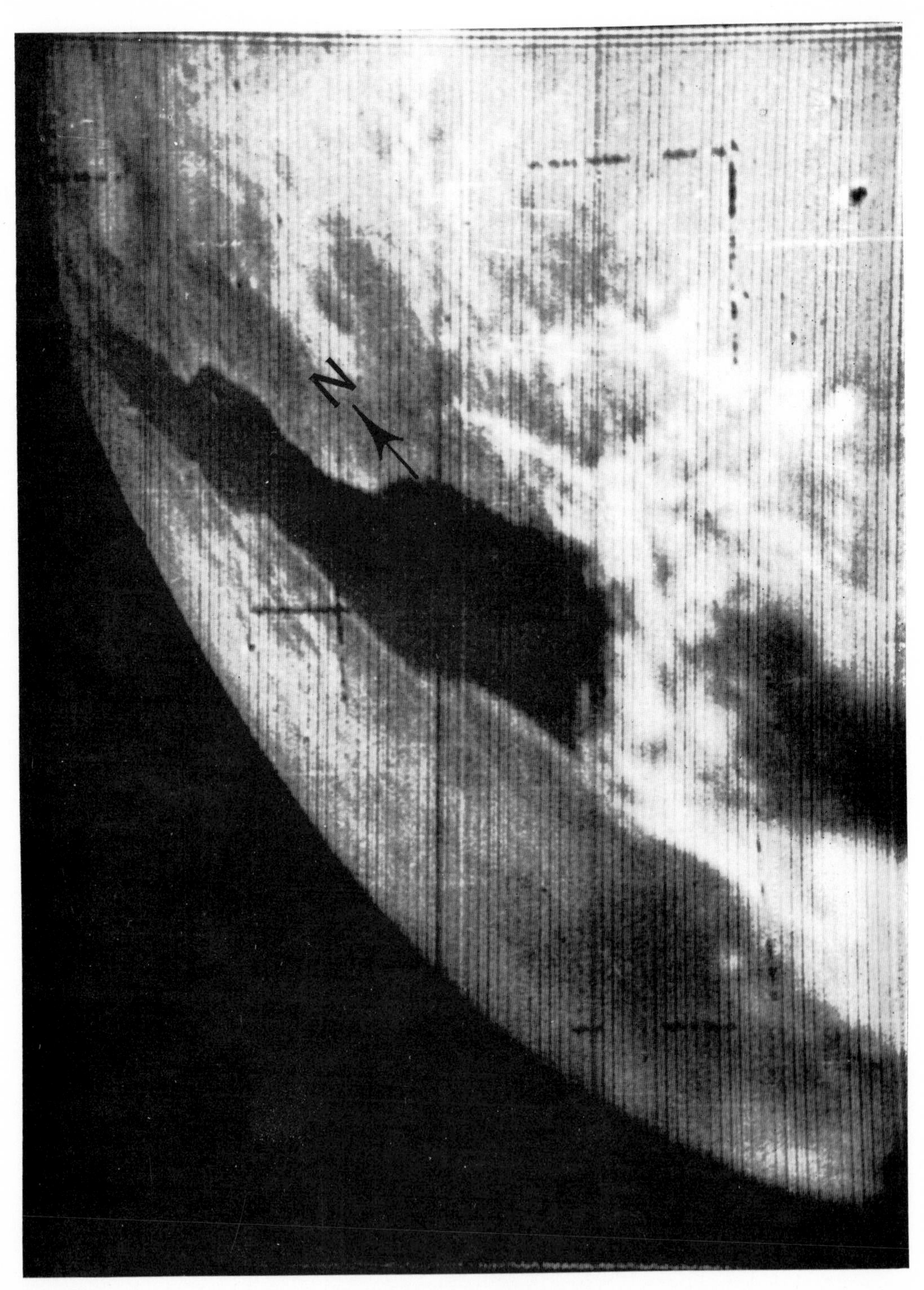

PLATE 5*b*.—Cf. legend for Pl. 5*a*

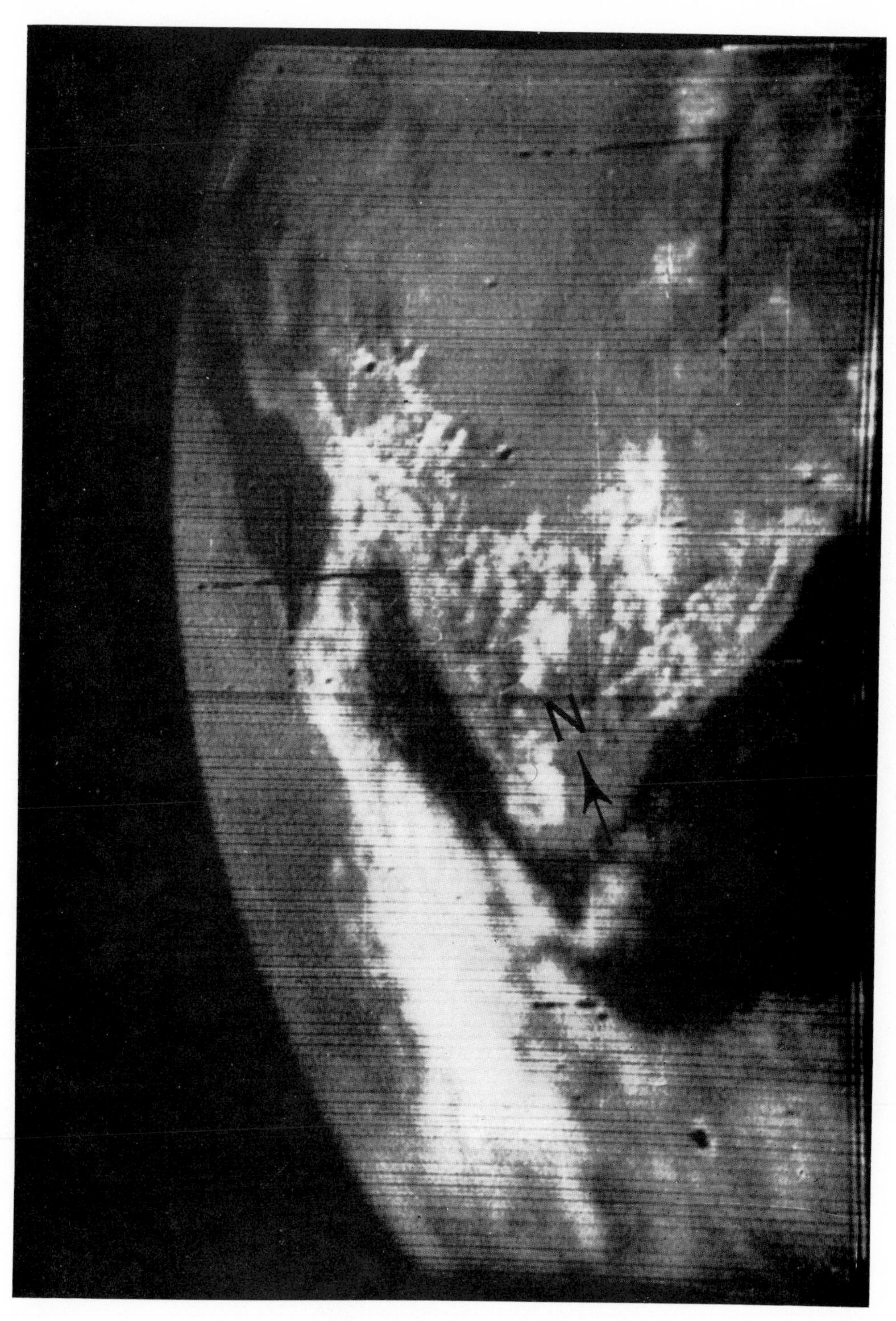

PLATE 5*c*.—Cf. legend for Pl. 5*a*

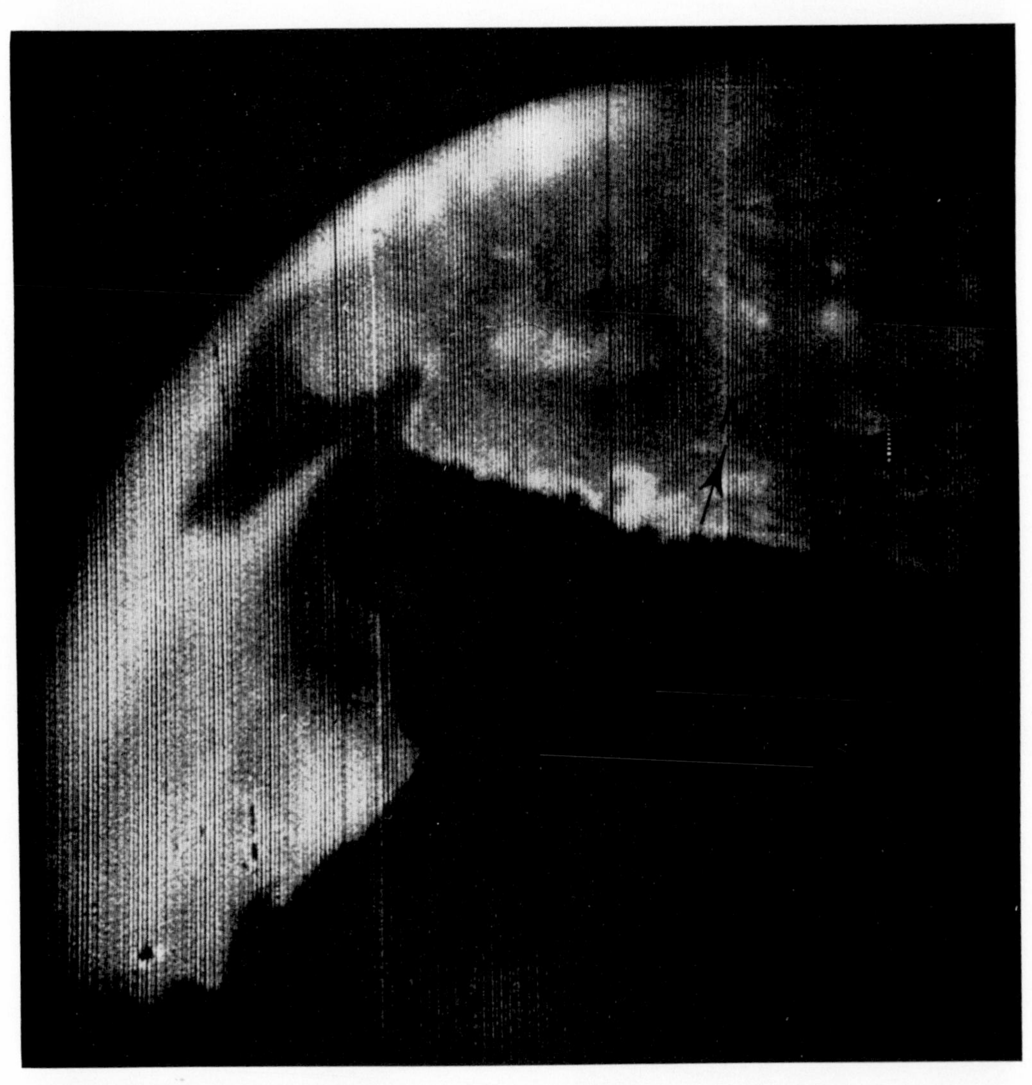

PLATE 6.—Land bordering on the Persian Gulf and the Arabian Sea. April 4, 1960; 09.30 G.C.T.

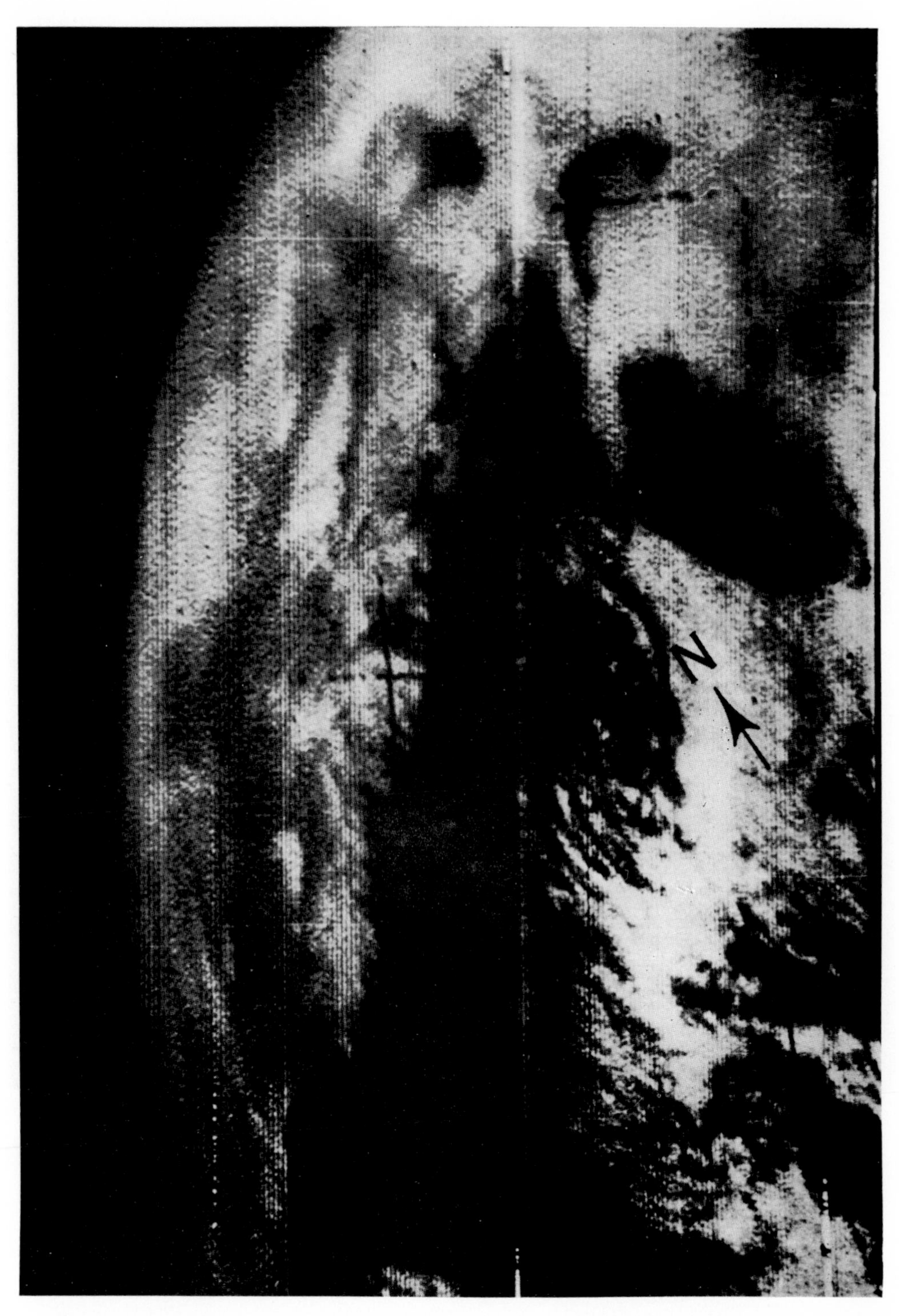

PLATE 7.—Hokkaido (Japan) and Maritime Territory of Siberia, separated by Gulf of Tartary and Sea of Japan. April 2, 1960; 03.00 G.C.T.

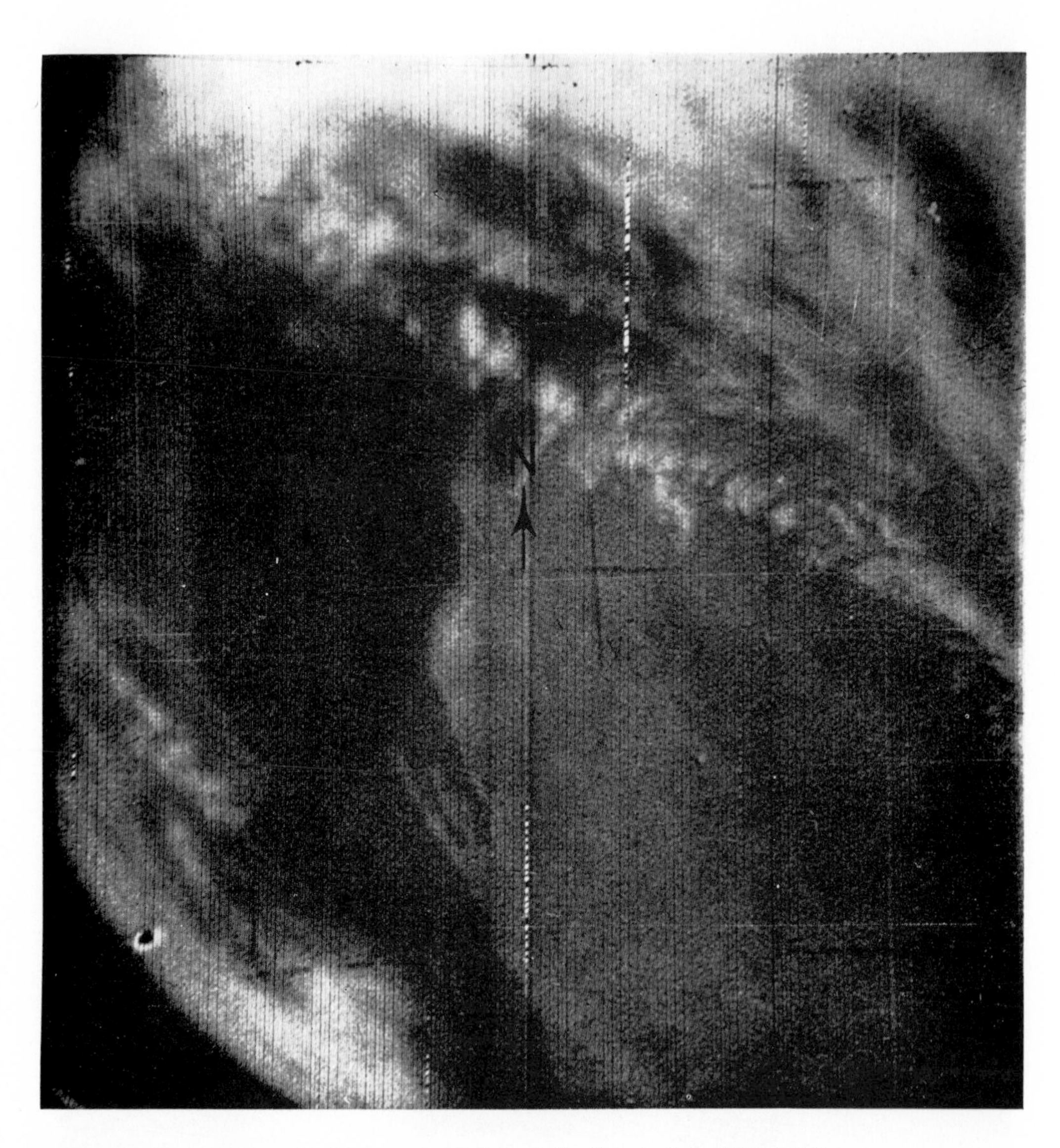

PLATE 8.—The west coast of Australia. May 5, 1960; 07.00 G.C.T.

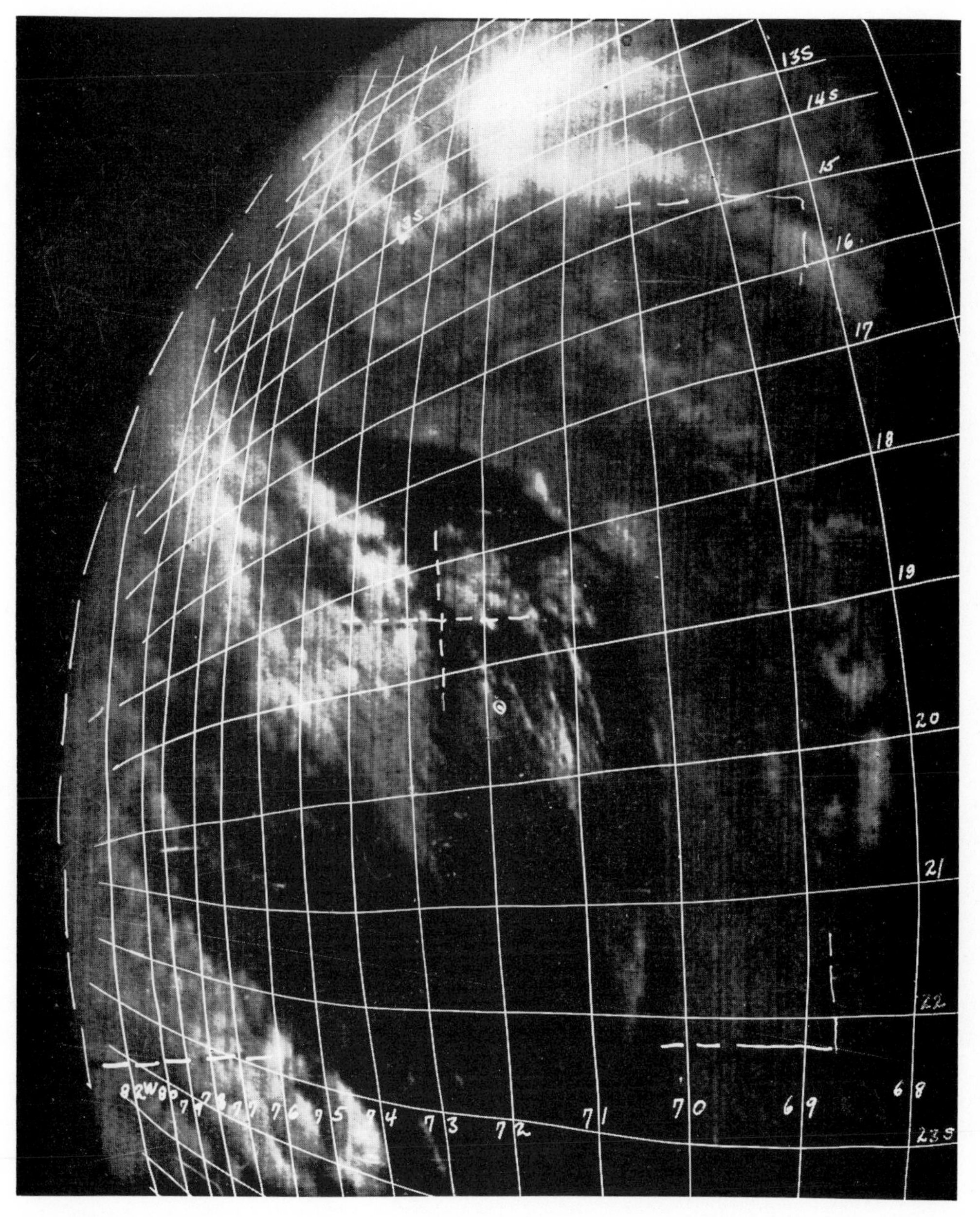

PLATE 9.—The west coast of South America, with superimposed latitude-longitude grid. April 15, 1960; 18.10 G.C.T.

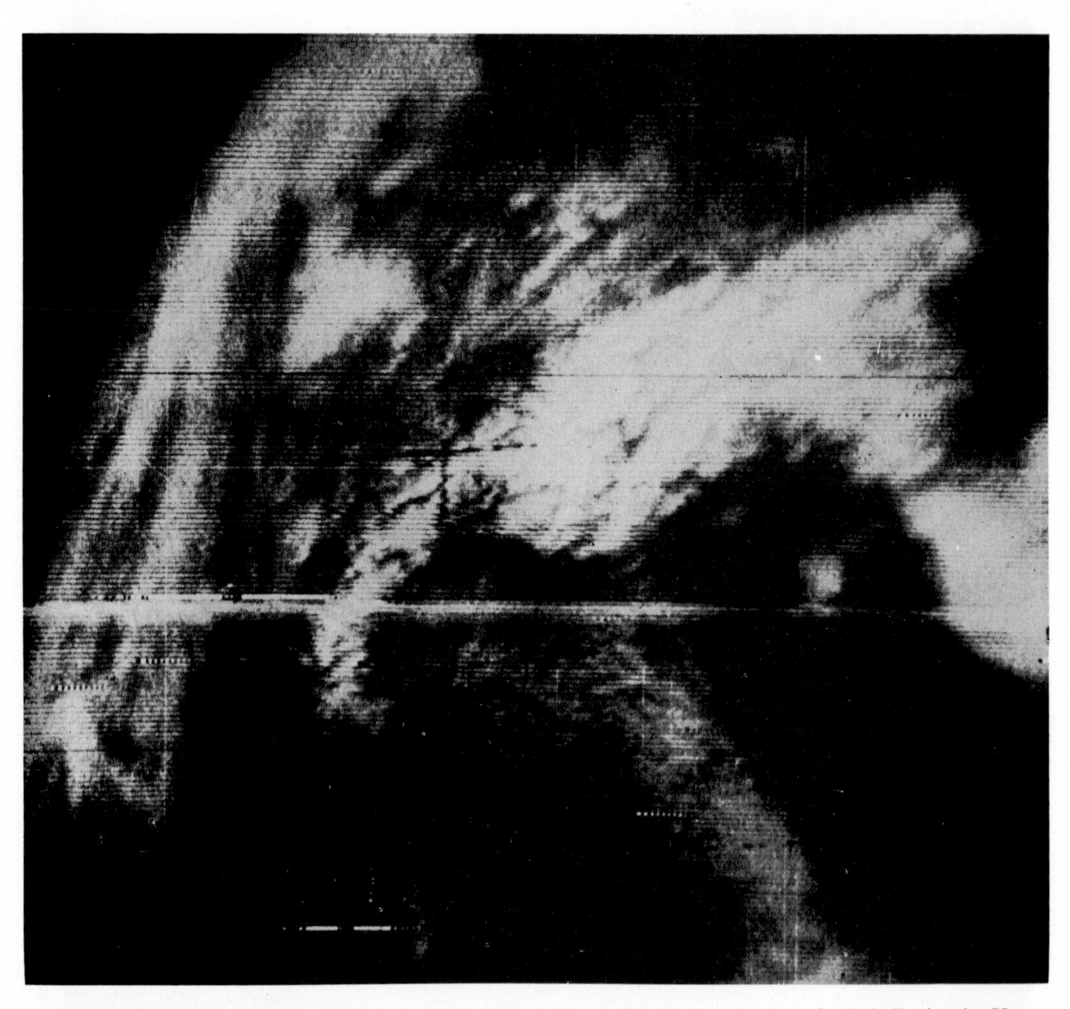

PLATE 10*a*.—The snow-covered Alps, April 12, 1960, Italy north of Rome, Corsica, the Pola Peninsula, Yugoslavia, and adjacent areas. The narrow lanes in the Alps, repeated from day to day, are low-lying valleys and lakes (cf. text). April 2, 1960; 11.10 G.C.T.

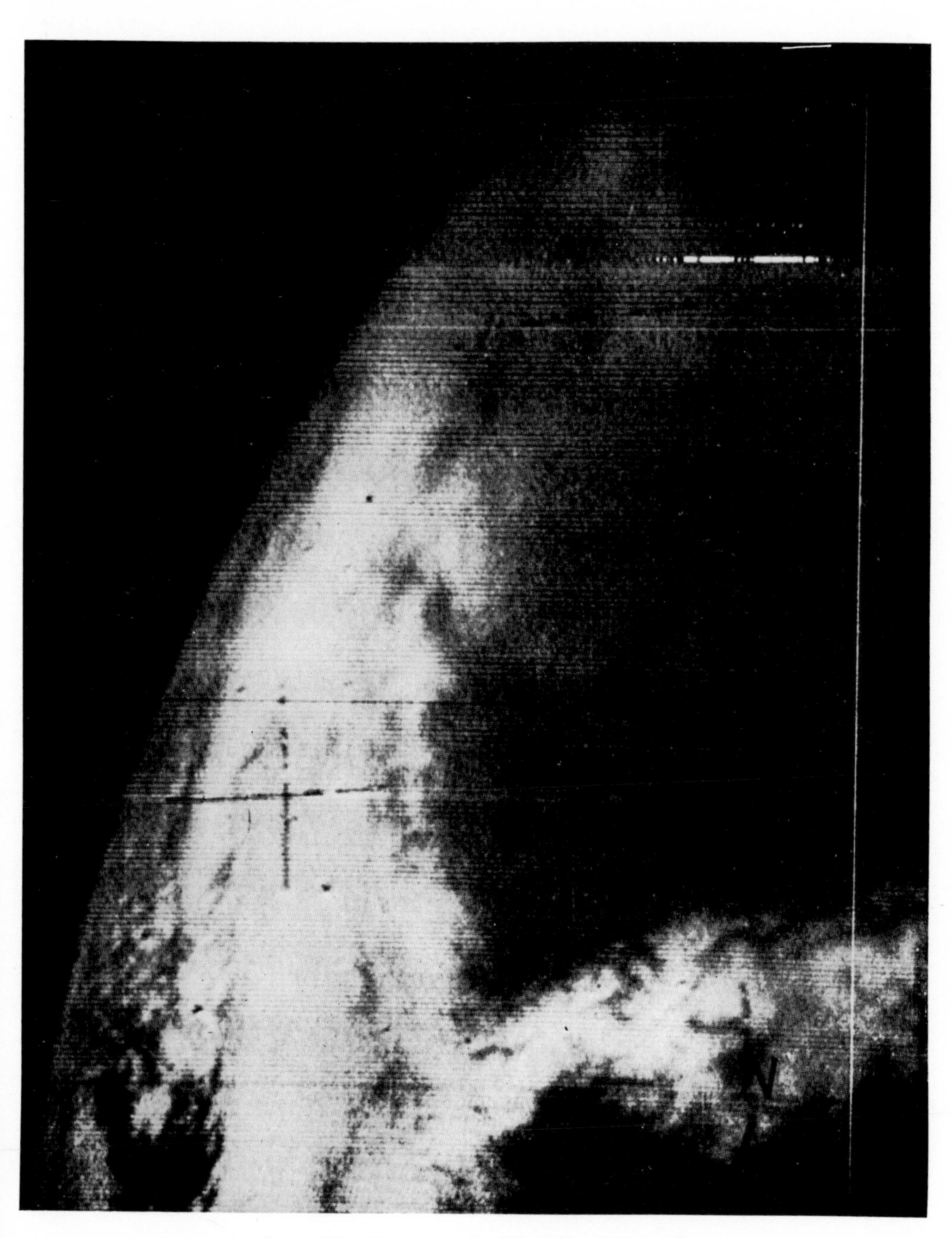

PLATE 10*b*.—Same area, April 3, 1960; 10.20 G.C.T.

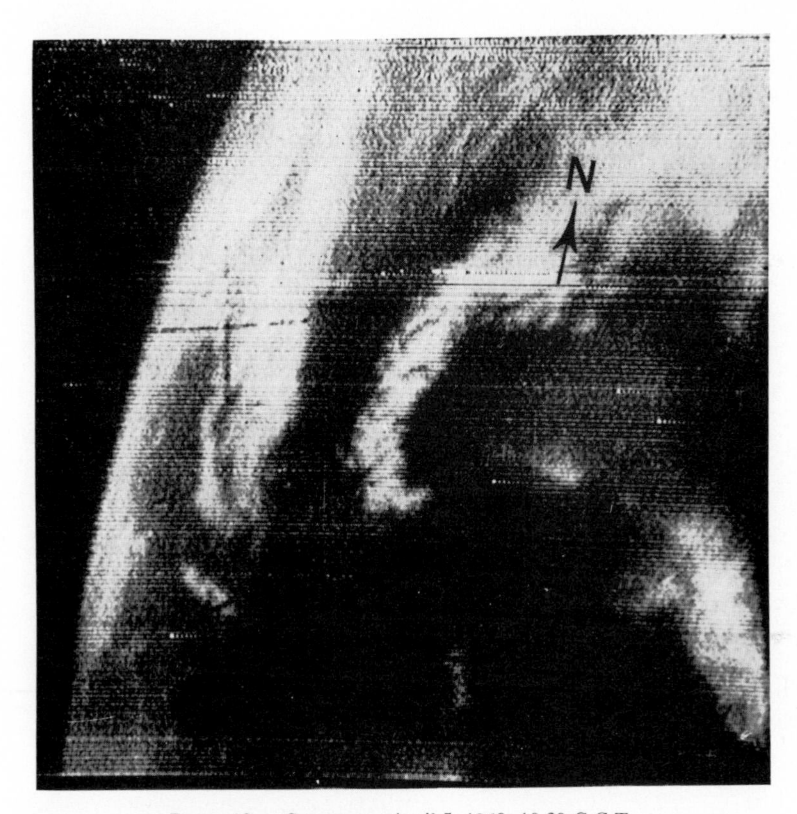

PLATE 10*c*.—Same area, April 5, 1960; 10.20 G.C.T.

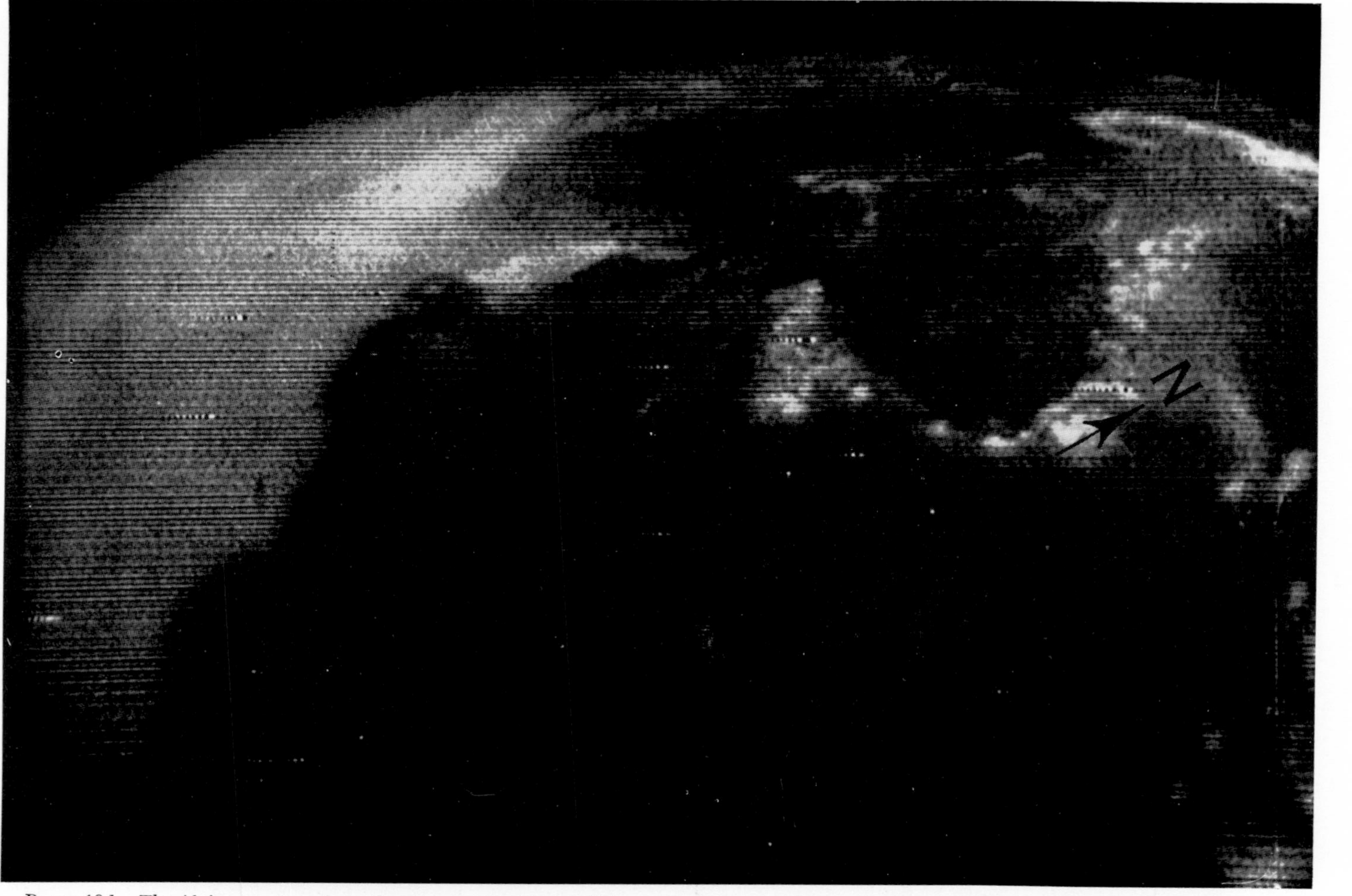

PLATE 10*d*.—The Alpine arc, 1 cm from the horizon; all of Italy, Sicily, and Sardinia, with the coastlines of Algeria, Tunisia, and Libya. April 7, 1960; 10.20 G.C.T.

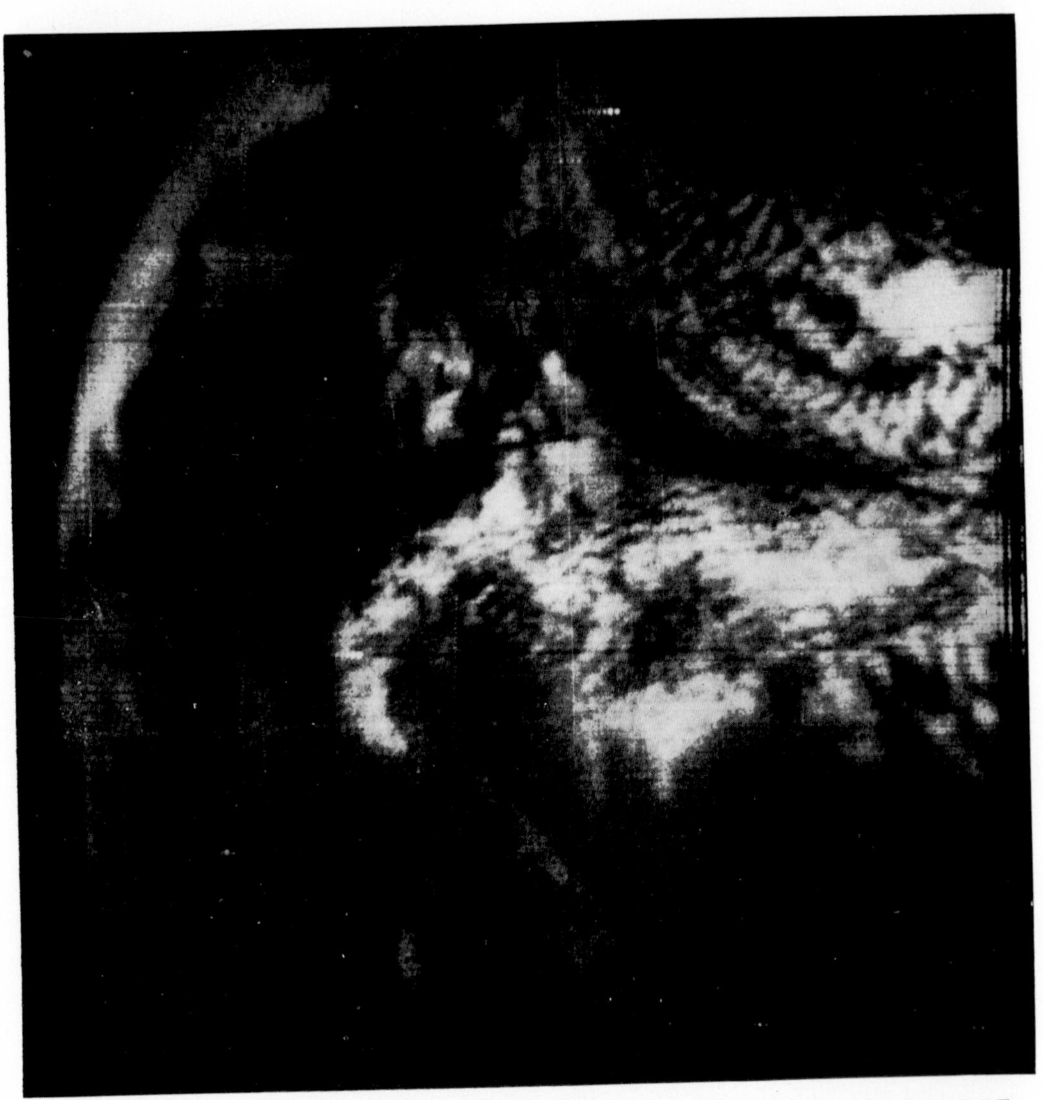

PLATE 10*e*.—Alpine arc with cloud over Germany, Austria, and Czechoslovakia. May 26, 1960; 14.10 G.C.T.

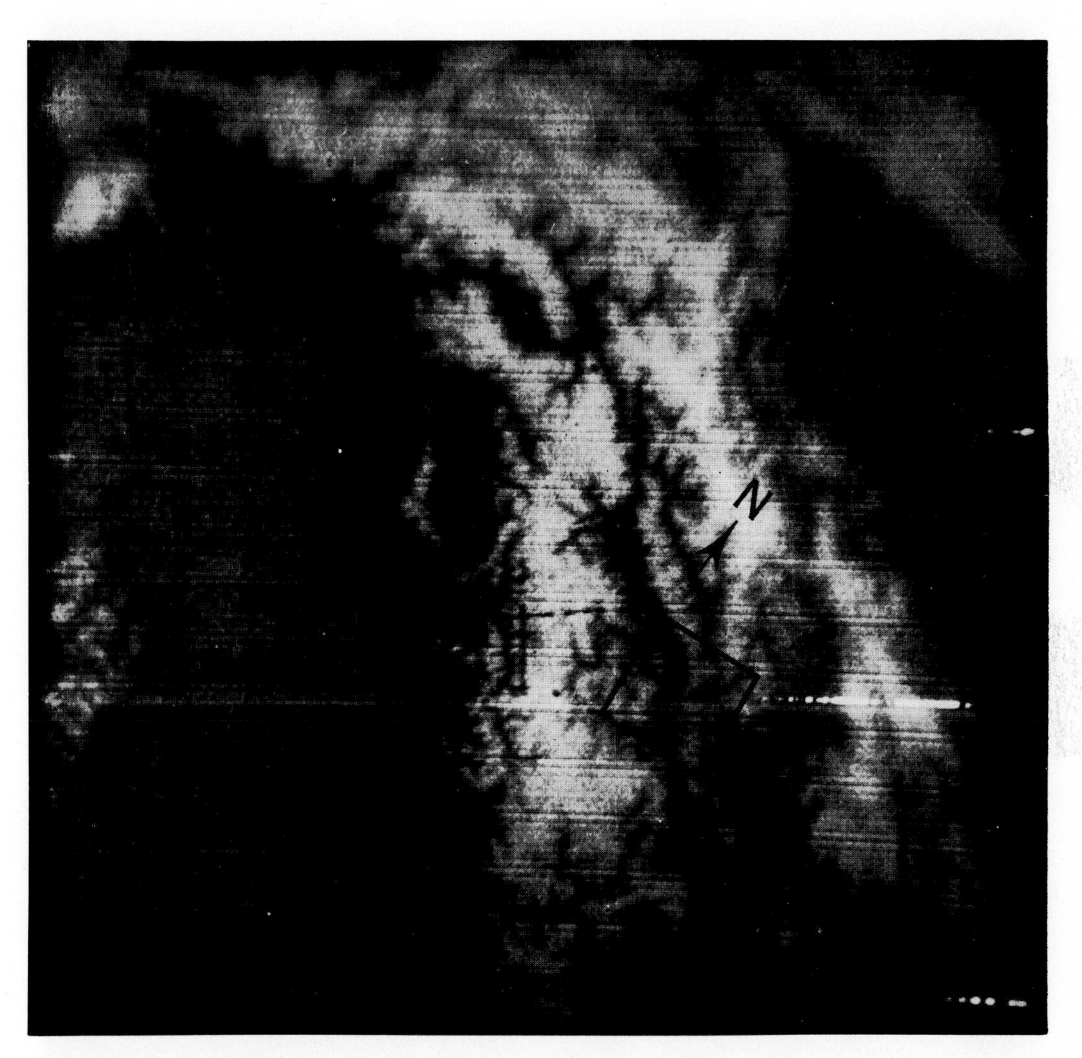

PLATE 11*a*.—Wide-angle view of snow-covered Himalayas in Kashmir. Black square is shown on larger scale on Pl. 11*b*. May 13, 1960; 10.10 G.C.T.

PLATE 11*b*.—Narrow-angle picture of snow-covered Ladakh Range (*upper central part*), (dark) Indus Valley (*left of center*), (dark) lake near Shushal (*right margin*). May 13, 1960; 10.10 G.C.T.

PLATE 12.—Florida Peninsula and Atlantic Coast to about Cape Lookout, South Carolina. Specular reflection of sun in coastal waters near Brunswick, Georgia. Gulf of Mexico, at lower left, is darker than the Peninsula. Ribbon-like cloud cover is near the Gulf Stream. May 16, 1960; 19.10 G.C.T.

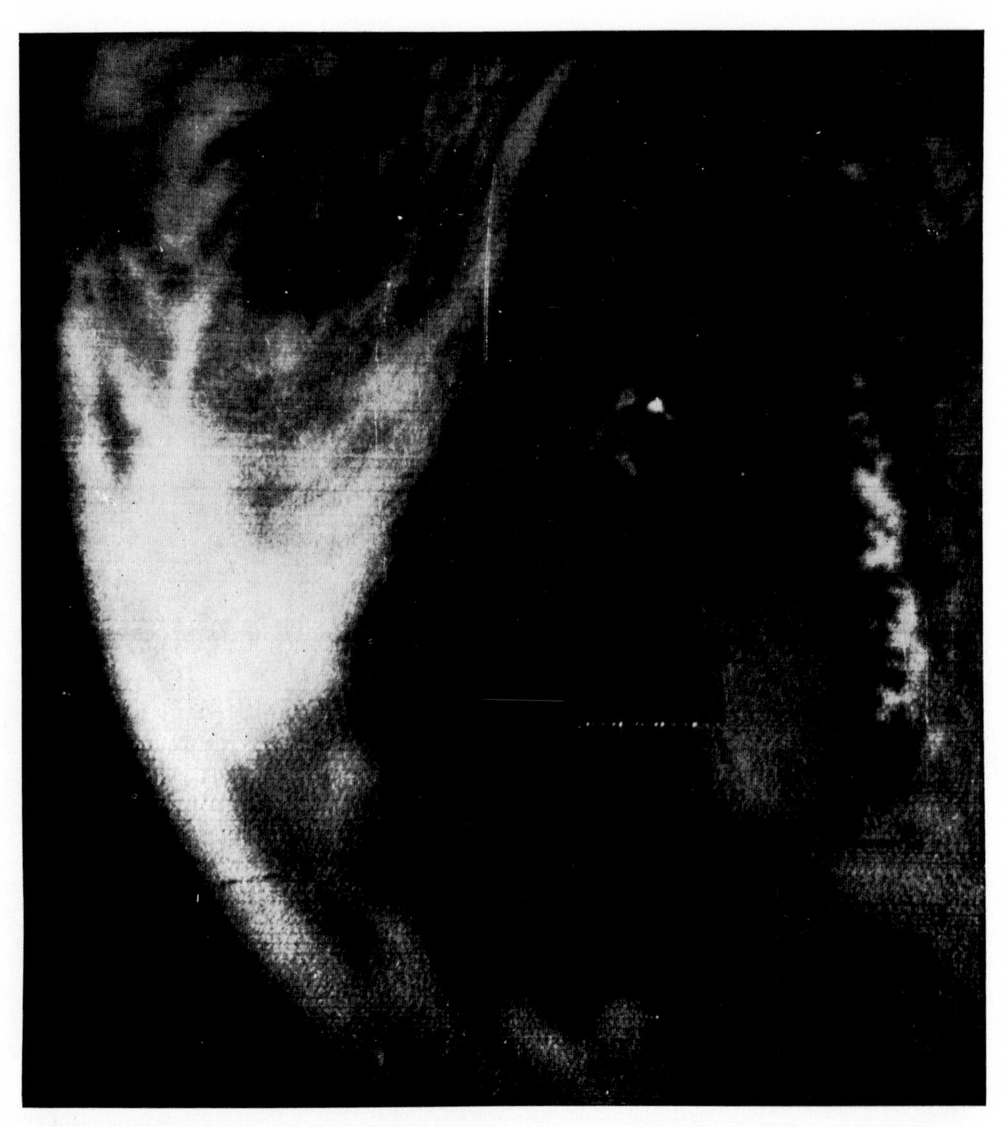

PLATE 13.—San Francisco Bay area (*center*), with specular reflection of sun on Sacramento River estuary, northeast of San Francisco peninsula. Coastline shown up to Eureka, California, with Los Angeles basin at bottom. Cloud-covered Sierra Nevada at right, clouds over Pacific at left. Mono Lake is black dot, 1 cm from right edge, at center. May 10 1960; 21.30 G.C.T.

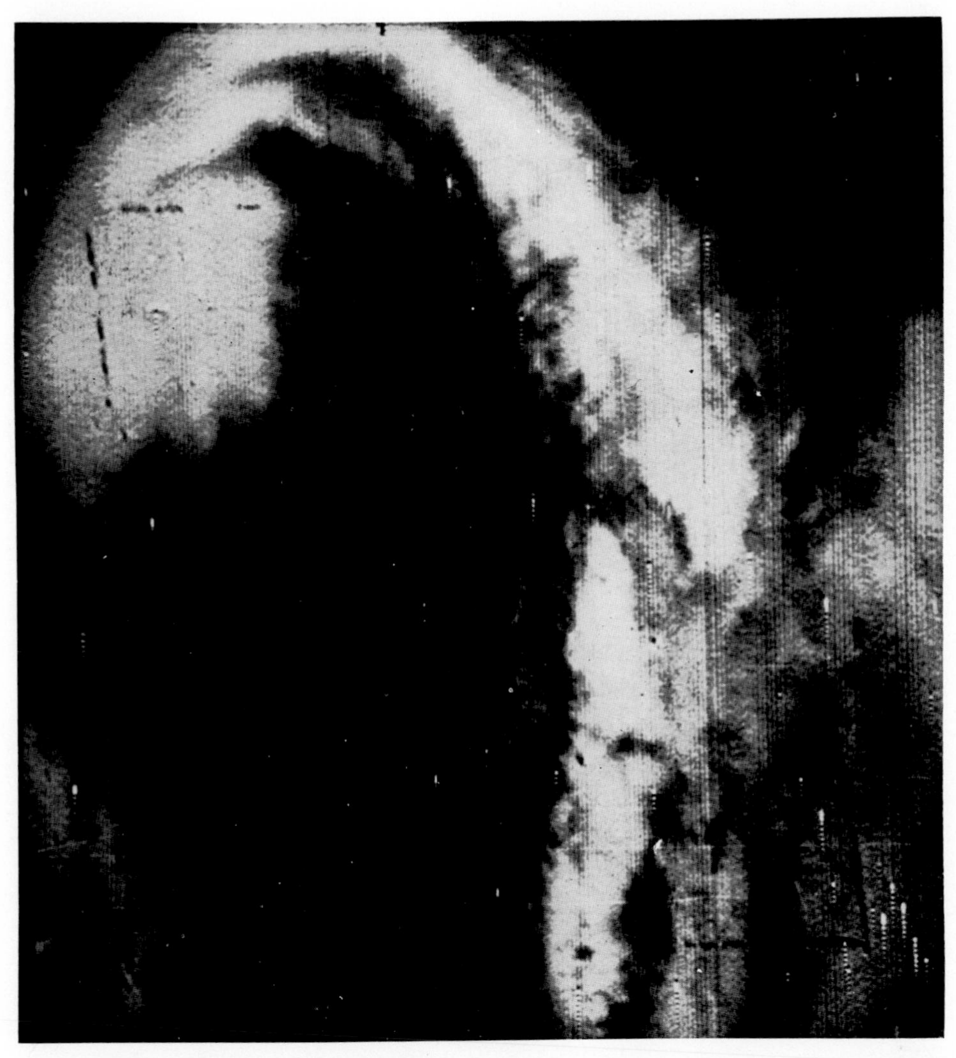

PLATE 14*a*.—A cyclonic and trailing cold frontal cloud system over the central United States. April 1, 1960; 20.30 G.C.T.

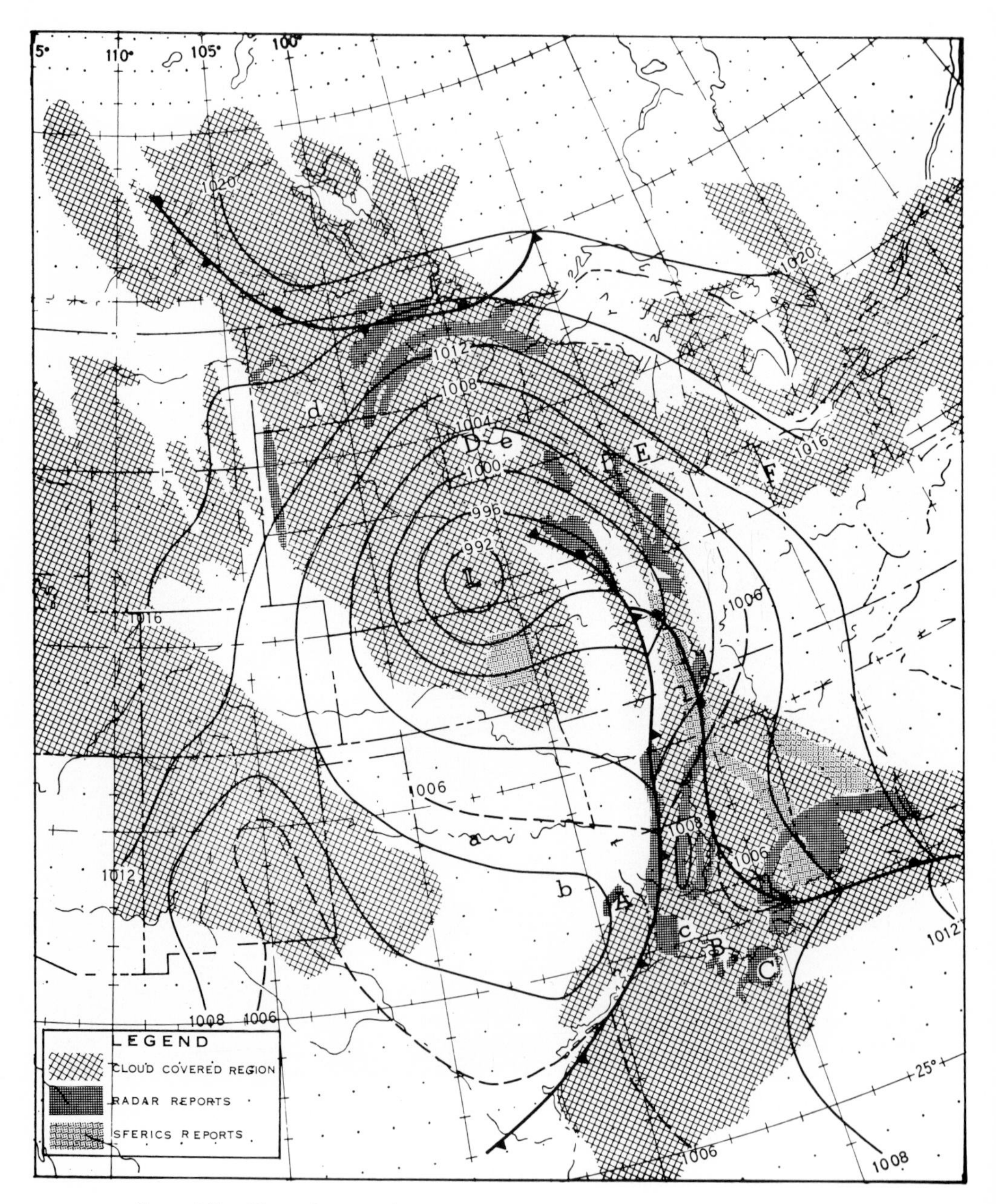

PLATE 14*b*.—The surface weather map near the time of the satellite picture in Pl. 14*a*

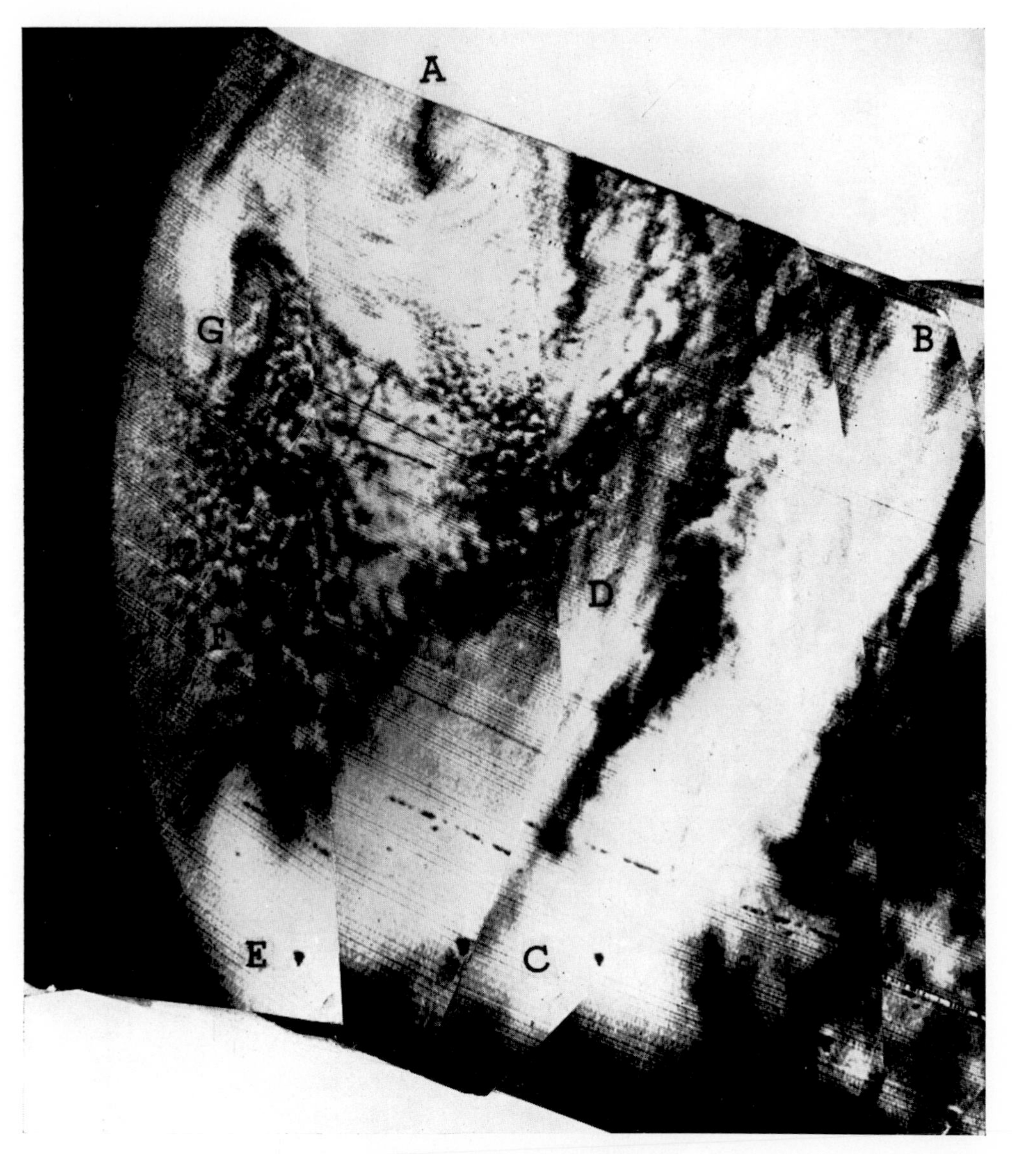

PLATE 15.—A composite picture of an occluded cyclonic cloud system in the Gulf of Alaska. April 1, 1960; 22.00 G.C.T.

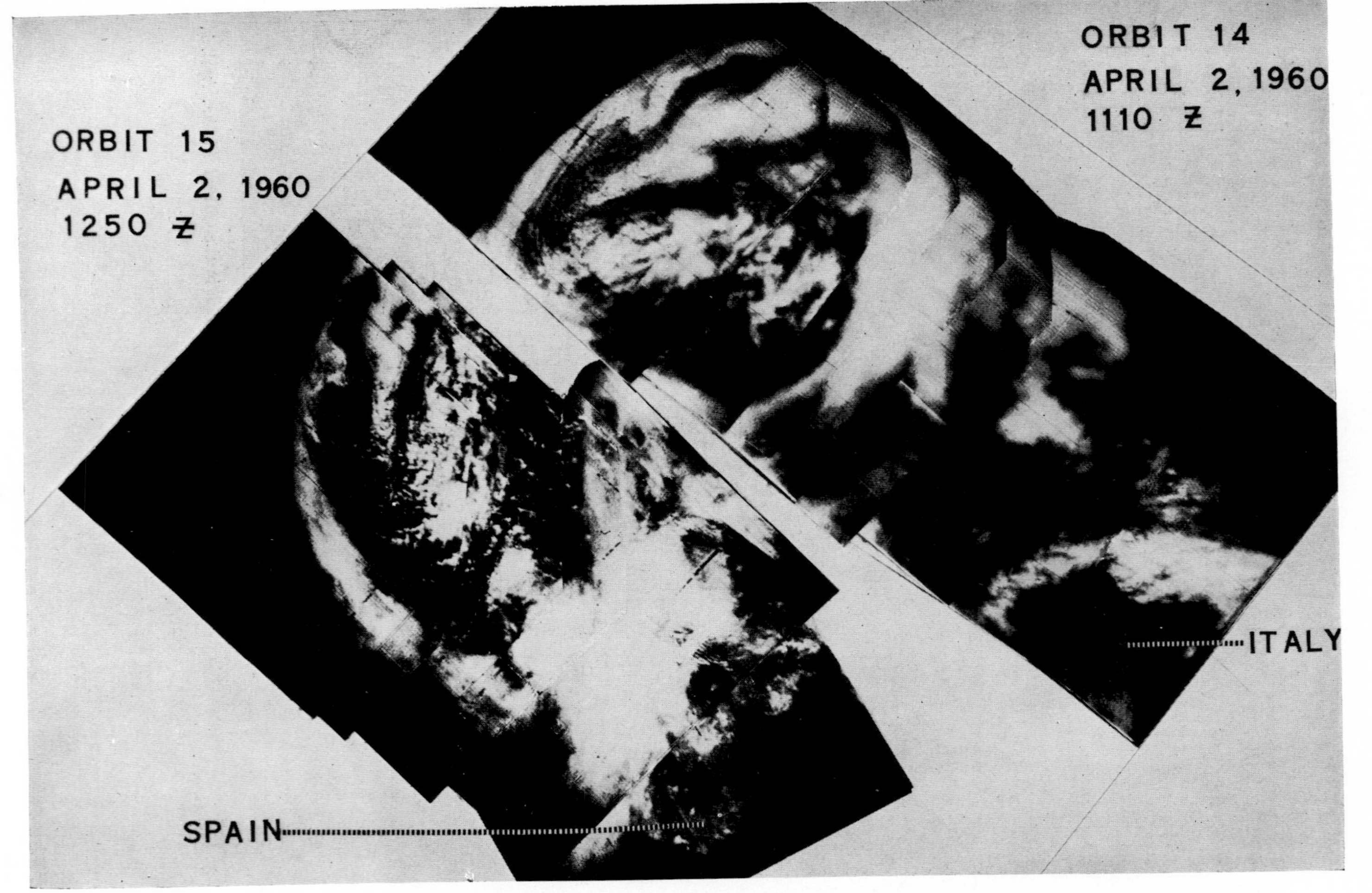

Plate 16.—A composite of satellite cloud pictures over an old occluded cyclone in the Atlantic Ocean

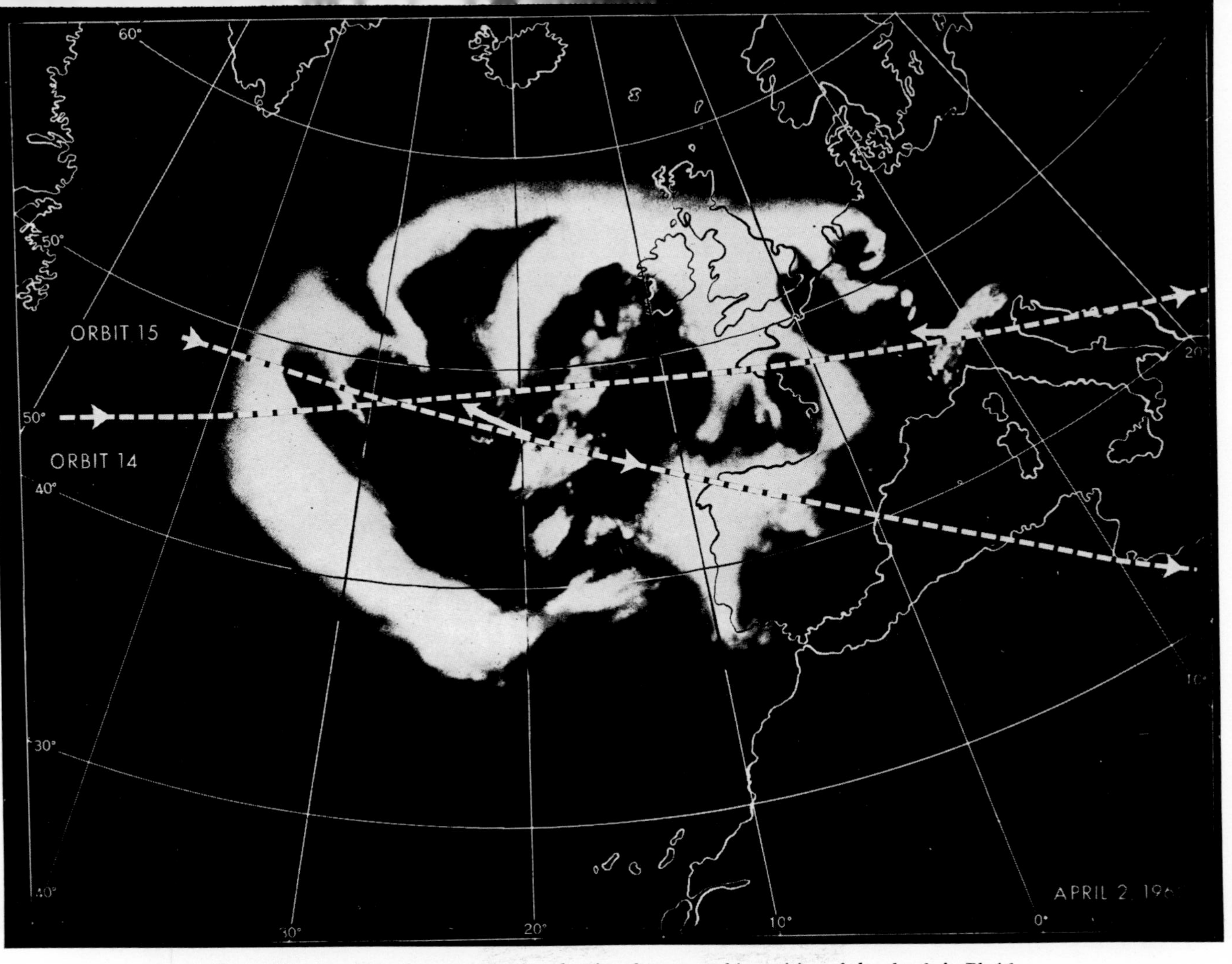

PLATE 17.—A schematic representation showing the geographic position of the clouds in Pl. 16

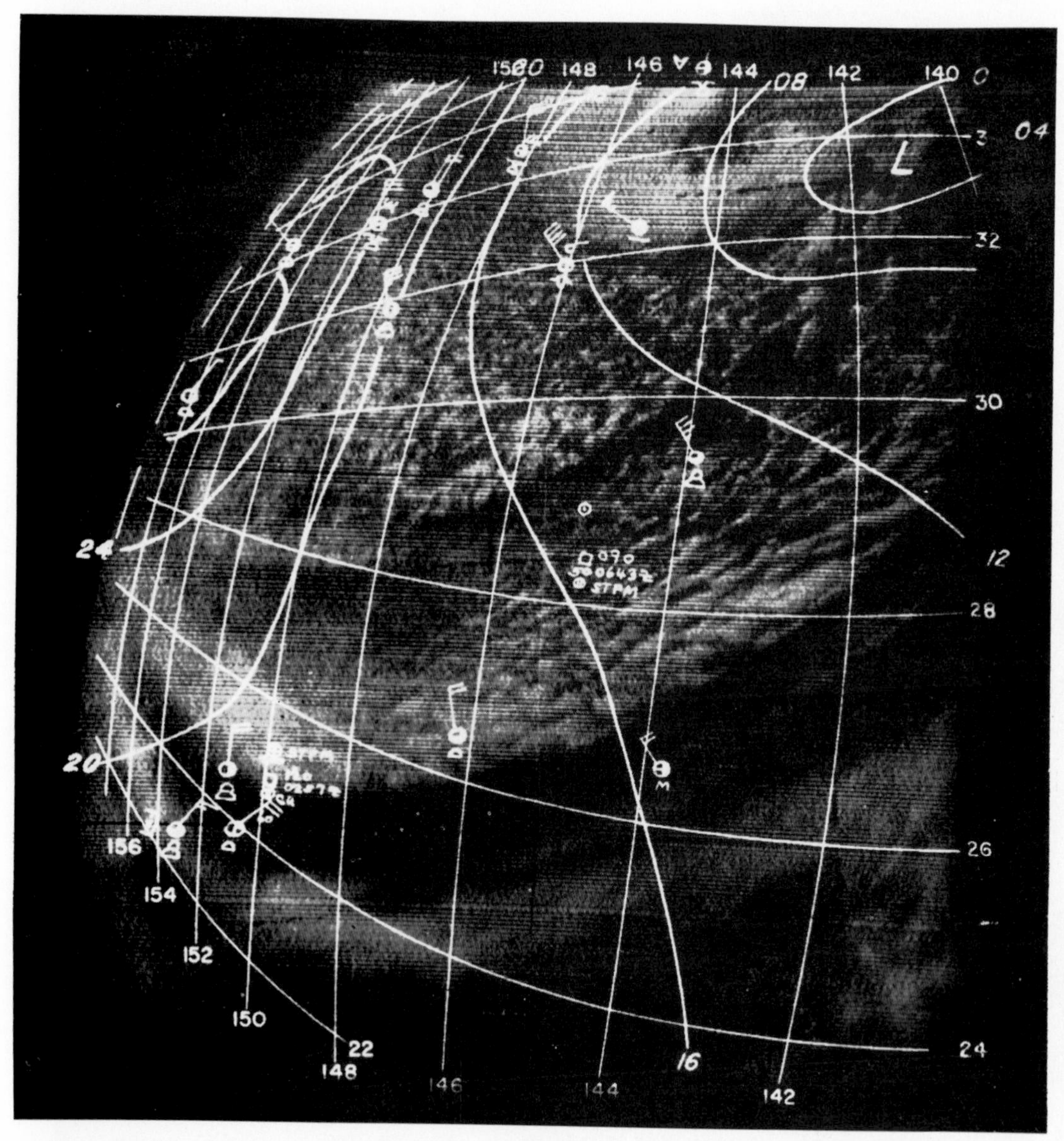

PLATE 18.—The cloud pattern associated with a cyclone lying between the United States and Hawaii. Latitude-longitude lines, surface isobars, and weather reports have been superimposed. April 4, 1960; 22.40 G.C.T.

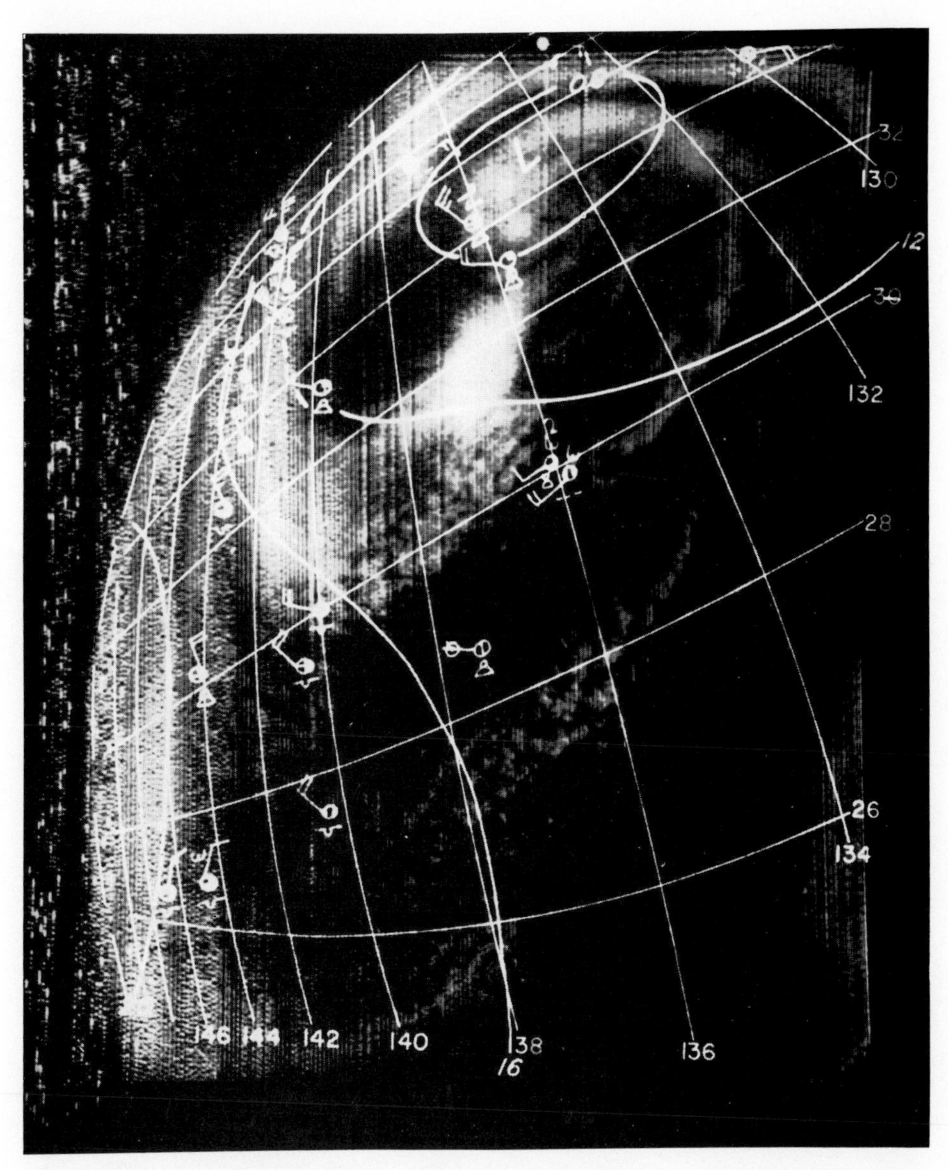

Plate 19.—A cloud picture of the same storm as the one in Pl. 18, but photographed about 24 hours later. April 5, 1960; 22.00 G.C.T.

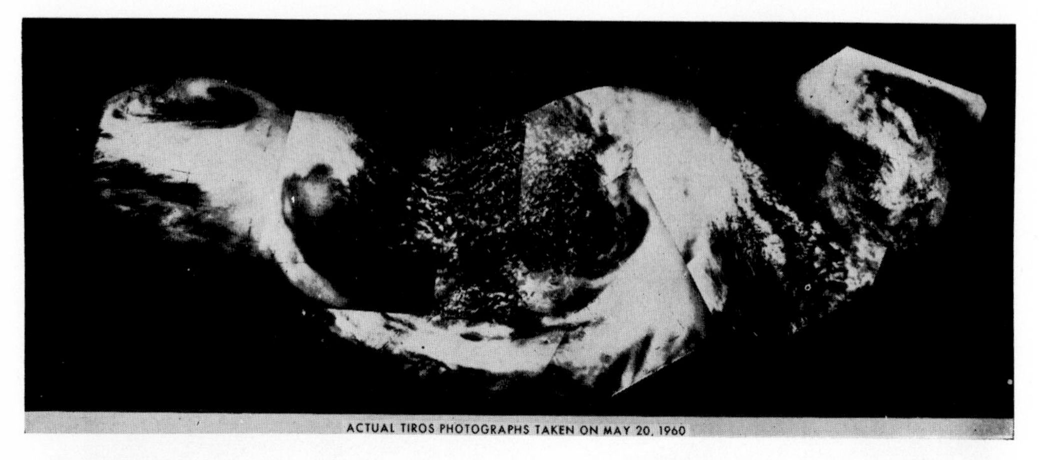

PLATE 20*a*.—A composite of satellite pictures from two orbits of TIROS I, showing two cyclonic cloud systems, joined by a frontal cloud band.

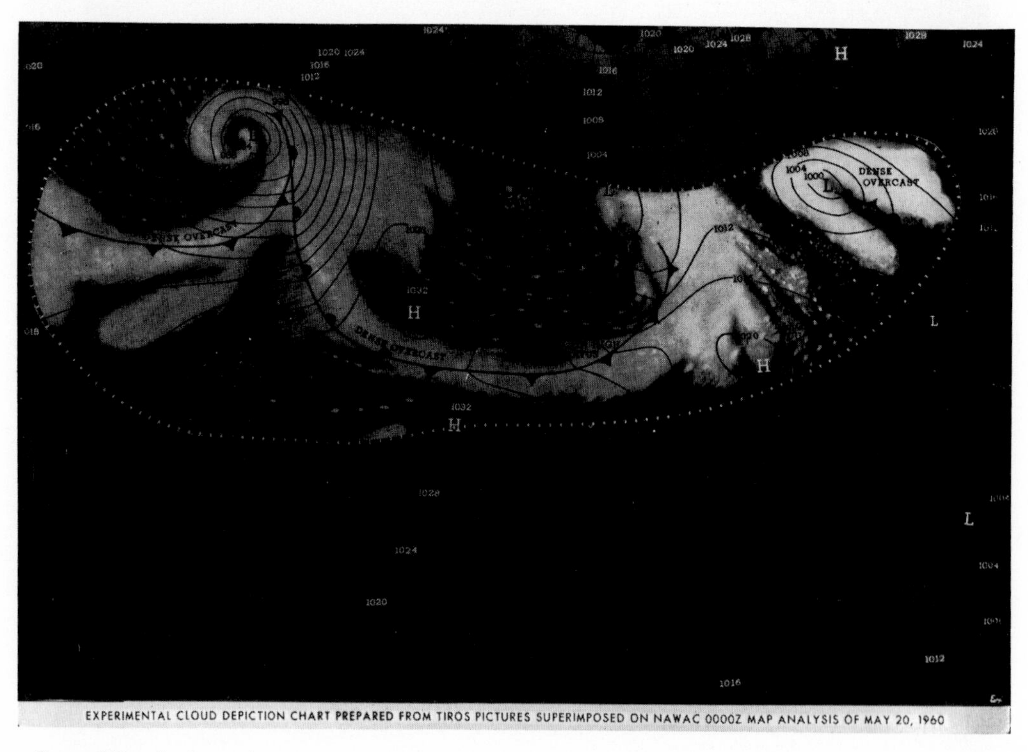

PLATE 20*b*.—A schematic representation of the cloud pictures, with the surface weather map superimposed

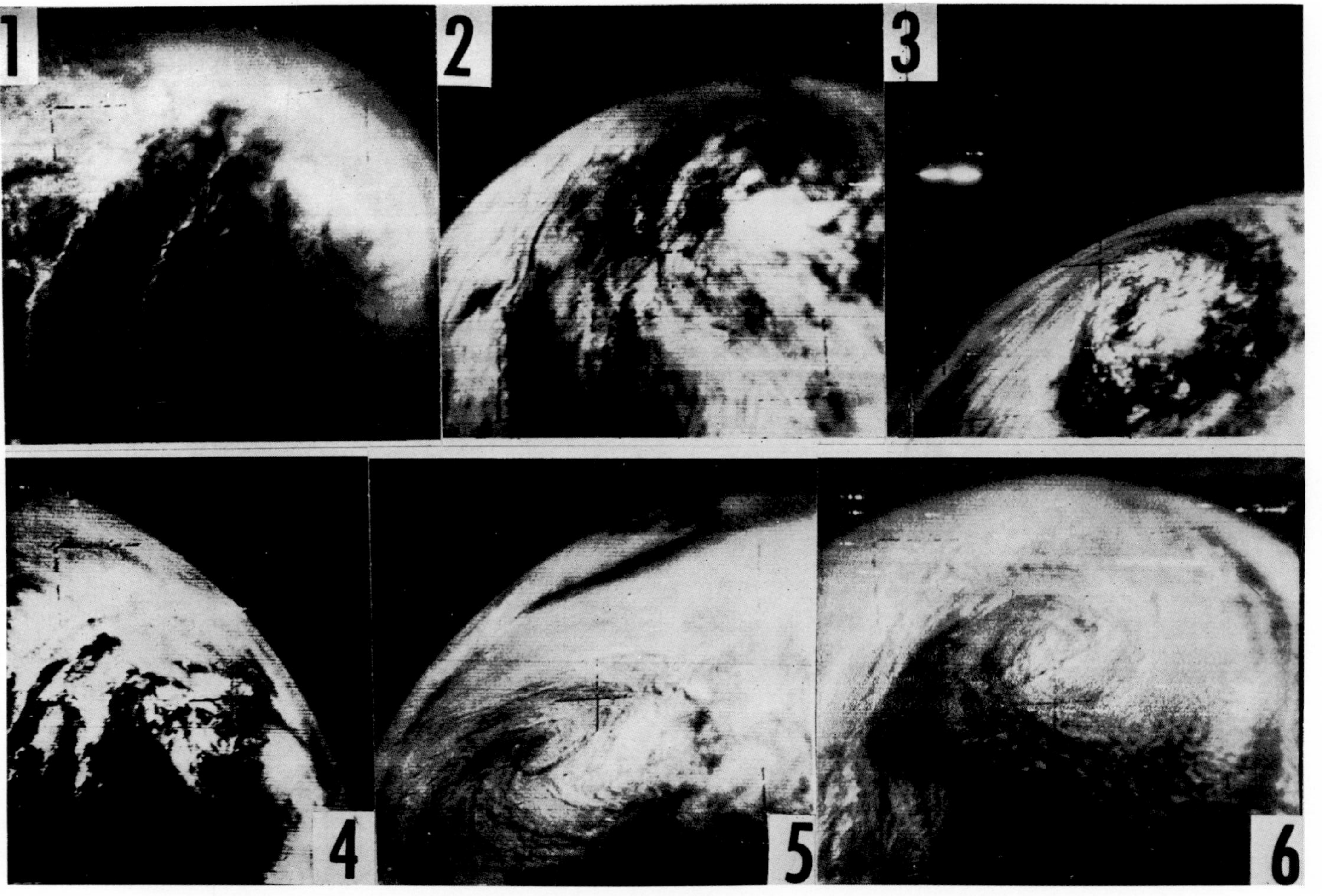

PLATE 21.—Six different cyclonic cloud patterns photographed in the Southern Hemisphere (see Pl. 3)

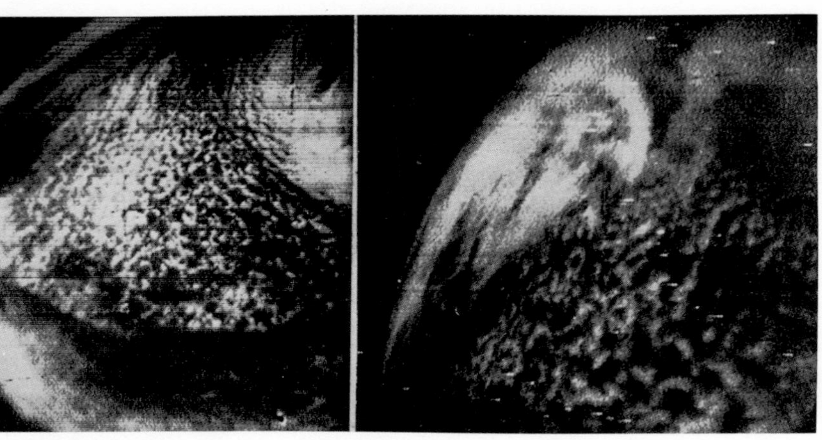

a — Orbit 7, April 1, 1960, 23.41 G.M.T. Picture Center 44 N., 166 W.

b — Orbit 50, April 4, 1960, 22.50 G.M.T. Picture Center 32 N., 149 W.

c — Orbit 46, April 4, 1960, 16.12 G.M.T. Picture Center 37 N., 57 W.

Plate 22.—Three pictures showing cellular patterns of clouds photographed by TIROS I

Report on TIROS I (Washington: National Aeronautics and Space Administration).

OLIVER, V. J. 1960 "TIROS Pictures a Pacific Frontal Storm," *Weatherwise*, Vol. **13**, No. 5 (October), cover.

RUSSELL, H. N., DUGAN, R. S., and STEWART, J. Q. 1945 *Astronomy*, Vol. **1**: *The Solar System* (Boston: Ginn & Co.), Frontispiece.

STAFF OF THE METEOROLOGICAL SATELLITE LABORATORY (U.S. WEATHER BUREAU) 1961 "Some Meteorological Results from TIROS I," *Report on TIROS I* (Washington: National Aeronautics and Space Administration).

STERNBERG, S., *et al.* 1960 "Roundup of TIROS I," *Astronautics*, **5**, 32–44 ff.

STROUD, W. G. 1960 "Our Earth as a Satellite Sees It," *Nat. Geog. Mag.*, **118**, 293–302.

WEXLER, H. 1957 "The Satellite and Meteorology," *J. Astronautics*, **4**, 1–6.

WEXLER, H., and FRITZ, S. 1960 "TIROS Reveals Cloud Formations," *Science*, **131**, 1708–1710.

WINSTON, J. S., and TOURVILLE, L. 1961 "An Occluded Cyclone over the Gulf of Alaska," *Report on TIROS I* (Washington: National Aeronautics and Space Administration).

WINSTON, J. S. 1960 "Satellite Pictures of a Cut-off Cyclone over the Eastern Pacific," *Monthly Weather Rev.*, **88**, 295–314.

CHAPTER 2

The Trans-Neptunian Planet Search

By CLYDE W. TOMBAUGH
New Mexico State University

1. EARLY INVESTIGATIONS OF TRANS-NEPTUNIAN PLANETS

THE discovery of Uranus by Sir William Herschel on March 13, 1781, was unexpected. Apparently, little thought was entertained on the possible existence of planets beyond the orbit of Saturn, which was the medieval boundary of the solar system. This is reflected in Herschel's interpretation of the object after he found it, thinking it was a comet. The discovery of Uranus was particularly important because it directed thought to regions beyond known boundaries. It paved the way for the dramatic mathematical prediction of the existence of Neptune by Le-Verrier and Adams and its detection by Galle at Berlin in 1846.

In both these discoveries, a perceptible disk was a criterion of identity. This thought was carried over by later investigators on the problem of a planet beyond the orbit of Neptune. But, as it turned out, the only detecting criterion was the slow, apparent motion through the star field.

As early as 1834, Hansen expressed the opinion, in correspondence with the elder Bouvard, that a single planet would not account for the increasing residuals in the longitude of Uranus. This thought was also the basis of Percival Lowell's theoretical investigations.

The theoretical aspects relating to this subject (such as orbits, perturbations, and masses) are dealt with in Chapter 3, and the present discussion will therefore be confined to the observational techniques and coverages in searches for bodies beyond the orbit of Saturn rather than that of Neptune. This nearer limit was chosen because it entailed no extra work.

The first serious search for a planet beyond the orbit of Neptune was probably by Todd (1880) with the 26-inch refractor at the United States Naval Observatory. Apparently, he searched along the invariable plane to a distance of 1° to either side. Extrapolating beyond Uranus and Neptune, he had in mind a planet at a mean distance of 52 units, with a diameter of 80,000 km subtending an angle of 2″.1. On thirty clear, moonless nights, between November 3, 1877, and March 5, 1878, he searched a strip from longitude 146°.8 to longitude 186°.1, employing magnifying powers of 400–600× and looking for an object with a perceptible disk. He suspected many objects, which were reobserved on the following nights and again several weeks later. He had much confidence in the negative results of his search.

Todd's search was carried out before the time of effective astronomical photography. To check stars to the thirteenth magnitude for planetary motion visually was quite out of the question, compared with the relatively few stars of the eighth magnitude used in the search for Neptune.

Attempts at prediction of a trans-Neptunian planet were made by Forbes (1880) and by Pickering (1928) on the basis of the effects of unseen planets on comets. Pickering's investigations led to a search at the Mount Wilson Observatory in 1919 by Humason, who took some photographs around the predicted position with a 10-inch astrographic telescope. This search met with no success. Later, in 1930, after the discovery of Pluto at Flagstaff, the planet was found on some of these plates, when it was known precisely where to look. Apparently, the images of Pluto were outside the area of close scrutiny, and there was locally a minor defect.

Among other astronomers who investigated the planetary residuals theoretically were Tisserand, Gaillot, and Lau.

2. EARLY WORK AT LOWELL OBSERVATORY

Percival Lowell became interested in the problem of a trans-Neptunian planet early in the century. He made theoretical studies culminating in his memoir on the motions of Uranus and Neptune (Lowell, 1915), which, as early as 1905, led him to initiate observational searches. He recognized that the photographic method, using short-focus telescopes, was the only one practicable. The first search was made with a 5-inch $F/7$ Brashear lens, with the plates taken 5° apart along the invariable plane. About 200 plates were taken by E. C. Slipher in 1906 and 1907 and 50 plates by K. P. Williams in 1907. The plates were exposed for 3 hours and reached sixteenth magnitude. However, only the central 5 cm of the plates had sharp images, and the magnitude loss in the main outer portions was considerable.

At first, Lowell examined these plates with a hand magnifier, with one plate of the pair above the other; but this was slow and not thorough. Later he used a modified Hartmann comparator and still later procured a Zeiss blink comparator. This instrument was found to be superior to any other means in detecting planetary motion on a plate crowded with star images.

Next, an extensive series of plates was taken with the 42-inch reflector by C. O. Lampland and E. C. Slipher. Since the exposures were less than 10 minutes in length, plates could be taken in rapid succession, which partly compensated for the small field. These plates recorded stars of seventeenth magnitude.

From 1914 to 1916 a 9-inch Brashear telescope, loaned by the Sproul Observatory, was used to carry on the search. A large number of plates were made with it by T. B. Gill and E. A. Edwards, supervised by Lampland. These plates were only partially examined by Lowell. Later, Lampland resumed the work and listed 515 asteroids and 700 variable stars (Putnam and Slipher, 1932).

Lowell died in November, 1916. No further search work was done until the completion of the 13-inch telescope in 1929.

3. THE LARGE SURVEY

3.1. The Telescope

In 1925 Mr. Guy Lowell, then trustee of the Lowell Observatory, purchased glass disks for a 13-inch objective. Dr. A. Lawrence Lowell, president of Harvard University, brother of Percival Lowell, provided the funds to build the telescope and mounting. The lens was figured by Mr. C. A. R. Lundin, and much of the mounting was made in the observatory shop. The instrument was ready for tests early in 1929, and the author was appointed at the same time to carry out the survey. It was discovered that the field of good images, though somewhat curved, was large enough to use 14 × 17-inch plates. Several plateholders of this size were built to curve the plates concave by about $\frac{1}{8}$ inch. The curvature could be adjusted by a central screw and four corner screws on the back. A special testing table was built to measure and adjust each plate to the standard curvature before it was placed in the telescope for exposure. The Zeiss blink-microscope-comparator was modified to accommodate 14 × 17-inch plates.

The focal length of the 13-inch objective is 66.6 inches (169 cm), giving a scale of 122″/mm, or 1.16 inches or 2.95 cm per degree. Thus the plates covered an area of sky nearly 12° × 14°; the net area was 162 square degrees. One hour's exposure recorded stars to $17\frac{1}{2}$ mag. in the central portion of the plates. There was a loss of 1.8 mag. from the center to the extreme corners, but the loss became serious only near the ends of the plate. The overlap between plates was 1° or more at the side and a few degrees at the ends; the data indicate that, on the average, the searched areas north of −30° declination reached the seventeenth magnitude.

The number of stars recorded per plate was appalling. In the Gemini region of the Milky Way, there were 400,000 star images per plate, while in the Scorpio-Sagittarius region it reached 1,000,000. In the thinnest regions, near the galactic poles, there were 40,000 stars per plate.

A $7\frac{1}{3}$-inch refractor, magnifying 171×, was attached to the 13-inch telescope tube for guiding.

3.2. The Method of Search

Observation was started April 1, 1929, and the Gemini region, in which Lowell's Planet X was suspected, was well past opposition. A series of plates taken there and farther east along the ecliptic was blinked a few months later, during the rainy season. It then became evident that it would be very difficult to distinguish between asteroids near their stationary points and distant planets.

This experience called for strict adherence to photography near the opposition point, where the angular motions of the asteroids would be much larger. The angular distances of the stationary points from opposition are 64° for Jupiter and 36° for Mars. Accordingly, 25° was taken as the upper limit permissible in the search program. In addition, of course, the pairs of plates had to be well matched.

The opposition point moves eastward about 30° per month. Since moonlight made searching impossible, each 30° interval had to be photographed within 2 weeks. The observing program was so planned that most regions were photographed within 15° of the opposition point. In this way, the numerous asteroids were at maximum apparent retrogression, and their short trailed images during each 1-hour exposure revealed their true identity at a glance. On the scale of the plates, they moved, on the average, about 7 mm/day.

One of the advantages of photographing at the opposition point is that all planets exhibit apparent westward motion, in amounts directly related to their distance from the sun. When the Trojan asteroids were encountered, their much shorter trails served to identify them quickly. A rough determination of the distance of any suspected planet could thus be made and proved extremely useful in several instances when regions had to be rephotographed for final check. Other advantages in photographing at the opposition point were as follows:

a) The opposition point is on the meridian at midnight. Long-exposure photographs taken near the meridian suffer least from atmospheric absorption and differential refraction; the latter effect leads to distortion of stellar images near the plate edges. When the search reached the low southern declinations, differential refraction was appreciable during 1 hour's exposure, even when the region was on the meridian. Consequently, the plates of those regions had to be duplicated to within the exact minute in hour angle, in order to match the distorted shapes of the images. Differential refraction also dilutes the images and diminishes the limiting magnitude. At more northern declinations, plates duplicated at hour angles differing by 1 hour were acceptable, if they were taken in the vicinity of the meridian. This permitted some flexibility in the rigorous observing schedule.

b) With optimum observing conditions around midnight, there was enough time to photograph two strips each year and each plate field in triplicate. By taking two belts, one north of the ecliptic and one south, the overlap required in allowing for the travel of hypothetical planets from year to year was kept to a minimum. This, in turn, *made it possible to extend the planet search to within Saturn's orbit with fairly thorough coverage for asteroids whose inclinations did not exceed 15°*. For more distant objects those with even higher inclinations would have been detected.

c) By photographing at the opposition point, all planets exterior to the earth's orbit are illuminated at full phase, and they are, moreover, at their minimum distance from the earth. Both permit the smallest possible bodies to fall within photographic range. However, for a body as distant as Pluto, this advantage amounts to only 0.1 mag.

d) The telescope was free early in the evening and before dawn to follow comets in their most frequently observed positions.

In September, 1929, the photographic work caught up with the opposition point on the ecliptic. Blink examination of the plates was then resumed with greater care and thoroughness.

3.3. Checking "Planet Suspects"

No matter how much care was bestowed on processing and handling the plates, a number of defects could not be avoided. When a possibly spurious image appeared on one plate of a pair within several millimeters from a similar object on the other and the "motion" so defined was westward, this was regarded as constituting a "planet suspect." The suspects which were obviously particles of dirt were dismissed at once. Other bits of foreign material, which did not exhibit a "hard" or sharp outline, were found to exhibit an "imbedded" appearance when the light behind the plate was moved sideways.

A third type of spurious image could not so readily be checked off. These were real silver deposits in the emulsion, having the same softness of outline and the same shape as stellar images. Nearly all of these appeared to be of the sixteenth and seventeenth magnitudes. The 5-inch Cogshall camera, which had been attached to the 13-inch telescope for the purpose of checking brighter objects, could furnish no help here. In such cases the region was rephotographed as soon as possible. However, lack of time did not permit blink examination of more than a fraction of the plates during the dark-of-the-moon period; instead, most of the examination has to be done during the following full-moon period. Consequently, suspects were often 2 or 3 weeks old before they could be rephotographed. This resulted in a greater uncertainty about where to look for the image. The repeat plate frequently contained one or more similar spurious images, so that this means of checking was time-consuming and unsatisfactory. The problem was solved by adopting the procedure of taking at least *three* good plates of each region within the same dark lunation. This procedure was followed throughout the whole search program with very few exceptions.

The experience with faint planet suspects showed the advantage of taking duplicate plates of each region about 2 nights apart. The third plate was therefore taken 2 nights after the second one. This interval on the scale of the 13-inch plates was found optimum for the blink examination. The interval was long enough to permit a detectable shift of a planet at opposition as far out as 400 astronomical units (ten times the distance of Pluto), yet short enough to disregard any pair of images with a separation of more than a few millimeters, the maximum shift for asteroids at the distance of Saturn. This greatly reduced the number of planet suspects because the number of spurious images is proportional to the plate area. As in later years the search progressed to great distances from the ecliptic, the optimum time interval was increased proportional to the secant of the ecliptic latitude.

Taking all three plates within 4 days or, at most, within a week had the further advantage of making it feasible to blink the first and third plates together if they were more perfectly matched than the other pairs. If unfavorable observing conditions interfered with the ideal interval between the first and second plates, this could generally be obtained for the second and third plates.

The possibility of the second image of a suspect being lost in the image of a brighter star was always checked on the extra plate. Thousands of such cases were checked. In the regions of the Milky Way about the galactic center, hundreds of these one-image suspects were found to be very faint variable stars, probably cluster types and eclipsing stars, whose minima were below the threshold of the plate. A great majority of the faint suspects were nothing else than accidental groupings of silver grains in the emulsion of the plates. It is estimated that about 20,000 planet suspects have been checked in the course of the search work over 14 years.

Considerable interest once developed when a fairly promising suspect of the sixteenth magnitude was found. The shift in position indicated a distance at about Uranus' orbit. Immediately more plates were taken, but the object failed to reappear. The original pair of plates had been taken near the fringe area with respect to the opposition point, because of unfavorable observing conditions. If the object was real, it was probably an unusually near asteroid, having a small angular distance between opposition and its stationary point.

Thus it was possible to extend the search with thoroughness to the limit of the plates. Only *negatives* were blinked. Examination of positive copies would have greatly increased the difficulty with defects.

3.4. The Discovery of Pluto

The first blink examination of plates taken at opposition began with those of the constellation of Aquarius in September, 1929. A pair of 14 × 17-inch plates in these non–Milky Way regions could be blinked in 2 days of hard work at the rate of examining 30,000 stars daily. As the search progressed eastward, month by month, through the constellations of Pisces, Aries, and western Taurus, all the plates taken in each dark lunation were examined by the end of the next bright lunation. But in eastern Taurus the number of star images per plate increased to 300,000. This increased the time of blink examination about fourfold. By the time that the examination of the Taurus plates was completed, the Gemini plates had been taken. The two regions of η and 36 Geminorum looked so forbidding, with nearly 400,000 stars per plate, that they were at first laid aside. Thus the author placed a pair of δ Geminorum plates on the blink comparator. Each plate contained 160,000 star images, or 1000 stars per square degree. It so happened that the first plate had been taken on a night—January 21, 1930—when the seeing was poor. The third plate was taken on January 29. The 6-day interval was three times longer than desired.

When one-quarter of the δ Geminorum pair had been examined, Pluto was discovered on February 18, 1930. The images were $3\frac{1}{2}$ mm apart. Since the plates had been taken strictly at opposition, this shift definitely indicated that the object lay beyond the orbit of Neptune. The images certainly looked real; they were nearly in the center of the plates. Their reality was confirmed on the 5-inch Cogshall plates taken concurrently on January 23 and 29, and even on the poor plates of January 21.

Clear skies on February 19 permitted the author to secure a fourth plate of the δ Geminorum region. The image was quickly found in the predicted place. The next night the observatory staff examined the object visually with the 24-inch telescope, to see whether it exhibited a disk. The seeing was quite steady, but no disk could be found. This was a disappointment. Percival Lowell had expected his Planet X to be of the same type as Neptune, which would make it as bright as the thirteenth magnitude and its disk diameter $1''$. Now there was perhaps some suspicion that the main body was still to be found.

A few nights later, Dr. Lampland, using the 42-inch reflector, compared the color photographically with that of Neptune. It confirmed the visual impression that the light of Pluto was yellowish and not bluish like that of Neptune. Pluto was $\frac{2}{3}$ mag. brighter visually than photographically. One-hour exposure plates with the 42-inch reflector by Lampland failed to reveal any satellites; consequently, there was no immediate means of determining Pluto's mass.

E. C. Slipher continued to examine the new planet visually with the 24-inch refractor on nights of very steady seeing, but he could not detect a disk. He set up a test target with illuminated graduated apertures about 1 mile away and observed them with the 24-inch refractor. As a result of these comparisons, he concluded that the new planet, under such faint illumination, could have a disk nearly $0''.5$ in diameter and not be detected by the best observations. Lampland started a long series of short-exposure plates of Pluto with the 42-inch reflector for positions. He secured a plate in every possible month until his death in December, 1951.

The discovery was announced on March 13, 1930. Numerous suggestions for names were received, among which Minerva and Pluto led. It seemed unwise to use Minerva, since one of the asteroids had been named after her. After careful consideration by the Lowell Observatory staff and the trustee, the name of Pluto was proposed and accepted.

After the discovery of Pluto, it was decided that the planet search should be extended on around the ecliptic and over wide areas of the sky (Slipher, 1938).

3.5. Pattern of Sky Coverage

The first year's plates taken in proper season consisted of a single strip, centered along the ecliptic. Thereafter, the search had been streamlined so that two strips around the sky could be observed at one time. They were parallel to the ecliptic strip, one north and one south of it, with sufficient overlap on the ends to prevent undue loss of magnitude and also to cover annual motion of intra-Neptunian objects with some inclination and as close as Saturn. This photographic program could be carried out, but the examination of the plates fell somewhat behind. By September, 1932, a solid belt 30°–35° wide (consisting of three adjacent strips), entirely around the sky, had been blink-examined.

In the third year unusually bad observing weather in the later winter months of 1932 made it necessary to abandon the second southern strip. In succeeding years it was possible again to photograph two strips at a time. These were at greater and greater distances from the ecliptic.

Such a vigorous observing program put a strain on the available observing hours with the telescope when too many nights of a lunation were not fully satisfactory. Often the check plate was taken by "forcing" the conditions. If there were passing patches of haze, the intensity and duration of the absorption was estimated, and the exposure was prolonged, sometimes up to 50–100 per cent. Yet the star images were usually well matched for blinking.

On many nights of excellent transparency, the seeing was bad enough to soften, expand, and dim the star images. Again, satisfactory check plates were obtained by prolonging the exposure. In a few instances blinkable pairs were obtained by duplicating a region on a night with similar seeing.

It was found that *the time required for thorough blink examination of a given area on well-matched plates was roughly proportional to the number of star images when they exceed about 400 per square inch,* but when the number of star images was less, the rate of sky coverage was proportional to the plate area. The reason for this appears to be that a roughly constant number of spurious aggregations of grains attract attention, whether there are any stars on the plate or not.

Figure 1 shows the area of the sky blink-examined by 1945. In the declination zone $-50°$ to $-40°$, differential refraction and poor seeing

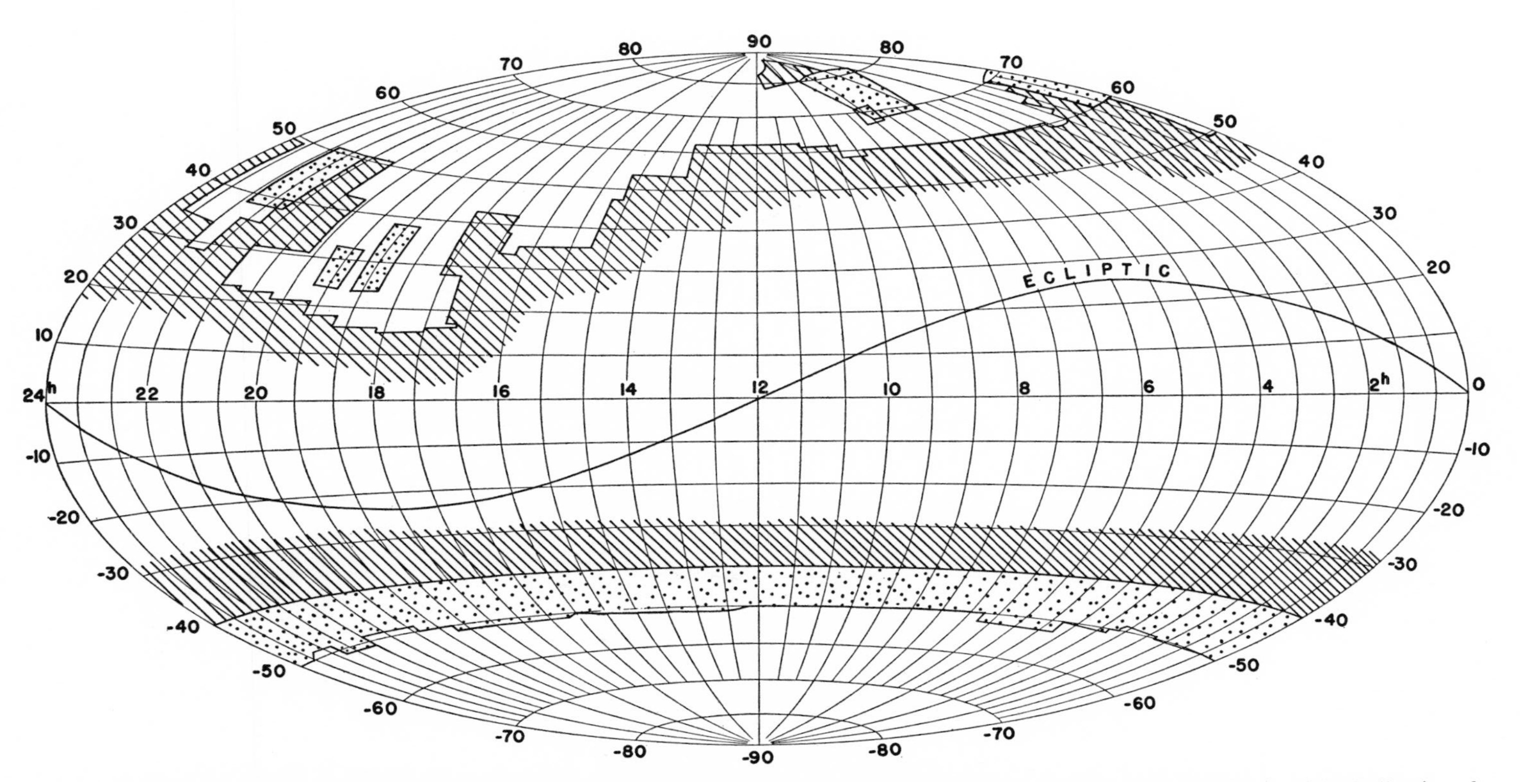

Fig. 1.—Area covered by trans-Neptunian planet search, Lowell Observatory, 1929–1945. *Dark shading:* limiting magnitude 16–17. *Light shading* (mostly between −40° and −50° declination): limiting magnitude 14–15.

caused such a loss in magnitude that the 5-inch Cogshall camera was practically as efficient as the 13-inch telescope. Less than 100 stars per square degree were recorded in the non–Milky Way regions of this zone. As an economy measure, only 14^h–23^h R.A. were photographed here with the 13-inch, this region containing the rich southern Milky Way. The remainder of the zone was photographed on 5-inch Cogshall plates only and reached 14–15 mag.

During the years 1933–1935, F. K. Edmondson carried on the photographic work during the author's absence. Thereafter, the tempo of the planet search was stepped up, aided by two grants from the Penrose Fund of the American Philosophical Society. With two workers, great progress was made in the two years from 1936 to 1938. During this period, H. Giclas made most of the exposures at the telescope, which permitted the author to devote his full time to the blink examination.

3.6. Search Belt to Eighteenth Magnitude

Beginning in November, 1939, the author commenced a series of $2\frac{1}{2}$-hour exposures with the 13-inch telescope along the ecliptic in regions not very crowded with stars. Seven plate regions in a continuous belt from 1^h to 5^h25^m R.A. were examined to the extreme limit of the plates. On one of these was a Selected Area, from which stars as faint as 18.6 mag. were identified. Over 1500 very faint planet suspects of the seventeenth and eighteenth magnitude were checked with good third plates. Four plates of $2\frac{1}{2}$ hours' exposures were made of the Pluto region, centered on η Cancri. A pair was very carefully blinked for possible companion planets, but nothing of promise was found. In May, 1940, a series of 2-hour exposures was made, centered on α and 28 Librae. These also were critically examined to the limit of the plates, probably eighteenth magnitude.

Thus a total of $7\frac{1}{2}$ hours of R.A. along the ecliptic, containing an area of 1530 square degrees, was examined to the eighteenth magnitude. From this sampling, it seems improbable that any zone of asteroids brighter than the eighteenth magnitude exists beyond Saturn.

This series of plates averaged 111 trailed asteroid images per pair. Their long trails aided in their detection. Several were found to vary in light in different portions of their trails.

3.7. Completion of Main Program

By May, 1943, the entire sky visible from Flagstaff, from 50° S. to the North Pole, had been photographed for blink examination, with three or more plates of each region and exposure time of 1 hour. From July, 1943, to August, 1945, no search work was done because of participation in war work. From August to November, 1945, blink examination was resumed on some plates in northern areas. Since then, no further searching has been done. It is hoped that the remaining unexamined photographed areas (see Fig. 1) can be blinked at a future time.

3.8. By-Products of Planet Search

The by-products are outside the scope of this volume and are therefore mentioned only briefly. They include the discovery of a new globular cluster (Lampland and Tombaugh, 1932); several galactic clusters, of which five were published (Tombaugh, 1938, 1941); and a cloud of 1800 galaxies (Tombaugh, 1937). Only one comet was found, of 9–10 mag., which is perhaps not surprising because the plates were taken at opposition. The comet was not reported because the plates were 1 year old when examined. On the other hand, asteroids were numerous, though not so numerous as they would have been if they had been photographed untrailed. For the average daily motion, the trailing caused a loss of about $2\frac{1}{2}$ mag. In all, 3969 asteroid images were marked on the plates, as well as 75 doubtful cases. Approximate positions (to 1′) and magnitudes were published of 744 asteroid images (Lampland and Newman, 1932; Lampland and Giclas, 1935; Lampland and Tombaugh, 1935); among these are 145 new objects, or 39 per cent. This indicates that some 775 new asteroids were recorded in the planet search.

Variable stars were numerous, of course; 1807 of them were marked on the search plates, probably mostly of short period because of the 2-day interval. A few hundred showed light-variations exceeding 1 mag.; the maximum range was 3 magnitudes. These objects were probably eclipsing binaries. A total of 29,548 extragalactic nebulae was counted, with a few thousand large enough to permit classification in Hubble's system.

3.9. General Comments and Statistics of the Survey

3.91. *Area and overlap.*—A total of 338 pairs of 13-inch telescope plates and 24 pairs of 5-inch Cogshall camera plates was examined with the blink comparator. In order to keep the work to a minimum, all unrequired overlap was ruled off. Even then, a total of 45,117 square degrees was blinked, of which 3511 were on 5-inch Cogshall plates of the southernmost strip. Thus over 90,000 square degrees of plate surface were critically scrutinized. This amounted to 75.4 square meters, or 810 square feet.

The total area of the sky is equal to 41,253 square degrees. Three-fourths of the sky, or about 30,000 square degrees, were covered in the planet search. Hence about 15,000 square degrees of examined sky consisted of overlap, to allow for motion of planetary bodies during time intervals between adjacent plate regions.

3.92. *Number of stars.*—The estimated number of stars in the examined areas total 44,675,000 (± a million), or an average of 1000 stars per square degree. Thus the author has examined about 90,000,000 star images of 30,000,000 different stars.

3.93. *Time required for examination.*—The blink examination of a pair of the richest Milky Way plates in the vicinity of the galactic center required 3 weeks of hard, steady work. The great richness in star images made it necessary to use an oblong diaphragm, 20 × 4 mm, in the eyepiece of the microscope. Wider, oblong diaphragms were used on less populous regions. Generally, the blink examination of 30,000–60,000 stars for planetary motion was a good day's work. The author estimates that he sat at the blink comparator 7000 hours, examining search plates and checking planet suspects. Three to 6 hours a day were all that one could blink with efficiency.

3.94. *Uniformity of the search.*—The lack of uniformity in the plate series falls into three categories:

a) Although the great majority of plates were 1-hour exposures, the limiting magnitude varied by 0.5 mag. in some cases because of differences between emulsions; the brand of plates had to be changed three times during the work. For instance, a particular brand and type were used satisfactorily for a few years; later shipments showed lower sensitivity. Several brands were tried. Also some very fast plates had to be rejected because of the larger number of spurious images.

b) Somewhat different exposure times were used. In regions of low ecliptic latitude two dozen or more pairs of plates in non–Milky Way regions were taken with exposures of 75, 90, 100, and 120 minutes, in addition to the special series of long exposures on the ecliptic. These plates reached magnitudes 17–18. No plates were used in the blink examination that had exposures of less than 1 hour.

c) There was bound to be some lack of uniformity from small differences in sky transparency and plate development. But the matching of plates indicates that the variation of the combined effect will not exceed 0.1–0.2 mag. The duplication of plates at the same hour angles eliminates, within a pair, appreciable differences due to atmospheric absorption.

However, the loss in magnitude from atmospheric absorption, seeing, and differential refraction increased rapidly south of −30°. In the zone between −35° and −40° the plates scarcely reach sixteenth magnitude.

In the zone between $-40°$ and $-50°$, the magnitude limit drops from 15 to 14. In Figure 1 the few lightly shaded regions in the north were blinked very hurriedly, but they are 50° or so from the ecliptic. All of the heavily shaded sky area is good to magnitudes 16–17.

3.95. *Dimensions of discoverable planets.*—In all the planet search work, only one new planet was found beyond the orbit of Saturn. No zones of large asteroids were found. No equivalents of the Trojan asteroids around the equilateral points of Saturn, Uranus, or Neptune were detected. However, they would have to be very large to be visible at the distances of Uranus and Neptune.

It is of interest to inquire into the sizes of discernible bodies at the magnitude limit of the planet search. This requires an assumption regarding the photographic albedo of the body.

For a planet with a circular orbit, observed at opposition, the distance from the sun is a and from the earth is $a - 1$. If the radius of the planet is R and its albedo A, the opposition brightness will be proportional to $AR^2/a^2(a - 1)^2$. For constant limiting magnitude and a given albedo, the radius R is therefore related to a by

$$\log R = \log a\,(a - 1) + \text{const.}\,,$$

or, for $a > 2$, very nearly

$$\log R = 2\,\log\,(a - \tfrac{1}{2}) + \text{const.}\,,$$

or, for $a > 10$, roughly

$$\log R = 2\,\log\,a + \text{const.} \tag{1}$$

The approximate equation (1) was used to compute the relations shown in Figure 2. The constant was evaluated from the moon, for which $2R =$ 3476 km, the visual opposition magnitude -12.55, the color index 0.84, and, hence, the photographic opposition magnitude -11.7. The full moon seen from 1 astronomical unit would be $(390)^2$ times, or 12.95 mag., fainter than this, or $+1.25$ mag. At 10 a.u. from the earth and the sun, the moon would appear at 11.25 mag.; at 40 a.u., 17.3 mag. Since a factor of 10 in a corresponds to a 10-mag. difference at opposition, the lines in Figure 2 are readily drawn. For a photographic albedo different from that of the moon, the magnitudes need a corresponding correction, a factor of 10 corresponding to $2\frac{1}{2}$ mag. If only visual albedos are known, the color index needs to be considered also. Figure 2 shows that, e.g., for an eighteenth-magnitude object, having the lunar albedo, the diameter is 160 km at the distance of Saturn, 660 km at the distance of Uranus, 2500 km at the distance of Pluto, and 15,000 km at 100 a.u.

The planet search work was done with such care and thoroughness that the author believes that no unknown distant planets brighter than the sixteenth magnitude exist and that any planet between magnitudes 16 and 17 had a good chance to be discovered.

4. POSSIBLE EXTENSION TO THE TWENTIETH MAGNITUDE

Just as the trans-Neptunian survey had begun, the Schmidt telescope was invented. This instrument offers advantages for planet searches un-

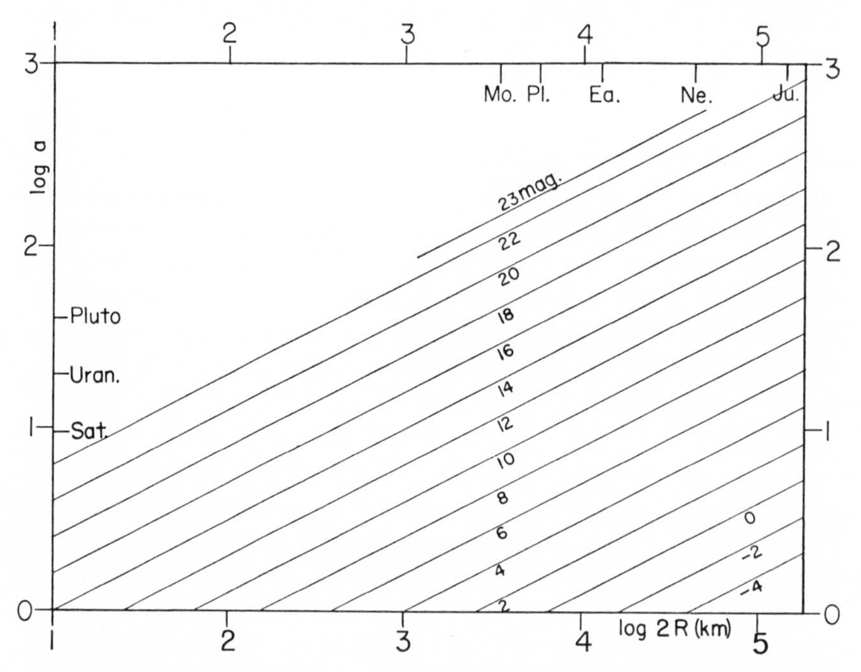

FIG. 2.—Photographic opposition magnitude for a planet of given diameter $2R$ (in km) and distance from the sun a, if the photographic albedo equals that of the moon, 0.05. For a greater albedo the magnitude is correspondingly greater.

matched by photographic refractors. It concentrates the light of the entire photographic spectrum into a small image, which increases the speed appreciably; the aperture can be made large without introducing undue absorption by the optics; and much faster F-ratios can be employed.

An extension of the survey to the twentieth magnitude is quite within reach of the 48-inch Schmidt telescope at Mount Palomar. Such an extension would, at an assumed photographic albedo of $2\frac{1}{2}$ times that of the moon, 0.13, reach the diameter of 40 km at Saturn's distance and 600 km near Pluto. A planet of the size of the earth or larger, if placed on the outer fringe of the planetary system, would be expected to possess a dense atmosphere and a high photographic albedo—say 0.8. This would increase the brightness 3 mag. over the graphs in Figure 2. At the twentieth magnitude it would detect a body having the earth's diameter at 290 a.u., Neptune's diameter at 540 units, and Jupiter's diameter at 950 units.

A new problem would arise, however, because of the enormous numbers of star images on the plates. Particularly in the Milky Way proper, searching beyond the seventeenth magnitude would appear prohibitive; the work of blink examination would increase roughly in proportion to the number of stars. Since the star density in the Milky Way between 17 and 20 mag. would increase 10–20 fold, the work would become truly prohibitive.

It is obvious, then, that a planet search to the twentieth magnitude must be confined to the non–Milky Way regions. At galactic latitude 30°, the number of stars per square degree down to the twentieth magnitude slightly exceeds that for the seventeenth magnitude along the galactic equator. Even this is nearly prohibitive. Half the sky area is within 30° of the galactic equator. At 60° galactic latitude the number of stars per square degree to the twentieth magnitude in yellow light would be less by another factor of 3 or so. No point on the ecliptic is more than about 60° from the galactic equator.

Dark interstellar clouds make it feasible to search nearer the Milky Way on the west sides of the ecliptic intersections than on the east sides. In all, about 220° of the ecliptic could be explored. If the search extended 7° to either side, with 1° overlap, approximately 5000 square degrees of sky would be involved. These would average roughly 5000 stars per square degree. Hence a total of 25,000,000 stars must be blink-examined for planetary motion. About 400 photographs would be required to cover in triplicate the area outlined. Allowing time for resetting the telescope on new regions and reloading, three to six photographs per hour could be taken, depending on the telescope. Approximately 75–140 hours of telescope time would be required. The examination and checking of planet suspects would take much more time. Three man-years would be required to accomplish the search outlined above.

It would be desirable to explore the entire ecliptic circle or the invariable plane to the twentieth magnitude. The only feasible way of doing this is to set up perennial patrol areas in Aquarius and eastern Cancer, which would be photographed each year at opposition season to catch possible objects as they emerge from the Milky Way. Each of the two patrol areas would be two photographs wide, to take care of annual motion for objects more distant than 12 a.u. The patrol area could be three to five photographs high in declination to cover inclinations of 7° and 11° respectively.

Although the area covered by the proposed twentieth-magnitude search is a small fraction of that covered with the 13-inch Lawrence Lowell telescope, the sampling would be large enough to yield very significant conclusions regarding the population and character of the outer portions of the solar system.

Addendum: Searches for small natural earth satellites of the earth and the moon.—The writer, in collaboration with Messrs. J. C. Robinson, B. A. Smith, A. S. Murrell, C. F. Capen, G. E. Thomas, C. Knuckles, and T. C. Bruce, undertook a systematic search for small earth satellites from the Lowell Observatory and other stations in the United States and from Quito, Ecuador, situated on the equator. The period was December, 1953, to October, 1958, during which interval observations were made on 80 per cent of all moonless nights. Both visual and photographic searches were made. The Lowell survey lasted about $2\frac{1}{2}$ years and used an $8\frac{1}{2}$-inch Schmidt camera of 13 inches focal length having a circular field of 13° diameter, as well as two K-24 aerial cameras (3 inches aperture, 7 inches focal length, 30° usable field). These instruments were mounted on an axis that was driven up to 350° per hour to guide on satellites in various distance zones from the earth. The inner boundary of the space effectively covered was 1300 miles or 2100 km above the surface of the earth. In all, 281 distance zones were searched, centered on either the equator or the ecliptic, depending on the distance from the earth. At the Lowell Observatory 5700 Schmidt films were taken, covering 800,000 square degrees, and 7750 K-24 film frames, covering some 7 million square degrees. The average stellar magnitude limit covered by the two instruments for satellite bodies was 14 and 12, respectively.

At Quito two 4-inch-aperture Taylor-Hobson lenses were used, with 6- and 8-inch focus, respectively, which were better suited than the K-24 cameras. These were attached to a 12-inch Fecker telescope, which was driven up to 1600° per hour, and which, in a stationary position, was also used visually, with a wide-angle eyepiece. The site was at 0°6′ south and 78°.44 west longitude, elevation 9300 feet (2800 meters); it was in operation from August, 1956, to October, 1958. About 1000 negatives were obtained with each of these cameras, making a grand total of 15,567 for the two stations.

The results of these surveys have been published elsewhere in full (Tombaugh, Robinson, Smith, and Murrell, 1959). In that publication the effective coverages of the observations were analyzed and represented in compact form in eighteen separate graphs and six tables. In addition, several diagrams and photographs illustrated the geometry involved, the instruments used, and samples of search films. Equations were given for calculating photographic drive rates and visual coverages. Also included were a summary of Quito observing weather and five graphs compiled for quick interpolation of satellite periods, angular velocities, stellar magnitudes, and sizes as functions of distance.

Several fairly promising, but faint, satellite suspects were encountered. Vigorous attempts were made to recover them, but without success. It was concluded that all or nearly all these objects were probably spurious, although some might conceivably have been satellites of high eccentricity or, more likely, very small asteroids passing the earth. The region surrounding the earth is therefore remarkably free of natural satellites. For bodies of the same albedo as the moon (7 per cent), seen at full phase, the detection limit with the 8½-inch Lowell Schmidt telescope, driven for circular motion at various distances, is about 1 foot at 4000 miles, about 10 feet at 40,000 miles, and about 50 feet (15 meters) at the distance of the moon. The present survey does not reach this ideal limit at all distances but is not far removed from it. The details may be seen from the full report.

Searches for *lunar* satellites are best made during total lunar eclipses. This has been done before, at Harvard in 1888 and by Barnard at the Lick Observatory in 1895. The latter found no satellites brighter than the twelfth magnitude. Tombaugh, Smith, and Capen observed the November 18, 1956, eclipse at the Lowell Observatory with the 13-inch Lowell telescope, the 5-inch Cogshall camera, and the 8½-inch Schmidt, mentioned above. This eclipse happened to be an unusually dark one and occurred in a very favorable position in the sky (near the zenith), with the moon near the perigree. Guiding was performed on the moon itself. The limiting magnitudes of the three cameras were 17½, 15½, and 14 mag., respectively. Allowing for the expected orbital motion of any satellites, it was concluded that practically all of lunar satellite space, except for the relatively small sectors in front of and behind the moon, was explored down to the thirteenth magnitude, which means a diameter for possible satellites of 100 feet, if the albedo is assumed to be 7 per cent. Sectors nearer "greatest elongation" were explored to progressively fainter limits for bodies from 100 to 15 feet in diameter, depending on the trailing loss, which decreases toward greatest elongation.

Over 500 lunar satellite suspects were marked; none of these has survived the process of plate-to-plate checking for confirmation.

This search was supported by the Office of Ordnance Research, U.S. Army, and by the Army Ballistic Missile Agency.

REFERENCES

Forbes, G. 1880 *Proc. Roy. Soc. Edinburgh.*, **10**, 429; **11**, 89.

Lampland, C. O., and Giclas, H. L. 1935 *A.N.*, **254**, 305.

Lampland, C. O., and Newman, K. A. 1932 *A.N.*, **247**, 69.

Lampland, C. O., and Tombaugh, C. W. 1932 *A.N.*, **246**, 171.

1935 *Ibid.*, **254**, 175.

Lowell, P. 1915 *Mem. Lowell Obs.*, No. 1.

Pickering, W. H. 1928 *Pop. Astr.*, **36**, 417.

Putnam, R. L., and Slipher, V. M. 1932 *Scient. Monthly*, **34**, 5.

Slipher, V. M. 1938 *Proc. Amer. Phil. Soc.*, **79**, 435.

Todd, D. P. 1880 *Amer. J. Sci.*, **20**, 232.

Tombaugh, C. W. 1937 *Pub. A.S.P.*, **49**, 259.

1938 *Ibid.*, **50**, 171.

1941 *Ibid.*, **53**, 219.

Tombaugh, C. W., Robinson, J. C., Smith, B. A., and Murrell, A. S. 1959 *The Search for Small Natural Earth Satellites* (University Park, N.M.: New Mexico State University).

CHAPTER 3

Orbits and Masses of Planets and Satellites

By DIRK BROUWER
Yale University Observatory

G. M. CLEMENCE
U.S. Nautical Almanac Office

1. INTRODUCTION

THE standard procedure followed in studying the motion of a body in the solar system consists of constructing an ephemeris based upon a gravitational theory and making a comparison between the ephemeris and observations. The problem of determining a first orbit for a newly discovered object is not considered here; for it see Stracke (1929), Crawford (1930), and Herget (1948).

The gravitational theory of a planet gives its heliocentric motion, which is determined by the orbital elements and the adopted masses of the perturbing planets. The discussion of a series of observations may then yield improved values for the elements and masses. The extent to which the discussion of a series of observations will be successful depends primarily on three conditions: the freedom of the observations from systematic errors, the adequacy of the theory on which the ephemeris is based, and the adequacy of the reduction of the observations.

The problem is complicated by the circumstance that the observations are made from an observatory on the surface of the Earth. The calculation of a geocentric ephemeris thus introduces the orbital motion of the Earth about the Sun into the problem. Now the tables of the Sun represent primarily the heliocentric motion of the center of mass of the Earth-Moon system; but, in order to furnish the geocentric position of the Sun, allowance must be made for the *lunar inequality* produced by the motion of the center of the Earth about the center of mass of the Earth-Moon system. This orbit is a reflection in miniature of the motion of the Moon about the Earth. The reduction factor is $\mu/(1 + \mu)$, if μ is the ratio between the Moon's mass and the Earth's mass. The lunar inequality thus introduces into the problem both the Moon's mass and the scale of the Moon's orbit in terms of the scale of the Earth's orbit.

Allowance must further be made for the fact that the observer moves about the axis of rotation of the Earth. This is done by applying a correction for geocentric parallax to the observed position, thereby reducing it to a geocentric position. This correction requires the introduction of the solar parallax and can be made if the distance of the observed body, expressed in astronomical units, is sufficiently well known. In very close approaches to the Earth, as in the case of Eros and similar objects, this may call for special caution; as a rule, the provisional orbit gives the geocentric distance with an adequate number of significant figures for applying the correction for geocentric parallax.

Finally, the positions of planets are always measured with reference to a fundamental star system. Meridian-circle and equatorial-micrometer observations require the reduction to apparent place by the application of precession, nutation, and aberration. In the simpler case of photographic observations it is customary to measure the position of an object in the solar system with reference to mean places of stars. This eliminates the application of nutation, but the so-called planetary aberration still enters into the reduction, though in a different form.

The study of the motions of the planets is therefore intricately entangled with the study of all the Earth's motions. The only way to make progress is by gradually improving the entire system of constants that characterize the solar system.

This is precisely the manner in which the subject has been developed since Newton's time. As the observational accuracy improved, the theories of the motions of the Sun, Moon, and planets had to meet higher requirements, and the methods of reduction of the observations had to be refined. Impressive advances were made during the nineteenth century. F. W. Bessel introduced the method of applying adequate corrections to meridian-circle observations to allow for imperfect adjustments of the instrument, and he contributed the modern method of rigorously applying corrections for precession, nutation, and aberration. These methods have become standard in all important series of meridian-circle observations since the middle of the nineteenth century. The monumental studies of the orbits of the principal planets by Laplace and Leverrier, successively, were important contributions on the theoretical side. Yet, when Simon Newcomb became superintendent of the *American Ephemeris and Nautical Almanac* in 1877, the year of Leverrier's death, a survey of the then existing conditions convinced him of the need to begin a new investigation of the orbits of the principal bodies in the solar system and the construction of a new fundamental system of star positions and proper motions. As Newcomb (1882) pointed out:

> When Leverrier commenced his work, the most striking feature which presented itself was the imperfections of the tables by Lindenau and Bouvard [based on Laplace's planetary theories]. . . . But there was another object, the desirableness of which was not immediately felt, but which must be more and more felt in the not distant future, namely, the attainment of uniformity in adopted astronomical data. So far was Leverrier from aiming at this object in its entirety, that his tables do not, in all cases, embody his final results. The consequence is, that notwithstanding that his work makes a greater epoch in astronomy than any of his immediate successors can hope to make, it does not wholly supply the wants of science in the immediate future.

Thus Newcomb embarked upon a tremendously ambitious program that occupied him beyond his retirement from the Nautical Almanac Office in 1897. The program of work that he mapped out included a discussion of all the observations of value of the positions of the Sun, Moon, planets, and the brighter stars made at the leading observatories of the world since 1750. Concerning the theoretical work to be included, Newcomb (1882, p. xi) wrote:

> The theories of the four inner planets naturally claim the first attention as embodying most of the fundamental elements of astronomy. This branch of the work includes not only the masses of the planets and the elements of the respective orbits, but the constants connected with the rotation of the Earth on its axis, namely, the annual precession, the obliquity of the ecliptic and its secular variation, the position of the equinox among the stars, and, indirectly, the positions of the fundamental stars. To these may be added the solar parallax and the mass of the Moon, as well as a number of quantities connected with those already mentioned.

A summary of the investigations of the motions of the four inner planets was published as a supplement to the *American Ephemeris and Nautical Almanac* for 1897 (Newcomb, 1895*a*). This was the first thoroughly uniform treatment of this problem, leading to a system of fundamental constants of astronomy that was consistent and in agreement with observational evidence.

2. THE SYSTEM OF ASTRONOMICAL CONSTANTS

By the "system of astronomical constants" is meant the constants of astronomy that are required for the interpretation of observations of the positions of celestial objects. For the most part, the constants are those that relate to the size, shape, constitution, and motions of the Earth. They are listed in Table 1. Numerous theoretical relations exist among them, so that when the values of some of them are known, others can be calculated. A certain minimum number of them can be selected, from which all the others can be calculated and which are independent in the sense that none of them are derivable from one another; these are called the *fundamental constants*, and the remaining ones are called *derived constants*. The choice of constants to be called "fundamental" is not a unique one; it is made in such a way as to simplify the theoretical relations with the derived constants as much as possible.

The system of fundamental astronomical constants adopted by the Paris Conference on Fundamental Stars in 1896 agrees in many respects with the values arrived at by Newcomb. However, a few minor inconsistencies were introduced into the compromise that took place at this conference. This not entirely perfect, but otherwise very satisfactory, system of fundamental constants has been in use for the construction of the national ephemerides by the *American Ephemeris* since 1900 and by the British *Nautical Almanac* and the *Connaissance des Temps* since 1901.

An attempt at constructing a thoroughly consistent system of astronomical constants was made by de Sitter (1938, edited and completed by D. Brouwer). Several improvements can now be made in the treatment of the problem.

1. De Sitter used "uniformly accelerated time" as the independent variable in dealing with secular changes in mean motions and related matters. It is now generally agreed that this choice was unfortunate and that the use of "ephemeris time" with an invariable unit is preferable. Clemence (1948*a*) has discussed the modifications that must be made in the treatment of the mean motions. This does not affect the following discussion.

2. One of the fundamental constants used by de Sitter is the dynamical oblateness, $H = (C - A)/C$. This arises in the expression for the precession constant P, according to Newcomb's definition,

$$P = \left(A' + B' \frac{\mu}{1+\mu} \right) H , \qquad (1)$$

in which μ is the ratio between the mass of the Moon and that of the Earth, while A' and B' are constants obtained from the lunar theory, the numerical values of which are known with all the precision required (see

Spencer Jones, **2**, 14). If de Sitter's values are denoted by subscripts zero, and

$$P = P_0(1+\delta)\,, \qquad \mu^{-1} = \mu_0^{-1}(1+z)\,, \qquad H = H_0(1+w)\,,$$

it is found with the values used by de Sitter that

$$1+w = (1+\delta)(1+0.67473\,z-0.211\,z^2)\,, \tag{2}$$

or, sufficient for most purposes,

$$w = \delta + 0.67473\,z\,. \tag{3}$$

The inconvenience of using H as a fundamental constant is that it must be derived from P and μ according to equation (1). The uncertainty in δ (order, 4×10^{-5}) is one decimal place beyond that in z (order 5×10^{-4}). Consequently, if H is treated as a fundamental constant, it is necessary in any adjustment of the data to provide that relation (2), or its approximate equivalent (3), is satisfied within the range permitted by the uncertainty in δ. If P is introduced as a fundamental constant, the symbol w can still be retained in the geodetic relations, provided that it is used merely as a symbol of an intermediary nature. For further comments, see under 4 below.

3. The difference between the observed value of the constant of nutation N and its computed value on the theory of a rigid Earth was the most serious discordance in the system of constants since Jackson (1930*a*) called attention to it. Jeffreys (1948*a*; also **2**, 45) has shown that the fluidity of the core of the Earth reduces the theoretical value of N. As long as the amount of this reduction cannot be accurately computed, the nutation constant must be treated as an independent constant. Whereas formerly both the ratio P/N and the lunar inequality,

$$L = \frac{\mu}{1+\mu}\,\frac{\pi_\odot}{\sin \pi_{☾}}\,,$$

were used for determining the Moon's mass, only the latter is now available.

4. De Sitter used the theory of hydrostatic equilibrium to obtain the relations among the constants of geodesy and especially to derive the oblateness of figure of the Earth from the constant of precession. This procedure has been criticized by Jeffreys (1937, 1941, 1948*a*) on the grounds that there are widespread observed variations in gravity from any formula consistent with the hydrostatic state. On the other hand, Jeffreys (1948*a*) has shown that there is close agreement between the values for the oblateness obtained with and without the assumption of

hydrostatic equilibrium. As was remarked by Spencer Jones (**2**, 18), the assumption of hydrostatic equilibrium may somewhat underestimate the real uncertainty in the determination of the flattening, but de Sitter's discussion suggests that the underestimation is not serious.

In a revision of de Sitter's article it would be desirable to follow Jeffreys' suggestion and use the correction to the oblateness rather than de Sitter's w in the correction factors. This is not the proper place for such a revision; we merely present a list of constants derived from the adoption of a set of fundamental constants resulting from an adjustment of the data that, in our view, corresponds to the best information now available. The derived quantities are based on de Sitter's formulae.

We have not included any information based on the motions of artificial satellites, which have indicated a substantial increase in the reciprocal of

TABLE 1

Revised "Fundamental" Constants

$R_1 = R_{10}(1+u)$	$u = (-0.441 \pm 0.123) \times 10^{-4}$
$g_1 = g_{10}(1+v)$	$v = (-0.138 \pm 0.014) \times 10^{-4}$
$P = P_0(1+\delta)$	$\delta = (+0.721 \pm 0.26) \times 10^{-4}$
$\kappa = \kappa_0 + \chi \times 10^{-3}$	$\chi = (+1.80 \pm 1.20) \times 10^{-4}$
$\lambda_1 = \lambda_{10} + \psi$	$\psi = (-2.40 \pm 1.10) \times 10^{-4}$
$\pi_\odot = \pi_{\odot 0}(1+x)$	$x = (-5.225 \pm 0.67) \times 10^{-4}$
$c = c_0(1+y)$	$y = (-0.033 \pm 0.133) \times 10^{-4}$
$\mu^{-1} = \mu_0^{-1}(1+z)$	$z = (-20.115 \pm 3.5) \times 10^{-4}$

the oblateness. Such information is increasing at so rapid a rate that any discussion of it made in 1961 would shortly be out of date, and it seems better to defer it until later.

The list of fundamental and derived constants is given in Table 1; the subscript zero designates the values in de Sitter's article. Particulars on the derivation of the new values are given in Section 2.1., and the resulting numerical values are given in Table 2. Errors are p.e. here and elsewhere, unless stated otherwise.

In this list R_1 denotes the radius of the Earth at the latitude whose sine is $\frac{1}{3}\sqrt{3}$; g_1 is the acceleration of gravity at this radius; κ is a measure of the depression of the equipotential surface below the surface of the ellipsoid, this depression being expressed by $-\kappa b \sin^2 2\varphi'$, where b is the equatorial radius, φ' is the geocentric latitude, and λ_1 is a constant describing the inner constitution of the Earth (de Sitter, 1924). The adopted values of δ and z correspond to

$$H = H_0(1+w), \quad w = (-12.86 \pm 2.37) \times 10^{-4}.$$

TABLE 2*

FUNDAMENTAL AND DERIVED CONSTANTS

Quantity	Revised Value	Conventional Value	Authority
	Fundamental Constants		
Mean radius of the earth, meters	$R_1 = 6\ 370\ 979$	6 371 269	Int. Ellipsoid
Acceleration of gravity in mean latitude (cm/sec²)	$g_1 = 979.756\ 5 \pm 14$	979.768	Stockholm 1930
Constant of precession (1900.0)	$P = 5\ 493''.553 \pm 0.143$	....	....
Constants depending on the constitution of the Earth	$\kappa = 0.000\ 000\ 68 \pm 12$	$\kappa = 0$	Int. Ellipsoid
	$\lambda_1 = 0.000\ 16 \pm 11$	....	....
Solar parallax	$\pi_\odot = 8''.798\ 4 \pm 6$	8″800	Paris, 1896
Velocity of light (km/sec)	$c = 299\ 773 \pm 4$	299 860	Newcomb
Reciprocal of Moon's mass	$\mu^{-1} = 81.366 \pm 29$	81.53	Hinks
	Derived Constants		
Dynamical flattening, (C−A)/C	$H = 0.003\ 275\ 205 \pm 777$	....	....
$\omega^2 R_1^3 / f M_1$	$\rho_1 = 0.003\ 449\ 83 \pm 4$	....	....
$\frac{3}{2} C / M_1\, b^2$	$q = 0.499\ 809 \pm 121$	....	....
$\frac{3}{2}(C-A)/M_1\, b^2$	$qH = J = 0.001\ 636\ 98 \pm 71$	....	....
Oblateness	$\epsilon = 0.003\ 365\ 50 \pm 72$	....	....
Reciprocal of oblateness	$\epsilon^{-1} = 297.133 \pm 63$	297.0	Int. Ellipsoid
Equatorial radius (meters)	$b = 6\ 378\ 098 \pm 78$	6 378 388	Int. Ellipsoid
Acceleration of gravity on the equator (cm/sec²)	$g_0 = 978.038\ 6 \pm 14$	978.049	Stockholm, 1930
Coefficient of $\sin^2 \varphi$ in gravity	$\beta = 0.005\ 290\ 38 \pm 75$	0.005 288 4	Stockholm, 1930
Coefficient of $\sin^2 2\varphi$ in gravity	$\gamma = -0.000\ 007\ 88 \pm 36$	−0.000 005 9	Stockholm, 1930
Astronomical unit (km)	$149\ 525\ 000 \pm 10\ 000$	....	....
Constant of aberration	$k = 20''.486\ 8 \pm 14$	20″47	Paris, 1896
Light-time	$\tau = 498^s.793 \pm 35$	498ˢ53	Paris, 1896
	$= 0^d.005\ 773\ 06 \pm 40$	....	....
Mass-ratio, Sun/(Earth and Moon)	$m^{-1} = 328\ 427 \pm 67$	329 390	Newcomb
Constant of sin of lunar parallax	$\pi'_{☾} = 3\ 422''.459 \pm 15$	3 422″54	Brown
Parallactic inequality	$P = 125''.050 \pm 9$	125″154	Brown
Constant of lunar inequality	$L = 6''.437\ 8 \pm 23$	6″425	Newcomb
Lunar inequality in Sun's longitude	$L_s = 6''.466\ 8 \pm 23$	6″454	Newcomb
Constant of nutation (observed value)	$N = 9''.207$	9″210	Newcomb

* The computed value of the constant of nutation for a rigid Earth is $N = 9''.224\ 5$.

TABLE 2—*Continued*

Quantity	Revised Value	Conventional Value	Authority
	Centennial Precessions per Tropical Century at 1900.0		
Lunisolar precession	$p_0=5\,039''.740$	$5\,037''.08$	Newcomb
Geodesic precession	$p_g=1''.917$	....	....
Planetary precession × cos Θ	$\lambda \cos \Theta=11''.392$	$11''.44$	Newcomb
General precession in longitude	$p=5\,026''.431$	$5\,025''.64$	Newcomb
	$m=4\,609''.240$	$4\,608''.51$	Newcomb
	$n=2\,004''.980$	$2\,004''.68$	Newcomb
$\Theta_{1900}=23°27'8''.26$ (Newcomb)	$\cos \Theta=0.917\,391\,8$	$\sin \Theta=0.397\,985\,4$	....

2.1. Comments on the Adopted Values of the Fundamental Constants

The geodetic constants, R_1 and g_1, are derived from the discussion by Jeffreys (1948*a*). The oblateness derived from the adopted values of the fundamental constants, especially P and μ^{-1}, is

$$\epsilon = 0.003\,365\,50 \;.$$

Following Jeffreys (1948*a*), we define

$$e' = 10^7 \times (\epsilon - 0.003\,352\,3) = +132 \;.$$

If this equation is used in Jeffreys' combination of data with the longitude terms omitted (1948*a*, p. 244), with the standard error 31, the solution becomes

$$a' = +98 \pm 78 \,, \qquad e' = +135 \pm 19 \;.$$

Then the equatorial radius in kilometers is found to be

$$b = 6\,378\,000 + a' \,, \qquad b = 6\,378\,098 \pm 78 \,,$$

to which the adopted mean radius corresponds. It will be noted that the oblateness derived on the basis of the hydrostatic theory is extremely close to the result of the combined solution without the longitude terms. The difference from Jeffreys' results is due to the difference in the adopted mass of the Moon.

The precession constant P is identical with the value adopted by Spencer Jones (**2**, p. 16). The resulting general precession in longitude agrees very nearly with the value obtained by Morgan and Oort (1951). Rabe's (1950) masses for the inner planets are used for the calculation of the planetary precession; the resulting general precession in longitude exceeds Newcomb's value used in star catalogues by 0ʺ.791 per tropical century.

The values of κ and λ_1 adopted are according to calculations by Bullard (1948), based on the density distribution in the Earth's interior by Bullen (1940, 1942, and **2**, p. 96).

For the solar parallax we have adopted the value derived by Rabe (1950). It depends on the gravitational determination of the mass of the Earth-Moon system in terms of the Sun's mass. This method appears to be less likely to be affected by possible systematic errors than the direct geometrical determination. Several recent compilations (Jones, 1941; Clemence, 1948*a*; Jeffreys, 1948*a*), published before Rabe's result appeared, used the value $\pi_\odot = 8''.790$ which Spencer Jones obtained by the geometrical method from the Eros observations of 1930–1931. It is therefore of some additional interest to present the results based on Rabe's value. We have changed Rabe's result from $8''.798\ 35 \pm 0''.000\ 39$ (p.e.) to $8''.798\ 4 \pm 0''.000\ 6$ (p.e.). The increase in probable error is arbitrary.

For the velocity of light we use the value $c = 299\ 773 \pm 4$ (p.e.) obtained by Dorsey (1944) from a discussion of all determinations. Since then, measurements with the use of short-wave radio frequencies have given substantially larger results which indicate $c = 299\ 793 \pm 1$ (p.e.). We prefer to retain the older value until the systematic difference between these results is explained. Only the constant of aberration and the light-time are affected.

The mass of the Moon was obtained from $L = 6''.437\ 8 \pm 0''.002\ 2$ (p.e.). The numerical value is identical with that obtained by Jeffreys (1942), a revision of the discussion by Spencer Jones (1941). It agrees well with a weighted mean of the various determinations of L. The probable error was derived from the discordance among them, which exceeds the internal probable errors of the individual determinations. The value of μ^{-1} that corresponds to our adopted value of L depends, of course, on the adopted value of the solar parallax.

The mass of the Moon enters so importantly into the system of constants that a few further comments on the problem of its determination are called for. From the observations of Eros at the opposition in 1931, Spencer Jones (1941) had derived for the lunar inequality $L = 6''.439\ 0 \pm 0''.001\ 5$ (p.e.). A revision of the discussion by Jeffreys (1942) led to the result $L = 6''.437\ 8 \pm 0''.001\ 7$ (s.e.). Later, Delano (1950) made a solution from the same data by a different method and obtained $L = 6''.442\ 9 \pm 0''.001\ 5$ (p.e.).

The procedure adopted by Spencer Jones involved the application of an empirical curve or formula to correct the ephemeris at times when the lunar equation vanishes. Thus the errors in the geocentric path of Eros and the lack of homogeneity of the star places are allowed for in a single empirical correction. Delano remarks that the observed deviations from the definitive ephemeris exceed the uncertainty in the star positions, and he proceeds to make an improvement of the orbit in an effort to reduce the residuals. He applies no further empirical corrections.

Regardless of the relative merits of the two procedures, it is important to note that the difference between the two results is three or four times the probable errors of the determinations. A list of all the available determinations shows that the discordance among the results is considerably greater than the internal uncertainties indicate. The available determina-

Determination	L	Weight
1. Newcomb, Sun, 1895	$6''456 \pm 0''012$	1
2. Gill, Victoria, 1897	6.414 ± 0.009	3
3. Hinks, Eros, 1909	6.4305 ± 0.0031	10
4. Morgan and Scott, Sun, 1939	6.450 ± 0.010	2
5*A*. Spencer Jones, Eros, revised 1941	6.4378 ± 0.0012	40
5*B*. Delano, Eros, 1950	6.4429 ± 0.0015	30

tions with the probable errors and assigned weights are given in the accompanying table. Three solutions were made:

(i) The result 5*A* was used with weight 40; with the solution

$$L = 6''436\,0 \pm 0''002\,3 \text{ (p.e.)}; \qquad \text{p.e. (weight 1)} = 0''017\,3 .$$

(ii) The result 5*B* was used with weight 30; yielding the solution

$$L = 6''438\,9 \pm 0''003\,0 \text{ (p.e.)}; \qquad \text{p.e. (weight 1)} = 0''020\,2 .$$

(iii) Both 5*A* and 5*B* were used, each with half the weight given:

$$L = 6''437\,3 \pm 0''002\,4 \text{ (p.e.)}; \qquad \text{p.e. (weight 1)} = 0''017\,2 .$$

Experiments with different weights indicated that any reasonable assignment of weights will yield a final result not far different from the adopted value. The adopted probable error also appears justified.

Brouwer and Ashbrook (1951) have suggested the use of the minor planet 619 Triberga for obtaining the mass of the Moon. The sidereal period of this planet is 4.001 1 years, and three times its synodic period is 3.999 6 years = 49.47 synodic months. With an interval of 4 years, the planet returns to the same field of the sky, with opposite phase of the lunar equation. The evaluation of the lunar equation will depend on the measurement of a maximum displacement of the order of 9″, a much smaller amount than in the case of a favorable opposition of Eros. Numerous observations and an extremely accurate orbit will be required, in order to make a significant contribution to the problem of determining the Moon's mass.

2.2. Unit of Distance

The astronomical unit of distance is derived from the adopted units of mass and time, which are the mass of the Sun and, until recently, the mean solar day. Owing to the varying length of the mean solar day, it is preferable to substitute the *ephemeris day*, which is equal to the average value of the mean solar day over the last 3 centuries (**2**, 37–38) and has in fact, under the name of *mean solar day*, often been used in practice. If m is the ratio of the mass of any planet to the mass of the Sun, n the observed angular mean motion of the planet expressed in radians per day, and k the Gaussian constant of gravitation, being 0.017 202 098 95 exactly (cf. Sec. 4.2), then a in the equation

$$n^2 a^3 = k^2 (1 + m) \tag{4}$$

is expressed in astronomical units. Equation (4) may be regarded as the *definition* of the astronomical unit. In dimensional analysis of equations in celestial mechanics it is necessary to assign to k^2 the dimensions $L^3T^{-2}M^{-1}$.

It should be remembered that the astronomical unit is not identical with the mean distance of the Earth from the Sun. For the Earth, equation (4) yields $a = 1.000\,000\,03$ a.u., and this is modified by the presence of disturbing planets, so that an ellipse having a semimajor axis of 1.000 000 2 a.u. gives a still better approximation to the actual orbit of the Earth than the value derived from equation (4). As a matter of historical interest it may be noted that the adopted value of k results from Gauss's use of a value for the mass of the Earth that is now known to be about 7 per cent too small.

It might have seemed logical to define the astronomical unit in such a way that equation (4) applied to the Earth would yield $a = 1$ precisely; this procedure would have required changing k each time that it was desired to introduce an improved value of m or n for the Earth, and a change in k would have required the value of a for all other planets and comets to be changed, which would have led to confusion. It is better to proceed in the way that has actually been followed, keeping k fixed and thus allowing a to be determined for any object independently of assumed values of m and n for the Earth.

In order to obtain the value of the astronomical unit in meters, it is necessary to know the *solar parallax*, which is defined as the angle subtended by the Earth's equatorial radius at a distance of 1 a.u., and the value of the equatorial radius in meters; the result is given in Table 2.

3. GENERAL CHARACTER OF PLANETARY MOTIONS

By the analysis of observational material alone, without recourse to gravitational theory, Kepler was able to infer that the planets move in ellipses around the Sun, with the Sun at one focus. The precision of observations has increased so much since Kepler's day that the departures from elliptic motion are now easily observed, and, if it is attempted to represent the actual motion of a planet by means of an ellipse whose elements are continuously changing, it can be seen that the changes are related to the positions of other planets in their orbits. If it were possible to extend the observations over some thousands of years, it might even be inferred, still without recourse to theory, that the changes in the elements are not exclusively periodic but are progressive as well. It is, however, doubtful whether, with the material presently at our disposal, consisting of 10 000–20 000 observations of each planet from Mercury to Saturn, inclusive, extending over 200 years, it would be possible to represent the motions by means of purely formal expressions that would suffice to predict future positions with much precision or for more than a very short interval of time; it would be found that purely formal representations of the motions would take very different forms at different epochs, and there is no assurance that any process of successive approximations would converge to a definite final form, unless the time interval covered by actual observations is supposed to increase without limit.

The position is greatly altered by the introduction of Newton's law of gravitation, as slightly modified by general relativity. It becomes possible, at least in principle, to construct mathematical curves that give the positions of the planets in three-dimensional inertial space as functions of the time and to fit these curves to existing observations, whence they may be used to calculate the configuration of the solar system at any past or future time. The theoretical limitations on the process are of two kinds. In the first place, the observations are not perfectly precise, so that it is not possible to calculate the actual curves of motion but only approximations to them that necessarily depart further and further from the actual motions at more and more remote times; however, good estimates of the probable magnitude of such departures can be made. In the second place, the motions are certainly not strictly in accordance with the adopted law of gravitation. Although observations have not yet shown any certain discrepancies, it is clear that the presence of interplanetary matter, for example, must have some effect. And the consequences of general relativity have not yet been fully understood for any system of moving bodies containing more than one finite mass. We are, however, as yet very far from having to consider these theoretical limitations. The effective limitations on celestial mechanics are of a more practical kind, arising from the difficulty of solving the differential equations of motion derived from Newton's law of gravitation.

If x, y, and z are rectangular co-ordinates of a planet referred to the Sun as origin and to axes having directions fixed in an inertial frame and if t is the time reckoned from an arbitrary fundamental epoch and m the mass of the planet in units of the Sun's mass, the equations of motion according to the inverse-square law are (cf. Moulton, 1914, p. 271)

$$\frac{d^2x}{dt^2}+\mu\frac{x}{r^3}=\frac{\partial R}{\partial x}, \qquad \frac{d^2y}{dt^2}+\mu\frac{y}{r^3}=\frac{\partial R}{\partial y}, \qquad \frac{d^2z}{dt^2}+\mu\frac{z}{r^3}=\frac{\partial R}{\partial z}, \tag{5}$$

where

$$\mu=k^2(1+m), \qquad r^2=x^2+y^2+z^2,$$

$$R=\Sigma k^2 m'\left\{[(x'-x)^2+(y'-y)^2+(z'-z)^2]^{-1/2}-\frac{x'x+y'y+z'z}{r'^3}\right\},$$

where the primed quantities refer to any disturbing planet, the sign of summation indicates summation over all disturbing planets, and k is the Gaussian constant.

Under general relativity the rigorous equations of motion, if expressed in rectangular co-ordinates having the conventional Euclidean properties in the strict sense, take an extremely complicated form. To a degree of precision considerably greater than that of observation, they may, however, be expressed as follows:

$$\begin{aligned}
\frac{d^2x}{dt^2}+\mu\frac{x}{r^3}(1-A)-B\frac{dx}{dt}&=\frac{\partial R}{\partial x},\\
\frac{d^2y}{dt^2}+\mu\frac{y}{r^3}(1-A)-B\frac{dy}{dt}&=\frac{\partial R}{\partial y},\\
\frac{d^2z}{dt^2}+\mu\frac{z}{r^3}(1-A)-B\frac{dz}{dt}&=\frac{\partial R}{\partial z},
\end{aligned} \tag{6}$$

where

$$A=\frac{2\mu}{c^2 r}-\frac{2v^2}{c^2}+\frac{2[x(dx/dt)+y(dy/dt)+z(dz/dt)]^2}{c^2r^2}+\frac{[x(dx/dt)+y(dy/dt)+z(dz/dt)]^2}{r^2[c^2-(2\mu/r)]},$$

$$B = \frac{2\mu [x (dx/dt) + y (dy/dt) + z (dz/dt)]}{r^3 [c^2 - (2\mu/r)]},$$

$$v^2 = \left(\frac{dx}{dt}\right)^2 + \left(\frac{dy}{dt}\right)^2 + \left(\frac{dz}{dt}\right)^2,$$

and c is the velocity of light.

It can be shown that if in the non-relativistic equations (5) the right-hand members are put equal to zero, the solution is a non-rotating ellipse, having the Sun at one focus. It can also be shown (Eddington, 1923; Chazy, 1928) that if in the relativistic equations (6) the right-hand members are put equal to zero, the solution is the ellipse of the former case, which is, however, not fixed but is rotating in its own plane in the direction of the planet's motion; the amount of rotation during one revolution of the planet is

$$\frac{12\pi^2 a^2}{c^2 T^2 (1 - e^2)}, \tag{7}$$

where a is the semimajor axis of the ellipse, T is the period of revolution of the planet, and e is the eccentricity of the ellipse.

It follows that the solution of equations (6) may be obtained by first solving equations (5) and then adding a correction given by equation (7). Since this correction is very small, this is the method adopted in practice. The correction (7) is of practical importance only for Mercury, Earth, and Mars (Clemence, 1947) but, with the lapse of many centuries, will become appreciable for the outer planets as well, since the angular velocity of the apse is very nearly proportional to $a^{-5/2}$.

The presence of the small factor m', which is always less than 0.001, on the right side of equations (5) renders it possible to solve them by successive approximations. In the first approximation R may be calculated on the assumption that all the planets move in exact ellipses; in the second approximation the results of the first are used to obtain a better value of R; and so on. Probably three approximations always suffice to give results as accurate as observations, and, excepting the mutual action of Jupiter and Saturn, two are generally adequate. The methods used for the solution are of two general kinds—numerical and analytic.

3.1. Numerical Solutions

Equations (5) may be solved by using any of the well-known methods for the solution of ordinary differential equations by step-by-step numerical integration. Until the advent of automatic electronic calculators the applications were limited mainly to the orbits of comets and minor planets over a few revolutions; in these cases it was possible to assume that the co-ordinates of all disturbing planets were known in advance, since the masses of comets and minor planets exert no observable influence on the motions of the planets, and hence there are only three simultaneous equations to be solved at each step. For the mutual perturbations of two planets it is necessary to solve six equations, three for each planet.

The only perfectly general application of the method thus far made to the principal planets is that by Eckert, Brouwer, and Clemence (1951), where the mutual actions of the five outer planets were rigorously calculated, giving fifteen simultaneous equations of the second order to be solved at every step and resulting in the rectangular heliocentric co-ordinates of the five outer planets at 40-day intervals from 1653 to 2060. This work involved more than five million multiplications and divisions and seven million additions and subtractions of large numbers, usually to fourteen decimal places.

Numerical integrations of the equations of motion are affected by errors originating in the dropped decimal places, which set an important limitation on the practical applications. It has been shown by Brouwer (1937) that, whereas the space-curve traced by any object can be determined by numerical integration with an error proportional to the square root of the number of steps of integration, the position of the object along the curve is affected by an error proportional to the three-halves power of the number of steps. Thus, in practice, the greater the number of intended steps of integration, the greater the number of significant figures that must be used in order to prevent the errors from exceeding a preassigned quantity; if the ratio of the number of steps in two different integrations is 4.64, then, in order to reach the same precision, one more significant figure is required in the more extended calculation. Another disadvantage of numerical integration is that, in order to calculate the position of an object at a remote time, it is necessary to calculate first the positions for all intervening times.

3.2. Analytic Solutions

In the analytic solutions of equations (5), the disturbing function R and its derivatives are expanded into infinite trigonometric series which are integrated term by term. If we consider only two planets, then in the first approximation R may be expressed as a function of the elements of two ellipses and the time, and we may use as elements a and a'; e and e'; J, the mutual inclination; θ, the longitude of the node of one orbit on the other; ω and ω', the angular distances of the perihelia from this node: and two more angular variables that give the positions of the planets in their elliptic orbits as functions of the time, say, the mean anomalies g and g'. If n and n' are the mean motions in a unit of time and c and c' are the values of the mean anomalies at the epoch, then $g = nt + c$ and $g' = n't + c'$. With this choice of elements and if $\alpha = a/a'$, the general form of R and its derivatives is

$$\frac{1}{a'}\Sigma F_{i,j,k,l,m}(\alpha, e, e', J)\cos(ig + jg' + k\omega + l\omega' + m\theta), \qquad (8)$$

where i, j, k, l, and m are integers, positive, negative, or zero, and for any particular values of them F may be expressed as a fourfold power series in α, e, e', and $\sin J$, which can be evaluated numerically with all the precision desired. The development in powers of α is not made explicitly but with the aid of the well-known Laplacian coefficients. A perfectly general solution would require that ω, ω', and θ be retained as variables throughout the course of the work, but these angles have periods that are of the order of 10^5 times as long as the periods of g and g'. Until now, no practicable way of obtaining a solution has been found when all five angular arguments are allowed to appear in literal form, owing to the very slow convergence of the series. Fortunately, it is not necessary for any practical purpose to have analytic expressions for the precise co-ordinates of the planets that are valid for more than a few hundred years. We are therefore permitted in the first approximation to give the three angular variables of long period the constant numerical values that they have at the epoch, and then formula (8) may be reduced to the form

$$\Sigma C_{i,j}(\alpha, e, e', J)\cos(ig + jg') + \Sigma S_{i,j}(\alpha, e, e', J)\sin(ig + jg'). \qquad (9)$$

The effect of this simplification is to produce some terms that, upon integration, take the form

$$C_i t \cos ig + S_i t \sin ig$$

in the first approximation, and, in the second approximation,

$$C_i t^2 \cos ig + S_i t^2 \sin ig .$$

These are known as *secular terms*, and they are equivalent to a representation of the periodic terms of very long period as power series in the time. The results can be valid only for such values of the time as render the series convergent, which may be of the order of 10^3 years, if the developments are carried through the third approximation.

Leaving aside the secular terms, the integration of terms of the form (9) with respect to the time introduces $in + jn'$ as a divisor of the coefficients C and S. When i and j have the same sign, the coefficients are reduced in size by the integration; but when they have opposite signs, the coefficients may be greatly increased. Such cases occur when n and n' are nearly in the ratio of two small integers, in other words, when two planets have motions that are nearly in resonance. In the second approximation, combinations of the motions of three planets appear, giving rise upon integration to the divisors $in + jn' + kn''$, which also may have small values, producing terms in the co-ordinates having large coefficients and long periods. An important property of expression (9) is that the lowest power of e or e' occurring in a coefficient is $|i + j|$; although values of i and j can be found that will make $in + jn'$ as small as we please, large values of i and j are associated with high powers of e and e', so that no practical difficulties seem to occur.

Nothing is certainly known about the mathematical convergence of the expressions for the co-ordinates obtained by integration of expressions of the form (8) or (9). With the methods so far used in studying the subject, the question is made to depend on the properties of the number n/n', which, being necessarily determined by observation, cannot be established exactly (Clemence, 1951).

Attempts at perfectly general analytic solutions of the equations of motion have been made by Poincaré (1892–1899), Gyldén (1893, 1908), Hill (1897, 1905, 1907), and Brouwer (1955). It is as yet too early to say whether success can be attained.

For the principal planets the most extensive applications of the classical planetary theory in which the form (9) is used have been made by Leverrier, Hill, and Newcomb; but as yet there is no single example in which the motion of a planet has been calculated as accurately as it has been observed. It remains to be seen whether automatic electronic calculating machines can be successfully applied to these problems and whether the difficulties of arranging the work to be carried out automatically are too great or whether some entirely new method that better utilizes the peculiar capabilities of the machines can be developed.

3.3. Perturbations of Long Period

The most celebrated perturbation of long period and the first to receive a dynamical explanation by Laplace and others is the great inequality in the motion of Jupiter and Saturn (Brown and Shook, 1933, p. 205). If we use the initial letters of the planets to denote their mean anomalies, the argument of the great inequality is $5S - 2J$, with a period of about 900 years; it displaces Saturn in longitude as much as $0\overset{\circ}{.}8$ from its elliptic position, and Jupiter 0.4 of this amount. The associated terms with arguments $10S - 4J$ and $15S - 6J$ are of appreciable magnitude, although the last is factored by the ninth power of the eccentricities (Hill, 1891).

Of comparable importance is the great inequality in the motion of Uranus and Neptune (Newcomb, 1874, p. 55), having the argument $2N - 1U$ and period more than 4000 years. This term, depending on the first power of the eccentricities, gives considerable difficulty in the development of the mutual perturbations.

In the motion of Mars there is an inequality with a period of 1800 years (Newcomb, 1895*b*, p. 73), having the argument $3J - 8M + 4E$ and a coefficient of about a minute of arc. The term in the longitude of the Earth having the same argument has a coefficient one-tenth as large, and in the motion of Jupiter it is entirely negligible.

Other appreciable terms of long period, of interest because they depend on high powers of the eccentricities, are those with arguments $13E - 8V$ and $15M - 8E$ (Newcomb, 1898).

3.4. Secular Perturbations

Although we have been forced in practice to develop the secular perturbations in powers of the time when we aim at the highest precision, we can nevertheless learn something of a more general nature about planetary motions by limiting consideration to the secular perturbations alone, that is, by limiting the disturbing function R to those terms of expression (8) where i and j are both zero. In this way it becomes possible to obtain a first approximation to a solution for eight of the principal planets simultaneously, in which the eccentricities, perihelia, inclinations, and nodes are expressed as sums of terms that are strictly periodic functions of the time. It has not yet been possible to include Pluto, the difficulty being that the orbits of Neptune and Pluto intersect if the perihelia and nodes are allowed to vary without restriction. The most recent application of the method is by Brouwer and van Woerkom (1950), who have refined it to include the principal effect of the great inequality between Jupiter and Saturn. The relatively minor effect of this refinement, together with the smallness of the neglected mass of Pluto, gives some ground for belief that the mathematical expressions have some physical significance for at least several million years, although they may have no validity at times as remote as a thousand million years.

Letting e be the eccentricity, π the longitude of perihelion measured from a fixed direction in inertial space, γ the inclination to the invariable plane of the solar system, and θ the longitude of the ascending node on the invariable plane, measured from a fixed direction, then for any planet j the results can be expressed in the form

$$e_j \sin \pi_j = \Sigma M_{i,j} \sin (a_i t + c_i), \quad e_j \cos \pi_j = \Sigma M_{i,j} \cos (a_i t + c_i),$$
$$\sin \gamma_j \sin \theta_j = \Sigma N_{i,j} \sin (b_i t + d_i), \quad \sin \gamma_j \cos \theta_j = \Sigma N_{i,j} \cos (b_i t + d_i), \tag{10}$$

where M, N, a, b, c, and d are constants and the summation extends over seven values of i for the inclinations and nodes and ten for the eccentricities and perihelia, two of which were introduced by the great inequality. The periods of the arguments range from 25 000 to 2 000 000 years and the M and N from nearly 0.2 to negligible values. For any value of i it turns out that a is nearly equal to $-b$, so that a single contribution to the motion causes the node to regress at nearly the same rate as the perihelion advances, but, because of the different sizes of the M and N, the net motions are quite different in some cases. Assuming the validity of expressions (10), one can fix maximum values for e and $\sin \gamma$, which are simply the sums of the M_i and the N_i without regard to sign. In case one M or N is greater in absolute magnitude than the sum of the others without regard to sign, then the excess is the minimum value of e or $\sin \gamma$; but when this is not the case, no minimum value can be assigned; e or $\sin \gamma$ may then become indefinitely small, and at such times π or θ will pass through a semicircumference with great speed. It may be noted, as an example, that 530 000 years ago the eccentricity of the Earth's orbit became so small that it is not known whether the perihelion advanced or regressed through half a circumference. In cases where no minimum value can be assigned to e or γ, no mean period can be assigned to π or θ. Table 3 gives the maximum and minimum values of e and $\sin \gamma$ and the mean periods of π and θ, in cases where they exist.

Figure 1 shows the Earth's eccentricity as the radius and the longitude of perihelion as the position angle for the next million years, with the perihelion referred to the fixed equinox of 1950.0. The points marked on the curve are at intervals of 10 000 years, every tenth one being numbered so that 0 corresponds to 1900 and number 10 to a million years later. Figures 2, 3, and 4 are entirely similar, but for Jupiter the interval shown is 650 000 years instead of a million, and for Uranus the shortest segments of the curve represent 50 000 years instead of 10 000. All four figures are based on special calculations by van Woerkom, using the data of Brouwer and van Woerkom (1950). It can be seen that the perihelia of the Earth and Jupiter do not always advance but occasionally regress for considerable periods of time, during which the curves are convex toward the origin; the same is true for Venus and Neptune. The motions of the nodes

TABLE 3*

SECULAR ELEMENTS OF THE PLANETS

PLANET	e Max.	e Min.	PERIOD OF π (MILLENNIA)	SIN γ Max.	SIN γ Min.	PERIOD OF θ (MILLENNIA)
Mercury.......	0.241	0.109	220	0.170	0.079	250
Venus.........	.074	....	....	.059	....	....
Earth.........	.067	....	....	.051	....	....
Mars..........	.141	.004	72	.102	....	....
Jupiter........	.062	.027	300	.008	.004	50
Saturn........	.086	.012	47	.018	.014	50
Uranus........	.067	....	....	.019	.016	450
Neptune.......	0.013	0.005	2000	0.014	0.010	1900

* e is the eccentricity, π the longitude of perihelion, γ the inclination to the invariable plane, θ the longitude of the node on the invariable plane.

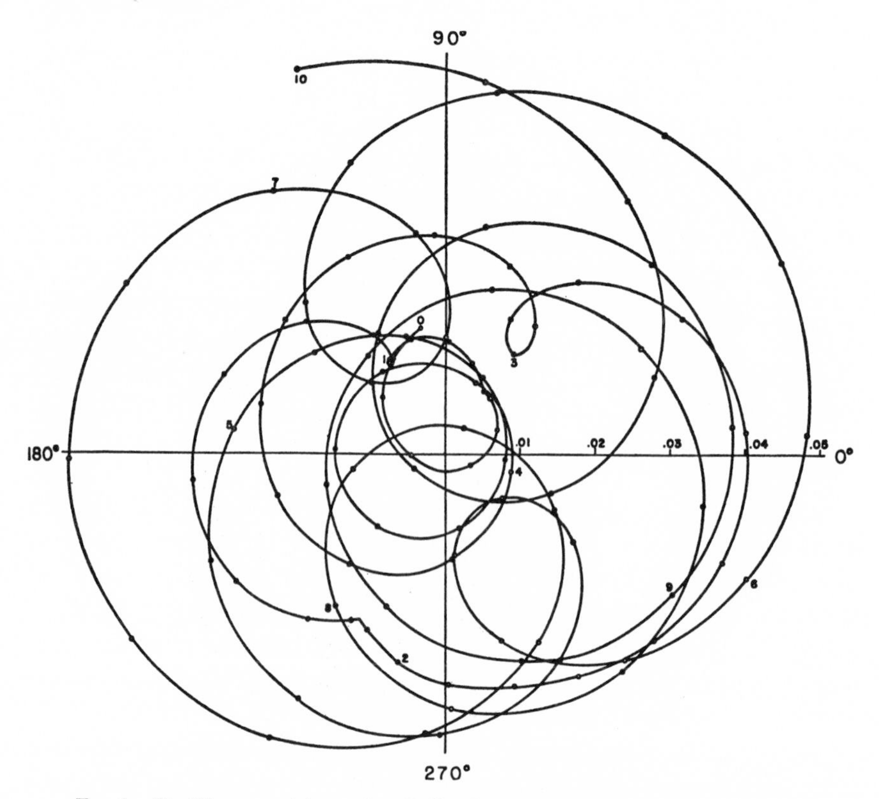

FIG. 1.—Earth's eccentricity and perihelion for 1 000 000 years, starting at 1900

on the invariable plane exhibit more uniformity; the diagram for the Earth resembles all the others to a considerable degree.

The character of the changes in the elements here considered depends very largely on the masses of the planets; in some instances small changes in the masses would produce very considerable modifications in the curves. This fact indicates nothing about the stability of the solar system, because we are dealing only with parameters that partially describe the motions of the planets and not with the motions themselves; we might equally well have chosen different parameters having entirely different properties. With the choice we have made, the fifth parameter required for a complete description of the space curve traced by a planet is the major axis $2a$. A well known theorem states that, with the time as independent variable, there are no secular perturbations of $1/a$ in the first and second approximations to the perturbations.

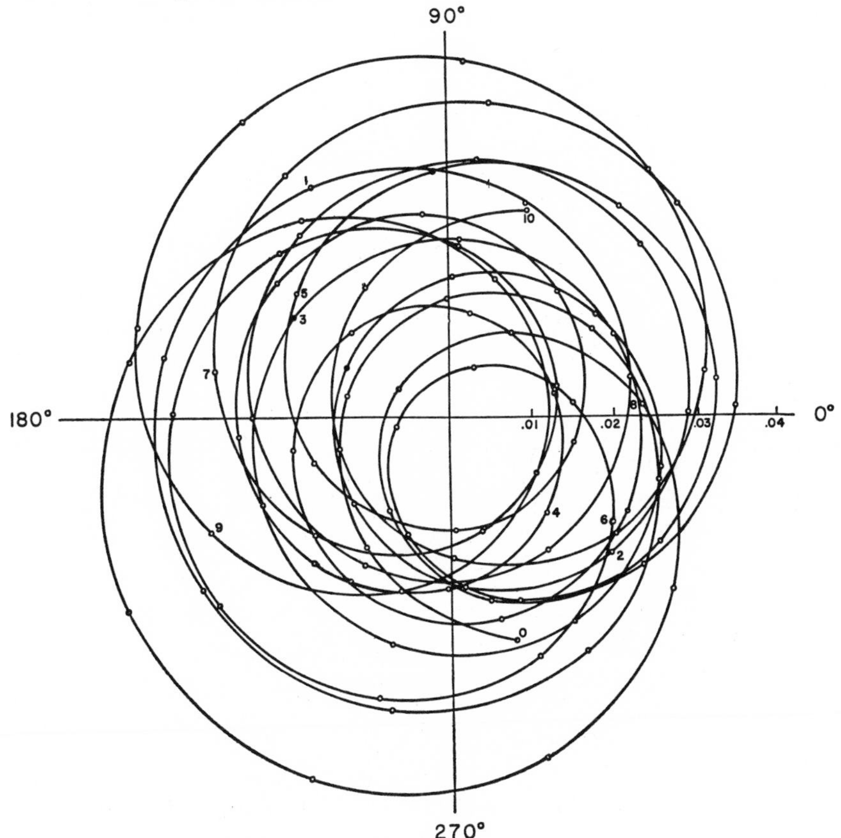

FIG. 2.—Earth's inclination and longitude of node referred to the invariable plane for 1 000 000 years, starting at 1900.

Evidently, the question whether the solar system is stable depends on the precise meaning attached to stability, but under any ordinary meaning of the word considerable interest attaches to the secular perturbations of the major axes. Laplace was the first to notice that when the perturbations of the elements a, e, π, γ, and θ are developed in the classical form, where the secular perturbations are represented by allowing powers of the time to appear as factors of the coefficients of periodic terms, there are no secular perturbations of a when only the first power of the disturbing masses is considered. Lagrange (1776) furnished a rigorous proof, and Poisson (1809) proved a famous theorem since known as *the theorem on the invariability of the major axes*, which asserts that no secular perturbations of a exist when the *second* power of the disturbing masses is included. Haretu (1878) showed that secular perturbations of the third order are, however, present, but Eginitis (1889) concluded that these terms, like the secular terms of the first order in the other elements, are representations of truly periodic terms of very long period. Nothing is known about terms of higher order than the third.

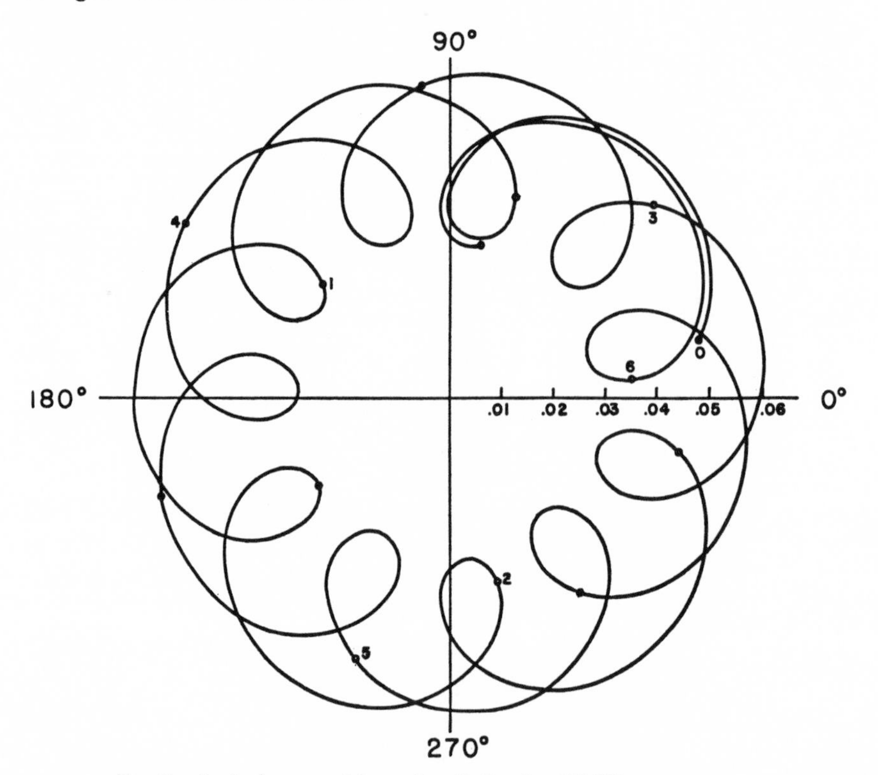

Fig. 3.—Jupiter's eccentricity and perihelion for 650 000 years

The importance of these results with respect to stability has been much exaggerated. The separation of the perturbations of the co-ordinates into perturbations of a, e, π, γ, and θ is an artificial one, convenient for calculation and description but having no particular physical significance (Brown and Shook, 1933). All that can be said definitely is that, with this choice of parameters and with the highly artificial methods of development that have been used, the element a is free from certain secular perturbations that affect the other elements. If, on the other hand, we choose to deal with the co-ordinates themselves, then, using similar methods of development, we find that the longitude, latitude, and radius vector all contain secular perturbations of all orders. We must have a more general form of planetary theory than any yet devised before anything definite can be proved about the stability of the solar system, except perhaps in a very restricted sense.

In the meantime it may be remarked that the measure of time used in the study of planetary motions is furnished by those motions themselves, the second being defined as 1/31 556 925.974 7 of the tropical year for 1900

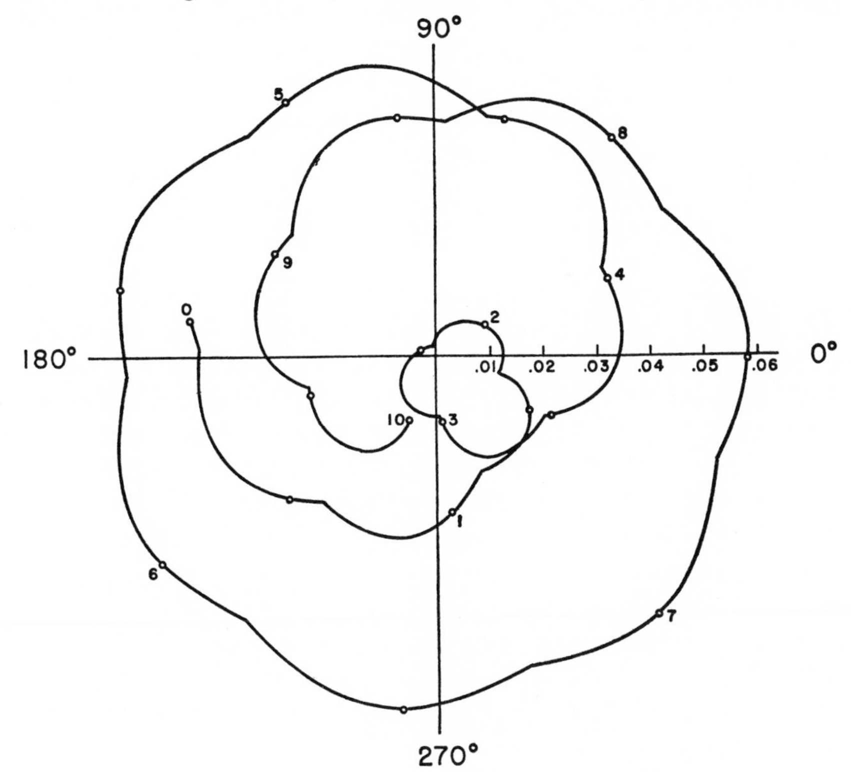

FIG. 4.—Uranus' eccentricity and perihelion for 1 000 000 years

January 0. It is possible to conceive of other measures of time (Clemence, 1948*b*, 1952) accelerated with respect to astronomical time, and in such measures of time the solar system would be indefinitely expanding or shrinking. But whether such time scales exist in nature is not yet known.

4. DETERMINATION OF PLANETARY MASSES

4.1. Determination of Planetary Masses from Planetary Motions

The mass of a planet can be determined from its action on other planets, including minor planets, and, in cases where no satellite is available, no other method is available in general. Comets are not suitable; the changes in the physical structure of a comet are associated with unpredictable changes in its motion, which observationally become entangled with the perturbations produced by the planets.

The calculated perturbations of one planet by another are proportional in amplitude to the mass of the disturbing planet. The discordances between observed and calculated positions of the disturbed planet can be correlated with the phase of the calculated perturbations and a correction to the calculated amplitude deduced, whence the correction to the mass used in the calculated perturbations follows immediately. It is necessary in all cases to include the mass correction as one of the unknowns in a least-squares solution, which also includes the elements of the orbit and occasionally also other unknowns, such as the elements of the Earth's orbit. And, of course, it is necessary that the mass sought be large enough to produce observable displacements in the position of the disturbed planet.

The *secular* perturbations of the four inner planets, as well as those of certain minor planets, can also be used to obtain information on some of the planetary masses. The discordance between the observed and calculated centennial variations of the eccentricity, perihelion, inclination, and node may be used as the right-hand members of equations of conditions, in which corrections to the disturbing masses appear as unknowns, the coefficients being proportional to the contributions of the several planets to the perturbation concerned. For the perihelia and nodes, the correction to the constant of precession must also be introduced as an additional unknown. The four inner planets give rise to sixteen equations of condition which may be solved by least squares. The principle of the method has been utilized by Leverrier, by Newcomb, and by others. The latest application is by Brouwer (1950), who introduced new refinements but was limited to the motions of Mercury and the Earth, modern data for Venus and Mars not yet being available.

4.2. Determination of Planetary Masses from the Scale of Satellite Orbits

If one assumes for a moment that the disturbing effects may be ignored, Kepler's third law would apply to a satellite's motion and yield

$$a^3 n^2 = k^2 (m_p + m_s) .$$

The Gaussian constant k has been defined (Sec. 2.2) in such a manner that a body with negligible mass revolving about the Sun in an orbit with semimajor axis equal to 1 a.u. would have the mean daily motion k,

$$\begin{aligned} k &= 3548''.187\,607 \\ &= 0^\circ.985\,607\,668\,6 \\ &= 0^{\text{rad}}.017\,202\,098\,95 . \end{aligned}$$

Thus, if a is expressed in astronomical units, there results

$$\frac{m_p + m_s}{M_\odot} = a^3 \left(\frac{n}{k}\right)^2 . \tag{11}$$

The observed sidereal motion, n_{obs}, is defined as the coefficient of t in the linear function of the time that results if the true longitude derived from the observations is freed from all the periodic terms and the precession. The constant a_{obs} is obtained by subtracting from the radius vector the elliptic terms and the periodic perturbations. Both n_{obs} and a_{obs} so obtained contain contributions due to the presence of disturbing forces. Allowance is made for these effects by modifying the formula expressing Kepler's third law to

$$\frac{m_p + m_s}{M_\odot} = a_{\text{obs}}^3 \left(\frac{n_{\text{obs}}}{k}\right)^2 \left(1 - \frac{\sigma}{2}\right), \tag{12}$$

in which, in the notation of Section 5.1,

$$\sigma = \frac{2JR^2}{a^2} + \frac{KR^4}{a^4} - \left(\frac{n'}{n}\right)^2 - \sum \frac{m_{s'}}{m_p}\, a^2 \frac{\partial A_0}{\partial a} . \tag{13}$$

The symbol A_0 stands for the constant term of $(a^2 + a_{s'}^2 - 2aa_{s'} \cos \psi)^{-1/2}$ developed as a Fourier series in ψ. The formula given corresponds to that obtained by H. Struve (1888, p. 116) or by Tisserand (1896, p. 18), and was used earlier by Laplace in the theory of Jupiter's Galilean satellites. It is assumed that σ is a small quantity, so that its square can be ignored.

A simple verification of the formula is obtained for the case in which the perturbations are to be considered due to the oblateness of the primary alone. Let the satellite's orbital plane coincide with the equatorial plane of the primary. The equations of motion then are

$$\frac{d^2x}{dt^2} = -\frac{\mu x}{r^3}\left(1 + \frac{JR^2}{r^2} + \frac{1}{2}\frac{KR^4}{r^4}\right),$$

$$\frac{d^2y}{dt^2} = -\frac{\mu y}{r^3}\left(1 + \frac{JR^2}{r^2} + \frac{1}{2}\frac{KR^4}{r^4}\right).$$

These equations are satisfied by the circular solution

$$r = a_0\,, \qquad x = a_0 \cos(n_0 t + \text{const.})\,, \qquad y = a_0 \sin(n_0 t + \text{const.})\,,$$

provided that

$$a_0^3 n_0^2 = \mu\left(1 + \frac{JR^2}{a_0^2} + \frac{1}{2}\frac{KR^4}{a_0^4}\right). \tag{14}$$

This is seen to be a special case of the general result, equations (12) and (13).

Kepler's third law with its appropriate correction factor is thus available for evaluating the masses in terms of the Sun's mass of all planets possessing satellites. The only exception is the mass of the Earth-Moon system, since the Moon's mean distance in astronomical units cannot be obtained from observations of the Moon alone. For all other satellites the necessary information concerning their orbits is derived from measurements of the position of the satellite relative to the center of the disk of the primary. The older measurements were made visually with a micrometer attached to an equatorial telescope or with a heliometer. Measurements on photographic plates have replaced the visual measurements to a large extent. Reference is made to chapter 18 for the limits now attainable with larger telescopes.

Whatever observing method may be used, the planetary mass ultimately depends on evaluation of the scale value, i.e., the number of seconds of arc that corresponds to 1 mm measured on the photographic plate or the value of one revolution of the micrometer screw of the filar micrometer or heliometer. Any error in the adopted scale value produces a systematic error in the deduced mass,

$$\frac{\Delta m}{m} = -\frac{3\Delta s}{s}.$$

A further important source of systematic errors may be present if a satellite is measured directly with reference to an extended planetary disk. Bessel (1831) referred the positions of Titan with the Königsberg heliometer to two opposite points of the open ring or to two opposite points of the planet's disk. A similar method has been used in the older micrometer measurements. H. Struve at Pulkovo introduced in 1884 the *relative* micrometer measurements of Saturn's satellites. No settings are made on the planet's disk or on the ring. Gill used the same method with the Cape heliometer on Jupiter's Galilean satellites a few years later. Similarly, in the many series of photographic plates used by de Sitter (1931) for Jupiter's satellites, settings on the satellite images only, not on the planet, are made. The disadvantage of having to solve simultaneously for the unknowns pertaining to the orbits of at least two satellites is more than canceled by the greater freedom from systematic errors and the appreciable reduction in the accidental errors. Of course, the method can be used only if at least two satellites not too different in magnitude are available.

4.3. Masses of the Planets

4.31. *Mercury.*—The conventionally adopted mass of Mercury, obtained by Newcomb (1895*a*, p. 175) from a general adjustment of quantities affecting the motions of the four inner planets, is

$$m^{-1} = 6\ 000\ 000\ .$$

Of the various separate determinations discussed by Newcomb (1895*a*, pp. 104–106), the only one that can be considered free from systematic error is his own, from Mercury's periodic perturbations on Venus, which is

$$m = \frac{1 \pm 0.32}{7\ 210\ 000}.$$

This result is now superseded by that of Duncombe (not yet published), who has rediscussed Newcomb's material, with the addition of all meridian observations of Venus made since, up to 1948, and finds

$$m = \frac{1 \pm 0.077}{5\ 880\ 000}.$$

Only two other determinations made since Newcomb's time are of any importance. Brouwer (1950, p. 15) finds, from the secular perturbations of Mercury and the Earth, a result which, taken together with Rabe's (1950) value of the mass of the Earth, gives

$$m = \frac{1 \pm 0.055}{6\ 480\ 000}.$$

Rabe (1950), from the perturbations of Eros by Mercury, finds

$$m^{-1} = 6\ 120\ 000 \pm 43\ 000\ ,$$

or nearly

$$m = \frac{1 \pm 0.007}{6\ 120\ 000}. \tag{15}$$

In spite of the disparity among the probable errors, the agreement among these determinations inspires more confidence than any one of them would deserve by itself. The only practicable way of increasing the weight of Rabe's result very much for many years to come would be by including the observations of Eros previous to 1926 in the determination.

4.32 *Venus.*—Newcomb's conventionally adopted value is

$$m^{-1} = 408\ 000\ .$$

As in the case of Mercury, the most precise single determination is that of Rabe (1950), who finds

$$m^{-1} = 408\ 645 \pm 208\ .$$

Three other determinations made since Newcomb's time may be regarded as free from systematic error. Morgan and Scott (1939) obtain, from periodic perturbations of the Earth by Venus in 1901–1936.

$$m^{-1} = 407\ 000 \pm 500\ ;$$

we have derived the probable error formally from the discordances between their four separate results, using their weights; probably the actual uncertainty is somewhat larger.

Clemence (1943), from periodic perturbations of Mercury by Venus, obtains

$$m^{-1} = 409\ 300 \pm 1\ 400\ .$$

Brouwer's (1950) result, obtained in the same way as the one quoted for Mercury, is

$$m^{-1} = 408\ 000 \pm 800\ .$$

The weighted mean of these four results, doubling the probable error of the 1939 result, is

$$m^{-1} = 408\ 600 \pm 200\ . \qquad (16)$$

Clemence and Scott (1942) have shown that the mass of Venus cannot be satisfactorily determined from its action on Mars until the theory of the motion of Mars is revised.

4.33. *The Earth-Moon system.*—The mass of the Earth cannot be determined from the distance and period of the Moon, because the distance of the Moon in astronomical units cannot be measured directly; direct measures of the lunar parallax give a result that can be expressed in astronomical units only if the mass of the Earth is already known. Neither can the mass be determined free from systematic error from the periodic perturbations of Venus and Mars by the Earth, because these perturbations are strongly correlated with the shape of the visible disks of the planets, which produce systematic errors in the observations. Something can be done with the secular perturbations, but the most precise determination appears to be from the perturbations of Eros by the Earth; the close approaches of Eros to the Earth produce very considerable perturbations and also permit very precise observations of them. Rabe's (1950) result is, for the Earth-Moon system,

$$m^{-1} = 328\,452 \pm 43\ .$$

Noteboom (1921), using observations of Eros in 1893–1914, all earlier than those used by Rabe, has obtained

$$m^{-1} = 328\,370 \pm 102\ .$$

Witt (1933), using observations of 1893–1931, obtained

$$m^{-1} = 328\,390 \pm 103\ .$$

The accordance of these determinations is remarkable, although the third is not independent of the other two, being based on partly overlapping material.

As already quoted, we are using

$$m^{-1} = 328\,427 \pm 67\ ,$$

which corresponds to our adopted value of the solar parallax $\pi_\odot = 8''.798\,4$ by the relation (de Sitter, 1938)

$$\pi_\odot^3 = 2.236\,92 \times 10^8 \frac{m}{1+m}\,. \tag{17}$$

The constant contains the factor $(1+\mu)^{-1}$ and thus depends on the adopted value of the Moon's mass in terms of the Earth's mass, for which we use $\mu^{-1} = 81.366 \pm 0.029$. Nevertheless, the constant in equation (17) is known with considerably greater precision than either $\pi_\odot$ or m; hence the formula may be used to determine either $\pi_\odot$ or m if the other is known.

Previous to the determinations cited, it was generally thought that $\pi_\odot$ could be measured directly with greater precision than that of the determination from m. In our judgment the position is now reversed; it is difficult to imagine any source of systematic error in the quoted determinations of m or to find any other reason for doubting the formal probable errors. Hence we regard the three values given above as summing up our present knowledge of the subject. It must be said, however, that our experience with all the other constants of astronomy has shown hardly a single instance where formal probable errors are to be believed. As a general rule, we have found that they must be doubled, in some cases trebled, in order to bring determinations of the same constant by independent methods into satisfactory agreement. Therefore, it should not be hastily assumed that the final word has been said on the mass of the Earth.

4.34. *Mars and the minor planets.*—The two satellites of Mars have been assiduously observed at the United States Naval Observatory in Washington, beginning with the observations by the discoverer, Asaph Hall, in 1877. The satellites have also been observed at Pulkovo, the Lick Observatory, the Yerkes Observatory, and recently at the McDonald Observatory (chap. 18). The measurements of the two satellites give accordant results. The value obtained by Hall (1878),

$$m^{-1} = 3\ 093\ 500\ ,$$

was adopted by Newcomb for the tables of the four inner planets and has been generally accepted.

Van den Bosch (1927), from a discussion of 27 determinations from the satellites during the years 1877–1909, obtains

$$m^{-1} = 3\ 088\ 000 \pm 5\ 000\ ,$$

which has been confirmed by Clemence, who finds

$$m^{-1} = 3\ 088\ 000 \pm 3\ 000$$

from an unpublished discussion including more recent observational material than was available to van den Bosch.

The determination of the mass of Mars from its action on other planets cannot compete in accuracy with that from the scale of the satellite orbits. This statement also applies to the result obtained by Rabe (1950) from his discussion of the observations of Eros.

The effect of the minor planets as a whole is to produce a small contribution to the motion of the perihelion of Mars, and a few attempts have been made to determine the combined mass of the minor planets by attributing the excess of the observed motion over the calculated one to this cause. The results so far obtained have been entirely spurious, owing to important defects in the theory of the motion of Mars. The future will show whether the probable error of a result so obtained is any smaller than the result itself; probably the most that can be hoped for is to determine an upper limit for the mass of the minor planets by this method.

4.35 *Jupiter.*—A compilation by Newcomb (1895*b*, p. 398) of various determinations gives for the Jupiter system

$$m^{-1} = 1\ 047.35 \pm 0.045\ .$$

This is obtained from the accompanying table.

		Wt.
All observations of satellites...........	1 047.82	1
Observations of comets...............	1 047.23	11
Observations of planets...............	1 047.38	32

De Sitter (1908, p. 718) rejects the older observations of the satellites and instead uses results of Gill's and Cookson's heliometer observations:

		Wt.
Heliometer observations of satellites....	1 047.42	20

Rejecting also the value derived from observations of comets, he takes, as the final result,

$$m^{-1} = 1\ 047.40 \pm 0.03\ .$$

Clemence, from Newcomb's material with revised weights, obtains

$$m^{-1} = 1\ 047.39 \pm 0.03\ . \qquad (18)$$

Hill (1873) remarked that thirteen minor planets in the region of the 2:1 resonance with Jupiter have principal Jupiter perturbations in longitude with coefficients between 1°4 and 9° and periods between 60 and 120 years and should therefore be suitable for an accurate evaluation of Jupiter's mass. Many of these planets have now been observed through an entire period of these perturbations. As a result of Clemence's initiative, they are now being observed at the United States Naval Observatory; the old observations have been collected and are being reduced to a uniform system. Clemence intends to undertake the calculation of orbits of the necessary precision with the aid of electronic calculators.

It has frequently been suggested that the faint outer satellites of Jupiter —VIII, IX, XI, and XII—might yield a mass determination more accurate than by any other method. De Sitter (1931) emphasizes the difficulty of such an undertaking. It appears, however, that the method is more promising now than it was a quarter of a century ago. A first requirement would be the calculation of ephemerides of high precision; rapid calculators are now available to perform the necessary numerical integrations. The observations of these satellites have been made almost exclusively with reflectors having small fields and reduced with star positions from the Astrographic Catalogues. The positions relative to Jupiter have then been obtained by using the planet's ephemeris positions. The method may be improved by making relative measurements of the satellites on plates taken with powerful wide-angle astrographs. Since the maximum angular distances of the satellites from the planet are approximately 2°, a field 5°

square would be sufficient. The project would be one of considerable magnitude.

4.36. *Saturn.*—The conventionally adopted value of the mass of Saturn is that obtained by Bessel (1833) from observations of Titan:

$$m^{-1} = 3\ 501.6 \pm 0.8 \ .$$

Convincing evidence has accumulated to indicate that this result is affected by a systematic error. The determinations from the scale of the satellite orbits made since Bessel's time have almost invariably indicated a smaller value for the reciprocal of the mass. A recent rediscussion by Jeffreys (1954) of the results obtained by G. Struve (1924–1937) yields

$$m^{-1} = 3\ 494.8 \pm 1.3 \ , \tag{19}$$

in which the internal probable error has been increased to allow for the uncertainty in the calibration of the micrometer screw. A similar value had been obtained earlier by H. Struve (1888).

Since Hill (1895) obtained for the mass of Saturn from Saturn's perturbations of Jupiter a result that confirmed Bessel's value,

$$m^{-1} = 3\ 502.2 \pm 0.53 \ ,$$

there has existed a contradiction between the mass of Saturn derived from the scale of the satellite orbits and that derived from the planet's action on Jupiter. This has recently been reduced by Hertz (1953), who finds, from a discussion similar to that by Hill,

$$m^{-1} = 3\ 497.64 \pm 0.27 \ . \tag{20}$$

Hertz used observations that have accumulated since Hill made his discussion. They are generally of higher systematic accuracy than the older observations available to Hill. Moreover, Hertz was able to use the numerical integration of the motions of the five outer planets. On several counts, therefore, Hertz's determination is entitled to greater confidence than Hill's earlier result. It would, nevertheless, be interesting to rediscuss the material used by Hill and to combine all the observations since 1750 into a single solution. Since much of the weight would be contributed by the great inequality, the precision of the determination should be considerably increased.

A new determination of the mass of Saturn from the satellites is in progress, based on a series of plates taken by Alden with the Yale telescope in the Southern Hemisphere. The plates were remeasured by F. H. Hollander, who gave particular attention to determining the scale value from star positions measured on the same plates. The reductions are now being carried out by M. S. Davis at Yale University.

4.37. *Uranus.*—The discussion by van den Bosch (1927) shows that appreciable systematic errors are present in the results from the individual satellites, as appears from the weighted means for m^{-1}:

Oberon	22 685 ± 21	Umbriel	22 919 ± 172
Titania	22 831 ± 17	Ariel	23 073 ± 97

Harris (1950) has discussed a series of photographic observations with the 82-inch reflector of the McDonald Observatory. The planet was not measured, and the scale was determined from reference stars. For the reciprocal of the mass, Harris finds

$$m^{-1} = 22\ 934 \pm 6\ . \tag{21}$$

This result is, strictly speaking, provisional, since it was obtained with a provisional value for Miranda, but the ensuing uncertainty cannot affect the result by more than one ten-thousandth of its amount.

Hill (1898) finds, from the perturbations by Uranus in the motion of Saturn,

$$m^{-1} = 23\ 239 \pm 89\ .$$

The slowness of the motion of Uranus renders the result vulnerable to the effects of systematic errors that vary with the time and also to imperfections in the theory of Saturn. The weight could be increased considerably by inclusion of observations since 1890 and use of the numerical integration orbit, similar to Hertz's evaluation of the mass of Saturn. It appears, however, that this method cannot compete with the photographic method from the scale of the satellite orbits.

4.38. *Neptune.*—The older value of the mass of Neptune,

$$m^{-1} = 19\ 700\ ,$$

was obtained by Newcomb (1874, p. 173) from the perturbations by Neptune on Uranus. The observations of Uranus from 1690 to 1872 were used for the purpose. This value of Neptune's mass was used by Hill in his theory of Jupiter and Saturn and by Newcomb in his second theory of Uranus. However, at the end of the work, Newcomb increased the mass to

$$m^{-1} = 19\ 314\ ,$$

a result that he had obtained from a discussion of the observations of Uranus from 1781 to 1896 and from observations of Triton. Gaillot (1910) gives

$$m^{-1} = 19\ 094 \pm 22\ ,$$

obtained from observations of Uranus from 1690 to 1903. The probable error was obtained by van den Bosch (1927, p. 94).

A discussion of the observations of Triton by Eichelberger and Newton (1926) yields the following values:

Visual observations.	19 176 ±25
Photographic observations.	19 655 ±36

The same remarks apply as with the mass of Uranus: results by the perturbation method are liable to be affected by the presence of slowly changing systematic errors, especially in the older meridian-circle observations. The large difference between the scale of the satellite orbit from visual and photographic observations is alarming. Perhaps the photographic observations are more seriously affected by systematic effects, since no effective precautions were taken to overcome the effect of a magnitude error due to the difference of about 5 mag. between Neptune and Triton.

Alden (1940, 1943) made a series of photographic observations of the satellite with the Yale refractor, using an objective grating giving a difference of 5 mag. between central image and first-order diffraction images. Thus the first-order diffraction images of the planet and the central image of the satellite are nearly equalized. These measurements are now being reduced at the Yale Observatory for a determination of the mass of the planet and for correcting the elements of the orbit of the satellite.

Van Biesbroeck (1951) found, from the orbit of Nereid, the provisional result $m^{-1} = 18\ 730$ and remarked that in a few years it would be possible to obtain both period and major axis more accurately and hence a reliable mass of the planet. Later (Van Biesbroeck, 1957) the more definitive value of

$$m^{-1} = 18\ 889 \pm 62 \tag{22}$$

was obtained.

4.39. *Pluto.*—The perturbations on Uranus and Neptune by Pluto furnish the only means of determining its mass, and those of Uranus are so small that they contribute little weight to the result. The conventionally adopted value is

$$m^{-1} = 360\ 000\ .$$

Brouwer (Eckert, Brouwer, and Clemence, 1951, p. xxiii) has shown that the determination from the motion of Neptune depends strongly on two prediscovery observations of Neptune in 1795 and on the latitude of Neptune, which is considerably perturbed by Pluto on account of the considerable motion of Pluto in latitude since the discovery of Neptune;

it is not possible to satisfy these observational data with a value of the mass of Pluto that is very different from the conventional one.

The determination from Uranus was somewhat strengthened recently by Brouwer (1955) by working in prediscovery observations of this planet. He finds, from a solution based on the longitude and latitude residuals of both Uranus and Neptune,

$$m^{-1} = 400\,000\,(1 \pm 0.10)$$

but remarks that the latitude residuals of Uranus show an unexplained systematic trend that cannot be removed by an adjustment of either the mass of Pluto or the orbital elements of Uranus. Thus the significance of the very satisfactory representation of the residuals in latitude of Neptune becomes doubtful.

A solution based on the longitudes alone gives

$$m^{-1} = 450\,000\,(1 \pm 0.20)\,.$$

Much of the weight depends on the observations by Lalande in 1795. Moreover, as Jackson (1930*b*) has shown, a wide range in the elements of Neptune is possible without requiring the presence of improbably large systematic errors in the observations since 1851. Thus the formal probable error is likely to be considerably smaller than the true uncertainty. The conclusion is that at present no reliable gravitationally determined mass of Pluto is available.

5. SATELLITE SYSTEMS

The relative motion of a satellite about its primary may, in a first approximation, be represented by an elliptic orbit. The motion would continue in an elliptic orbit with constant elements if the only force acting on a satellite were an attraction to the center of the primary varying inversely as the square of the distance. For some satellites in the solar system this is such a good approximation that a fixed ellipse may serve for a number of revolutions without producing observable deviations from the actual motion. For other satellites a fixed ellipse can represent the motion for only a small fraction of a revolution. All depends on the magnitude of the disturbing forces in comparison with the attraction by the primary. Four different perturbing causes may be distinguished: (A) The oblateness of the primary and, for Saturn's satellites, the attraction by the rings. In the Moon's motion there are also minute observable effects due to the lack of sphericity of the Moon. All other satellites may be treated as point masses. (B) The attraction by the Sun. (C) The attractions by other satellites revolving about the same planet. (D) The attractions by other planets in the solar system. These are important only in the case of the Moon's motion.

5.1. The Disturbing functions

Let x, y and z be the rectangular co-ordinates of a satellite in a co-ordinate system having its origin at the center of the primary. If the perturbing forces are ignored, the equations of motion are

$$\frac{d^2x}{dt^2}=\frac{\partial F_0}{\partial x}; \qquad \frac{d^2y}{dt^2}=\frac{\partial F_0}{\partial y}; \qquad \frac{d^2z}{dt^2}=\frac{\partial F_0}{\partial z}; \tag{23}$$

if

$$F_0=\frac{\mu}{r}; \qquad r^2=x^2+y^2+z^2; \qquad \mu=k^2(m_p+m_s),$$

where m_p = mass of primary, m_s = mass of satellite, and k = Gaussian constant.

The oblateness of the primary requires the addition to the force function, F_0, of the disturbing function,

$$R_A=\frac{\mu}{r}\left[-\frac{2}{3}\frac{JR^2}{r^2}P_2(\sin\delta)+\frac{4}{15}\frac{KR^4}{r^4}P_4(\sin\delta)+\ldots\right]. \tag{24}$$

The notation is essentially the same as in **2**, 8. To avoid confusion with a, the semimajor axis of the orbit of the satellite, the planet's equatorial radius is represented here by R; P_2 and P_4 are Legendre polynomials; δ is the angle between the radius vector and the equatorial plane of the primary. The following Legendre polynomials will be needed in this section:

$$P_1(x)=+x, \qquad P_2(x)=-\tfrac{1}{2}+\tfrac{3}{2}x^2,$$

$$P_3(x)=-\tfrac{3}{2}x+\tfrac{5}{2}x^3, \qquad P_4(x)=+\tfrac{3}{8}-\tfrac{15}{4}x^2+\tfrac{35}{8}x^4.$$

J and K are constants characteristic of the primary. If ξ, η, ζ are rectangular co-ordinates in a co-ordinate system having the ζ-axis along the axis of rotation of the primary,

$$J=\frac{3}{2m_pR^2}\int\left[\tfrac{1}{2}(\xi^2+\eta^2)-\zeta^2\right]dm=\frac{3}{2}\frac{C-A}{m_pR^2}, \tag{25}$$

$$K=\frac{5}{m_pR^4}\int\left[\tfrac{9}{32}(\xi^2+\eta^2)^2-\tfrac{9}{4}(\xi^2+\eta^2)\zeta^2+\tfrac{3}{4}\zeta^4\right]dm. \tag{26}$$

The notation is the same as that used in **2** and also by Tisserand (1896), where derivations are found of formulae used in this section. Different notations are used by other authors. H. Struve (1888, p. 63) and Jeffreys (1953, 1954) use

$$k = JR^2\,, \qquad l = \tfrac{2}{5}KR^4\,;$$

Brouwer (1946) uses

$$k_2 = \tfrac{1}{3}JR^2\,, \qquad k_4 = \tfrac{1}{10}KR^4\,.$$

The integrals J and K vanish for a body with spherical symmetry, in which case the moments of inertia C (about the axis of rotation) and A (about an axis in the equatorial plane) are equal.

The magnitude of the disturbing forces derived from the disturbing function R_A, relative to the attraction by the primary, is of the order JR^2/a^2, if a is the semimajor axis of the satellite orbit.

The disturbing function due to the attraction by the Sun may be written

$$R_B = \frac{\mu}{r}\frac{M_\odot}{m_p + m_s}\left[\frac{r^3}{r'^3}P_2(\cos S) + \frac{m_p - m_s}{m_p + m_s}\frac{r^4}{r'^4}P_3(\cos S) + \ldots\right], \tag{27}$$

in which r' is the radius vector of the primary in its relative orbit about the Sun and S is the angle between the directions to the Sun and the satellite as seen from the center of the primary. The factor $(m_p - m_s)/(m_p + m_s)$ is taken into account in the theory of the Moon's motion (Brown, 1896, p. 7); in all other satellite theories it may be replaced by unity.

The order of magnitude of the solar perturbations on a satellite (in terms of the attraction by the primary) is indicated by the ratio

$$\frac{M_\odot}{m_p + m_s}\frac{a^3}{a'^3} = \frac{n'^2}{n^2} \quad \text{(very nearly)}\,,$$

if a' is the semimajor axis of the orbit of the primary about the Sun and n and n' are the mean daily motions of the satellite and the primary, respectively. Thus the ratio $(n'/n)^2$ may be considered a reliable indicator of the magnitude of the solar perturbations on a satellite orbit.

Let $m_{s'}$ represent the mass of any other satellite revolving about the same planet. The disturbing function due to the attraction of such satellites, then, is

$$R_C = \Sigma k^2 m_{s'}\left(\frac{1}{\Delta_{s'}} - \frac{r\cos S_{s'}}{r_{s'}^2}\right), \tag{28}$$

in which

$$\Delta_{s'}^2 = (x - x_{s'})^2 + (y - y_{s'})^2 + (z - z_{s'})^2$$
$$= r^2 + r_{s'}^2 - 2rr_{s'}\cos S_{s'}\,,$$

and $S_{s'}$ is the angle at the primary between the radii vectores of the satellites s and s'. The perturbing forces are of the order $m_{s'}/m_p$ compared with the attraction by the primary. In the present condition of the subject they are significant only among Jupiter's Galilean satellites and among the satellites of Saturn, outward to Iapetus.

The perturbation theories that deal with the mutual perturbations among satellites are in many respects similar to planetary theories. However, in planetary theories it is permissible to allow expansions in powers of the time in the coefficients of periodic terms, in view of the slowness of the motions of perihelia and nodes. In satellite systems these motions are generally so rapid that such expansions would be hopelessly inadequate within a few months or a few years.

In dealing with the mutual perturbations among satellites, we are not interested primarily in the numerous periodic terms that characterize a planetary theory. In most cases the periodic terms have such small coefficients that they are unobservable. In special circumstances, however, the coefficients of periodic terms are magnified many fold in the integration due to resonance. This feature makes it possible to concentrate particularly on a limited class of terms in the development of the perturbations. A further simplifying circumstance is the smallness of the mutual inclinations or the eccentricities, or both, of the orbits of most satellites for which mutual perturbations are to be treated.

Perturbations by other planets are important in the theory of the Moon's motion. They are of two types: the direct perturbations and the indirect perturbations. An example of an indirect planetary perturbation is the famous secular acceleration in the Moon's mean longitude. This is caused by the secular diminution of the eccentricity of the Earth's orbit, which is a consequence of the attractions by the planets on the Earth (see **2**, 24). For further details and references Brown's *Introductory Treatise on the Lunar Theory* (Brown, 1896) may be consulted.

There is no doubt that if exhaustive analytical theories for the outer satellites of Jupiter and Saturn were to be constructed, it would be necessary to treat the perturbations by the other planets. For the present, the first need would be further perfection of these general theories.

The thirty-one satellites now known in the solar system present a great variety of perturbation problems. Table 4 gives a list of pertinent data concerning the satellites. The letters A, B, and C given in the final column indicate the type of perturbation that is predominant. Though the perturbations of types A, B, and D exist for every satellite and the perturbations of type C whenever a planet has more than a single satellite, it

TABLE 4

ELEMENTS AND MASSES OF SATELLITES

	PERIOD	a/R	n'/n	m_s/m_p	TYPE
	The Earth: R = 1.000, m = 1.000				
Moon	27^d32	60.27	+0.074 80	$(1.229\,0 \pm 0.000\,4) \times 10^{-2}$	B
	Mars: R = 0.532, m = 0.107 66 ± 0.000 10				
Phobos	0.318 9	2.755	+0.000 46	...	A
Deimos	1.262	6.919	+0.001 84	...	A
	Jupiter: R = 11.194, m = 317.360 ± 0.009				
V	0.498 2	2.539	+0.000 11	...	A
(I) Io	1.769	5.905	+0.000 41	$(3.81 \pm 0.30) \times 10^{-5}$	C, A
(II) Europa	3.551	9.396	+0.000 82	$(2.48 \pm 0.05) \times 10^{-5}$	C
(III) Ganymede	7.155	14.99	+0.001 65	$(8.17 \pm 0.10) \times 10^{-5}$	C
(IV) Callisto	16.69	26.36	+0.003 85	$(5.09 \pm 0.40) \times 10^{-5}$	C, B
VI	250.6	160.7	+0.057 9	...	B
VII	260.1	164.4	+0.060 0	...	B
X	260	164	+0.060 0	...	B
XII	617	290	−0.142	...	B
XI	692	313	−0.160	...	B
VIII	735	326	−0.171	...	B
IX	758	332	−0.175	...	B
	Saturn: R = 9.351, m = 95.031 ± 0.007				
Mimas	0.942 4	3.111	+0.000 09	$(6.69 \pm 0.13) \times 10^{-8}$	A, C
Enceladus	1.370	3.991	+0.000 13	$(1.27 \pm 0.36) \times 10^{-7}$	A, C
Tethys	1.888	4.939	+0.000 17	$(1.14 \pm 0.02) \times 10^{-6}$	A, C
Dione	2.737	6.327	+0.000 25	$(1.82 \pm 0.04) \times 10^{-6}$	A, C
Rhea	4.518	8.835	+0.000 42	...	C
Titan	15.95	20.48	+0.001 48	$(2.411 \pm 0.013) \times 10^{-4}$	A, B
Hyperion	21.28	24.83	+0.001 98	...	C
Iapetus	79.33	59.67	+0.007 37	...	A, B, C
Phoebe	550.4	216.8	−0.051 16	...	B
	Uranus: R = 3.72, m = 14.497 ± 0.004				
Miranda	1.413	5.494	+0.000 05	...	A
Ariel	2.520	8.079	+0.000 08	...	A
Umbriel	4.144	11.25	+0.000 14	...	A
Titania	8.706	18.46	+0.000 28	...	A
Oberon	13.46	24.69	+0.000 44	...	A
	Neptune: R = 3.50, m = 17.19 ± 0.27				
Triton	5.877	15.85	−0.000 09	$(1.32 \pm 0.23) \times 10^{-3}$	A
Nereid	359.4	249.5	+0.005 97	...	B

is the rule, rather than the exception, that the perturbations of one type determine the character of the motion.

5.2. Dynamical Constants of Planets Derived from Satellite Motions

If the only perturbing action on a satellite is that arising from the oblateness of the primary, the motions of the pericenter and node are given by the following expressions (Brouwer, 1946):

$$\frac{d\varpi}{dt} = +n\left(\frac{JR^2}{a^2} - \frac{1}{2}\frac{J^2R^4}{a^4} + \frac{KR^4}{a^4}\right),$$
$$\frac{d\theta}{dt} = -n\left(\frac{JR^2}{a^2} - \frac{3}{2}\frac{J^2R^4}{a^4} + \frac{KR^4}{a^4}\right), \qquad (29)$$

which are valid if the square of the eccentricity and the square of the inclination of the orbit with the equatorial plane can be ignored. This applies with few exceptions to all the inner satellites. The principal exception to be treated in this section is the case of Triton, with inclination $\gamma = 160°$ with respect to the planet's equator. The formula for the motion of the node θ on the equatorial plane of the planet then becomes

$$\frac{d\theta}{dt} = -n\left(\frac{JR^2}{a^2} - \frac{3}{2}\frac{J^2R^4}{a^4} + \frac{KR^4}{a^4}\right)\cos\gamma\,. \qquad (30)$$

Attention is called to the notation γ, θ for inclination and node referred to the planet's equator, while we shall use the conventional notations I and Ω for inclination and node referred to the planet's orbital plane.

In all known cases of satellite motion, the term having J^2 as a factor is so small that the observed motions of ϖ and θ of a single satellite will not permit the evaluation of both J and K. This can be accomplished only if either or both motions can be determined from observation for two or more satellites belonging to the same system and if in at least one of these the K-term contributes significantly. After J and K have been determined, the planet's oblateness, ϵ, may be obtained from the relation (de Sitter, 1938, p. 220)

$$\epsilon = (J + \tfrac{1}{2}\phi_1)(1 + J) - \tfrac{1}{6}K\,, \qquad (31)$$

in which

$$\phi_1 = \frac{\omega^2 R_1^3}{f m_p},$$

and f is the constant of gravitation (see **2**, 9, where the symbol ρ is used instead of ϕ). In this expression R_1 is the mean radius, for which it is sufficient to write

$$R_1^3 = R^3(1 - \epsilon)$$
$$= R^3(1 - J - \tfrac{1}{2}\phi_1)\,.$$

Hence, if we put

$$\phi = \frac{\omega^2 R^3}{f m_p},$$

then

$$\phi_1 = \frac{\phi(1-J)}{1+\frac{1}{2}\phi}.$$

It is simplest to compute ϕ by using ω_*, R_*, and m_*, expressed in terms of $\omega_\oplus$, $R_\oplus$, and $m_\oplus$. Then

$$\phi = P_*^{-2} R_*^3 m_*^{-1} \phi_\oplus,$$

in which P is the period of rotation. We shall use

$$\begin{aligned} P_\oplus &= 23^{\text{h}}.934\,5, \\ R_{1\oplus} &= 8''.798\,4 \text{ at distance 1 a.u.} \\ &= 4.265\,6 \times 10^{-5} \text{ a.u.}, \\ m_\oplus^{-1} &= 332\,463, \\ \phi_\oplus &= 0.003\,461\,41. \end{aligned}$$

5.21. *The satellites of Mars*.—The principal perturbations are those arising from the oblateness of the planet. Woolard (1944) in a discussion of the data ascribes to this cause the annual motions of $-158^\circ.5 \pm 0^\circ.5$ in the node of Phobos, and $-6^\circ.279\,5 \pm 0^\circ.000\,7$ in the node of Deimos. From the latter he derives $J = 0.002\,920$, obtained with $R = 4''.680$ at distance 1 a.u., which corresponds to $R_* = 0.531\,9$.

The first-order formulae give

$$\begin{aligned} J &= 0.002\,920 - 0.005\,8\,\frac{\delta R}{R}, \\ \phi_1 &= 0.004\,548 + 0.013\,6\,\frac{\delta R}{R}, \\ \epsilon &= 0.005\,209 + 0.001\,0\,\frac{\delta R}{R}, \\ \epsilon^{-1} &= 192.0 - 36.1\,\frac{\delta R}{R}, \\ \frac{\epsilon}{\phi_1} &= 1.145 - 3.22\,\frac{\delta R}{R}. \end{aligned}$$

The observational uncertainty in J and ϕ_1 is less than one unit in the last place given for the assumed value of R. The uncertainty in the observed motion of the node of Phobos is too great to warrant an evaluation of K.

5.22. *The fifth satellite of Jupiter*.—This satellite is so close to the planet that the only observable perturbations are those due to the oblateness. In a discussion by H. Struve (1906) the motions both of the node and of the pericenter are determined. A discussion of all available observations by van Woerkom (1950) shows that the eccentricity is too small for a reliable determination of the motion of the pericenter. The motion of the node is well determined. In good agreement with H. Struve's value, van Woerkom finds

$$\frac{d\theta}{dt} = -914^\circ\!.62 \pm 0^\circ\!.36 \text{ per Julian year}$$

$$= - \quad 2^\circ\!.504\ 1 \pm 0^\circ\!.001\ 0 \text{ per mean solar day}$$

$$= (- \quad 0.003\ 465\ 3 \pm 0.000\ 001\ 4)\, n\,,$$

$$n = 722^\circ\!.631\ 73 \text{ per mean solar day}\,.$$

Thus

$$\frac{JR^2}{a^2} - \frac{3}{2}\frac{J^2R^4}{a^4} + \frac{KR^4}{a^4} = 0.003\ 465\ 3 \pm 0.000\ 001\ 4\,. \qquad (32)$$

With the equatorial radius of Jupiter, 18″.930 at mean distance 5.202 8 a.u. and the radius of the circular orbit from equation (14), we have

$$R = 98''\!.489\,, \qquad a = 250''\!.06\,,$$

both reduced to the distance 1 a.u., whence

$$\frac{R}{a} = 0.393\ 86\,.$$

From Jupiter's Galilean satellites, in which the effects of the K-terms in the motions of the pericenters and nodes are almost negligibly small, de Sitter's final discussion of the observations (de Sitter, 1931) gives

$$J = 0.022\ 06 \pm 0.000\ 22\,.$$

The equation obtained from the motion of the node of the fifth satellite then gives

$$K = 0.002\ 53 \pm 0.001\ 41\,.$$

With $P = 9^h52^m$,

$$\phi_1 = 0.084\ 25\,,$$

and, consequently,

$$\epsilon = 0.065\ 18\,, \qquad \epsilon^{-1} = 15.34\,,$$

$$\frac{\epsilon}{\phi_1} = 0.773\ 7\,, \qquad \frac{K}{J^2} = 5.20\,, \qquad \kappa = +0.000\ 51\,.$$

The motions of the pericenters and nodes of the satellites yield the function JR^2/a^2; here a is well determined from the mean motion and the mass of the planet. Thus the value of J depends on the value adopted for R. This also applies to ϕ_1, but a change in R affects J and ϕ_1 in opposite directions. Consequently, the value of $J + \frac{1}{2}\phi_1$ and therefore of ϵ is less sensitive than either J or ϕ_1 to such a change. The ratio K/J^2 is independent of the adopted equatorial radius.

The equations that determine J from the motions of the Galilean satellites contain, as other unknowns, corrections to the masses of the satellites that are notoriously difficult to determine. This accounts for the wide range in values of J obtained even in modern times, such as 0.021 865 by de Sitter (1918, p. 49) and 0.022 273 by Sampson (1921, p. 175). In de Sitter's last discussion (1931) an extensive collection of new observational material of high accuracy was fully used for the first time.

5.23. *The inner satellites of Saturn.*—The observational material on these satellites was summarized by G. Struve (1924–1933), who had continued the excellent series of observations begun by his father, H. Struve (1888). Perturbations of types B and C are present in these satellites, but if these contributions are subtracted from the observed motions of the nodes and pericenters, the amounts due to the oblateness of the planet can be derived. There is no difficulty about the separation of the unknowns such as there is in the system of Jupiter's Galilean satellites. A recent discussion by Jeffreys (1954) yields the representation of the secular motions given in the accompanying table.

Satellite		Annual Motion		O—C
		Obs.	Calc.	
Mimas	Apse	365°60	365°86	−0°26 ±0°13
	Node	365.23	364.93	+ .30 ± .13
Enceladus	Apse	....	152.52	....
Tethys	Node	72.227	72.231	− .004 ± .044
Dione	Apse	30.75	30.72	+ .03 ± .28
Rhea	Node	10.20	10.05	+ .15 ± .05
Titan	Apse	0.5012	0.4994	+ .001 8 ± .004 7
	Node	0.492	0.4994	−0.007 ±0.020

For an adopted equatorial radius of the planet, the values of J and K obtained from this solution are well determined; the contributions due to the K-terms are 5°74 for Mimas and 0°32 for Tethys. With $R = 8''625$ at Saturn's mean distance of 9.538 85 a.u., Jeffreys finds

$$J = 0.02501 \pm 0.00003 , \qquad K = 0.00386 \pm 0.00026 .$$

To the adopted value of R correspond, with $P = 10^h24^m5$,

$$\phi_1 = 0.14231 , \qquad \epsilon = 0.09792 , \qquad \epsilon^{-1} = 10.21 ,$$

$$\frac{\epsilon}{\phi_1} = 0.6881 , \qquad \frac{K}{J^2} = 6.17 , \qquad \kappa = +0.00160 .$$

Some of these quantities differ slightly from those given by Jeffreys. This is due to the fact that, in the formulae used, the third powers of J, ϕ_1, and ϵ have been ignored; expressions that are equivalent to the second order may give slightly different numerical results.

Jeffreys' solution was originally made without taking into account the terms having J^2 as a factor (Jeffreys, 1953). The inclusion of these terms materially affects the value of K that is obtained from the solution. Also, the computations were first carried out with H. Struve's value, $R = 8\overset{''}{.}750$, for the equatorial radius. This changes J and K from 0.024 30 and 0.003 64 to the values given above and illustrates how strongly the results, with the exception of the ratio K/J^2, depend on the adopted planetary radius.

For a comparison with proposed models for Saturn, see Jeffreys' (1954) discussion. In the case of Saturn the effect of the attraction of the rings on the inner satellites is included. For the body of the planet alone, the ratio K/J^2 may be somewhat smaller than the ratio obtained from the satellite motions, but no reliable estimate of the necessary correction is available.

5.24. *The satellites of Uranus.*—In comparison with our present knowledge of the orbits of the satellites of Jupiter and Saturn, only very meager information is available concerning the satellites of Uranus. Newcomb (1875) concluded from a discussion of observations of the satellites with the 26-inch equatorial of the United States Naval Observatory that "there is but slight evidence of any real eccentricity of the orbits and no evidence of any mutual inclinations." This applied to the four satellites, Ariel, Umbriel, Titania, and Oberon. A new satellite, Miranda, was discovered by Kuiper (1949*b*) in 1948 at a distance from Uranus about two-thirds that of Ariel. This satellite might be included in Newcomb's statement on the orbits.

An important step forward was made with a series of photographic observations with the 82-inch reflector of the McDonald Observatory. The measurements and discussion by Harris (1950) yield the eccentricities given in the accompanying table, suggesting that, at least for Ariel,

Miranda	$\ll 0.01$
Ariel	$.002\ 8 \pm 0.000\ 5$
Umbriel	$.003\ 5 \pm\ .000\ 4$
Titania	$.002\ 4 \pm\ .000\ 2$
Oberon	$0.000\ 7 \pm 0.000\ 2$

Umbriel, and Titania, real eccentricities have at last been obtained. These values also indicate that the eccentricities 0.008 ± 0.001 for both Ariel and Umbriel, obtained by Bergstrand (1904, 1909), are spurious and merely reflect the insufficient accuracy of the observations. Then the motion of the pericenter of Ariel found by Bergstrand, $+15°$ per annum, must also be spurious. An additional series of observations of the same accuracy as those discussed by Harris can furnish reliable secular motions.

Only the perturbations due to the oblateness of the planet are significant for all five of the satellites of Uranus, except possibly for resonance phenomena. The ratio R/a for Miranda is intermediate between the values for Tethys and Dione among Saturn's satellites. Hence the evaluation of K will require secular motions of high accuracy. It would therefore be helpful if the motions of the nodes could also be determined.

The smallness of the mutual inclinations of the orbits and the fact that the orbits of the satellites as viewed from the Earth are so wide open most of the time have so far prevented their evaluation. Most favorable for determining the orientations of the satellite orbits are the times when the orbits are seen edge-on. These occur at intervals of 42 years when Uranus crosses the intersections of its orbital plane with its equatorial plane. The next crossing will occur early in the year 1966.

5.25. *The satellites of Neptune.*—Triton, the bright satellite, was discovered by Lassell in 1846, the year of the optical discovery of Neptune. The orbit is so nearly circular that no eccentricity has been established. About 40 years after the discovery, the observations had clearly shown a gradual change in the orientation of the orbital plane. Tisserand (1888) and Newcomb (1888) independently ascribed the observed change to the effect of the oblateness of the planet. If the orbital plane makes an angle γ with the plane of Neptune's equator, the pole of the orbit will describe a circle with spherical radius γ about the pole of rotation of Neptune. If a sufficient fraction of a revolution of the node has been observed, the center of the circle can be determined.

From a discussion of the observations in the interval 1848–1923, Eichelberger and Newton (1926) find, for the inclination, $\gamma = 159\overset{\circ}{.}9 \pm 2\overset{\circ}{.}3$, and, for the period of revolution of the pole of the orbit around the pole of Neptune's equator, 585 ± 66 years.

The difficulty of the problem is that both the radius and the center of the small circle described by the pole of the orbit must be determined from an arc that, at the time of the discussion by Eichelberger and Newton, extended over only 46°. Actually, the situation was even less favorable: the observations before 1863 are weak and had to be rejected, leaving an arc of only 37°. The reduction of a series of photographic observations by Alden (1940, 1943), now in progress at the Yale Observatory, should add appreciably to the weight of the determinations. So also would a further series of observations at the present time.

Well determined from the observations is the rate with which the pole of the orbit moves among the stars, $\sin\gamma\ d\theta/dt$. If P_θ is the period of revolution of the pole of the orbit, the function $P_\theta/\sin\gamma$ is equally well determined. H. Struve (1894) finds $P_\theta/\sin\gamma = 1734$ years. The results obtained by Eichelberger and Newton correspond to 1704 years. The probable error of the latter may be estimated to be approximately 5 years.

Formula (30) is simplified to

$$-\frac{1}{n}\frac{d\theta}{dt}=\frac{JR^2}{a^2}\cos\gamma$$

and may be written

$$\frac{1}{2}\frac{JR^2}{a^2}\sin 2\gamma = -\frac{\sin\gamma}{n}\frac{d\theta}{dt}$$

$$= +0^\circ\!.211\,3 \pm 0^\circ\!.000\,6\ ,$$

the right-hand member being well determined from the observations. Kuiper's (1949*c*) value of Neptune's radius,

$$R_* = 3.50 \pm 0.03\ ,$$

and

$$\sin 2\gamma = 0.646 \pm 0.062\ ,$$

then yield

$$J = 0.007\,4 \pm 0.000\,7\ .$$

Neptune's period of rotation was determined by Moore and Menzel (1928) from spectroscopic observations. They find

$$P = 15^{\mathrm{h}}\!.8 \pm 1^{\mathrm{h}}\!.0\ ,$$

which in their discussion appears to depend on their assumed value of the radius but which is really independent of it. With these values of R, P, and J,

$$\phi_1 = 0.019\,4 \pm 0.003\,1\ , \qquad \epsilon = 0.017\,1 \pm 0.001\,8\ ,$$

$$\epsilon^{-1} = 58.5 \pm 6.2\ , \qquad \frac{\epsilon}{\phi_1} = 0.88 \pm 0.13\ .$$

The second satellite, Nereid, was discovered by Kuiper (1949*a*). The orbit by Van Biesbroeck (1957) shows that the satellite has a direct motion and that the orbit is inclined 27°.8 to the Earth's equator. The ratio R/a is 0.004, and $n'/n = 0.006\,0$. Thus both the perturbations due to the figure of the planet and the solar perturbations are extremely small and the secular motions of pericenter and node correspondingly slow.

5.3. Solar Perturbations on Satellites

In many respects the Moon's motion is a typical example of the motion of a satellite perturbed by the solar attraction. As was remarked in Section 5.1., the ratio $(n'/n)^2$ is a convenient indicator of the magnitude of the solar perturbations. For most satellites the ratio n'/n is considerably smaller than the Moon's 0.074 8. Examples are Jupiter IV, 0.003 9; Iapetus, 0.007 4; and still smaller values for other satellites (see Table 4). In all such cases it is sufficient to employ the abbreviated equations for the variation of arbitrary constants, develop the disturbing function in powers of the eccentricities and inclinations, and obtain the perturbations of the first order in the elements. Examples of the use of this method are given by H. Struve (1888) and by Tisserand (1896). No unknown parameters enter into the problem, such as J and K in the perturbations due to the figure of the primary and the satellite masses in the mutual perturbations among the satellites. This method is still applicable if the mutual perturbations among the satellites must be treated simultaneously, but it is inadequate if the value of n'/n is too large. The only known examples are shown in the accompanying table. With such large values of n'/n, the

	n'/n
Moon	+0.074 8
Phoebe	− .051
Jupiter VI, VII, X	+ .060
Jupiter VIII, IX, XI, XII	−0.14 to −0.17

first-order perturbations can give only a very rough approximation to the solution; it is necessary to employ methods that permit an orderly and certain development of the successive approximations. For the various methods that have been used in the construction of theories of the Moon's motion, we refer to the treatises by Brown (1896) and Tisserand (1894), but special mention must be made of the work by Delaunay (1860, 1867).

Delaunay's solution of the "main problem" of the lunar theory consists of a purely literal development. The longitude, the parallax, and the latitude are found in the following form:

$$\text{Longitude} = nt + \text{const.} + \Sigma A_{p_1, p_2, p_3, p_4} \sin (p_1 D + p_2 l + p_3 l' + p_4 F) ,$$

$$\text{Parallax} = \text{const.} + \Sigma B_{p_1, p_2, p_3, p_4} \cos (p_1 D + p_2 l + p_3 l' + p_4 F) ,$$

$$\text{Latitude} = \Sigma C_{p_1, p_2, p_3, p_4} \sin (p_1 D + p_2 l + p_3 l' + p_4 F) .$$

The coefficients A, B, C are all expanded in powers of the parameters n'/n, e, e', $\sin \frac{1}{2}I$, and a/a', where e is the mean eccentricity of the Moon's orbit; e' is the eccentricity of the Earth's orbit, treated as a constant; I is the mean inclination of the Moon's orbit with the plane of the ecliptic; and a/a' is the ratio of the semimajor axes.

The arguments D, l, l', and F have the following meaning: let λ be the Moon's mean longitude, λ' the Sun's mean longitude; then

$$D = \lambda - \lambda', \quad \text{the mean elongation of the Moon from the Sun},$$

$$l = \lambda - \varpi, \quad \text{the Moon's mean anomaly},$$

$$l' = \lambda - \varpi', \quad \text{the Sun's mean anomaly},$$

$$F = \lambda - \Omega, \quad \text{the mean argument of the latitude}.$$

These four arguments are linear functions of the time; the mean motions of λ and λ' are n and n', respectively, and are accurately known. Delaunay's theory gives the values of $d\varpi/dt$ and $d\Omega/dt$ as functions of the parameters of the problem. For example, the first three terms in the expressions are

$$\frac{d\varpi}{dt} = n\left[+\frac{3}{4}\left(\frac{n'}{n}\right)^2 + \frac{225}{32}\left(\frac{n'}{n}\right)^3 + \frac{4071}{128}\left(\frac{n'}{n}\right)^4 + \dots\right],$$

$$\frac{d\Omega}{dt} = n\left[-\frac{3}{4}\left(\frac{n'}{n}\right)^2 + \frac{9}{32}\left(\frac{n'}{n}\right)^3 + \frac{273}{128}\left(\frac{n'}{n}\right)^4 + \dots\right].$$

Delaunay's results were obtained to the ninth power of n'/n for $d\varpi/dt$ and to the seventh power for $d\Omega/dt$. In addition, he derived numerous smaller contributions that have powers of e^2, e'^2, $\sin^2 \frac{1}{2}I$, and a/a' as factors. Similar expressions were obtained for the coefficients of the numerous periodic terms in the three co-ordinates of the Moon—the longitude, the parallax, and the latitude.

An unfortunate property of the solution is the extremely slow convergence of some of the series, not only for the motion of the perigee, but also for the coefficients of many of the periodic terms. This slow convergence is found almost exclusively in series progressing in powers of n'/n. The convergence with respect to other parameters is, in general, satisfactory. For this reason, Hill (1878) proposed a method for dealing with the main problem of the lunar theory in which developments in powers of n'/n are avoided altogether. A fixed numerical value for this parameter is used throughout; the other parameters are retained in literal form. Hill developed in detail only some important parts of the theory. A complete theory of the Moon's motion based on these principles was carried to completion by Brown (1897–1908). The *Tables of the Motion of the Moon* constructed from this theory (Brown, 1919) have been used for the calculation of the lunar ephemeris since 1923 and represent a tremendous advance over previous lunar tables.

Aside from the drawback of the slow convergence, Delaunay's theory is the most perfect analytical solution of any problem in celestial mechanics. Since it is a purely literal development, the solar perturbations for any satellite can be obtained from his results by the substitution of appropriate numerical values for the parameters. For satellites with n'/n appreciably smaller than that for the Moon, 0.074 8, the use of Delaunay's theory is not particularly valuable. In view of the lower accuracy needed in applications to satellites other than the Moon, only a few terms suffice, and these can readily be obtained in other ways. Thus the interesting cases are the satellites with larger values of n'/n, listed above.

Delaunay's theory has actually been applied to the motions of Jupiter VI and VII by Ross (1907*a*, *b*). All perturbations exceeding 3″ in the geocentric motion are included. Ross remarks that the large eccentricity and especially the high inclination with Jupiter's orbit render it doubtful whether the periodic perturbations can be obtained from Delaunay's expressions with sufficient accuracy for an adequate representation of the observed motion. An application to Phoebe was also made by Ross (1905) and recently revised by Zadunaiski (1954). Ross correctly adopted the theoretical values for the motions of the pericenter and node obtained from Delaunay's expressions, augmented by the small contributions due to the figure of Saturn and its rings. Zadunaiski obtains these secular motions from the observations. This is no improvement: the uncertainties of the solution are many times the uncertainties of Delaunay's expressions converted into numbers.

If only limited accuracy is obtainable from Delaunay's theory for these satellites, its applicability to satellites such as Jupiter VIII must necessarily be questionable. The large eccentricity and inclination cause slow convergence according to powers of these parameters, in addition to the slow convergence according to powers of n'/n. Brown (1923) showed that the portion depending on n'/n alone of the motion of the pericenter of Jupiter VIII may be obtained from Delaunay's expression with an error of 1 per cent of its amount by including terms up to the eleventh power of n'/n. The error was reduced by a factor of one-fiftieth by making an estimate of the remainder. The errors were obtained by comparison with a direct calculation by the method that was used so effectively by Hill (1877) for the Moon's motion. Stating the error as a fraction of the motion rather overemphasizes the inadequacy of Delaunay's developments. Actually, the calculation with extrapolation yields the part depending on n'/n of the motion of the pericenter of Jupiter VIII with an error of only 0°01 per century. This indicates that Delaunay's expressions may be used to advantage for obtaining the periods of the fundamental arguments even for Jupiter VIII, IX, XI, and XII. The usefulness of Delaunay's expressions for evaluating the coefficients of the periodic terms is more questionable. Brown expresses the opinion that the chief assistance which can be furnished by estimates from Delaunay is the knowledge of the orders of magnitude of the various terms.

Having given up the attempt to use Delaunay's expression, Brown developed a purely numerical theory of Jupiter VIII (Brown, 1930; Brown and Brouwer, 1937), in which the perturbed true longitude is used as an independent variable. The constants of the eccentricity and inclination used in the theory are only rough approximations. This and the inconvenience of the independent variable chosen detract considerably from the usefulness of the theory for ephemeris computation.

Beginning with the work by Cowell, Crommelin, and Davidson (1909), Cowell's method of numerical integration has been used for the construction of ephemerides. Further numerical integration on this orbit was done by Miss Boeva (1933, 1936) and by Grosch (1948). Of interest are graphs and tables by Grosch of elliptic elements with an interval of 80 days over a period of 6000 days. Numerical integrations have also served for ephemeris computations for the other satellites in this class.

5.4. The Orbits of Iapetus and of the Satellites of Mars

5.41. *The orbit of Iapetus.*—It was remarked earlier that, as a rule, in the motions of satellites one type of perturbation predominates. The outstanding exception is the orbit of Iapetus. The solar perturbations, if acting alone, would cause the pole of Iapetus' orbit to describe a circle about the pole of the orbit of Saturn. The computations give

$$\frac{1}{n}\frac{d\Omega}{dt} = -C_{\odot} \cos I = -4.0973 \times 10^{-5} \cos I .$$

The oblateness of the planet and the attractions of the satellites, if acting alone, would cause the pole of Iapetus' orbit to describe a circle about the pole of rotation of Saturn. If the contribution by Titan is stated separately and those by the other satellites ignored, this may be written

$$\frac{1}{n}\frac{d\theta}{dt} = -(C_p + C_{Ti}) \cos \gamma = -[7.025 \times 10^{-6} + 0.11176 m_{Ti}] \cos \gamma .$$

With $m_{Ti}^{-1} = 4147$, the result is

$$\frac{1}{n}\frac{d\theta}{dt} = -3.3975 \times 10^{-5} \cos \gamma .$$

It is seen that the two contributions are of the same order of magnitude. The plane of Saturn's equator makes an angle of 26°.74 with Saturn's orbital plane. It is therefore obvious that the orbit of Iapetus cannot remain close to either of the two planes. The equations of the variation of the elements yield the following results:

a) Throughout the motion,

$$C_{\odot} \cos^2 I + (C_p + C_{Ti}) \cos^2 \gamma = \text{Constant} .$$

b) The pole of the orbit describes a nearly circular path about a mean pole, which lies on the great circle passing through the poles of Saturn's orbital plane and of Saturn's equator. This mean pole is defined by the relations

$$C_{\odot} \sin 2 I_L = (C_p + C_{Ti}) \sin 2\gamma_L ,$$

$$I_L + \gamma_L = A = 26^{\circ}.74 .$$

The problem was solved first by Laplace. The plane defined by the mean pole is called the *proper* or *Laplacian plane.* It passes through the line of intersection of Saturn's orbital and equatorial planes and makes the angle I_L with the former, γ_L with the latter. An extensive treatment of the theory is given by Tisserand (1896, p. 91).

The observational record of Iapetus begins with some observations by William Herschel in 1787; the modern series of higher precision begins with H. Struve in 1885. From the path described by the pole so far, it is not yet possible to determine the center of the curve with great precision. Knowing the center of the curve would yield the ratio $(C_p + C_{Ti})/C_{\odot}$ and thus an evaluation of the mass of Titan. H. Struve found for this ratio 0.750. Jeffreys (1953) finds for the ratio 0.814 ± 0.010, corresponding to $m_{Ti}^{-1} = 4\ 243 \pm 63$. This method promises eventually to give a very strong determination of Titan's mass. It takes the pole of Iapetus about three thousand years to describe its curve with mean radius about 8°. The problem resembles that of the path of the pole of Triton, but in the case of Iapetus it is known that the center of the curve must lie on the great-circle arc passing through the poles of Saturn's orbital and equatorial planes, while the center of the path of Triton's pole must be obtained from the observations without any such aid. Thus, though the period of revolution of the pole of Iapetus is five times the corresponding period for Triton, the center of the path is better determined from the available observations than the length of the period might indicate.

5.42. *The satellites of Mars.*—The same theory applies to the motions of the poles of the orbital planes of the satellites of Mars. Only the solar attraction and the effect of the planet's oblateness are to be considered. For the inner satellite, Phobos, the effect of the oblateness is overwhelmingly much greater than the solar effect, as indicated by the ratio $C_p/C_\odot = 2\,350$. For the outer satellite, Deimos, the ratio is 23.76. The radii of the paths described by the poles are $1\overset{\circ}{.}13$ and $1\overset{\circ}{.}77$, respectively, the periods $2\overset{y}{.}262 \pm 0\overset{y}{.}007$ and $55\overset{y}{.}014 \pm 0\overset{y}{.}006$, according to Woolard (1944), who gives a detailed discussion of the problem. From the two satellites Burton (1929) finds $25\overset{\circ}{.}20$ for the angle between the Martian orbital and equatorial planes, and for the angle γ_L between the Laplacian planes and the Martian equator, $0\overset{\circ}{.}01$ for Phobos, $0\overset{\circ}{.}92$ for Deimos.

5.5. Mutual Perturbations among Satellites

5.51. *Jupiter's Galilean satellites.*—The effect of the oblateness of the planet is considerable on Satellite I ($0\overset{\circ}{.}13$ per day in pericenter and node), and the solar attraction produces significant perturbations in the motion of Satellite IV. Nevertheless, the mutual perturbations play the dominant role in this system. Of particular interest is the resonance relationship among Satellites I, II, and III. The mean longitudes of these satellites satisfy the relations

$$\lambda_1 = (4 - \kappa)\,\tau + \lambda_{10}\,, \qquad \lambda_2 = (2 - \kappa)\,\tau + \lambda_{20}\,, \qquad \lambda_3 = (1 - \kappa)\,\tau + \lambda_{30}$$

in which

$$\tau = (n_2 - n_3)(t - t_0)\,,$$

n_1, n_2, and n_3 being the mean daily motions of the satellites and κ a constant. The numerical values are

$$\kappa = +0.014\,483\,925\,,$$

$$n_2 - n_3 = 51\overset{\circ}{.}057\,115\,5\,.$$

Thus the mean motions are nearly commensurable in pairs, and, in addition, the relation

$$n_1 - 3n_2 + 2n_3 = 0$$

is rigorously satisfied. The observations further show that the epoch t_0 can be so chosen that

$$\lambda_{10} = \lambda_{30} = \lambda_{20} + 180°\,.$$

These relationships suggest that the orbits are close to a periodic solution. De Sitter (1918) computed such a periodic solution, in which the terms

$$\delta\psi_1 = +0\overset{\circ}{.}463 \sin(2\lambda_1 - 2\lambda_2)\,, \qquad \delta\psi_2 = -1\overset{\circ}{.}073 \sin(\lambda_1 - \lambda_2)\,,$$

$$\delta\psi_3 = -0\overset{\circ}{.}068 \sin(\lambda_2 - \lambda_3)\,,$$

appear as the principal terms in the longitudes. These terms are included in the intermediate orbit that de Sitter uses as the basis of his theory of the motions of the Galilean satellites. In other theories, such as those by Laplace, Souillart (see Tisserand 1896), and Sampson (1921), they are obtained among the perturbations. It is remarkable that the resonance magnification can produce such large perturbations by the small satellite masses. In terms of Jupiter's mass they are

$$\text{I},\ 0.381\times 10^{-4}\ \pm 30;\qquad \text{II},\ 0.248\times 10^{-4}\ \pm 5;\qquad \text{III},\ 0.817\times 10^{-4}\ \pm 10,$$

according to the latest determination by de Sitter (1931). The mass of the fourth satellite is $(0.509 \pm 0.040) \times 10^{-4}$.

The actual orbits do not correspond exactly to a periodic solution but may be obtained from the periodic solution by modifying the co-ordinates

TABLE 5

SECULAR ELEMENTS OF THE GALILEAN SATELLITES

Satellite	Proper Eccentricity	Proper Inclination
I..........	0°000 6±0°000 4	0°031 7±0°001 4
II..........	.007 5± .000 8	.466 8± .000 9
III.........	.079 6± .001 3	.178 8± .001 0
IV.........	0.421 8±0.000 5	0.245 2±0.000 9

and velocity components at a chosen epoch. The modifications so introduced correspond to eccentricities and inclinations. At this point it is important to include the fourth satellite because the eccentricities and longitudes of the pericenters of the four satellites are connected by a system of secular variations and likewise the inclinations and the longitudes of the nodes. According to de Sitter (1931), the proper eccentricities and proper inclinations, the latter referred to the plane of Jupiter's equator, are those listed in Table 5.

The terms so introduced into the motions of the satellites have periods nearly equal to the periods of revolution. The longitudes of Satellites I, II, and III also contain terms with periods ranging from 405 to 486 days, i.e., *long* compared with the periods of revolution. They are associated with the equations of the centers.

The *libration*, so named by Laplace, represents a variation of the periodic solution that could be present even if all proper eccentricities and proper inclinations were zero. In addition to the periodic solution in which the relation $\lambda_1 - 3\lambda_2 + 2\lambda_3 = 180°$ is satisfied exactly, the problem admits an infinite number of solutions, in which the relation changes into

$$\lambda_1 - 3\lambda_2 + 2\lambda_3 = 180° + V .$$

The angle V oscillates about zero with an amplitude that must be determined from observation. The period of the libration, about 6 years, is a function of the masses of the satellites. The longitudes of the three satellites contain terms having the same period, while the ratios of the amplitudes depend on known functions of the masses.

Although the theory of the libration has been known since Laplace first developed it, various efforts to detect these terms in the observations have failed to show their presence conclusively. It is certain that the amplitude is very small.

One reason for the difficulty in determining the amplitude of the libration is of an observational nature: various important periods in the longitudes are somewhat longer than 400 days, the synodic period of Jupiter. If observations are concentrated near the opposition dates of Jupiter, a term with a period of 480 days will appear in the residuals in longitude to have a period of nearly 6 years. De Sitter overcame this drawback, at least in part, by planning two series of photographic observations in each opposition: the mean date of the first about 2 months before opposition, the mean date of the second about 2 months after opposition. In eclipse observations the separation of the periods is even more difficult. A complication of a different nature is the slow convergence of the analytical developments in powers of the masses. The combination of these various circumstances accounts for the slowness with which the constants of the problem and especially the masses of the satellites have been improved by the comparison of theory with observation. The motions of the Galilean satellites constitute one of the most fascinating theoretical problems in the solar system; the further improvement of the constants remains as a challenge to future workers.

5.52. *Librations among the satellites of Saturn.*—Among Saturn's satellites there are three pairs for which the mean motions are nearly commensurable. All three of these systems exhibit librations.

Mimas—Tethys: The mean motions are very nearly in the ratio 2:1. The inclinations of both satellites with respect to Saturn's equator are large enough for the longitudes of the node to be well determined. The observations show that the angle $2\,\lambda_{Mi} - 4\,\lambda_{Te} + \theta_{Mi} + \theta_{Te}$ oscillates about zero. The period of this libration is $70^{y}\!.78 \pm 0^{y}\!.08$; its amplitude is exceptionally large; the angle varies from $-\,95^{\circ}\!.3$ to $+\,95^{\circ}\!.3$. Both Mimas and Tethys have terms in their mean longitudes with the period of the libration. The ratio of the coefficients, H and H', of the principal libration terms in the longitudes yields, according to the theory by H. Struve (1898),

$$\frac{m_{Mi}}{m_{Te}} = 2\,\frac{a_{Mi}}{a_{Te}}\,\frac{H'}{H}.$$

Since $H = 44^\circ\!.390 \pm 0^\circ\!.082$, $H' = 2^\circ\!.065 \pm 0^\circ\!.019$, and $a/a' = 0.629\,6$,

$$\frac{m_{Te}}{m_{Mi}} = 17.07 \pm 0.16\,.$$

The period of the libration furnishes a second equation for the masses. Jeffreys (1953), in a revision of the discussion by G. Struve (1924–1937, Heft 4), finds

$$m_{Te} = \frac{1}{876\,400}\,(1 \pm 0.018) = 1.141 \times 10^{-6}\,(1 \pm 0.018)\,,$$

$$m_{Mi} = \frac{1}{1\,496\,000}\,(1 \pm 0.020) = 6.69 \times 10^{-8}\,(1 \pm 0.020)\,.$$

All masses in this section are expressed in terms of Saturn's mass.

Enceladus—Dione: As in the system Mimas—Tethys, the mean motions are very nearly in ratio 2:1; the inclinations with respect to Saturn's equatorial plane are very small; the eccentricities are small (0.004 and 0.002, respectively) but permit the evaluation of the longitudes of the pericenters from the observations. The observations show that $\lambda_{En} - 2\,\lambda_{Di} + \varpi_{En}$ remains very near zero. The theory was first developed by Woltjer (1922*a*) with the simplification that results if the mass of Enceladus and the eccentricity of Dione are ignored. The period of the libration and the observed eccentricity then yield the mass of Dione for which Woltjer's discussion gives $m_{Di} = 1/530\,000$. A detailed account of the theory without these simplifications has been given by Jeffreys (1953). From the data given by G. Struve (1924–1937, Heft 4) he obtains:

$$m_{En} = \frac{1}{7\,900\,000}\,(1 \pm 0.28) = (1.27 \pm 0.36) \times 10^{-7}\,,$$

$$m_{Di} = \frac{1}{548\,000}\,(1 \pm 0.023) = (1.825 \pm 0.041) \times 10^{-6}\,.$$

The coefficients of the libration terms in the mean longitudes are very small, $0^\circ\!.024$ in Enceladus and $0^\circ\!.015$ in Dione; both longitudes contain an additional term of almost equal amplitude arising from the term in the disturbing function with argument $\lambda_{En} - 2\lambda_{Di} + \varpi_{Di} = \lambda_{En} - 2\lambda_{Di} + \varpi_{En} - (\varpi_{En} - \varpi_{Di})$.

Titan—Hyperion: The ratio of the mean motions in this case is very nearly 4:3. The libration was discovered by Newcomb (1891) after Hall (1884) had shown that the pericenter of Hyperion has a retrograde motion of about 20° per annum. The angle $4\lambda_{Hy} - 3\lambda_{Ti} - \varpi_{Hy}$ oscillates about 180°, with an amplitude of about 36°.

The problem is one of the most difficult in celestial mechanics. The large ratio of the semimajor axes, 0.825, causes the series representing the disturbing function to converge very slowly. For this reason Woltjer (1928), who developed a complete theory of the motion, uses numerical methods for obtaining the basic series. A significant feature of the problem is the relatively large eccentricity of the orbit of Titan, 0.028 7. Thus the terms with argument $4\lambda_{Hy} - 3\lambda_{Ti} - \varpi_{Ti}$ in the disturbing function, which have Titan's eccentricity as a factor, play an important role in the theory. Woltjer includes only the first power of e_{Ti} in his theory. Jeffreys (1953) remarks that extending the theory to the second power in e_{Ti} and possibly higher terms in m_{Ti} would be an undertaking comparable with the lunar theory.

The mass of Hyperion is so small that no effect of its action on Titan is observed; the large perturbations by Titan on Hyperion offer, in principle at least, an excellent possibility for obtaining the mass of Titan with high accuracy.

The most direct approach is that by Eichelberger (1911), who computed special perturbations in the elements for a 6-month period and compared with observations by A. Hall. The result is $m_{Ti}^{-1} = 4\,172 \pm 39$.

Other determinations of the mass of Titan from the motion of Hyperion, such as from the period of libration, the motion of the pericenter, and the motion of the node are not so reliable, on account of incompleteness of the theory. This also applies to the method used by Hill (1888), who computed a periodic solution by numerical integration. A new computation by this method by Brouwer (1924) with improved basic data yields $m_{Ti}^{-1} = 4\,143$. A critical discussion by Jeffreys (1953), in which determinations from the effect of Titan on other satellites are also included, yields $m_{Ti}^{-1} = 4\,147 \pm 22$. This is the best result available at present. With rapid calculating machines now available, Eichelberger's method could well be repeated over a longer period and should be capable of yielding a much reduced uncertainty.

5.53. *The orbit of Rhea.*—The mean motion is not near a low-order commensurability with Titan or any other satellite. The eccentricity is very small, 0.001. The mean longitude of the pericenter is the same as that of Titan, with annual motion 0°.5. As Woltjer (1922*b*) has shown, the eccentricity and longitude of the pericenter must be interpreted as a perturbation by Titan, of the nature of a forced oscillation. The constant of integration corresponding to the eccentricity and longitude of the pericenter appears in the free oscillation superposed on the forced oscillation. G. Struve (1924–1937, Heft 4) finds 38 years for the period of the free oscillation and 0.000 30 for the coefficient in the eccentricity. This solution is of the nature of the ordinary solution of the secular variations of the planetary problem. The mass of Rhea is practically indeterminate from its contribution to the secular variations of other satellites in the Saturn system. Jeffreys (1953) finds $m_{Rh} = (0.4 \pm 2.6) \times 10^{-6}$. H. Struve estimates the mass $m_{Rh}^{-1} = 250\ 000$, which is within the Jeffreys limits.

While the systems Mimas—Tethys, Enceladus—Dione, and Titan—Hyperion offer illustrations of librational motions in the vicinity of periodic orbits in resonance regions, this is not the case with the orbit of Rhea. It is therefore misleading to refer, as G. Struve (1924–1937, Heft 1, p. 9) did, to the oscillation in Rhea's eccentricity and longitude of the pericenter as a libration. The only unusual feature about Rhea's orbit is that the amplitude of the free oscillation in the eccentricity-pericenter solution is smaller than the amplitude of the forced oscillation.

5.6. The Mass of Triton

From photographic material, Alden (1940, 1943) derived the ratio between the masses of Triton and Neptune by finding the center of mass from a comparison with an accurate ephemeris. The method had previously been used by Nicholson, van Maanen, and Willis (1931), but Alden's measurements have many times the weight of the earlier ones. A weighted mean of all the data gives

$$\frac{m_s}{m_p + m_s} = 0.001\ 32 \pm 0.000\ 23\,.$$

Thus

$$m_{\text{Triton}} = (0.022\ 7 \pm 0.004\ 0)\ m_{\text{Earth}}$$

$$= (1.85 \pm 0.32)\ m_{\text{Moon}}\,.$$

5.7. Secular Accelerations and Retardations

The secular retardation in the Earth's rotation and the related secular acceleration in the longitudes of the Sun and Moon have been discussed in **2**, chapters 1 and 2. The discussion by Spencer Jones (1939) established that the modern observations give, for the quadratic terms in the mean longitudes ascribed to tidal friction,

$$\text{Moon:} \quad (+5''.22 + s)\,(T/\text{century})^2\,,$$

$$\text{Sun:} \quad (+1''.23 \pm 0''.04 + 0.074\ 8\,s)\,(T/\text{century})^2\,.$$

The unknown s must be determined by making use of ancient records of eclipses and occultations.

The quadratic term in the Sun's mean longitude is entirely due to the retardation in the Earth's rate of rotation. By merely multiplying the term in the Sun's mean longitude by the ratio $n/n_{\odot}$, the corresponding term in the mean longitude of any other celestial body with the exception of the Moon can be obtained. For the Moon, there results

$$(+16\overset{''}{.}44 \pm 0\overset{''}{.}54 + s)(T/\text{century})^2 .$$

as the acceleration due to the secular retardation of the Earth's rate of rotation, leaving, as a true retardation,

$$\text{Moon:} \qquad (-11\overset{''}{.}22 \pm 0\overset{''}{.}54)(T/\text{century})^2 .$$

Quadratic terms in the mean longitudes of various satellites are given in Table 6.

TABLE 6

QUADRATIC TERMS IN SATELLITE LONGITUDES

Satellite	$\Delta\lambda$	$\Delta n/n$ per period	Reference
Moon........	$-3\overset{\circ}{.}117\times10^{-7}\, t_y^2$	$(-9.69\pm0.47)\times10^{-12}$	
Phobos.......	$+1.882\times10^{-3}\, t_y^2$	$(+7.97\pm0.72)\times10^{-12}$	Sharpless (1945)
Deimos......	$-2.66\ \times10^{-4}\, t_y^2$	$(-1.76\pm1.09)\times10^{-11}$	*Ibid.*
J V..........	$+3.5\ \ \times10^{-4}\, t_y^2$	$(+3.62\pm1.45)\times10^{-12}$	Van Woerkom (1950)
J I..........	$+1.2\ \ \times10^{-5}\, t_y^2$	$+1.6\ \times10^{-12}$	De Sitter (1928)
J II.........	$+5.0\ \ \times10^{-6}\, t_y^2$	$+2.6\ \times10^{-12}$	*Ibid.*
J III........	$+1.5\ \ \times10^{-6}\, t_y^2$	$+3.2\ \times10^{-12}$	*Ibid.*

A common remark to these results is that in the discussions that produced these data, with the exception of the values for the Moon and Jupiter V, the reduction from mean solar to ephemeris time has not been adequately taken into account. The discussion of Jupiter's Galilean satellites by de Sitter included this reduction, but his treatment will require some revision. It is not likely, however, that this refinement will materially alter the results for any of the satellites listed. The probable errors given indicate that, with the exception of Phobos and the Moon, the quantities derived from observation are of the same order of magnitude as their probable errors. Nevertheless, the uniformity of $\Delta n/n$ per period is remarkable.

The interpretation of the results presents considerable difficulty. Jeffreys (1952, p. 240) comments that the positive value for Phobos would be expected if the acceleration is due to tidal friction, since Phobos revolves more rapidly than Mars rotates. On the same interpretation, the sign of the more doubtful result for Deimos is what we should expect, but a negative sign for Jupiter V would be expected, since this satellite revolves less rapidly than the planet rotates. Kerr and Whipple (1954) have discussed the problem and examined the possible interpretation of the acceleration in the mean longitude of Jupiter V by assuming a resisting medium. They conclude that such a resisting medium must have a space density at least 10^3 times greater than can be expected from other considerations. The first need is evidently for more accurate observational data.

It is important to note that the true secular term in the Moon's mean longitude, $-11''.22\, T^2$, is obtained from the observations of the last 3 centuries, without reference to the ancient data, and is not affected by the irregular fluctuations in the Earth's rate of rotation. The problem of evaluating the retardation in the Earth's rate of rotation may be reduced to solving the single unknown s, provided that it is justified to assume that the effect of tidal friction has remained essentially constant during the last 25 centuries. On this point geological evidence is reassuring.

The essential difficulty is that, on the assumption of random cumulative changes in the Earth's rate of rotation (Brouwer, 1952*a*, *b*), the uncertainty in the evaluation of s from observations is much greater than the uncertainty of the ancient observations alone would contribute. With van Woerkom's (1953) mean error, reduced to probable error, Brouwer's result is

$$s = -3''.00 \pm 6''.40 \,,$$

yielding

$$\text{Moon:} \quad (+2''.22 \pm 6''.40)\,(T/\text{century})^2 \,,$$

$$\text{Sun:} \quad (+1.01 \pm 0.48)\,(T/\text{century})^2 \,.$$

Jeffreys (**2**, 54) remarks that the theoretical ratio of about 7 to 1 between the coefficients in the Moon and Sun may be obtained by using 7″ for the Moon and 1″ for the Sun, agreeing with both results within the uncertainties. This choice would violate the relations involving the unknown s. In order to obtain the ratio 7 to 1, it would be necessary to put $s = +7''.12$, yielding

$$\text{Moon:} \quad +12''.34T^2 \,,$$

$$\text{Sun:} \quad +\ 1.76T^2 \,,$$

which differ by slightly more than the mean errors from the solution obtained from the ancient observations.

Concerning the remark by Jeffreys (**2**, 55) that he mistrusts the method of least squares in this problem, it may be said that, regardless of the method used, the coefficient of T^2 in the solution is almost exclusively determined by the ancient residuals. However, the mean errors were obtained by van Woerkom by evaluating spurious T^2 terms in artificially constructed sequences. This procedure established the lack of determinateness in the solution of the secular accelerations from observations and is independent of the least-squares solution.

REFERENCES

ALDEN, H. L.	1940	*A.J.*, **49**, 70.
	1943	*Ibid.*, **50**, 110.
BERGSTRAND, O.	1904	*Uppsala Nova Acta*, No. 20.
	1909	*Ark. f. mat., astr., fys.*, Vol. **6**, No. 6.
BESSEL, F. W.	1831	*A.N.*, **9**, 1, 381.
	1833	*Ibid.*, **11**, 17.
BIESBROECK, G. VAN	1951	*A.J.*, **56**, 110.
	1955	*Ibid.*, **60**, 38, 57, and 275.
	1957	*Ibid.*, **62**, 272.
BOEVA, H.	1933	*Bull. Inst. Astr. Leningrad*, No. 32.
	1936	*A.J., U.S.S.R.*, **12**, 476.
BOSCH, C. A. VAN DEN	1927	*De Massa's van de Groote Planeten* (Dissertation, University of Utrecht)
BROUWER, D.	1924	*B.A.N.*, **2**, 119.
	1937	*A.J.*, **46**, 149.
	1946	*Ibid.*, **51**, 110.
	1950	*Bull. Astr.*, Vol. **15**, Fasc. 3.
	1952*a*	*Proc. Nat. Acad. Sci.*, **38**, 1.
	1952*b*	*A.J.*, **57**, 125,
	1955	*M.N.*, Vol. **115** (George Darwin lecture).
BROUWER, D., and ASHBROOK, J.	1951	*A.J.*, **56**, 57.
BROUWER, D., and WOERKOM, A. J. J. VAN	1950	*Astr. Papers Amer. Ephem. Naut. Almanac*, Vol. **13**, Part 2.
BROWN, E. W.	1896	*An Introductory Treatise on the Lunar Theory* (Cambridge: Cambridge University Press).
	1897–1908	*Mem. R.A.S.*, **53**, 39, 163; **54**, 1; **57**, 51; **59**, 1.
	1919	*Tables of the Motion of the Moon* (with the assistance of H. B. HEDRICK) (New Haven: Yale University Press).
	1923	*A.J.*, **35**, 1.
	1930	*Trans. Yale U. Obs.*, No. 6, Part 4.
BROWN, E. W., and BROUWER, D.	1937	*Trans. Yale U. Obs.*, No. 6, Part 8.

Brown, E. W., and Shook, C. 1933 *Planetary Theory* (Cambridge: Cambridge University Press).

Bullard, E. C. 1948 *M.N., Geophys. Suppl.*, **5**, 186.

Bullen, K. E. 1940 *Bull. Seism. Soc. America*, **30**, 235.

1942 *Ibid.*, **32**, 19.

Burton, H. E. 1929 *A.J.*, **39**, 155.

Chazy, J. 1928 *La Théorie de la relativité et la mécanique céleste* (Paris: Gauthier-Villars).

Clemence, G. M. 1943 *Astr. Papers Amer. Ephem. Naut. Almanac*, Vol. **11**, Part 1, p. 46.

1947 *Rev. Mod. Phys.*, **19**, 361.

1948*a* *A.J.*, **53**, 169.

1948*b* *A.S.P. Leaflets*, No. 235.

1951 *M.N.*, **111**, 234.

1952 *Amer. Scientist*, **40**, 260; *A.S.P. Leaflets*, Nos. 283, 284.

Clemence, G. M., and Scott, F. P. 1942 *A.J.*, **49**, 188.

Cowell, P. H., Crommelin, A. C. D., and Davidson, C. 1909 *M.N.*, **69**, 42.

Crawford, R. T. 1930 *Determination of Orbits of Comets and Asteroids* (New York: McGraw-Hill Book Co.).

Delano, F. 1950 *A.J.*, **55**, 129.

Delaunay, C. 1860 *Mém. Acad. Sci. Paris*, Vol. **28**.

1867 *Ibid.*, Vol. **29**.

Dorsey, N. E. 1944 *Trans. Amer. Phil. Soc.*, N.S., Vol. **34**, Part 1.

Eckert, W. J., Brouwer, D., and Clemence, G. M. 1951 *Astr. Papers Amer. Ephem. Naut. Almanac*, Vol. **12**.

Eddington, A. S. 1923 *The Mathematical Theory of Relativity* (Cambridge: Cambridge University Press).

Eginitis, D. 1889 *Ann. Obs. Paris*, Vol. **19**.

Eichelberger, W. S. 1911 *Pub. U.S. Naval Obs.*, 2d ser., **6**, Appendix I, B, 1–17.

Eichelberger, W. S., and Newton, A. 1926 *Astr. Papers Amer. Ephem. Naut. Almanac*, No. 9, Part 3.

Gaillot, M. A. 1910 *Mém. Obs. Paris*, **28**, 83.

Grosch, H. R. J. 1948 *A.J.*, **53**, 180.

Gyldén, H. 1893, 1908 *Traité analytique des orbites absolues des huit planètes principales* (Stockholm: F. & G. Beijer).

HALL, A. 1878 *Observations and Orbits of the Satellites of Mars with Data for Ephemerides in 1879* (Washington: Government Printing Office).

1884 *M.N.*, **44**, 361.

HARETU, S. C. 1878 Thesis submitted to the Sorbonne.

HARRIS, D. 1950 Dissertation, University of Chicago (unpublished).

HERGET, P. 1948 *The Computation of Orbits* (Cincinnati: Published privately).

HERTZ, H. G. 1953 *Astr. Papers Amer. Ephem. Naut. Almanac*, Vol. **13**, Part 1.

HILL, G. W. 1873 *Mem. Amer. Acad. Arts and Sci.*, N.S., **9**, 417; *Collected Mathematical Works* (privately printed), **1**, 105.

1877 *Collected Mathematical Works*, **1**, 243.

1878 *Amer. J. Math.*, **1**, 5, 129, 245; *Collected Mathematical Works*, **1**, 284.

1888 *A.J.*, **8**, 57; *Collected Mathematical Works*, **2**, 135.

1891 *A.J.*, **11**, 49.

1895 *Astr. Papers Amer. Ephem. Naut. Almanac*, Vol. **7**, Part 1.

1897 *A.J.*, **17**, 81; *Collected Mathematical Works*, **4**, 123.

1898 *Astr. Papers Amer. Ephem. Naut. Almanac*, Vol. **7**, Part 2.

1905 *A.J.*, **25**, 1; *Collected Mathematical Works*, **4**, 320.

1907 *Collected Mathematical Works*, **4**, 345.

JACKSON, J. 1930*a* *M.N.*, **90**, 733.

1930*b* *Ibid.*, p. 728.

JEFFREYS, H. 1937 *M.N., Geophys. Suppl.*, **4**, 1.

1941 *M.N.*, **101**, 34.

1942 *Ibid.*, **102**, 194.

1948*a* *M.N., Geophys. Suppl.*, **5**, 219.

1948*b* *M.N.*, **108**, 206.

1952 *The Earth* (3d ed.; Cambridge: Cambridge University Press).

1953 *M.N.*, **113**, 81.

1954 *Ibid.*, **114**, 433.

JONES, H. SPENCER 1939 *M.N.*, **99**, 541.

1941 *Mem. R.A.S.*, Vol. **66**, Part 2.

KERR, F. J., and WHIPPLE, F. L. 1954 *A.J.*, **59**, 124.

KUIPER, G. P. 1949*a* *Harvard Ann. Card*, No. 994.

1949*b* *Pub. A.S.P.*, **61**, 129.

1949*c* *Ap. J.*, **110**, 93.

1950 *Pub. A.S.P.*, **62**, 133.

LAGRANGE, J. L. 1776 *Mém. Acad. Berlin; Œuvres*, **4**, 255.

MOORE, J. H., and MENZEL, D. H. 1928 *Pub. A.S.P.*, **40,** 234.

MORGAN, H. R., and OORT, J. H. 1951 *B.A.N.*, **11,** 379.

MORGAN, H. R., and SCOTT, F. P. 1939 *A.J.*, **47,** 197.

MOULTON, F. R. 1914 *An Introduction to Celestial Mechanics* (2d ed.; New York: Macmillan Co.).

NEWCOMB, S. 1874 *Smithsonian Contr. Knowledge*, **19,** 173.

1875 *Washington Obs. 1873*, Appendix I.

1882 *Astr. Papers Amer. Ephem. Naut. Almanac*, **1,** ix.

1888 *A.J.*, **8,** 143.

1891 *Astr. Papers Amer. Ephem. Naut. Almanac*, Vol. **3,** Part 3.

1895*a* *The Elements of the Four Inner Planets and the Fundamental Constants of Astronomy* (Suppl. to Amer. Ephem. for 1897).

1895*b* *Astr. Papers Amer. Ephem. Naut. Almanac*, **5,** 398.

1898 *Ibid.*, **6,** 19.

NICHOLSON, S. B., MAANEN, A. VAN, and WILLIS, H. C. 1931 *Pub. A.S.P.*, **43,** 261.

NOTEBOOM, E. 1921 *A.N.*, **214,** 169.

POINCARÉ, H. 1892–1899 *Les Méthodes nouvelles de la mécanique céleste* (Paris: Gauthier-Villars).

POISSON, S. D. 1809 *J. École Polytechnique*, **15,** 1.

RABE, E. 1950 *A.J.*, **55,** 112.

ROSS, F. E. 1905 *Ann. Harvard Coll. Obs.*, Vol. **53,** No. 6.

1907*a* *Lick Obs. Bull.*, **4,** 110.

1907*b* *A.N.*, **174,** 359.

SAMPSON, R. A. 1921 *Mem. R.A.S.*, Vol. **63.**

SHARPLESS, B. P. 1945 *A.J.*, **51,** 185.

SITTER, W. DE 1908 *Proc. K. Acad. Wetensch. Amsterdam*, p. 713.

1918 *Ann. Sternw. Leiden*, Vol. **12,** Part 1.

1924 *B.A.N.*, **2,** 99.

1928 *Ibid.*, Vol. **16,** Part 3.

1931 *M.N.*, **91,** 706.

1938 *B.A.N.*, **8,** 213 (revised and edited by D. Brouwer).

STRACKE, J. 1929 *Bahnbestimmung der Planeten und Kometen* (Berlin: J. Springer).

STRUVE, G. 1924–1937 *Pub. Berlin-Babelsberg*, Vol. **6,** Hefte 1–5.

STRUVE, H. 1888 *Obs. Pulkovo*, Suppl. 1.
1894 *St. Petersburg Mem.*, Vol. **42**, No. 4.
1906 *Sitz. Berlin*, p. 760.
TISSERAND, F. 1888 *C.R.*, **107**, 804.
1894 *Traité de mécanique céleste* (Paris: Gauthier-Villars), Vol. **3**.
1896 *Ibid.*, Vol. **4**.
WITT, G. 1933 *Astr. Abh. Ergänzungsh.*, A.N., **9**, No. 1.
WOERKOM, A. J. J. VAN 1950 *Astr. Papers Amer. Ephem. Naut. Almanac*, Vol. **13**, Part 1.
1953 *A.J.*, **58**, 10.
WOLTJER, J., JR. 1922*a* *B.A.N.*, **1**, 23, 36.
1922*b* *Ibid.*, p. 175.
1928 *Ann. Sternw. Leiden*, Vol. **16**, Part 3.
WOOLARD, E. W. 1944 *A.J.*, **51**, 33.
ZADUNAISKI, P. E. 1954 *A.J.*, **59**, 1.

This chapter was written mainly in 1955. Some values have since been brought up to date, but no major revisions have been made by the authors.

CHAPTER 4

*The Stability of the Solar System**

By YUSUKE HAGIHARA
Professor Emeritus, University of Tokyo

WILL the present configuration of the solar system be preserved without radical changes for a long interval of time? What can be said about the arrangement of the planetary orbits in the distant past? Were the satellite systems initially very different? A student of the solar system will further note the presence of gaps in the distribution of semimajor axes of the asteroids and in Saturn's rings. Are these gaps due to gravitational causes? Questions such as these constitute aspects of the problem of the stability of the solar system.

The existence of a solution for the three-body problem was proved by Sundman, but the solution does not have a form that permits dealing with questions of stability. At present, celestial mechanics enables us to compute the positions of the planets and the satellites within the accuracy of modern observations for a long, but limited, interval of time. The question of the stability of the solar system is closely related to the form of the solution and to the behavior of the series employed. The problem can be put as follows: *What is the interval of time, at the end of which the configuration deviates from the present by a given small amount?* Present mathematics hardly permits this question to be answered satisfactorily for the actual solar system. We must limit ourselves to a description of the present status of the solution of this difficult problem.

Newton's laws of motion are assumed. Einstein's relativistic theory of gravitation will not be treated here nor such mechanisms as encounters with stars nor the dissipation of energy and momentum during the passage through a hypothetical resisting medium. The basis of our discussion is the gravitational theory of Newton and the Galilean notion of space and time.

* Condensed from *Stability in Celestial Mechanics* (Hagihara, 1957).

1. THEORY OF PERTURBATION

The motion of planets around the sun is described by a system of differential equations in rectangular co-ordinates referred to the center of the sun:

$$\frac{d^2x_i}{dt^2}+k^2(m_0+m_i)\frac{x_i}{r_i^3}=\frac{\partial R_i}{\partial x_i},$$

$$\frac{d^2y_i}{dt^2}+k^2(m_0+m_i)\frac{y_i}{r_i^3}=\frac{\partial R_i}{\partial y_i},\quad (i=1,2,\ldots,\text{N}),\qquad (1)$$

$$\frac{d^2z_i}{dt^2}+k^2(m_0+m_i)\frac{z_i}{r_i^3}=\frac{\partial R_i}{\partial z_i},$$

where x_i, y_i, and z_i are the rectangular co-ordinates of the ith planet referred to the sun, N is the number of the planets considered, k^2 the constant of gravitation, m_0 the mass of the sun, m_i the mass of the ith planet, $r_i^2 = x_i^2 + y_i^2 + z_i^2$, and

$$R_i=k^2\sum_j m_j\left(\frac{1}{\Delta_{ij}}-\frac{x_ix_j+y_iy_j+z_iz_j}{r_j^3}\right),$$
$$\Delta_{ij}^2=(x_i-x_j)^2+(y_i-y_j)^2+(z_i-z_j)^2\,.\qquad (2)$$

The function R_i is called the *disturbing function*. If R_i is zero, that is, if we neglect the action of other planets on the motion of the ith planet, then the motion of the planet is Keplerian; the orbit is then a conic section with the sun in one of its foci. In such a case the orbit of the ith planet is described as a function of time t with six constants of integration. They are the semimajor axes a, the eccentricity e, the inclination i of the orbit to the reference plane, which is usually taken to be the ecliptic; the longitude of the perihelion, $\tilde{\omega}$, the longitude of the ascending node Ω, and the mean longitude λ of the epoch. The mean anomaly l is related to the mean longitude λ by $\lambda = l - \tilde{\omega} = nt + \epsilon$, where n is the mean motion and

$$n_i^2a_i^3=k^2(m_0+m_i)\,\mu_i\,,$$

for the ith planet.

If $R_i \neq 0$, the "constants" obtained above for $R_i = 0$ may now be regarded as *slowly variable quantities*, since $m_i \ll m_0$. Then system (1) is found to be equivalent to

$$\frac{da}{dt}=2\sqrt{\left(\frac{a}{\mu}\right)}\frac{\partial R}{\partial \epsilon},$$

$$\frac{de}{dt}=-\frac{\cot\phi}{\sqrt{(\mu a)}}\frac{\partial R}{\partial \varpi}-\frac{\tan\phi/2\,\cos\phi}{\sqrt{(\mu a)}}\frac{\partial R}{\partial \epsilon},$$

$$\frac{di}{dt} = -\frac{1}{\sqrt{(\mu a)}\cos\varphi\sin i}\frac{\partial R}{\partial\Omega} - \frac{\tan i/2}{\sqrt{(\mu a)}\cos\phi}\left(\frac{\partial R}{\partial\varpi} + \frac{\partial R}{\partial\epsilon}\right),$$

$$\frac{d\Omega}{dt} = \frac{1}{\sqrt{(\mu a)}\cos\phi\sin i}\frac{\partial R}{\partial i}, \tag{3}$$

$$\frac{d\varpi}{dt} = \frac{\cot\phi}{\sqrt{(\mu a)}}\frac{\partial R}{\partial e} + \frac{\tan i/2}{\sqrt{(\mu a)}\cos\phi}\frac{\partial R}{\partial i},$$

$$\frac{d\epsilon}{dt} = -2\sqrt{\left(\frac{a}{\mu}\right)}\frac{\partial R}{\partial a} + \frac{\tan\phi/2\cos\phi}{\sqrt{(\mu a)}}\frac{\partial R}{\partial e} + \frac{\tan i/2}{\sqrt{(\mu a)}\cos\phi}\frac{\partial R}{\partial i},$$

where $e = \sin\phi$ and the suffix i is omitted. In system (3) it is supposed that the co-ordinates of various planets in the expression R are replaced by the solution of the system (1) as functions of t with the integration constants obtained by integrating system (1) with $R = 0$. As the eccentricities and the inclinations of the actual planets are small, each term of the sum in equations (2) can be expanded in the following remarkable form (if we omit the suffixes i for the disturbed planet and put primes instead of the suffixes j in equations [2] for the quantities referred to the disturbing planet):

$$R = \Sigma C\cos D\,,$$

$$C = A\,e^{q_1}e'^{q_2}(\tan i)^{q_3}(\tan i')^{q_4}\,, \tag{4}$$

$$D = (j_1 + j_3 + j_5)\,\lambda + (j_2 + j_4 - j_5)\,\lambda' - j_1\varpi - j_2\varpi' - j_3\Omega - j_4\Omega'\,,$$

where A is a homogeneous function of degree -1 in a and a'; $\lambda = nt + \epsilon$, $\lambda' = n't - \epsilon'$, and q_1, q_2, q_3, and q_4 are positive integers or zero; j_1, j_2, j_3, j_4, and j_5 are positive or negative integers or zero; and $q_1 - |j_1|$, $q_2 - |j_2|$, $q_3 - |j_3|$, and $q_4 - |j_4|$ are even positive integers or zero; and, finally, $j_3 + j_4$ and hence also $q_3 + q_4$ are always even. Moreover, R can be expanded in positive integral powers and products of $e\cos\tilde{\omega}$, $e\sin\tilde{\omega}$, $e'\cos\tilde{\omega}'$, $e'\sin\tilde{\omega}'$, $\tan i\cos\Omega$, $\tan i\sin\Omega$, $\tan i'\cos\Omega'$, and $\tan i'\sin\Omega'$. The sum $q_1 + q_2 + q_3 + q_4$ of each term $C\cos D$ in equations (4) is called the *degree* of the term. Brown has called this a d'Alembert series.

In the right-hand members of system (3), n is supposed to appear only in the arguments of the trigonometric functions in this expansion of R, and n in their coefficients should be replaced by a such that $\mu = n^2a^3$. Further, for the convenience of the integration, it is supposed that n in the mean longitude does not vary when partially differentiated with respect to a but that nt should be replaced by $\int n\,dt$. The differential equations (3) are called the formulae for the *variation of elements* (see bibliography at end of section).

The right-hand members of system (3) are small, of the order of the disturbing masses, and, accordingly, the variation of the elements is generally very small. The ordinary method of integration rests entirely on this remarkable feature. Thus we expand the elements according to the disturbing masses,

$$a = a_0 + \delta_1 a_0 + \delta_2 a_0 + \ldots, \qquad a' = a'_0 + \delta_1 a'_0 + \delta_2 a'_0 + \ldots,$$

$$\ldots, \qquad \ldots,$$

$$\Omega = \Omega_0 + \delta_1 \Omega_0 + \delta_2 \Omega_0 + \ldots, \qquad \Omega' = \Omega'_0 + \delta_1 \Omega'_0 + \delta_2 \Omega'_0 + \ldots,$$

$$\ldots, \qquad \ldots,$$

$$n = n_0 + \delta_1 n_0 + \delta_2 n_0 + \ldots, \qquad n' = n'_0 + \delta_1 n'_0 + \delta_2 n'_0 + \ldots,$$

$$R = R_0 + \delta_1 R_0 + \delta_2 R_0 + \ldots,$$

$$R_0 = \Sigma C_0 \cos D_0 ,$$

$$D_0 = q(n_0 t + \epsilon_0) + q'(n'_0 t + \epsilon'_0) - j_1 \varpi - j_2 \varpi' - j_3 \Omega_0 - j_4 \Omega'_0 ,$$

$$q_1 = j_1 + j_3 + j_5 , \qquad q' = j_2 + j_4 - j_5 .$$

In the first approximation, which is called the *perturbation of the first order*, we get by formal term-by-term integration with respect to t, insofar as t appears explicitly on the right-hand sides of system (3),

$$\delta_1 a_0 = 2\sqrt{\frac{a_0}{\mu}} \cdot \sum \frac{qC_0 \cos D_0}{qn_0 + q'n'_0},$$

$$\ldots, \tag{5}$$

$$\delta_1 \Omega_0 = \frac{1}{\sqrt{(\mu a_0)}\sqrt{(1-e_0^2)}\sin i_0} \cdot \sum \frac{(\partial C_0/\partial i_0)\sin D_0}{qn_0 + q'n'_0} .$$

In this manner each term in the expansion of the right-handed members of series (5) corresponds to the successive term in the expansion of R_0. This is called the *inequality*, with the period given by $2\pi/(qn_0 + q'n'_0)$. The period is short, comparable to the period of Keplerian revolution of the planets, provided that $qn_0 + q'n_0$ is not small.

Suppose that $q = q' = 0$; then this method of integration fails, and the result is

$$\delta_1 a_0 = 0 ,$$

$$\ldots,$$

$$\delta_1 \Omega_0 = \frac{t}{\sqrt{(\mu a_0)}\sqrt{(1-e_0^2)}\sin i_0} \sum{}' \frac{\partial C_0}{\partial i'_0} \cos \langle D_0 \rangle , \tag{6}$$

$$\ldots,$$

$$\langle D_0 \rangle = j_1 \varpi_0 + i_2 \varpi'_0 + j_3 \Omega_0 + j_4 \Omega'_0 ,$$

where Σ' denotes the sum over such terms as $q = q' = 0$. Terms proportional to t now appear, and the eccentricities and the inclinations either increase or decrease with increasing value of t. Such terms are called *secular*. At first sight, the secular terms might appear to disturb the stability of the solar system, by allowing, for example, an inner planet at its aphelion to approach the outer planet at its perihelion, thus causing large mutual perturbations.

Now suppose that $qn_0 + q_0'n_0' = 0$ but $q \neq 0$, $q' \neq 0$. Then a similar situation occurs, and terms proportional to t make their appearance. This happens when the mean motions n_0 and n_0' in the first approximation are mutually commensurable. There are cases in which $qn_0 + q'n_0'$ is very small, though not strictly zero, with comparatively small integral values of $|q|$ and $|q'|$. Such a denominator in equations (5) is called a *small divisor*, and the amplitude of the corresponding inequality becomes very large, and terms with large coefficients appear in the formal series obtained by the formal integration of the differential equations. This spoils even the formal convergence of the series. However, the series obtained in this manner is only a formal solution, and, as is described in Section 6, it diverges and therefore is not a true solution of the equations. The problem of stability is closely related to the form of the solution of the planetary motion in the uniformly convergent expression. For the case considered, the solution by this method of successive approximation, which is sometimes called the "Lagrangian method," is simply not adequate, even for the formal integration of the problem.

So far we have discussed perturbations of the first order. Next we consider *second-order perturbations*, that is, the second approximation to the formal solution of the system (3). Let E represent any one of the six elements of the disturbed planet and E' any one of the six elements of the disturbing planet, then the equations for the variation of elements take the form

$$\frac{dE}{dt} = m'F(\lambda, \lambda', a, a', \ldots, \Omega, \Omega', \ldots),$$

with

$$\lambda = \int n\,dt + \epsilon, \qquad \lambda' = \int n'\,dt + \epsilon,$$

$$E = E_0 + \delta_1 E_0 + \delta_2 E_0 + \ldots, \qquad E' = E_0' + \delta_1 E_0' + \delta_2 E_0' + \ldots,$$

$$\lambda = \lambda_0 + \delta_1 \lambda_0 + \delta_2 \lambda_0 + \ldots, \qquad \lambda' = \lambda_0' + \delta_1 \lambda_0' + \delta_2 \lambda_0' + \ldots,$$

and the second-order terms of E are determined by

$$\frac{d\,\delta_2 E}{dt} = m'\left(\frac{\partial F_0}{\partial \lambda_0}\,\delta_1\lambda_0 + \frac{\partial F_0}{\partial \lambda_0'}\,\delta_1\lambda_0' + \frac{\partial F_0}{\partial E_0}\,\delta_1 E_0 + \ldots\right).$$

The function F_0 is derived from R_0 and is of a form similar to R_0. The formal integration proceeds as in the case of the first-order calculation. Similarly, we can proceed to the computation of the perturbation of any higher order.

The final form of the formal solution thus obtained can be proved by the method of mathematical induction (Poincaré, 1905) to be made up of the terms

$$m'^{s} A \frac{t^{p}}{(qn_0 + q'n_0')^{r}} e_0^{q_1} e_0'^{q_2} (\tan i_0)^{q_3} (\tan i_0')^{q_4} \times \frac{\sin}{\cos}(q\lambda_0 + q'\lambda_0' - j_1\varpi_0 - j_2\varpi_0' - j_3\Omega_0 - j_4\Omega_0'), \tag{7}$$

where s, p, r, q_1, q_2, q_3, and q_4 are positive integers or zero; j_1, j_2, j_3, j_4, and j_5 are positive or negative integers or zero, such that $j_1 + j_3 + j_5 = q$, $j_2' + j_4' - j_5' = q'$; and A is a function of a and a'. When $p \neq 0$, but $q \neq 0$ and $q' \neq 0$, the term is called *mixed secular*. When $p \neq 0$ and $q = q' = 0$, the term is called *purely secular;* $q_1 + q_2 + q_3 + q_4$ is called the *degree*, s the *order*, $s - p$ the *rank*, $s - p/2 - r/2$ the *class* of the term. We get the following theorems:

(i) There is no term of negative rank in the development of the elements $\lambda_0 - n_0 t$ and $\lambda_0' - n_0' t$.

(ii) There is no mixed secular term of zero rank.

(iii) There is no term of zero rank in the development of a and a' or n and n'.

It should be emphasized again that this solution is purely formal and, as has been proved by Poincaré (cf. Sec. 6), the series is not uniformly convergent and hence does not represent the solution in the mathematically rigorous sense of the word. However, the formal solution can be used for representing the solution to the degree of accuracy of present-day observational technique when the series are cut off at the terms of some higher order and degree (Sec. 6). It has been remarked by Poincaré that these formal series are semiconvergent, such as the Stirling series employed in mathematical statistics. The question may be raised how far this will hold true when the degree of required accuracy is elevated.

This method can easily be extended to the case where there are N planets. In this case the argument of the trigonometric function (generalized from expression [7]) contains 3N sets of angular variables, $\lambda_1, \tilde{\omega}_1, \Omega_1$; $\lambda_2, \tilde{\omega}_2, \Omega_2$; . . . ; $\lambda_N, \tilde{\omega}_N, \Omega_N$; and its coefficient 3N sets of linear variables a_1, e_1, i_1; a_2, e_2, i_2; . . . ; a_N, e_N, i_N.

REFERENCES

The fundamental treatises on the perturbation theory are as follows:

INTRODUCTORY

CHARLIER, C. L. 1927 *Die Mechanik des Himmels* (2d ed.; Leipzig: W. de Gruyter), Vols. **1** and **2**.

CHEYNE, C. H. H. 1862 *Elementary Treatise on the Planetary Theory* (London: Macmillan & Co., Ltd.).

DZIOBEK, O. 1888 *Mathematische Theorie der Planetenbewegung* (Leipzig: Barth).

FINLAY-FREUNDLICH, E. 1958 *Celestial Mechanics* (London: Pergamon Press).

MOULTON, F. R. 1914 *An Introduction to Celestial Mechanics* (2d ed.; New York: Macmillan Co.).

PLUMMER, H. C. 1918 *An Introductory Treatise on Dynamical Astronomy* (Cambridge: Cambridge University Press).

SMART, W. M. 1953 *Celestial Mechanics* (London: Longmans, Green & Co.).

TISSERAND, F. 1889–1896 *Traité de mécanique céleste* (Paris: Gauthier-Villars), Vols. **1–4**.

ADVANCED OR SPECIALIZED

ANDOYER, H. 1923–1926 *Cours de mécanique céleste* (Paris: Gauthier-Villars), Vols. **1** and **2**.

BROWN, E. W. 1896 *An Introductory Treatise on the Lunar Theory* (Cambridge: Cambridge University Press).

BROWN, E. W., and SHOOK, C. A. 1933 *Planetary Theory* (Cambridge: Cambridge University Press).

HAPPEL, H. 1941 *Das Dreikörperproblem* (Leipzig: K. F. Koehler).

NEWCOMB, S. 1874 *Smithsonian Contr. to Knowledge*, No. 281.

POINCARÉ, H. 1905–1919 *Leçons de mécanique céleste* (Paris: Gauthier-Villars), Vols. **1** and **2**.

SIEGEL, C. L. 1956 *Vorlesungen über Himmelsmechanik* (Berlin: Springer).

WINTNER, A. 1947 *Analytical Foundations of Celestial Mechanics* (Princeton: Princeton University Press).

2. POISSON'S THEOREM ON THE INVARIABILITY OF SEMIMAJOR AXES

The first question to be answered regarding the stability of the solar system is whether or not the dimension of a planetary orbit changes progressively; in other words, whether or not the expression for any of the semimajor axes has pure or mixed secular terms, implying that a planet may approach another planet, or the sun, or even leave the solar system. Laplace, Lagrange, and Poisson discussed the problem, but Tisserand (1876) first proved that there is no secular term in the expression for any of the semimajor axes in the first and second orders. Haretu (1885) discovered a secular term in the third order, and Eginitis (1889) computed the term for the Earth to be $-2.065 \times 10^{-17}\, at$, in which t is expressed in days. Andoyer (1902), Poincaré (1897*a*, 1905, chap. xi), Sundman (1940), and Hagihara (1944) proved the theorem in a more elegant manner by using canonical elements. The following is a summary of the treatment by Hagihara (1944).

We take Jacobi's relative canonical co-ordinates (Jacobi, 1842, 1844) for describing the motion of planets around the Sun, replacing the actual planets according to Bertrand (1852) and Radau (1866) by N fictitious planets with masses

$$m'_i = \frac{M_{i-1}}{M_i}\, m_i \qquad (i = 1, 2, \ldots, \mathrm{N}),$$
$$M_0 = m_0, \qquad M_i = m_0 + \sum_{j=1}^{i} m_j, \tag{8}$$

and with co-ordinates x'_i, y'_i, z'_i, such that

$$x_1 = x'_1, \qquad x_2 = x'_2 + \frac{m_1}{M_1}\, x'_1, \qquad x_3 = x'_3 + \frac{m_2}{M_2}\, x'_2 + \frac{m_1}{M_1}\, x'_1, \ldots,$$

$$y_1 = y'_1, \qquad y_2 = y'_2 + \frac{m_1}{M_1}\, y'_1, \qquad y_3 = y'_3 + \frac{m_2}{M_2}\, y'_2 + \frac{m_1}{M_1}\, y'_1, \ldots,$$

$$z_1 = z'_1, \qquad z_2 = z'_2 + \frac{m_1}{M_1}\, z'_1, \qquad z_3 = z'_3 + \frac{m_2}{M_2}\, z'_2 + \frac{m_1}{M_1}\, z'_1, \ldots.$$

Thus the first planet is referred to the Sun, the second planet to the center of mass of the Sun and the first planet, the third planet to the center of mass of the Sun and the first two planets, etc. For each of the fictitious planets we use Poincaré's set of canonical variables (Poincaré, 1897*b*, 1905, chap. iv):

$$\Lambda = \beta \sqrt{a}, \qquad \Gamma = \beta \sqrt{a}\,[1 - \sqrt{(1 - e^2)}],$$
$$Z = \beta \sqrt{[a(1 - e^2)]}\,(1 - \cos i), \tag{9}$$
$$\lambda = l + \varpi, \qquad \gamma = -\varpi, \qquad z = -\Omega,$$

with

$$\beta_i = k^2 m'_i m_0 m_i \qquad (i = 1, 2, \ldots, \mathrm{N}).$$

Then the differential equations for the variations of elements take the canonical form

$$\frac{d\Lambda_i}{dt}=\frac{\partial F}{\partial \lambda_i}, \qquad \frac{d\Gamma_i}{dt}=\frac{\partial F}{\partial \gamma_i}, \qquad \frac{dZ_i}{dt}=\frac{\partial F}{\partial z_i},$$
$$\frac{d\lambda_i}{dt}=-\frac{\partial F}{\partial \Lambda}, \qquad \frac{d\gamma_i}{dt}=-\frac{\partial F}{\partial \Gamma}, \qquad \frac{dz_i}{dt}=-\frac{\partial F}{\partial Z_i} \tag{10}$$
$$(i=1,\,2,\,\ldots,\,\mathrm{N}),$$

where

$$F=F_0+\mu F_1\,, \qquad F_0=\sum_{i=1}^{\mathrm{N}}\frac{\beta_i^4}{2\,m_i'\Lambda_i}, \quad \text{and}$$

$$F_1=\sum_{\substack{i_1,\,j_1,\,i_2,\,\ldots,\,i_{\mathrm{N}},\,j_{\mathrm{N}}\\ s_1,\,s_2,\,\ldots,\,s_{\mathrm{N}}}} A_{i_1j_1i_2\ldots i_{\mathrm{N}}j_{\mathrm{N}}}\Gamma_1^{i_1}Z_1^{j_1}\Gamma_2^{i_2}\ldots\Gamma_{\mathrm{N}}^{i_{\mathrm{N}}}Z_{\mathrm{N}}^{j_{\mathrm{N}}} \tag{11}$$
$$\times\cos\,(s_1\lambda_1+\ldots+s_{\mathrm{N}}\lambda_{\mathrm{N}}+p_1\gamma_1+q_1z_1+\ldots+p_{\mathrm{N}}\gamma_{\mathrm{N}}+q_{\mathrm{N}}z_{\mathrm{N}})\,.$$

Here μ denotes a small quantity of the order of the disturbing mass m_i', and

$$\lambda_i=\int n_i dt+\epsilon_i\,, \qquad n_i=\frac{\partial F}{\partial \Lambda_i} \qquad (i=1,\,2,\,\ldots,\,\mathrm{N})\,.$$

The summation is extended to integers such that

$$2i_a-|p_a|=2\times\text{integer}, \quad 2j_a-|q_a|=2\times\text{integer} \qquad (a=1,\,2,\,\ldots,\,\mathrm{N}),$$
$$s_1+s_2+\ldots+s_{\mathrm{N}}+p_1+\ldots+p_{\mathrm{N}}+q_1+\ldots+q_{\mathrm{N}}=0\,,$$
$$2i_1+2j_1+\ldots+2i_{\mathrm{N}}+2j_{\mathrm{N}}\geq|s_1+s_2+\ldots+s_{\mathrm{N}}|\,.$$

If we can prove that no purely secular term appears as far as a certain order in m_i', then, by referring to equations (8), it is obvious that there appears no purely secular term as far as the same order in m_i. Hence it is sufficient to prove the theorem with respect to m_i', that is, to μ. Put

$$F_1=V'+\mu V''+\mu^2 V''+\mu^3 V'''+\ldots+\mu^n V^{(n+1)}+\ldots,$$
$$\Lambda=\Lambda^{(0)}+\delta_1\Lambda+\delta_2\Lambda+\ldots,$$
$$\lambda=\lambda^{(0)}+\delta_1\lambda+\delta_2\lambda+\ldots,$$

indicating the orders of the terms explicitly. Then the variation of the semimajor axes in the first and second orders is expressed by

$$\frac{d\,\delta_1\Lambda_i}{dt}=\mu\frac{\partial V'}{\partial \lambda_i} \qquad (i=1,\,2,\,\ldots,\,\mathrm{N}), \tag{12}$$

$$\frac{d\,\delta_2\Lambda_i}{dt} = \mu^2 \frac{\partial V''}{\partial \lambda_i} + \mu \cdot \sum \frac{\partial^2 V'}{\partial \lambda_i \partial \Lambda_j} \delta_1\Lambda_j + \mu \cdot \sum \frac{\partial^2 V'}{\partial \lambda_i \partial \lambda_j} \delta_1\lambda_j$$
$$+ \mu \sum_j \left(\frac{\partial^2 V'}{\partial \lambda_i \partial \Phi_j} \delta_1\Phi_j + \frac{\partial^2 V'}{\partial \lambda_i \partial \phi_j} \delta_1\phi_j \right), \qquad (13)$$

where Φ_j and ϕ_j are an arbitrary pair of canonically conjugate variables other than Λ_j and λ_j. From the terms of zero order we know that there is no term with any of the λ_i's in the first equation of system (10). Hence $\Lambda_i = \Lambda_i^{(0)}$ is constant in the zero order.

The terms containing λ_i are always periodic, as the form of equations (11) shows. Hence $d\delta_1\Lambda_i/dt$ has no constant term independent of any λ_i on the right-hand side of equations (12). Thus there is no secular, pure or mixed, term in $\delta_1\Lambda_i$, and the rank of terms in $\delta_1\Lambda_i$ is 1.

TABLE 1

COMPUTED SECULAR VARIATION PER CENTURY OF MEAN CENTENNIAL MOTION (NEWCOMB)

PERTURBED BY	PLANET			
	Mercury	Venus	Earth	Mars
Venus......	−0.″0426	....	−0.″0104	+0.″0010
Earth......	− .″0029	+0.″0128	....	+ .″0119
Mars.......	+ .″0003	− .″0001	− .″0012	....
Jupiter.....	− .″0039	− .″0046	− .″0308	+ .″0004
Saturn......	−0.″0004	+0.″0015	+0.″0021	+0.″0036
Total...	+0.″0495	+0.″0096	−0.″0403	+0.″0169

It can be proved by generalizing this process that $\delta_2\Lambda_i$ has no purely secular terms arising from the second-order equation (13) but has mixed terms of rank 1. The first and the second terms of equation (13) can, however, give rise to the factor D^{-1}, where D is a denominator of the form $s_1n_1 + s_2n_2 + \ldots + s_Nn_N$ with integral coefficients $s_1, s_2, \ldots, s_N$. It can easily be proved that, by the integration of a mixed secular term, the power of t is unaltered and that the sum of the exponent of t and the power of D^{-1} is increased by 1 for all terms arising from the integration. Thus it is shown that the class of a term in $\delta_2\Lambda_i$ is at least $\frac{1}{2}$.

The proof of Poisson's theorem and that of the theorems on ranks and classes given in Section 1 are complete for the second-order terms. We can proceed similarly to any higher-order terms (Hagihara, 1944).

Newcomb (1895, p. 187) has computed for the inner planets the secular variation of the mean centennial motions, as shown in Table 1. Thus the mean centennial motion of the earth, being $100 \times 360 \times 3600''$, decreases by $0''.04$ per century, or 1 part in 3×10^{11} per year, mainly due to the perturbation by Jupiter. Hence the sidereal year becomes longer by 1.1×10^{-7} days per century and the tropical year longer by 6.16×10^{-6} days per century.

Accordingly, if other effects are neglected, about 10^{11} years would be needed for the mean motion to change by appreciable amounts. Brown (1928; Brown and Shook, 1933) estimated the validity of deductions about earlier configurations of the solar system to be restricted to 10^8 years, but the computations leading to this estimate have not been published.

REFERENCES

ANDOYER, H. 1902 *Ann. Obs. Paris*, Mem. No. 23.

BERTRAND, J. 1852 *J. math. pures appl.*, Vol. **17**.

BROWN, E. W. 1928 *Bull. Amer. Math. Soc.*, **34**, 265.

BROWN, E. W., and SHOOK, C. A. 1933 *Planetary Theory* (Cambridge: Cambridge University Press), p. 248.

EGINITIS, D. 1889 *Ann. Obs. Paris*, Mem. No. 19.

HAGIHARA, Y. 1944 *Japan. J. Astr. Geophys.*, **21**, 9.

HARETU, SPIRU C. 1885 *Ann. Obs. Paris*, Mem. No. 18.

JACOBI, C. G. J. 1842 *A.N.*, Vol. **20**.

1844 *J. reine angew. Math.*, Vol. **26**, see also TISSERAND, F., 1889, *Traité de mécanique céleste* (Paris: Gauthier-Villars), Vol. **1**.

NEWCOMB, S. 1895 *The Elements of the Four Inner Planets and the Fundamental Constants of Astronomy* (Washington: Government Printing Office).

POINCARÉ, H. 1897*a* *Bull. Astr.*, **14**, 261; cf. HILL, G. W., 1904, *A.J.*, **22**, 27.

1897*b* *Bull. Astr.*, **14**, 241.

1905 *Leçons de mécanique céleste* (Paris: Gauthier-Villars), Vol. **1**.

RADAU, R. 1866 *Ann. École Norm. Sup.*, Vol. **5**.

SUNDMAN, K. 1940 *Festschrift für Elis Strömgren* (Copenhagen: E. Munksgaard), p. 263.

TISSERAND, F. 1876 *Mém. Acad. Sci. Toulouse*, Ser. 7, Vol. **7**.

3. THE LAPLACE-LAGRANGE THEORY OF SECULAR PERTURBATION IN THE ECCENTRICITIES AND INCLINATIONS

With the method of integration described in Section 1, a series was obtained with pure and mixed secular terms, proportional to t and higher powers of t. If this type of series represented the actual solution, the present configuration of the solar system would not be preserved. Laplace and

Lagrange have avoided secular terms, although only formally, with their theory of secular perturbations. The secular terms are treated separately and the corresponding equations are integrated in periodic functions with periods corresponding to the reciprocals of the disturbing masses.

Consider the first-order perturbation. Then, as Section 1 shows, the semimajor axes are constant. We put

$$e_i \sin \varpi_i = h_i\,, \qquad \tan i_i \sin \Omega_i = p_i\,,$$
$$e_i \cos \varpi_i = k_i\,, \qquad \tan i_i \cos \Omega_i = q_i \qquad (i = 1, 2, \ldots, \mathrm{N})\,.$$

The differential equations for the secular inequalities, then, are

$$\begin{aligned} \frac{dh_i}{dt} &= \frac{1}{\sqrt{(\mu a_i)}}\frac{\partial R_i}{\partial k_i}\,, & \frac{dp_i}{dt} &= \frac{1}{\sqrt{(\mu a_i)}}\frac{\partial R_i}{\partial q_i}\,, \\ \frac{dk_i}{dt} &= \frac{-1}{\sqrt{(\mu a_i)}}\frac{\partial R_i}{\partial h_i}\,, & \frac{dq_i}{dt} &= \frac{-1}{\sqrt{(\mu a_i)}}\frac{\partial R_i}{\partial p_i}\,, \end{aligned} \tag{14}$$

where only secular terms in R_i are considered. The secular part of the disturbing function is developed in positive integral powers of h_i, k_i, p_i, and q_i, according to the property of the disturbing function described in Section 1. Then, if we take terms to the second degree, the solution, obtained by making use of the secular determinant has the form

$$\begin{aligned} a_i\sqrt{(m_i n_i)}\; e_i \sin \varpi_i &= \sum_{j=1}^{\mathrm{N}} M_j^{(i)} \sin (g_j t + \beta_j)\,, \\ a_i\sqrt{(m_i n_i)}\; e_i \cos \varpi_i &= \sum_{j=1}^{\mathrm{N}} M_j^{(i)} \cos (g_j t + \beta_j)\,, \end{aligned} \tag{15}$$

where the g_j's are the roots of the secular determinant and one of the $M_j^{(i)}$—say $M_j^{(1)}$—and β_j (with $j = 1, 2, \ldots, \mathrm{N}$) are the necessary integration constants, 2N in number. The ratios $M_j^{(i)}/M_j^{(1)}$ (with $i = 1, 2, \ldots, \mathrm{N}$) are determined in the course of the solution with each of the roots g_j of the secular determinant. We get similar expressions for the inclinations and the nodes, with f_j replacing g_j.

The roots of the secular determinant are all real, but they may be multiple. If all the roots are simple, the solution can be expressed formally in purely trigonometric form. The non-existence of multiple roots has not been ascertained except for $\mathrm{N} = 2$ and $\mathrm{N} = 3$ (Seeliger, 1879). The form of the solution in the case when there is a pair of multiple roots has been studied by Charlier (1900*a*). In this case a term proportional to t appears, and formal stability is lost. On the other hand, if the series for the secular part of R_i is cut off after the second-degree terms, the following integrals result:

$$\sum_{i=1}^{N} m_i n_i a_i^2 e_i^2 = \text{Constant} , \qquad \sum_{i=1}^{N} m_i n_i a_i^2 \tan^2 i_i = \text{Constant} . \quad (16)$$

These are the famous integrals of Laplace and Lagrange. From them we find that the eccentricities and the inclinations will remain small, since the constants on the right-hand sides are small with the present values of e and i. This circumstance alone establishes the stability of the solar system with regard to e and i within the accuracy of equations (16). This accuracy is limited by the neglect of terms higher than the second degree in the deviation of both equations (15) and equations (16). This implies an uncertainty of about 1 part in 400, being the value of e_2 for Jupiter. Because the value of the product mna^2 for the Earth is only 1/700 of that of Jupiter, the contribution to the sums in equations (16) by the terrestrial planets is unimportant, and no conclusions can be derived for these planets from equations (16). Equations (15) may, of course, be applied to each planet separately. In addition, equations (16) are limited by the assumption that $a_i =$ constant, and hence their validity is in no case greater than that of the theorem on the constancy of a (see Sec. 2).

Equations (15) can also be written as follows:

$$a_i \sqrt{(m_i n_i)}\, e_i \sin(\varpi_i - g_k t - \beta_k) = \sum_{j,\, j\neq k}{}' M_j^{(i)} \sin[(g_j - g_k)t + \beta_j - \beta_k] ,$$

$$a_i \sqrt{(m_i n_i)}\, e_i \cos(\varpi_i - g_k t - \beta_k)$$
$$= M_k^{(i)} + \sum_{j,\, j\neq k}{}' M_j^{(i)} \cos[(g_j - g_k)t + \beta_j - \beta_k] .$$

If the sum of the absolute values of $M_1^{(i)}, M_2^{(i)}, \ldots, M_{k-1}^{(i)}, M_{k+1}^{(i)}, \ldots, M_N^{(i)}$ is smaller than $|M_k^{(i)}|$ for one of the planets, then $\cos(\tilde{\omega}_i - g_k t - \beta_k)$ cannot become zero, and $\tilde{\omega}_i$ cannot pass through $g_k t + \beta_k \pm \pi/2$, so that $\tilde{\omega}_i$ makes a periodic oscillation around $(g_k t + \beta_k)$ or $(g_k t + \beta_k + \pi)$. In this case the perihelion makes a revolution around the sun, and g_k is called *the mean motion of the perihelion* of that planet. The above is sometimes called the Lagrangian condition and is sufficient, but not necessary, for the existence of this mean motion. Stockwell (1872) formulated this problem when he computed the numerical values of the secular inequalities by this method. He showed that the Lagrangian condition was not satisfied in the case of the perihelia of Venus and the Earth and, in the case of the nodes of Venus, the Earth, and Mars. Since then the question has been studied by several writers (Gyldén 1872, 1893, 1894, 1895; Cavallini 1895). Bohl

(1909) succeeded in proving the existence of mean motions of the perihelion and the node in the case $N = 3$. Weyl (1916) proved the theorem for $N = 4$ on the basis of an extension of Kronecker's theorem on the theory of numbers (Bernstein, 1912; Bohr, 1913; Weyl, 1914); he later extended it (Weyl, 1938, 1939) to *any* value of N on the concept of probability distribution and the theorem of random walks by Pearson and Kluyver (Watson, 1922) and obtained a general formula for the mean motions of the perihelion and node. Wintner (1929–1940) discussed the question of the asymptotic probability distribution, using almost-periodic functions and the ergodic theorem (described in Sec. 9).

For the asteroids the stability of the eccentricities and inclinations cannot be inferred from equations (16), as their masses are vanishingly small. We must instead return to equations (14) applied to an asteroid and substitute the known solution (15) for the major planets. The integration is straightforward. The locus in the (h, k)-plane for the asteroid moves along a circle with a radius which Hirayama (1922, 1927, 1933) has called the *proper eccentricity*. He has applied this concept to a systematic study which led to his discovery of the asteroid families.

The above integration fails, however, for the nodes at $a = 1.951$. However, if we include higher-degree secular terms in the expansion of the disturbing function, the integration can be carried out with elliptic functions (Charlier, 1900*b;* Trousset, 1917). Further, LeVerrier noticed that between Jupiter and Saturn there is a place where asteroidal inclinations would increase indefinitely. This feature also disappears when higher-degree terms are taken into account, as has been done by Tisserand (1882) and Idman (1900).

For the discussion of secular perturbations of the eccentricities we have considered terms up to the second degree in the expansion of the disturbing function. Harzer (1895) applied the theory to the eight principal planets by also taking higher-degree terms into account. The quantities h_i, k_i, p_i, and q_i are expressed in trigonometric series with arguments $g_jt + \beta_j$ and $f_jt + \delta_j$. Hence, using the ordinary method of successive approximation, we find corresponding to a term such as $h_i^\alpha k_j^\beta$ on the right-hand sides of the differential equations (14) a term with argument $\alpha(g_it + \beta_i) + \beta(g_jt + \beta_j)$. The formal integration may give a very small divisor $\alpha g_i + \beta g_j$, because g_i and g_j are of the order of disturbing masses. This difficulty has been resolved by Hagihara (1928, 1930*a*, 1940) on the basis of the so-called Peano-Baker matrix method of integration of differential equations with periodic coefficients; no small divisor then occurs in the solution which can be represented by purely periodic functions with arguments proportional to t. Analogous discussions have been applied by the same authors to the secular perturbations of the inclinations. However, Kosai (1954) pointed out the importance of mixed terms in e and i and supplemented Hagihara's work; the formal results were still expressible in sums of periodic terms.

The foregoing result can be looked upon as supplementing the theory of Newcomb (1874), which expresses the equations of planetary motion by series of purely periodic terms having linear functions of time as the arguments (cf. Sec. 6) and which therefore establishes the formal stability of the solar system. "Formal" designates here that the series are not uniformly convergent.

The actual numerical computation of the values of the secular variations of the major planets was carried out by LeVerrier and later by Stockwell (1872) and was recently revised by Brouwer and van Woerkom (1950) with allowance for Harzer's (1895) work (see Figs. 1–4 of chap. 3).

Gauss (1876) invented a new method of computing the secular variation of elements. He found that the secular variations due to another planet are equal to those produced by an elliptical ring of the same mass, with a density distribution corresponding to the time spent by the planet in different sectors of its orbit, and that these variations may thus be computed with inclusion of higher-degree terms. The analysis was revised by Hill (1901, 1907), and numerical values were calculated by Doolittle (1912). If these theoretical results are compared with the observational data compiled by Newcomb, certain differences appear; they occur in the motions of the perihelion of Mercury, the node of Venus, and the perihelion of Mars, as well as in the variation of the eccentricity of Mercury (Newcomb, 1895; Hagihara, 1930*b*). The first discordance found its interpretation by Einstein's theory of general relativity and is still its principal astronomical test. The other discordances are smaller, and a full discussion of them lies outside the scope of this chapter (cf. chap. 3).

REFERENCES

BERNSTEIN, S. 1912 *Math. Ann.*, **71**, 417.

BOHL, P. 1909 *J. reine angew. Math.*, **135**, 189.

BOHR, H. 1913 *Acta Math.*, **36**, 202.

BROUWER, D., and VAN WOERKOM, A. J. J. 1950 *Astr. Papers, Amer. Ephem. Naut. Almanac*, Vol. **13**, No. 2.

CAVALLINI, C. B. S. 1895 *Medd. Lund Astr. Obs.*, No. 19.

CHARLIER, C. V. L. 1900*a* *Öfversigt Vetensk.-Akad., Förhandl.*, **57**, 1083.

1900*b* *Bull. Astr.*, **17**, 209.

DOOLITTLE, E. 1912 *Trans. Amer. Phil. Soc.*, **22**, 2.

GAUSS, C. F. 1876 *Werke* (Göttingen: K. Gesell. Wiss.), **3**, 331. See also HILL, G. W., 1906, *Collected Mathematical Works* (Washington: Carnegie Institution), **2**, **1**.

GYLDÉN, H. 1872 *Svensk. Vetensk.-Akad., Handl.*, Vol. **11**.

1893 *Traité analytique des orbites absolues des huit planètes principales* (Stockholm: Beyers), Vol. **1**.

1894 *Bull. Akad. Sci. St. Petersbourg*, Sér. 5, Vol. **1**.

1895 *Astr. Iaktt. Unders. Stockholm Obs.*, Vol. **5**, Cf. LEVI-CIVITA, T., 1911, *Ann. École Norm. Sup.*, Ser. 4, **28**, 325; TREVISANI, L., 1912, *Atti R. Ist. Veneto Sci.*, **71**, 1089.

HAGIHARA, Y. 1928 *Proc. Phys.-Math. Soc. Japan*, Ser. 3, **10**, 1, 34, 87, and 127.

1930*a* *Ibid.*, **12**, 22.

1930*b* *J. Astr. Soc. Japan*, **1**, 1 (in Japanese).

1940 *Festschrift für Elis Strömgren* (Copenhagen: E. Munksgaard), p. 58.

HARZER, P. H. 1895 *Die säkularen Veränderungen der Bahnen der grossen Planeten, Preisschrift* (Leipzig: S. Hirzel).

HILL, G. W. 1901 *Amer. J. Math.*, **23**, 317.

1907 *Collected Mathematical Works* (Washington: Carnegie Institution), **4**, 219.

HIRAYAMA, K. 1922 *Japan. J. Astr. Geophys.*, **1**, 55.

1927 *Ibid.*, **5**, 137.

1933 *Proc. Imp. Acad. Japan*, **9**, 482.

IDMAN, A. 1900 *Öfversigt Vetensk.-Akad., Förhandl.*, **57**, 977.

KOZAI, Y. 1954 *Pub. Astr. Soc. Japan*, **6**, 41.

NEWCOMB, S. 1874 *Smithsonian Contr. to Knowledge*, No. 281.

1895 *The Elements of the Four Inner Planets and the Fundamental Constants of Astronomy* (Washington: Government Printing Office).

SEELIGER, H. VON 1879 *A.N.*, **93**, 353. For $N > 3$ refer to G. DARBOUX's article in *Œuvres de Lagrange* (Paris: Gauthier-Villars, 1888), **11**, 492–497 (note 8).

STOCKWELL, J. N. 1872 *Smithsonian Contr. to Knowledge*, No. 232.

TISSERAND, F. 1882 *Ann. Obs. Paris*, Vol. **16**.

TROUSSET, J. 1917 *Ann. Obs. Bordeaux*, Vol. **16**.

WATSON, G. N. 1922 *A Treatise on the Theory of Bessel Functions* (Cambridge: Cambridge University Press), p. 419.

WEYL, H. 1914 *Nachr. Gesell. Wiss. Göttingen*, p. 234.

1916 *Math. Ann.*, **77**, 313.

1938 *Amer. J. Math.*, **60**, 889.

1939 *Ibid.*, **61**, 143.

WINTNER, A. 1929 *Math. Zs.*, **30**, 290.
1932 *J. London Math. Soc.*, **7**, 242.
1933*a* *Math. Zs.*, **36**, 618.
1933*b* *Amer. J. Math.*, **55**, 309, 335, 603, 606.
1934*a* *Ibid.*, **56**, 401.
1934*b* *Math. Zs.*, **37**, 479.
1934*c* *Monatsh. Math. Phys.*, **41**, 1.
1935 *Amer. J. Math.*, **57**, 821, 839.
1938 *Asymptotic Distribution and Infinite Convolution* (Princeton: Princeton University Press).
1940 *Proc. Nat. Acad. Sci.*, **26**, 126. See also HEAVILAND, E. K., 1933, *Proc. Nat. Acad. Sci.*, **19**, 549; HEAVILAND, E. K., and WINTNER, A., 1934, *Amer. J. Math.*, **56**, 1 and 17; VAN KAMPEN, E. R., and WINTNER, A., 1937, *Amer. J. Math.*, **59**, 175; KERSHNER, R. B., and WINTNER, A., 1936, *Amer. J. Math.*, **58**, 91.

4. THEORY OF LIBRATION

The method used in the solution of equations (5) breaks down for $qn_0 + q'n_0' = 0$. Now consider a system consisting of an asteroid and Jupiter, both moving around the Sun. When the mean motions of the asteroids are plotted, *gaps* are found around the mean motions commensurable with that of Jupiter and a clustering tendency slightly different at mean motions. It seems that the motions of asteroids at the commensurability points are unstable and that such mean motions pass to nearby values.

Klose (1923) has ascribed this tendency to the presence of small divisors in the ordinary formal expansion of the co-ordinates. It should be borne in mind, however, that the solution discussed in Section 1 is not uniformly convergent, and hence we cannot draw any definite conclusions about the behavior for long time intervals from the ordinary formal expansions of the perturbation theory. Hirayama (1918, 1928) once considered the possible existence of a resisting medium for the explanation of the gaps, but he later withdrew the suggestion. The *families* of asteroids which he discovered are considered in Volume **5**.

Brown (1928) concluded that near a commensurability point the mean motion of an asteroid varies rapidly and makes a libration about this point. He studied in detail the resonance phenomena in simplified dynamical systems such as a coupled pair of pendulums (Brown, 1932). Wilkens (1933), Okyay (1935), and Urban (1935) discussed the cases of multiple commensurability among asteroids, with respect to Jupiter and Saturn.

Similar gaps or density minima exist in the rings of Saturn (cf. chap. 15). The Cassini division and smaller irregularities appear related to commensurabilities between the mean motion of the ring particles and that of Mimas or Enceladus.

Contrary to the case of the asteroids, among the satellites and planets cases exist where the mean motions are almost exactly commensurable. Jupiter I, II, and III have mean daily motions such that (de Sitter, 1908, 1909, 1918, 1928)

$$n_1 = 203\overset{\circ}{.}48895528\,, \qquad n_2 = 101\overset{\circ}{.}37472396\,, \qquad n_3 = 50\overset{\circ}{.}31760833\,,$$

$$n_1 - 2n_2 = 0\overset{\circ}{.}73950736\,, \qquad n_2 - 2n_3 = 0\overset{\circ}{.}73950730\,,$$

$$n_1 - 3n_2 + 2n_3 = 0\overset{\circ}{.}00000006\,.$$

The Saturn satellites, Mimas and Tethys, have mean motions in the ratio 2:1, Enceladus and Dione also 2:1, and Titan and Hyperion 4:3. Jupiter and Saturn have mean motions nearly in the ratio 5:2, Uranus and Neptune 2:1. In all these cases the co-ordinates will show terms with small divisors, and the method of solution described in Section 1 fails. For Jupiter and Saturn and for Uranus and Neptune, for example, the computations are based on the methods of Hansen and Newcomb; the solutions are semiconvergent in the sense of Poincaré (1893).

The question thus arises whether the existence of a near-commensurability of the mean motions formally endangers the stability of the system. This question may be dealt with by the method of Delaunay (1860, 1867) or its equivalent by von Zeipel (1916*a*, *b*, 1917, 1918) or that of Brown (1936, 1937; Brown and Shook, 1933, chap. vi). Delaunay's method consists of first taking the constant term and one periodic term out of the expansion of the disturbing function and solving the equations of motion with the necessary integration constants. The integration is carried out with contact transformations, and one such step is called an *operation*. The periodic term is eliminated by the operation. A second periodic term is treated in a similar manner, leading to a canonical system of equations for determining the variation of the constants in the foregoing step. The operation is repeated several times, and the final solution is obtained in a purely trigonometric series with linear functions of the time as arguments. The behavior of the equations of motion in the presence of small divisors will be described here on the basis of a general theory of libration by Hagihara (1944), which followed Delaunay's method as extended by the work of Poincaré (1902) and Andoyer (1903) on the Hecuba group of asteroids.

The equations of motion are used in the form (10) for an asteroid (without primes) and Jupiter (with primes). Consider the contact transformation,

$$v=\lambda-t\,,\qquad \sigma=-s\lambda+s'\gamma+(s+s')t\,,$$
$$\tau=-s\lambda+s'z+(s+s')t\,,\qquad U=\Lambda+sS+sT\,, \tag{17a}$$
$$S=\frac{1}{s'}\Gamma\,,\qquad T=\frac{1}{s'}Z\,,\qquad R'=F-\Gamma-Z+\Lambda\,.$$

The mean motions of the asteroid and Jupiter are supposed to be nearly commensurable in the ratio $n/n' = (s + s')/s$, where s and s' are integers. The quantity v denotes the difference of the mean longitudes of two bodies. Then the equations of motion become

$$\frac{dU}{dt}=\frac{\partial R'}{\partial v}\,,\qquad \frac{dS}{dt}=\frac{\partial R'}{\partial \sigma}\,,\qquad \frac{dT}{dt}=\frac{\partial R'}{\partial \tau}\,,$$
$$\frac{dv}{dt}=-\frac{\partial R'}{\partial U}\,,\qquad \frac{d\sigma}{dt}=-\frac{\partial R'}{\partial S}\,,\qquad \frac{d\tau}{dt}=-\frac{\partial R'}{\partial T}\,, \tag{17}$$

where S and T are of the order of the squares of the eccentricity and the orbital inclination, respectively; $d\sigma/dt$ and $d\tau/dt$ are small; and $d\lambda/dt$ is nearly equal to $(s + s')/s$. We put

$$x=\sqrt{(2S)}\cos\sigma\,,\qquad \xi=\sqrt{(2T)}\cos\tau\,,$$
$$y=\sqrt{(2S)}\sin\sigma\,,\qquad \eta=\sqrt{(2T)}\sin\tau\,. \tag{18a}$$

Then x, y, ξ, and η are small, and the canonical equations (17) become

$$\frac{dU}{dt}=\frac{\partial R'}{\partial v}\,,\qquad \frac{dx}{dt}=\frac{\partial R'}{\partial y}\,,\qquad \frac{d\xi}{dt}=\frac{\partial R'}{\partial \eta}\,,$$
$$\frac{dv}{dt}=-\frac{\partial R'}{\partial U}\,,\qquad \frac{dy}{dt}=-\frac{\partial R'}{\partial x}\,,\qquad \frac{d\eta}{dt}=-\frac{\partial R'}{\partial \xi}\,. \tag{18}$$

Here R' is expanded in a trigonometric series with integral multiples of v as arguments and with integral power series of x, y, ξ, and η, multiplied by a function of $\Lambda = U - sS - sT$, as coefficients. As this function of Λ can be expanded in an ascending power series of $s(S + T)$ or of $\frac{1}{2}s\,(x^2 + y^2 + \xi^2 + \eta^2)$, the Hamiltonian function R' is expressed in an integral power series of $\cos pv$, $\sin pv$, x, y, ξ, and η with functions of U as the coefficients, where p denotes an integer.

Delaunay's method neglects all short-period terms in R' and considers the constant terms and one of the long-period terms, with integral multiples of σ and τ as the arguments. Further, the terms of higher degree in x, y, ξ, and η and in the orbital inclination are neglected. Now consider only the part R of R', where R does not change when y is replaced by $-y$. Hence it is of the form

$$R = A_0 + A_1 x^{s'} + A_2 (x^2 + y^2) + A_3 x^{2+s'} + A_4 (x^2 + y^2)^2 . \quad (19)$$

The coefficients A_0, A_1, A_2, A_3, and A_4 are functions of $\sqrt{\Lambda}$ and $\sqrt{\Lambda'}$.

As R contains neither t nor v, we get integrals with arbitrary constants h and c: $R = h$, and $U = c$. We shall now discuss these integrals. If R_0 is defined by putting $m' = 0$,

$$R_0 = \frac{1}{2\Lambda^2} + \Lambda - s'S = \frac{1}{2U^2} + U + \left(\frac{1}{U^3} - s - s'\right) \frac{x^2 + y^2}{2} . \quad (20)$$

The double points of this curve are given by

$$\frac{\partial R_0}{\partial x} = \frac{\partial R_0}{\partial y} = 0 .$$

There are two solutions, $x = y = 0$, and the intersections A_0 and B_0 of the circle

$$\frac{s}{(U - sS)^3} - s - s' = 0$$

with the x-axis (cf. Fig. 1). The two points A_0 and B_0 correspond to the case when the mean motion of the asteroid is exactly in the ratio $(s + s')/s$

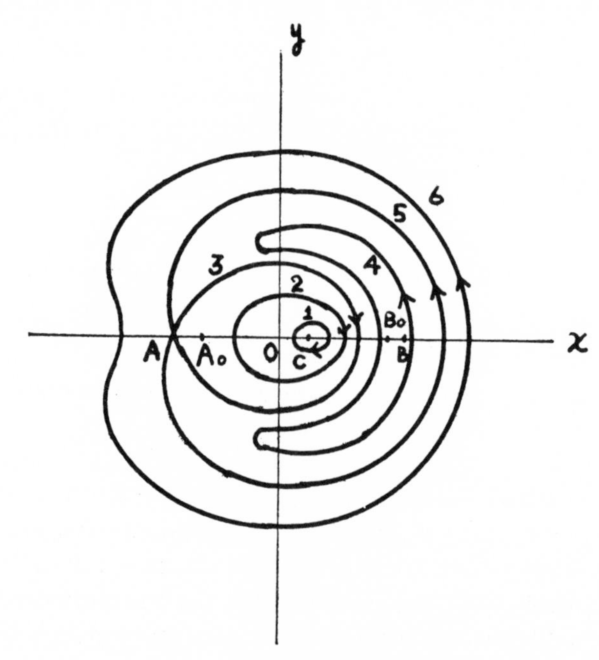

FIG. 1.—Classes of periodic orbits

to that of Jupiter. When $m' \neq 0$, the points O, A_0, and B_0 of Figure 1 displace to C, A, and B, respectively. The curve $R = h$ is shown and represents the orbit in the xy-plane. The oval near O corresponds to a periodic orbit with the eccentricity zero. Thus it represents a periodic solution of the first sort (in the sense of Poincaré, 1892), as is shown in Section 7. The periodic orbits in the neighborhood of A_0 and B_0 correspond to the periodic solutions of the second sort, the eccentricities of the generating solutions being nearly equal to $\sqrt{(2s'S)}$. The radius vector of Figure 1 is $\sqrt{(2S)}$, and the argument is σ. Curves *1* and *4* in Figure 1 represent librations, where σ is the corresponding critical argument,

$$\sigma = -s\lambda + s'\gamma + (s+s')t = (s+s')\lambda' - s\lambda - s'\varpi .$$

Curve *5* is asymptotic to A. Curve *6* makes a complete revolution about the origin.

The most important case is $s' = 1$. We then have two integrals,

$$R = A_0 + A_1 x + A_2(x^2+y^2) + A_3 x^3 + A_4(x^2+y^2)^2 = h ,$$

$$U = \Lambda + \frac{s}{2}(x^2+y^2) = c .$$

Let $\Lambda = \Lambda_0 - \zeta$, with $\zeta \ll \Lambda_0$. The quantity ζ denotes the variable part of Λ. Substituting this value of Λ in the integrals R and U and expanding, since $\Lambda_0 = U = c$, we get

$$\frac{3}{2c^4}(\zeta - a)^2 - b + A_1(2S)^{1/2}\cos\sigma + A_2(2S) + A_3(2S)^{3/2}\cos^3\sigma + A_4(2S)^2 = 0 , \tag{21}$$

$$\zeta = sS = \tfrac{1}{2}s(2S) , \tag{22}$$

with properly chosen constants a and b.

We draw a diagram with ζ and $\sqrt{(2S)}$ as the two rectangular co-ordinates. Curve (22) is a parabola, and curve (21) is contained between two curves of the fourth degree:

$$\frac{3}{2c^4}(\zeta - a)^2 - b \pm A_1(2S)^{1/2} + A_2(2S) \pm A_3(2S)^{3/2} + A_4(2S)^2 = 0 , \tag{23}$$

because $-1 \leq \cos\sigma \leq 1$. Curves (23) are nearly parabolic in the region not far from the co-ordinate origin (see Fig. 2). Also shown are examples of curves (23) corresponding to $\sigma = 0$ and $\sigma = \pi$, with $a > 0$ and $b/A_1 > 0$. The part of the parabola (22) lying between the intersections with curve (23) expresses the dependence of the eccentricity upon the semimajor axes satisfying the periodic solution; it is shown heavy in Figure 2. These figures are similar to those by Hirayama (1918, 1928), although we have taken the terms of higher degree into account. For the Hecuba group ($s = 1$), Wilkens (1927) has carried out a numerical integration of the differential equations, but his statement concerning the absence of periodicity is based on a misunderstanding. This matter was cleared up by Hirayama (1928), who previously (1918) had developed a general theory of libration. In a subsequent paper Wilkens (1930) conceded the correctness of Hirayama's analysis. In the case of the Hilda group ($s = 2$), Hirayama and Akiyama (1937, 1938) have made numerical computations similar to Wilkens but by the method of special perturbations. The points representing the computed values of a and e lie indeed very near the curves drawn by Hirayama. Similar studies were made for the Thule group ($s = 3$) (Ura and Takenouchi, 1951; Kozai, 1952, 1953).

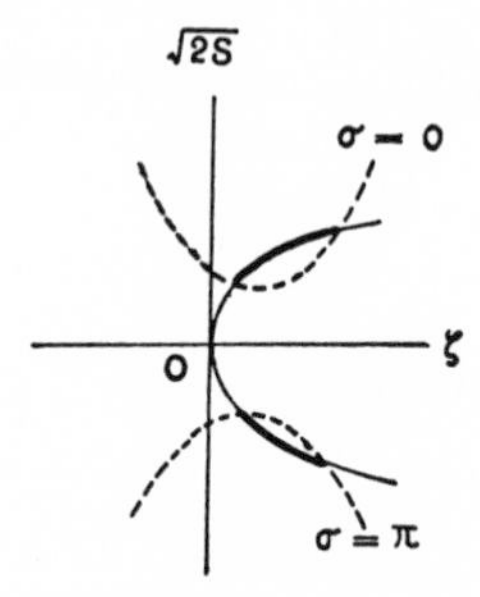

FIG. 2.—Dependence of eccentricity on semimajor axis in periodic solution

Next we consider the time variation of σ (see eq. [17*a*]). To this end we return to the equations of motion (18). We now put (Hagihara, 1944)

$$\frac{1}{2\,U^2} + U - h = U' \,,$$

$$2\beta = \frac{s}{U^3} - (s+1) \,, \qquad \gamma = \frac{s}{2}\frac{\partial A_1}{\partial U} \,,$$

$$\delta = \frac{3\,s^2}{8\,U^4} - \frac{3}{2}\frac{\partial A_3}{\partial U} \,,$$

and

$$p = x + \sqrt{(-1)}\,y \,, \qquad q = x - \sqrt{(-1)}\,y \,.$$

Then the equations determining σ or x and y as functions of t take the form

$$\frac{dp}{dt} = -2\sqrt{(-1)}\,\frac{\partial \Psi}{\partial q} \,, \qquad \frac{dq}{dt} = 2\sqrt{(-1)}\,\frac{\partial \Psi}{\partial p} \,,$$

where

$$\Psi = U' - \tfrac{1}{2} A_1 (p+q) + \beta p q + \tfrac{1}{2}\gamma p q (p+q) + \delta p^2 q^2 = 0 \,.$$

The biquadradic form Ψ is symmetric with respect to p and q and may be written as

$$\Psi = A q^2 + B q + C \qquad \text{or} \qquad \Psi = A' p^2 + B' p + C' .$$

Then, as $\Psi = 0$,

$$\frac{\partial \Psi}{\partial q} = \sqrt{(B^2 - 4AC)}\,, \qquad \frac{\partial \Psi}{\partial p} = \sqrt{(B'^2 - 4A'C')}\,.$$

Put

$$G(p) = 16AC - 4B^2\,,$$

then we have

$$(dt)^2 = \frac{(dp)^2}{G(p)} = \frac{(dq)^2}{G(q)}\,.$$

These equations define the time variations of p and q and therefore of x and y and hence of σ (eq. [18*a*]). We are thus led to the study of the inversion of the elliptic integral $\int dp/\sqrt{[G(p)]}$. As $G(p)$ is a quartic equation for p, which can be written as

$$G(p) = a_0 p^4 + 4a_1 p^3 + 6a_2 p^2 + 4a_3 p + a_4\,, \qquad a_0 = -\gamma^2\,,$$

the solution is obtained by a suitable choice of constants using Weierstrass's elliptic functions with the two invariants g_2 and g_3, in the form

$$p = -\frac{a_1}{a_0} + \zeta(u + \epsilon) - \zeta u - \zeta\epsilon\,,$$

$$u = \pm\sqrt{(a_0)}\,(t - t_0)\,,$$

where ζ is the Weierstrass elliptic function and ϵ is defined by the Weierstrass elliptic function

$$\wp(\epsilon) = \frac{a_1^2 - a_0 a_2}{a_0}\,.$$

The behavior of the motion can now be discussed by the sign of the discriminant $-g_2^3 - 27g_3^2$, and those of a_0, $a_1^2 - a_0 a_2$, and $12\,(a_1^2 - a_0 a_2)^2 - a_0^4 g_2$. It is found that *the motion is periodic, corresponding to either a libration or a revolution, and in certain limiting cases asymptotic; but the values of x and y never increase indefinitely so as to endanger the stability.* A motion is usually said to be a libration when the angular variable representing the motion oscillates within a limited interval. This variable is called the *critical argument.* In our case the variable u makes a complete revolution, while the variable σ makes a libration. It is noted that whether a motion is a libration or a revolution depends on the variable used (Hagihara, 1944).

The discussion of the case $s = 1$, $s' = 0$—that is, of the *Trojan group*—can be made in a similar manner by using elliptic functions. In the more complete Trojan theory by Brown (1923, 1925) he proved that the long-period terms, such as the libration caused by the Sun and Jupiter, do not increase in amplitude even by the action of other planets such as Saturn. Brown's result proves the stability of the libration against the action of other planets.

We now turn to satellite systems. The theory can be applied to the motion of Hyperion under the action of Titan (Woltjer, 1917, 1918, 1919, 1921, 1922*a*, 1928), for which $s = 3$, $s' = 1$. Application of the theory to this case shows that the critical argument is

$$\sigma = -3\lambda + \gamma + 4t .$$

For the system Mimas and Tethys ($s = s' = 1$) the critical argument is similarly found to be

$$\tau = -2\lambda + 2(z - z') + 4t ,$$

while for the system Enceladus and Dione ($s = s' = 1$), for which Woltjer (1922*b*) says there is a libration, the critical argument is

$$\sigma = -\lambda + \gamma + 2t .$$

The most important case is the system of the first three Galilean satellites of Jupiter. The critical argument now is

$$\vartheta = \lambda_1 - 3\lambda_2 + 2\lambda_3 .$$

This combination of the three longitudes can occur only by combining two terms in the disturbing function, such as

$$F = A_0 + A_1 m_2 m_3 \cos(2\lambda_3 - \lambda_2) + A_2 m_1 m_3 \cos(4\lambda_3 - \lambda_1) + A_3 m_1 m_2 \cos(2\lambda_2 - \lambda_1) ,$$

where A_0, A_1, A_2, and A_3 contain Γ_1, Γ_2, Γ_3, and γ_1, γ_2, γ_3, besides Λ_1, Λ_2, Λ_3. The equations of motion are

$$\frac{d\Lambda_j}{dt} = \frac{\partial F}{\partial \lambda_j}, \qquad \frac{d\Gamma_j}{dt} = \frac{\partial F}{\partial \gamma_j},$$

$$\frac{d\lambda_j}{dt} = -\frac{\partial F}{\partial \Lambda_j}, \qquad \frac{d\gamma_j}{dt} = -\frac{\partial F}{\partial \Gamma_j} \qquad (j = 1, 2, 3) .$$

From these equations one finds

$$\frac{d^2\lambda_j}{dt^2} = B_j \sin \vartheta \qquad (j = 1, 2, 3) .$$

Hence

$$\frac{d^2\vartheta}{dt^2} + H^2 \sin \vartheta = 0 . \qquad (24)$$

In this approximation the coefficients B_j and H^2 are constants. Equation (24) describes the motion of a spherical pendulum. When we consider higher-degree terms and the effect of other satellites and planets, periodic terms make their appearance on the right-hand side. If one of the periods of these periodic terms becomes nearly commensurable with $2\pi/H$, then the motion becomes quite complicated but interesting. This general problem is usually called "Duffing's problem." It has been studied by Iglisch (1930, 1932, 1935, 1936) and others (Stoker, 1950) on the basis of the theory of non-linear integral equations developed by Schmidt (1908).

Thus the presence of small divisors in the ordinary perturbation theory does not make the situation worse as to the stability of the motion. We return to this matter in the next section. According to Brown (1928; Brown and Shook, 1933, p. 249), "so long as the past history of the solar system was supposed to be confined within an interval of 10^8 years, deductions as to its initial configuration from its present configuration appeared to have some degree of value," even by taking such resonance phenomena into account.

The theory of libration described above can be applied to the motion of an artificial satellite with critical inclination arc cos $1/\sqrt{5}$, for which one gets a small divisor by Brouwer's (1959) method of solution (Hagihara, 1961; see also Kozai, 1961, and Hori, 1960).

REFERENCES

ANDOYER, H. 1903 *Bull. Astr.*, **20**, 321.

BROUWER, D. 1959 *A.J.*, **64**, 378.

BROWN, E. W. 1923 *Trans. Yale U. Obs.*, Vol. **3**, Part 1.

1925 *Ibid.*, Part 3.

1928 *Bull. Amer. Math. Soc.*, **34**, 265.

1932 *Elements of the Theory of Resonance* (Cambridge: Cambridge University Press). Cf. KRYLOV, N., and BOGOLIUBOV, N., 1934, *Application des méthodes de la mécanique non-linéaire à la théorie des perturbations des systèmes canoniques* (Kiev: Acad. Sci. Ukraine).

1936 *M.N.*, **97**, 56, 62, 116.

1937 *Ibid.*, **98**, 170. Cf. BIRKHOFF, G. D., 1927, *Amer. J. Math.*, **49**, 1; 1929, *Dynamical Systems* (New York: American Math. Soc.), chap. 3, as discussed in our Sec. 6.

BROWN, E. W., and SHOOK, C. A. 1933 *Planetary Theory* (Cambridge: Cambridge University Press).

DELAUNAY, C. 1860 *Théorie du mouvement de la lune* ("Mém. Acad. Sci. Inst. Imp. France," Vol. **28**).
1867 *Ibid.*, Vol. **29**

HAGIHARA, Y. 1944 *Japan. J. Astr. Geophys.*, **21**, 29.
1961 *Smithsonian Contr. to Astrophysics*, Vol. **5**, No. 6.

HIRAYAMA, K. 1918 *Coll. Sci. Tokyo Imp. U.*, Vol. **41**.
1928 *A.J.*, **38**, 147. Cf. HAGIHARA, Y., 1928, *Proc. Phys.-Math. Soc. Japan*, **10**, 15; 1928, *Japan. J. Astr. Geophys.*, **5**, 1.

HIRAYAMA, K., and AKIYAMA, K. 1937 *Proc. Imp. Acad. Japan*, **13**, 191.
1938 *Japan. J. Astr. Geophys.*, **15**, 137.

HORI ,G. 1960 *A.J.*, **65**, 291.

IGLISCH, R. 1930 *Monatsh. Math. Phys.*, **37**, 325.
1932 *Ibid.*, **39**, 173.
1935 *Math. Ann.*, **111**, 568.
1936 *Ibid.*, **112**, 221.

KLOSE, A. 1923 *A.N.*, **218**, 401.

KOZAI, Y. 1952 *Pub. Astr. Soc. Japan*, **3**, 84.
1953 *Ibid.*, **4**, 172.
1961 *Smithsonian Contr. to Astrophysics*, Vol. **5**, No. 6.

OKYAY, T. 1935 *A.N.*, **255**, 277.

POINCARÉ, H. 1892 *Les méthodes nouvelles de mécanique céleste* (Paris: Gauthier-Villars), Vol. **1**, chap. 3. See also our Sec. 7.
1893 *Ibid.*, Vol. **2**.
1902 *Bull. Astr.*, **19**, 289.

SCHMIDT, E. 1908 *Math. Ann.*, **65**, 370.

SITTER, W. DE 1908 *Proc. Acad. Sci. Amsterdam*, Vol. **10**.
1909 *Ibid.*, Vol. **11**.
1918 *Ann. Leiden Obs.*, Vol. **12**, Part 1.
1928 *Ibid.*, Vol. **16**, Part 2; see also SAMPSON, R. A., 1921, *Mem. R.A. Soc.*, Vol. **63**.

STOKER, J. J. 1950 *Non-linear Vibrations* (New York: Interscience Publishers), chap. 4.

URA, T., and TAKENOUCHI, T. 1951 *Pub. Astr. Soc. Japan*, **2**, 93.

URBAN, K. 1935 *A.N.*, **256**, 109.

WILKENS, A. 1927 *Sitz. Bayer. Akad. Wiss.*, p. 147.
1930 *A.N.*, **240**, 201.
1933 *Sitz. Bayer. Acad. Wiss.*, p. 71.

WOLTJER, J. 1917 *Proc. Acad. Sci. Amsterdam*, **19**, 1225.
1918 *Investigations in the Theory of Hyperion* (Dissertation, Leiden).
1919 *Proc. Acad. Sci. Amsterdam*, **21**, 886, 1166.
1921 *M.N.*, **81**, 603.
1922*a* *B.A.N.*, **1**, 13, 31, 96, 176.

1922b *Ibid.*, **1**, 23, 36.
1928 *Ann. Leiden Obs.*, Vol. **16**, Part 2.
ZEIPEL, H. VON 1916a *Ark. f. mat., astr., fys.*, Vol. **11**, No. 1.
1916b *Ibid.*, Vol. **11**, No. 7.
1917 *Ibid.*, Vol. **12**, No. 9.
1918 *Ibid.*, Vol. **13**, No. 3.

5. STABILITY OF SATELLITE SYSTEMS AND THE RINGS OF SATURN

The motion of our satellite, the Moon, was successfully studied by Hill (1873, 1886, 1905) on the basis of his ingenious theory involving the periodic solution of the problem. Suppose that the Sun at infinity moves around the Earth as origin in a circle at a uniform rate and that the Moon moves in the rotating co-ordinate plane with the x-axis always pointing toward the Sun. In this rotating system Hill has chosen as the intermediary orbit for the Moon a periodic orbit with two arbitrary integration constants. From this intermediary orbit the actual orbit has been obtained by the superposition of small periodic terms and small empirical secular terms as may be needed and which may be due largely to tidal friction. Thus the principal feature on the stability of the motion of the Moon can be judged by the behavior of the intermediary orbit, provided that the amplitudes of the superimposed periodic terms always remain within certain narrow limits.

Let x and y be the rectangular axes, the x-axis always pointing toward the Sun. Let the mean motion of the Sun around the Earth as origin be n' in a stationary frame. In this plane problem the equations of motion of the Moon are

$$\begin{aligned} &\frac{d^2x}{dt^2} - 2n'\frac{dy}{dt} + \left(\frac{\mu}{r^3} - 3n'^2\right)x = 0\,, \\ &\frac{d^2y}{dt^2} + 2n'\frac{dx}{dt} + \frac{\mu}{r^3}\,y = 0\,, \end{aligned} \tag{25}$$

where μ is the sum of the masses of Moon and Earth multiplied by the constant of gravitation. Strictly speaking, it has also been assumed that the lunar mass is infinitesimal compared with that of the Earth. The appropriate Jacobi integral is readily found:

$$\frac{1}{2}\left[\left(\frac{dx}{dt}\right)^2 + \left(\frac{dy}{dt}\right)^2\right] = \frac{\mu}{r} + \frac{3}{2}\,n'^2x^2 - C\,,$$

with the constant of integration C. The left-hand member is positive or zero. Hence the motion occurs within the region where

$$\frac{\mu}{(x^2+y^2)^{1/2}} + \frac{3}{2}\,n'^2x^2 - C \geq 0 \tag{26}$$

and is thus bounded by a so-called *zero-velocity curve*. The behavior of this curve is shown in Figure 3, *a*, *b*, and *c*, according as $(2C)^{3/2}$ is larger than, equal to, or smaller than $9\mu n'$. In our case $(2C)^{3/2} > 9\mu n'$ and $OX_1 = OX_2 = \sqrt{(2C/3n'^2)} = 500.4992$; $OA_1 = 109.694$; $OA_2 = 109.655$; and $OB_1 = OB_2 = 104.408$, the unit being the equatorial radius of the Earth. The Moon is at present within the oval of zero velocity, $OA_1B_1A_2B_2$ of Figure 3, *a*. Since it cannot cross the boundary of the oval, it must always remain inside. This proves, within the approximations of the problem introduced above, the stability of the motion of the Moon.

Similar arguments were applied by Hagihara (1952) to other satellite systems, with a slight change of units. Again, the Sun is assumed to have uniform circular motion at infinity, and the problem is assumed to be

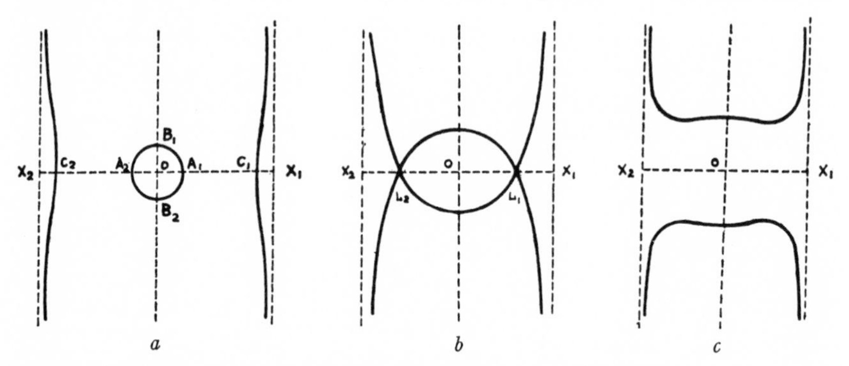

FIG. 3.—Zero-velocity curves for the intermediary orbit in the three-body problem: (*a*) $(2C)^{3/2} > 9\mu n'$; (*b*) $(2C)^{3/2} = 9\mu n'$; (*c*) $(2C)^{3/2} < 9\mu n'$.

plane. The satellite mass is considered infinitesimal. For all satellites of the solar system with the exception of Jupiter VIII, IX, XI, and XII, the situation is then similar to that of the Moon. These four exceptional satellites form a group, all at nearly the same mean distance from Jupiter, all having retrograde motions. For this group $(2C)^{3/2} < 9\mu n'$. This is the case shown in Figure 3, *c*, for which the criterion does not disclose stability of the motion. At about half the mean distance of VIII, IX, XI, and XII, Jupiter has a second group of satellites, VI, VII, and X. These satellites have direct motions and are stable on the basis of the Jacobi integral. For Phoebe of Saturn and Triton of Neptune we get $(2C)^{3/2} > 9\mu n'$, and the present orbits are inside the oval of Figure 3, *a*, even though the motions are retrograde (Hagihara, 1952). Brown (1930; Brown and Brouwer, 1937) and Kovalevsky (1959) have studied the motion of the eighth satellite of Jupiter but have found no peculiarity indicating instability. Nicholson (1944) and Hori (1957*a*, 1958) have studied the ninth satellite. The stability of the motion of a retrograde planet has been proved by Jackson

(1913), Moulton (1914), and Hirayama (1927) by means of the usual Laplace theory of secular perturbations.

In some of the satellite systems, pairs occur whose mean motions are almost exactly commensurable. Such a case has been treated in Section 4 on the basis of the perturbation theory of Section 2 (small divisors). But the question naturally arises whether the stability is affected when other disturbing bodies, such as the other satellites and the Sun, and the figure of the planet are taken into account. Usually, the stability of the motion is discussed for the three-body problem without taking such effects into consideration (Brown, 1901). As we have remarked, in the case of the Trojan asteroids, Brown has proved that the amplitude of the long-period libration terms does not grow without limit by the action of planets other than Jupiter (Brown, 1923, 1925; Brouwer, 1933; Eckert, 1933; Aoki, 1955).

Hagihara (1927) has discussed the stability of a satellite system composed of several satellites two of which have nearly commensurable mean motions. He took the effects of the solar perturbation and of the figure of the planet into account. The method is based on the theory of the characteristic exponents defined in Section 7. Starting with an intermediary non-periodic orbit with a moving pericenter for each of the two satellites with nearly commensurable mean motions, the differential equations for the disturbed motion are formed. If these equations in the first approximation are regular and if all n characteristic exponents have negative real parts, then the undisturbed motion is stable, and the disturbed motion approaches the undisturbed asymptotically. On this criterion the stability of the motion of the satellite system is proved for a special configuration of the pericenters of the satellite pair.

Saturn has several satellite pairs with such commensurability relations (see Sec. 4). The actual configurations of the orbits are such that stability exists in each case. For Titan and Hyperion the conjunction always occurs in the aposaturnium of Hyperion (Newcomb, 1891; Woltjer, 1922, 1928; Brouwer, 1924). For Enceladus and Dione the conjunction occurs at the perisaturnium of Enceladus. For Mimas and Tethys the relation exists for the *node* instead of the perisaturnium. The conjunction of the two satellites occurs near the mid-point between their two ascending nodes on Saturn's equator.

The most interesting case is that of Jupiter I, II, and III. As the intermediary orbit for the three satellites, de Sitter (1908, 1909, 1918, 1928) adopted a periodic solution with the same perijove motion. The stability of this solution was proved by the method of characteristic exponents. The perijove of II lies in the direction opposite to those of I and III. If I and III were in conjunction at a certain epoch and II in opposition, all in their apojoves, then the motion would be unstable. But with I and III in conjunction and II in opposition, all in their perijoves, the motion will be stable. The actual case is the second.

Even if there is no commensurability relation between the mean motions, the stability is of interest, such as for Rhea under the action of Titan. Hagihara (1927, p. 81, 1940) has taken an intermediary nonperiodic orbit with moving perisaturnium and proved the stability of the motion by a matrix method of integrating a system of differential equations for disturbed motion.

Saturn's rings show one gap and some irregularities in the radial distribution of the ring material (see chap. 15). One may suspect that the "gaps" lie near the points of commensurability with the mean motion of Mimas or that of Enceladus. Indeed, Cassini's division occurs near the commensurability points 2 to 1 with Mimas and 3 to 1 with Enceladus. The stability of motion of small bodies at the commensurability point has been discussed by Goldsbrough (1921, 1922, 1924; see also Pendse, 1933, 1935, 1937). Goldsbrough considered the small bodies situated at the vertices of an equilateral polygon with Saturn at the center and rotating around Saturn. This configuration is what was called by Dziobek (1900) a *central figure* (Andoyer, 1906; Meyer, 1933). Goldsbrough considered a small deviation of a body from this equilibrium position and discussed the solution of the equation of the disturbed motion under the action of Mimas from the point of view of stability. He proved the instability of the motion around the two principal commensurability points. Hagihara (1940) has applied Peano-Baker's matrix method for discussing this problem and obtained greater mathematical generality of the solution. Lichtenstein (1923, 1924, 1932) has based the discussion on the theory of nonlinear integro-differential equations. He has proved the existence of periodic solutions and found that the deviation from the equilibrium position propagates in wave form.

We may now summarize the main conclusions of this section. The stability of a satellite system is, if anything, strengthened by the presence of a commensurability relation in the mean motions, provided that certain conditions on the relative configuration are fulfilled (as, for example, the case of J I, II, III). But if the system contains a great number of members and the positions of the pericenters and the longitudes in the orbits are arbitrary (as in the asteroid ring and the ring of Saturn), then gaps will arise in the distribution of the mean motions or distances, apparently due to the disturbing action of close encounters with small neighboring masses (Hagihara, 1961).

REFERENCES

ANDOYER, H. 1906 *Bull. Astr.*, **23**, 50.

AOKI, S. 1955 *Pub. Astr. Soc. Japan*, **7**, 105–113; *Tokyo Astr. Rep.*, No. 124.

BROUWER, D. 1924 *B.A.N.*, **2**, 119.

1933 *Trans. Yale U. Obs.*, Vol. **6**, Part 7.

BROWN, E. W. 1901 *Bull. Amer. Math. Soc.*, **8**, 103.

1923 *Trans. Yale U. Obs.*, Vol. **3**, Part 1.

1925 *Ibid.*, Part 3.

1930 *Ibid.*, Vol. **6**, Part 4.

BROWN, E. W., and BROUWER, D. 1937 *Trans. Yale U. Obs.*, Vol. **6**, Part 8.

DZIOBEK, O. 1900 *A.N.*, **152**, 33.

ECKERT, W. J. 1933 *Trans. Yale U. Obs.*, Vol. **6**, Part 6.

GOLDSBROUGH, G. R. 1921 *Phil. Trans. A*, **222**, 101.

1922 *Proc. R. Soc. London, A*, **101**, 280.

1924 *Ibid.*, **106**, 526.

HAGIHARA, Y. 1927 *Japan. J. Astr. Geophys.*, **5**, 1.

1940 *Festschrift für Elis Strömgren* (Copenhagen: E. Munksgaard), p. 58.

1952 *Proc. Imp. Acad. Japan*, **28**, 182. Data for Jupiter XII have been taken from NICHOLSON, S. B., 1952, *Pub. A.S.P.*, Leaflet No. 275.

1961 *Smithsonian Contr. to Astrophysics*, Vol. **5**, No. 6.

HILL, G. W. 1873 *Amer. J. Math.*, **5**, 129, 245.

1886 *Acta Math.*, **8**, 1.

1905 *Collected Mathematical Works* (Washington: Carnegie Institution), **1**, 243, 284.

HIRAYAMA K. 1927 *Proc. Imp. Acad. Japan*, **3**, 9.

HORI, G. 1957*a* *Pub. Astr. Soc. Japan*, **9**, 51.

1957*b* *Proc. Japan. Acad.*, **33**, 395.

1958 *Ibid.*, **34**, 263.

JACKSON, J. 1913 *M.N.*, **74**, 62.

KOVALEVSKY, J. 1959 *Bull. Astr.*, **23**, 11.

LICHTENSTEIN, L. 1923 *Math. Zs.*, **17**, 62.

1924 *Probleme der Astronomie: Festschrift für Hugo von Seeliger* (Berlin: J. Springer), p. 200.

1932 *Ann. Scuola Norm. Sup., Pisa*, Ser. 2, **1**, 173.

MEYER, O. 1933 *Ann. Obs. Bordeaux*, **17**.

MOULTON, F. R. 1914 *M.N.*, **75**, 40.

NEWCOMB, S. 1891 *Astr. Papers, Amer. Ephem. Naut. Almanac*, Vol. **3**, Part 3.

NICHOLSON, S. B. 1944 *Ap. J.*, **100**, 57.

PENDSE, C. G. 1933 *Phil. Mag.*, Ser. 2, **16**, 575.

1935 *Phil. Trans. A*, **234**, 145.

1937 *Phil. Mag.*, Ser. 2, **23**, 425.

ROY, A. F., and OVENDEN, M. T.	1954	*M.N.*, **114**, 232.
	1955	*Ibid.*, **115**, 296.
SITTER, W. DE	1908	*Proc. Acad. Sci., Amsterdam*, Vol. **10**.
	1909	*Ibid.*, Vol. **11**.
	1918	*Ann. Sterrew. Leiden*, Vol. **12**, Part 1.
	1928	*Ibid.*, Vol. **16**, Part 2.
WOLTJER, J.	1922	Dissertation, Leiden.
	1928	*Ann. Sterrew. Leiden*, Vol. **16**, Part 2.

6. THE INTEGRALS OF PLANETARY THEORY

If we could express in a mathematically rigorous sense the exact solution of the differential equations of planetary theory in periodic functions of the time t, or in the quasi-periodic functions of Bohl and Esclangon, or in the almost-periodic functions of Harald Bohr, then the stability of the planetary system would follow at once. During the last century and since then, the usual procedure has been to express the solution as a trigonometric series whose arguments are linear functions of the time, without examining the convergence of the assumed expansion.

Newcomb (1871, 1874) obtained the integrals of planetary theory in such a form by extending the Laplace-Lagrange theory of secular perturbations (Delaunay, 1860, 1867; von Zeipel, 1916, 1917, 1918; Brown and Shook, 1933; Brown, 1936, 1937). The planetary co-ordinates are expressed in Section 1 as sums of terms of the form

$$m'h \frac{\cos}{\sin} (i_1\lambda_1 + \ldots + i_N\lambda_N + j_1\varpi_1 + \ldots + j_N\varpi_N + k_1\Omega_1 + \ldots + k_N\Omega_N) ,$$

where

$$\lambda_i = n_i t + \epsilon_i \qquad (i = 1, 2, \ldots, N) ,$$

$$h = A e_1^{j_1} \ldots e_N^{j_N} (\tan i_1)^{k_1} \ldots (\tan i_N)^{k_N} (1 + A_1 e_1^2 + A_2 e_2^2 + \ldots) ,$$

$$(i_1 + \ldots + i_N + j_1 + \ldots + j_N + k_1 + \ldots + k_N = 0) ,$$

and $A, A_1, A_2, \ldots$, are functions of the semimajor axes. By the Laplace-Lagrange theory of secular perturbations we have

$$e_i \frac{\sin}{\cos} \varpi_i = \sum_j N_j^i \frac{\sin}{\cos} (g_j t + \beta_j) ,$$

$$\tan I_i \frac{\sin}{\cos} \Omega_i = \sum_j M_j^i \frac{\sin}{\cos} (f_j t + \delta_j) \quad (i = 1, 2, \ldots, N) ,$$

where I is the inclination. Substituting these, we get the expression for the co-ordinates in the form

$$\sum m' A' \begin{matrix} \sin \\ \cos \end{matrix} (i_1\lambda_1 + \ldots + i_N\lambda_N + j_1\zeta_1 + \ldots + j_N\zeta_N + j'_1\zeta'_1 + \ldots + j'_N\zeta'_N),$$

$$(i_1 + \ldots + i_N + j_1 + \ldots + j_N + j' + \ldots + j'_N = 0),$$

where N is the number of planets and A' is a function of N_i, M_i, and the semimajor axes. Thus there are 3N arguments, all linear functions of t:

$$g_i t + \beta_i = \zeta_i, \qquad f_i t + \delta_i = \zeta'_i,$$

$$\lambda_i = m_i t + \epsilon_i \qquad (i = 1, 2, \ldots, N).$$

When there are commensurability relations among n's, then Section 4 shows that the co-ordinates can nevertheless be expressed in this form with suitable change in the variables. Even when there are commensurability relations among g_i's and f_i's, Newcomb's theory remains valid, as has been proved by Hagihara (1928, 1930). Delaunay (1860, 1867) obtained the integrals of the planetary and lunar theory in purely periodic form by his ingenious method of transformations described in Section 4. Von Zeipel's method (1916, 1917, 1918) and Brown's new method (1936, 1937) are related to this procedure.

Lindstedt (1882, 1883*a*, *b*, *c*, 1884) has reduced the discussion of the formal solution to that of the differential equation

$$\frac{d^2x}{dt^2} + n^2 x = \Psi_0 + \Psi_1 x + \Psi_2 x^2 + \ldots,$$

with the constants n, α, β, and A. In order to obtain a first approximation, this equation is replaced by

$$\frac{d^2x}{dt^2} + n^2(1-\sigma)x = -n^2\sigma x - 2Ax\cos(\alpha t + \beta),$$

in which A is supposed small, $x = \eta_0 \cos w$ and $w = n\sqrt{(1-\sigma)}t + \tilde{\omega}$, η_0 and $\tilde{\omega}$ being constants. The quantity σ is so chosen that there is no term $\cos w$ in the right-hand member. By successive approximations the solution is obtained in the form

$$x = \sum_{j=-\infty}^{\infty} p_j \cos[j(\alpha t + \beta) + w].$$

Bruns (1883*a*, *b*), Tisserand (1892*a*, *b*), Stieltjes (1884), Bohlin (1888), Charlier (1904), and others have modified or generalized the process. Gyldén (1881, 1882, 1887, 1891) considered an equation of the type

$$\frac{d^2x}{dt^2} + [n^2 + 2A\cos(\alpha t + \beta)]x = U,$$

where

$$U = \sum_i A_i \cos(\alpha_i t + \beta_i),$$

a known function of t. Gyldén solved the equation by using Hermite's solution of Lamé's equation,

$$\frac{d^2x}{dt^2} - (2k^2 \mathrm{sn}^2 u - h)\, x = 0,$$

and expanding it in ϑ-series of the theory of elliptic functions, i.e., in purely trigonometric forms. Poincaré (1893*a*, *b*) discussed all these solutions and proved that none of them is uniformly convergent. He published (1897) a beautiful method for solving the equations of the three-body problem in trigonometric series of such a type.

In the problem of three bodies the method was applied by von Zeipel (1916, 1917, 1918) to the motion of asteroids. Bohlin (1907, 1908, 1921, 1923, 1925, 1931, 1935) tried to obtain the solution in three independent variables instead of the single variable t by the transformation by which Sundman (1908) succeeded in proving the existence of the solution. Let n_1, n_2, and n_3 correspond to the mean motions; a, b, and c, the constants corresponding to the semimajor axes; and r_1, r_2, and r_3, the mutual distances. By the transformation $n_1\, dt = r_1\, du/a$, $n_2\, dt = r_2\, dv/b$, $n_3\, dt = r_3\, dw/c$, Bohlin obtained the solution in the form

$$\sum_{-\infty}^{\infty} A_{hkl}\, e^{\sqrt{(-1)}\,(hu+kv+lw)},$$

and found five classes of expansions, corresponding to the roots of the quintic equation determining the three equilibrium points in Euler's particular solution of the straight-line case of the three-body problem. Hill (1886, 1905) in his lunar theory reduced the problem to a linear differential equation now called by his name,

$$\frac{d^2x}{dt^2} + (a_0 + a_1 \cos t + a_2 \cos 2t + \ldots)\, x = 0.$$

When $a_2 = a_3 = \ldots = 0$, then the equation reduces to the type called the "Mathieu equation," which has become important in the modern theory of solids and metals. Whittaker (1914) invented a method of successive approximation similar to Lindstedt's and solved it in purely trigonometric series subject to certain condition among the constants. Ince (1915, 1924) generalized the method to solve Hill's equation, and Goldsbrough (1921, 1922, 1924) applied it to Saturn's rings (cf. Sec. 5). Hagihara (1941) applied Peano-Baker's matrix method (Baker, 1915) of solution to several problems in celestial mechanics. The latter method is, in reality, founded on Lie's theory of continuous groups. The region of validity of periodic solutions of such equations can be defined by either method. Whittaker (1916) studied his "Adelphic" integral for a dynamical system with two degrees of freedom in a canonical form for which the Hamiltonian function begins with a linear term $s_1q_1 - s_2q_2$. He obtained different forms of the solution, depending on whether the ratio s_1/s_2 was rational or irrational. These solutions are all expanded in convergent series. By contrast the solution expressed in the form of a *single* equation is not uniformly convergent, as was proved by Poincaré (1882, 1883, 1884, 1885, 1889, 1892, 1893*a*, *b*, 1895, 1896, 1898).

On the convergence of the series used in perturbation theory, Bruns (1884) discovered a remarkable theorem. The perturbation of the first order, as defined in Section 1, is

$$E = \sum{}' \frac{B_{ii'jj'}}{in + i'n'} \sin (i\lambda + i'\lambda' + D_{jj'}) + Ct + E_0 ,$$

$$\lambda = nt + \epsilon , \qquad \lambda' = n't + \epsilon' ,$$

$$D_{jj'} = jg + j'g' \neq 90° ,$$

where Σ' denotes the sum over i, i', j, j' such that $in + i'n' \neq 0$; C and E_0 are constants; and $B_{ii'jj'}$ is expanded in integral power series of the eccentricities and the inclinations. In order to discuss the convergence of this series, we consider the case when $|in + i'n'|$ is small. Let the series be

$$\sum \frac{K_{ii'}\kappa^r}{i - i'\nu} \sin (i\lambda - i'\lambda' + D) \qquad (|\kappa| < 1) ,$$

with $\nu = n'/n$; $r = |i - i'|$; and $i, i' = 1, 2, \ldots$. If ν is rational, then the terms for which $i - i'\nu = 0$ are included in the term Ct. If ν is irrational, we consider a series with $\nu < 1$,

$$S = \sum \left| \frac{K_{ii'}\kappa^r}{i - i'\nu} \right| .$$

The series $\Sigma K_{ii'}\kappa^r$ is absolutely convergent with the radius of convergence less than 1. Given a real number ν_0 and a positive or negative number δ, there exists an infinite number of points in the interval from $\nu = \nu_0$ to $\nu = \nu_0 + \delta$ at which the above series converges, however small one may choose δ, and also an infinite number of points at which the series diverges. That is, the points at which the series converges and the points at which the series diverges lie everywhere densely in the domain of ν. Gyldén (1888) considered this question from the point of view of probability distributions. Recently Petersson (1924) and Siegel (1941) extended the proof by basing it on modern mathematics. The final conclusion is that the series obtained so far for the integrals of planetary theory are not uniformly convergent and hence that the stability of motion cannot be decided by the form of these integrals, even though they look like sums of periodic terms. Possibly the form of the solution may be discussed by a closer examination of the properties explored by Painlevé (1895) of the singular points in the three-body problem, apart from Taylor series expansions in a regularizing variable (Sundman, 1907, 1908, 1913).

For how long will the expansion, cut off after a certain number of terms, be valid within a given tolerance? Consider a canonical system of differential equations,

$$\frac{dq_i}{dt} = \frac{\partial H}{\partial p_i}, \qquad \frac{dp_i}{dt} = -\frac{\partial H}{\partial q_i} \qquad (i = 1, 2, \ldots, \text{N}),$$
$$H = H_2 + H_3 + \ldots, \qquad H_2 = \sum_{j=1}^{N} \lambda_j p_j q_j. \tag{27}$$

By using canonical transformations, the solution of the transformed equation, up to the terms of degree s, can be obtained in the form

$$p_i = A_i e^{-\gamma_i t}, \qquad g_i = B_i e^{\gamma_i t},$$
$$\gamma_i = \frac{\partial H(A_1, B_1, \ldots, A_{\text{N}}, B_{\text{N}})}{\partial c_i} \qquad (i = 1, 2, \ldots, \text{N}),$$

with integration constants A_i and B_i, where the integrals $p_i q_i = c_i =$ constant are obtained ($i = 1, 2, \ldots, \text{N}$). Put

$$\epsilon^2 = \sum_{j=1}^{\text{N}} p_j q_j$$

and let the values of ϵ, p_j, and q_j for $t = t_0$ be, respectively, ϵ_0, p_j^0, and q_j^0. Then it can be shown that the error introduced by cutting off at terms of degree s, upon integration of the periodic series, does not exceed

$$A\epsilon_0^{s+1} + B\epsilon_0^{s+2} |t - t_0| + C\epsilon_0^{s+3} |t - t_0|^2 \tag{28}$$

during the time interval

$$|t - t_0| \leq \frac{1}{2\text{N} sQ\epsilon_0^s}, \tag{28$'$}$$

with suitably chosen finite positive constants A, B, and C. The notion of *complete stability* by Birkhoff (1927, 1929) is based on this result. The above discussion can easily be extended to the case when the differential equations contain t explicitly in the form of periodic functions. Hagihara (1931, 1944) generalized the discussion to the canonical equations treated by Poincaré with the Hamiltonian H ($x_1, y_1, x_2 \ldots, y_m; \xi_1, \eta_1, \xi_2, \ldots, \eta_n$), by assuming that we have a solution $x_i = y_i = 0$ ($i = 1, 2, \ldots, m$), $\xi_j = A_j$, and $\eta_j = B_j$ ($j = 1, 2, \ldots, n$); and that the expansion of H in powers of x_i and y_i begin with quadratic terms with constant coefficients. The quantities x_i and y_i correspond to the semimajor axes and the longitudes in the planetary theory and ξ_j and η_j to the perihelia and the nodes.

Siegel (1952, 1954) discussed the existence of a normal form, such as equation (27), of an analytic differential equation in the neighborhood of an equilibrium point. He proved for $N = 2$ and for λ_1 and λ_2, both purely imaginary, that the canonical transformation of a Hamiltonian system to a normal form is, in general, divergent but that, in this case, the equilibrium is not necessarily unstable.

REFERENCES

BAKER, H. F.	1915	*Phil. Trans. A*, **216**, 129.
BIRKHOFF, G. D.	1927	*Amer. J. Math.*, **49**, 1.
	1929	*Dynamical Systems* (New York: American Math. Soc.), chap. iii.
BOHLIN, K.	1888	*Bihang Svensk. Vetensk.-Akad., Handl.*, **14**, 5.
	1907	*Astr. Iaktt. Unders. Stockholm Obs.*, Vol. **8**.
	1908	*Ibid.*, Vol. **9**.
	1921	*Ark. f. mat., astr., fys.*, Vol. **16**.
	1923	*Astr. Iaktt. Unders. Stockholm Obs.*, Vol. **10**.
	1925	*Ark. f. mat., astr., fys.*, Vol. **18**.
	1931	*Astr. Iaktt. Unders. Stockholm Obs.*, Vol. **11**.
	1935	*Ark. f. mat., astr., fys.*, Vol. **25**.
BROWN, E. W.	1936	*M.N.*, **97**, 56, 62, 116.
	1937	*Ibid.*, **98**, 170.
BROWN, E. W., and SHOOK, C. A.	1933	*Planetary Theory* (Cambridge: Cambridge University Press), chap. vi.
BRUNS, H.	1883*a*	*A.N.*, **106**, 193.
	1883*b*	*Ibid.*, **107**, 129. Cf. CALLANDREAU, O., 1883. *A.N.*, **107**, 33.
	1884	*Ibid.*, **109**, 215.
CHARLIER, C. V. L.	1904	*Ark. f. mat., astr., fys.*, **1**, 449.
DELAUNAY, A.	1860	*Théorie du mouvement de la lune* ("Mém. Acad. Sci. Inst. Imp. France," Vol. **28**).
	1867	*Ibid.*, Vol. **29**.
GOLDSBROUGH, G. R.	1921	*Phil. Trans. A*, **222**, 101.
	1922	*Proc. R. Soc. London*, **101**, 280.
	1924	*Ibid.*, **106**, 526.

Gyldén, H. 1881 *Vierteljahrsschr. Astr. Gesell.*, **16**, 296.
1882 *A.N.*, **103**, 321.
1887 *Acta Math.*, **9**, 185.
1888 *Öfversigt Svensk. Vetensk.-Akad., Förhandl.*, **45**, 77, 349.
1891 *Acta Math.*, **15**, 65. Cf. Tisserand, F., 1888, *Ann. Fac. Sci. Toulouse*, **2**, 1.

Hagihara, Y. 1928 *Phys.-Math. Soc. Japan*, Ser. 3, **10**, 1, 34, 87, 127.
1930 *Ibid.*, **12**, 22.
1931 *Proc. Imp. Acad. Japan*, **7**, 44.
1941 *Festschrift für Elis Strömgren* (Copenhagen: E. Munksgaard), p. 58.
1944 *Proc. Imp. Acad. Japan*, **20**, 617, 622.

Hill, G. W. 1886 *Acta Math.*, **8**, 1.
1905 *Collected Mathematical Works* (Washington: Carnegie Institution), **1**, 243.

Ince, E. L. 1915 *M.N.*, **75**, 436.
1924 *Proc. London Math. Soc.*, **23**, 56.

Lindstedt, A. 1882 *A.N.*, **103**, 211, 257.
1883*a* *Mém. Acad. Sci. St. Pétersbourg*, Ser. 7, Vol. **31**, No. 4.
1883*b* *A.N.*, **105**, 91.
1883*c* *C.R.*, **97**, 1276.
1884 *An. École Norm. Sup.*, Ser. 3, **1**, 85. Cf. von Zeipel, H., 1898, *Bihang Svensk. Vetensk.-Akad., Handl.*, **24**, 1, and 1901, *ibid.*, **26**, 1; Tisserand, F., 1889, *C.R.*, **98**, 1207, and 1888, *Ann. Fac. Sci. Toulouse*, **2**, 1; Andoyer, H., 1890, *Ann. Fac. Sci. Toulouse*, **4**, 1.

Newcomb, S. 1871 *J. math. pures appl.*, Ser. 2, **16**, 321.
1874 *Smithsonian Contr. to Knowledge*, No. 281.

Painlevé, P. 1895 *Leçons sur la théorie analytique des équations différentielles* (Paris: Hermann).

Petersson, H. 1924 *Abh. Math. Sem., Hamburg U.*, **3**, 324. Cf. Wintner, A., 1929, *Math. Zs.*, **31**, 434.

Poincaré, H. 1882 *C.R.*, **95**, 766.
1883 *Ibid.*, **97**, 1471.
1884 *Bull. Astr.*, **1**, 319.
1885 *C.R.*, **101**, 1131.
1889 *Ibid.*, **108**, 21.
1892 *Ibid.*, **114**, 1305.
1893*a* *Acta Math.*, **13**, 1.
1893*b* *Méthodes nouvelles de la méchanique céleste* (Paris: Gauthier-Villars), Vol. **2**.
1895 *C.R.*, **120**, 57.
1896 *Ibid.*, **122**, 497, 557. See also Poincaré, 1893*a* and *b*.

	1897	*Bull. Astr.*, **14**, 241. Cf. HAGIHARA, Y., 1931, *Proc. Phys.-Math. Soc. Japan*, **13**, 39; WHITTAKER, E. T., 1902, *Proc. London Math. Soc.*, **34**, 206.
	1898	*Bull. Astr.*, **15**, 289.
SIEGEL, C. L.	1941	*Ann. Math. Princeton U.*, **42**, 806.
	1952	*Nachr. Gesell. Wiss. Göttingen*, p. 21.
	1954	*Math. Ann.*, **128**, 144.
STIELTJES, T. J.	1884	*A.N.*, **109**, 145, 261.
SUNDMAN, K. F.	1907	*Acta Soc. Sci. Fenn.*, Vol. **34**.
	1908	*Ibid.*, Vol. **35**.
	1913	*Acta Math.*, **36**, 105.
TISSERAND, F.	1892*a*	*C.R.*, **114**, 441.
	1892*b*	*Bull. Astr.*, **9**, 102.
WHITTAKER, E. T.	1914	*Proc. Edinburgh Math. Soc.*, **32**, 75.
	1916	*Proc. R. Soc. Edinburgh*, **37**, 95. Cf. CHERRY, T. M., 1926, *Proc. London Math. Soc.*, **26**, 211, and 1927, *ibid.*, **27**, 151.
ZEIPEL, H. VON	1916	*Ark. f. mat., astr., fys.*, Vol. **11**, Nos. 1 and 7.
	1917	*Ibid.*, Vol. **12**, No. 9.
	1918	*Ibid.*, Vol. **13**, No. 3.

7. PERIODIC SOLUTIONS AND CHARACTERISTIC EXPONENTS

Stability could be proved if the planetary configuration has the property of recurrence or periodicity in time in a mathematically rigorous sense, rather than as represented by a semiconvergent series, as discussed in previous sections. Such is the approach of Hill and Poincaré, who proved the existence of periodic solutions. Unfortunately, these periodic solutions do not apply exactly to existing systems, although quite nearly so, but only to idealized configurations. Moreover, the deviation of an actual orbit from such an intermediary orbit, though it may be small at present, might increase with time because of the non-uniform convergence or even the divergence of the series employed.

As an intermediary orbit for the motion of the Moon, Hill (1886, 1905) took a periodic orbit for a particle of vanishingly small mass. As stated in Section 5, this orbit was situated in a rotating plane with the x-axis pointing to the Sun, while the latter moves around the Earth in the reference plane in a circular orbit of infinite radius with mean motion n'. The periodic orbit selected was a particular solution of the equations of motion for which the quantity corresponding to the eccentricity of the Moon's orbit in the ordinary Keplerian motion is zero. This intermediary orbit Hill called the *variational curve*. For the Moon this orbit starts at the epoch $t = t_0$ from a point on the x-axis in a direction perpendicular to this axis and then cuts the other side of this axis, also perpendicularly, after a half-period $\pi/(n - n')$. The co-ordinates of the Moon in the intermediary orbit can be represented by a series of periodic terms,

$$\sum_{i=0}^{\infty} A_{2i+1} \frac{\cos}{\sin} (2i+1)\,\xi\,, \qquad \xi = (n-n')(t-t_0)\,, \tag{29}$$

where the constants A_{2i+1} (with $i = 0, 1, 2, \ldots$) can be expressed in terms of a certain integration constant a, resembling the semimajor axes in the Keplerian motion and the constant $m = n'/(n - n')$. The coefficients can be calculated according to Hill's scheme and are real, as was proved by Poincaré (1909). Wintner (1925, 1926*b*, *c*, 1938*a*, *b*) proved that the series are convergent.

In order to come closer to the actual orbit of the Moon, Hill considered an orbit varied from the periodic intermediary orbit, by taking the eccentricity or, more rigorously, the quantity corresponding to the eccentricity in the Keplerian motion into account. By supposing that the above deviation is small, he formulated differential equations giving the time dependence of this deviation. Consider the equations

$$\frac{dx_i}{dt} = f_i(x_1, x_2, \ldots, x_N;\, t)\,, \tag{30}$$

where $i = 1, 2, \ldots, N$, in which f_i is analytic in $x_1, x_2, \ldots, x_N$ and periodic in t. Suppose that we know a particular solution $x_i = X_i(t)$, where $i = 1, 2, \ldots, N$, with period T. We vary the initial condition and denote the solution by $x_i = X_i(t) + \xi_i(t)$, where $i = 1, 2, \ldots, N$. If we assume that $\xi_i(t)$, $i = 1, 2, \ldots, N$, remains small for all values of t, then the solution can be approximately represented by the solution of linear differential equations with periodic coefficients.

$$\frac{d\xi_i}{dt} = \sum_{j=1}^{N} \left(\frac{\partial f_i}{\partial x_j}\right)_{x_i = X_i(t)} \xi_j(t) \qquad (i = 1, 2, \ldots, N)\,, \tag{31}$$

where the periodic function $x_i = X_i(t)$ $(i = 1, 2, \ldots, N)$ is substituted for x_i in $(\partial f_i/\partial x_j)$. Poincaré has called equations (31) the "variational equations." By denoting by δp the normal deviation from the generating periodic intermediary orbit, Hill obtained the linear differential equation called by his name, referred to in Section 6,

$$\frac{d^2 \delta p}{dt^2} + (n-n')^2 \left(\sum_{j=-\infty}^{\infty} \Theta_j \cos 2j\xi\right) \delta p = 0\,,$$

where the Θ_j's are constants which in turn contain the constants in the generating periodic solution, with $\Theta_j = \Theta_{-j}$ at least to the order of magnitude m^{2j}, m being 0.08085 for the Moon. Hill solved this equation by referring to a determinant (called also by his name) with an infinite number of rows and columns. This infinite determinant has been proved convergent by Poincaré (1893, 1900). In this manner Hill could calculate the motion of the perigee and obtain agreement with observation, something that had been attempted in vain for two centuries. Adams, followed by Cowell, computed the motion of the node by similar methods, independently of Hill. The entire lunar theory was elaborated by Brown.

Poincaré took a keen interest in Hill's ingenious concept of periodic solutions and proved the existence of such solutions in both the perturbation and the three-body problems.

Let μ be, as before, a small parameter of the order of magnitude of the disturbing mass, and consider the differential equations

$$\frac{dx_i}{dt} = X_i\,(x_1, x_2, \ldots, x_N; t; \mu) \qquad (i = 1, 2, \ldots, N)\,,$$

where X_i is analytic in $x_1, x_2, \ldots, x_N$, t, and μ, and periodic with respect to t. Suppose that this system admits a periodic solution

$$x_i = \phi_i\,(t; \mu = 0) \qquad (i = 1, 2, \ldots, N)\,, \quad (32)$$

with period T such that $\phi_i(t) = \phi_i(t + T)$ for all values of t. Poincaré's formulation of the problem of the existence of periodic solutions, then, is: Do periodic solutions exist in the neighborhood of the generating periodic solution ($\mu = 0$), even when μ is not zero but small? Poincaré proved that this is the case, as μ can take at least an enumerable set of values near $\mu = 0$, on the basis of his lemma on the existence of solutions of differential equations containing a parameter μ.

In the planetary problem of three bodies Poincaré (1889, 1891) classified the *periodic solutions of the first genus* (for which the nearby periodic solutions are of period equal, or nearly equal, to that of the generating periodic solution) into three sorts. Consider two planets of small mass revolving around the Sun. When the generating periodic solution for the motion of the two planets is such that they revolve around the Sun with uniform circular motion and with noncommensurable mean motions, then the nearby periodic solutions with small orbital eccentricities are such that the perihelia rotate around the Sun. These *periodic solutions* are said to be *of the first sort.* Hill's intermediary orbit is a special case. For *periodic solutions of the second sort* the orbits of the two planets have finite eccentricities and a special configuration of the perihelia; further, the mean motions are commensurable, and the ratio of the eccentricities is related to the ratio of the major axes. The nearby periodic solutions are such that their perihelia are at rest. The discussion of the stability of the periodic solutions of the first sort has been revised by Hagihara (1927; see also Brendel, 1898, 1903, 1910; Levi-Civita, 1901; Poincaré, 1902; Hölder, 1931) and Wintner (1930, 1931*a*, 1936) by an analytic continuation of the periodic solutions of the second sort. A periodic solution of the first sort is a degenerate case of the second sort and is at the bifurcation point at which the exchange of stability takes place. Inclusion of finite orbital inclinations leads to *periodic solutions of the third sort*. When the period of a nearby periodic solution is equal, or nearly equal, to an integral multiple of the period of the generating solution, then the *periodic solution* is *of the second genus*. Further, when the quantities kept constant in the generating solution become variable and periodic, then the solution is said to be of the *second species*, in contrast to those of the *first species*, for which the variables of the generating and the nearby periodic solutions are the same.

The periodic solutions of the second sort can be applied for establishing the stability of the motion of characteristic asteroids for which the mean motions are nearly commensurable with that of Jupiter. Such periodic solutions have been obtained by Hill (1902) for the Hestia group (de Sitter, 1908, p. 47; Klose, 1926), with a ratio of the mean motions $n:n' = 3:1$; by Poincaré (1902), Andoyer (1903), and Schwarzschild (1903) for the Hecuba group, $n:n' = 2:1$ (see also Simonin, 1897, 1912; Popoff, 1912; Wilkens, 1913; von Ziepel, 1916, 1917, 1918; Heinrich, 1922, 1925*a*, *b;* Rasdolsky, 1926); by Dziewulski (1909) for the Hilda group, $n:n' = 3:2$ (Hopkins, 1915); by Woltjer (1917, 1918, 1919, 1928) for the Thule group, $n:n' = 4:3$; and by Heinrich (1912*a*, *b*) for $n:n' = 5:3$; by Wilkens (1910) for $n:n' = (p + 2):p$; and by Kępiński (1925) and Woltjer (1923) for $n:n' = (p + 1):p$; and, finally, by Brown (1911) and others (Kępiński, 1913) for the Trojan asteroids, for which $n = n'$. The existence of the gaps in the distribution of the asteroids, where the mean motions would be nearly commensurable with that of Jupiter, might be explained on the basis of the small probability for the perihelion to satisfy the relation which should be fulfilled for a stable periodic solution of the second sort. Indeed, in the actual case of the three innermost Galilean satellites of Jupiter the mean motions are almost commensurable in the double ratio 4:2:1; moreover, a relation $n_1 - 3n_2 + 2n_3 = 0$ is almost exactly satisfied, and the relative position of the perijoves are such as has been proved by de Sitter (1908, 1909) to be stable. Darwin (1897) computed several cases of periodic solutions and discussed their stability in the restricted three-body problem for two finite masses in the ratio 1 to 10, while Strömgren (1934) obtained an extensive range of periodic solutions for masses in the ratio 1 to 1 and studied the whole continuum of the various classes. Wintner (1931*b*; Martin, 1931) discussed Strömgren's principle of classification on a topological basis. The existence of periodic solutions can be proved by using the theory of non-linear integral equations or the theory of an infinite system of non-linear algebraic equations founded on Schmidt's theory of non-linear integral equations (Hölder, 1929, 1931; Wintner, 1926*a*, *b*). Lichtenstein (1920, 1925) applied the theory to the motion of Saturn's rings.

Poincaré's criterion for the stability of periodic solutions is based on the use of characteristic exponents. Consider equations (31), in which the coefficients on the right-hand sides are periodic in t with the period T of the generating periodic solution. Here it is supposed that the $\xi_i(t)$'s are all small for all values of t. By the theory of Floquet on the solution of linear differential equations with periodic coefficients, the solution is of the form

$$\xi_i = e^{\alpha_k t} S_{ik} \qquad (i, k = 1, 2, \ldots, \text{N}),$$

where the S_{ik}'s are periodic functions with period T. Poincaré (1889, 1891) called the α_k's the *characteristic exponents*. It is apparent that, if all α_k's have negative real parts, then the varied motion asymptotically approaches the generating solution as $t \to +\infty$, and that, on the contrary, if all α_k's have positive real parts, the deviations ξ_i increase with time and the solution becomes invalid. Thus, in order that the generating periodic solution may be stable for both $t \to +\infty$ and $t \to -\infty$, the characteristic exponents α_k should all be purely imaginary. If t does not appear explicitly in the variational equations, at least one of the characteristic exponents is zero. The same is true if there exists a uniform integral. In the canonical system of equations the characteristic exponents can be grouped into pairs of equal absolute value but opposite signs.

Let the original variational equations, including all higher-degree terms, be

$$\frac{d\xi_i}{dt} = \Xi_i \qquad (i = 1, 2, \ldots, \text{N}),$$

where the Ξ_i's are periodic in t and expanded in positive integral powers of ξ_i. Poincaré (1889, 1891) attempted to prove that the ξ_i's can be expanded in powers of $A_1 e^{\alpha_1 t}, A_2 e^{\alpha_2 t}, \ldots, A_\text{N} e^{\alpha_\text{N} t}$ with periodic functions of t as the coefficients, if the ξ_i's are sufficiently small and provided that a certain linear combination of the a_k is not small. The result was an asymptotic solution of the original differential equations (30). However, Poincaré's arguments are unsatisfactory. Wintner (1926*a*, *b*) has proved the existence of asymptotic solutions on the basis of the theory of non-linear infinite analysis. Moulton *et al.* (1920) have computed asymptotic orbits in certain cases of the problem of three bodies.

A non-astronomical application of asymptotic solutions has been made by Hagihara (1948), who applied Peano-Baker's matrix method of solution to the theory of the motion of electrons in a magnetron under the action of a periodic electric potential superposed on electrostatic and magnetic fields. The characteristic exponents are complex, and the solution is not uniformly convergent. This is due to the drift of electrons from the heated cathode to the split-anodes and is, in fact, the cause of the emission of microwaves from the magnetron.

REFERENCES

ANDOYER, H. 1903 *Bull. Astr.*, **10**, 321.

BRENDEL, M. 1898 *Theorie der kleinen Planeten* (Abh. k. Gesell. Wiss. Göttingen, math.-phys. Kl., N.F.), Vol. **1**, No. 2.
1903 *Ibid.*, Vol. **6**, No. 4.
1910 *Ibid.*, No. 5.

BROWN, E. W. 1911 *M.N.*, **71**, 438.

DARWIN, G. H. 1897 *Acta Math.*, **21**, 99.

DZIEWULSKI, W. 1909 *A.N.*, **183**, 65.

HAGIHARA, Y. 1927 *Japan. J. Astr. Geophys.*, **5**, 1.
1948 *J. Phys. Soc., Japan*, **3**, 70.

HEINRICH, W. W. 1912*a* *A.N.*, **192**, 325.
1912*b* *Rozpřavy české Akad.*, Vol. **21**.
1922 *Bull. Astr.*, Ser. 2, **2**, 425.
1925*a* *M.N.*, **85**, 625.
1925*b* *Pub. Inst. Astr. U. Prague*, No. 2.

HILL, G. W. 1886 *Acta Math.*, **8**, 1.
1902 *A.J.*, **22**, 93, 117.
1905 *Collected Mathematical Works* (Washington: Carnegie Institution), **1**, 243.

HÖLDER, E. 1929 *Math. Zs.*, **31**, 197.
1931 *Ber. math.-phys. Kl. sächs. Akad. Wiss.*, **83**, 179.

HOPKINS, L. A. 1915 *A.J.*, **29**, 81.

KĘPIŃSKI, F. 1913 *A.N.*, **194**, 49.
1925 *Acta Astr.*, *A*, Vol. **1**.

KLOSE, A. 1926 *Mitt. Inst. Theoret. Ap.*, No. 1.

LEVI-CIVITA, T. 1901 *Ann. math. pura appl.*, Ser. 3, **5**, 221.

LICHTENSTEIN, L. 1920 *Math. Zs.*, **17**, 62.

	1925	*Probleme der Astronomie: Festschrift für Hugo von Seeliger* (Berlin: J. Springer), p. 200.
MARTIN, M. H.	1931	*Amer. J. Math.*, **53,** 259.
MOULTON, F. R.	1920	*Periodic Orbits* (Washington: Carnegie Institution).
POINCARÉ, H.	1889	*Acta Math.*, **13,** 1.
	1891	*Méthodes nouvelles de mécanique céleste* (Paris: Gauthier-Villars), Vol. **1.**
	1893	*Ibid.*, Vol. **2,** chap. xvii.
	1900	*Bull. Astr.*, **17,** 134.
	1902	*Ibid.*, **19,** 177 and 289.
	1909	*Leçons de mécanique céleste* (Paris: Gauthier-Villars), **2,** 34.
POPOFF, K.	1912	Thesis, Paris.
RASDOLSKY, A.	1926	*A.N.*, **227,** 315.
SCHWARZSCHILD, K.	1903	*A.N.*, **160,** 385.
SIMONIN, M.	1897	*Ann. Obs. Nice*, Vol. **6.**
	1912	Thesis, Paris.
SITTER, W. DE	1908	*Proc. Acad. Sci. Amsterdam*, Vol. **10.**
	1909	*Ibid.*, Vol. **11.**
STRÖMGREN, E.	1934	*Bull. Astr.*, **9,** 87. Cf. COE, C. J., 1932, *Trans. Amer. Math. Soc.*, **34,** 811; MOULTON, 1920; MURRAY, F. J., 1926, *Trans. Amer. Math. Soc.*, **28,** 74 and 109; PERRON, O., 1936, *Math. Ann.*, **113,** 95, and *Monatsh. Math. Phys.*, **43,** 81; WEISS, G., 1937, *Math. Zs.*, **43,** 446.
WILKENS, A.	1910	*Astr. Abh.*, No. 17.
	1913	*A.N.*, **195,** 385.
WINTNER, A.	1925	*A.N.*, **224,** 7.
	1926*a*	*Math. Ann.*, **95,** 544.
	1926*b*	*Ibid.*, **96,** 284.
	1926*c*	*Math. Zs.*, **24,** 259.
	1930	*Ber. math.-phys. Kl. sächs Akad. Wiss.*, **82,** 3.
	1931*a*	*Math. Zs.*, **34,** 350.
	1931*b*	*Ibid.*, p. 322.
	1936	*Proc. Nat. Acad. Sci.*, **22,** 435.
	1938*a*	*Amer. J. Math.*, **59,** 795.
	1938*b*	*Ibid.*, **60,** 937. Cf. HOPF, E., 1929, *Sitz. preuss. Akad. Wiss.*, **23,** 401; HÖLDER, 1929.
WOLTJER, J., JR.	1917	*Proc. Acad. Sci. Amsterdam*, Vol. **19.**
	1918	Dissertation, Leiden.
	1919	*Proc. Acad. Sci. Amsterdam*, Vol. **21.**
	1923	*B.A.N.*, **1,** No. 38, 219.
	1928	*Ann. Leiden Obs.*, Vol. **16,** Part 3.
ZEIPEL, H. VON	1916	*Ark. f. math., astr. fys.*, Vol. **11,** Nos. 1 and 7.
	1917	*Ibid.*, Vol. **12,** No. 9.
	1918	*Ibid.*, Vol. **13,** No. 3.

8. SECULAR STABILITY AND TEMPORARY STABILITY

Now that we have shown that the stability of the solar system cannot be decided in the rigorous mathematical sense by the straightforward method of examining the formal solutions so far obtained, we shall have to seek for a criterion without trying to solve the equations of motion. In this fresh start we propose to analyze the notion of stability *ab initio*.

Consider a dynamical system with a dissipative force derived from a potential f, for which the Lagrangian equations of motion are

$$\frac{d}{dt}\left(\frac{\partial T}{\partial \dot{q}_i}\right)-\frac{\partial T}{\partial q_i}=\frac{\partial U}{\partial q_i}-\frac{\partial f}{\partial \dot{q}_i} \qquad (i=1,2,\ldots,\mathrm{N}).$$

Suppose that $q_1 = q_2 = \ldots = q_\mathrm{N} = 0$ is a point of equilibrium and consider the behavior of motion in this neighborhood. We easily obtain the relation

$$\frac{d(T-U)}{dt}=-2f.$$

This shows that the energy $T - U$ always decreases, if $f > 0$. If f does not depend on $\dot{q}$, then the dissipation is *incomplete*, and the energy remains constant; but if f depends on $\dot{q}$, then the dissipation is *complete*, and $T - U$ decreases. The equations of motion can be solved by substituting

$$q_i=\lambda_i e^{rt} \qquad (i=1,2,\ldots,\mathrm{N}),$$

with constant λ_i. If the real part of $r = \rho$, the necessary and sufficient condition for the stability of the motion is $\rho \leq 0$. If the equilibrium is stable, then the force function U is maximum with or without the dissipative force. If $\rho > 0$, the equilibrium will be unstable even with dissipative force. Hence the addition of dissipative force does not alter the stability character for the state of the equilibrium of a conservative dynamical system. This is Thomson and Tait's (1876) theorem (Levi-Civita, 1926; Rayleigh, 1945). It should be remarked that the state may change from unstable to stable when we add purely gyroscopic forces $Q_1, Q_2, \ldots, Q_\mathrm{N}$, such that $Q_1\dot{q}_1 + Q_2\dot{q}_2 + \ldots + Q_\mathrm{N}\dot{q}_\mathrm{N} = 0$; but a stable state remains stable. Thus, if we suppose that

$$\frac{\partial U}{\partial q_1}=\frac{\partial U}{\partial q_2}=\ldots=\frac{\partial U}{\partial q_\mathrm{N}}=0$$

defines a point of equilibrium, then (i) the system is stable for U = maximum when the reacting force is derived entirely from U; (ii) U = maximum is the necessary and sufficient condition for stability when we add dissipative force; (iii) U = maximum is the necessary and sufficient condition for stability when complete dissipative and gyroscopic forces are present besides U; and (iv) U = maximum is a sufficient, but not a necessary, condition for stability when gyroscopic and incomplete dissipative forces are present besides U. A dynamical system is said to be *secularly stable* when it is stable with or without complete dissipative force. When it is stable only without dissipative or with incomplete dissipative force, it is said to be *temporarily stable* (Stapper, 1911; Klein and Sommerfeld, 1923). If we do not consider any dissipative force, the condition U = maximum is sufficient, but not necessary, for stability, and it is difficult to give a necessary condition. Poincaré discussed the stability only on the basis of small oscillations from the equilibrium point, using characteristic exponents; this method did not allow him to derive a necessary condition. The stability discussed on the basis of characteristic exponents is of such nature.

For a rotating mass, U will have the form

$$W + \tfrac{1}{2} J \omega^2 ,$$

where W denotes the force function for the mutual attraction of the constituent particles, ω the angular velocity, and J the moment of inertia. The condition U = maximum is sometimes called *Lejeune-Dirichlet's condition*. Poincaré (1887, 1902) adopted U = maximum as the criterion of stability in his theory of the equilibrium figures of a rotating fluid mass. The same criterion was adopted by Robin in considering stability problems in thermodynamics and by Lamb (1908) in hydrodynamics (cf. Painlevé, 1904; Duhem, 1911; Jouguet, 1929). If a dynamical system is stable in Dirichlet's sense, then the velocity decreases slowly if complete dissipative forces are applied. Hadamard (1897, 1898) proved that a necessary condition for secular stability in Dirichlet's sense is W = maximum.

Let the force function be $F(x_1, x_2, \ldots, x_N; \alpha)$ with a parameter α, and suppose that the dynamical system is in equilibrium, such that

$$\frac{\partial F}{\partial x_1} = \frac{\partial F}{\partial x_2} = \ldots = \frac{\partial F}{\partial x_N} = 0 .$$

This system may admit several roots. As we vary the parameter α, the roots vary continuously, and we get linear series of solutions corresponding to each of the roots. Two or more roots may coincide for a certain value of α if and only if the Hessian of F with regard to $x_1, x_2, \ldots, x_N$ vanishes. When the Hessian is zero and changes its sign at $\alpha = \alpha_1$, then two or more linear series pass through $\alpha = \alpha_1$, provided that the solutions are real in the neighborhood of α_1. The point $\alpha = \alpha_1$ is called a *bifurcation point* of the linear series. If the solution changes from real to imaginary in passing through $\alpha = \alpha_1$, the point is called a *limiting point*. By passing through a bifurcation point, the stability character may be exchanged. Poincaré (1887, 1902) applied this idea to the equilibrium figures of a rotating fluid mass. Similar consideration can be applied to the three-body problem. Hölder (1929, 1931) discussed it, using the theory of non-linear integral equations, which had been applied to the equilibrium figures by Lichtenstein (1920, 1925). It should be mentioned that the establishment of equilibrium configurations requires the study of the first-order variations, while the criterion of stability necessitates the study of second-order variations, so that the judgment of stability is a complicated problem in itself.

The notion of stability has been refined by Liapounoff (1897, 1907). He considered the case in which the stability character cannot be decided by the method of characteristic exponents. He proceeded as follows.

Suppose that we have an undisturbed motion represented by $q_i = f_i(t)$, where $i = 1, 2, \ldots, m$, and consider two disturbed motions q_{i0}, $\bar{q}_{i0}$ with initial conditions

$$q_{i0} = f_i(t_0) + \epsilon_i\,, \qquad \bar{q}_{i0} = f_i(t_0) + \bar{\epsilon}_i \qquad (i = 1, 2, \ldots, m)\,. \quad (33)$$

Here ϵ_i and $\bar{\epsilon}_i$ are constants called the *perturbations*. Let Q_i, where $i = 1, 2, \ldots, n$, be a set of given real continuous functions of q_i and $\bar{q}_i$, where $Q_i = F_i$ for the undisturbed motion. The quantities Q_i are functions of ϵ_i, $\bar{\epsilon}_j$, and t, and $Q_i - F_i = 0$ for all values of t for $\epsilon_i = \bar{\epsilon}_j = 0$, where $i, j = 1, 2, \ldots, m$. Further, let L_k, where $k = 1, 2, \ldots, n$, be a set of given positive numbers. If we can choose for all values of L_k, however small, two sets of positive numbers E_i and $\bar{E}_j$, where $i, j = 1, 2, \ldots, m$, such that, when

$$|\epsilon_1| < E_i\,, \qquad |\bar{\epsilon}_j| < \bar{E}_j \qquad (i, j = 1, 2, \ldots, m)\,,$$

we have

$$|Q_k - F_k| < L_k \qquad (k = 1, 2, \ldots, n) \quad (34)$$

for all values of $t > t_0$, then the undisturbed motion is said to be stable *with respect to* Q_i. Otherwise it is unstable with respect to Q_i.

The nature of the stability depends on the solution of the differential equations satisfied by $Q_i - F_i = x_i$ $(i = 1, 2, \ldots, n)$, viz.,

$$\frac{dx_i}{dt} = X_i \qquad (i = 1, 2, \ldots, n)\,.$$

The right-hand members are known functions of $x_1, x_2, \ldots, x_n$, and t; they reduce to zero for $x_1 = x_2 = \ldots = x_n = 0$ and can be expanded in absolutely convergent integral power series of $x_1, x_2, \ldots, x_n$, provided that

$|x_i| \leq A_i$ $(i = 1, 2, \ldots, n)$, where the A_i's are small positive constants. If there exist constants λ_1 and λ_2 such that a function $x = x(t)$ has the property that $xe^{\lambda t}$ tends to zero at $\lambda = \lambda_1$ and to infinity at $\lambda = \lambda_2$ as $t \to \infty$, then there exists a real number λ_0 such that $xe^{\lambda t}$ tends either to infinity or to zero at $\lambda = \lambda_0 + \epsilon$ as $t \to \infty$, according as ϵ is positive or negative, or vice versa, provided that $|\epsilon|$ is sufficiently small. Such a number λ_0 is called the *characteristic number* of x. It can be shown that if the solution obtained in the first approximation to the given differential equations is regular and has its characteristic numbers $\lambda_1, \lambda_2, \ldots, \lambda_n$ and if we take k $(\leq n)$ arbitrary constants $\alpha_1, \alpha_2, \ldots, \alpha_k$, then the given differential equations can be satisfied formally by a series

$$\sum L^{(m_1, m_2, \ldots, m_n)} \alpha_1^{m_1} \alpha_2^{m_2} \ldots \alpha_k^{m_k} \exp\left(-\sum_{i=1}^{k} m_i \lambda_i t\right), \qquad (35)$$

and that there exists a positive constant α such that the series is absolutely convergent for all positive values of t when $\lambda_1, \lambda_2, \ldots, \lambda_n$ are positive for $|x_i| \leq \alpha$ $(i = 1, 2, \ldots, n)$. Thus the undisturbed motion is stable in this case. When there are k $(\leq n)$ positive characteristic numbers, then the undisturbed motion is conditionally stable. The function represented by the uniformly convergent series (35) is a quasi-periodic function in the sense of Bohl (1893, 1906; Esclangon, 1904, 1917; Hagihara, 1947) and is of the form discussed in Section 6 if the characteristic numbers are all purely imaginary.

If a function V of $x_1, x_2, \ldots, x_n$ becomes zero only when $x_1 = x_2, \ldots = x_n = 0$, then V is called a *definite* function. If the equations for a disturbed motion are such that it is possible to find a definite function V, of which the total differential coefficient,

$$V' = \frac{\partial V}{\partial x_1} X_1 + \frac{\partial V}{\partial x_2} X_2 + \ldots + \frac{\partial V}{\partial x_n} X_n + \frac{\partial V}{\partial t},$$

satisfies $VV' < 0$ or vanishes identically, then the disturbed motion is stable. The proof of this theorem is connected with the theorem of Lejeune-Dirichlet on the stability of motion in the case when the energy integral exists. There are several recent publications by Soviet mathematicians in this field (Andronov and Witt, 1933; Moisseiev, 1936, 1937; Artemiev, 1938; Malkin, 1938).

Liapounoff (1897, 1907) applied the theory to systems of differential equations with constant coefficients and to a system with periodic coefficients. Bohl (1904, 1906, 1910; cf. Horn, 1903, 1906) has extended the theory to the case with quasi-periodic coefficients (Hagihara, 1944). Similar arguments (Favard, 1927, 1932; Bochner, 1929, 1930, 1931, 1933; Cameron, 1936, 1938) have been applied to differential equations with almost-periodic coefficients in the sense of Harald Bohr. These theories might be applied to the perturbation theory of celestial mechanics, and it would then be possible to prove that the solution of the equations of motion could be expressed in quasi- or almost-periodic functions, provided that the given differential equations or their solutions could be shown to satisfy the restrictions stated in the existence and convergence theorems for such functions. However, the question that must be settled at the outset is whether the differential equations in celestial mechanics actually satisfy such restrictions.

REFERENCES

ANDRONOV, A., and WITT, A. 1933 *Phys. Zs. U.S.S.R.*, **4**, 606.

ARTEMIEV, N. 1938 *Compositio Math.*, **6**, 78.

BOCHNER, S. 1929 *Math. Ann.*, **102**, 489.
1930 *Ibid.*, **103**, 588.
1931 *Ibid.*, **104**, 579.
1933 *J. London Math. Soc.*, **8**, 283.

BOHL, P. 1893 Dissertation, Dorpat.
1904 *J. reine angew. Math.*, **127**, 179.
1906 *Ibid.*, **131**, 268.
1910 *Bull. Soc. Math. France*, **38**, 5.

CAMERON, R. H. 1936 *Ann. Math. Princeton*, **37**, 29.
1938 *Acta Math.*, **69**, 21.

DUHEM, H. 1911 *Traité d'énergétique* (Paris: Gauthier-Villars), Vol. **2**, chaps. 17 and 18.

ESCLANGON, E. 1904 Thesis, Paris.
1917 *Ann. Obs. Bordeaux*, **16**, 53.

FAVARD, J. 1927 *Acta. Math.*, **51**, 31.
1932 *Functions presque-périodiques* (Paris: Gauthier-Villars).

HADAMARD, J. 1897 *J. math. pures et appl.*, Ser. 5, Vol. **3**.
1898 *Ibid.*, **4**, 27.

HAGIHARA, Y. 1944 *Proc. Imp. Acad. Tokyo*, **20**, 617.
1947 *Foundations of Celestial Mechanics* (Tokyo: Kamade Publisher), **1***A*, 114.

HÖLDER, E. 1929 *Math. Zs.*, **31**, 307.
1931 *Ber. math.-phys. Kl. sächs. Akad. Wiss.*, **83**, 179

HORN, A. 1903 *J. reine angew. Math.*, **126**, 194.
1906 *Ibid.*, **131**, 224.

JOUGUET, M. E. 1929 *J. École Polytech.*, Ser. 2, Vol. **27**.

KLEIN, F., and SOMMERFELD, A. 1923 *Theorie des Kreisels* (Leipzig: Teubner), **2**, 342.

LAMB, H. 1908 *Proc. R. Soc. London, A*, **80**, 168.

LEVI-CIVITA, T. 1926 *Meccanica razionale* (Bologna: Zanichelli), Vol. **2**, Part 1, chap. 6.

LIAPOUNOFF, A. 1897 *J. math. pures appl.*, Ser. 5, **3**, 81.

1907 *Ann. Fac. Sci. Toulouse*, **9**, 203.

LICHTENSTEIN, L. 1920 *Math. Zs.*, **17**, 62.

1925 *Seeliger Festschrift* (Berlin: J. Springer), p. 200.

MALKIN, I. G. 1938 *Rec. Math.*, **3**, 47.

MOISSEIEV, N. 1936 *Math. Ann.*, **113**, 452.

1937 *Math. Zs.*, **42**, 513.

PAINLEVÉ, H. 1904 *C.R.*, **138**, 1555.

POINCARÉ, H. 1887 *Acta Math.*, **7**, 1.

1902 *Leçons sur les figures d'équilibre d'une masse fluide* (Paris: Gauthier-Villars).

RAYLEIGH, J. W. S. 1945 *Theory of Sound* (2d ed.; New York: Dover Publications), **1**, 81.

STAPPER, M. 1911 *Ann. Obs. Bordeaux*, Vol. **14**.

THOMSON, W., and TAIT, P. G. 1876 *Natural Philosophy* (Oxford: Clarendon Press), **1**, 391.

URA, T., and HIRASAWA, Y. 1954 *Proc. Japan Acad.*, **30**, 726.

9. TOPOLOGICAL THEORY; ERGODIC THEOREM; ALMOST-PERIODIC MOTION AND STABILITY

Another approach to the problem of stability was initiated by Poincaré. This was to study the behavior of the trajectories qualitatively or topologically, apart from the metrical properties that were discussed in the preceding sections; that is, (i) whether a trajectory is closed and surrounded by other nearby closed trajectories, in which case the solution is periodic in the sense of Poincaré; (ii) whether a nearby trajectory recedes with time from the original, in which case it is unstable; or (iii) whether all trajectories are confined inside a closed curve and fill up the region everywhere densely, in which case it is stable in the sense of Poisson.

Suppose that we have a differential equation

$$\frac{dx}{X}=\frac{dy}{Y},$$

where X and Y are polynomials in x and y. The solution-curves in the two-dimensional representation are called the *characteristics* by Poincaré (1879, 1881, 1882, 1885, 1886, 1895, 1928) (see Fig. 4). Except for singular points, only one characteristic passes through each point. Poincaré gave several kinds of singular points: a saddle point (Fig. 4, *b*) through which two characteristics pass; a node (Fig. 4, *c*) through which infinitely many characteristics pass; and a focus (Fig. 4, *d*) toward which or out of which a characteristic winds in a spiral form. A characteristic which returns to its initial point after traveling a finite length of path is called a *cycle*, and it divides the surface into two regions. An arc of a curve which intersects any characteristic at a non-zero angle is called an *arc without contact*. Take such an arc AB and a point M_0 on AB. Follow a characteristic starting at M_0. When it intersects AB again at M_1, M_1 is called a *consequent* of M_0, and M_0 an *antecedent* of M_1. If there is a limit H of the successive consequents of M_0, then the characteristic through H is a cycle and, indeed, a limiting cycle. If a closed characteristic which does not pass through any singular point is not a limiting cycle, then there exists an annular region

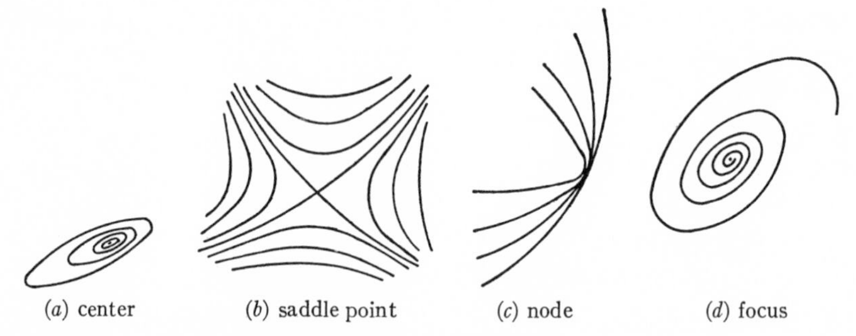

(*a*) center (*b*) saddle point (*c*) node (*d*) focus

FIG. 4.—Types of singular points according to Poincaré

inside which every characteristic is a closed curve (Birkhoff, 1920, 1932*a*, 1932*b*, 1935*a*, 1935*b*, 1936*a*, 1936*b*; Dulac, 1937). When a characteristic starting at a point inside a small circle returns to the inside of this circle an infinite number of times, however small we may choose the radius of the circle, then the characteristic is stable, otherwise unstable. Instability is general and stability exceptional (Levi-Civita, 1901). A singular point surrounded by a set of non-intersecting single cycles is called a *center* (Fig. 4, *a*) and is stable (Bendixon, 1900). The necessary and sufficient condition to be satisfied by X and Y in order that $x = y = 0$ should be a center is expressed by an infinite number of equations, but they can be satisfied when X and Y and the characteristics satisfy a certain symmetry condition. Such is the case of Hill's variational curve. The work of Poincaré on the curves defined by differential equations has recently become useful in the study of non-linear vibrations in physics and engineering (Krylov and Bogoliubov, 1943; Lefschetz, 1946, 1950; Andronov and Chaiken, 1949; McLachlin, 1950; Stoker, 1950).

It is known that, according to Maupertuis's principle, a dynamical trajectory is defined by

$$\delta \int_{t_0}^{t} 2T dt = 0 , \qquad \text{or} \qquad \delta \int_{s_0}^{s_1} 2 (U + h) \, d s = 0 ,$$

when there exists the energy integral $H = h$. If there is no external force acting on the particle, the trajectory is a geodesic on a surface defined by

$$d s^2 = \sum_{i,\, j} g_{ij} d x_i d x_j , \qquad \text{where} \qquad 2T = \sum_{i,\, j} g_{ij} \left(\frac{d x_i}{dt}\right)\left(\frac{d x_j}{dt}\right).$$

Thus the problem is reduced to the study of geodesics on a surface so defined as to correspond to the kinetic energy. Hadamard (1898) first studied the geodesics on a surface of negative curvature, and Poincaré (1905) on a surface of positive curvature. Consider a closed geodesic G_0 and a nearby geodesic G on a surface of positive curvature and let the arc length u be measured along G_0 and the length v on the normal to G_0 from its foot on G_0. Denote the square of the surface line element by $ds^2 = dv^2 + C^2 du^2$, where we take $C = 1$ and $\partial C/\partial v = 0$ for $v = 0$. By supposing v small, we have

$$\frac{d^2 v}{d u^2} = \left(\frac{\partial^2 C}{\partial v^2}\right)_{v=0} v .$$

The coefficient of v on the right-hand side is periodic with the period U of G_0. The solutions are, according to the theory of Floquet,

$$v_1 = e^{\alpha u} \phi (u) , \qquad v_2 = e^{-\alpha u} \psi (u) ,$$

where $\phi(u)$ and $\psi(u)$ are periodic functions of u with period U and α is the characteristic exponent. If α is imaginary, then G_0 is stable; and if α is real and not zero, then G_0 is unstable. The successive intersections of G_0 and G are called the *foci*. If $\phi(u)$ never becomes zero in the unstable case, no point on G_0 can be a focus, and its instability is of the *first kind*. If $\phi(u)$ becomes zero in the unstable case, the instability is of the *second kind*. The successive foci of a point lying between a zero point of $\phi(u)$ and a zero point of $\psi(u)$ tend to a limiting point, which is called the *limiting focus*, and the instability is of the *third kind*. The longest and the shortest principal ellipses on an ellipsoid are stable, but the intermediate principal ellipse is an unstable solution of the second kind, while the ellipse on the neck of a hyperboloid of one sheet is an unstable solution of the first kind.

When two closed geodesics coincide and then become imaginary, one of them is stable and the other unstable (cf. Sec. 8). Thus the difference between the numbers of stable and unstable simple closed geodesics is constant on a surface of positive curvature. The work of Poincaré has been extended by Birkhoff (1917) for the study of dynamical trajectories; by Denjoy (1932) on the surface of an anchor ring (Ura and Hirasawa 1954); and by Morse (1924) and others on a surface of any genus by making use of the theory of Fuchsian groups, automorphic functions, universal covering surfaces, and the fundamental groups in Poincaré's topological theory (Poincaré, 1895, 1899*a*, 1900, 1902*a*, *b*, 1904).

Let the polar co-ordinates of a point M be r, θ and draw two concentric circles $C:r = a$ and $C':r = b$, $a > b$. Consider a one-to-one continuous transformation T which transforms any point (r, θ) of the ring bounded by C and C' to a point (r_1, θ_1) of the ring and any point (a, α) on C to a point (a, α_1), $\alpha_1 < \alpha$, on C and any point (b, α') on C' to a point (b, α_1'), $\alpha_1' > \alpha_1$, on C', or vice versa, and suppose that T has an integral invariant extended over the ring region such that there exists a function $f(r, \theta) > 0$ with the property

$$\iint f(r,\theta)\, r\,dr\,d\theta = \iint f(r_1,\theta_1)\, r_1 dr_1 d\theta_1 \,.$$

Then there exist at least two points, and hence infinitely many points, invariant under T inside the ring region. This theorem is called the *geometric theorem* of Poincaré (1912*a*) but has been proved by Birkhoff (1912*a*). It occupies a central position with regard to the fixed-point theorems in the modern development of topology by Kerékjártó (1923), L. E. J. Brouwer (1910), Alexander (1923), Alexandrov and Hopf (1935), Veblen (1922), Lefschetz (1930), and Nielsen (1927). Poincaré (1912*a*) applied this theorem for proving the existence of periodic solutions in the restricted problem of three bodies, and Birkhoff (1912*a*, 1928*a*) for proving the existence of periodic trajectories in general dynamical problems of two degrees of freedom.

The trajectories in a dynamical system of two degrees of freedom, such as the restricted problem of three bodies (cf. Fig. 5), can, since the Jacobi integral exists, be represented in a three-dimensional manifold of motion. Various possible states of motion can be considered as being generated by a transformation T from a state of motion at a certain epoch. An analytic surface S in this manifold, which is cut always in the same sense and at least once in a fixed interval of time by every trajectory and is bounded by a set of closed trajectories, is called a *surface of section*, according to Birkhoff. Poincaré's theory on the consequents is applied to the transformation T of the surface of section S into itself. When a point on S coincides with its consequent under T, then the trajectory represents a periodic solution. The equations can be transformed into

$$\frac{dx}{at} = X\,, \qquad \frac{dy}{at} = Y\,, \qquad \frac{du}{at} = U\,,$$

and it can be proved that there exists a relation

$$\frac{\partial X}{\partial x}+\frac{\partial Y}{\partial y}+\frac{\partial U}{\partial u}=0\,,$$

so that the trajectories can be represented by streamlines of molecules of an incompressible fluid confined in a vessel of finite dimension and hence that there exists an integral invariant $\int\int\int dxdydu$ = constant. Thus Poincaré's geometric theorem is applied, and we can prove the existence of periodic solutions of the problem. Birkhoff (1914) has studied in detail the topological character of the trajectories inside the zero-velocity oval

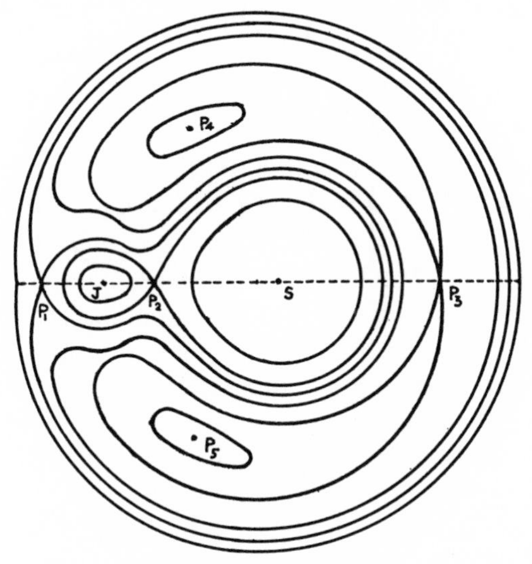

FIG. 5.—Representative zero-velocity curves for x, y-plane for the restricted three-body problem.

around either of the two finite masses in the restricted problem of three bodies, and Koopman (1927) of those outside the large oval of zero velocity around the point at infinity. Birkhoff and Lewis (1933; Lewis, 1934) have further extended the geometric theorem of Poincaré in various aspects. Chazy (1922, 1929), by extending the method used by Sundman (1907, 1908, 1913) of proving the existence of a solution in the three-body problem at finite distances, has made a topological study of the trajectories at infinity (Levi-Civita, 1903, 1906, 1919; Bisconcini, 1906; Block, 1909; Chazy, 1918; Freundlich, 1918; Siegel, 1941). Hotelling (1925) dis-discussed the manifold of motion on a surface of arbitrary genus. Morse (1935) went further and built up his theory of calculus of variations in the large.

If the successive consequents of a point on the surface of section S corresponding to a trajectory varied from the original periodic solution recede farther and farther from the invariant point corresponding to the original periodic solution by the repetition of T, then the original periodic state of motion is unstable, and if the consequent approaches the invariant point, then the original state of motion is stable. These states of motion are called *asymptotic to the periodic state of motion*. Thus the behavior as to the stability of the periodic orbits depends on the nature of the invariant point under the surface transformation T. In the neighborhood of an invariant point the transformation T can be written (Levi-Civita, 1901) with the curvilinear isothermal co-ordinates u and v on S as

$$u_1 = au + bv + \dots , \qquad v_1 = cu + dv + \dots , \qquad ad - bc > 0 .$$[1]

If the two roots ρ_1 and ρ_2 of the characteristic equation formed with these coefficients a, b, c, d are not both equal to 1, then the invariant point $u = v = 0$ is called *simple*, otherwise *multiple* (Birkhoff, 1912*a*). A simple point with $0 < \rho_1 < 1 < \rho_2$ is called *directly unstable*, and a simple point with $\rho_1 < -1 < \rho_2$ *inversely unstable*, and any other simple point $ad - bc > 0$ is called *stable*. If a one-to-one analytic transformation T can be generated by a deformation of an analytic closed surface S of genus p into itself, then the difference between the number of directly unstable invariant points and the number of other invariant points is $2p - 2$. Each type of invariant point corresponds to a type of periodic orbit. Birkhoff (1925, 1927) further studied this type of surface transformation in detail by making use of formal series of expansions.

Another approach to the stability of motion has recently been developed on the idea of the regional transitivity inaugurated by Poincaré. His recurrence theorem states that, if the motion of a point P in a Euclidean space is restricted to remain inside a finite region D and if there exists an integral invariant of volume, then there exist trajectories passing through the interior of a sphere of radius r in the region D an infinite number of times, however small we may choose r, and the trajectories passing through the sphere a finite number of times only are exceptional. The ratio of the probability of the occurrence of these two types of trajectories is in the ratio of the powers of incommensurable and commensurable numbers. In such a case the motion is called by Poincaré (1889, 1899*b*) *stable in the sense of Poisson*. Sundman's work on the possibility of analytic continuation beyond the binary collisions enables us to apply this theorem to the three-body problem. The necessary and sufficient condition for a triple collision,

[1] The ellipses stand for higher degree terms in u and v. We do not consider them in the discussion here.

which is excepted in this application, is that the total angular momentum of the three bodies is zero. The idea of the recurrence property appeared in Poincaré's works on the three-body problem, but it was stated explicitly for the first time in his lectures on probability (Poincaré, 1912*b*) when he discussed the problem of the so-called *battages des cartes*, which was further developed by Markov (1912) and Hostinsky (1931, 1932) as Markov chains. Schwarzschild (1896) has applied this recurrence theorem for denying the ultimate capture of a comet by Jupiter within the scope of conservative dynamics. Von Zeipel (1905), recently followed by Maitre (1937), discussed it in detail, especially applied to the problem of capture of Comet Lexell, and proved that the captured orbit of a comet is unstable and that the orbit of a comet which was hyperbolic for $t = -\infty$ should tend to be hyperbolic for $t = +\infty$ unless it be quite exceptional. Recently, however, Merman seems to have proved the existence of capture orbits, notwithstanding either this theory or the work of Chazy (1922; Leimanis and Minorsky, 1958).

Consider a vessel containing N molecules. Then the motion of the N molecules is described by curves on a $(6N - 1)$-dimensional manifold imbedded in the $6N$-dimensional phase space of the N molecules, as there exists the integral of energy. There exists the integral invariant of the volume. Take a volume σ of Σ and consider the path of a molecule. In a steady state the mean sojourn time of the molecule inside σ—that is, the sum of the time intervals during which the path of the molecule lies inside σ divided by the whole time interval we consider—tends to the ratio of the volume σ to the whole volume of the vessel, as the whole time interval we consider tends to infinity. In other words, the time mean of the sojourn time inside σ is in the limit equal to the space mean of the volume. This is the ergodic hypothesis of Ehrenfest (1909) and was made the foundation of the gas-kinetic theory, although there were severe criticisms against it by Loschmidt, Zermelo, and von Mises (1920, 1922, 1931; Hagihara 1933), in connection with the theory of Brownian motion by Einstein and Smoluchowsky and the theory of stochastic phenomena by Kolmogorov (1933) and Khintchine (1933, 1949; see also Feller, 1950; Bartlett, 1955).

The entity of the images of a point P on Σ by the interation of T is called a *complete sequence*, and if $T^{\alpha}(P) = T^{\beta}(P)$, $\alpha < \beta$, then it is a *periodic point group*. The complete sequence, together with the ω limit point

$$\lim_{\kappa\to\infty} T^{\kappa}(P)$$

and the α limit point

$$\lim_{\kappa\to\infty} T^{-\kappa}(P)\,,$$

is called a *complete group*. If $T(E)$ of a point set E coincides with E, then E is an *invariant point set.* A complete group is a *closed invariant set.* When a complete group does not contain in itself a closed subset proper to its type, it is called a *minimal set*, and a complete group in a minimal set, a *recurrent* point group, according to Birkhoff (1926). The necessary and sufficient condition for a point group to be recurrent is that we can choose a positive integer m large enough that k successive points $T^m(P)$, $T^{m+1}(P)$, . . . , $T^{m+k-1}(P)$ should lie within a distance ϵ from any arbitrary image of P for any small positive number ϵ. If we cannot find δ such that a point P within a distance δ from a minimal set E always remains within the distance δ by the iteration of T and T^{-1}, then the recurrent point group is unstable, and the complete group generated from P is asymptotic to E; otherwise it is stable. In the neighborhood of an unstable invariant point O there are two point sets with O as the α limit point and two point sets with O as the ω limit point. If these two point sets have a common point, the latter point is said to be *doubly asymptotic* (Poincaré, 1889, 1899*b*). Birkhoff (1926) further developed the concept of wandering regions and the central motion on the basis of Cantor's transfinite number in the point-set theory. The concept of recurrent motion is wider than that of periodic motion, and the concept of central motion is wider than that of recurrent motion. A point which is both the α limit point and the ω limit point of the complete sequence of itself is called *pseudo-recurrent* (Birkhoff, 1912*b*, 1917, 1928*b*). A pseudo-recurrent point is a central point, and a recurrent point is pseudo-recurrent. But a central motion is not necessarily recurrent.

Fabre (1937, 1938) has proposed taking a recurrent motion as an intermediary orbit in planetary theory by calling it an absolute orbit in a more refined sense than that of Gyldén. His absolute orbit is represented by quasi-periodic functions with two arguments, not necessarily of time, but of an angular variable λ so related to t, as $t = g\lambda +$ a quasi-periodic function of λ, with a constant g. The actual motion of a planet is considered as a small deviation from this absolute orbit, but the series representing this deviation turns out to be divergent.

If one of two connected regions without a common point can become an image of the other by the iteration of T, then T is called *transitive*, otherwise *intransitive*. When there exists no invariant set E measurable in the sense of Lebesgue such that $0 < m(E) < m(\Sigma)$, T is called *metrically transitive*. If E is metrically transitive and its measure is conserved under T, then the measure $m(E)$ of the set E composed of the pseudo-recurrent points and their limit points is equal to $m(\Sigma)$, and it is stable in the sense of Poisson, apart from the sets of measure zero. This is a refined form of Poincaré's recurrence theorem. For any function $f(P)$ of a point of Σ, the ratio

$$\int_{\Sigma} f(P)\, dm / m(\Sigma)$$

tends to a constant independent of P by the iteration of T or T^{-1} almost everywhere, that is, apart from sets of measure zero. From this theorem follows the concept of ergodicity that the time means are equal to the corresponding space means almost everywhere. Von Neumann (1932) has proved this mean ergodic theorem in Hilbert space on the basis of the idea of convergency in the mean. He proved that a dynamical system is quasi-ergodic when and only when a measurable set A, invariant under T apart from sets of measure zero, is reduced to 0 or Σ apart from zero sets. The necessary and sufficient condition for quasi-ergodicity is the metrical transitivity. If the set of points P of the intersection of a trajectory with the surface of section S fills up the whole S, then the measurable set of the complete trajectory in Σ has measure zero or $m(\Sigma)$, and is called *strongly transitive* by Birkhoff (1931), i.e., metrically transitive. Thus he proved that the limit exists of the ratio of the sojourn time $\bar{t}$ in the volume σ to the whole time t elapsed since the initial epoch, such that

$$p = \lim_{t\to\infty} \frac{\bar{t}}{t} = 1$$

is the probability of the sojourn of the particle in the volume σ. This theorem is called the *individual ergodic theorem*. Hopf (1932*a*, *b*, 1937) generalized the idea of ergodicity to the sense of weak convergence and developed the notion of mixture and statistical regularity. The mixture includes metrical transitivity. Wiener (1939; Wiener and Wintner, 1941) discussed homogeneous chaos and applied it to the theory of turbulence.

The notion of stability in the sense of Liapounoff has been made definite by basing it on the theory of the almost-periodic functions of Harald Bohr (1924, 1925, 1926, 1932*a;* Besicovitch, 1932). Let $f(t)$ be a real or complex function defined and continuous for all real values of t. Suppose that we can find a positive number $l(\epsilon)$ for a given $\epsilon > 0$, however small ϵ may be, such that there exists at least one number $\tau(\epsilon)$ in any interval of length l for which we have

$$f(t+\tau) - f(t) \leq \epsilon$$

for all values of t. Then $f(t)$ is said to be *almost-periodic*. For any almost-periodic function $f(t)$ there exists a definite and finite limit,

$$\lim_{T\to\infty} \frac{1}{T}\int_0^T f(t)\, dt = M[f(t)],$$

and an almost-periodic function can be represented uniquely by a Fourier series,

$$f(t) \sim \sum_n A_n e^{\sqrt{(-1)}\lambda_n t}$$

with mutually independent numbers $\lambda_n (n = 1, 2, 3, \ldots)$ such that

$$\lim_{N\to\infty} M\left[\left| f(t) - \sum_{n=1}^{N} A_n e^{\sqrt{(-1)}\lambda_n t}\right|^2\right] = 0\,.$$

A quasi-periodic function of Bohr is included in this class of functions, and its Fourier series is represented by

$$\Sigma A_{n_1 n_2, \ldots n_m} \exp\left[\sqrt{(-1)}\,(n_1 a_1 + n_2 a_2 + \ldots + n_m a_m)\, t\right],$$

where $a_1, a_2, \ldots, a_m$ are linearly independent constants; $n_1, n_2, \ldots, n_m$ are mutually independent integers; and the sum is extended over all positive and negative integral values, including zero. This expression is what we have seen in equation (35) of Section 8.

The almost-periodic motion was first studied by Muckenhoupt (1929) for a vibrating plate and was extended by Bohr (1930, 1932*b*) and Jessen (Bohr and Jessen, 1932; Jessen and Tornehave, 1952) on a circle and by Fenchel and Jessen (1935) on a plane domain. Fenchel and Jessen generalized the almost-periodicity to that in modulo λ, such that $|f(t + \tau) - f(t) - n(\tau, t)\lambda| \leq \epsilon$ instead of $|f(t + \tau) - f(t)| \leq \epsilon$, and discussed the almost-periodic motions in a plane domain G bounded by a finite number of circles by showing them to correspond to the almost-periodic motions in modulo the fundamental group of G on the universal covering surface. The work was extended further to the motion on any topological surface.

Consider a motion described by $x = f(t, a)$, starting from $x = a$ at $t = t_0$, and another motion described by $\bar{x} = f(t, \bar{a})$ starting from $\bar{x} = \bar{a}$ at $t = t_0$ for $|t - t_0| \leq S$, and suppose that $|\bar{x} - x| < \epsilon$ for $|\bar{a} - a| < \delta = \delta(\epsilon, S, a)$. If $\delta(\epsilon, S, a) = \delta(\epsilon)$, then the motion is said to be *S-stable* by Franklin (1929) and is, if the domain of the trajectories is bounded and closed, stable in the sense of Liapounoff. The necessary and sufficient condition for a stable motion to be almost-periodic is that it should be a recurrent motion S-stable. Markov (1933) has related the motion of stability in the sense of Liapounoff to the almost-periodic motion, that is, any motion contained in a bounded and closed point set is almost-periodic, if it is stable in the sense of Liapounoff. Bohr and Fenchel (1936) called a motion stable in the sense of Liapounoff *strongly stable* and a motion stable in the sense of Poisson, *weakly stable*. If we choose $\delta = \delta(l, \epsilon, \tau_0)$ for any value of τ_0 such that $|x(\tau_0 + t) - x(\tau + t)| \leq \epsilon$ is derivable from $|x(\tau_0) -$

$x(\tau)| \leq \delta$ in $-l > t \leq +l$ and if the point set $x(t_n)$ converges to $x(t_0)$ for $n \to \infty$, then the motion $x(t)$ is stable in the sense of Poisson. A motion which is weakly stable and bounded is necessarily periodic in a plane. The second condition is satisfied by an almost-periodic motion. Hence an almost-periodic motion on a plane satisfying the first condition is necessarily periodic (Hagihara, 1942).

We do not know as yet whether the motions of the planets and the satellites in the solar system satisfy any of these classes of orbits, although they have a much wider domain of applicability than periodic orbits. But I should like to emphasize that our practical experience is more advanced than its mathematical verification. For example, the non-convergent expressions for the planetary perturbations are fairly successful in predicting the positions of planets within the accuracy of observation, while the rigorous expression obtained by Sundman is hardly of use for practical purposes (Belorizky, 1931, 1932, 1933). The discussion in Sections 1–5 is sufficient to establish the stability of the solar system for practical purposes, if we remark that forces other than the purely Newtonian, such as frictional, tidal, or electromagnetic, will be operating during the long future history of the solar system.

REFERENCES

Alexander, J. W. 1923 *Trans. Amer. Math. Soc.*, **25**, 173.

Alexandrov, P. S., and Hopf, H. 1935 *Topologie* (Berlin: J. Springer), chap. **14**.

Andronov, A. A., and Chaikin, C. E. 1949 *Theory of Oscillations* (Princeton: Princeton University Press).

Bartlett, M. S. 1955 *An Introduction to Stochastic Processes with Special Reference to Methods and Applications* (Cambridge: Cambridge University Press).

Belorizky, D. 1931 *C.R.*, **193**, 314, 766, 1321.
1932 *Ibid.*, **194**, 769, 1449.
1933 Thesis, Paris.

Bendixon, I. 1900 *Acta Math.*, **24**, 1.

Besicovitch, A. S. 1932 *Almost-Periodic Functions* (Cambridge: Cambridge University Press).

Birkhoff, G. D. 1912*a* *Trans. Amer. Math. Soc.*, **14**, 14.
1912*b* *Bull. Soc. Math. France*, **40**, 305.
1914 *R.C. circ. mat. Palermo*, **39**, 1.
1917 *Trans. Amer. Math. Soc.*, **18**, 199.
1920 *Acta Math.*, **43**, 1.
1925 *Ibid.*, **47**, 297.
1926 *Nachr. Gesell. Wiss. Göttingen.*
1927 *Acta Math.*, **50**, 359.

Birkhoff, G. D. 1928*a* *Acta Szeged*, **4**, 6.
1928*b* *Dynamical Systems* (New York: American Math. Soc.), chap. vii.
1931 *Proc. Nat. Acad. Sci.*, **17**, 650.
1932*a* *Ann. Inst. Poincaré*, **2**, 369.
1932*b* *Bull. Soc. Math. France*, **60**, 1.
1935*a* *Mém. Ponteficio Acad. Sci.*, Ser. 3, Vol. **1**.
1935*b* *Ann. Scuola Norm. Sup. Pisa*, Ser. 2, **4**, 267.
1936*a* *Ibid.*, **5**, 9.
1936*b* *J. math. pures appl.*, Ser. 9, **15**, 339.

Birkhoff, G. D., and Lewis, D. C. 1933 *Ann. math. pura appl.*, Ser. 4, **12**, 117.

Bisconcini, G. 1906 *Acta Math.*, **30**, 49.

Block, H. 1909 *Medd. Lunds Astr. Obs.*, Ser. 2, No. 6.

Bohr, H. 1924 *Acta Math.*, **45**, 29.
1925 *Ibid.*, **46**, 101.
1926 *Ibid.*, **47**, 239.
1930 *Danske Vidensk. Selsk.*, Vol. **10**, No. 10.
1932*a* *Fast-periodische Funktionen* (Berlin: J. Springer).
1932*b* *Commentarii Math. Helvetici*, **4**, 51.

Bohr, H., and Fenchel, W. 1936 *Dansk. Vidensk. Selsk.*, Vol. **14**, No. 1.

Bohr, H., and Jessen, B. 1932 *Ann. Scuola Norm. Sup. Pisa*, Ser. 2, **1**, 385.

Brouwer, L. E. J. 1910 *Math. Ann.*, **69**, 176.

Chazy, J. 1918 *Bull. Astr.*, **35**, 321.
1922 *Ann. École Norm. Sup.*, Ser. 3, **39**, 29.
1929 *J. math. pures appl.*, Ser. 9, **8**, 353.

Denjoy, A. 1932 *J. math. pures appl.*, Ser. 9, **11**, 333.

Dulac, H. 1937 *C.R.*, **204**, 1703.

Ehrenfest, P., and Ehrenfest, T. 1909 *Begriffliche Grundlagen der statistischen Aufassung in der Mechanik* (Enzykl. Math. Wiss., IV, 2, II [Berlin: Teubner]).

Fabre, J. 1937 *Bull. Astr.*, Ser. 2, **10**, 297.
1938 *Ibid.*, **11**, 17.

Feller, W. 1950 *An Introduction to Probability Theory and Its Applications* (New York: John Wiley & Sons).

Fenchel, W., and Jessen, B. 1935 *Dansk. Vidensk. Selsk.*, Vol. **13**, No. 6.

Franklin, P. 1929 *Math. Zs.*, **30**, 325.

Freundlich, E. 1918 *Sitz. preuss. Akad. Wiss.*, p. 168.

Hadamard, J. 1898 *J. math. pures appl.*, Ser. 5, **4**, 27.

Hagihara, Y. 1933 *J. Phys. Math. Soc. Japan*, **7**, 253, 321.
1942 *Ibid.*, **16**, 52.

Hopf, E. 1932*a* *Proc. Nat. Acad. Sci.*, **18**, 93, 204, 333.
1932*b* *Sitz. preuss. Akad. Wiss.*, p. 182.
1937 *Ergodentheorie* (Berlin: J. Springer).

HOSTINSKY, B. 1931 *Méthodes générales du calcul des probabilités* (Paris: Gauthier-Villars).

1932 *Ann. Inst. Poincaré*, **3**, 1.

HOTELLING, J. 1925 *Trans. Amer. Math. Soc.*, **27**, 329.

JESSEN, B., and TORNEHAVE, H. 1952 *Acta Math.*, **77**, 137.

KERÉKJÁRTÓ, T. 1923 *Vorlesungen über Topologie* (Berlin: J. Springer), Vol. **1**.

KHINTCHINE, A. 1933 *Asymptotische Gesetze der Wahrscheinlichkeitsrechnung* (Berlin: J. Springer).

1949 *Mathematical Foundation of Statistical Mechanics* (New York: Dover Publications).

KOLMOGOROV, A. N. 1933 *Grundlagen der Wahrscheinlichkeitsrechnung* (Berlin: J. Springer).

KOOPMAN, B. O. 1927 *Trans. Amer. Math. Soc.*, **29**, 287.

KRYLOV, N., and BOGOLIUBOV, N. 1943 *Non-linear Mechanics* (Princeton: Princeton University Press).

LEFSHETZ, S. 1930 *Topology* (New York: American Math. Soc.), chap. 6.

1946 *Lectures on Differential Equations* (Princeton: Princeton University Press).

1950 *Contributions to the Theory of Non-linear Oscillations* (Princeton: Princeton University Press).

LEIMANIS, H., and MINORSKY, N. 1958 *Dynamics and Non-linear Mechanics* (New York: John Wiley & Sons).

LEVI-CIVITA, T. 1901 *Ann. math. pura appl.*, Ser. 3, **5**, 221.

1903 *Ibid.*, **9**, 1.

1906 *Acta Math.*, **30**, 305.

1919 *Ibid.*, **42**, 99.

LEWIS, D. C. 1934 *Amer. J. Math.*, **56**, 25.

MCLACHLIN, N. W. 1950 *Non-linear Differential Equations* (Oxford: Oxford University Press).

MAITRE, V. 1937 *J. Observateurs*, **20**, 121.

MARKOV, A. 1912 *Wahrscheinlichkeitsrechnung* (Leipzig: Teubner).

1933 *Math. Zs.*, **36**, 708.

MISES, R. VON 1920 *Phys. Zs.*, **21**, 225, 256.

1922 *Naturwiss.*, **10**, 25.

1931 *Wahrscheinlichkeitsrechnung* (Leipzig: Barth), p. 16.

MORSE, M. 1924 *Trans. Amer. Math. Soc.*, **26**, 25.

1935 *Calculus of Variations in the Large* (New York: American Math. Soc.).

MUCKENHOUPT, C. F. 1929 *J. Math. Phys. M.I.T.*, **8**, 163.

NEUMANN, J. VON 1932 *Proc. Nat. Acad. Sci.*, **18**, 70, 263.

NIELSEN, J. 1927 *Acta Math.*, **50**, 189.
POINCARÉ, H. 1879 Thesis, Paris.
1881 *J. math. pures appl.*, Ser. 3, **7**, 375.
1882 *Ibid.*, Ser. 3, **8**, 251.
1885 *Ibid.*, Ser. 4, **1**, 167.
1886 *Ibid.*, **2**, 151.
1889 *Acta Math.*, **13**, 1.
1895 *J. École Polytech.*, Ser. 2, **1**, 1.
1899*a* *R.C. circ. mat. Palermo*, **13**, 285.
1899*b* *Méthodes nouvelles de mécanique céleste* (Paris: Gauthier-Villars), Vol. **3**.
1900 *Proc. London Math. Soc.*, **32**, 277.
1902*a* *Bull. Soc. Math. France*, **30**, 49.
1902*b* *J. math. pures appl.*, Ser. 5, **8**, 169.
1904 *R.C. circ. mat. Palermo*, **18**, 45.
1905 *Trans. Amer. Math. Soc.*, **6**, 237.
1912*a* *R.C. circ. mat. Palermo*, **33**, 375.
1912*b* *Calcul des probabilités* (2d ed.; Paris: Gauthier-Villars), chap. xvi.
1928 *Œuvres de H. Poincaré* (Paris: Gauthier-Villars), Vol. **1**.
SCHWARZSCHILD, K. 1896 *A.N.*, **141**, 1.
SIEGEL, C. L. 1941 *Ann. Math. Princeton*, **42**, 127.
STOKER, J. J. 1950 *Non-linear Vibrations* (New York: Interscience Publishers).
SUNDMAN, K. F. 1907 *Acta Soc. Sci. Fenn.*, Vol. **34**.
1908 *Ibid.*, Vol. **35**.
1913 *Acta Math.*, **36**, 105.
URA, T., and HIRASAWA, Y. 1954 *Proc. Japan. Acad.*, **30**, 726.
VEBLEN, O. 1922 *Analysis Situs* (New York: American Math. Soc.)
WIENER, N. 1939 *Duke Math. J.*, **5**, 1.
WIENER, N., and WINTNER, A. 1941 *Amer. J. Math.*, **63**, 416.
ZEIPEL, H. VON 1905 *Bull. Astr.*, **22**, 449.

CHAPTER 5

Planetary Interiors

By RUPERT WILDT
Yale University Observatory

1. SURVEY OF THE PROBLEM

ASTRONOMICAL observations have furnished information about the size, shape, mass, and rate of rotation of the several planets. The task is to interpret these data in terms of the physical state and chemical composition of the bulk of planetary matter. Some guidance, naturally, is afforded by what has been learned from sounding the depths of the Earth. Further suggestions derive from cosmogonic speculation. At present, however, it is impossible to formulate a wholly deductive theory of planetary constitution. The principal mode of attack has been to construct mathematical models of planetary configurations, to vary their assumed composition, and to match their bulk properties with those of the astronomical bodies. The theory of the interior of the Sun and stars owes much of its success to the fact that stellar matter is completely volatilized and amenable to treatment as an ideal gas. The planets, on the other hand, rank among the cold cosmic bodies, and undoubtedly the bulk of their matter is in the condensed—i.e., liquid and solid—state. It is precisely for this reason that the study of the internal constitution of the planets has made scant progress.

In practice, the analysis of planetary configurations has profited from a partition unattainable in the stellar case. When dealing with gaseous stars, the analysis of the mechanical and thermal equilibrium must be carried out simultaneously. The density gradient inside a self-luminous star is inextricably bound up with the temperature gradient sustained by the internal generation of energy, because both the elasticity and the opacity of a gas depend strongly on temperature. But for liquids and solids the effect of temperature changes is, on the whole, rather small and may be ignored in zeroth approximation. Therefore, the equation of state approximately reduces to a pressure-density relation. By combining any such relation with the equation of hydrostatic equilibrium, the internal density distribution for a "cold" body is found by quadratures. Afterward, for any given distribution of matter endowed with sources of radioactive heat, the internal temperatures can be evaluated according to the theory of heat conduction. A further approximation to the density distribution could be obtained if the thermal variation of the compressibility were known; and the procedure could be repeated until the results converged. The scheme is obvious but rather academic. Its execution has not progressed beyond the first step, and even these results are still *sub judice*. They are bound to remain indeterminate until doubts regarding the equation of state have been settled.

The plan of this chapter is to present, first, a summary of a few astronomical data and certain general considerations concerning the equilibrium configurations of planets. These theoretical results are formal in the sense that they hold irrespective of the material of which the planets are built. The following section deals with physicochemical theories and experimental data relevant to the equation of state of planetary matter. Mention is also made of seismological facts and interpretations that cast light upon the behavior of matter inside the Earth. The importance of these terrestrial data lies in their predominantly empirical nature. Much confidence in theoretical deductions is needed when dealing with the other planets. The third and fourth sections describe particular models of the terrestrial and Jovian planets. Inevitably, these models will require revision as knowledge of the equation of state advances. Section 2, giving a glimpse of some problems of solid-state physics, is perhaps the most important part of this chapter; for it indicates the limitations that the present state of physical theory imposes upon the analysis of planetary constitution. It is not without interest to note that only thirty years ago a monograph on the physical nature of the planets (Graff, 1929) could afford to dismiss the subject of planetary interiors in a few lines.

1.1. Astronomical Evidence

Many geometrical and mechanical constants characterizing the planets individually are not known with great precision. There is a notorious lack of agreement among the numerical values quoted in different treatises. However, the point to be made here cannot be obscured by these uncertainties. Table 1 is intended to illustrate the prime astronomical result bearing upon planetary interiors, namely, that, according to their mean density, the planets fall into two groups, known as the "terrestrial" and

"Jovian" planets. The dividing line may be drawn at about 2.5 gm cm^{-3}, i.e., near the density of the terrestrial rocks. Pluto has been omitted from Table 1, since its mass is not reliably known (Brouwer, 1955, and chap. 3). The Moon, Jupiter I, and Jupiter II are doubtless bodies of the terrestrial type, but Mimas, Enceladus, and Tethys must be classified as Jovian. The status of the other satellites is less certain. At any rate, the grouping observed among the planets probably prevails among the satellites too.

TABLE 1

PRINCIPAL PLANETS AND SATELLITES

SOME PHYSICAL DATA

Planet or Satellite	Mass	Mean Radius	Mean Density (gm/cc)
Mercury......	0.0543	0.38	5.46
Venus........	0.8136	0.961	5.06
Earth........	1.0000	1.000	5.52
Mars.........	0.1069	0.523	4.12
Jupiter.......	318.35	10.97	1.35
Saturn........	95.3	9.03	0.71
Uranus.......	14.54	3.72	1.56
Neptune......	17.2	3.38	2.47
Moon........	0.0123	0.273	3.33
Jupiter I......	0.0121	0.255	4.03
Jupiter II.....	0.0079	0.226	3.78
Jupiter III....	0.0260	0.394	2.35
Jupiter IV....	0.0162	0.350	2.06
Mimas.......	0.000006	0.04:	0.5:
Enceladus.....	0.000014	0.05:	0.7:
Tethys.......	0.000109	0.08	1.2
Dione........	0.000176	0.07:	2.8:
Rhea.........	0.00038	0.102	2.0:
Titan.........	0.0235	0.377	2.42

Mass of the Earth: 5.975×10^{27} gm
Mean radius of the Earth: 6.37123×10^{8} cm

A conspicuous difference in respect of polar flattening accentuates the distinction between terrestrial and Jovian planets. The geometric ellipticity of a planetary disk, as measured with a filar micrometer (either directly at the telescope or on photographs), is affected by systematic errors of physiologic and photographic origin, which are not well understood. At first sight, therefore, it might appear that the geometric ellipticities are inferior to the so-called dynamic ones, whose derivation is as follows. The oblateness of a spinning body is intimately related to its internal density distribution. If the latter is known, the geometric ellipticity of the body's surface can be predicted accurately (see Sec. 1.2); but the converse is not true. Another clue to the internal density distribution of a planet is given by the perturbations in the motions of its satellites. Again, if the former is known, the latter can be predicted. Hence, for any postulated density distribution inside a planet, the observed perturbations in the satellite motions imply a definite "dynamic" ellipticity of the planet's surface. It would be preferable to label the results of such computations "hypothetical" ellipticities. This usage would guard against the possible misunderstanding that the hypothetical ellipticities are a substitute for the geometric ones. Only strictly empirical data can yield unequivocal information about the internal constitution of the planets. Such are the geometric ellipticity and the higher terms in the expansion of the external potential of a planet, which are the product of exhaustive analysis of the satellite motions (see chap. 3). The correct procedure, therefore, is to vary the density distribution, as suggested by physical considerations, until the corresponding hypothetical ellipticity matches the geometric one; and this density distribution simultaneously ought to account for the satellite motions.

1.2. Equilibrium Configurations of Planets

The basic problem in the analysis of the internal constitution of the planets may now be stated as follows: Given a material of known compressibility, shaped by gravitation into a body of prescribed mass and angular velocity, what are the mean radius, the polar flattening, and the moment of inertia of the resulting equilibrium configuration? The answer, in a certain sense, is supplied by the well-known theories of the figure of the Earth. Even in their modern versions (Darwin, 1899; de Sitter, 1924), the emphasis has been on the determination of the figure when the density distribution along some radius is known. DeMarcus (1958) has recast the theory into a form more appropriate for solving the problem of the figure of a planet when only the equation of state of the material is known.

In hydrostatic equilibrium the equipotential surfaces are also surfaces of constant density, assumed to be nearly ellipsoids of revolution, whose equation is

$$\xi = r\left[1 - \frac{4\epsilon^2}{45} - \frac{2}{3}\left(\epsilon + \frac{23\epsilon^2}{42} + \frac{4\kappa}{7}\right) P_2(\cos\phi) + \frac{8}{35}\left(\frac{3}{2}\epsilon^2 + 4\kappa\right) P_4(\cos\phi)\right]. \quad (1)$$

Here ξ is length of the radius vector from the center of the planet, ϕ the co-latitude, r the mean radius, ϵ the ellipticity, and κ a measure of the departure from the ellipsoid. The mean radius r is defined as the radius of a sphere which contains precisely the same volume as the surface of constant density. Replacing the gravitational potential, $V(\xi, \phi)$, by $V(r)$ and dropping second-order terms in the latter, the equations of hydrostatic equilibrium are, with ω the angular velocity,

$$\frac{dP}{dr} = -\left[g(r) - \tfrac{2}{3}\omega^2 r\right] \rho(P) \qquad (2)$$

and

$$g(r) = \frac{GM(r)}{r^2} = \frac{4\pi G}{r^2}\int_0^r \rho(x)\, x^2 dx\,. \qquad (3)$$

Here $M(r)$ is the mass inside the sphere of radius r; $M(R) = M$ is the total mass of the planet, R being the mean radius of the free surface of the configuration. The term $(\frac{2}{3})\omega^2 r$ in equation (2) is the average value over a sphere of radius r of the radial component of the centrifugal force.

The assumption of hydrostatic stress is, of course, an approximation; it is indeed probably rather poor throughout the shallow crust of a planet, where the mechanical strength of the solid material is of the same order of magnitude as the mean pressure. But the latter increases steadily with depth, while the strength (or maximum stress difference sustained) probably decreases; and a depth is soon reached where the approximation becomes satisfactory. In the Earth, from about 100 km down, the stress must be nearly hydrostatic, in the sense that the stress differences are always a minute fraction of the mean pressure. Evidently the assumption of hydrostatic stress will fail in sufficiently small bodies like the asteroids.

In order to build model planets of a material whose equation of state, $\rho(P)$, is known, equations (2) and (3) are solved as though the configuration had spherical symmetry, the oblateness being allowed for in the definition of the parameter r. The resulting density distribution, $\rho(r)$, is correct to the first order in ω^2. Thereafter, the quantities ϵ and κ are evaluated, by successive approximation, from two non-linear integral equations containing various integrals over $\rho(r)$. If the density distribution is known to the first order in ω^2, ϵ and κ are obtained correct to the second order; and with these values the density distribution can be recalculated to the second order by retaining the second-order terms in the expression for $V(r)$.

The coefficients J and K in the conventional expansion of the external potential of the planet, viz.,

$$V_e = -\frac{GM}{\xi}\left[1 - \frac{2}{3} J \frac{a^2}{\xi^2} P_2(\cos\phi) + \frac{4}{15} K \frac{a^4}{\xi^4} P_4(\cos\phi) - \dots\right], \qquad (4)$$

where a is the equatorial radius of the planet, are related to values of ϵ and κ at the surface (subscript s) by

$$\epsilon_s - J = \frac{1}{2}\epsilon_s^2 - \frac{1}{7}\epsilon_s\left(\frac{\omega^2 R^3}{GM}\right) - \frac{4}{7}\kappa_s + \frac{1}{2}\left(\frac{\omega^2 R^3}{GM}\right) \tag{5}$$

and

$$K = \frac{24}{7}\kappa_s + 3\epsilon_s^2 - \frac{15}{7}\epsilon_s\left(\frac{\omega^2 R^3}{GM}\right). \tag{6}$$

If the coefficients J and K can be deduced from the motions of a satellite, equations (5) and (6) afford a test of the tentative density distribution, $\rho(r)$, or, by implication, of the hypothetical equation of state, $\rho(P)$.

Instead of the ellipticity ϵ, de Sitter's variable ϵ' may be introduced, which is defined by

$$\epsilon' = \epsilon - \frac{5}{42}\epsilon^2 + \frac{4}{7}\kappa . \tag{7}$$

Denoting, as before, by the subscript s the values that these variables take at the surface of the configuration, de Sitter has set an upper limit to κ_s, namely,

$$|\kappa_s| < \frac{5}{16} m\epsilon_s - \frac{1}{4}\epsilon_s^2 , \tag{8}$$

with $m = (\omega^2 R^3/GM)$. However, according to Jeffreys (1953), the inequality (8) is not general, and in the same paper Jeffreys has corrected several other errors in de Sitter's formulae. Following Radau, the logarithmic derivative of ϵ' is introduced as an auxiliary variable,

$$\eta = \frac{d \ln \epsilon'}{d \ln r} .$$

It satisfies the equation

$$(1+\eta)^{1/2} = \frac{5}{r^5 \bar{\rho}(r)} \int_0^r F(\eta)\, \rho(x)\, x^4 dx , \tag{9}$$

in which $\bar{\rho}(r)$ is the average density of matter interior to the sphere of radius r,

$$\bar{\rho}(r) = \frac{3}{r^3} \int_0^r \rho(x)\, x^2 dx , \tag{10}$$

and $F(\eta)$ stands for

$$\frac{1 + \eta/2 - \eta^2/10}{\sqrt{(1+\eta)}} = F(\eta) . \tag{11}$$

Equation (9) has to be integrated numerically, the solution of equations (2) and (3) being substituted under the integral in equation (10); the boundary condition is that η must vanish at the center of the planet. Carrying the integration of equation (9) up to the surface, $r = R$, the parameter η_s is obtained, and the ellipticity of the planet's surface can be found by solving the equation

$$\eta_s \epsilon_s' = \frac{5}{2} m - 2\epsilon_s' + \frac{10}{21} m^2 + \frac{4}{7} \epsilon_s^2 - \frac{6}{7} m \epsilon_s \,, \tag{12}$$

which is the boundary condition for η at the surface. In the Radau-Darwin approximation, $F(\eta)$ in equation (9) is replaced by unity. Hence, letting $r = R$,

$$(1 + \eta_s)^{1/2} = \frac{5}{R^5 \bar{\rho}(R)} \int_0^R \rho(r) r^4 dr \,. \tag{13}$$

After integrating by parts and substituting equation (10) for $\bar{\rho}(r)$, it follows that

$$\frac{2}{5}(1 + \eta_s)^{1/2} = 1 - \frac{3}{R^5 \rho(R)} \int_0^R \rho(r) r^4 dr \,. \tag{14}$$

Since the moment of inertia of a spherical configuration about its axis of rotation is

$$I = \frac{8\pi}{3} \int_0^R \rho(r) r^4 dr = \frac{2M}{R^3 \bar{\rho}(R)} \int_0^R \rho(r) r^4 dr \,, \tag{15}$$

there results from equations (14) and (15) the approximate relation

$$\frac{2}{5}(1 + \eta_s)^{1/2} = 1 - \frac{3}{2} \frac{I}{MR^2} \,. \tag{16}$$

Now, if in equation (12) the squares of ϵ_s and m and the cross-term are neglected, this equation reduces to

$$\eta_s = \frac{5}{2} \frac{m}{\epsilon_s} - 2 \,. \tag{17}$$

By substitution of equation (17) in equation (16), we obtain

$$\frac{I}{MR^2} = \frac{2}{3}\left[1 - \frac{2}{5}\left(\frac{5}{2}\frac{m}{\epsilon_s} - 1\right)^{1/2}\right] ; \tag{18}$$

accordingly, the moment of inertia of a planet can be estimated from the observed value of m/ϵ_s (Jeffreys, 1924). For a homogeneous sphere, i.e., $\rho(r)$ = const. = $\bar{\rho}(R)$, the left-hand side of equation (18) becomes $\frac{2}{5}$. For configurations whose mass is strongly concentrated toward the center, the Radau-Darwin approximation breaks down; it always becomes inadequate for $I/MR^2 < 0.3$ and is quite poor for the Jovian planets (Jeffreys, 1951).

1.3. Unstable Planetary Configurations

For a planet of spherical symmetry and for any specified function $\rho(P)$, the solution $P(r)$ of equations (2) and (3) of the preceding section determines uniquely the march of pressure in terms of the pressure P_c prevailing at the center of the configuration. Hence the entire configuration is determined if P_c is known, including the total mass given by the equation

$$M = 4\pi \int_0^R \rho \left[P(r) \right] r^2 dr . \qquad (19)$$

It must not be assumed, however, that P_c is uniquely determined by M. The existence of multiple solutions, $P_c(M)$, and of corresponding multiple solutions, $R(M)$, was first demonstrated by Ramsey (1950*a*) for a special density-pressure relation, namely, $\rho = \rho_0 =$ const. for $0 < P < P^*$, and $\rho = \lambda\rho_0$ for $P > P^*$, where $\lambda > \frac{3}{2}$ is a constant; that is to say, an incompressible material is supposed to undergo a discontinuous change in density, or a phase transition (see Sec. 2.11), at a critical pressure P^*. With such a material there exist for some values of M three possible configurations, of which either the one with largest P_c or the one with smallest P_c is the most stable; planets with very small cores (consisting of the high-pressure phase) are the least stable of these three configurations. Schatzman (1951) has extended Ramsey's analysis to planetary models composed of an incompressible material undergoing two, or even three, successive phase transitions at distinct critical pressures. Such have been observed in the laboratory for many substances, of which ice is an outstanding example. The discriminants for instability established by Schatzman are naturally more complicated than Ramsey's simple inequality, $\lambda > \frac{3}{2}$. The conclusion is that the existence of further phase transitions does not eliminate the instabilities deriving from the first one; thus Ramsey's principal result is upheld, and for purposes of general orientation it will suffice to discuss the case of material with a single phase transition. It may also be pointed out that Schatzman's general analysis is relevant to the study by Datta (1954) of a possible mechanism for the ejection of the Moon from a primordial Earth-Moon body, originally proposed by Bullen (1951).

The assumption that the material be incompressible, except at the critical pressure, does not seem to be essential. Lighthill (1950) has proved the following theorem. If $\rho(P)$ is continuous for $0 < P < P^*$, but takes a value $\lambda\rho^*$ (where $\rho^* = \rho[P^*]$) for P just greater than P^*, then in the dependence of the total mass M on the central pressure P_c the derivative dM/dP_c is changed discontinuously by a factor $(3 - 2\lambda)/\lambda^2$ at $P = P^*$. Moreover, the derivative, with respect to P_c, of any other bulk characteristic (such as the radius R) changes by the same factor, which is positive if λ does not exceed $\frac{3}{2}$. But for $\lambda > \frac{3}{2}$ this factor becomes negative, so that

dM/dP_c changes sign. Since M cannot decrease with P_c indefinitely, the behavior of $M(P_c)$ in a certain critical range, $M_1 < M < M_0$, must be qualitatively as shown in Figure 1.

A configuration is stable against small radial perturbations if dM/dP_c is positive, and it is unstable if dM/dP_c is negative. The configurations A and C in Figure 1 are therefore stable, and B is unstable. Now A has no core, because $P_c < P^*$, and C has a large core. All configurations B located on the branch YX' have smaller cores than the configurations C located on the branch $X'Y'$. For this reason, Ramsey has termed the phenomenon under discussion "the instability of small cores." It is further illustrated by the corresponding energy relation $U(R)$, where U is the total energy (sum of elastic energy stored in the compressed material and of potential energy of the configuration), as shown schematically in Figure 2. The letters A, B, etc., refer to the same equilibrium configurations as those in Figure 1. The top and bottom curves in Figure 2 refer to the masses M_0 and M_1, which represent limiting cases. For either mass a maximum and a minimum in the energy diagram coalesce to give a point of inflection; hence there is only one stable configuration for these masses. For mass M_0 the stable configuration is Y', which has a core; for M_1 the stable configuration is X, which has no core. The total energy of the stable configuration C with a core will, in general, differ from that of the stable configuration A without a core; this difference, ΔU, attains its greatest values near M_0 and M_1 and changes sign at some intermediate mass. Ramsey has estimated that transitions between the two stable configurations would release energies sufficient to expel a sizable fraction of the parent body's mass.

The planetary transitions envisaged by Ramsey would occur if the planet's mass M crosses the critical mass range (M_0, M_1) in either direction. Secular changes of M could result from accretion or loss of mass by dissipation into space; or internal cooling could advance a planet into the critical range, because M_0 and M_1, strictly speaking, are functions of the internal temperatures. Ramsey believes that the asteroids and meteorites are fragments of an exploded planet of the terrestrial type, and he attributes the instability in question to the transition of silicates into a metallic phase at the pressure prevailing in the Earth at a depth of 2900 km. Attractive as this hypothesis appears, Ramsey's ideas regarding the chemical composition of the terrestrial planets, so far, have not been accepted by most geophysicists and astronomers. It is appropriate, therefore, to distinguish clearly between the general theory of planetary instability, whose

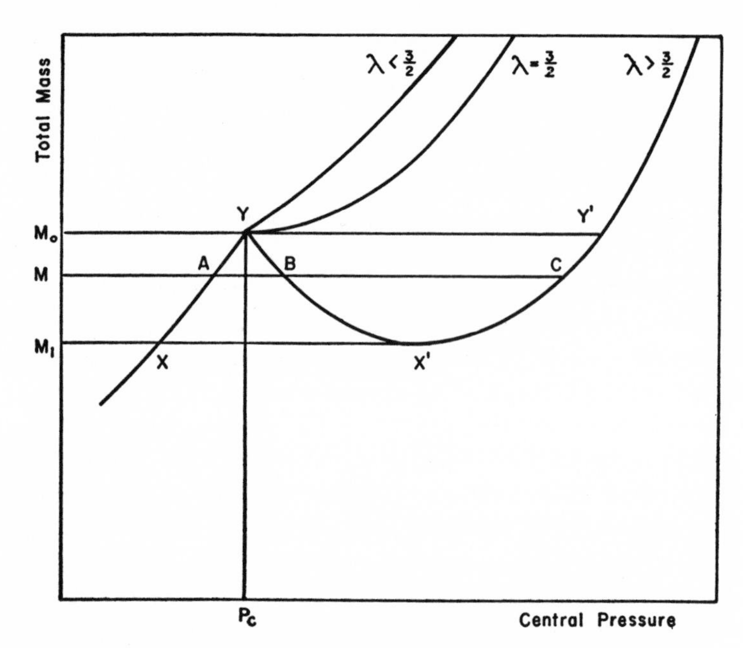

FIG. 1.—Instability of small planetary cores (Ramsey)

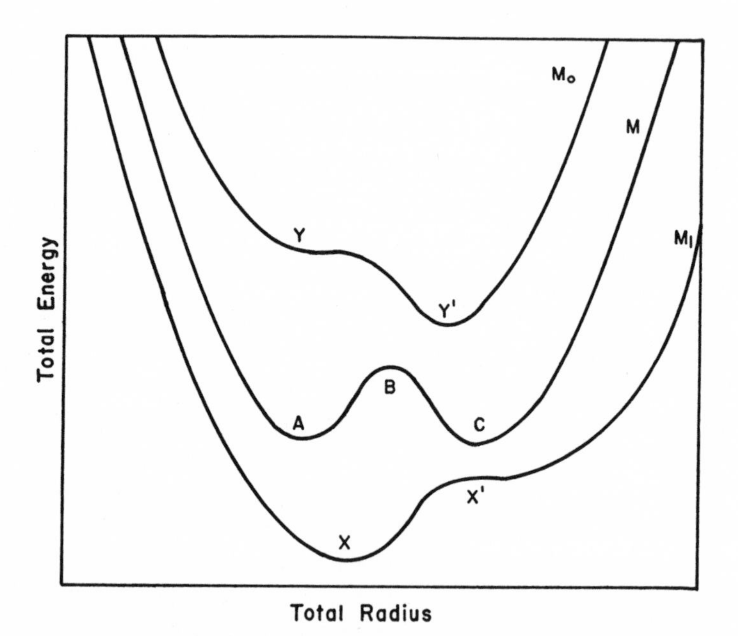

FIG. 2.—Instability of small planetary cores (Ramsey)

merits are undisputed, and the special hypothesis underlying application of this theory to the terrestrial planets (see Sec. 3.2).

Recently Lighthill's theorem has been generalized for a slowly rotating planet (DeMarcus, 1954). With the same notation as above, the discriminant for instability is

$$\lambda > \frac{3}{2} - \frac{\omega^2}{4\pi G \rho^*}.$$

A planet rotating with the angular velocity of Jupiter, in order to exhibit the effects in question, requires only a 37 per cent discontinuity in density if the mantle density is about 0.3 gm cm^{-3} (molecular hydrogen), instead of a 50 per cent discontinuity for a non-rotating planet.

1.4. Estimates of the State Variables

It is convenient to mention here sundry integral theorems on the equilibrium of planets. They relate certain bulk properties to the mean values of the state variables, like pressure and temperature, or to their values at the center. Deeper insight into the structure of the planets must depend on the elaborate studies sketched in Section 1.2. This analysis will profit from prior knowledge of the order of magnitude of the state variables. The integral theorems furnish this information.

The internal pressures can be estimated from two integral theorems well known in the theory of gaseous stars (Chandrasekhar, 1939). Their application to the internal constitution of non-gaseous planets is justified if the bulk of the planetary matter does not have sufficient strength to support tangential stresses. The physical content of the first three theorems stated below (eqs. [20], [21], and [24]) is the following: Consider a spherical equilibrium configuration of mass M and radius R with a density distribution, arbitrary except for the rather weak condition that $\bar{\rho}(r)$, the average density of the material interior to r, does not increase outward. From the given configuration two other configurations of uniform density can then be constructed, one with a constant density equal to $\bar{\rho}$, the mean density of the given configuration, and the other with a constant density equal to ρ_c, the density at the center of the given configuration. Intuition will suggest that the properties of the given configuration or, at any rate, some of them, are intermediate between those of the two constructs. The theorems to be stated simply express that certain physical variables of the given configuration in fact have values respectively greater than those for the configuration of uniform density with $\rho = \bar{\rho}$, and respectively less than those for the configuration of uniform density with $\rho = \rho_c$.

Of chief interest is a pair of inequalities bracketing the central pressure,

$$\frac{3}{8\pi}\frac{GM^2}{R^4} \leqslant P_c \leqslant \frac{3}{8\pi}\frac{GM^2}{R^4}\left(\frac{\rho_c}{\bar{\rho}}\right)^{4/3}. \tag{20}$$

Similarly, the average pressure throughout the configuration is bounded by

$$\frac{3}{20\pi}\frac{GM^2}{R^4} \leqslant \bar{P} \leqslant \frac{3}{20\pi}\frac{GM^2}{R^4}\left(\frac{\rho_c}{\bar{\rho}}\right)^{4/3}, \tag{21}$$

where $\bar{P}$ is defined by

$$M\bar{P} = \int_0^R P dM(r), \tag{22}$$

$M(r)$ being the mass contained in the sphere of radius r. The lower limits of P_c and $\bar{P}$, according to equations (20) and (21), are listed in Table 2 for the principal planets. Theorems (20) and (21) hold irrespective of the

TABLE 2

LOWER LIMITS TO INTERNAL PRESSURES AND TEMPERATURES

Planet	$P_c>$ (Atm.)	$\bar{P}>$ (Atm.)	$T_c>$ (° K)	$\bar{T}>$ (° K)
Mercury..........	3.0×10^5	1.2×10^5	650	400
Venus..........	1.2×10^6	4.8×10^5	1900	1200
Earth..........	1.7×10^6	6.8×10^5	2400	1500
Mars..........	1.6×10^5	7.0×10^4	1600	300
Jupiter..........	1.0×10^7	4.1×10^6	68,000	42,000
Saturn..........	1.9×10^6	7.7×10^5	24,000	16,000
Uranus..........	1.8×10^6	7.5×10^5	11,000	7000
Neptune..........	3.6×10^6	1.5×10^5	12,000	8000

equation of state of the material. Additional theorems concern the internal temperatures, but their proof requires the further assumption that the material behaves like an ideal gas. The average temperature, defined by

$$M\bar{T} = \int_0^R T dM(r), \tag{23}$$

is restricted by the inequalities

$$\frac{1}{5}\frac{\mu m_0}{k}\frac{GM}{R} \leqslant \bar{T} \leqslant \frac{1}{5}\frac{\mu m_0}{k}\frac{GM}{R}\left(\frac{\rho_c}{\bar{\rho}}\right)^{1/3}, \tag{24}$$

where μ is the dimensionless molecular weight on the chemical scale, m_0 the corresponding mass unit (1.66×10^{-24} gm), and k the Boltzmann constant. An important theorem, due to Chandrasekhar (1938), establishes an absolute minimum of the central temperature of a gaseous configuration having constant molecular weight throughout, namely,

$$(T_c)_{\min} = \frac{1}{2} \frac{\mu m_0}{k} \frac{GM}{R} Q_{\min} . \tag{25}$$

With decreasing mass, the dimensionless function $Q_{\min}$ approaches the value of 0.640, which is appropriate for masses of planetary order. Lower limits of T_c and $\bar{T}$ computed from equations (23) and (25) with $\mu = 1.008$, are also listed in Table 2. These figures are instructive, though it is not suggested that they are relevant to the present internal state of the planets.

If the pressure-density relation, $P = P(\rho)$, were known for the material constituting the innermost core of a planet, one could substitute $P(\rho_c)$ for P_c in equation (20) and thereby bracket ρ_c. It may happen, of course, that the inequalities for ρ_c are not satisfied by a given combination of M and $\bar{\rho}$, which would indicate that such a planet cannot have a core obeying the equation of state to be tested. As an example, consider the pressure-density relation of a (non-relativistic) Fermi-Dirac gas, i.e.,

$$P = K \left(\frac{Z}{A}\right)^{5/3} \rho^{5/3} , \tag{26}$$

with $K = 9.913 \times 10^{12}$ dynes cm^{-2}. Here Z is the (dimensionless) atomic number and A the (dimensionless) atomic weight; ρ has the dimension gm cm^{-3}. The only influence of the chemical composition is through the ratio (A/Z), which is about 2 for most substances, but 1 for hydrogen. By substitution of equation (26) for $P_c(\rho_c)$ in equation (20) it is found that

$$\beta^{1/5} \bar{\rho}^{4/5} \leqslant \rho_c \leqslant \beta \quad \text{and} \quad \beta = \frac{4\pi}{3} \left(\frac{G}{2K}\right)^3 \left(\frac{A}{Z}\right)^5 M^2 . \tag{27}$$

The outer members of this inequality constitute a necessary, though not sufficient, condition for the attainment of complete degeneracy at the center of a planet, namely,

$$\bar{\rho} \leqslant \beta = \frac{4\pi}{3} \left(\frac{G}{2K}\right)^3 \left(\frac{A}{Z}\right)^5 M^2 , \tag{28}$$

or numerically, with $A/Z = 2$,

$$\beta = 1.824 \times 10^{-4} \left(\frac{M}{M_E}\right)^2 \text{gm cm}^{-3} ,$$

where M_E is the mass of the Earth. Jupiter ($\beta = 18.5$) and Saturn ($\beta = 1.66$) are the only planets satisfying equation (28) with $A/Z = 2$. On the other hand, it has been claimed that an equation of state based on the Thomas-Fermi model (see Sec. 2.21), namely,

$$P = K' Z^{4/3} A^{-2} \rho^2 , \qquad K' = 5.07 \times 10^{12} \text{ dynes cm}^{-2} , \tag{29}$$

should be adequate to represent the behavior of matter at depths greater than a few hundred kilometers below the surface of the planets (Scholte, 1947). In fact, equation (29) cannot hold, even at the center of the planet, unless Z is improbably large. For, by the same argument as above, the analogue of equations (27) is

$$\left(\frac{\bar{\rho}\gamma M}{R}\right)^{1/2} \leqslant \rho_c \leqslant \left(\frac{1}{\bar{\rho}}\right)^{1/2} \left(\frac{\gamma M}{R}\right)^{3/2}, \qquad \gamma = \frac{GA^2}{2K'Z^{4/3}}. \tag{30}$$

Hence

$$\bar{\rho} \leqslant \frac{\gamma M}{R}, \tag{31}$$

and

$$R \geqslant \left(\frac{3K'}{2\pi G}\right)^{1/2} \frac{Z}{A} \left(\frac{1}{Z}\right)^{1/3}, \tag{32a}$$

or, numerically,

$$\frac{R}{R_E} \geqslant 9.595 \frac{Z}{A} \left(\frac{1}{Z}\right)^{1/3}, \tag{32b}$$

with R_E being the radius of the Earth. For hydrogen this inequality is satisfied by Jupiter, but not by Saturn; if equation (20) is generalized for a slowly rotating planet (DeMarcus, 1958), then Saturn too will satisfy the corresponding generalization of inequality (32) for hydrogen. For heavy elements ($A/Z = 2$), it follows from inequality (32*b*) that

$$Z \geqslant 110.4 \left(\frac{R_E}{R}\right)^3 .$$

It would be unrealistic, therefore, to expect planetary models built on equations of state like (26) or (29) to have a general bearing on the internal constitution of the planets.

Another integral theorem, noted by Jeffreys (1924), qualifies the density, ρ_s, of the material composing the planet's solid surface. We put equation (15) into the form

$$\frac{I}{MR^2} = \frac{2\rho_s}{\bar{\rho}(R)} \int_0^R \frac{\rho(r)}{\rho_s} \frac{r^4 dr}{R^5}, \tag{33}$$

where $\bar{\rho}(R)$ now refers to the average density of the material inside the *solid* surface, of radius R. Since, further,

$$\int_0^R \frac{\rho(r)}{\rho_s} \frac{r^4 dr}{R^5} \geqslant \frac{1}{5}, \tag{34}$$

if $\rho(r)/\rho_s \geqslant 1$ everywhere, it is seen that

$$\rho_s \leqslant \frac{5}{2} \frac{I}{MR^2} \bar{\rho}(R) . \tag{35}$$

This inequality provides useful information only if I/MR^2 is much smaller than $\frac{2}{5}$, i.e., for the Jovian planets, which also have low mean densities. It would be permissible to identify $\bar{\rho}(R)$ with the mean density $\bar{\rho}$ of the Jovian planets, given in Table 1, unless their solid surface were located at extremely great depth below the cloud layer defining the visible surface. This point requires special examination (see Sec. 4.2).

The same type of argument was used by Ramsey (1951) to set a lower limit to the central density, ρ_c. It is assumed that the central density is the largest in the planet. Evidently the moment of inertia is larger than it would be if the planet consisted of a central core of radius a and density ρ_c, surrounded by an envelope of negligible mass; that is,

$$I > \frac{8\pi}{15} \rho_c a^5 , \qquad \rho_c a^3 = \bar{\rho} R^3 ,$$

from which it follows that

$$\rho_c \geqslant \left(\frac{5}{2} \frac{I}{MR^2}\right)^{-3/2} \bar{\rho} . \tag{36}$$

2. SOME PROPERTIES OF MATTER UNDER HIGH PRESSURE

2.1. Moderate Pressures (Experimental Range)

For chemically pure substances, copious experimental data are now available, extending up to pressures of 100,000 atm. in some cases. But the behavior of mixtures of several chemical compounds shows features that as yet defy prediction from general principles, though they admit of some a posteriori classification by the theory of phase equilibria. *Phase* is the generic term comprising, in a chemically homogeneous system, the distinct states of aggregation as well as allotropic modifications, e.g., different crystal structures of the solid. In systems composed of more than one chemical substance, the term *phase* denotes the spatially separate domains of uniform chemical composition; there are usually one gaseous and several (immiscible) liquid and solid phases. In a gravitational field, coexisting phases will tend to become stratified according to the sequence of their specific gravities. A non-trivial example is a refrigerated mixture of hydrogen and helium in the interval between their respective critical temperatures (33° and 5° K); subjected to progressive compression, this mixture eventually will decompose into a liquid phase consisting of hydrogen and a denser phase of gaseous helium, with the former floating on top of the latter! This case is rather an exceptional one, insofar as knowledge of the properties of the components sufficed to foresee the odd behavior of the composite system. Generally speaking, neither chemical thermodynamics nor the quantum theory of the liquid and solid states is of much use for predicting in detail the properties of mixed materials inside the planets from measurements made on their components at much lower pressures. This is strikingly illustrated by the discovery (Krichevsky, 1940) that a homogeneous gaseous system on compression may split into distinct gaseous phases. A mixture of two-thirds of NH_3 and one-third of H_2 by volume, at 413° K and 5000 kg cm^{-2}, decomposes spontaneously into two phases containing 76 and 33 per cent of NH_3, respectively. Immiscible gaseous phases also exist in a mixture of NH_3 and CH_4 under high compression, and in the ternary system N_2-NH_3-CH_4 also (Krichevsky and Ziclis, 1943). Although it had been foreseen by van der Waals that two gaseous phases might coexist, in a binary system, at temperatures higher than the critical temperature of the less volatile component, the possibility of this very phenomenon was denied by other investigators until fairly recently.

2.11. *Thermodynamics of phase transitions.*—For any particular phase the internal energy is a continuous function of pressure and temperature (and possibly of other state parameters, e.g., magnetic-field strength), which is known as the *caloric equation of state* of the individual phase. Phase transitions are accompanied by evolution, or absorption, of heat. This so-called *latent heat*, L, and the change in volume, ΔV, of one mole of substance undergoing a phase transition are related to the temperature derivative of the equilibrium pressure, P_e, by the equation

$$L = T \frac{dP_e}{dT} \Delta V . \tag{37}$$

The special form of equation (37) applying to the process of evaporation is known as *Clapeyron's equation*,

$$L_v = T \frac{dP_s}{dT} (V_s - V_l) , \tag{38}$$

with L_v the latent heat of evaporation, P_s and V_s the pressure and molar volume of the saturated vapor, and V_l the molar volume of the coexistent liquid. Along the vapor-tension curve, $P_s(T)$, both L_v and the difference $V_s - V_l$ decrease monotonically; and they vanish together at the critical point, the upper terminal of $P_s(T)$. Another version of equation (37) is

$$L_m = T \frac{dP_m}{aT} (V_{\text{fluid}} - V_{\text{solid}}) , \tag{39}$$

which applies to the process of melting (or freezing), L_m being the latent heat of melting and P_m the melting pressure. Figure 3 shows the vapor-tension and melting-curves of a number of common gases and of water; to avoid confusion, the phase boundaries of the numerous allotropic modifications of ice have been omitted (see Fig. 4 instead), and only their triple points on the melting-curve have been indicated. The dashed parts of the melting-curves are extrapolations based on a formula proposed by Simon (1929),

$$\log_{10}(P_m + a) = b \log_{10} T + c , \tag{40}$$

in which a, b, and c are three constants obtained by fitting the experimental melting data. Equation (40) is flexible enough to represent the

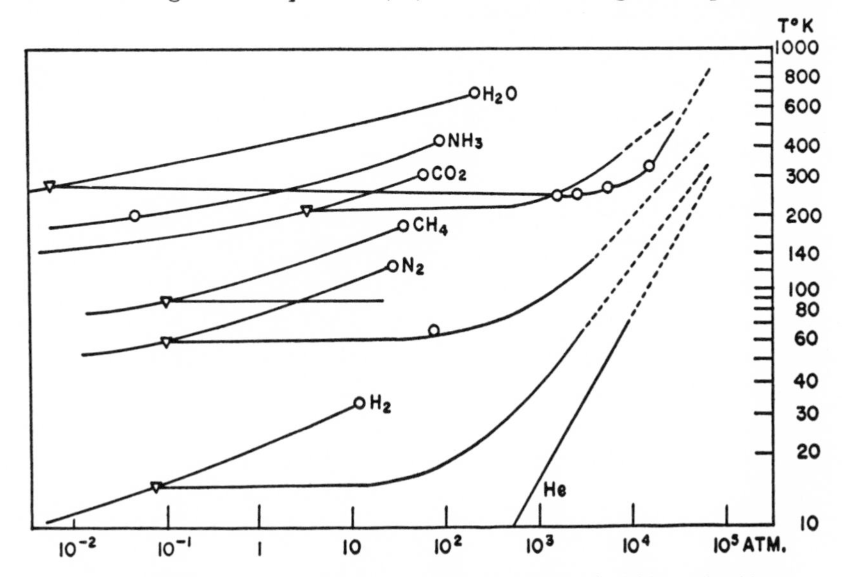

FIG. 3.—Phase diagrams of hydrogen, helium, and some hydrides and oxides

melting-curves of all substances so far tested. More recently, it has been replaced (Simon, 1937) by the expression

$$P_m = a\left[\left(\frac{T}{T_0}\right)^c - 1\right], \tag{41}$$

where T_0 is the normal melting point of the substance, and only two constants, a and c, are obtained by fitting the experimental melting line; a has been called internal pressure,[1] the c is about 1.5–2 for most substances (metals excepted). For helium the quantity T_0 needed reinterpretation, but the use of the internal pressure a as a parameter automatically took

[1] *Internal pressure* = $(\partial U/\partial V)_{T=\text{const}}$. In the particular case of a fluid obeying the van der Waals equation of state, $(P + a/V^2)(V - b) = RT$, the internal pressure has the same value as the *pressure of cohesion*, a/V^2.

account of the quantum effects here playing a role. For alkali metals the constant c ranges from 3.8 to 4.8. Equation (41) has now received some theoretical justification (de Boer, 1952; Salter, 1954). It is believed, therefore, that the "reduced" formula (41) for the temperature dependence of the melting pressure has the same kind of validity as that of the well-known "reduced" vapor-pressure formula. Hence the fact that helium can be solidified at temperatures of about fifty times the "normal" melting temperature by exerting pressures of about five hundred times the internal pressure (Holland, Huggill, and Jones, 1951) may be taken to apply to

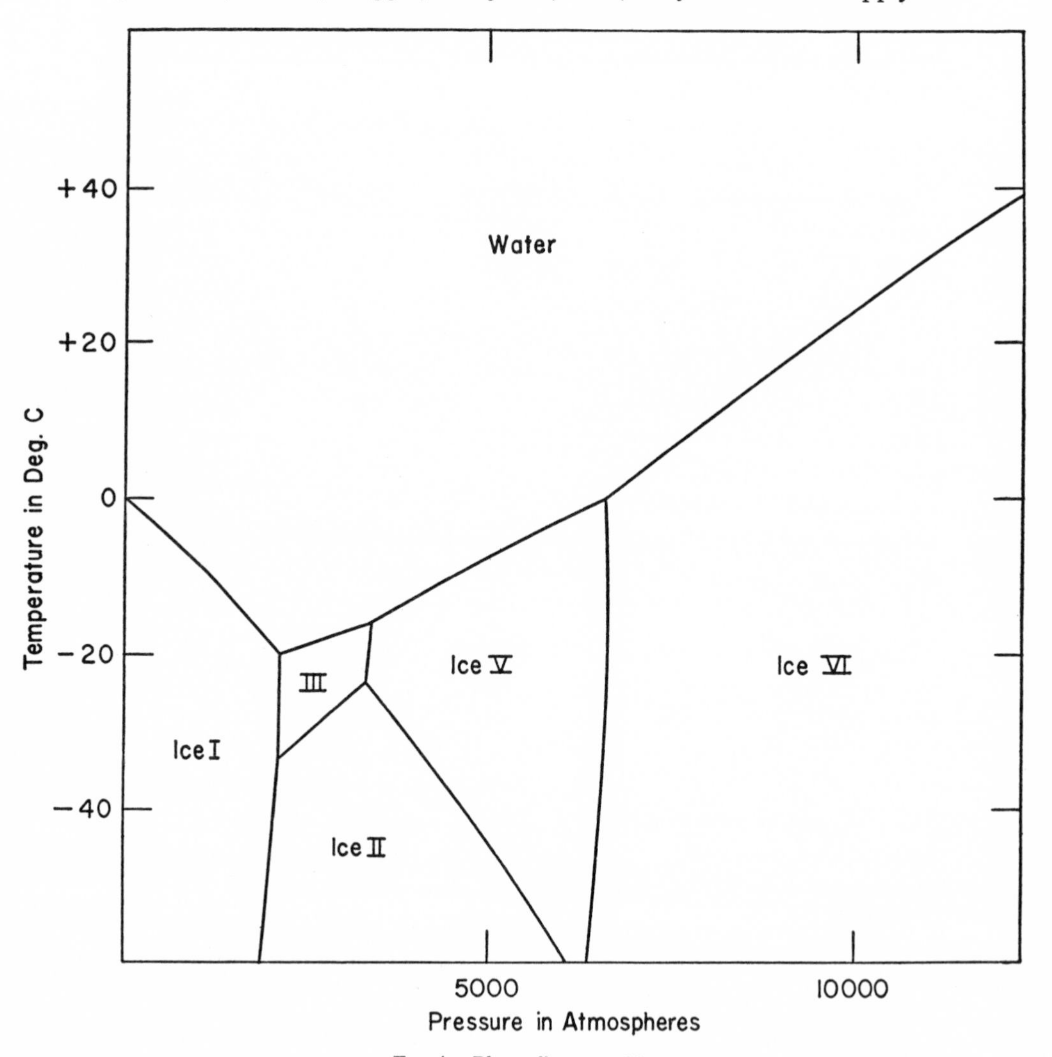

Fig. 4.—Phase diagram of ice

other substances under similar conditions (Simon, 1952). This statement is based on the assumption that the particular conditions of temperature and pressure are not so extreme as to produce changes on a subatomic scale (e.g., changes in electronic structure or pressure ionization); for such changes cannot be taken care of by equation (41).

Speculation continues regarding the ultimate behavior of the melting line at pressures far beyond the range ever likely to be realized in the laboratory. It is now regarded as certain that the melting pressure will rise indefinitely with increasing temperature; that is to say, the melting line cannot level off or reapproach the pressure axis, as Schames and Tammann had surmised. What remains in doubt is whether the melting line will just "fade out" or terminate in a critical point fluid-solid. The approach to such a point would be revealed by a simultaneous decrease along the melting line of the difference in molar entropy and volume of the coexistent phases. This effect is conspicuously absent from the recent experimental data for the most favorable model substance, i.e., helium (Dugdale and Simon, 1953); in fact, the differences between the phases become more pronounced at higher melting temperatures. Eventually, one would expect, this trend would be reversed. At first sight, it may seem difficult to conceive of a discontinuity of states at pressures under which the ordinary crystal structure breaks down completely. But reflection will show that, even at densities such that the atomic structures of neighboring atoms overlap completely and the electrons are effectively free, the nuclei may still tend toward an ordered distribution representing a minimum of potential energy—provided, of course, that the temperature is low enough, or, more precisely, under the condition that the zero-point vibrations of the nuclei shall not be so strong as to prevent the formation of a "lattice." Violation of this condition leads to "melting"; that is to say, the "nuclear crystal" will become disordered. Critchfield (1942) has given estimates of this kind of melting point for superdense hydrogen.

The pressure dependence of the melting point has geophysical and cosmogonic implications. In a chemically homogenous fluid shell of planetary dimensions, stirred by convection and cooling by radiation from the surface, both the melting point and the temperature increase with depth. If the former rises faster than the latter, freezing will first occur at great depth and then spread upward; and on the opposite assumption, solidification would commence at the free surface. This argument was given by Kelvin. The inference from modern laboratory data is that in the Earth solidification advanced from the bottom upward (Adams, 1924). Interest also attaches to the conclusion (Dugdale and Simon, 1953) that it is impossible to melt a normal substance by adiabatic, reversible compression, as such a compression would lead further into the domain of stability of the solid. The term "normal" applies to substances having $dP_m/dT > 0$. Bridgman has shown that, with increasing pressure, all abnormal substances, e.g., ice I, undergo polymorphic transitions and then behave normally.

2.12. *Compressibility and viscosity of fluids.*—Diverse interpolatory formulae have been designed to represent the *thermal* equation of state of an imperfect gas. Their value for purposes of extrapolation is quite limited. From the theoretical standpoint, the most appropriate form is a power series in the reciprocal molar volume, V, namely,

$$PV = RT\left(1 + \frac{b}{V} + \frac{c}{V^2} + \frac{d}{V^3} + \ldots\right), \tag{42}$$

in which R denotes the ideal gas constant and $b, c, \ldots,$ are known as the second, third, etc., *virial coefficients.* Following Amagat, experimenters have adopted the custom of using for each gas a unit, V_0, of molar volume such that $V_0 = 1$ at $0°$ C and $P = 1$ atm. The Amagat unit varies from one gas to another according to the extent of its deviation from perfection at standard temperature and pressure; but this variation is very small, and, unless extreme precision is demanded, the Amagat unit may be taken to correspond to 2.24×10^4 cm^3. The dimensionless quantity,

$$\frac{V_0}{V} = \frac{\text{Density at } (T, P)}{\text{Density at } (0° \text{C}, 1 \text{ atm.})},$$

is often called the *Amagat density*. By substituting equation (42) in the equation of hydrostatic equilibrium, the density distribution in an isothermal atmosphere can readily be found; it should be noted that the virial coefficients are temperature functions only. The evaluation of the density distribution in an adiabatic atmosphere poses a problem inherently more difficult, since it involves the dependence on temperature and pressure of C_p/C_v (the ratio of the specific heat capacities at constant pressure and volume) or, generally speaking, the *caloric* equation of state, i.e., the full dependence of the internal energy on P and T. Except for the noble gases, C_p/C_v is never a constant. It should be noted, moreover, that, for diatomic and polyatomic molecules, C_p/C_v is temperature-dependent even in the ideally gaseous state, i.e., with pressure tending to zero. This effect may be quite large in planetary atmospheres, as for molecular hydrogen in the tropospheres of Jupiter and Saturn.

The best empirical representations of the equation of state of liquids are, perhaps, the Tait equation and its modification by Kirkwood, which afford an excellent representation of the equation of state of water up to 25,000 atm. (Richardson, Arons, and Halverson, 1947). Next to their compressibility, the viscosity of liquids is their most important mechanical characteristic in the present context. No useful predictions regarding the viscosity of superdense gases and liquids appear to have been made from the rigorous kinetic theories of transport processes. The Eyring theory of absolute reaction rates (Glasstone, Laidler, and Eyring, 1941) has been moderately successful in predicting the transport coefficients in the liquid phase and explaining them on a simple pictorial basis. Furthermore, Andrade (1934) has put forward a theory of viscosity designed to apply to liquids near their freezing point. It gives satisfactory account of the viscosity of liquid metals (Andrade, 1952) and even of liquid helium. But it fails when applied to liquids under a pressure exceeding a few thousand atmospheres, the predicted viscosities falling short of the observed ones. These deviations are the result of incipient deformation of the molecules proper, with which the Andrade theory cannot be expected to cope. Examination of Bridgman's measurements of viscosity under very high pressure has led to the hypothesis (Kirsch, 1935) that the melting-curve may be a line of constant viscosity in the coexistent liquid phase. This interesting suggestion does not appear to have been followed up.

Finally, an intriguing speculation by Landau and Zeldovich (1943) may be noted. Their surmise is that a metallic liquid, in a certain pressure range, might undergo a phase transition to a dielectric liquid, before evaporating. In this case a second triple point (liquid metal–dielectric liquid–vapor) must exist at a temperature and pressure higher than that of the familiar triple point (solid metal–liquid metal–vapor). A self-gravitating sphere of such a liquid metal conceivably could contain a dielectric liquid core.

The distinction between liquids and solids seems obvious enough, but its usefulness is restricted to such clear-cut cases as crystals and their melts. The inadequacy of this classification may be illustrated by recent experimental work (Kuhn and Vielhauer, 1953) on the velocities of ultrasonic waves in a vitreous material and their variation over a wide range of temperatures: propagation of transversal waves was observed up to temperatures 30°–40° above the melting point, i.e., in a medium described as unquestionably liquid. The study of the deformation and flow of materials, known as *rheology*, has many applications to geophysics and the cosmogony of the planets. Attention may therefore be called to a comprehensive, yet brief, monograph dealing with principles of theoretical rheology (Reiner, 1949). A more-up-to-date treatment by the same author is given in the *Handbuch der Physik* (Reiner, 1958).

2.13. *Compressibility of solids.*—Deep inside the planets all materials undergo finite deformations so large as to render inapplicable the classical (infinitesimal) theory of elasticity. Among modern theories of finite strain, the one developed by Murnaghan (1937) is particularly useful in problems concerning the compressibility at the highest pressures accessible to experimentation. Murnaghan's theory imposes no restriction on the magnitude of strain. But, because of its thermodynamic character, it cannot supply values for the various parameters distinguishing any particular material, and those must be found by other means. Hence, in practice, limitations arise from ignorance of the coefficients of higher order than the second appearing in the development of the strain energy as function of the strain. The complete form of Murnaghan's theory is required for discussion of the effect of pressure upon individual elastic constants. But the relation between hydrostatic pressure and volume of an isotropic body may be obtained directly from the thermodynamic relation

$$P = -\left(\frac{\partial F}{\partial V}\right)_T , \tag{43}$$

where F is the Helmholtz free energy.

Now in the case of hydrostatic strain of a medium of quasi-isotropic (or cubic) symmetry, the strain degenerates into a single component,

$$\epsilon = \frac{\partial u}{\partial x} - \frac{1}{2}\left(\frac{\partial u}{\partial x}\right)^2 = \frac{\partial v}{\partial y} - \frac{1}{2}\left(\frac{\partial v}{\partial y}\right)^2 = \frac{\partial w}{\partial z} - \frac{1}{2}\left(\frac{\partial w}{\partial z}\right)^2 . \tag{44}$$

Here, u, v, and w denote the components of displacement, and x, y, and z are the Cartesian co-ordinates after displacement. The relation between strain ϵ and volume V is

$$\frac{V_0}{V} = \frac{\rho}{\rho_0} = (1 - 2\epsilon)^{3/2} = (1 + 2f)^{3/2} , \tag{45}$$

with the subscript zero referring to zero pressure; the auxiliary variable f is positive for compression. It is further assumed that for hydrostatic pressure the strain energy may be expanded in the form

$$F = af^2 + bf^3 + cf^4 + \ldots . \tag{46}$$

By definition, F includes an indeterminate absolute constant; moreover, on the right-hand side of equation (46), a term proportional to F is eliminated by choosing as the initial configuration a state devoid of any strain.

The constants $a, b, c, \ldots$, are functions of temperature. Substituting equation (46) in equation (43), the pressure is found to be

$$P=-\left(\frac{df}{dV}\right)\left(\frac{dF}{df}\right)_T=(3V_0)^{-1}(1+2f)^{5/2}(2af+3bf^2+\ldots) \quad (47)$$

or, with a notation introduced by Birch (1947),

$$P=3K_0 f(1+2f)^{5/2}(1-2\xi f+\ldots), \quad (48)$$

where K_0 is the isothermal incompressibility (or bulk modulus). Conversely,

$$\frac{1}{K_{T,0}}=-\frac{1}{V_0}\left(\frac{\partial V}{\partial P}\right)_{T,0}=\frac{1}{\rho_0}\left(\frac{\partial \rho}{\partial P}\right)_{T,0}=\beta_{T,0} \quad (49)$$

is the isothermal compressibility. The convergence of the series in equation (48) depends on the amount of compression, i.e., f, and on the magnitude of the coefficients of the higher powers of f. Both K_0 and ξ are functions of temperature alone. If the coefficients beyond ξ are neglected, there result the following important relations (all for isothermal compression):

$$P=3K_0 f(1+2f)^{5/2}(1-2\xi f), \quad (50)$$

$$K=K_0(1+2f)^{5/2}[1+7f-2\xi(2+9f)], \quad (51)$$

and

$$\left(\frac{dK}{dP}\right)_T=\frac{12+49f-2\xi(2+32f+81f^2)}{3[1+7f-2\xi f(2+9f)]}. \quad (52)$$

Finally, the explicit form of the isothermal equation of state is

$$P=\frac{3}{2}K_0\left[\left(\frac{\rho}{\rho_0}\right)^{7/3}-\left(\frac{\rho}{\rho_0}\right)^{5/3}\right]\left\{1-\xi\left[\left(\frac{\rho}{\rho_0}\right)^{2/3}-1\right]+\ldots\right\}. \quad (53)$$

This expression is often referred to as the "Murnaghan-Birch equation of state."

Reviewing experimental results for compression of metals under pressures up to 100,000 atm., Birch (1952) finds that they are best represented by an equation of type (53) in which ξ is nearly zero; the same conclusion holds for the cubic compounds garnet, periclase, fluorite, and lithium fluoride. A fair representation is obtained even for corundum, beryl, and topaz, to which the theory does not strictly apply because of their lower symmetry. After analysis of Bridgman's data for silica and olivine, Verhoogen (1953) has urged that the initial value of dK/dP is certainly not always 4, as equation (52) would predict for $\xi = 0$; moreover, he has called attention to the fact that, for sodium chloride, $(dK/dP)_{P=0}$ depends on temperature, which likewise would contravene equation (52) only if $\xi = 0$. Such effects indicate that the higher terms in the expansion of F are not negligible for some non-metals. But Verhoogen's criticism, aimed at Birch's use of equation (52) with $\xi = 0$, is clearly not pertinent to the unrestricted form of Murnaghan's theory of isotropic compressibility. Birch has also reviewed the problems of thermal expansion, adiabatic compressibility, and change of compressibility with temperature, all of which are involved in the interpretation of seismic data and thus bear upon our concept of the Earth's interior. These phenomena, at present, do not require discussion in regard to the constitution of other planets, as their internal temperatures are still a matter of conjecture. To sum up, the Murnaghan-Birch theory of finite strain provides a geometrically correct framework into which experimental data on isothermal compressibility can be fitted successfully.

In the range of pressures from 10^5 to 10^7 atm., seismic data are the only source of information regarding the compressibility of matter. Bullen (1946) has generalized from seismic experience and enunciated a pressure-compressibility hypothesis. He suggests that at pressures exceeding a few hundred thousand atmospheres the compressibility is independent of chemical composition and that the bulk modulus becomes a universal function,

$$K = 2.25 + 2.86P + 0.16P^2 , \tag{54}$$

with K and P in megabars. Ramsey (1950*b*) has rejected formula (54) as an arbitrary assumption and replaced it by the expression

$$K = K_0 + \alpha P , \tag{55}$$

which is supposed to be asymptotically valid at high pressures. His argument derives from a heuristic hypothesis in the atomistic theory of solids, used by Born and Lennard-Jones. According to this hypothesis, the repulsive potential is inversely proportional to a power n of the interatomic distance; hence the coefficient α is explicitly related to n, viz.,

$$\alpha = \tfrac{1}{3} (n + 3) . \tag{56}$$

Consequently, α must depend on chemical composition, if weakly. In fact, Ramsey finds α equal to 3.7 in the mantle and 3.8 in the core of the Earth; and this close coincidence he accepts as prima facie evidence of the chemical identity of core and mantle. The same discussion leads to values of K_0 in the mantle and core that differ widely (0.5 and 2.1 megabars, respectively). Ramsey regards this discrepancy as a conclusive refutation of Bullen's hypothesis. Both Ramsey's and Bullen's treatments fail at low pressure, so that formulae (54) and (55) cannot be compared directly, as can equation (53), with experiment. Evidently, equations (54) and (55) are not in agreement with the measurements of large compressions. Birch has emphasized, moreover, that in neither case has the observed variation of the seismic velocities, V_S and V_P, been used directly to find dK/dP; instead, the density distribution in the Earth's interior has first been calculated with the aid of various assumptions and then K from the well-known relation

$$K_s = \rho \left(V_P^2 - \tfrac{4}{3} V_s^2 \right), \tag{57}$$

where K_s is the adiabatic compressibility; and this was then found to be approximately linear with pressure.

2.2. Extreme Pressures (Theoretical Deductions)

At densities exceeding a certain critical value, depending on the composition of the material, the structures of neighboring atoms begin to overlap, and the outer electrons behave as if free, irrespective of the temperature. This phenomenon is now commonly called *pressure ionization*. The breakdown of atomic structure under increasing compressions is a gradual one, and there finally results a gas of free electrons and bare nuclei. To call this ultimate state a "gas" is a linguistic convention. The term does not imply a thermodynamic distinction from a corresponding solid; rather it points to the absence of structural differentiation on the scale of interatomic distances in ordinary solids. The electrons are free in the sense that the conductive electrons in a metal are free, and their progressive liberation is accompanied by conspicuous changes in the physical characteristics of the material: dielectric solids eventually turn into good conductors, i.e., approach the metallic state. For instance, under a pressure of 280,000 atm., produced by explosion shocks, the electric conductivity of such non-metals as iodine and red phosphorus increases by factors greater than 10^6 and 10^4, respectively (Alder and Christian, 1956).

It is by no means clear whether matter in the state of pressure ionization represents a new thermodynamic phase or possibly several phases. The criterion is, of course, the appearance of discontinuities in the pressure-density and the pressure-energy relations during isothermal changes of state. The perfection of the shock experiments just mentioned, to the degree where the new technique would yield precision values of the compressibilities, is awaited with much interest. All phase transitions in solids so far investigated, by *static* measurements of the compression, have proved to be associated with major lattice rearrangements; the only exception is, perhaps, a discontinuity in the pressure-density relation of cesium, discovered by Bridgman, which has tentatively been attributed to a rearrangement of the electronic structure of individual atoms. In no case has there been a *conclusive* theoretical demonstration that density jumps exist in the pressure range realized at depths of more than 1000 km in the Earth or at corresponding depths in other planets. This statement, strictly speaking, applies even to solid hydrogen, though there is some reason to believe that this element will undergo at least one phase transition, namely, from a dense molecular phase to a metallic one more than 50 per cent denser; such a large discontinuity, if real, would still be exceptional, percentagewise. For atoms with several electrons, phase transitions might occur with each successive ionization, but no detailed theoretical deduction of such a sequence has ever been undertaken. The standard theoretical treatment of elements heavier than hydrogen (see Sec. 2.21) amounts to a deliberate smoothing-out of the spatial distribution of electrons and thereby obviates the very possibility of deciding whether or not phase transitions exist. In numerous papers Ramsey has repeated the apodictic statement that such ought to occur, but there is no cogent theoretical reasoning to support this doctrine. The controversial point is whether pressure ionization produces a sequence of transitions with *major* discontinuities in the density, as Ramsey has it. It still appears to be an open question whether progressive pressure ionization is a more or less discontinuous process, attended by *minor* changes in density, or even a continuous one, without any phase transitions.

Once the ionization of the innermost electronic shell is complete, the equation of state of the "completely degenerate" electron gas, obeying Fermi-Dirac statistics, assumes the rather simple form

$$P\,(\text{dynes cm}^{-2}) = K_1\,[\rho\,(\text{gm cm}^{-3})\,]^{5/3}\,, \tag{58}$$

$$K_1 = \frac{h^2}{5m_e}\left(\frac{3}{8\pi}\right)^{2/3}\left(\frac{Z}{A}\frac{1}{m_0}\right)^{5/3} = 9.913 \times 10^{12}\,,$$

where h is the Planck constant, m_e the electron mass, Z the atomic number of the element, and A its (dimensionless) atomic weight, while $m_0 = 1.66 \times 10^{-24}$ gm is the mass unit corresponding to $A = 1$. Equation (58) is valid up to densities of 10^5–10^6 gm cm^{-3}. Relativistic effects come into play at still higher densities, which are irrelevant in the present context because they are encountered only in certain white dwarfs. It has already been shown that even at the centers of the terrestrial planets equation (58) cannot hold (Sec. 1.4). The boundaries of its domain of validity have been discussed and graphed by Wares (1944) and Wrubel (1958) without calling attention to the fact that at densities less than 100 gm cm^{-3} and temperatures lower than 10^4 °K phase equilibria play a dominant role.

At much higher temperatures, undoubtedly, a degenerate (Fermi) gas decompressed isothermally would, by a *continuous* change of state, end up as an ideal (Maxwell) gas, and probably a fair representation of the intermediate stages could be achieved by the so-called "partially degenerate" equation of state which Wares (1944) used. At temperatures lower than 1000° K, however, isothermal decompression of a high-density sample of matter is a *discontinuous* process, since ultimately it must run across the melting line and, possibly, across other phase boundaries before that.

2.21. *Pressure ionization of heavy elements.*—Several investigators (Slater and Krutter, 1935; Jensen, 1938; and others) have used the Thomas-Fermi model of the atom for approximate computations of potential fields and charge densities in metals as functions of the lattice spacing (i.e., the density). They made two simplifying assumptions: (1) the temperature of the electrons and nuclei was taken to be absolute zero, and (2) the electron was assumed to exert no force on itself, but only on other electrons (neglect of exchange effects). With these simplifications, a set of universal potential functions is found, applicable to all atomic numbers by a simple change in scale of linear dimensions. A refined theory (Dirac, 1930) including the exchange effects does not lead to the similarity transformations just mentioned, and it becomes necessary to obtain separate solutions for each atomic number Z. Feynman, Metropolis, and Teller (1949) have reconsidered the Thomas-Fermi-Dirac (T.F.D.) model, including the exchange effects, and more recently March (1955) and Gilvarry (1957) have given asymptotic solutions to the T.F.D. equation.

Attempts were also made to extend the theory to simple molecules of high symmetry, like CH_4 and NH_3 (Sen, 1938). In a suitable mixture of ammonia and hydrogen ($NH_3 + \frac{1}{2}H_2$) pressure ionization should lead to the formation of metallic ammonium ($NH_4^+ + e^-$), the stability of which has long been a matter of speculation among chemists, because in chemical combination the ammonium ion, NH_4^+ mimics the ions of the alkali metals. Bernal and Massey (1954) estimate that the transition from a mixture of solid NH_3 and H_2 to ammonium metal might occur at fairly low pressures somewhere between 50,000 and 250,000 atm.

In Figures 5, 6, and 7 Bridgman's experimental data for the elements and a few compounds are shown on the left side, and the pressure-density relations derived from the T.F.D. model (Elsasser, 1951) on the right side. The theoretical curves are devoid of any discontinuities, i.e., successive phase transitions; for these cannot properly be represented by the model. The theory clearly fails when the curves are extrapolated downward to normal densities. To make them join the experimental data, a point of inflection is needed somewhere. The theory is asymptotically correct for pressures exceeding, say, 10 megabars (= 10^{13} dynes cm^{-2}). Bridging the gap between 10 megabars and the terminal points of the experimental curves introduces an element of arbitrariness in the range of geophysical interest. In order to minimize this arbitrariness, Knopoff and Uffen (1954) have fitted to the experimental data an equation of state of the Murnaghan-Birch type, which appears to hold up to half a megabar at least (Duvall and Zwolinski, 1955), and then have interpolated between the

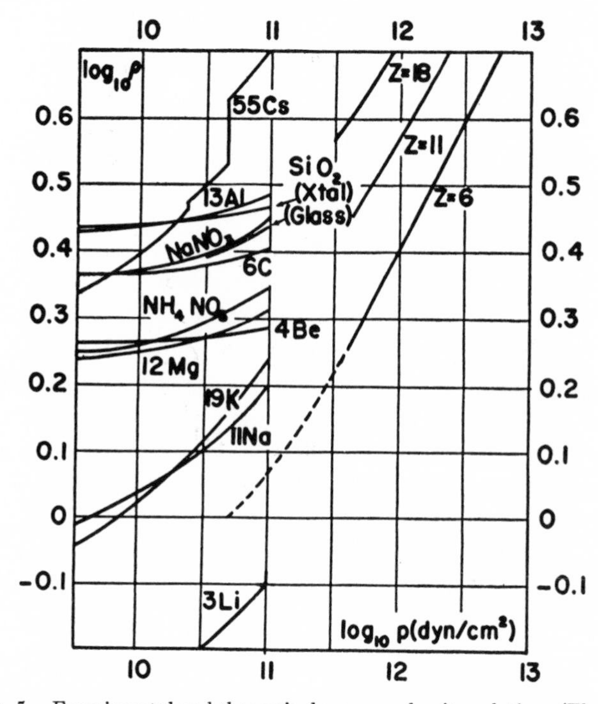

FIG. 5.—Experimental and theoretical pressure-density relations (Elsasser)

Murnaghan-Birch curve and the T.F.D. solution. This procedure is applicable even to compounds and mixtures of such, because, to a good approximation, in the state of pressure ionization the atomic volumes of the several components behave additively. Knopoff and Uffen first assign to each compound a "representative atomic number," Z, which is a suitable weighted mean of the atomic numbers of the constituent atoms. The T.F.D. solution for a pure element having this Z is then taken to be the pressure-density relation of the compound; no difficulty arises if Z assumes non-integral values. Knopoff and Uffen find that for fayalite (Fe_2SiO_4) and forsterite (Mg_2SiO_4) the Murnaghan-Birch curves approach the T.F.D. curves asymptotically at pressures of about 100 megabars. For SiO_2 the interpolation is somewhat unsatisfactory, presumably because the higher terms in the Murnaghan-Birch expansion are not negligible.

All these computations refer to the pressure-density relation at absolute zero. An extension of the theory to non-zero temperature was given by Feynman, Metropolis, and Teller (1949), but detailed and systematic predictions based on this extension are not yet available. Knopoff and Uffen

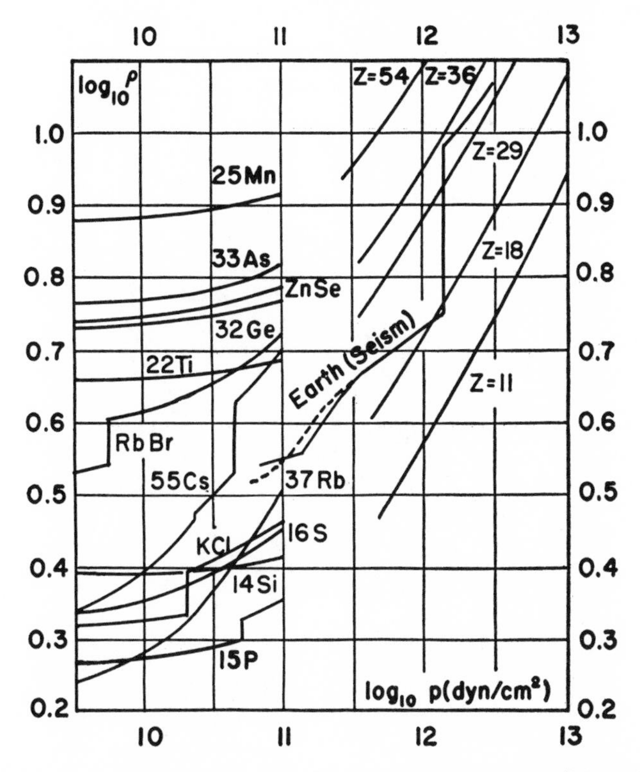

Fig. 6.—Experimental and theoretical pressure-density relations (Elsasser)

estimate that the temperature corrections to their pressure-density relations are, at most, 8 per cent at temperatures less than 10^4 °K. The Feynman-Metropolis-Teller treatment, however, does not shed any light on the possible existence of several solid phases or on the transition from the solid to the fluid state, if such discontinuity persists at pressures of 100 megabars and more. As pointed out before, the ultimate behavior of the melting-curves is still in doubt, and the theoretical delineation of phase boundaries in the intermediate range of densities relevant to the planets remains a challenging problem.

2.22. *Solid hydrogen and helium.*—The isothermal compressibility at 4.2° K of solid hydrogen and helium is known from experiment up to pressures of 20,000 atm. (Stewart, 1956). These empirical data have proved the inadequacy of all theoretical equations of state proposed prior to 1956, which were reviewed critically by DeMarcus (1958). For pressures exceeding 20,000 atm., the experimental data have to be extrapolated, either analytically, as in the case of helium (DeMarcus, 1958), or without the

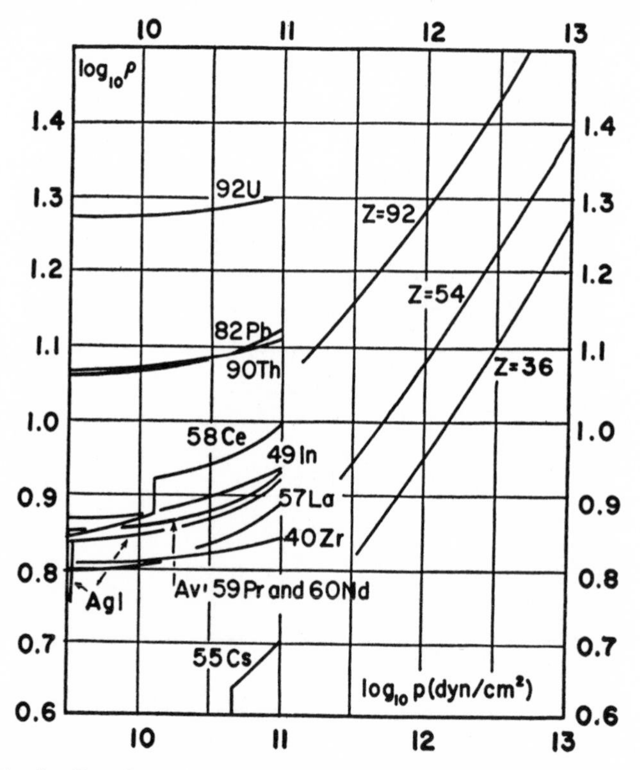

FIG. 7.—Experimental and theoretical pressure-density relations (Elsasser)

benefit of a definite analytic form as guide. The latter course was preferred by DeMarcus (1958) for molecular hydrogen, although Stewart had already fitted a Murnaghan-Birch type of equation to his experimental data. The possible existence of a metallic phase of hydrogen was first studied by Wigner and Huntington (1935). Later on, Kronig, de Boer, and Korringa (1946) gave detailed formulae for computing the internal energy of metallic hydrogen as a function of the density. Because of certain objections to their procedure, DeMarcus (1958) has reverted to the methods employed by Wigner and Huntington. The results of his computations, for pressures greater than 300,000 atm., are very close to the predictions made by Kronig, de Boer, and Korringa if the zero-point energy is added to their results. Excerpts from the pressure-density relations obtained by DeMarcus are given in Table 3.

In order to find the phase boundaries of solid hydrogen, one would have

TABLE 3

PRESSURE DENSITY RELATIONS FOR SOLID HYDROGEN AND HELIUM

$P\times10^{-12}$ Dynes/Cm²	Gm/Cm³	$P\times10^{-12}$ Dynes/Cm²	Gm/Cm³
Molecular Hydrogen		Metallic Hydrogen	
0.0	0.089	0.052	0.5487
0.0002	0.0961	0.182	0.6001
0.0004	0.1010	0.355	0.6582
0.0006	0.1052	0.586	0.7239
0.0008	0.1090	0.893	0.7988
0.0010	0.1123	1.292	0.8843
0.002	0.1252	1.834	0.9824
0.003	0.134	2.547	1.0957
0.005	0.147	3.495	1.2270
0.007	0.158	4.766	1.3802
0.009	0.167	6.476	1.5601
0.011	0.175	8.795	1.7725
0.013	0.182	11.963	2.0254
0.015	0.188	16.336	2.3287
0.02	0.201	22.437	2.6958
0.03	0.220	31.046	3.1442
0.04	0.236		
0.05	0.248		
0.06	0.261	Helium	
0.07	0.270		
0.08	0.279		
0.09	0.287	0.0002	0.234
0.1	0.295	0.0010	0.323
0.2	0.356	0.0040	0.431
0.3	0.401	0.0100	0.534
0.4	0.439	0.0200	0.626
0.5	0.471	0.04	0.742
0.6	0.500	0.08	0.884
0.7	0.528	0.10	0.936
0.8	0.553	0.20	1.15
0.9	0.576	0.40	1.40
1.0	0.601	0.80	1.77
1.5	0.704	1.00	1.89
2.0	0.797	2.00	2.45
2.5	0.886	4.00	3.16
3.0	0.970	8.00	4.15
3.5	1.047	10.00	4.56
4.0	1.114	20.00	6.21
		30.00	7.58
		40.00	8.64
		80.00	11.41

to know the free energies of all the phases involved. The prediction of phase transitions, for a given temperature, is then accomplished by plotting the free enthalpies of the several phases, per mole of H atoms, versus pressure; any point where these graphs intersect marks the pressure of a phase transition, at the given temperature. Unfortunately, it is by no means certain that at low temperatures the common molecular and metallic phases of hydrogen are the only stable modifications. Wigner and Huntington have conjectured the existence of an intermediate layer-like lattice. This possibility had been overlooked in all work on hydrogen planets up until now. Ignoring this contingency, DeMarcus finds the transition from the molecular to the metallic phase to occur at a pressure of 3.5×10^6 atm., or at 1.95×10^6 atm. for an alternate extrapolation of Stewart's data. Clearly, the predicted transition pressure is quite sensitive to the preferred value of the density of the molecular phase. Moreover, he reaches the reassuring conclusion that the existence of a layer-like modification of hydrogen would not invalidate his estimates regarding the composition of cold planets of predominantly hydrogenic nature.

3. TERRESTRIAL PLANETS

3.1 The Earth as Prototype

The internal constitution of the Earth has been treated at length in Volume **2**, chapter 3, to which reference is made for the main conclusions. This account does not refer to the speculations of Kuhn and Rittmann (1941) about the chemical homogeneity of our globe. Although their ideas have failed to be generally accepted, they have nevertheless influenced much recent thinking about the interior of the terrestrial planets. Therefore, a review of these speculations is in order here. Briefly, Kuhn and Rittmann suggested that the Earth consists of a thin solid crust, a shallow magmatic zone of molten silicates underneath, and otherwise of practically unchanged "solar matter," rich in hydrogen. They denied the existence of chemical discontinuities in the deep interior and explicitly rejected the conception of an iron core. Three main points of their reasoning require comment.

1. The chief argument for the chemical homogeneity of the Earth's interior is its high viscosity, which, so it is claimed, would have prevented gravitational separation of immiscible phases even during periods of the order of 10^9 years. On the other hand, complete gravitational separation into a silicate mantle and an iron core could have occurred in bodies of a few hundred kilometers' diameter (Kuhn, 1942). Urey (1952) has discussed various mechanisms for the fall of the liquid iron phase toward the

Earth's center, without reaching a verdict; but he has not acknowledged Kuhn and Rittmann's objections as fatal. Birch (1952), who questioned their estimate of the viscosity, regards the thesis of the impossibility of the segregation of an iron core as wholly unproved.

2. Having dispensed with the hypothesis of a liquid iron core, Kuhn and Rittmann still had to account for the sudden disappearance of transversal seismic waves at a depth of 2900 km. The explanation they urged makes use of the relaxation time, τ, of a viscous medium, which is a measure of the rate at which tangential stresses die out after a sudden deformation has been applied. Their suggestion was essentially to replace the discontinuity in density between mantle and core by a very rapid change in relaxation time. Against disturbances having periods small compared with τ the medium will behave as if it were an elastic solid, while toward those having periods large compared with τ it will display a fluid character. Accordingly, Kuhn and Rittmann had to assume that the relaxation time of the material inside the Earth decreases steadily with depth and at 2900 km reaches a critical value of the order of 10 seconds. They did not advance a detailed explanation of what causes this decrease in the relaxation time, but they expressed belief that an increase in the hydrogen content with depth would be sufficient. Subsequently, Kuhn (1948) elaborated this point by a more detailed study of the chemical equilibria (possible reduction of silica by hydrogen and formation of silicon hydride and of free silicon dissolved in iron). Also important are their conclusion that the "apparent" radius of the Earth's core ought to depend on the period of the seismic waves and the recent announcement (Kuhn and Vielhauer, 1953) that this dispersion effect has now been verified by Altenburg. This claim deserves the attention of the seismologists. Jeffreys (1952) found that the introduction of elasticoviscosity per se does not help in explaining the seismic data and cited evidence contradicting the 1941 version of Kuhn and Rittmann's arguments.

3. Although Kuhn and Rittmann admitted continuous changes in chemical composition with depth, i.e., varying hydrogen content, they denied the existence of a discontinuity in the density distribution at the depth of 2900 km. This was reintroduced, on different grounds, by Kronig, de Boer, and Korringa (1946); they proposed to regard the so-called boundary of the Earth's core as a manifestation of the phase transition in solid hydrogen (see Sec. 2.22), with the additional assumption that the adjacent phases are "loaded" with heavier atoms, dispersed in such concentrations among the molecules and atoms of hydrogen as to give the required density of about 8 gm cm^{-3}. This density would imply a hydrogen

content of the order of 10 per cent by weight in the atomic phase at the critical depth. No reasons are given why the atomic phase should have a smaller relaxation time than the molecular one. By ignoring this requirement, which is crucial to the interpretation of the seismic data on the hypothesis of Kuhn and Rittmann, Kronig *et al.* have rather obscured the point originally at issue. At the same time, however, they paved the way for a hypothesis that is fundamentally at variance with Kuhn and Rittmann's insistence on a *continuous* density distribution throughout the interior of the Earth.

This new hypothesis is due to Ramsey (1948), who accepted the conventional interpretation of the seismic data, i.e., a major discontinuity in density at 2900 km below the surface, and advanced a novel explanation for it, namely, a hypothetical phase transition in "olivine." The inner core of the Earth he attributed tentatively to a second-phase transition in the same material. If it were to apply solely to the Earth, the merits of this hypothesis would be of little concern in the context of this chapter. But its avowed purpose is to provide a uniformitarian principle for explaining the internal constitution of the terrestrial planets. It is simply postulated that the pressure-density relation revealed by analysis of the seismic data holds for all of them. However, in order to represent the mean densities of these planets, Ramsey was forced to alter Bullen's density distribution for the Earth by an arbitrary increase in density with depth, supposed to result from a concentration of heavy elements, particularly iron orthosilicate, toward the center. It may be remarked that, strictly, such a concentration is inconsistent with Ramsey's acceptance of Kuhn and Rittmann's thesis that segregation of an iron core is impossible. Details of this modification of Bullen's pressure-density relation are given in a second paper, analyzing the nature of the Earth's core (Ramsey, 1949).

Enough has been said to relate the point of departure of Ramsey's ideas to the speculations of Kuhn and Rittmann. This is not the place to consider *cosmogonic* arguments for or against the existence of iron cores in the terrestrial planets. Opinion regarding their reality is still sharply divided.[2]

[2] The following comment (Birch, 1952) characterizes aptly, if somewhat caustically, the uncertainty of our knowledge: "Unwary readers should take warning that ordinary language undergoes modification to a high-pressure form when applied to the interior of the Earth: a few samples of equivalents follow:

"High-pressure form:	Ordinary meaning:
certain	dubious
undoubtedly	perhaps
positive proof	vague suggestion
unanswerable argument	trivial objection
pure iron	uncertain mixture of all the elements"

It is convenient to describe, first, Ramsey's models and the similar ones of Bullen and Low (Sec. 3.2) and thereafter to deal with several aspects of the problem of iron cores (Sec. 3.3) This division of the subject matter is unavoidably somewhat arbitrary.

3.2. Models with Phase Transitions

If it could be shown from quantum mechanics that a phase transition in olivine does occur under the precise pressure at the core boundary and that it is associated with a change in density of the right order, then indeed Ramsey's hypothesis would be irresistible. What he has shown, in fact, is only that the work done on crossing the core boundary, i.e., the product of pressure and change in specific volume computed from seismic data, is roughly equivalent to the energy difference between the ground state and the first excited state of a quartz crystal. Now this computed work might as well be associated with a smaller change in volume at a higher pressure; and, until either of these can be predicted separately, the geophysical argument lacks cogency. Moreover, if the predicted equilibrium pressure of the phase transition should exceed that at the center of the Earth, the argument would be void. Too little attention has been paid to the further requirement that the denser phase must be liquid, while the coexistent less dense phase is solid. Hence Ramsey's hypothesis implies the existence of a triple point on the phase diagram of olivine, the three coexistent phases being the ordinary solid, a dense liquid, and a solid of still higher density. It may also be asked whether the dense liquid is separated from the ordinary liquid by another phase boundary. If so, there would be a second triple point (ordinary solid–ordinary liquid–high-density liquid) at a pressure inferior to the pressure at the core boundary, while the first-mentioned triple point would occur at a superior pressure. These are thermodynamic corollaries to Ramsey's main thesis that have escaped notice in the literature.

In the absence of the requisite physical theory, Ramsey attached great significance to the astronomical test of his hypothesis, which seemed to reproduce the mean densities of the terrestrial planets and the Moon within the range of accuracy of the masses and diameters then regarded as best. A more recent upward revision of the mass of Mercury has marred this simple picture, and Mercury's mean density is now hopelessly out of line with Ramsey's hypothesis. It has been suggested that this divergence does not necessarily argue against the doctrine of a common primitive composition of all terrestrial planets; for Mercury is the planet closest to the Sun, and the temperature at the subsolar point of its surface now suffices to melt lead. In the past, Mercury's mean distance from the Sun may have been even less. Therefore, it is entirely plausible that the initial density of this planet could have been raised by preferential evaporation. But, if this is admitted, Ramsey's astronomical test loses its force.

As far as the Earth is concerned, Ramsey's views have been broadly indorsed by Bullen (1949*a*, *b*, 1950*a*, 1957), with the minor modification that he prefers for the inner core a distinct chemical composition rather than a second phase transition in olivine. Bullen and Low (1952) have

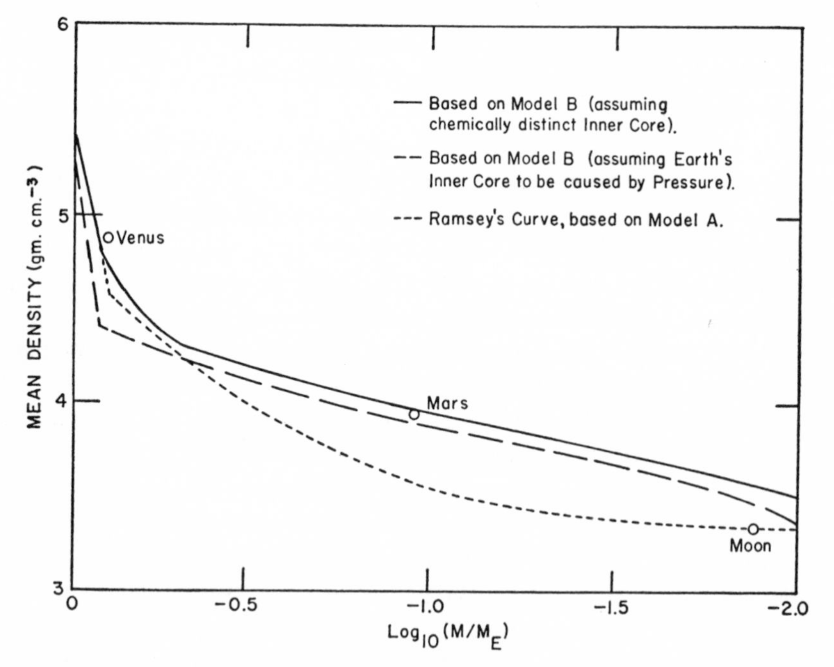

FIG. 8.—Variation of mean density with mass of planets of terrestrial type (Bullen and Low)

critically examined Ramsey's density calculations for the terrestrial planets and have repeated them, using for the pressure-density relation the new Model B of the Earth (Bullen, 1950*b*); Ramsey's work was based on the earlier Model A (Bullen, 1947). The results of these computations are illustrated by Figure 8, which is adapted from the paper by Bullen and Low. The graph representing Model A does not include the *ad hoc* corrections applied by Ramsey, which were intended to rectify the 10 per cent discrepancy between the calculated and observed values of the mean density of Mars. Although Model B fits Mars much better, the authors make reservations as to what may be the best value for the diameter of this planet. In respect of Venus, they hold that the presence of chemically distinct inner cores in the planets is indicated when Model B is used. Model A fits the Moon best, but it is far from certain that the Moon is a typical terrestrial planet. Mercury's image point is not shown, for reasons already stated. The modern data for Venus satisfy the theoretical curve less well than do the data used by Bullen and Low.

A phase transition of the non-controversial kind, in olivine, has been proposed as explanation of the pronounced steepening of the density gradient inside the Earth at depths from about 200 to 900 km (Bernal, 1936). This speculation has quite recently received definitive experimental support (Ringwood, 1957, 1958*a*, 1959), and it is certain that the pressures and temperatures required to change olivine into an allotropic modification of the spinel type do occur in the upper 1000 km of the mantle. Because of solid-solution effects, the transition will occur over a considerable depth range. It is also clear that the thermodynamic treatment of the mantle as binary system Mg_2SiO_4-Fe_2SiO_4 is unsatisfactory; even the representation as ternary system Mg_2SiO_4-Fe_2SiO_4-SiO_2 is probably an oversimplification (Meijering and Rooymans, 1958). In a qualitative way, at any rate, Bernal's conjecture may now be regarded as verified by experiment. The phase transition is attended by changes in density of the chief constituents of olivine, namely, forsterite (Mg_2SiO_4) and fayalite (Fe_2SiO_4), of about 11 and 12 per cent, respectively (Ringwood, 1957, 1958*a*). This moderate compression, caused by a well-understood rearrangement of lattice structure, stands in marked contrast to the hypothetical enormous changes in density invoked by Ramsey and others in order to account for the discontinuity at the core boundary.

3.3. Iron Cores

Not much can be said about particular models with iron cores. Brown (1950) computed the mean densities of the terrestrial planets from pressure-density relations for iron, oxygen, and silicon that were interpolated between Bridgman's experimental data and the Thomas-Fermi curves; but he did not attempt to evaluate the internal density by distribution of these bodies. The idea of applying the seismic pressure-density relation to the other terrestrial planets goes back to Jeffreys (1937), who attempted to test Bernal's conjecture described in the preceding section. Since the experimental verification of the latter (Sec. 3.2), Jeffreys' density distributions for Venus and Mars command new interest; but his model for Mercury is obsolete because of the subsequent revision of its mass. Attention may also be called to a paper by Ringwood (1960), entitled "The Chemical Evolution and Densities of the Planets."

The reasons for the widely accepted identification of the Earth's core with iron—or an iron-rich alloy rather—are simple. In the first place, it has been tempting to regard the iron phase of meteorites as the vestige of a planet whose inside was bared by an early catastrophe. Moreover, iron is the most abundant heavy metal in the Sun and stars, while none of the heavier elements (excepting cobalt and nickel) has a cosmic abundance approaching that of iron within a factor of one-hundredth. A transition from rock to iron provides a natural explanation of the sharpness of the interface at the depth of 2900 km, and near the melting point of olivine the iron in contact with it would be liquid. The dependence on pressure of the melting point of iron can now be estimated with fair accuracy (Simon, 1953). If the core of the Earth were pure iron, the equilibrium temperature at the depth of 5000 km (surface of the inner core) would have to be about 3900° K, on the assumption that the inner core is solid; and at the center of the Earth the temperature would have to exceed 4900° K in order to maintain the core liquid throughout. A plausible case has been made for the spontaneous formation of a solid inner core in a homogeneous iron core originally liquid (Jacobs, 1954). No precise predictions of melting points have been given as yet for iron alloyed with moderate amounts of other metals. Among these the chief candidates have always been nickel and cobalt, but recently silicon has come under discussion (MacDonald and Knopoff, 1958; Ringwood, 1960).

According to Urey (1952), the Earth's mantle contains pockets of liquid iron even now, and the movement of this liquid phase toward the core has caused a steady decrease in the moment of inertia during the geologic ages (cf. Spencer Jones, Vol. **2**, chap. 1). The existence of heavy cores in Venus and Mercury is obvious from their mean densities. It may prove feasible to deduce the strengths of the magnetic moments of these two planets from the perturbations that such would exert on terrestrial magnetic storms (Houtgast, 1955). Since the dipole field of the Earth is now generally believed to arise from dynamo effects in the liquid core, there would appear to be some hope of obtaining essentially novel information about the physical state of the cores of Venus and Mercury.

4. THE JOVIAN PLANETS

The belief that the Jovian planets are gaseous throughout has proved difficult to shake. Jeffreys (1923) was the first to emphasize that the interiors of these bodies cannot resemble those of hot gaseous stars. He considered the amount of heat lost by a planet whose interior is in complete thermal communication with the surface, and he found it sufficient to solidify the planet on any reasonable hypothesis concerning the original temperature. Jeffreys (1924) inferred, moreover, from the inequality (35) derived in Section 1.4, that the density of the surface material must be less than 0.8 gm cm^{-3} on Jupiter and less than 0.4 gm cm^{-3} on Saturn. However, he did not accept these figures at their face value because they seemed extraordinary for a solid material. Rather he suggested that the solid cores of Jupiter and Saturn are covered by vast, tenuous atmospheres, which would not appreciably contribute to the total mass of these planets and thereby account for the low mean densities computed from the apparent diameters of the cloudy atmospheres. If this view were correct, the surface densities deduced from inequality (35) would be spurious indeed. Fessenkov (1924) likewise noticed that the surface density of Jupiter and Saturn must be lower than that of any known liquid or solid, hydrogen and helium excepted; but he did not formulate any clear picture of the physical nature of the surface strata, which he preferred to think of as gaseous. Eventually it became clear that a literal interpretation of Jeffreys' original inference is inescapable (Wildt, 1934). Accordingly, the following physical model was proposed. A heavy core ($\rho_1 = 5.5$ gm cm^{-3}) is surrounded by a high-pressure modification of ice ($\rho_2 = 1.0$ gm cm^{-3}), which in turn is covered by a mantle of solidified hydrogen and helium ($\rho_3 = 0.35$ gm cm^{-3}). The unknowns of the problem—namely, the radii of the two surfaces of discontinuity—are determined by two observables—the mean density and the moment of inertia of the planet. The chief result was the realization that the outer layers of Jupiter and Saturn must consist of solid hydrogen (and possibly some helium), all other light materials being ruled out by their higher density. Slightly different values for the densities of the three phases were adopted later (Wildt, 1938). In particular, attention was called to the remarkable compressibility of the ordinary, molecular phase of hydrogen and to the replacement of the latter, in the deeper layers of the mantle, by the metallic phase predicted by Wigner and Huntington; Kothari (1938) also remarked upon the importance of this phase transition. It is worth noting that the densities ρ_3 and ρ_2, given above, agree closely with what we now know to be the average densities in the shells of molecular and metallic hydrogen. This modern reinterpretation of the masses of shells 2 and 3 would raise the abundance of hydrogen (by mass) to more than 50 per cent; and, despite the assumption of constant density inside each shell, this simplest model is seen to be compatible with a predominantly hydrogenic constitution of Jupiter and Saturn. The hypothesis of the intermediate ice shell reflects the state of knowledge of the cosmic abundances twenty years ago. Our views on the chemical composition of the Sun and stars have since undergone a major change, which was bound to accord hydrogen an even more prominent place in the chemistry of the planets.

The 1934 model has been criticized for ignoring the compressibility of the three phases. This defect has largely been corrected by recent work. DeMarcus (1951), Fessenkov and Massevich (1951), and Ramsey (1951) have arrived independently at essentially the same conclusions. Their concurrence is not surprising, for they have relied on almost identical assumptions regarding the equation of solid hydrogen. A good deal of earlier work along similar lines was based on physical approximations so unrealistic that the conclusions fail to carry any conviction. This statement also applies, as far as the Jovian planets are concerned, to an otherwise interesting paper by Brown (1950).

4.1. Model Planets of Pure Hydrogen

DeMarcus (1951) and Ramsey (1951) have computed the masses and radii of spherical (non-rotating) configurations of pure hydrogen. The underlying physical theory of solid hydrogen has been reviewed in Section 2.22. Ramsey's publication contains only a graph of the mean density of these planetary models versus the logarithm of their mass. DeMarcus has given a mass-radius diagram, on logarithmic scales, which is reproduced in Figure 9. Dots indicate the configurations of planetary size found by direct integration. The dots representing white dwarf configurations, on the top of the diagram, were taken from Chandrasekhar's theory (it is immaterial here that white dwarfs of exclusively hydrogenic nature are now believed to be impossible for other physical reasons). Figure 9 illustrates the observation, originally made by Russell (1935), that the white dwarfs and planets form a single family of cold cosmic bodies, it being assumed that they are chemically homogeneous and that the chemical composition is the same for all. Since among the planets the radius increases with mass, while the opposite is true of the white dwarfs, Russell predicted the existence of a maximum radius for a cold body, which DeMarcus has located for hydrogen bodies to be at approximately 79,000 km. Earlier illustrations of Russell's idea, by Kothari (1936) and others, had only qualitative significance because the pressure-density relations they used did not afford an adequate representation of the compressibility of planetary matter.

The complex features of the lower part of Figure 9 are a manifestation of the phase transition. The central pressure, P_c, in a small hydrogen planet is too low to support a metallic core. A core is possible only if the mass of the planet exceeds 70 M_E, M_E being the mass of the Earth. The locus of configurations that are wholly molecular in composition is the lowest branch of the mass-radius diagram, which terminates in a cusp (not shown on Fig. 9); at this point P_c reaches the critical pressure of the phase transition. Between 70 M_E and 77 M_E the mass-radius diagram has three distinct branches. The branch to the right has already been identified. The other two branches represent configurations having equal mass, but cores of different size. For reasons given at length in Section 1.3, the equilibrium configurations having the smaller cores are unstable and so would not occur in nature. Finally, a hydrogen planet always has a core if its mass exceeds 77 M_E, and the fraction that is metallic rapidly increases with the

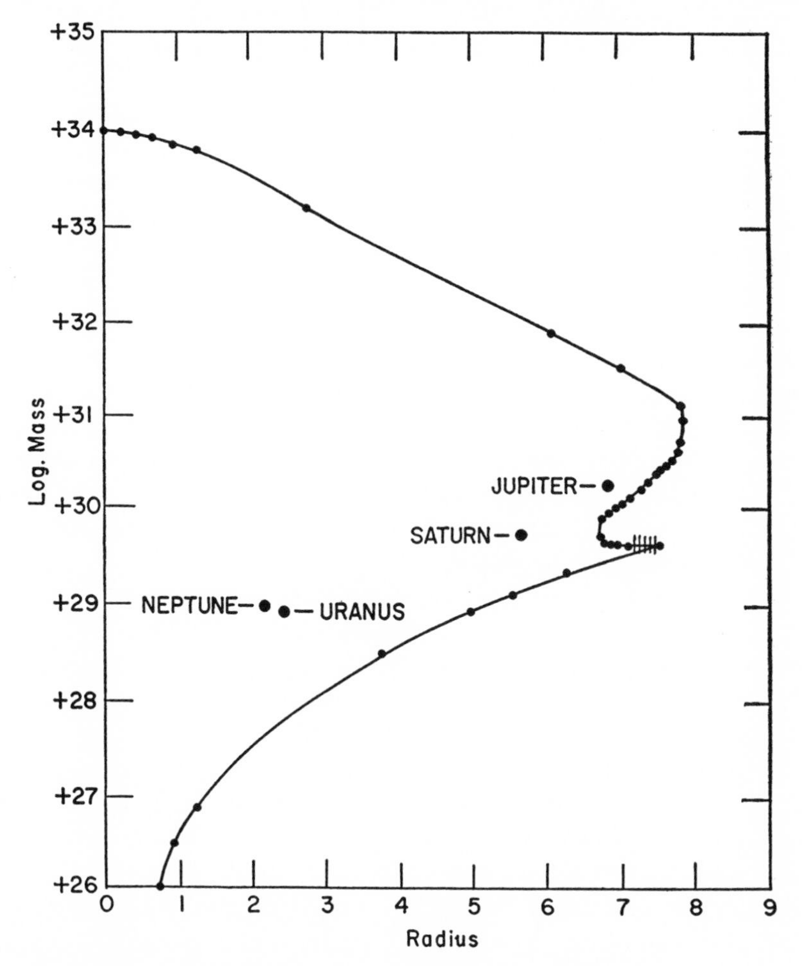

FIG. 9.—Mass-radius diagram for spherical configurations of solid hydrogen (DeMarcus). *Ordinates:* $\log_{10} M$; *abscissae:* radii in units of 10^9 cm. The vertical bars indicate an area where complicated features cannot adequately be represented on this scale.

total mass of the planet. All the numerical values just given relate to DeMarcus' integrations and are about 20 per cent smaller than those quoted by Ramsey.

A later study by Abrikosov (1954) of spherical configurations of pure hydrogen is based on the following equation of state, with ρ being the density in gm cm^{-3}:

$P = 2.37 \exp(-5.88\rho^{-1/3}) \times 10^9$ atm., for the molecular phase;

$P = (1.06\rho^{5/3} - 0.887\rho^{4/3}) \times 10^7$ atm., for the metallic phase;

Transition pressure $P = 2.4 \times 10^6$ atm., with change in density from 0.62 to 1.12 gm cm^{-3}.

Abrikosov omitted the detailed derivation of these equations of state, which will be published elsewhere. His mass-radius diagram, of which Figure 10 is an adaptation, differs markedly from Figure 9, as it must because of the alternate equations of state he preferred; the dashed curve represents the radii of the metallic cores.

The Jovian planets, whose positions have been entered in Figures 9 and 10 in order to facilitate comparison with the models, lie to the left of the mass-radius graph; that is to say, the Jovian planets fall short of the size they would have if consisting of pure hydrogen. Since this is the least dense and most compressible solid material known, it is physically impossible to realize more extended configurations of solid matter, and the entire field to the right of the mass-radius graph is "forbidden territory." The horizontal distance of the planetary symbols from the mass-radius graph is some measure of the proportion of elements heavier than hydrogen accumulated in the Jovian planets, which show a distinct grouping. Uranus and Neptune must be *much* poorer in hydrogen than Jupiter and Saturn. The course of the mass-radius graph is particularly sensitive to the critical pressure of the phase transition. Unfortunately, precisely this quantity is left rather indeterminate by the physical theory and has to be fixed somewhat arbitrarily. A revision of this quantity might be required for either of two reasons or their combination. In the first place, if the estimate of the constant part of the free energy of the metallic phase is in error and the pressure-density relation of the molecular phase is correct, then the critical pressure of the phase transition has been overestimated or underestimated, according to the sign of the error. On the other hand, if the constant part of the free energy of the metallic phase is correct but the compressibility of the molecular phase is erroneous, then the critical pressure of the phase transitions would be too low or too high, depending on whether the adopted compressibility was too low or too high. The consequences are fairly clear in the first case, where an increase in the critical pressure would lower the mean density of a purely hydrogenic planet of prescribed mass. For the second alternative, or a combination of the two cases, it does not seem possible to foresee the results without detailed calculations. For composite models the consequences of changing the critical pressure are wholly unpredictable, as the outcome might strongly depend on the assumed distribution of the heavier materials. Incidentally, for the molecular phase the problem of the constant part of the free energy does not arise, because it is known experimentally. Attention must again be called to the possible existence of an intermediate phase (layer-lattice?). Even in the absence of such a phase, the mass-radius graph is much modified by DeMarcus' (1958) revision of the critical pressure. It is now less certain whether the cusp of the mass-radius graph exists, because of the greater compressibility of the molecular phase.

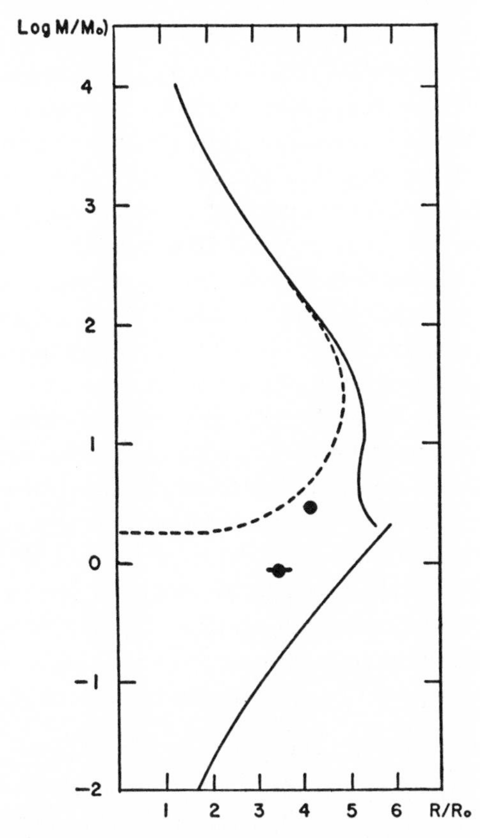

FIG. 10.—Mass-radius diagram for spherical configurations of solid hydrogen (Abrikosov). The radii and masses are expressed in units of 1.66×10^9 cm and 6.48×10^{29} gm.

Despite these unresolved difficulties, it cannot be doubted any longer that Jupiter and Saturn consist predominantly of solid hydrogen. Leaving all cosmogonic speculations aside, the argument relies on experimental evidence, sound physical theory, and very plausible conjectures:

A. Over the pressure range accessible to experimentation, hydrogen is the least dense of all solids, by a factor of 2 or more.

B. For the domain of pressure ionization, theory predicts that hydrogen is less dense, by a factor of 2, than any other element.

C. It seems plausible, therefore, that in the intermediate pressure range also hydrogen is less dense, by a factor of 2, than any other materials, helium included.

D. It is assumed, moreover, that Jupiter and Saturn are "cold" bodies, in the sense that it is permissible to ignore the effects of temperature on the pressure-density relation. The temperature gradients in these bodies have been considered by DeMarcus (1958).

It is noted that Jupiter and Saturn differ in mean density by a factor of less than 2 from model planets of the same mass that consist of pure hydrogen. Now consider a sequence of models of increasing mass built of a fictitious material, having a pressure-density relation such that at any pressure the corresponding density is precisely twice that of solid hydrogen. It can then be shown that the mass-radius graph of the fictitious models lies wholly within (i.e., on Figs. 9 and 10 to the left of) the mass-radius graph of the hydrogen models. The *abscissae* of the graph representing the fictitious models (i.e., their radii) are obtained by scaling down in the ratio 1:2 the abscissae of the mass relation of the pure hydrogen models, and its *ordinates* (i.e., the masses) by applying an analogous scale reduction in the ratio 1:4; this precept derives from the fact that the two mass-radius relations here considered are connected by a homology transformation. On the premises stated above, it is clear that the area between the hydrogen-curve and the fictitious one can be populated only by configurations containing some hydrogen and that, qualitatively, these must be the richer in solid hydrogen, the closer their representative points are to the boundary defined by the pure hydrogen configurations. The points for Jupiter and Saturn, in fact, fall on the strip between the two curves, and their location indicates that either planet contains substantial amounts of hydrogen.

4.2. The Fluid Shell of Hydrogen Planets

It remains to be shown that the mass of the atmosphere is negligible compared with the total mass of the model planets described in the preceding section. A physical limit to the depth of an isothermal atmosphere in the hypercritical state is imposed by the spontaneous solidification of a gas at the melting pressure of the solid phase (Wildt, 1934). As illustrated by Figure 3 (see Sec. 2.11), on a P-T diagram the melting line separates the domain of stability of the gas from that of the solid. Descent into an isothermal atmosphere leads through a sequence of states represented, on Figure 3, by a line parallel to the P-axis, and any such parallel must somewhere cross the melting line; solidification of the gas under its own weight occurs at this intersection. If the compressibility of the gaseous phase is known, the P-T diagram could be overlaid with a grid of lines of constant density, and, after integration of the equation of hydrostatic equilibrium, these lines could be labeled according to their depth below the fiducial level. It was found impractical to prepare such a diagram, but this sketch of its construction may clarify the physical argument. On the assumption of ideal compressibility of the atmosphere, it was concluded that the melting pressure of H_2 would be reached at a depth, below the visible surface of Jupiter and Saturn, not exceeding twenty times the scale height of an H_2 atmosphere at 150° K, which amounts to less than 1 per cent of the radii of these planets (Wildt, 1934). The assumption of ideal compressibility tends to underestimate the depth, but it was clear that the true depth could not differ from the estimate given by a factor as large as 10.

An exact treatment, taking full account of the imperfect compressibility, is now available (DeMarcus, 1957); since this publication is only an abstract, some details of the procedure will be added here. The Amagat densities (cf. Sec. 2.13) were taken from an exhaustive discussion of the equation of state of the hydrogen molecule (Woolley, Scott, and Brickwedde, 1948). Since the Amagat densities tabulated by these authors do not go beyond 500, the isothermals were extrapolated up to the melting line by the formula $PV/RT = \exp(a\rho - b\rho^2)$. The constants a and b, which are functions only of the temperature, were determined by the conditions that the formula should give the correct value of PV/RT and of its partial derivative with respect to the Amagat density at the terminal point of the experimental isothermal. The extrapolation was stopped on reaching the melting pressure, P_m, computed from the Simon equation

$$\log_{10}(237.1 + P_m \text{ kg cm}^{-2}) = 1.85904 \log_{10}(T_m \text{ °K}) + 0.24731 ,$$

as compiled by Woolley, Scott, and Brickwedde (1948); this equation represents the measured melting data, up to 5000 kg cm^{-2}, within the spread of their experimental errors. As a check on the extrapolated molar volumes of the fluid phase, molar volumes of solid molecular hydrogen were computed for the melting pressure corresponding to the same temperature as the fluid isothermal. The difference in molar volume of the solid and fluid phase, i.e., the expansion caused by melting, was found to be nearly constant along the melting line, which is roughly what one would expect on general grounds. The extrapolations of the two molar volumes lead to concordant results. For the integration of the equation of hydrostatic equilibrium along each isothermal, the linear co-ordinate h was then replaced by the dimensionless ratio h/H, where $H = kT/gm$ is the scale height of the H_2 atmosphere. It should be noted that the conversion factor from h/H to kilometers depends on both temperature and surface gravity; for $T = 150°$ K approximate values of H are 25 km (Jupiter) and 60 km (Saturn). By integration of the density-depth graph (see Fig. 11) it can then be established that even under the most unfavorable circumstances the total mass of isothermal H_2 atmospheres on Jupiter and Saturn is still less than one-tenth the mass of the Earth. Hence the atmospheric mass of the model planets is indeed negligible.

An interesting feature of Figure 11 is the bottom layer of nearly constant density ($0.1 < \rho < 0.2$). Because of its density and attendant viscosity, this layer is likely to resemble dynamically the ocean rather than

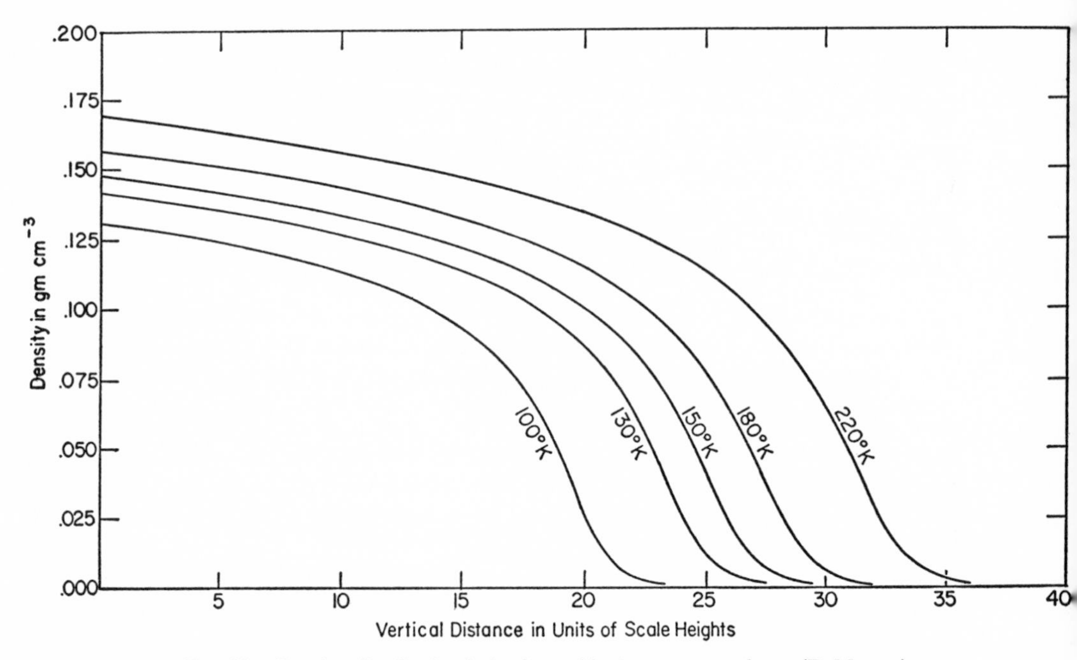

FIG. 11.—Density distribution in isothermal hydrogen atmospheres (DeMarcus)

the terrestrial atmosphere.[3] Tidal friction in the hydrogen oceans on Jupiter and Saturn might manifest itself in the motions of their inner satellites; for Jupiter V the sign of the observed acceleration is wrong for a tidal explanation. Although its density is high as compressed gases go, this fluid shell would seem to lack the buoyancy required to keep the Great Red Spot afloat. All available evidence, which was summarized by Peek (1939), seemed to support the view that this curious object is a solid body that is not rigidly attached to the core of Jupiter but has drifted back and forth across the planet's face during the last hundred years (Wildt, 1939). Granting this premise, it might be objected that the observed drift argues strongly against the assumption of a shallow isothermal atmosphere. Now a much deeper H_2 atmosphere could, in fact, be built on an adiabatic model, but its density could not very well exceed the density of the coexistent solid phase, i.e., about 0.35 gm cm^{-3}. Thus the difficulty regarding the buoyancy remains unresolved. But if the Red Spot is not a floating solid, how is one to reconcile its indubitable permanency with its astounding mobility?

The adiabatic lapse rate is so large that red heat (1000° K) would be reached at a depth of about 500 km on Jupiter and 1000 km on Saturn, the zero level corresponding to $T = 150°$ K and $P = 1$ atm. These figures are based on an unpublished investigation of the lapse rate in H_2 atmospheres that allows rigorously for the variation of C_p/C_v with temperature. This effect, which is quite large because of the outsize rotational quanta of H_2, was overlooked in earlier work on adiabatic models (Peek, 1937). In order to extend the adiabatic relations to even greater depths, it would be necessary to take account of the dependence of C_p/C_v on pressure, too. Explora-

[3] Huygens anticipated, by more than two hundred and fifty years, the concept of this ocean shading imperceptibly into the atmosphere. In his discourse on other planets and their inhabitants he remarked, in the words of an anonymous translator of the *Cosmotheoros*, as follows:

"... The Air I confess may be much thicker and heavier than ours, and so, without disadvantage to its Transparency, be fitter for the volatile Animals. There may be too many Fluids ranged over one another in rows as it were. The Sea perhaps may have such a fluid lying on it, which tho' ten times lighter than Water, may be a hundred times heavier than Air; whose utmost Extent may not be so large as to cover the higher places of their Earth.

"... But since 'tis certain that the Earth and Jupiter have their Water and Clouds, there is no reason why the other Planets should be without them. I can't say that they are exactly of the same nature with our Water; but that they should be liquid their use requires, as their beauty does that they should be clear. For this Water of ours, in Jupiter and Saturn, would be frozen up instantly by reason of the vast distance of the Sun. Every Planet therefore must have its Waters of such a temper, as to be proportion'd to its heat. Jupiter's and Saturn's must be of such a nature as not to be liable to Frost. ..."

tory work along these lines has been undertaken by DeMarcus (1958); tentative adiabatics of fluid hydrogen extending close to the melting-curve are shown in Figure 3.5 of his paper. DeMarcus admits that in the Jovian planets substantial portions of the molecular phase might be fluid, perhaps even up to the point of stability of the metallic phase. But, he finds, heating of dense hydrogen in the fluid phase to hypercritical temperatures would not render it markedly less dense than the corresponding solid at absolute zero. Consequently, ignorance in regard to the depth of the fluid shell on Jupiter and Saturn does not preclude the construction of models intended merely to represent the internal density distribution (cf. Sec. 4.3).

General interest is now focused on the internal temperatures of the Jovian planets, because the polarization of Jupiter's radio emission (chaps. 12, 13, 14) indicates the existence of a magnetic field, which might result from dynamic effects in a liquid core of high electrical conductivity. Metallic hydrogen would meet the requirements as an excellent conductor (Critchfield, 1942), but nothing is known about its melting point in the relevant pressure range. The decision as to fluidity versus solidity of the molecular mantle depends on the unknown heat flow from the interior and the thermal conductivity. Although a pure crystal of H_2 is a much better conductor of heat than rocks and other compact non-metals (DeMarcus, 1958), the behavior of the material forming the mantles of Jupiter and Saturn is quite a different matter, for the effects of impurities on the thermal conductivity of dielectrics are not well understood.

4.3. Composite Models

At the outset, one has to decide on a pressure-density relation likely to represent the behavior of a homogeneous mixture of two or more materials. Additivity of the molar volumes of the components is assumed, though it cannot hold except as a first approximation. Ramsey (1951) and Miles and Ramsey (1952) constructed models of Jupiter and Saturn on two extreme assumptions, namely, (1) a homogeneous mixture of hydrogen and helium and (2) a small, *incompressible* core of high density imbedded in the *compressible* phases of pure hydrogen; some intermediate models, with small incompressible cores imbedded in a homogeneous mixture of hydrogen and helium, were also studied. The second-order hydrostatic theory of a rotating planet was applied by Message (1955) to the models proposed by Miles and Ramsey. Using the general nature of the dependence of the coefficient of the fourth harmonic on that of the second, as computed from the models, and the value of the latter coefficient derived from observations of the great satellite motions, Message estimated the secular motion of the node of the fifth satellite and found it close to the observed value. Nevertheless, a critical examination of the pressure-density relations for hydrogen and helium that Miles and Ramsey used revealed that they require considerable modification (DeMarcus, 1958).

DeMarcus has built new models based on his interpolatory equations for helium and molecular hydrogen and his recalculation of the properties of metallic hydrogen. The final models (DeMarcus, 1958) are detailed in Tables 4 and 5. The primary aim was to achieve a reasonable representa-

TABLE 4

CALCULATED PROPERTIES OF THE INTERIOR OF JUPITER (AFTER DEMARCUS)

r/R	ρ (Gm/Cm³)	M (10^{27} Gm)	P (10^{12} Dynes/Cm²)
1.000	0.00016	1902.0	0
0.994	0.138	1900.31	0.0068
0.98942	0.185	1897.15	0.0200
0.98942	0.197	1897.15	0.0200
0.86	0.593	1707.76	1.06
0.802	0.777	1591.13	1.93
0.802	1.08	1591.13	1.93
0.7	1.56	1270.27	5.07
0.6	2.12	940.33	9.52
0.5	2.66	630.96	15.3
0.4	3.14	379.85	21.9
0.3	3.58	203.05	29.2
0.2	4.08	100.37	37.6
0.15	4.40	72.19	43.2
0.1	19.09	33.98	63.5
0.05	27.90	5.19	96.3
0	30.84	0	110

tion of the observed coefficients of the external potentials of Jupiter and Saturn, and the success may be judged by Table 6.

In regard to the physical interpretation of his models, DeMarcus has been quite cautious. In particular, no claim is advanced that the planets are, in fact, solid throughout. Rather it is suggested that the equations of state of a fluid molecular phase of hydrogen and that of a melt of the metallic phase do not differ sufficiently from the adopted pressure-density relations to vitiate either the matching of the data in Table 6 or the attempted abundance estimates. DeMarcus has been careful to point out that one could not deduce the chemical composition of Jupiter and Saturn even if the precise values of pressure, density, and temperature at every level surface were known and, in addition, if the exact equations of state, $\rho = f(P, T)$, for hydrogen and helium were available. But, he concludes, with only a rough knowledge of the internal densities and a rather crude estimate of the equation of state of molecular hydrogen, one can give a fair estimate of a lower limit to the hydrogen content if certain plausible assumptions are made. Proceeding from a stated set of assumptions, the weight fractions of hydrogen are shown to exceed 0.78 and 0.63 for Jupiter and Saturn, so that the ratio of the total number of hydrogen atoms to the total number of helium atoms in these bodies must be greater than 14 and 11, respectively. These lower limits are quite insensitive to the relaxation, essentially one at a time, of the assumptions mentioned above, excepting effects of temperature on the equation of state. Even these would not markedly alter the abundance estimates if the internal temperatures of Jupiter and Saturn were only a few thousand degrees. A surprising result

TABLE 5

CALCULATED PROPERTIES OF THE INTERIOR OF SATURN (AFTER DEMARCUS)

r/R	ρ (Gm/Cm^3)	M (10^{27} Gm)	P (10^{12} Dynes/Cm^2)
1.000	0.00016	569.00	0.000001
0.990	0.092	568.20	0.00186
0.970	0.185	561.44	0.0200
0.970	0.197	561.44	0.0200
0.95	0.236	551.86	0.0478
0.90	0.293	524.45	0.137
0.80	0.397	464.55	0.396
0.70	0.498	404.36	0.775
0.60	0.611	348.36	1.32
0.5227	0.719	309.78	1.93
0.5227	0.999	309.78	1.93
0.40	1.289	238.64	3.99
0.30	4.155	173.00	8.74
0.20	9.445	75.476	24.0
0.10	13.92	11.696	45.2
0	15.62	0	55.5

TABLE 6

COEFFICIENTS OF EXTERNAL POTENTIAL

	JUPITER		SATURN	
	Model Planet	Observed*	Model Planet	Observed*
ϵ_s	0.0645	0.0652	0.0988	0.0978
κ_s	.00049	.00052	.00195	.00159
J	.02130	.02206	.02594	.02501
K	0.00250	0.00253	0.00588	0.00386

* Observed values deduced from satellite motions, after Brouwer and Clemence.

of this analysis was the stability of the hydrogen abundances with respect to large variations in the equation of state of molecular hydrogen. DeMarcus traced the origin of this phenomenon to the consequences of the phase transition. If the densities of hydrogen are lowered from the values listed in Table 3, the pressure at which the phase transition occurs will be lowered too. Accordingly, the densities in the outer shell of the planetary models, where the molecular phase is stable, are lowered, and at the same time the volume of that shell is reduced, because of the stability of the metallic form at lower pressures. The cancellation of the two effects is almost complete. This discussion by DeMarcus of the abundance problem, incidentally, provides an explanation of why his estimates, after all, differ so little from the earlier ones obtained by Ramsey and Miles.

Uranus and Neptune are more than ten times as dense as hydrogen configurations of the same mass. Obviously, the pressure density of hydrogen is not a proper point of departure for an inquiry into the internal constitution of these bodies. Brown (1950) has suggested that they are composed mainly of ice and of solid methane and ammonia. Rough calculations of the 0° K isotherm of NH_3 have been made (Bernal and Massey, 1954), and work on models for these planets is under way (DeMarcus, private communication).

REFERENCES*

ABRIKOSOV, A. A. 1954 *Voprosi Cosmogonii*, **3**, 11–19 (Moscow: Akademia Nauk).

ADAMS, L. H. 1924 *Washington Acad. J.*, **14**, 459.

ALDER, B. J., and CHRISTIANS, R. H. 1956 *Phys. Rev.*, **104**, 550.

ANDRADE, E. N. DAC. 1934 *Phil. Mag.*, **17**, 497.

BERNAL, J. D. 1936 *Observatory*, **59**, 268.

BERNAL, M. J. M., and MASSEY, H. S. W. 1954 *M.N.*, **114**, 172.

BIRCH, F. 1947 *Phys. Rev.*, **71**, 809.

1952 *J. Geophys. Res.*, **57**, 227.

BOER, J. DE 1952 *Proc. R. Soc. London, A*, **215**, 4.

BOER, J. DE, and BLAISSE, B. S. 1948 *Physica*, **14**, 149.

BOER, J. DE, and MICHELS, A. 1938 *Physica*, **5**, 945.

BROUWER, D. 1955 *M.N.*, **115**, 221.

BROWN, H. 1950 *Ap. J.*, **111**, 641.

BULLARD, E. C. 1948 *M.N., Geophys. Suppl.*, **5**, 186.

* This list is not an exhaustive bibliography. Some effort has been made to list papers published after the first draft of this chapter was completed (1955), but it has not been possible to include all important recent literature in the revision.

BULLEN, K. E. 1946 *Nature*, **157**, 405.

1947 *Theory of Seismology* (Cambridge: Cambridge University Press).

1949*a* *M.N.*, **109**, 457.

1949*b* *Ibid.*, p. 688.

1950*a* *Ibid.*, **110**, 256.

1950*b* *M.N., Geophys. Suppl.*, **6**, 50.

1951 *Nature*, **167**, 29.

1957 *M.N., Geophys. Suppl.*, **7**, 271.

BULLEN, K. E., and LOW, A. H. 1952 *M.N.*, **112**, 637.

CHANDRASEKHAR, S. 1938 *Ap. J.*, **87**, 535.

1939 *An Introduction to the Study of Stellar Structure* (Chicago: University of Chicago Press).

CRITCHFIELD, C. N. 1942 *Ap. J.*, **96**, 1.

DARWIN, G. 1899 *M.N.*, **60**, 82.

DATTA, A. N. 1954 *M.N., Geophys. Suppl.*, **6**, 535.

DEMARCUS, W. C. 1951 Thesis, Yale University.

1954 *A.J.*, **59**, 116.

1957 *Les Molécules dans les astres* (Colloque International d'Astrophysique, Liège), p. 182.

1958 *A.J.*, **63**, 2.

DIRAC, P. A. M. 1930 *Proc. Cambridge Phil. Soc.*, **26**, 376.

DUGDALE, J. S., and SIMON, F. E. 1953 *Proc. R. Soc. London, A*, **218**, 291.

DUVALL, G. E., and ZWOLINSKI, B. J. 1955 *J. Acoust. Soc. America*, **27**, 1054.

ELSASSER, W. M. 1951 *Science*, **113**, 105.

FESSENKOV, V. G. 1924 *Russ. Astr. J.*, **1**, 102.

FESSENKOV, V. G., and MASSEVICH, A. G. 1951 *Russ. Astr. J.*, **28**, 317.

FEYNMAN, R. P., METROPOLIS, N., and TELLER, E. 1949 *Phys. Rev.*, **75**, 1561.

GILVARRY, J. J. 1957 *J. Chem. Phys.*, **27**, 150.

GLASSTONE, S., LAIDLER, K. J., and EYRING, H. 1941 *Theory of Rate Processes* (New York: McGraw-Hill Book Co., Inc.).

GRAFF, K. 1929 *Hdb. d. Ap.* (Berlin: J. Springer), Vol. **4**.

HOLLAND, F. A., HUGGILL, J. A. W., and JONES, G. O. 1951 *Proc. R. Soc. London, A*, **207**, 268.

HOUTGAST, J. 1955 *Nature*, **175**, 678.

JACOBS, J. A. 1954 *Nature*, **173**, 258.

JEFFREYS, H. 1923 *M.N.*, **83**, 350.

1924 *Ibid.*, **84**, 534.

1937 *M.N., Geophys. Suppl.*, **4**, 62.

	1951	*M.N.*, **111**, 410.
	1952	*The Earth* (3d. ed.; Cambridge: Cambridge University Press).
	1953	*M.N.*, **113**, 97.
JENSEN, H.	1938	*Zs. f. Phys.*, **111**, 373.
KIRSCH, G.	1935	*Anzeiger Akad. Wiss. Wien, math. naturw. Kl.*, **72**, 95.
KNOPOFF, L., and UFFEN, R. J.	1954	*J. Geophys. Res.*, **59**, 471.
KRICHEVSKY, I.	1940	*Acta Physicochem. U.R.S.S.*, **12**, 480.
KRICHEVSKY, I., and ZICLIS, D.	1943	*Acta Physicochem. U.R.S.S.*, **18**, 264.
KRONIG, R., BOER, J. DE, and KORRINGA, J.	1946	*Physica*, **12**, 245.
KUHN, W.	1942	*Naturwiss.*, **30**, 689.
	1946	*Experientia*, **2**, 391.
	1948	*Ibid.*, **4**, 23.
KUHN, W., and RITTMANN, A.	1941	*Geol. Rundschau*, **32**, 215.
KUHN, W., and VIELHAUER, S.	1953	*Zs. f. phys. Chem.*, **202**, 124, and *Geochim. et Cosmochim. Acta*, **3**, 169.
LANDAU, L., and ZELDOVICH, J.	1943	*Acta Physicochem. U.R.S.S.*, **18**, 194.
MACDONALD, G. J. F., and KNOPOFF, L.	1958	*Geophys. J., R.A.S.*, **1**, 284.
MARCH, H. N., MEIJERING, J. L., and ROOYMANS, C. J. M.	1955	*Proc. Phys. Soc. London, A*, **68**, 726.
	1958	*Proc. Kon. Nederl. Akad. Wetensch. B*, **61**, 333.
MESSAGE, P. J.	1955	*M.N.*, **115**, 550.
MILES B., and RAMSEY, W. H.	1952	*M.N.*, **112**, 234.
MURNAGHAN, F. D.	1937	*Amer. J. Math.*, **59**, 239.
PEEK, B. M.	1937	*M.N.*, **97**, 574.
	1939	*J. Brit. Astr. Assoc.*, **50**, 4.
RAMSEY, W. H.	1948	*M.N.*, **108**, 406.
	1949	*M.N., Geophys. Suppl.*, **5**, 409.
	1950*a*	*M.N.*, **110**, 325.
	1950*b*	*M.N., Geophys. Suppl.*, **6**, 42.
	1951	*M.N.*, **111**, 427.
REINER, M.	1949	*Twelve Lectures on Theoretical Rheology* (Amsterdam: North-Holland Publishing Co.).
	1958	"Rheology." In *Encyclopedia of Physics* (Berlin: J. Springer), Vol. **6**.
RICHARDSON, J. M., ARONS, A. B., and HALVERSON, R. R.	1947	*J. Chem. Phys.*, **15**, 785.

RINGWOOD, A. E. 1957 *Geochim. et Cosmochim. Acta*, **13**, 303.
1958*a* *Ibid.*, **15**, 18.
1958*b* *Ibid.*, p. 195.
1959 *Ibid.*, **16**, 192.
1960 *Ibid.*, **20**, 241.

RUSSELL, H. N. 1935 *Observatory*, **58**, 259.

SALTER, L. 1954 *Phil. Mag.*, **45**, 369.

SCHATZMAN, E. 1951 *Bull. Acad. R. Belgique* (5), **37**, 599.

SCHOLTE, J. G. 1947 *M.N.*, **107**, 237.

SEN, P. 1938 *Zs. f. Ap.*, **16**, 297.

SIMON, F. 1929 *Zs. f. phys. Chem.* (*B*), **6**, 331.
1937 *Trans. Faraday Soc.*, **33**, 65.
1952 *L. Farkas Memorial Volume* ("Research Council of Israel Special Publications," No. 1 [Jerusalem]), p. 37.
1953 *Nature*, **172**, 746.

SITTER, W. DE 1924 *B.A.N.*, **2**, 97.

SLATER, J. C., and KRUTTER, H. M. 1935 *Phys. Rev.*, **47**, 559.

STEWART, J. W. 1956 *J. Phys. Chem. Solids*, **1**, 146.

UREY, H. 1952 *The Planets, Their Origin and Development* (New Haven: Yale University Press).

VERHOOGEN, J. 1953 *J. Geophys. Res.*, **58**, 337.

WARES, G. 1944 *Ap. J.*, **100**, 158.

WIGNER, E., and HUNTINGTON, H. B. 1935 *J. Chem. Phys.*, **3**, 764.

WILDT, R. 1934 *Veröff. U. Sternw. Göttingen*, No. 40.
1938 *Ap. J.*, **87**, 508.
1939 *Proc. Amer. Phil. Soc.*, **81**, 135.

WOOLLEY, H. W., SCOTT, R. B., and BRICKWEDDE, F. G. 1948 *J. Res. Nat. Bur. Stand.*, **41**, 379.

WRUBEL, M. H. 1958 "Stellar Interiors." In *Encyclopedia of Physics* (Berlin: J. Springer), Vol. **51**.

CHAPTER 6

Photometry of the Moon

By M. MINNAERT
Utrecht Observatory

THE study of the photometric properties of a planetary surface is a special method of investigation, almost without parallel in astrophysics. From the variation of the brightness under varying conditions of illumination, information concerning the geometrical and physical properties of the surface may be derived which cannot be discovered by any other method. Full information is obtained only if photometric observations are available in different wave lengths and in both planes of polarization. In *detailed photometry* one studies a planetary area so small that geometrically it may be considered to be placed at one point on the planetary sphere; but nevertheless one will integrate over many samples of surface material and slopes and get only mean values. *Integrated photometry* studies the total emission of a celestial body; it gives less information but still allows useful conclusions. No other celestial body can be studied photometrically in such detail as the moon, and in no other case is there such an opportunity for the application of photometric methods in all their variety. Monographs on the photometry of planets and satellites are due to Schoenberg (1929) and Sharonov (1954*a*, 1958).

1. INTEGRATED PHOTOMETRY

The apparent magnitude of the full moon has been measured several times. We mention only the more recent results; the first two values were discussed by Kuiper; all are reduced to the international photovisual system:

Pettit (1926), $-12^{m}.75$ (radiometric determination)
Calder (1937), -12.69 (photoelectric determination)
Nikonova (1949), -12.67 (photoelectric determination)

The difference between the magnitudes of the full moon and of the sun is important, since it determines the lunar albedo and may be used to connect the sun with the system of stellar magnitudes. Two modern determinations are found:

Rougier (1933), −14.29
Nikonova (1949), −14.02

Finally, it may be useful to quote the value of the "lunar constant"; this means the brightness of the full moon expressed in lux, as observed outside the atmosphere, the moon being at its apogee and the earth at its mean distance from the sun. Sytinskaya (1957*b*) published a survey of 36 determinations by different authors, yielding, as a mean value, 0.342 ± 0.011.

TABLE 1

THE INTEGRATED PHASE-CURVE OF THE MOON

a	Rougier	Bullrich	a	Rougier	Bullrich
0°........	100	100	90°.......	8.0	6.7
10........	72.5	73.2	100........	5.6	4.7
20........	57.8	56.0	110........	3.9	3.6
30........	43.7	42.3	120........	2.5	2.4
40........	33.9	32.0	130........	1.6	1.2
50........	26.3	23.3	140........	0.9	0.9
60........	20.8	16.7	150........	0.4	0.4
70........	16.6	12.4	160........	0.002	0.002
80........	11.5	8.7	....	....	....

In all these photometric determinations, the difficult extrapolation to phase angle zero gives an inevitable uncertainty.

The variation of the total radiation of the moon in the course of a lunation has been studied by several astrophysicists. Russell (1916) gave a critical survey of the results then available and published a curve which has been widely used. More recently, a careful determination was made by Rougier (1933; cf. Table 1 of this chapter). He pointed a potassium photocell without a telescope toward the moon and matched the signal with that of an incandescent lamp, placed at a variable distance. His curve has been so reduced as to make the total radiation at full moon equal to 100. A curious asymmetry is found between the waxing and the waning crescent, the latter radiating more, though the area of the dark plains at that time is relatively greater. Between full moon and quarter phases the waxing moon is slightly brighter. Rougier's curve refers to a mean wave length of 4450 A.

Another modern determination was made by Bullrich (1948) by not very precise visual measurements with a Bechstein photometer; the deviations from Russell's curve are in the same sense but somewhat larger than those found by Rougier. On the whole, the two modern curves agree very well (Table 1). Whether the remaining differences are due to measuring errors or to the wave-length difference is uncertain; the latter would imply that the moon becomes slightly bluer away from full moon, the ratio increasing to about 0.2 mag. The influence of wave length, if any, is thus small.

2. METHODS OF DETAILED LUNAR PHOTOMETRY

Lunar photometry is a comparatively recent branch of astrophysics. The early visual estimates, interesting even now in view of possible changes on the moon, were often made on the scale of Schroeter (1791–1802), of which 1 step, according to Markov (1952), corresponds to 0.15 mag. A more precise relation between the radiance ρ and the step number N is found to be $\rho = 0.047 + 0.136N$, the value of 1 step thus varying from 0.45 to 0.08 stellar magnitudes between $N = 1$ and $N = 9.5$ (Sytinskaya, 1953*b*). The early photographic work of Pickering (1882) has suffered from incorrect calibration.

The first rather reliable visual measurements were made by Wisliscenus (1895, published by Wirtz in 1915) and by Barabashev (1922); the method has been used even in recent times (Sytinskaya and Sharonov, 1952). It is well known that the eye is superior to any instrumental receiver in distinguishing small detail.

Most investigations, however, make use of *photographic* photometry and measure the transmission of the plate with a Hartmann photometer or microphotometer. The radiance range to be measured is so considerable that diaphragms have to be introduced during part of the lunation. Special corrections have to be introduced for the veil which is always found around lunar photographs and which may be due to the bright sky around the moon, to photographic effects, or to diffracted or multiply reflected light inside the telescope (this last source of error was especially emphasized by Markov, 1950*b*). Probably the most systematic attempt at such corrections was made by Fessenkov, Parenago, and Staude (1928), though the results convey the impression that they have been overcorrected.

Since the introduction of the multiplier to astronomical photometry, lunar measurements have also been made with this invaluable instrument. Markov (1948) studied, first, the integrated radiance of successive annular regions of the lunar surface, by putting an iris diaphragm in the focal plane and screening off successive rings. Afterward he (1950*b*) followed the radiance of eighteen points of the lunar surface during the whole lunation, taking some G- and K-type stars as standards and determining the extinction for each night. Markov points out several sources of error and concludes that photoelectric photometry should be made for areas not greater than $5'' \times 5''$.

Special difficulties arise if the "absolute" photometric radiance is desired. This involves intercomparisons during a whole lunation or, better, direct comparisons of moon and sun. The first is often done by photographing extra-focal images of stars or by obtaining a double series of lunar photographs on consecutive days of the lunation and developing the plates in overlapping pairs. In this way all records are interconnected

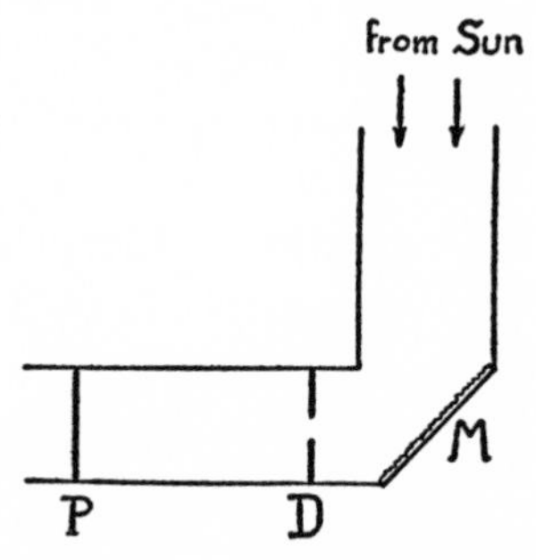

FIG. 1.—Method of standardization and comparison with the sun (Sytinskaya, 1949). M = magnesia screen; D = diaphragm; P = photographic plate.

(Barabashev, 1927; Fedoretz, 1952). Other authors use a standard lamp for comparison but then have to determine the atmospheric extinction for each night.

Several methods for solar comparisons have been discussed by Sharonov (1936) and Mrs. Sytinskaya (1948, 1949). Especially useful is a MgO screen illuminated by the sun, which is an almost ideal isotropic diffuser if prepared properly (chemically pure magnesium, layer thickness at least 0.3 mm on a reflecting metal or on chalk). Other "white" screens may be used, such as chalk, ground opal glass, or white marble, provided that they are calibrated afterward. Such a screen, perpendicularly illuminated by the sun, has along the normal a radiance of $\pi(16')^2/\pi = 1/46{,}400$ times the mean radiance of the solar disk. Using three different forms of this method on thirteen points of the full moon, Mrs. Sytinskaya reached a precision of 5–7 per cent in an absolute comparison with the sun. Her results may be used for secondary calibrations. We especially note her simple and convenient measuring device, explained in Figure 1.

Van Diggelen (1959) applied a different principle. He measured on photographs the radiance at between 500 and 2500 points, depending on the phase, spaced at regular intervals in two co-ordinates. The sum of the readings was assumed to be equal to the integrated radiance for that phase of the lunation according to Rougier (1933), which determined the reduction factor for each photograph. It would be useful if all investigations on detailed photometry could similarly be compared to the well-established integrated phase-curve; this may be illustrated by the discussions of Fessenkov and Parenago (1929).

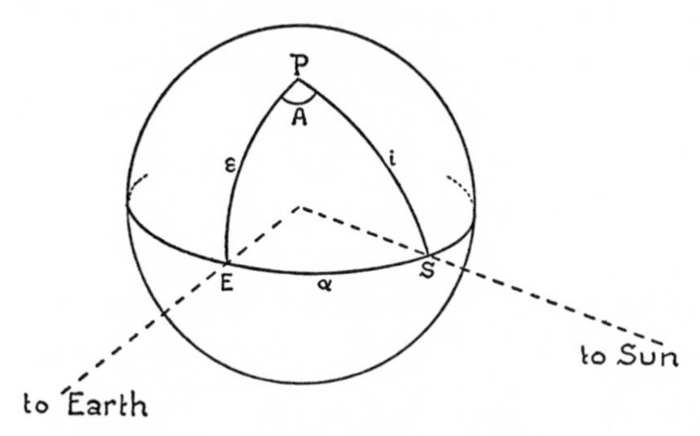

FIG. 2.—Photometric co-ordinates

Radiance measurements of lunar objects have usually been expressed in terms of the radiance of an ideal white screen. Two numbers are commonly used:

$$\text{The radiance coefficient } r = \frac{B}{B_0} = \frac{\text{Radiance observed}}{\text{Radiance of white screen under same inclination}},$$

$$\text{The radiance factor } \rho = \frac{B}{B_n} = \frac{\text{Radiance observed}}{\text{Radiance of white screen, normal to incident rays}}.$$

Evidently, $\rho = r \cos i$. The numbers r and ρ can be directly compared with those of terrestrial objects, whereas the albedo must be derived by complicated and often uncertain reductions.

Consider the moon as a sphere, with the radius through E directed to the earth, and the radius through S to the sun (see Fig. 2). For a surface element P on the sphere, the angle of incidence of the light is $SP = i$, the angle of observation is $EP = \epsilon$, the phase angle $ES = \alpha$. Now

$$\rho = \rho_0 f(i, \epsilon, \alpha),$$

and the function f is normalized by such a factor that $f = 1$ when $i = \epsilon = 0$; thus ρ_0 is the value of ρ for an object near the center of the full moon.

We shall call this the *normal albedo* and call $f(i, \epsilon, \alpha)$ the *photometric function* of the material. Both concepts are widely used in lunar photometry and are quite simple for practical purposes.

Instead of the variables i, ϵ, α, we could have chosen i, ϵ, A, where $A = EPS$ is the azimuth difference between the plane of incidence and the plane of observation. The choice of α has the advantage that it is the same for all points on a lunar photograph and that it is the variable used in integral lunar photometry. It is true that the azimuth A is more fundamental in the formulae for diffuse reflection; but, since the photometric features of the moon are found to be determined primarily by shadows of surface detail and not by scattering, it seems reasonable to keep to the more practical co-ordinates.

3. RESULTS OF DETAILED PHOTOMETRY

Data on the photometry of lunar objects are available in the following publications:

Wisliscenus-Wirtz........	1915	20 objects	Whole lunation	Visual
Götz....................	1919	55 objects	Full moon	Photographic
Rosenberg..............	1921	55 objects	Full moon	Photographic
Barabashev.............	1922	7 objects	Whole lunation	Visual
Öpik....................	1924	64 objects	Whole lunation	Photographic
Schoenberg..............	1925	Few objects	Whole lunation	Visual
Barabashev.............	1927	31 objects	Whole lunation	Photographic
Fessenkov, Parenago, and Staude..............	1928	Whole surface	Whole lunation	Photographic
Fessenkov and Parenago..	1929	Limb	Whole lunation	Photographic
Bennett................	1938	59 objects	Whole lunation	Photographic
Radlova................	1941	35 objects	Full moon	Visual
Radlova................	1943	97 objects	Full moon	Photographic
Graff....................	1949	79 objects	Full moon	Photographic
Sytinskaya..............	1949	13 objects	Full moon	Photographic
Markov..................	1950*b*	18 objects	Whole lunation	Photoelectric
Fedoretz................	1952	168 objects	Whole lunation	Photographic
Sytinskaya and Sharonov.	1952	90 objects	Whole lunation	Visual
Van Diggelen	1959	45 crater bottoms	Whole lunation	Photographic

Fedoretz' results, especially, are a rich source of information, the more valuable because he has obtained measurements very near full moon, by observing just before a lunar eclipse (i and ϵ differing by at most $1^\circ.5$). Even between $\alpha = 4^\circ.5$ and $\alpha = 1^\circ.5$ the mean increase of radiance is still of the order of 15 per cent.

4. THE PHOTOMETRIC FUNCTION

4.1. Lunar Formations

A full description of the photometric properties of a lunar formation as functions of i, ϵ, and A cannot be obtained from measurements of its radiance during the lunation because, for a given object, i and A do not vary independently and ϵ always has the same value. Progress can be made by assuming that there are a number of similar objects, distributed over the entire lunar surface, which follow the same photometric function, though the albedo may differ. It is then possible to combine these areas as if the entire sphere were covered by the same material.

4.11. *Principles of symmetry.*—Assuming that considerable areas, distributed all over the lunar surface, are covered by a homogeneous layer, we may predict from geometry certain relations between the radiance at different points of the disk (see Fig. 3).

a) At any phase, points symmetric with respect to the radiance equator have the same radiance. As an example, we compare in Figure 4 some

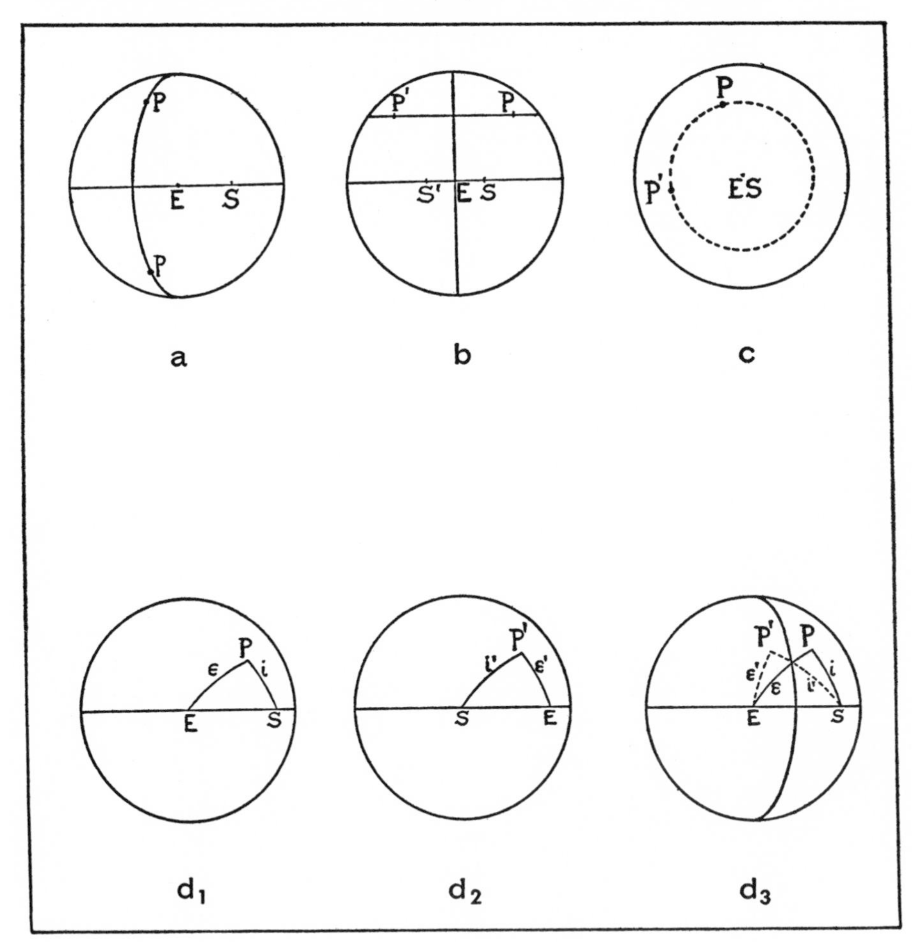

FIG. 3.—Symmetry properties

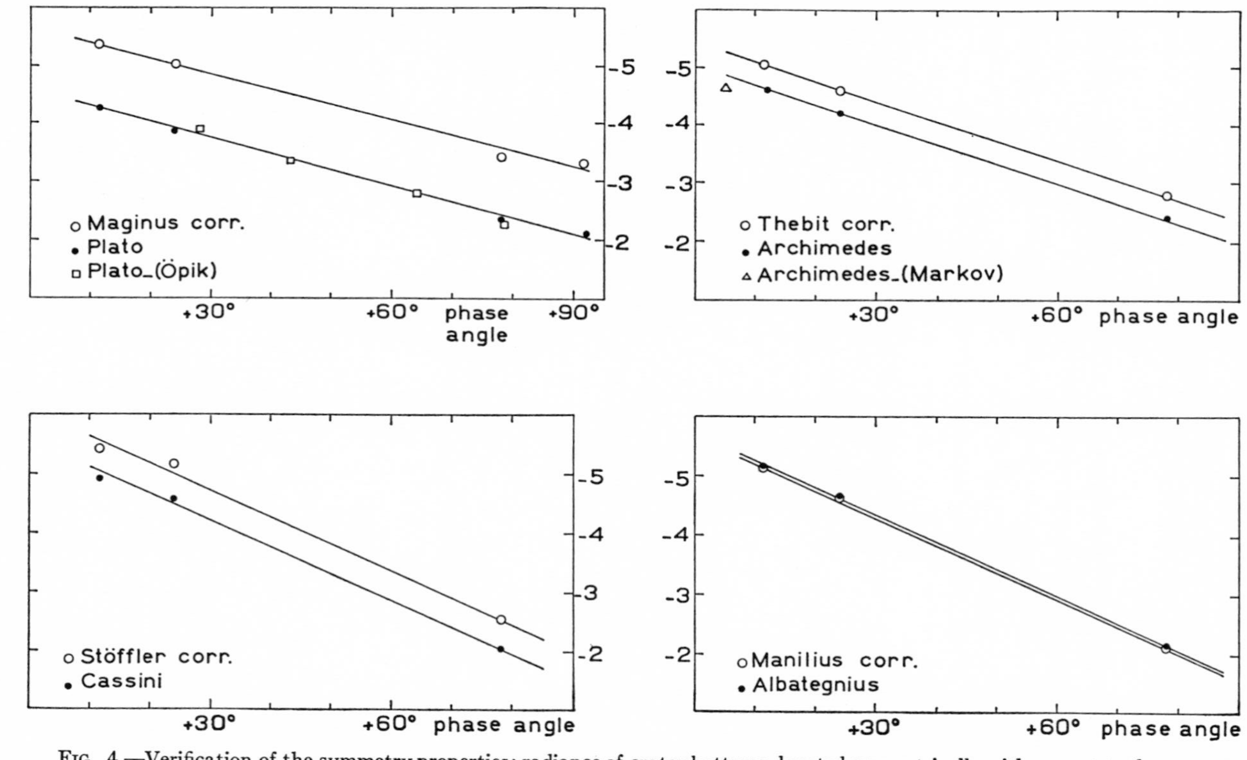

FIG. 4.—Verification of the symmetry properties: radiance of crater bottoms, located symmetrically with respect to the equator, one north (*full circle*) and one south (*open circle*). Corrections have been made for the small longitude differences (after van Diggelen, 1959). *Ordinates:* brightness in magnitudes, with average brightness at full moon being −5 mag.

crater bottoms in the northern hemisphere with crater bottoms in the southern hemisphere, symmetrically located. Slight corrections for the lack of perfect symmetry have been introduced.

b) Two points, symmetric with respect to the central meridian, will have the same radiance if photographed at phases symmetric with respect to full moon.

c) At full moon, points on a circle concentric with the limb have the same radiance.

d) Compare a surface area, irradiated at unit intensity under the angle i and observed to have a radiance B under the angle of observation ϵ. From the reversibility of the light-rays, connected with the second law of thermodynamics, it follows that the same surface area, irradiated by unit intensity under the angle $i' = \epsilon$ and observed under the angle $\epsilon' = i$, will have the radiance B', such that $B/B' = \cos i/\cos \epsilon = \cos i/\cos i'$ (Minnaert, 1942). Applying this to the moon, we see that for a given phase of the lunation, characterized by the selenographic longitude of the sun $\lambda_{\odot}$, the illuminated part of the disk is divided into two parts by the *mirror meridian* at longitude $\lambda_{\odot}/2$, located halfway between the limb and the terminator. Any point P in the terminator half may be compared with its "image" P' on the same parallel in the limb half, symmetric with respect to the mirror meridian. If a point and its image belong to the same type of lunar formation, their radiances are in the ratio

$$\frac{B}{B'} = \frac{\cos i'}{\cos i} = \frac{\cos \lambda}{\cos \lambda'},$$

as follows from the reciprocity principle and simple trigonometry. For any point on the lunar disk it is easy to express in its rectangular or polar co-ordinates the co-ordinates of the image and the ratio of the radiances, if the surface composition is uniform. These symmetry principles permit one to check whether the areas compared are indeed covered by the same types of material or, if this is known to be true, whether the measurements are consistent.

e) Consider the brightness distribution along the equator and take two points related by the reciprocity principle: i, ϵ and i', ϵ', such that $i' = \epsilon$ and $\epsilon' = i$. Let these points be at selenocentric distances β from the terminator and β from the limb, respectively. Their radiance values B_t and B_l will then be related by

$$\frac{B_l}{B_t} = \frac{\cos i'}{\cos i} = \frac{\cos (90° - \alpha - \beta)}{\cos i}.$$

Thus, if the observed fact is used that the brightness near the limb is nearly constant, the radiance near the terminator is found to be $B_t = c \cos i$. At the limb itself β vanishes and we have the relation

$$B_l^* = c \sin \alpha = -\left(\frac{dB}{d \cos i}\right)_t = -\left|\frac{dB}{d \sin \epsilon}\right| \sin \alpha .$$

Thus a simple relation exists between the *radiance* near the limb and the *radiance gradient* near the terminator; this is used on page 227. A similar relation exists along each parallel of latitude.

4.12. *Empirical results.*—From the available measurements some general results have been obtained.

a) Except for differences in the normal albedo, the radiance at full moon is the same for all points of the disk (Markov, 1924; Markov and Barabashev, 1926); this means that $f(i, \epsilon, \alpha) = 1$ if $i = \epsilon$. For all areas inside each of the concentric rings this was to be expected for reasons of symmetry. However, the extensive measures of Fessenkov, Parenago, and Staude (1928) and those of Fedoretz (1952), contrary to the conclusions of the authors themselves, appear to show a slight increase in the radiance toward the limb, at least for the mountain regions. The measures of Markov (1948) point in the same direction.

b) All formations reach their greatest radiance at full moon, independently of their position on the lunar disk. This important law was discovered for the maria by Barabashev (1922) and generalized to all lunar formations by Markov (1924) on the basis of rather primitive measurements, but has been confirmed since. The results of Fedoretz (1952) show that deviations from the law are infrequent and small, though for some areas the maximum may be displaced up to 13° after full moon; this occurs especially for craters surrounded by bright rays (van Diggelen, 1959, p. 45).

c) The bright limb of the moon shows nearly the same brightness over at least three-fourths of its arc (Schoenberg, 1925). This result has rarely been checked.

d) According to Öpik (1924), the photometric function $f(i, \epsilon, \alpha)$ is nearly the same for all types of formations. However, Fedoretz (1952)—who directly compares the maria and the neighboring crater bottoms, the dark and the bright spots on some crater bottoms, and the mountain regions—found clear differences for the crater rays and the bright craters, which showed a more pronounced maximum near full moon than did the other features.

e) Except for differences in albedo, the isophotes at any phase are meridians (see p. 226).

4.13. *Empirical formulae.*—Several authors have tried to group their measures in such a way that they could be represented by simple formulae of which the constants were determined empirically. Öpik (1924) expressed the radiance in stellar magnitudes and found practically the same expression for the terrae and the maria (except for the albedo difference, the constant term Δa):

$$B_m = a + \Delta a - 2.5\,k \log \cos i - k'\epsilon .$$

Here a and k are functions of the phase, while k' is practically constant for phase angles between 28° and 121°.

Schoenberg (1925) tried to express his results by an extension of the theoretical formula of Lommel-Seeliger. He wrote:

$$B = \text{Const.}\,\frac{\cos i}{\cos i + \frac{1}{3}\cos \epsilon}\,\psi(\alpha) .$$

However, this formula is not compatible with the reciprocity principle and is therefore unacceptable.

Fessenkov (1928) and Fessenkov, Parenago, and Staude (1928) have made similar attempts. They put

$$B = \text{Const.}\,\frac{\cos i\,(1 + \cos^2 \alpha/2)}{\cos i + 0.225\,(1 + \tan^2 \alpha)\cos \epsilon} .$$

This formula must be rejected for the same reason.

Tschunko (1949), by generalization of some results by Bennett, arrives at an empirical law:

$$B = \text{Const.}\,\frac{\tan^{-1} q}{\alpha + \tan^{-1} q}, \qquad \text{where} \qquad q = \frac{\sin \alpha}{(\cos \epsilon/\cos i) - \cos \alpha} .$$

Here the reciprocity principle is satisfied; the law means that the radiance of any lunar region increases linearly with the phase up to full moon and decreases again linearly until the terminator passes over the object (cf. Fig. 5). No additional functions or constants have to be introduced; even the phase dependence is described: for example, a point at the center will have a radiance $1 - \alpha/90°$; for a point at the limb, $\epsilon = 90°$, $q = -\tan \alpha$, $\tan^{-1} q = 180° - \alpha$, $B = \text{Const}\,[1 - (\alpha/180°)]$. However, as Tschunko recognizes, his law is only a first approximation, since for most lunar objects the $B(\alpha)$ lines are not straight but curve upward.

4.14. *Description by isophotes.*—Minnaert has tried to avoid aprioristic formulae and derived the photometric properties directly from the observations by obtaining systems of isophotes. The reduction is carried out in the following steps:

(i) From a series of standardized photographs at various α, derive ρ for a number of areas of the same type, each characterized by their i and ϵ. We assume that their photometric function is the same, while the albedos may be different.

(ii) Assume that the radiance at full moon is independent of the location on the disk, and determine the normal albedo ρ_0 for each of the areas

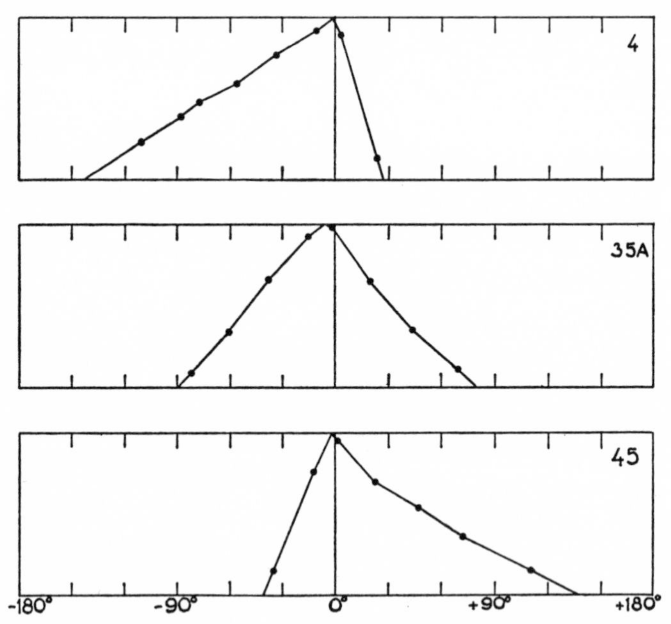

FIG. 5.—The empirical photometric function of lunar formations according to Bennett's observations (Tschunko, 1949).

measured, which at that moment is equal to the directly observed radiance factor.

(iii) For every day of the lunation, divide the radiance factor, measured on each area, by the corresponding albedo. The measures may now be considered as referring to one and the same surface material.

(iv) In order to diminish the effects of the remaining inhomogeneities and errors of measurement, make use of the existing symmetry properties and take mean values, combining (*a*) the north and south hemispheres; (*b*) the east and west sides at symmetric phases before and after full moon; (*c*) the limb half and the terminator half, combining each point with its image and reducing the radiance of the one to that of the other by multiplying by cos i/cos i' (see p. 221). By these reductions, the density of points increases by the factor 8, and mean values may be determined with increased accuracy. The data showed that, upon reduction, the values of the limb half agreed quite well with those of the terminator half.

(v) Draw a system of isophotes for each phase α.

Such a reduction has been carried out by Minnaert separately for the terrae and the maria, making use of the measurements of Fessenkov, Parenago, and Staude (1928); but it should be repeated with more modern data. Of the isophote diagrams, obtained for six different phase angles, two are reproduced in Figure 6. The general shape of the curves is considered fairly reliable but not the numerical values, owing to systematic errors in the measurements. Figure 6 suggests that the isophotes are

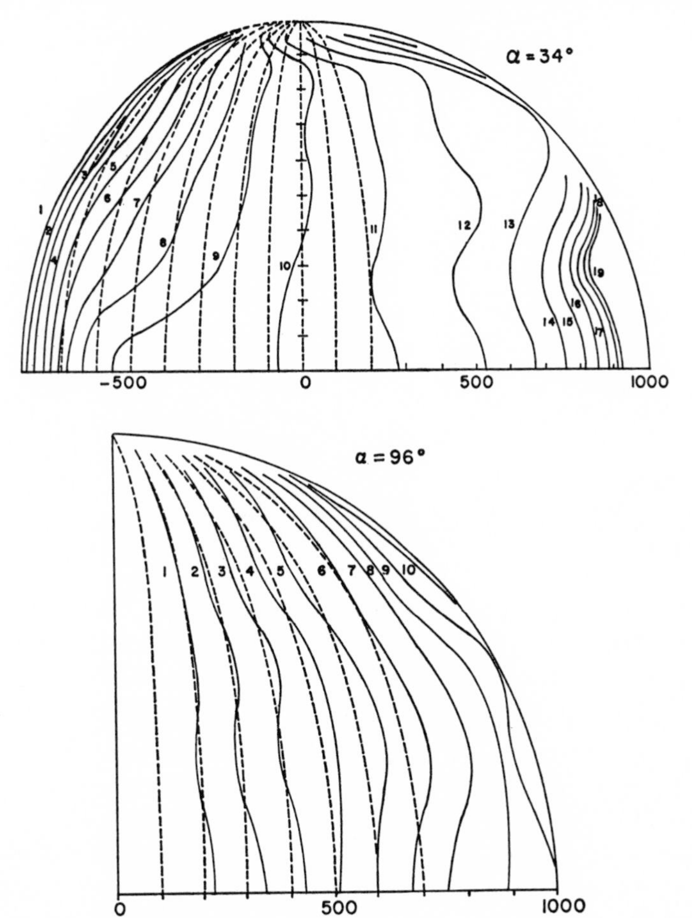

FIG. 6.—Isophotes of the moon, for two phases, $\alpha = 34°$ and $96°$ (Fessenkov *et al.*, 1928, and Minnaert).

approximately meridians, appearing as ellipses; the same is true for the other phase angles studied and according to most empirical formulae obtained thus far. Accepting this result, we may determine for each phase the mean radiance of each meridian. Therefore, while the system of curves obtained for the different phases best summarizes the photometric properties of the moon, a set of curves giving for each phase the mean radiance as a function of selenographic longitude or giving simply the radiance along the equator is almost equivalent.

TABLE 2*A*

VALUES OF RADIANCE COEFFICIENT r (UNIT 0.001) ALONG THE EQUATOR FOR TERRAE (ORLOVA, 1956)

ϵ*	i							
	0	10°	20°	30°	40°	50°	60°	70°
80°....	28	....	45	62	75	102	....	....
70....	36	43	51	74	87	120	190	362
60....	43	52	58	87	100	150	248	243
50....	51	62	67	103	124	193	171	156
40....	61	72	81	124	161	144	117	112
30....	72	87	109	143	124	100	90	89
20....	86	104	131	120	92	79	72	73
10....	102	131	113	98	74	65	58	61
0....	124	108	88	78	61	52	47	50
10....	102	84	72	64	53	45	40	45
20....	86	73	62	53	46	37	34	38
30....	72	60	53	42	37	32	30	35
40....	61	52	47	35	31	27	27	32
50....	51	42	37	30	27	25	26	30
60....	43	35	32	25	24	25	25	27
70....	36	25	25	22	22	25	25	27
80....	28	17	17	20	20	25	25	27

* The values of ϵ, top to bottom, correspond to increasing phase from just after first quarter to just before last quarter.

In order to derive such a system of curves, we make use of the photometric measurements of Fedoretz (1952) and those of Sytinskaya and Sharonov (1952), summarized and averaged by Orlova (1955, 1956). Her tables for the terrae and the maria are nearly the same. She reduced all values by a factor such that the mean value of ρ at full moon became 0.124 for the terrae and 0.081 for the maria and obtained in this way the absolute values of r, reproduced in our Table 2*A* for the terrae. By multiplying all values by cos i, we reconvert them into the directly observed values of ρ and obtain Table 2*B*, giving the best description of the radiance distribution over the lunar disk.

On the equator $|\alpha| = \epsilon - i$; thus one easily selects the ρ values beyond $i = 70°$ and $\epsilon = 80°$. The trend of the curves toward $\epsilon = 90°$ seems clear, and from the extrapolated limb radiance we may derive the slope near the terminator, according to the relation given on page 222. These parts of the curves are shown broken, and they fit the measured parts satisfactorily. The system of curves shows some deviations from reciprocity requirements: in general, the terminator half is too high with respect to the limb half, an error easily made because the observer is inclined to measure the illuminated areas and neglect the shadows. The amount of the deviation is shown for $\alpha = 40°$.

TABLE 2*B*

VALUES OF RADIANCE FACTOR ρ (UNIT 0.001) ALONG THE EQUATOR, FOR TERRAE (ORLOVA, 1956)

ϵ*	*i*							
	0	10°	20°	20°	40°	60°	60°	70°
80°....	28	....	42	54	57	66	....	....
70....	36	42	48	64	67	77	95	124
60....	43	51	54	75	77	97	124	83
50....	51	61	63	89	95	124	86	53
40....	61	71	76	107	123	93	58	38
30....	72	86	102	124	95	64	45	30
20....	86	102	123	104	70	51	36	25
10....	102	129	106	85	57	42	29	21
0....	124	106	83	67	47	33	24	17
10....	102	83	68	55	41	29	20	15
20....	86	72	58	46	35	24	17	13
30....	72	59	50	36	28	21	15	12
40....	61	51	44	30	24	17	14	11
50....	51	41	35	26	21	16	13	10
60....	43	34	30	22	18	16	12	9
70....	36	25	23	19	17	16	12	9
80....	28	17	16	17	15	16	12	9

* For values of ϵ see note to Table 2*A*.

On the whole, Figure 7 gives a satisfactory survey of the photometric properties of the moon. The value of the mathematical formulae quoted in Section 4.13 and of the models discussed in Sections 4.2 and 4.3 may be best appreciated by deriving from them similar graphs and comparing them with Figure 7.

It is emphasized that the symmetry conditions, valid for any spherical surface, determine the general nature of the curve system and that the special properties of the lunar surface are only expressed by (*a*) the remarkable distribution of the radiance over the full moon, almost independent of ϵ, and (*b*) the distribution of the radiance at the limb as a function of α (cf Fig. 8).[1] For the point where a curve begins is given by $\epsilon_t = 90 - \alpha$; its initial slope is proportional to $\cos i$; the proportionality constant is determined by the limb radiance (cf. p. 222); and the maximum value is the limb radiance itself. Given the limb radiance, it is thus only the curvature of each curve that to some extent is left undetermined.

Instead of the limb radiance as a function of α, we could also consider the integrated radiance (p. 217), which has a similar run; this, however, is so far less representative in that it is also affected by the accidental surface distribution of maria and terrae and by crater rays.

[1] This important curve has also been determined and published by Orlova (1958).

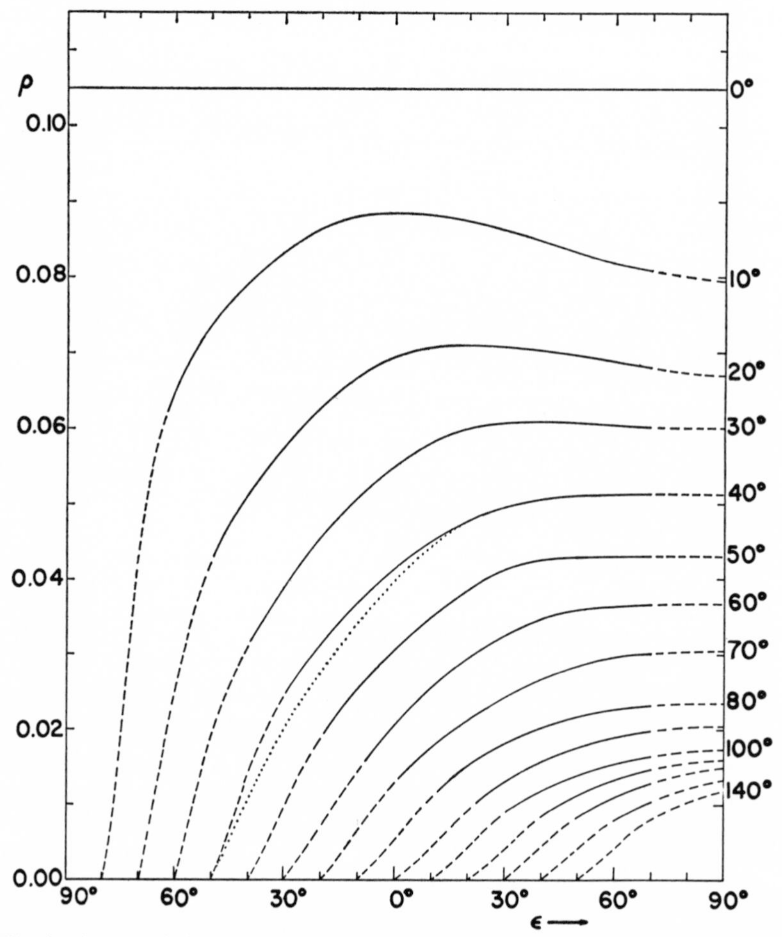

FIG. 7.—Survey of the photometric properties of the terrae of the moon (after Orlova, 1956, and Minnaert). The graphs give, for different phases α, the radiance factor ρ as a function of α along the equator. *Broken curves:* extrapolations; *dotted curve:* curve for $\alpha = 40°$ corrected according to the reciprocity principle.

The table of Orlova may be used also to construct the *indicatrix* of the lunar surface. This is a polar diagram representing r or ρ as a function of the direction of the scattered light. A set of such curves, relating to the coefficient r for the mountain regions, is reproduced in Figure 9 (Orlova, 1956). The preponderance of back radiation over forward radiation is striking and is found to increase together with the angle of incidence.

4.2. Comparison with Theoretical Models

Many attempts have been made to analyze the penetration of light-rays into a plane layer of a diffusing solid substance, taking into account both scattering and absorption. The older theories were oversimplified; much better results have been reached recently. We enumerate some of the laws proposed (Schoenberg, 1929).

Lambert: $\rho = \rho_0 \cos i$. Minnaert has shown that this law may be expected to apply near the terminator (cf. van Diggelen, 1951).

Lommel-Seeliger: $\rho = C \cos i/(\cos i + \lambda \cos \epsilon)$; for the derivation, only the primary scattering was taken into account, and the scattering of a volume element was assumed to be isotropic. In the derivation of this formula, the introduction of the coefficient λ is unjustified, the reciprocity principle requires $\lambda = 1$.

Lommel and Schoenberg (1929, p. 38) and Fessenkov (1916) have tried to include also the second-order scattering as well as anisotropic scattering (see Fig. 10). Their results are very complicated. Markov used Fessenkov's formula in simplified form:

$$\frac{\cos i}{\cos i + \cos \epsilon} (1 + \cos^2 \alpha) .$$

Chandrasekhar (1950, especially chap. vi) succeeded in calculating the effect of higher-order scattering, the elementary phase function in each volume element being either isotropic or proportional to $1 + \chi \cos \theta$, or to $1 + \cos^2 \theta$, or to still more complicated functions.

Pokrowski (1925) and Berry (1923) have introduced an entirely different model, by considering the scattering medium as a swarm of minute little mirrors, placed at random. Their formula is

$$\rho = b \cos \epsilon + \frac{a}{2} F(i, n) .$$

All these formulae fail to describe the lunar observations. This has led to the working hypothesis that the photometric properties of the moon are determined primarily by the *shadows* of millions of surface irregularities and that the precise form of the diffuse reflection law is relatively

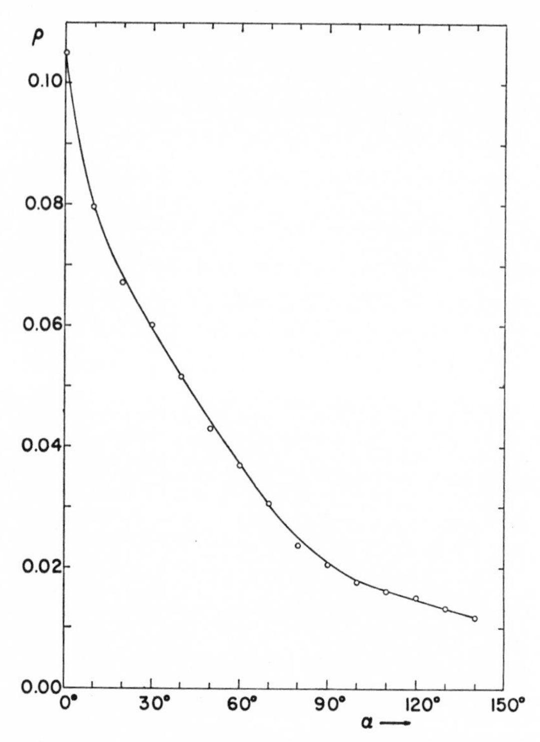

Fig. 8.—Radiance factor ρ near the illuminated limb, as a function of the phase α (based on Fig. 7).

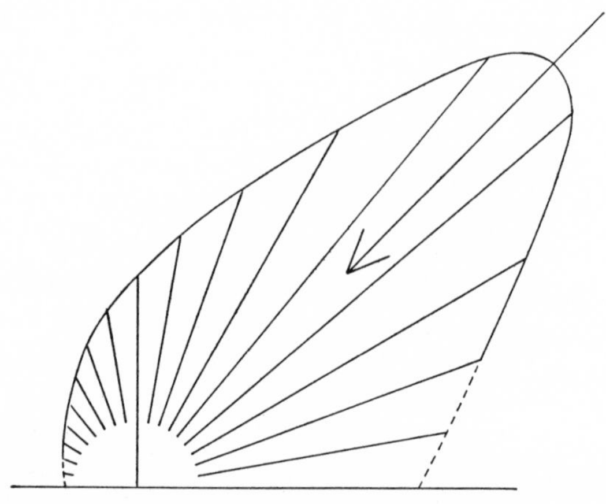

Fig. 9.—Indicatrix of the lunar surface (terrae) in the equatorial plane (Orlova, 1956)

unimportant compared with the effects of surface relief. The mean radiance under different conditions of illumination has been calculated for several model surfaces, among which the following are the most interesting: plane surface, covered with (*a*) cones (Schoenberg, 1925); (*b*) hemispherical elevations (Schoenberg, 1925; Markov and Barabashev, 1925, 1926; (*c*) spherical domes (Schoenberg, 1925); (*d*) hemispherical holes (Barabashev, 1924*a*; Schoenberg, 1925; Markov, 1952, p. 71); (*e*) cylindrical holes (Barabashev, 1924*a*; Schoenberg, 1925; Markov and Barabashev, 1925, 1926); (*f*) grooves, either parallel or in random orientation (Barabashev, 1922; criticized by Schoenberg, 1925, p. 7; Markov, 1923); (*g*) hemi-ellipsoidal excavations (Bennett, 1938, improved by van Diggelen, 1959, p. 70).

The photometric behavior of many of these theoretical models has been investigated only for special conditions of illumination and observation. The result appears to be that a model with hemispherical or hemi-ellipsoidal excavations gives the best representation of the observations. The maximum brightness at full moon, and quick decline toward small phase angles is at once understandable from such a model (Fig. 11).

A systematic method for arriving at a satisfactory model was followed by Tschunko (1949). He investigated the relative photometric contributions of surface elements inclined to the normal at given angles and de-

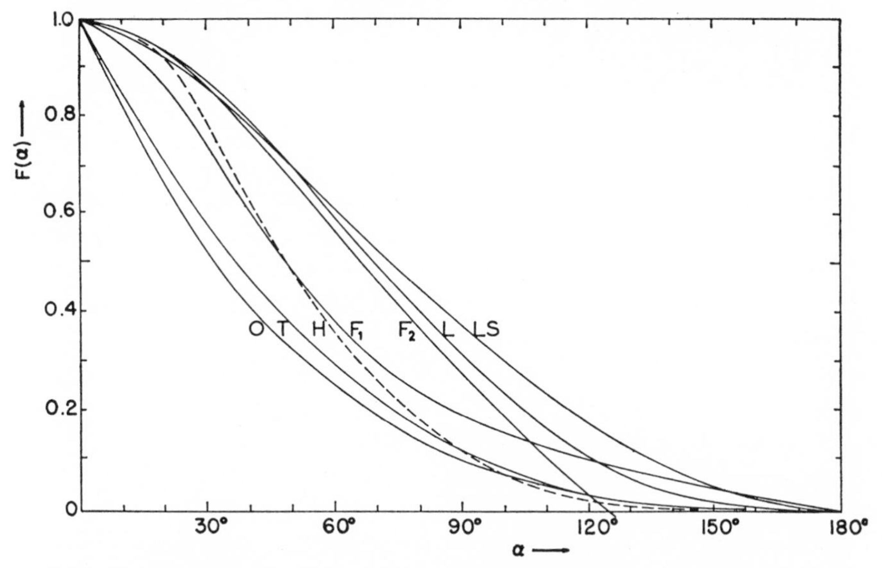

FIG. 10.—Integrated phase-curve of the moon, compared with different photometric formulae (Tschunko, 1949). L = Lambert; L-S = Lommel-Seeliger; F_1, F_2 = Fessenkov; T = Tschunko; O = observed; H = heat radiation of the moon.

duced the frequency distribution of the slopes. He was thus able to draw the profile of the typical lunar surface irregularity, which he found to approximate a hemispherical elevation. Other authors have found, however, that models with elevations do not have the required properties, and it is suggested that Tschunko's procedure might have been applied to excavations instead of elevations. The photometric function of the plane layer on which the elevations are built or in which the holes are excavated plays only a minor role, as may be seen from calculations by Markov and Barabashev (1925, 1926).

The photometric behavior is, of course, independent of the absolute size of the irregularities. Holes of 1 meter in diameter will give the same

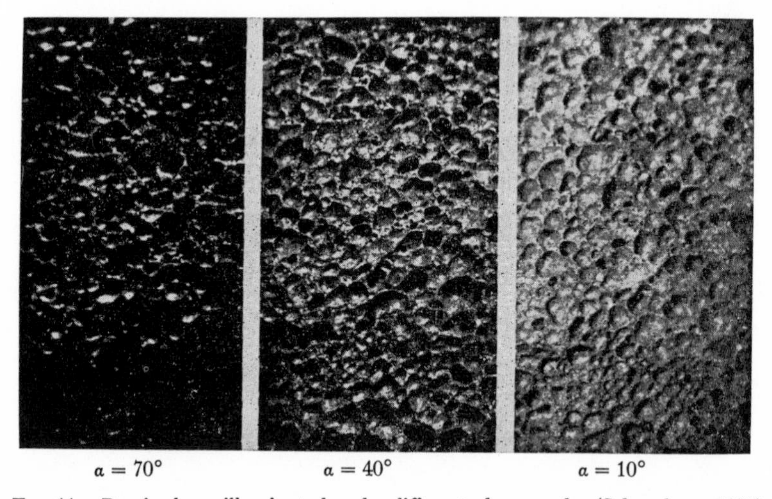

$\alpha = 70°$ $\alpha = 40°$ $\alpha = 10°$

FIG. 11.—Pumice-lava, illuminated under different phase angles (Schoenberg, 1925)

over-all photometric properties as holes of 1 mm, as long as the irregularities are much larger than the wave length of light. The calculations made thus far take into account only primary shadows and perspective effects but not reflection by illuminated parts on neighboring dark areas. If reflection is included, different photometric functions will correspond to the same surface model, depending on the albedo of the material. It is true that the effect of the slopes and the corresponding ratio between bright and dark areas are independent of the albedo; but the relative contribution by regions in the primary shadow will be important only for high albedo. Little attention has yet been given to this effect.

The interpretation of the photometric properties of local areas should lead also to an understanding of the phase-curve of the moon as a whole. This approach is of interest also for the study of Mercury, the asteroids, and the satellites, where only the integrated effects can be observed. Calculations of the integrated phase-curves for spheres covered with various types of surface irregularities would therefore be of interest. Figure 10 shows how far some theoretical laws are able to explain the integrated phase-curve.

A simplification was attempted by Zöllner (1865), who hoped that the integrated phase-curve of a corrugated sphere would be almost the same as that of a cylinder with similar corrugations and with its axis perpendicular to the plane containing the light-source, the object, and the observer. For the moon this assumption seems quite good. If the isophotes may be considered as meridians (p. 225), the integrated phase function will be the same as that of a strip along the equator, which in turn will be proportional to the phase function of a cylinder. With this assumption the integrated phase function has been computed for (*a*) a cylinder covered with contiguous prismatic ridges, parallel to the axis (Zöllner, 1865; errors in the integration corrected by Searle in 1884; and Seeliger in 1886; cf. Müller, 1897, p. 77); (*b*) the same if the tops of the ridges are truncated (Zöllner, 1865); (*c*) a cylinder covered with contiguous grooves with vertical walls and horizontal bottoms, the width being in a given ratio to the depth (Riyves, 1952).

In comparing the photometry of the moon with model *c* (on the assumption that there are smooth strips between the grooves), Riyves uses an interesting approach. He assumes that the walls of the grooves have hardly any influence when the phase angle exceeds 110°; for larger phase angles the radiation is due to the strips between the grooves, and the phase-curve may be assumed to correspond to that of a Lambert sphere. If this concept is now applied to the moon, Lambert's law yields the contribution of the smooth areas for the smaller phase angles. Subtracting this from the observed lunar radiation, we obtain the contribution of the "grooves" themselves, which may be compared with the theoretical model. Riyves' result is that the observed lunar phase-curve is satisfactorily explained if about two-thirds of the lunar surface is covered with grooves which are slightly deeper than wide. It is noted, however, that the model is an artificial one. The calculations will have to be repeated for various other models before the best schematic representation of the mean lunar surface is known.

4.3. Comparison with Terrestrial Materials

The empirical derivation of the phase-curve of a terrestrial sample is at present made more readily than the theoretical investigation of a model. Moreover, complicated and variable surface features are accessible experimentally, the influence of multiple reflections is automatically taken into account, and there is a greater chance that the natural conditions are simulated. The objects thus investigated are of two kinds: (*a*) simple geometrical forms and (*b*) natural samples. A brief survey of the available literature follows.

a) Hemispherical holes (Wilsing and Scheiner, 1921); parallel cylinders lying on a plane surface (Wilsing and Scheiner, 1921); cracks with sloping walls (Fedoretz, 1952); cylindrical holes (Fedoretz, 1952); cone-shaped pits (Fedoretz, 1952); spheres lying on a plane (van Diggelen, 1959); glass beads on a plane (van Diggelen, 1959); humps (van Diggelen, 1959); hemi-ellipsoidal holes (van Diggelen, 1959).

We recall the importance of the albedo of the material used. Albedo effects are found in the experiments by Wilsing and Scheiner (1921, p. 65). The albedo effect is apparent if white and nearly black models are compared: but for albedos not over 15 per cent the effects are negligible.

Van Diggelen's (1959) measurements on a very dark model, with hemi-ellipsoidal holes, gave excellent agreement with the photometric properties of the lunar surface; only it was necessary to subtract from the observed intensity the contribution due to the level areas between the holes. The surface of the moon is apparently an assembly of closely packed holes of all sizes, superposed and juxtaposed, excavated in a dark material. These model experiments confirm the results obtained by theoretical calculations (p. 231).

The model with opaque spheres, lying on a plane, does not match the photometric properties of the lunar surface (van Diggelen, 1959). The lunar surface is therefore not covered with a layer of stones. If the spheres are transparent, as with glass beads, special properties are encountered which are known from the use of projection screens. The brightness maximum at full phase is more pronounced here than for any other terrestrial material and matches that of the moon. It is not completely excluded that glass beads cover parts of the moon, such as crater rays, and that they affect their photometric behavior.

b) The natural objects investigated are very numerous. Of special interest are materials which may be found on the lunar surface. Most measurements have been made for the case where the planes of incidence and observation coincide ($A = 0°$ or $180°$). The studies include the following: quartz sand of different grain size, granite, gneiss, amber, selenite, apatite, hematite, diabase, fine gravel (Orlova, 1952); volcanic ashes (van Diggelen, 1959); rubber sponge, pumice, gravel (Rougier, 1933); river sand, sandstone, volcanic tuff (Fedoretz, 1952); and a layer of the lichen *Cladonia* (van Diggelen, 1959).

Attempts have been made to represent the photometric function of a material by analytical expressions. Sytinskaya (1946, 1949) investigated objects under full-phase conditions ($i = \epsilon$) and measured the brightness dependence on $i = \epsilon$ between the normal and the direction of illumination and observation. The radiance distribution is represented by $B = B_0 \times \cos^q i$, where q is a measure of the surface smoothness. It was then found that

$q > 1$ for polished surfaces;

$q = 1$ for an ideal orthotropic surface, following Lambert's law; for photometric screens, values of 0.85 to 0.92 are found;

$q < 1$ for most natural objects (rocks, 0.1–0.6; sand, 0.3–0.5; gravel and pitted surfaces, 0.1–0.2; vegetation, sometimes negative down to − 0.14, never above 0.4);

$q = 0$ for the moon.

None of the rocks or sands investigated had the low lunar value.

Sharonov (1940) introduced another classification scheme. The sample is illuminated under 45° with the normal, and the radiance is measured in the plane of incidence as a function of ϵ. This gives the *indicatrix.* The following classes of objects are considered: (i) the radiance is almost equal in all directions; (ii) it is greater around the direction of specular reflection; (iii) it is greater in the direction toward the source; (iv) the indicatrix has two maxima, one back to the source and one near specular reflection.

Examples of such measurements are given by Orlova (1952). They may be compared with the lunar indicatrix (p. 229 and Fig. 9). No terrestrial materials are known for which the indicatrix is so elongated toward the source as for the moon. Sharonov uses the name "specular factor" for the ratio between the radiance in the direction of specular reflection and that in the direction of the source. Van Diggelen's attempt to reproduce the optical properties of the lunar surface by a thick layer of the loosely ramified lichen *Cladonia rangiferina* was particularly successful. As a result of the smallness of the lunar gravitational force, very loose surface formations may be formed, with properties similar to those of our model. The true structure may be intermediate between this and the hemi-ellipsoidal cups.

It should be stressed that no model or terrestrial material can be considered as reproducing the photometric properties of the lunar surface in the absence of tests under many varying conditions of illumination, as determined by the values of i, ϵ and α. For such tests, van Diggelen selected four representative points of the moon, for which the radiance had been well measured over the lunation. This is certainly the minimum number of areas that should be compared.

5. THE NORMAL ALBEDO

5.1. Lunar Formations

The normal albedo, as defined on page 218, can be found only for objects near the center of the full moon. However, we have seen (p. 224) that at full moon the lunar formations probably have a radiance that is independent of their position on the disk ($f = 1$ when $i = \epsilon$). In any case, we may tentatively assume that the radiance ρ_0 of the individual lunar formations near full moon corresponds to their normal albedo.

Sytinskaya (1953*a*) has published a catalogue of 104 normal visual albedos, derived from the best available data and reduced to absolute values. She finds mean values of ρ_0 for different types of formations, distinguished mainly by topography and brightness, as follows:

Dark plains (maria)	0.065
Brighter plains (paludes)	.091
Mountain regions (terrae)	.105
Crater bottoms	.112
Bright rays	.131
Brightest spot (Aristarchus)	.176
Darkest spot (inside Oceanus Procellarum)	0.051

Her catalogue was extended by the photographic measurements of van Diggelen (1959, p. 52), reduced to the same scale. Ezerskaya and Ezerski (1959) showed that the measurements of Fedoretz (1952) may also be very well reduced to her system.

The values obtained by Sytinskaya correspond to full moon, but not to "true full moon." Fedoretz (1952), observing just before the partial phase of an eclipse, succeeded in obtaining results for a phase angle down to 1°5. The necessary correction for the finite value of the phase angle cannot be found from direct observation, but it is certainly not negligible, since the radiance increases so steeply near the opposition, and it will in general be different for the individual objects. Orlova (1954) derived the reduction from graphical extrapolation toward phase angle zero and published a catalogue of improved values of the albedos; the correction on the albedo may amount up to 0.027.

Lunar photographs seem to show two main types of surface: the dark maria and the brighter terrae. However, Sytinskaya (1953*a*), from the statistical distribution of ρ_0, has found all values between 0.07 and 0.13 to be almost equally frequent; with an extension up to 0.18 for small areas, probably to be attributed to crater rays at full moon (Fig. 12). Similarly, by a statistical treatment of the full-moon measures of Fessenkov, Parenago, and Staude (1928), Minnaert has found that the albedo of the terrae is greatest in extended terrae and smallest for regions dispersed between the maria; this suggests that there are many transitions between the two main types.

Roth (1949) has given a catalogue of 433 very bright objects on the full moon. Interesting are Graff's (1949) photometric data on crater rays of Tycho.

5.2. Terrestrial Materials

Measurements of the albedo of terrestrial rocks were made by Wilsing and Scheiner (1909, 1921), and several other series of such measurements have been published since. Of the recent investigations, we mention the work of Radlova (1943), of Sytinskaya (1949), and especially that of Sharonov (1954*b*), who used the extensive measurements of Budnikova and Borissova (1953). Assuming that the lunar surface rocks are abrased

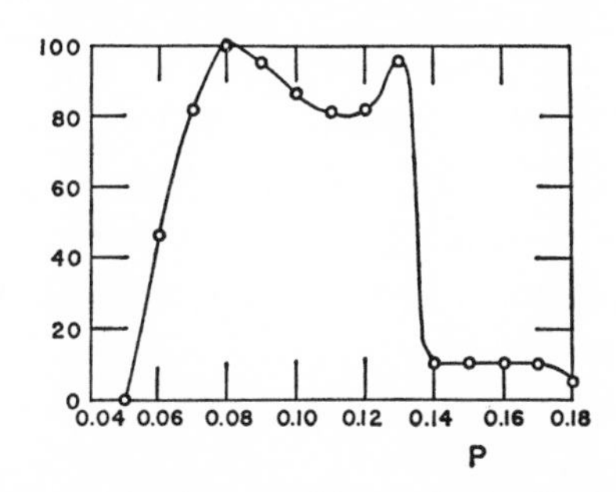

Fig. 12.—Statistical distribution of lunar albedos (Sytinskaya, 1953*a*)

by the frequent impact of small meteorites, Sharonov investigated only newly broken rock surfaces. Moreover, since the optical properties of any type of terrestrial rock vary from sample to sample, he measured a number of samples of each type and constructed frequency-curves for the albedo as well as for the color index. He subdivided his rocks into nine classes: carbonates, conglomerates, fine-grained rocks, basic magmatic rocks, neutral magmatic rocks, acidic magmatic rocks, metamorphic rocks, meteorites, and melted crusts of meteorites. For none of these classes was there a sufficient resemblance, in albedo and color, to the frequency-curves of the lunar formations—which show a much smaller variability in optical properties. This may simply be due to the fact that any observable area of the moon already contains numerous samples, of which the measures give the mean.

Certain other complications must be taken into account:

a) Direct comparisons between the apparent normal albedos of a lunar formation and a terrestrial object are invalidated if the first has an irregular surface while the second is smooth. Lunar material, if plane, would have a higher albedo than it now has. It is easily found that a flat surface, following Lambert's law, would have an albedo 50 per cent larger than that of a hemispherical hole. Tschunko (1949) has calculated for his model that the normal albedo of the moon would increase to about 0.20 if the surface were plane, with local variations between 0.10 and 0.40. The effect of the pits on the surface brightness should always be borne in mind.

b) The thermal data on the lunar surface indicate that it is covered by pulverized material in powder form. From terrestrial experience we know that pulverization gives rise to total reflection at the boundaries of the grains; this increases the albedo and decreases the color saturation, thus working in the opposite sense from the decrease by pitting mentioned under *a*. It is probably still justifiable to assume that lunar surface rocks are dark; however, quantitative comparisons become very difficult. It will be necessary to investigate the photometric and colorimetric properties of powders derived from pulverized rocks.

c) The comparison of the colors is examined in Section 6.

In summary, the outstanding result, reached by all these photometric investigations, is the great *similarity* of the photometric functions of all lunar formations, even when their albedos are as *different* as those of the maria and the terrae. It would be impossible to understand this result if the photometric properties were determined by diffusion and absorption within two different solid materials having smooth, plane surfaces. An explanation becomes possible if the photometric properties are due to microforms and their shadows. We must assume, then, that these microforms are the *same* all over the moon but that they have been molded from materials of *different* composition and albedo—in particular, different between the maria and the terrae. For dark materials (multiple reflections negligible), differences of a factor of 2 in the albedo are not expected to modify the photometric function appreciably. Such structures could be formed by the impact of meteorites and micrometeorites, striking the different parts of the surface equally and forming innumerable small pits. It would also explain why the polarization of the moon corresponds to that of a layer of volcanic ash (cf. chap. 9), while the microscopic photometric properties do not. Whether the polarization due to a layer of fine dust, as formed by meteoritic impact, is the same as that of volcanic ash remains to be checked.

In the case of the crater rays, especially interesting photometric properties appear, such as the increase in brightness near full moon. The same is found in a number of bright craters, which at full moon stand out as luminous specks. To the same category belong the bright sector on the bottom of Plato, the bright patch around Linné, the bright zone on the bottom of Ptolemy, etc. (Markov, 1952, p. 70). Wilsing and Scheiner (1921) have interpreted these features as surfaces covered by crystalline salt, which could have been deposited there by volcanic exhalations. This assumption does not explain the observations. Another hypothesis would be to assume a layer of glass beads (O'Keefe, 1957); however, van Diggelen's (1959, p. 93) laboratory measurements on such a model do not give very satisfactory agreement.

6. COLORS OF LUNAR FORMATIONS

Measurements of the color of the moon as a whole have usually been made by measuring its albedo in two spectral regions, yellow and blue, either by visual, photographic, or photoelectric means, and by determining the color index. The image of the moon was usually reduced optically to the size of a stellar image. We mention the following results:

	Color Index (Mag.)
Russell (1916)	+1.18
Rougier (1937)	+1.10
Radlova (1939)	+1.08
Radlova (1943)	+1.07

From these data we must subtract the color index of the sun, which, according to Russell, is +0.79 mag.; its uncertainty affects the result for the color index of the lunar surface, which is about 0.30 mag. It is clear that a direct comparison between sun and moon would have advantages; this has been made by Sytinskaya (1943) and Sharonov (1952, 1953, 1955). Sharonov's latest results (1955) give the difference, moon − sun, as +0.332 mag. Teyfel (1958*a*) finds a difference of +0.28 mag., which he considers to be on the low side. All authors confirm the reddish color.

First inspection, especially in a reflector, shows the lunar surface to be without striking color differences. Photographic tests include qualitative comparisons between exposures taken through two or three different filters; but such comparisons are questionable unless the plate gradations are known to be independent of color and exposure time. This remark applies, among others, to the colored map of the full moon by Miethe and Seegert (1911), derived from a comparison between photographs at 3300–3600 A and at 5900–6300 A. The color differences were displayed in a two-color print, the ultraviolet image being printed green and the orange image red. The authors give a list of "red" areas, among which Mare Humorum was the most pronounced. More recently, similar two-color prints have been published by Barabashev (see Sharonov 1958, p. 359). The main results were confirmed by Rosenberg (1921) when he compared his visual measurements with the photographic results of Götz. Wood

(1910, 1912) obtained photographs in the ultraviolet (with a silver filter transmitting at 3230 A), violet, and yellow, which disclosed a spot near Aristarchus, described by the author as almost black on the ultraviolet image and attributed by him to a surface layer of sulfur. However, Keenan's (1931) measurements yielded a difference of only 20 per cent between the reflecting power of this spot in the ultraviolet and in the infrared. Wright (1929), who obtained a series of fine lunar images in six colors, made further qualitative comparisons, while Barabashev (1952, 1953) made color photographs. Considering the lack of pronounced colors on the moon, it is understandable that Wright emphasized the need for *quantitative* measures.

In principle, the measurement consists in the determination of the radiance coefficient ρ for at least two colors; but the necessary precision can be reached only with difficulty. The first systematic investigation was begun in 1909 by Wilsing and Scheiner, first in five spectral regions and later (1921) in ten regions between 5410 and 6420 A. They found that the lunar surface has a yellowish ("brownish") hue, consistent with the integrated color index. Most terrestrial rocks investigated showed the same tinge; among these, volcanic ash and river sand closely match the lunar surface. The Potsdam measurements were later reduced and revised by Sharonov (1956).

The following color measurements on a number of specific objects have been made.

a) With a blue wedge: Fessenkov (1929), Radlova (1941, 1943). This method, though not exact, might be very sensitive to small color differences.

b) By photographic or visual photometry in several spectral regions, in general selected by color filters: Barabashev (1924*b*), 3 regions; Markov (1950*a*, *b*), photoelectrically in 2 regions; Ezerski and Fedoretz (1956), 7 regions; Barabashev and Chekirda (1956), 5 regions; Teyfel (1958*a*), 2 regions; Barabashev and Ezerski (1959), crater bottoms in 2 regions.

c) By photographic photometry across the spectrum: Teyfel (1958*c*), Vigroux (1956), Barabashev, Ezerski, and Fedoretz (1959), Polozhentseva (1960).

In general, the albedo is found to increase monotonically toward the red. Still most authors conclude, from their results, that a color index from two wave lengths only does not give a satisfactory description; this refers especially to the deviations which the curves for some objects show in the ultraviolet and infrared.

It is generally assumed that the colors do not vary with the phase. This assertion, based mainly on some results of Danjon (1937, p. 146), and recently confirmed by Polozhentseva, has not been tested in more detail.

The most important color catalogues of lunar objects, judging by the number of measured objects, are those of Radlova (1941, 1943) and Barabashev and Chekirda (1954). Very generally, the mean color index C is found to increase with the albedo (except for the bright crater bottom Aristarchus and some other details): the continents are redder than the maria. The ratio $\Delta C/\Delta \log I$ is found to be 0.5 (Teyfel, 1958*a*) or 0.6 (Radlova, 1943). Teyfel (1958*b*) explains this by assuming that the surface of the moon is covered by two substances, mixed in various proportions, for which the ratio $C/\log I$ differs.

An interesting question concerns the color differences between the individual objects. These are mostly determined from the ratio ρ_1/ρ_2 of the albedos in two colors, which the observer compares with a certain standard: either the value of that ratio at one definite point of the moon, or the mean value for the whole moon, or the value for a white screen, illuminated by the sun. The results show surprising discrepancies. On the one hand, Barabashev, Markov, Chekirda, Ezerski, and Fedoretz find considerable variations from one object to the other, of the order of 0.25 mag. on the international scale. In the careful relative measurements of Polozhentseva, contrasts of the order of 15 per cent are found. On the other hand, Sharonov, Radlova, Sytinskaya, and Vigroux find only slight differences, amounting perhaps to 0.05 mag.

Barabashev, Ezerski, and Fedoretz (1959) defend their standpoint by noting that the albedo may vary over $\Delta \log \rho = 0.55$, and they conclude that this will probably correspond to a color difference of $\Delta C = 0.55 \times 0.5 = 0.27$ mag., which is about what they had measured.

On the other hand, Radlova (1957) compared six color catalogues of lunar objects and showed that no pair of catalogues presenting a satisfactory agreement could be found. Radlova and Sharonov (1958) also inserted an artificial object inside the lunar image and found a color difference of 0.1 mag. to be immediately visible, while $\Delta C = 0.01$ mag. is about the threshold for color distinction; from the absence of directly visible color differences, they conclude that ΔC cannot much exceed 0.05 mag., which corresponds to $\Delta T_c = 300°$. However, these measurements apply to the color contrast between local details and their surroundings; it seems probable that the eye will be less sensitive if the areas compared are farther apart.

Sharonov (1958), by careful inspection of the lunar image, noticed a yellowish spot to the northeast of Aristarchus, the brownish hue of the central parts of Mare Serenitatis, etc. Firsoff (1958) makes similar observations. Kuiper (1956), with the 82-inch telescope, reports several limited regions which are distinctly yellower than the remainder of the moon, among these the same region near Aristarchus; numerous small white craters, located within maria; and parts of the shore lines of Sinus Roris. However, the much greater differences, found by some authors, will have to be attributed to errors in photographic photometry.

A special problem is the comparison of the color of some large craters with that of the bright crater rays ejected from them (Barabashev and Chekirda, 1955). In the vicinity of the crater the color is found to be the same as that of the crater bottom; but, farther on, it rather suddenly becomes bluer. For Copernicus this change occurs at distances of from 0–113 km from the crater; for Kepler at 133–235 km; for Tycho at 332–837 km. The authors suggest that acid exhalations of the crater have modified the material in the neighborhood.

For comparison with terrestrial rocks, it is useful to take into account Krinov's (1940) classification of minerals. He distinguishes five albedo classes, according to the coefficient r near 5500 A, and four color types according to the ratio r_{6500}/r_{4000}. Radlova (1943) finds fair agreement between lunar objects and meteorites, but this is not confirmed by the subsequent investigation of Sharonov (1954*b*). As long as the measurements of the moon are not in better agreement with each other, such comparisons have little value. Moreover, in all comparisons between the colors of the moon and terrestrial rocks, we have to keep in mind the following factors:

a) The pulverization of the lunar rocks will tend to deprive them of their color; this in itself is probably the explanation of the general grayness of the lunar surface.

b) Terrestrial rock samples are mostly modified by atmospheric effects, unless fresh surfaces are studied.

c) Lunar rocks may have been modified by exhalations, solar ultraviolet radiation, or cosmic rays.

d) Impurities may change considerably the color of a mineral.

Sharonov (1954*b*) tries to avoid these difficulties by studying many samples of each rock. Just as for the albedo, he finds a much smaller variation in the color on the moon than for the earth; we are inclined to attribute this also to the averaging effect in lunar observations. Larger color differences might be discovered by observations in the ultraviolet and infrared.

Still better is the method of Sytinskaya (1957*a*), who plots in a color-albedo diagram the representative points for a great number of terrestrial rocks, for meteorites (1953*c*) for the lunar surface. From this simultaneous comparison of two properties, she finds that the lunar rocks do not correspond to any of the groups of terrestrial minerals; in the diagram they occupy a small area only, quite outside the other fields. She concludes that the lunar surface must be continually pitted by micrometeorites and supposes that these small explosions melt the rocks locally and build a dark crust which would be the surface of the moon (1954). This does not seem acceptable; there is ample evidence that the exterior layers must be porous and covered by dust.

Barabashev and Chekirda (1959) have melted different kinds of rocks, in the air and in vacuum, and have shown that the optical properties of the melting crust are very different from these of the moon. Irradiation by ultraviolet rays or roentgen rays, bombardment by protons, had only a small influence. Of course, the density, heat conductivity, luminescence, reflection of radio waves, and fluorescence should also be taken into account.

Provisionally, we are inclined to picture the lunar surface as very porous (photometric properties), all the details being covered by a thin layer of dust (polarimetry).

A satisfactory identification of lunar surface rocks will probably require a simultaneous comparison of photometric properties (albedo and phase function), color properties, and the polarization-curves. Of these three, the color has probably the least significance, unless measurements in regions outside the visual spectrum should show more striking features.

7. LUNAR PHOTOMETRY AS AN AID TO TOPOGRAPHY

When the lunar plains are illuminated by a low sun, surface details and wavelike ridges become surprisingly conspicuous. The slightest surface slope results in a considerable deviation in brightness. In these conditions of illumination it would probably be legitimate to apply Lambert's cosine law. However, it is still simpler to determine the surface brightness of the surrounding mare as a function of the distance to the terminator and thus derive the relation between radiance and slope. A simple integration then yields the height at each point. By this method, van Diggelen (1951) was able to derive the profiles of the low ridges in Mare Imbrium, which proved not higher than 100–200 meters; altitude differences of 10–20 meters could already be detected.

This method has not yet been used widely but might be applied to other formations.

8. POSSIBLE FLUORESCENCE OF THE LUNAR SURFACE

Because of the absence of a lunar atmosphere, the solid surface of our satellite is exposed to the full intensity of ultraviolet and corpuscular radiations from the sun. Could this irradiation produce fluorescence phenomena? They are shown under similar conditions by many terrestrial rocks.

Two methods of investigation have been applied; both are said to have given positive results.

From our knowledge of the upper atmospheric layers, obtained by rocket techniques, it is possible to compute the brightness distribution in the shadow of the earth, taking into account the refraction and the absorption by ozone and by dust (Link, 1956, 1958). A comparison with photometric measurements during lunar eclipses shows that in 73 per cent of the cases *an excess of light* is observed in the penumbra, at a distance of a few minutes of arc from the boundary of the umbra. This is precisely the zone, illuminated by the chromosphere and the lower corona, from which ultraviolet and corpuscular radiation might be expected. The variations in the amount of these radiations would explain why the effect is not always equally well pronounced; moreover, the different parts of the lunar surface cannot be supposed to show equally strong fluorescence. Link (1947) has even tried to correlate the excess of brightness with the fluctuations in the solar constant. The effect was confirmed by the observations of Cimino (1953, 1955, 1958) and his collaborators, as also by Fortini (1954). As a striking example we mention that the brightness of Mare Crisium, compared photoelectrically with an adjacent area, was found to show an excess up to 2 magnitudes at certain stages of a lunar eclipse.

Fluorescent light is emitted in broad bands, characteristic for the irradiated solid. If such a fluorescence occurs, the spectrum of the moon should correspond to the spectrum of the sun on which a continuous radiation is superposed in some specific spectral regions. This means that the intensity distribution

$$\frac{i}{i_0} = f(\Delta\lambda)$$

within a Fraunhofer line should be modified into $(i + \epsilon)/(i_0 + \epsilon)$, which means that the line becomes shallower. The effect must be most pronounced inside very deep spectral lines. Dubois (1956) has shown by a laboratory experiment that this is really the case and has tried to discover the effect on the moon. On about half of the lunar areas investigated a positive result was obtained, especially on 87 per cent of the maria and on 100 per cent of the crater bottoms; a map of the fluorescent lunar regions has been drawn (Dubois, 1959).

Several fluorescence bands were found, different according to the region investigated (Dubois, 1956, 1959); the fluorescence there may reach 10–20 per cent of the intensity of the solar radiation. The effect was not always equally strong, which may be attributed to variations in the solar activity. Dubois emphasizes that a small amount of luminescent material mixed with ordinary rocks in, say, a proportion of 1 per cent could be detected, owing to the low albedo of the moon.

At about the same time, similar results were obtained by Kozyrev (1956). However, he finds fluorescence effects only on Aristarchus, especially after Full Moon; spectral maxima occur near λ 3900 and λ 4300. From a quantitative estimate, he concludes that the solar ultraviolet would not be strong enough to explain the observed fluorescence. However, a corpuscular stream of 5.10^3 protons cm^{-3} with a velocity of 1500 km/sec would be sufficient, which seems acceptable when the sun is disturbed. Also this author notices considerable variations with time.

It is clear that the results here reported are only preliminary. Further confirmation and detailed research would be most interesting.

REFERENCES[2]

Barabashev, N. P. 1922 *A.N.*, **217**, 445.
1924*a* *Ibid.*, **221**, 289.
1924*b* *Russ. A.J.*, **1**, 44 (No. III–IV).
1927 *Pub. Kharkov Obs.*, **1**, 35.
1952 *Astr. Tsirk. Akad. Nauk U.S.S.R.*, No. 127, p. 5.
1953 *Circ. Astr. Obs. Kharkov*, No. 12, p. 3.
Barabashev, N. P., and Chekirda, A. T. 1954 *Trudy Obs. Kharkov*, Vol. **3**, No. 11.
1955 *Circ. Astr. Obs. Kharkov*, No. 13.
1956 *Russ. A.J.*, **33**, 549; *Circ. Astr. Obs. Kharkov*, No. 15, p. 9.
1959 *Russ. A.J.*, **36**, 851.
Barabashev, N. P., and Ezerski, V. I. 1959 *Astr. Tsirk. Akad. Nauk U.S.S.R.*, No. 205, p. 9.
Barabashev, N. P., Ezerski, V. I., and Fedoretz, V. A. 1959 *Russ. A.J.*, **36**, 496.
Bennett, A. L. 1938 *Ap. J.*, **88**, 1.
Berry, E. M. 1923 *J. Opt. Soc. America*, **7**, 627.
Budnikova, N. A., and Borissova, A. P. 1953 *Vestnik U. Leningrad*, No. 8, p. 81.
Bullrich, K. 1948 *Ber. Deutsch. Wetterd.*, *U.S.-Zone*, No. 4.
Calder, W. A. 1937 *Harvard Ann.*, **105**, 445.
Chandrasekhar, S. 1950 *Radiative Transfer* (Oxford: Clarendon Press).

[2] Russian publications which the writer was able to consult are included; some additional papers are summarized in Sharonov (1954*a*, 1958) and in the monograph of V. P. Dzhapiashvili, *Bull. Ap. Obs. Abastumani*, **21**, 1957.

CIMINO, A. A.	1953	*Rend. Lincei*, Ser. VIII, **14**, 619.
	1955	*Ibid.*, Ser. VIII, **18**, 173.
	1958	*Ibid.*, Ser. VIII **25**, 58.
DANJON, A.	1937	*Ann. Obs. Strasbourg*, **3**, 139.
DIGGELEN, J. VAN	1951	*B.A.N.*, **11**, 283.
	1959	*Rech. Obs. Utrecht*, Vol. **14**, No. 2, and Dissertation, Utrecht.
DUBOIS, J.	1956*a*	*C.R. Soc. Franç. Phys.*, October 25.
	1956*b*	*Astronomie*, **70**, 225.
	1958	*Ibid.*, **72**, 267.
	1959	*Rozpr. Česk. Akad. Věd.*, *Rada M.P.V.*, Vol. **69**, No. 6.
EZERSKAYA, V. A., and EZERSKI, V. I.	1959	*Astr. Tsirk. Akad. Nauk U.S.S.R.*, No. 205, p. 11.
EZERSKI, V. I., and FEDORETZ, V. A.	1955	*Astr. Tsirk. Akad. Nauk U.S.S.R.*, No. 159, p. 18.
	1956	*Circ. Astr. Obs. Kharkov*, No. 15, p. 17.
FEDORETZ, V. A.	1952	*Pub. Kharkov Obs.*, Vol. **2**, and *Uch. Zap. Kharkov U.*, **42**, 49.
FESSENKOV, B.	1916	*Bull. Soc. Astr. Russ.*, Vol. **22**.
	1928	*Russ. A.J.*, **5**, 219.
	1929	*A.N.* **236**, 7.
FESSENKOV, B., and PARENAGO, P.	1929	*Russ. A.J.*, **6**, 279.
FESSENKOV, B., PARENAGO, P., and STAUDE, N.	1928	*Pub. Astr. Inst. Russ.*, **4**, 1.
FIRSOFF, V. A.	1958	*Sky and Telescope*, **17**, 329.
FORTINI, T.	1954	*Atti Lincei*, Ser. VIII, **17**, 209.
GÖTZ, P. W. P.	1919	*Veröff. Sternw. Oesterberg-Tübingen*, Vol. **1**, No. 2.
GRAFF, K.	1949	*Sitz. Österr. Akad. Wiss.*, Ser. II*a*, **157**, 19; also *Mitt. Sternw. Wien*, Vol. **4**, No. 6.
KEENAN, P. C.	1931	*Pub. A.S.P.*, **43**, 203.
KOZYREV, N. A.	1956	*Izvest. Krymsk. Astr. Obs.*, **16**, 148.
KRINOV, E. L.	1940	*Russ. A.J.*, **17**, 40 (No. 4).
	1947	*Spektralnaya otrazhatelnaya sposobnost prirodnych obrazovanii* (Akad. Nauk U.S.S.R.).
KUIPER, G. P.	1946	Personal communication.
LINK, FR.	1947	*Colloque C.N.R.S.*, *Lyon*, p. 308.
	1956	*Die Mondfinsternisse* (Leipzig: Akadem. Verlagsgesellsch.).
	1958	*Bull. Astr. Inst. Czech.*, **9**, 169.
LOMMEL, E.	1887	*Sitz. Akad. München*, **17**, 95.
	1889	*Wied. Ann. d. Phys.* **36**, 473.
MARKOV, A.	1923	*A.N.*, **221**, 65.
	1924	*Ibid.*
	1927*a*	*Ibid.*, **231**, 57.

	1927*b*	*Russ. A.J.*, **4**, 60.
	1948	*Ibid.*, **25**, 172.
	1950*a*	*Izves. Akad. Kazakhstan*, No. 90, p. 92.
	1950*b*	*Bull. Abastumani*, **11**, 107.
	1952	*Pulkovo Izvest.*, **19**, 64 (No. 149).
MARKOV, A., and BARABASHEV, N.	1925	*A.N.*, **226**, 129.
	1926	*Russ. A.J.*, **3**, 55.
MIETHE, A., and SEEGERT, B.	1911	*A.N.*, **188**, **9**, 239 and 371.
	1914	*Ibid.*, **198**, 121.
MINNAERT, M.	1941	*Ap. J.*, **93**, 403.
MÜLLER, G.	1897	*Die Photometrie der Gestirne* (Leipzig: Engelmann).
NIKONOVA, E K.	1949	*Izvest. Crimean Obs.*, **4**, 114.
O'KEEFE, J. A.	1957	*Ap. J.*, **126**, 466.
ÖPIK, E.	1924	*Pub. Astr. Obs. Tartu*, Vol. **26**, No. 1.
ORLOVA, N. S.	1952	*Uch. Zap. U. Leningrad*, No. 153, p. 166.
	1954	*Vestnik U. Leningrad*, **9**, 77.
	1955	*Astr. Tsirk. Akad. Nauk U.S.S.R.*, No. 156, p. 19.
	1956	*Russ. A. J.*, **33**, 93.
	1958	*Astr. Tsirk. Akad. Nauk U.S.S.R.*, No. 192, p. 20.
PETTIT, E.	1926	*Ap. J.*, **81**, 33.
PICKERING, E.	1882	*Selenograph. J.*[3]
POKROWSKI, G. J.	1925	*Zs. f. Phys.*, **32**, 563.
POLOZHENTSEVA, T. A.	1960	*Pulkovo Izvest.*, **21**, 180
RADLOVA, L. N.	1939	Quoted in SHARONOV (1958), p. 282.
	1941	*Uch. Zap. U. Leningrad*, No. 82, p. 99.
	1943	*Russ. A.J.*, **20**, 1 (No. 5–6).
	1957	*Astr. Tsirk. Akad. Nauk U.S.S.R.*, No. 179, p. 7.
RADLOVA, L. N., and SHARONOV, V. V.	1958	*Russ. A.J.*, **35**, 788.
RIYVES, B. G.	1952	*Pub. Astr. Obs. Tartu*, **32**, 129.
ROSENBERG, H.	1921	*A.N.*, **214**, 137.
ROTH, H.	1949	*Sitz. Österr. Akad. Wiss.*, IIa, **157**, 25; also *Mitt. Sternw. Wien*, Vol. **4**, No. 7.
ROUGIER, M. G.	1933	*Ann. Obs. Strasbourg*, **2**, 205.
	1937	*Ibid.*, **3**, 257.
RUSSELL, H. N.	1916	*Ap. J.*, **43**, 103.
SCHOENBERG, E.	1925	*Acta Soc. Fenn.*, Vol. 50.
	1929	*Hdb. d. Ap.*, **2**, Part 1, 72.
SCHRÖTER, J. H.	1791–1802	*Selenographische Fragmente* (Göttingen: Lilienthal).
SEARLE, A.	1884	*Proc. Amer. Acad. Arts and Sci.*, **19**, 310.
SEELIGER, H.	1886	*Vierteljahrsschr. Astr. Gesellsch.*, **21**, 216.
	1888	*Sitz. Akad. München*, **18**, 201.

[3] This paper is rather inaccessible. A good summary is found in G. Müller (1897, p. 345).

SHARONOV, V. V.	1936	*Uch. Zap. U. Leningrad*, No. 6, p. 26.
	1940	*Ibid.*, No. 53, p. 5.
	1952	*Ibid.*, No. 153, p. 114.
	1953	*Russ. A.J.*, **30**, 532.
	1954*a*	*Uspekhi Astr. Nauk*, **6**, 181.
	1954*b*	*Russ. A.J.*, **31**, 442.
	1955*a*	*Astr. Tsirk. Akad. Nauk U.S.S.R.*, No. 157, p. 19.
	1955*b*	*Vestnik U. Leningrad*, No. 11, p. 113.
	1956*a*	*Ibid.*, No. 1, p. 156.
	1956*b*	*Astr. Tsirk. Akad. Nauk U.S.S.R.*, No. 166, p. 9.
	1958	*Priroda Planet* (Moscow).
SYTINSKAYA, N. N.	1943	*Doklady Akad. Nauk U.S.S.R.*, **38**, 28.
	1946	*Bull. U. Leningrad*, No. 6, p. 16.
	1948	*Absolutnaya fotometria protyazhennykh nebesnykh obyektov* (Leningrad: Zhdanov University Press).
	1949	*Uch. Zap. U. Leningrad*, No. 116, pp. 123 and 138; also *Trudy Leningrad Obs.*, Vol. **13**.
	1953*a*	*Russ. A.J.*, **30**, 295.
	1953*b*	*Astr. Tsirk. Akad. Nauk U.S.S.R.*, No. 144, p. 11.
	1953*c*	*Ast. Tsirk. Akad. Nauk U.S.S.R.*, No. 136, p. 22.
	1954	*Trudy Obs. Stalinabad*, **20**, 106; see also *Astr. Tsirk. Akad. Nauk U.S.S.R.*, No. 153, p. 17.
	1957*a*	*Uch. Zap. U. Leningrad*, No. 190, p. 74; also *Trudy Leningrad Obs.*, Vol. 17.
	1957*b*	*Russ. A.J.*, **34**, 899.
	1959	*Ibid.*, **36**, 315.
SYTINSKAYA, N. N., and SHARONOV, V. V.	1952	*Uch. Zap. U. Leningrad*, No. 153, p. 114; also *Trudy Leningrad Obs.*, Vol. **16**,
TEYFEL, V. G.	1958*a*	*Astr. Tsirk. Akad. Nauk U.S.S.R.*, No. 192, p. 21.
	1958*b*	*Ibid.*, No. 194, p. 11.
	1958*c*	*Ibid.*, No. 196, p. 5.
TSCHUNKO, H. F. A.	1949	*Zs. f. Ap.*, **26**, 279; also *Mitt. Sternw. Heidelberg*, No. 74.
VIGROUX, E.	1956	*J. Obs.*, **39**, 134.
WILSING, J., and SCHEINER, J.	1909	*Pub. Astr. Obs. Potsdam*, Vol. **20**, No. 61.
	1921	*Ibid.*, Vol. **24**, No. 77.
WISLISCENUS-WIRTZ, C.	1915	*A.N.*, **201**, 289.
WOOD, R. W.	1910*a*	*Pop. Astr.* **18**, 67.
	1910*b*	*M.N.*, **70**, 226.
	1912	*Ap. J.*, **36**, 75.
WRIGHT, W. H.	1929	*Pub. A.S.P.*, **41**, 125.
ZÖLLNER, F.	1865	*Photometrische Untersuchungen* (Leipzig: Engelmann).

CHAPTER 7

Photometry of Lunar Eclipses

By D. BARBIER
Institut d'Astrophysique, Paris

1. INTRODUCTION

THE eclipsed moon does not disappear completely but remains faintly luminous. Usually, observers assign to the center of the shadow the color of red brick, while the edge is described as gray or greenish gray. The brightness of the moon at the center of the shadow, which varies from one eclipse to another and also with the wave length, is of the order of 2×10^{-5} of that of the full moon in green light.

The light of the eclipsed moon comes from the solar rays that are refracted and weakened in passing through the earth's atmosphere. Seen from a point on the moon at the center of the shadow, the earth appears as a disk whose maximum radius is $61'.5$. The sun's radius is $16'$, and twice the horizontal refraction at the earth's surface is about $73'$ under average conditions. As a result, the sun's limb is raised by refraction to $27'$ beyond that of the earth. Hence the moon remains illuminated by the sun even under the most unfavorable conditions. The terrestrial atmosphere is made luminous by the scattering of solar light. One may ask whether this light causes an appreciable illumination of the moon. The brightness of the twilight sky in the direction of the sun is of the order of the full moon's brightness or 5×10^{-6} times the brightness of the sun. This atmosphere, limited to 8-km altitude, the height of an equivalent homogeneous one, is seen from the moon as a ring of $60'$ radius and $0'.08$ width. Its area is 30 square minutes of arc, which corresponds to one-thirtieth of the solar disk. The light scattered by the atmosphere therefore produces an illumination $\frac{1}{30} \times 5 \times 10^{-6}$ or 2×10^{-7}, which is about one-hundredth of the brightness observed in green light. Only at short wave lengths can this light become appreciable in comparison with the direct light which is strongly absorbed by the earth's atmosphere for the central regions of the shadow. Link (1950) has made a more detailed computation of the influence of scattering.

Let us consider which regions of the atmosphere affect the brightness of a point on the moon just on the edge of the geometric shadow. The point of the sun farthest from the earth's rim is distant 32′ from it. In order to illuminate the moon, that point has to be raised by an angle ω, which appears, according to Table 1, to be the value of the double horizontal refraction for 7-km altitude. At the edge of the shadow, no ray reaching the moon has passed through the atmosphere at a lower elevation. A ray from the center of the sun has to be raised by 16′ and must

TABLE 1*

MEAN OPTICAL PROPERTIES OF EARTH'S ATMOSPHERE FOR HORIZONTAL RAYS WHICH TRAVERSE IT AT ALTITUDE h

h (km)	m	ω	h (km)	m	ω
0	78.4	73′.5	22	3.25	4′.03
2	60.6	55.5	24	2.38	2.95
4	47.5	44.9	26	1.76	2.18
6	37.0	36.4	28	1.28	1.62
8	28.4	30.3	30	0.94	1.16
10	21.4	25.1	32	0.69	0.85
12	15.6	19.6	34	0.51	0.62
14	11.4	14.2	36	0.38	0.46
16	8.40	10.3	38	0.28	0.33
18	6.09	7.56	40	0.21	0.24
20	4.45	5.56			

* m = air mass; ω = angle of refraction.

pass at 13 km above the earth's surface. The rays which cause the illumination of the moon at the edge of the shadow have therefore crossed the atmosphere at rather high altitudes, which are practically cloudless. The center of the shadow receives light from rays which have gone through the lower atmosphere, well known from direct observations, and the principal phenomenon affecting the illumination of the moon at this point is the obstruction caused by clouds on the terrestrial limb.

It is clear that the parts near the edge of the geometric shadow are the most interesting because they give information about the upper atmosphere. The results may be compared with theoretical values which for the cloudless upper atmosphere can safely be derived.

For any point the *density* of the shadow is defined as the logarithm of the ratio of the intensity, I_0, of the non-eclipsed full moon to the intensity, I, of the eclipsed moon,

$$\delta = \log I_0/I \,. \tag{1}$$

Here δ is usually expressed either in decimal logarithms or in magnitudes. This definition raises a problem which has apparently never been mentioned before. The non-eclipsed full moon, outside the penumbra, can be observed only at phase angles larger than $1^\circ\!.5$. But, according to photometric measures, especially Rougier's (1933), the moon's albedo changes for small angles very rapidly with the phase; if the moon could be observed at zero phase, it might perhaps be 10 per cent brighter than the full moon. Shadow densities which are of importance for the theory, i.e., densities for which I_0 and I correspond to the same phase, are not accessible to observation and are presumably greater than those given by the observers. The excess brightness of the moon in the penumbra observed by Link (1947*b*) is possibly due to this effect. The complete history of the physical observations of lunar eclipses and their interpretation will not be given. Early investigations are discussed by Dyson and Woolley (1937).

2. METHODS OF OBSERVATION

2.1. General Remarks

Some of the older observations measured the total magnitude of the eclipsed moon. This information is vague, and we now prefer measuring the shadow density at one or more well-defined points. The precision of the photometry is limited by the following circumstances:

a) The measurement must be made at a point whose position is sufficiently well defined with respect to both the lunar surface and the shadow of the earth during eclipse. As this shadow moves rapidly over the lunar surface, the measurement must be of short duration.

b) When filters are used, they should have a narrow pass band because the energy distribution in the spectrum varies greatly from point to point in the shadow and differs from the distribution in the spectrum of the non-eclipsed moon even in the gray regions. Measures with filters having wide pass bands will introduce serious errors if treated as monochromatic.

c) The difference in brightness between the eclipsed and the non-eclipsed moon is large and difficult to compensate for.

d) During the partial phases, the penumbral part is incomparably brighter than the eclipsed part. Scattering in the atmosphere and the instrument may vitiate the measures if not suppressed or allowed for.

We shall not attempt to examine or catalogue all the observations of lunar eclipses that have been made. Instead we take some typical examples, indicating the advantages or drawbacks of the methods used.

2.2. Visual Photometry

Danjon (1921) inaugurated a long series of observations, using a double-image photometer (Danjon, 1931; Rougier and Dubois, 1942, 1943; Dubois, 1949, 1950, 1952, 1953). Except for the 1921 series, all these measures were made with a "cat's-eye" photometer and three Wratten filters: red (W25A), green (W58), and blue (W44A). The measurement consists in placing the two images of the moon produced in the photometer in exterior contact and equalizing them near the contact points. One of the points thus equalized is at a distance d_1 from the center of the shadow

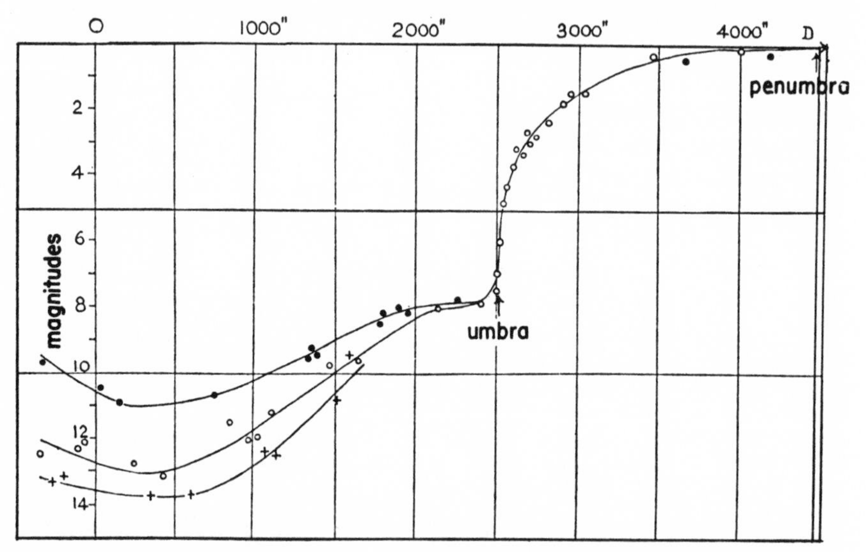

Fig. 1.—Light-curve for lunar eclipse obtained with a double-image photometer by Rougier and Dubois. • = red; o = green; and + = blue. Entry into the penumbra and into the geometrical shadow is indicated by arrows.

and the other at a distance d_2. The measure furnishes the value of $a_1 I(d_1)/a_2 I(d_2)$, where a_1 and a_2 are the albedos of the moon at the two points considered. The ratio a_1/a_2 is measured when the moon is not eclipsed, so that the value of $I(d_1)/I(d_2)$ is found. When the preceding limb, which was at distance d_1, reaches the distance d_2, the following limb is at a distance d_3, and another measure is made, giving $I(d_2)/I(d_3)$. Step by step, starting from the non-eclipsed moon, one reconstructs the whole curve of shadow density $I_0/I(d)$. Figure 1 shows the density-curve thus obtained at the eclipse of March 2–3, 1942.

This technique has the important advantage of completely eliminating the effect of atmospheric absorption and the influence of scattered light. Furthermore, the equalization of the images is instantaneous, and there is no uncertainty about the positions of the measured points in the shadow. Finally, the measures can be made with a small instrument. The only weak point in the method is the hypothesis that the shadow density depends on the distance to the center only and that it does not vary with time.

2.3. Photoelectric Photometry

In photoelectric photometry a fairly large telescope is needed, in order to observe sufficiently small lunar areas. Cuffey (1952) used the 36-inch Indiana reflector. As in the visual measures, one has the advantage of instantaneous operation, but it is impractical to measure the scattered light. For this, one would have to measure successively the moon and the sky, and too much time would be involved in resetting the telescope. It does not seem, therefore, that, for the present, photoelectric methods will supersede other techniques.

2.4. Photographic Photometry

The greatest advantage of photographic photometry is that the shadow density can be recorded simultaneously at numerous points of the lunar disk. The greatest difficulty, however, is the limited range of the plate, which cannot measure at one step intensity ratios larger than 20 or 30. Link (1948*b*) has used the method successfully and found not only that the density of the shadow depends on the distance to its center but that at a given point the density may vary by a factor of 2 in 10 minutes.

2.5. Photographic Spectrophotometry

Compared with the photographic method using filters, photographic spectrophotometry has the advantage of giving intensity measures at well-defined wave lengths. Furthermore, absorption bands of oxygen and ozone—the principal absorbing gases of the atmosphere—can be recognized in the spectrum. If the moon is projected on the slit with a short-focus objective, one can study on each spectrum several points of the shadow simultaneously, but the striae crossing the spectrum, corresponding to the accidental features of the lunar surface, make it necessary to obtain microphotometer tracings only of regions of nearly uniform structure. This requires a short slit, which is a disadvantage. This is the process used by Barbier, Chalonge, and Vigroux in 1942 and has led to the curves shown in Figure 2.

If the telescope has a long focal length, one can measure only one point at a time, but the measurement becomes easier. This method was used by Chalonge and his collaborators to secure the spectra studied by Vigroux (1954).

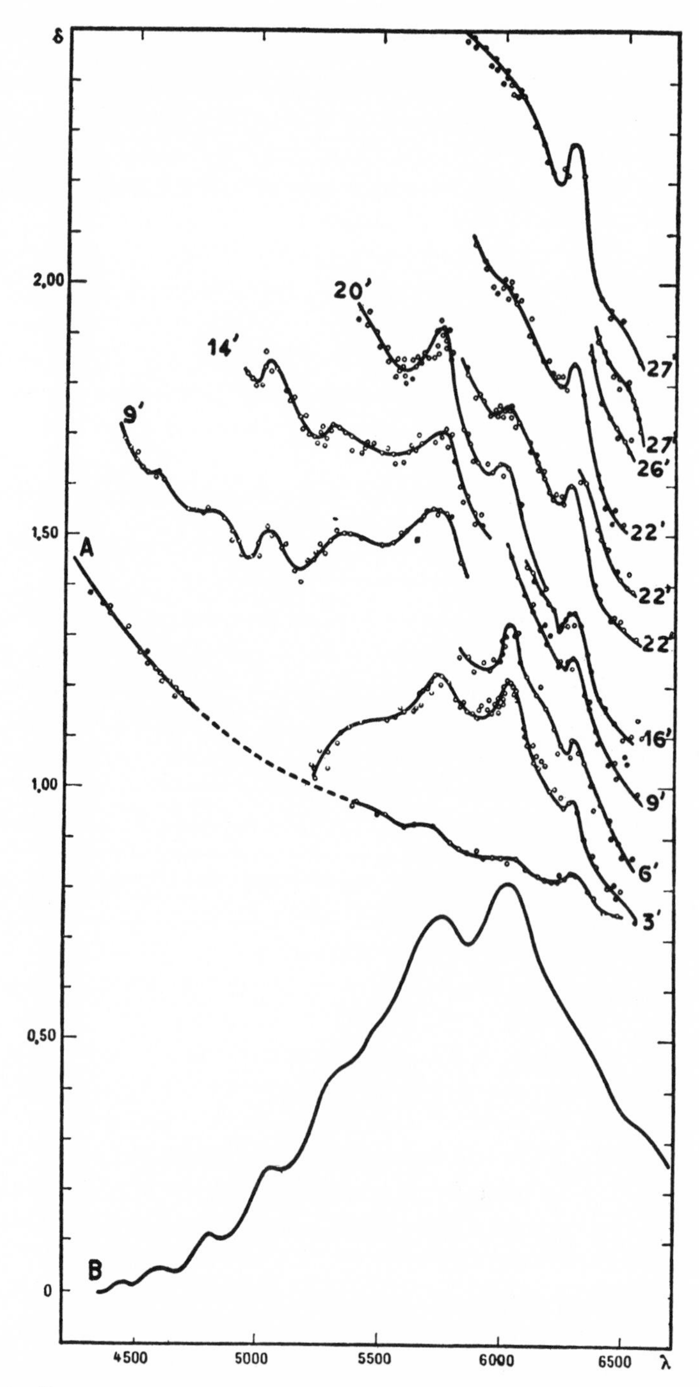

FIG. 2.—*Upper curves:* intensity distribution in spectrum of the eclipsed moon. Numerals indicate angular distance from edge of shadow. *A:* optical density of terrestrial atmosphere from measures of Sirius near horizon. *B:* optical density of 10-cm layer of pure ozone. Observations by Barbier, Chalonge, and Vigroux.

With an $F/3.5$ spectrograph, the exposure time for the edge of the shadow is about 2 minutes in the visible—the most interesting spectral region; it is much longer for the center, so there is some uncertainty as to the location of the measured points in the shadow, the latter having moved during the exposure. During the partial phases one cannot take into exact account scattered light from the atmosphere or in the instrument; Dufay (1931) estimated that, on his spectra obtained during the partial phase, the whole of the blue and 80 per cent of the red are due to scattered light.

Thus none of the methods is entirely satisfactory, and the method used will depend on the problem. The observational data will be discussed following a review of the theory of lunar eclipses.

TABLE 2*

NORMAL SHADOW DENSITIES AND MEAN AIR MASSES (CF. LINK)

γ'/λ	0.62	0.54	0.46	0.62	0.54	0.46
	Densities			Mean Air Masses		
0'....	2.35	2.48	2.71	7.3	5.7	4.1
2....	2.52	2.70	3.04	10.2	8.4	6.6
4....	2.63	2.85	3.28	13.0	11.0	8.6
6....	2.70	2.97	3.48	15.5	13.2	10.5
8....	2.75	3.07	3.67	17.8	15.3	12.4
10....	2.81	3.17	3.84	20.1	17.2	14.2
20....	3.04	3.59	4.64	30.0	26.8	22.9

* Only losses through refraction and molecular diffusion are taken into account. γ' is the distance from the edge of the geometrical shadow.

3. THEORY

The theory was established almost simultaneously by Link (1932), Fessenkov (1932, 1937), and de Saussure (1931). We shall follow Link's method, which is the most complete and the best adapted to computation with the help of the numerous tables given by this author. More recently Link has made new presentations with several additions (Link, 1948*a*, 1956). Some slight changes in Link's theory are introduced here, and in the comparison between theory and observation we have used an approach better adapted to spectra. Tables 1 and 2, taken from more complete tables given by Link (1932, 1948*a*) are valid for the mean lunar parallax of 57'. The results for points less than 20' from the edge of the shadow are hardly affected by variation of the parallax.

For a horizontal ray passing at a height h above the ground, let the refraction be ω and the air mass m. Let P_I (Fig. 3) be the plane of the

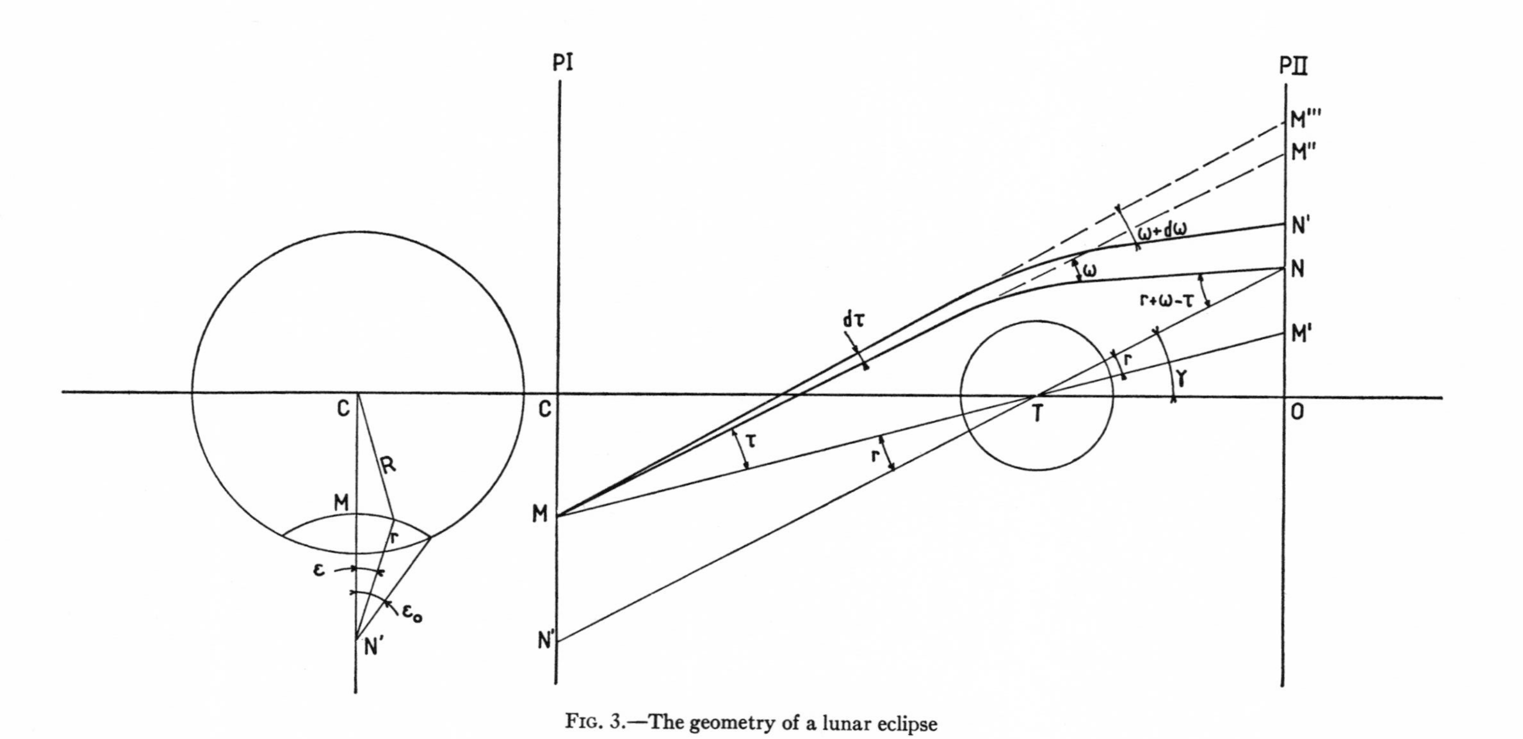

FIG. 3.—The geometry of a lunar eclipse

solar disk and P_{II} a parallel plane at the moon's distance on which the shadow-density distribution is to be found. C and T are the centers of sun and earth, O the center of the shadow.

Let the angular radius of the shadow be

$$\rho = p_M + p_S - R_S . \tag{2}$$

in which p_M and p_S are the parallaxes of the moon and the sun and R_S is the radius of the sun. The location of a point N of the shadow is characterized by its angular distance γ to the center or γ' to the edge, so that $\gamma' = \rho - \gamma$. The solar ray in the plane of the figure which arrives at N after passing at an altitude h above the ground[1] and undergoing a refraction ω intersects the plane P_I at M. This ray makes an angle τ with the direction MT. Let r be the angle between TN and the extension TM' of MT. From Figure 3, we see that

$$\tau L_S = a + h ,$$

$$(r + \omega - \tau) L_M = a + h ,$$

where L_S and L_M are the distances TC and TO and a is the radius of the earth.

Introducing the parallaxes of the sun and moon, we find

$$\begin{aligned} \tau &= \left(1 + \frac{h}{a}\right) p_S , \\ r &= \left(1 + \frac{h}{a}\right) (p_S + p_M) - \omega , \end{aligned} \tag{3}$$

where ω is a function of h.

All the points of the solar disk situated on a circle of radius r, center N' (*left side*), illuminate the point N after the rays have undergone the same refraction and absorption. Let $I(R)$ designate the solar brightness as a function of the distance to the center and e^{-D} the reduction of the radiation, including the loss by refraction, when passing through the earth's atmosphere. The density of the shadow will be

$$\delta = \log_e \frac{\int I(R)\, d\sigma}{\int I(R)\, e^{-D} d\sigma} , \tag{4}$$

in which $d\sigma = r dr d\epsilon$ is an element of the solar surface and D is independent of ϵ. The integrals extend over the entire solar disk. We have

$$R^2 = r^2 + \gamma^2 - 2 r \gamma \cos \epsilon ,$$

[1] We make no distinction here between h and h', the minimum altitude of the asymptotic rays. The difference is of no practical importance except at low altitudes. For these altitudes, Link takes account of the difference.

and, for a point on the solar limb,

$$R_S^2 = r^2 + \gamma^2 - 2\, r\gamma \cos \epsilon_0 \,.$$

The brightness of the sun's disk is given with sufficient precision by

$$I(R) = 1 - q + \frac{q}{R_S} \sqrt{(R_S^2 - R^2)}$$

$$= 1 - q + \frac{q}{R_S} \sqrt{(2\, r\gamma)}\, \sqrt{(\cos \epsilon - \cos \epsilon_0)}\,,$$

where q is the coefficient of darkening and a function of wave length. We then have

$$\int I(R)\, e^{-D} d\sigma = 2 \int_{\gamma - R_S}^{\gamma + R_S} e^{-D}\, r\, d\, r \int_0^{\epsilon_0} I(R)\, d\epsilon \,.$$

By numerical integration,

$$r \int_0^{\epsilon_0} I(R)\, d\epsilon = P(\gamma, r) - qQ(\gamma, r)\,.$$

The numerical values of P and Q as given in Link's tables should be multiplied by 2. We also have

$$\int I(R)\, d\sigma = \pi R_S^2 \left(1 - \frac{q}{3}\right),$$

giving

$$\delta = \log_e \pi R_S^2 \left(1 - \frac{q}{3}\right) - \log_e \int_{\gamma - R_S}^{\gamma + R_S} e^{-D} (P - qQ)\, d\, r\,,$$

or, since r is a function of h,

$$\delta = \log_e \pi R_S^2 \left(1 - \frac{q}{3}\right) - \log_e \int_{h_2}^{h_1} e^{-D} (P - qQ) \frac{d\, r}{d\, h}\, d\, h\,. \tag{5}$$

The limiting heights, h_1 and h_2, are found from the equations

$$\gamma + R_S = \left(1 + \frac{h_1}{a}\right)(p_S + p_M) - \omega(h_1)\,, \tag{6}$$

$$\gamma - R_S = \left(1 + \frac{h_2}{a}\right)(p_S + p_M) - \omega(h_2)\,. \tag{7}$$

The values of h_1 and h_2 are given in the accompanying table as functions

γ (min. of arc)	0	5	10	15	20
h_1 (km).....	37.8	20.7	16.2	13.7	11.9
h_2 (km).....	7.5	5.9	4.7	3.6	2.5

of the distance γ' from the edge of the shadow. Now we may calculate e^{-D}. Consider, first, the reduction through refraction of the light emitted by the point M between cones of aperture τ and $\tau + d\tau$. In the absence of an eclipse, this light will cover a ring on P_{II} between the circles $M'M''$ and $M'M'''$, of area

$$2\pi\,(L_S + L_M)^2\,\tau d\tau\ .$$

Taking refraction into account, the light will cover the ring between the circles of radius $M'N$ and $M'N'$, the area of which is

$$2\pi L_M^2\,r d r\ .$$

The reduction of light by refraction is

$$\Delta = \left(\frac{L_M}{L_S + L_M}\right)^2 \frac{r}{\tau}\frac{dr}{d\tau}$$

$$= \left(\frac{p_S}{p_S + p_M}\right)^2 \frac{r}{\tau}\frac{dr}{d\tau}.$$

But

$$\frac{d\tau}{dh} = \frac{p_S}{a}, \qquad \text{and} \qquad \frac{dr}{dh} = \frac{p_S + p_M}{a} - \frac{d\omega}{dh}.$$

So that

$$\Delta = \left[1 - \frac{\omega}{p_S + p_M}\left(1 - \frac{h}{a}\right)\right]\left(1 - \frac{a}{p_S + p_M}\frac{d\omega}{dh}\right). \tag{8}$$

The refraction ω changes little with wave length, so that Δ is nearly neutral. We write

$$D_1 = \log_e \Delta\ . \tag{9}$$

A second cause of diminution of light is extinction by Rayleigh scattering. We write

$$D_2 = mk\ , \tag{10}$$

the air mass, m, being expressed as the ratio to the vertical mass at the earth's surface under normal conditions. The absorption coefficient k varies very nearly as λ^{-4}.

A third cause of dimming is the absorption by atmospheric ozone. This has an absorption band—the Chappuis band—in the yellow-red part of the spectrum. Let D_3 be the density due to ozone absorption and possibly to other causes. Then

$$D = D_1 + D_2 + D_3\ ,$$

D_1, D_2, and D_3 being functions of h.

The shadow density may first be calculated by neglecting the ozone absorption. This absorption as seen on the moon is apt to vary along the earth's limb between wide limits on account of differences in the total amount of ozone and its distribution in altitude. The ozone absorption is therefore unknown in the regions of the atmosphere to be considered at the time of the eclipse. The integral equation (5) could, in principle, make known the function $D_3(h)$, but, as its solution would be laborious and the result affected by observational errors, we may be satisfied with the following procedure for an estimation of D_3.

Write equation (5) in the form

$$-\delta = \log_e \int_{h_1}^{h_2} e^{-D} F_\lambda(h)\, dh\,, \tag{11}$$

where

$$F_\lambda(h) = \frac{P - qQ}{\pi R_S^2 [1 - (q/3)]} \frac{dr}{dh}. \tag{12}$$

Figure 4 shows values of $F_\lambda(h)$ for different values of γ.

The first method of solution, used by Link, is to expand e^{-D_3} in powers of $m - \bar{m}$, m being the horizontal air mass at altitude h and $\bar{m}$ a mean value to be determined. Then

$$e^{-D_3} = e^{-D_3(\bar{m})} + (m - \bar{m}) \left(\frac{d\, e^{-D_3}}{dm} \right)_{\bar{m}} + \dots .$$

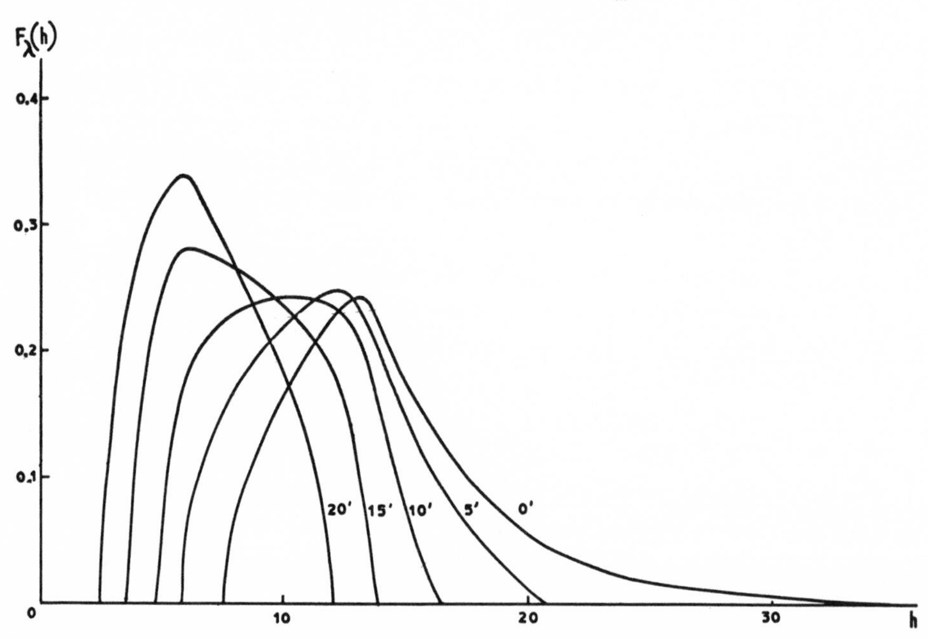

FIG. 4.—Values of $F_\lambda(h)$ for different distances from the geometrical shadow edge, measured for $\lambda = 6200$ A. Note that $F_\lambda(h)$ is practically independent of λ.

If we take

$$\bar{m} = \frac{\int_{h_1}^{h_2} m\, e^{-(D_1+D_2)} F_\lambda(h)\, dh}{\int_{h_1}^{h_2} e^{-(D_1+D_2)} F_\lambda(h)\, dh}, \tag{13}$$

we obtain, keeping only the first two terms of the expansion,

$$D_3(\bar{m}) = \delta - \int_{h_1}^{h_2} e^{-(D_1+D_2)} F_\lambda(h)\, dh. \tag{14}$$

This gives an estimate of the absorption by ozone for a horizontal ray as a function of the mean air mass or qualitatively as a function of the height. Then $D_3(\bar{m})$ is the residual between observation and a theory which neglects ozone absorption.

A second method consists in expanding e^{-D} in powers of $(h - \bar{h})$. We write

$$e^{-D} = e^{-D(\bar{h})} + (h - \bar{h}) \left(\frac{d\, e^{-D}}{dh} \right)_{\bar{h}} + \dots, \tag{15}$$

where

$$\bar{h} = \frac{\int_{h_1}^{h_2} h F_\lambda(h)\, dh}{\int_{h_1}^{h_2} F_\lambda(h)\, dh}.$$

The accompanying table gives the mean height $\bar{h}$ as a function of the distance γ' to the shadow edge. We then have

γ (min. of arc)	0	5	10	15	20
$\bar{h}$(km).....	14.4	11.9	9.9	8.2	6.7

$$D(\bar{h}) = \delta - \log_e \int_{h_1}^{h_2} F_\lambda(h)\, dh = \delta. \tag{16}$$

From the definition of $F_\lambda(h)$ the value of the integral is unity. This approximation cannot be very good because the variation of e^{-D} is far from linear with altitude; but it is sufficient to justify the process used by Barbier, Chalonge, and Vigroux (1942) and by Vigroux (1954) in interpreting the spectra. This is seen as follows.

Consider a small spectral interval in the region of ozone absorption.

The density of the shadow due to a mean air mass $\bar{m}$ and a mean optical thickness $\bar{\epsilon}$ of ozone is

$$\delta = k\bar{m} + k_{O_3}\bar{\epsilon} + \text{constant}\,; \qquad (17)$$

k_{O_3} is the absorption coefficient of ozone. The constant accounts for the weakening by refraction and a possible systematic error in the densities, which are difficult to measure in absolute value by spectrophotometry. Equations (16) and (10) lead to a form resembling equation (17):

$$km(\bar{h}) + k_{O_3}\epsilon(\bar{h}) + \text{constant} = \delta\,, \qquad (18)$$

so that the average values $\bar{m}$ and $\bar{\epsilon}$ determined by equation (17) can be equated to the values of $\bar{m}$ and ϵ corresponding to $\bar{h}$. The above result, of course, has a meaning only if the mean altitudes given by equation (15) are sufficiently precise. A verification is obtained from a discussion of the spectrophotometric observations (see Fig. 9). Another check is made by comparing the densities computed from Link's table for 0.46 μ with those computed from formula (16). The densities are given in decimal logarithms.

γ' (min. of arc)	0	5	10	15	20
δ (Link)....	2.71	3.38	3.84	4.24	4.64
δ (eq. [16])..	3.25	3.81	4.18	4.67	5.19

Formula (16) gives rise to a density error varying from 0.34 to 0.55. This error, which is large if the densities themselves are of interest, is equivalent to an underestimate of 2 km in height, which may be considered acceptable.

We have obtained the function $\epsilon(h)$ representing the reduced optical thickness of ozone encountered by a ray passing at a height h above the ground. What interests us most is the function representing the distribution of ozone as a function of the height z. We have (Fig. 5)

$$\epsilon(h) = \int_{-\infty}^{+\infty} f(z)\,dx\,,$$

but

$$x^2 = (a+z)^2 - (a+h)^2 \simeq 2a(z-h)\,;$$

hence

$$\epsilon(h) = \sqrt{(2a)}\int_h^{\infty} \frac{f(z)\,dz}{\sqrt{(z-h)}}\,,$$

an Abel equation, of which the solution is

$$f(z) = -\frac{1}{\pi\sqrt{2a}}\int_z^{\infty} \frac{d\epsilon}{dh}\,\frac{dh}{\sqrt{(h-z)}}\,. \qquad (19)$$

In practice the inversion of Abel's equation is not possible because, in the table of $\bar{h}$, the function $\epsilon(h)$ is not known beyond 14.4 km. Following Link (1932), it is preferable to try to represent the observations by a layer of ozone of uniform density f_0 between the limits H and $H + \Delta$, and zero density outside these limits, so that we have

$$\begin{aligned} \epsilon(h) &= 2 f_0 \sqrt{2a}\,[\sqrt{(H+\Delta-h)} - \sqrt{(H-h)}]\,, && \text{if } 0 < h < H\,, \\ \epsilon(h) &= 2 f_0 \sqrt{2a}\sqrt{(H+\Delta-h)}\,, && \text{if } H < h < H+\Delta\,, \\ \epsilon(h) &= 0 && \text{if } H+\Delta < h\,. \end{aligned} \quad (20)$$

The distribution of atmospheric ozone obtained by other methods leads to values of optical thickness for horizontal rays analogous to those found

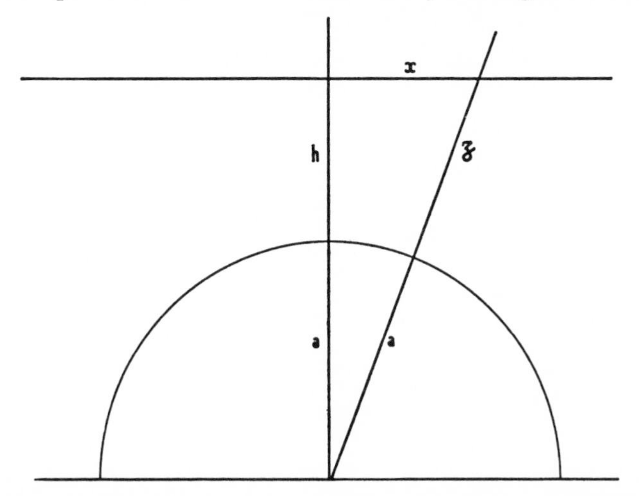

FIG. 5.—Representation of horizontal ray passing through layer at height h above the ground (see text).

by eclipse observers. Penndorf (1948) has made this comparison for the observations of Barbier, Chalonge, and Vigroux.

If h is much below the level H of the layer, the horizontal optical density of a thin layer is

$$\epsilon = f_0 \Delta \sqrt{\frac{2a}{H}}\,,$$

which is independent of h. The optical density measured vertically is

$$\epsilon_0 = f_0 \Delta\;;$$

so that we have, for the ratio of the horizontal and vertical densities, the simple expression

$$\frac{\epsilon}{\epsilon_0} = \sqrt{\frac{2a}{H}}\,. \quad (21)$$

Additions to the theory.—Link (1948*a*) has studied the effects of variations in atmospheric conditions on refraction and the resulting changes in the shadow density, extending the computations for the density up to the center of the shadow. Svetska (1949) has examined the influence of clouds on the central part of the shadow, and Link (1947*b*) has developed the theory of the penumbra.

The eclipse seen from the moon.—To interpret the results, we need to determine which parts of the terrestrial atmosphere are effective in illumi-

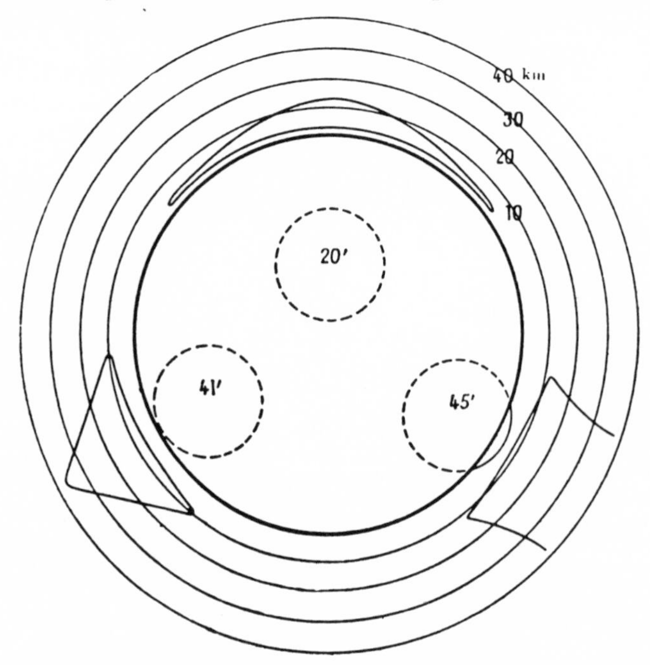

Fig. 6.—The eclipsed sun seen from the moon. Distances near the edge of the earth are multiplied by 100. The dotted circles indicate the true position of the sun (Link).

nating a given point on the moon. Seen from the moon, the center of the terrestrial disk is that point which has the moon at the zenith, which is found from the ephemeris. If we neglect parallax and refraction, the terrestrial limb is the great circle which has this point as its pole.

For a point on the moon at a distance γ from the center of the shadow at position angle P, the center of the sun is seen at a distance γ from the center of the earth's disk also in position angle P. Thus we can easily find the sun's position relative to points on the earth's limb. Link (1947*a*) gives further details. He has further determined the position of the solar disk as seen from the moon, refraction being taken into account. Figure 6 is taken from his publications (1932, 1948*a*).

4. INTERPRETATION OF THE OBSERVATIONS

4.1. Photometric Observations

Figure 7 represents the differences $O - C$ between the observed and computed densities, neglecting absorption by ozone; the observations are of the September 26, 1931, eclipse of the moon discussed by Link (1932). The measures in the blue are Link's (1931); the measures in green and red are by Danjon (1931). The differences $O - C$ are given as functions of the mean air masses as defined by Link (cf. eq. [13]).

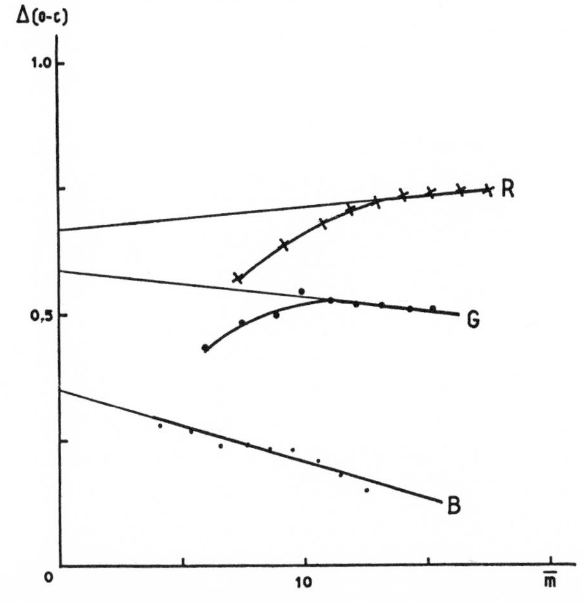

Fig. 7.—Difference between observed and calculated densities as a function of the mean air mass (Link). B = blue, G = green, R = red.

The observed densities are higher than the computed ones. The difference is greater in green light than in blue and still larger in the red. Further, the differences in the green and red become smaller when the mean air mass is less than 10. Link shows that this result can, for the most part, be explained by the existence of atmospheric ozone, which absorbs more in the red than in the green and for which absorption in the blue is negligible. The decrease for small air masses shows that the abundance of ozone decreases above 20 km. For this eclipse the numerical interpretation gives a horizontal thickness of about 8 cm of ozone, corresponding to forty times the normal vertical thickness, which is admissible.

The difference in the blue between the observed and calculated values cannot be attributed to atmospheric ozone, for which the absorption coefficient is very small in this spectral region. Link believes that it proves the presence of an absorbing layer high in the terrestrial atmosphere. Strictly speaking, it is possible to explain the slope of the line *B* (Fig. 7) by the uncertainty of some of the numerical data utilized, but the ordinate of 0.35 at the origin remains, and it is this value which represents the horizontal thickness of the layer. According to formula (21), if the layer is at 100-km altitude, the horizontal absorption is eleven times as great as at the zenith (eight times at 200 km). The density at the zenith would therefore be of the order of 0.03. The existence of such a layer has been suggested by Bauer and Danjon (1923) to account for their observations of the sun at the horizon; its density would be of the same order as the value deduced from lunar eclipses.

Link (1948*a*) has called attention to the following difficulties. Measures of absorption at the zenith are not consistent with a layer of optical density > 0.006. Furthermore, if such a scattering layer existed, the resulting twilight phenomena would be much more intense than those actually observed. It would be of interest to investigate more closely the causes of errors arising from the use of filters having wide pass bands.

The existence of a high absorbing layer is therefore not established with certainty. Paetzold (1952) concludes, however, that such a zone must exist, in order to account for the magnification of the terrestrial shadow. This magnification had been discovered long ago by Brosinsky and Hartmann, from a discussion of the times at which several lunar craters entered and left the shadow. It is also demonstrated by the photometric measures if we define the limit of the shadow by the location of the point of inflection in the light-curve (Fig. 1). Paetzold finds that the enlargement varies between 1.7 and 3 per cent, and he attributes this to a variation in thickness of an absorbing layer such as Link proposed, but he does not explain why observation of other kinds do not bring it out.

4.2. Spectrophotometric Observations

Ozone has absorption maxima at 6020, 5750, 5340, 5050, 4800, and 4600 A. Spectrophotometric observations (cf. Fig. 2) have established beyond doubt the presence of the absorption spectrum of ozone in the light of the eclipsed moon (Barbier, Chalonge, and Vigroux 1942). Also the bands B(6867 A) and α (6276 A) of oxygen have been observed. Vigroux (1954) established that the 5788 A band of oxygen can be detected during an eclipse, even though it is superimposed on the band 5750 A of ozone.

Figure 8, due to Barbier, Chalonge, and Vigroux (1942), shows the variations of intensity of the ozone bands and the α band of oxygen as a function of the air mass.

It is not possible to use all the observations obtained during one eclipse in order to deduce a single curve for the distribution of ozone with altitude. This was pointed out by Vigroux (1954), while Chalonge (1955) demonstrated it for the eclipse of January 29, 1953. At the beginning of the eclipse, the center of the moon was illuminated by rays having crossed the atmosphere over Canada and the United States; later they passed over Alaska, Mongolia, and finally Sumatra.

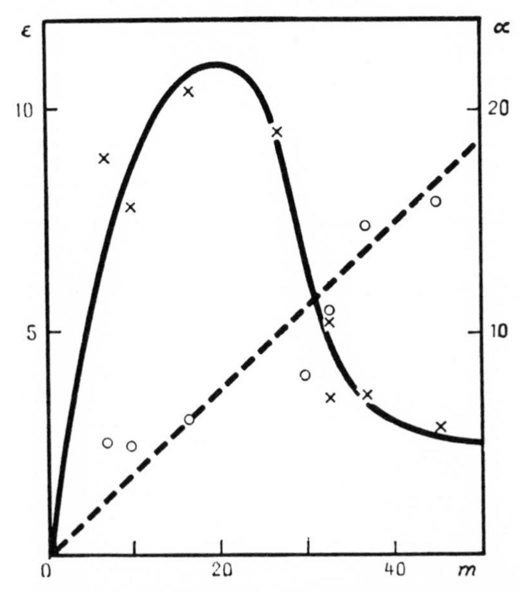

FIG. 8.—Variation of the optical thickness of ozone (*continuous line*) and of the α band of oxygen (*dotted line*), in terms of the air mass traversed, in centimeters (Barbier, Chalonge, and Vigroux).

As we have seen, equation (18) will furnish a mean value of the height of the rays from spectrophotometric data. Further, the theory gives a mean height from equation (15). Figure 9 shows the relation between the observed and computed values of $\bar{h}$, using observations by Barbier, Chalonge, and Vigroux (1942) and Vigroux (1954) made during totality. The agreement is very good except that the observed values tend to be 15 per cent greater than the computed ones. A single abnormal point relates to a 1942 observation for which the determination of the air mass was especially uncertain. The small systematic difference between the observed and computed altitudes is not serious, in view of the precision that one may expect.

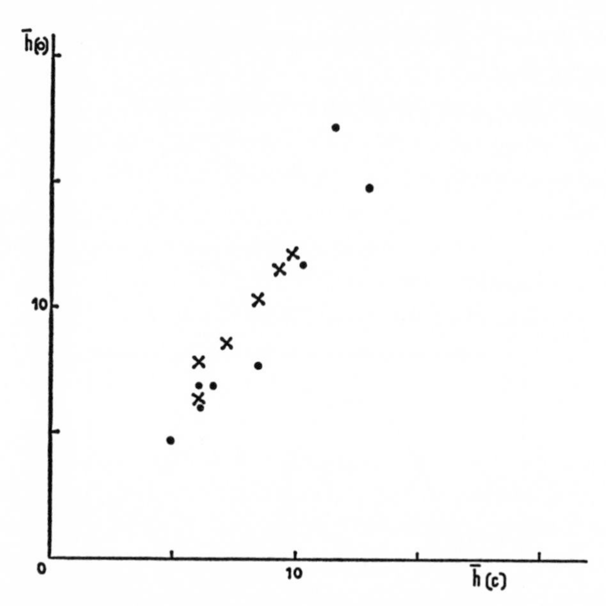

FIG. 9.—Relation between observed and calculated mean heights. • = Barbier, Chalonge, and Vigroux; x = Vigroux.

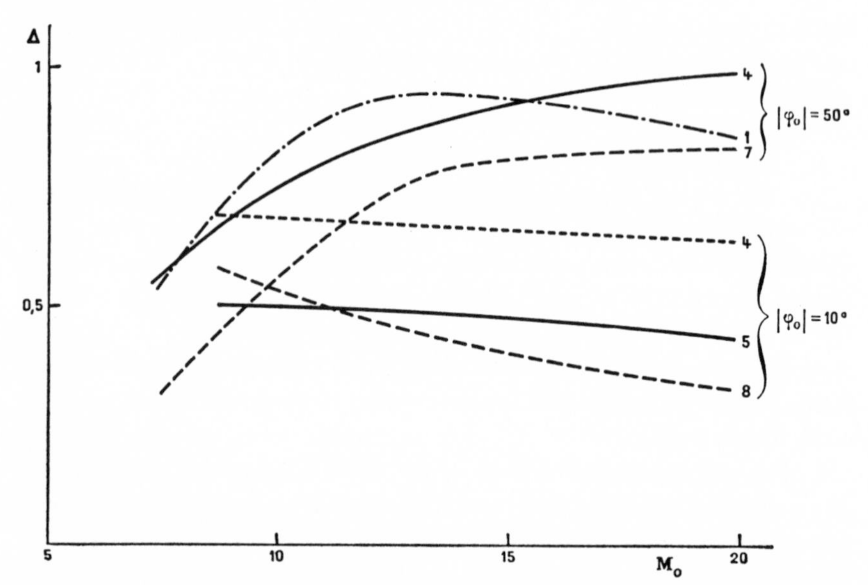

FIG. 10.—Difference between observed and calculated densities as a function of the mean air mass. The curves in the case when the effective rays traverse the atmosphere at high latitudes are distinctly different from the case of traversal at low latitudes.

4.3. Brightness of the Eclipse

From one eclipse to another there are differences in brightness and color. In 1920 Danjon established a 0–4 scale of luminosity which he used in classifying old eclipses. He found that during the 2 years following a solar minimum the eclipses are quite dark; during the next 4 years the brightness increases, after which it remains nearly constant until the following minimum, when a marked drop in brightness sets in. G. de Vaucouleurs confirmed this in 1944.

Danjon's classification does not make a distinction between the central part of the shadow, whose intensity depends on the degree of cloudiness above the terrestrial terminator (other factors being equal), and the outer parts, whose brightness depends on the thickness of the ozone layer. Danjon's classification refers essentially to the central part of the shadow, for Link (1946) has shown from photometric measures that, at 5′ from the edge of the shadow, the intensity is completely independent of the Danjon classification. He has also shown that the differences between observation and theory in terms of the mean air mass follow very different laws if the effective rays cross the atmosphere at low or high latitudes (Fig. 10). A possible explanation is that the ozone layer is at a greater height at low latitudes. Paetzold (1954) remarks, moreover, that the differences in color often noticed in the outer part of the shadow between beginning and end of the eclipse can be attributed to this.

5. CONCLUSIONS

On the whole, explanation of the phenomena presented by lunar eclipses appears to be in a satisfactory state. The most important open question is the reality of the high absorbing layer. The observations needed to settle this question are photometric measures of shadow densities near its limits in blue light, using a very narrow spectral range. It is also worthwhile continuing systematic photometric measures of eclipses in the central part of the shadow, in order to discover the real significance of Danjon's findings.

Lunar eclipses can provide improved knowledge of the upper atmosphere. So far, it is true, eclipses have not added new results; they have served only to check information obtained by other methods. The determination of the horizontal thickness of ozone will be useful only if it is combined with simultaneous measures in those regions of the earth that are effective in producing the brightness of the eclipsed moon. Apart from ozone, it is not impossible that other constituents of the high atmosphere might be identified at lunar eclipses by spectroscopic methods. It should be remembered that Moore and Brigham (1927) reported unidentified absorption lines near the edge of the shadow, which have not been seen again, perhaps because of insufficient dispersion. Continuous absorption in the infrared in the high atmosphere, if it existed, could also be detected by photometry. A first attempt by Bloch and Falgon (1949) shows that such photometry is still possible in the near infrared. All in all, it is possible that surprises may still come from the region 2 or 3 minutes of arc on either side of the geometric edge of the shadow. In this region the absorption effect in the higher atmosphere has the greatest influence.

REFERENCES

BARBIER, D., CHALONGE, D., and VIGROUX, E. 1942 *Ann. d'ap.*, **5**, 1.

BAUER, E., and DANJON, A. 1923 *C.R.*, **176**, 761.

BLOCH, M., and FALGON, R. 1949 *C.R.*, **228**, 1417.

BRIGHAM, L. A., and MOORE, J. H. 1927 *Pub. A.S.P.*, **39**, 223.

CHALONGE, D. 1955 *Astronomie*, **69**, 319.

CUFFEY, J. 1952 *Ap. J.*, **115**, 17.

DANJON, A. 1920 *C.R.*, **171**, 1127.
1921 *Ibid.*, **173**, 706.
1928 *Ann. Obs. Strasbourg*, **2**, Fasc. 1, 1.
1931 *C.R.*, **193**, 716.

DUBOIS, J. 1949 *Astronomie*, **63**, 165.
1950 *Ibid.*, **64**, 164, 458.
1952 *Ibid.*, **66**, 105.
1953 *Ibid.*, **67**, 44.

DUFAY, J. 1931 *C.R.*, **193**, 711.

DYSON, F., and WOOLLEY, R. v.d.R. 1937 *Eclipses of the Sun and Moon* (Oxford: Oxford University Press).

FESSENKOV, V. 1932 *Bull. Acad. Sci. U.S.S.R.*, Ser. 7, **1**, 9.
1937 *Astr. J. Soviet Union*, **147**, 505.

LINK, F. 1931 *C.R.*, **193**, 998.
1932 *Bull. Astr.*, Ser. II, **8**, 77.
1946 *Ann. d'ap.*, **9**, 227.
1947*a* *Bull. Astr.*, Ser. II, **13**, 171.
1947*b* *Časopis pro pěstovani matematiky a fysiky*, **72**, 65.
1948*a* *Ann. géophys.*, **4**, 47, 211.
1948*b* *Bull. Astr. Inst. Czechoslovakia*, **1**, 13.
1950 *Bull. Astr.*, Ser. II, **15**, 143.
1956 *Die Mondfinsternisse* (Liepzig: Akademische Verlagsgesellschaft Geest und Portig K.-G.).

LINK, F., and GUTH, V. 1940 *Zs. f. Ap.*, **20**, 1.
PAETZOLD, H. K. 1952 *Zs. f. Ap.*, **30**, 282.
1954 *Die Sterne*, **30**, 13.
PENNDORF, R. 1948 *J. Meteorol.*, **5**, 156.
ROUGIER, G. 1933 *Ann. Obs. Strasbourg*, Vol. **2**, Fasc. 3.
ROUGIER, G., and DUBOIS, J. 1942 *Astronomie*, **56**, 81, 173.
1943 *Ibid.*, **57**, 65, 129.
SAUSSURE, M. DE 1931 *Verh. Naturforsch. Gesell. Basel*, **41**, 211.
ŠVETSKA, Z. 1949 *Bull. Astr. Inst. Czechoslovakia*, **1**, 131.
VAUCOULEURS, G. DE 1944 *C.R.*, **218**, 655.
VIGROUX, E. 1954 *Ann. d'ap.*, **17**, 399.

CHAPTER 8

Photometry and Colorimetry of Planets and Satellites

By DANIEL L. HARRIS
*Yerkes Observatory**

1. INTRODUCTION

THE photometry of the planets and the satellites has been the subject of many investigations, both observational and theoretical. As good summaries of the earlier work are available in the *Handbuch der Astrophysik*, Volumes **2** and **7** (Berlin, 1929, 1936), we restrict the present discussion to the more recent observational and theoretical results, except where earlier data are needed for the treatment of specific problems. However, frequent reference will be made to Zöllner's (1865) excellent magnitude observations of the sun, moon, and planets made from 1862 to 1864 and to Müller's (1893) comprehensive investigations made from 1877 to 1891. These are of particular importance in connection with the possibility of long-period variations in the magnitudes of the planets.

For the purposes of this chapter, systematic observations of magnitudes and colors of planets and satellites have been made with the 82-inch telescope of the McDonald Observatory since 1950. These observations, here included in summary form, were made by Dr. Kuiper and the writer. We wish to acknowledge the assistance received from the Office of Naval Research, which supported the initial phases of this work under Contract No. N9onr-87100, starting July 1, 1950, under the title "The Determination of Photometric Constants of the Solar System."

2. STANDARD SYSTEM OF MAGNITUDES AND COLORS

As in most photometric investigations, it is desirable to reduce all the observations to a standard system so that the various results may be

* Now at T.E.N.S. Corporation, Bethesda, Md.

directly compared. In the present discussion we have adopted the U, B, V color system of Johnson and Morgan (1953) and the R, I color system of Hardie (1956).

The V (visual) magnitude has for solar radiation an effective wave length of about 0.554μ, which is very close to that of the average eye for the higher intensities. The zero point of the V magnitudes was fixed to obtain agreement with the *IPv* magnitudes of the North Polar Sequence.

The effective wave lengths for solar radiation of the U (ultraviolet), B (blue), V (visual), R (red), and I (infrared) magnitudes are given in Table 1. The zero points of the U and B magnitudes were so chosen that the colors $U - B$ and $B - V$ are 0.00 for unreddened stars of spectral type A0 V on the MK system of spectral classification (Johnson and Morgan, 1953). However, as all discussion of colors in this chapter is with reference to the sun, we can consider the zero points of the colors to be defined by

TABLE 1

EFFECTIVE WAVE LENGTHS OF PHOTOELECTRIC MAGNITUDES FOR SOLAR RADIATION

	U	B	V	R	I
$\lambda(\mu)$	0.353	0.448	0.554	0.690	0.820
$\lambda^{-1}(\mu)$	2.83	2.23	1.81	1.45	1.22

the mean values for stars of the same spectral type, G2 V, i.e., $U - B = 0.14$, $B - V = 0.63$, $V - R = 0.45$, and $R - I = 0.29$.

The choice of the V magnitude system was made because the important earlier observations were made visually. In future investigations the use of a magnitude with a shorter effective wave length would admit more accurate intercomparison of the major planets, in which the visual region is strongly affected by their atmospheric bands (e.g., Hardie and Giclas, 1955).

The relative response to solar radiation of the 1P21 photomultiplier and filter combinations used in establishing the U, B, V system (Johnson and Morgan, 1951) is shown in Figure 1. The plotted points include the effects of one air mass of extinction, two reflections by aluminized mirrors, and the solar absorption lines averaged over intervals of 100 A. The integrated effects of the absorption lines amount to about 0.66 mag. for U, 0.29 mag. for B, and 0.07 mag. for V.

Table 2 gives the V magnitudes and $B - V$, $U - B$ colors for a few of the bright stars employed by earlier investigators in their photometric work. These results have been compiled from published photoelectric

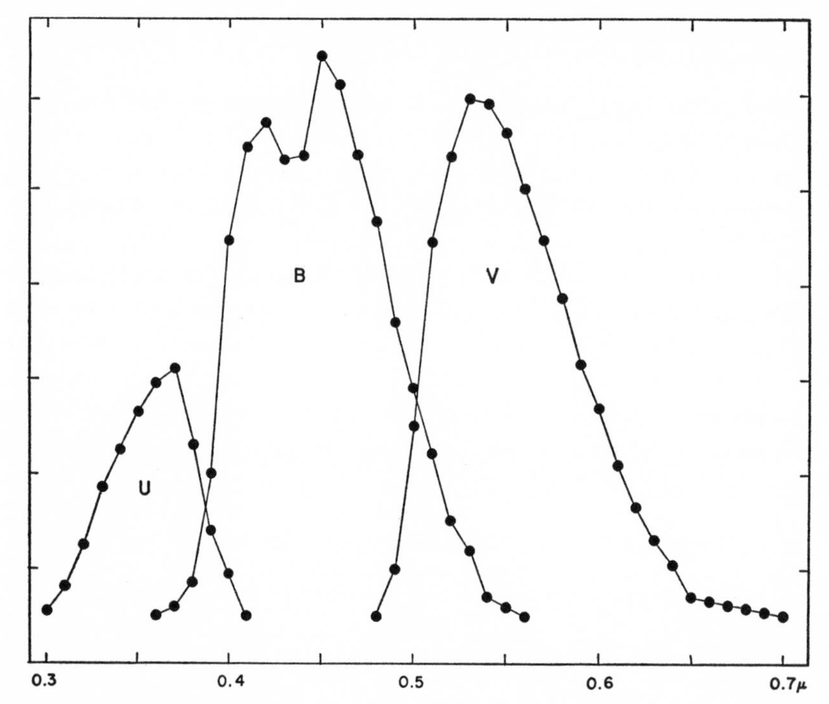

FIG. 1.—Relative wave-length response of photometric system to solar radiation. See also Table 1.

TABLE 2

MAGNITUDES AND COLORS OF COMPARISON STARS FOR SUN AND PLANETS

Star	V	$B-V$	$U-B$	MK
α CMa.........	-1^m42	$+0^m01$	-0^m08	A1 V
α Boo.........	−0.06	+1.23	+1.26	K2p III
α Lyr..........	+0.04	0.00	−0.01	A0 V
α Aur.........	+0.05	+0.80	+0.44	G8 III: +F
β Ori..........	+0.14v*	−0.04	−0.67	B8 Ia
α CMi.........	+0.37	+0.41	0.00	F5 IV–V
α Aql..........	+0.77	+0.22	+0.08	A7 IV, V
α Tau.........	+0.86v†	+1.53	+1.89	K5 III
β Gem.........	+1.16	+1.02	+0.84	K0 III
α Cyg.........	+1.26	+0.09	−0.25	A2 Ia
α Leo..........	+1.36	−0.11	−0.36	B7 V
α Gem AB......	+1.60	+0.02	0.00	Al V+A5 m
α Ari..........	+2.00	+1.15	+1.12	K2 III

* Variable; range 0.12 mag.

† Range 0.15 mag.

measures and from unpublished observations made at the McDonald and Lowell Observatories.

3. THE SUN

A determination of the albedos of the planets and satellites requires a knowledge of the apparent magnitude of the sun. Since Kuiper's (1938) discussion, three new determinations of the magnitude of the sun have been made, and the earlier visual data can be improved by a redetermination of the magnitudes of the comparison stars.

The visual determinations of the magnitude of the sun are those given by Russell (1916*a*) and Kuiper (1938). They are reduced to the V system in Table 3. Giving half-weight to Pickering's determination, which is of lower internal accuracy, we have, from the visual determinations, $V = -26.88$.

Woolley and Gascoigne (1948) have made a spectrophotometric determination of the difference in magnitude between the sun and Sirius, which

TABLE 3

VISUAL DETERMINATIONS OF MAGNITUDE OF SUN

Observer	V	Observer	V
Zöllner.....	−26.82	Ceraski....	−26.87
Fabry......	−26.90	Pickering...	−26.98

gives $V = -26.99$ for the sun. However, this is essentially a monochromatic magnitude, in that it does not include the effects of absorption lines in either the sun or Sirius. For the sun the effect amounts to 0.073 mag. (see Sec. 2), while for Sirius the effect is only about 0.01 mag. (e.g., Milford, 1950). Allowing for this difference, we obtain $V = -26.93$.

Nikonova (1949) made a photoelectric determination of the magnitude of the sun, using the moon as an intermediary. Her result appears to be quite reliable and gives $V = -26.80$.

Stebbins and Kron (1956) have recently made a determination of the magnitude of the sun, using a standard lamp as an intermediary. They conclude that $V = -26.73$, with an internal probable error of less than 0.02 mag.

Comparing the three recent determinations, which are of comparable internal accuracy, we see that the systematic errors affecting one or more of the values are far more important than the accidental errors of observation. If we give equal weight to the three modern determinations and to the mean of the visual determinations, we find $V = -26.83$. Double weight for the photoelectric determinations gives $V = -26.81$, which we have adopted.

We have omitted here the two photographic determinations by King and Birke and the radiometric determination by Pettit and Nicholson used by Kuiper (1938) in his discussion, because of uncertainties in the reduction to the V system.

We have not assigned a mean error to the magnitude of the sun, as it depends on the relative weights assigned to the individual determinations. However, the adopted value is uncertain by not more than ± 0.1 mag.

4. MAGNITUDES OF THE PLANETS

4.1. Introduction

The apparent magnitude of a planet is a function of the planet's distance from the sun, r, the planet's distance from the earth, d, and the solar phase angle, α, which is the angular distance at the planet between the earth and the sun. For Mercury, Venus, and the moon the solar phase angle varies from 0° to 180°; for Mars the angle can attain a value of about 47°, while for the more distant planets the angle reaches only a few degrees.

Neglecting possible variations in brightness due to rotation or intrinsic causes, the observed magnitude of a planet is given by

$$V = V(1, 0) + 5 \log rd + \Delta m(\alpha), \tag{1}$$

where $V(1, 0)$ is the magnitude of the planet reduced to unit distance from the sun and earth and phase angle zero and $\Delta m(\alpha)$ is the correction for the variation in magnitude of the planet with phase angle.

The quantity $\Delta m(\alpha)$ is composed of two parts; the first arises from the fact that the fraction of the illuminated disk visible from the earth varies with α, while the second, and much larger, effect is due to the properties of diffuse reflection from the planet's surface or atmosphere.

From the observational point of view, it is sufficient to express $\Delta m(\alpha)$ as a power series in α, retaining as many terms as are required adequately to represent the observations. For the outer planets it is generally sufficient to consider only the first term, and the coefficient of α so found is often referred to as the *phase coefficient.*

The magnitude, $V(1, 0)$, corresponds to the quantity g introduced by earlier investigators; the new symbol is used here to specify the spectral region to which the observations refer.

For some purposes it is convenient to use the *mean opposition magnitude*, V_0, which is related to $V(1, 0)$ by

$$V_0 = V(1, 0) + 5 \log a(a - 1), \tag{2}$$

where a is the semimajor axis of the planet's orbit.

If the brightness of the planet is *variable*, the problem of predicting the apparent magnitude at a given time is considerably more complicated. The variability may arise from the following causes: (*a*) the planet's surface is spotted, giving rise to a brightness variation with the rotation period; (*b*) changes in the planet's atmosphere or surface features, giving rise to variations of an irregular character. These may be short-lived or persist for considerable periods of time.

For planets exhibiting intrinsic variability we have tried to determine mean values of $V(1, 0)$ and $\Delta m(\alpha)$ and indicate graphically the variations from these mean values. In particular, it is noted that the variation with rotational phase is a function of solar phase angle as well. This follows not only from geometrical considerations but also because the properties of diffuse reflection must depend on the surface features.

In connection with the question of long-period variations in the magnitudes of the planets, as discussed by Becker (1933, 1948), it is necessary to utilize earlier observations along with the modern photoelectric determinations. Of the older observations, those of Zöllner (1865) and Müller (1893) are considered the most reliable, and we have reduced these results, using modern values of V for the comparison stars and modern phase functions. Because of the uncertainty in the magnitude scales and in the reductions to the standard system, we have omitted all photographic values, except the photovisual determinations by King (1923, 1930), which are considered systematically reliable.

As far as short-period variations are concerned, we have given most of the weight to the photoelectrical results because of their far greater precision.

4.2. Magnitudes of Individual Planets

4.21. *Mercury.*—The variation of Mercury with solar phase angle has been determined by Danjon (1949) over the range $3° < \alpha < 123°$. As the rotation period is equal to the planet's period of revolution around the sun, only small changes from the mean curve are observed at different librations. The corrected analytical expression for the phase function is (Danjon, 1954)

$$\Delta m\,(\alpha) = 3.80\left(\frac{\alpha}{100°}\right) - 2.73\left(\frac{\alpha}{100°}\right)^2 + 2.00\left(\frac{\alpha}{100°}\right)^3 . \tag{3}$$

Danjon's magnitude of Mercury at unit distance is $m_0 = -0.21$. This value may be reduced to the V system by means of an average correction determined from the comparison stars used by Danjon. This gives $V(1, 0) = -0.36$. Danjon also reduced Müller's observations, made in 1878–1888, to his own photometric system and to opposition with the aid of his own $\Delta m(\alpha)$-curve; the result is a value 0.10 mag. brighter, i.e., $V(1, 0) =$

−0.46. As Mercury has no appreciable atmosphere, no long-period variations in the brightness are expected.

4.22. *Venus.*—The variation of Venus with solar phase angle has been determined by Danjon (1949) over the range $0^\circ\!.9 < \alpha < 170^\circ\!.7$. His analytical expression for the phase function is

$$\Delta m(\alpha) = 0.09\left(\frac{\alpha}{100^\circ}\right) + 2.39\left(\frac{\alpha}{100^\circ}\right)^2 - 0.65\left(\frac{\alpha}{100^\circ}\right)^3. \qquad (4)$$

Danjon's magnitude of Venus, reduced to the V system as was done for Mercury, gives $V(1, 0) = -4.29$. Danjon's reduction of Zöllner's observations made in 1865 gives −4.28, while his reduction of Müller's observations between 1877 and 1890 gives −4.21. These older observations, while of lower weight, give no indication of a long-period variation in the brightness of Venus.

4.23. *The earth.*—The problem of the photometry of the earth has been considered by Danjon in Volume **2**, chapter 15. The analytical phase function was found to be

$$\Delta m(\alpha) = 1.30\left(\frac{\alpha}{100^\circ}\right) + 0.19\left(\frac{\alpha}{100^\circ}\right)^2 + 0.48\left(\frac{\alpha}{100^\circ}\right)^3. \qquad (5)$$

The magnitude difference, earth − sun, is 22.94 and shows an annual variation with a range of 0.51. With our adopted magnitude of the sun, −26.81, we obtain $\bar{V}(1, 0) = -3.87$.

As Danjon emphasizes (Vol. **2**, chap. 15, Sec. 3), these results are based on measures of the earth light obtained in France and do not necessarily apply to the earth as a whole. The magnitude of the earth, as seen from the sun, probably shows a diurnal variation due to land and oceans, as well as variations from the variable cloud coverage.

4.24. *Mars.*—The variation in the brightness of Mars with rotational period was first established by Guthnick and Prager (1918) and was recently confirmed by Johnson and Gardiner (1955) and by Kuiper and Harris. The variation is best defined by series of observations made near quadrature. Figure 2, *a*, shows the difference in red magnitude between Mars and β Geminorum observed by Guthnick and Prager in 1916, plotted against the longitude of the central meridian, λ. The 1954 observations of $\bar{V}(1, 0)$ made at the Lowell and McDonald Observatories, when the solar phase angles were greater than 30°, are shown in Figure 2, *b*, while the corresponding $B - V$ colors are shown in Figure 2, *c*. Three observations made at McDonald in 1952 are plotted as open circles.

As was shown by Johnson and Gardiner, the similarity in the variation of V and $B - V$ indicates that the change in blue and ultraviolet light is considerably less than in visual or red light. This must be largely due to the fact that the Martian atmosphere is nearly opaque to ultraviolet and blue light, where the disk of the planet shows little contrast with rotational phase, while in visual light the atmosphere is more transparent and the contrast of the surface gives rise to the rotational variation observed.

Johnson and Gardiner noted that the scatter of their observations about a mean curve was more than could be explained by observational errors.

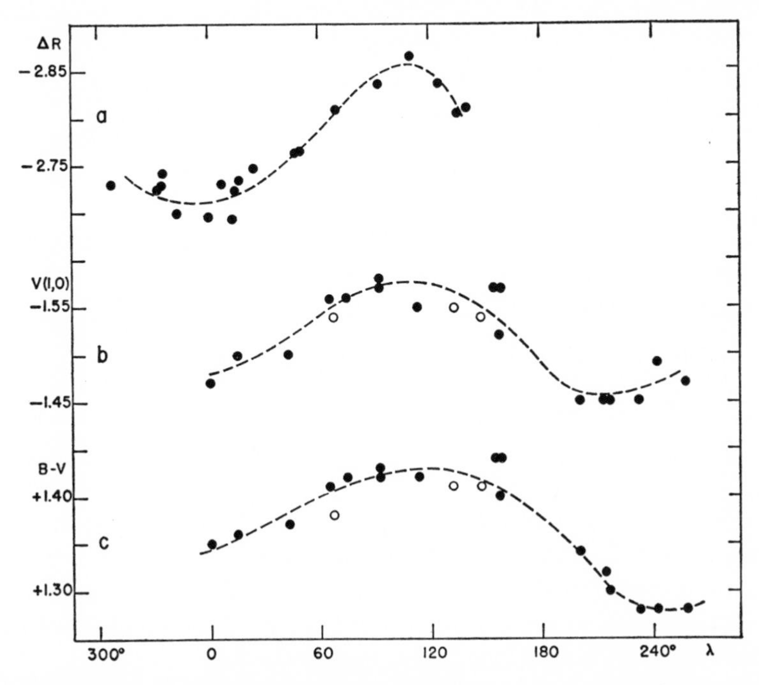

Fig. 2.—Variation of Mars with rotational phase angle. (*a*) In red light (Mars − β Gem) by Guthnick and Prager (1916); (*b*) in yellow light, *V* (1, 0), from Lowell and McDonald observations (1954); (*c*) in the *B* − *V* color. (See text.)

It is probable that variations in the transparency of the Martian atmosphere contributed to this scatter; if the atmosphere is clearer than normal, the observed magnitude will be fainter than average, while an increase in the "haze" will make the observed magnitudes brighter. An extreme example occurred on May 3–4, 1952, when the ultraviolet magnitude of Mars was 0.3 mag. brighter, the blue magnitude about 0.2 mag. brighter, and the visual magnitude 0.1 mag. brighter than average (see Sec. 6.2).

The problem of determining the solar phase variation over the 47° range in phase angle is complicated by the rotational variation and the changes in the Martian atmosphere; a linear relation satisfies the observations quite well. The extensive photoelectric series at the Lowell Observatory in 1954 gives a phase coefficient of 0.016 mag/degree in the visual region, and 0.017 mag/degree in the blue and ultraviolet. The long series of visual observations by Müller gave 0.015 mag/degree.

Becker (1933) suggested that the mean opposition magnitude of Mars varied with a range of about 0.48 mag., but no period was established. The older measures reduced to the V system are given in Table 4, where it is seen that the values are remarkably accordant and show no evidence of a long-period variation. The mean value of $V(1, 0)$ is -1.52 mag. The reduction to mean opposition is -0.49 mag., giving $\bar{V}_0 = -2.01$.

TABLE 4

VALUES OF V(1, 0) FOR MARS

Opposition	Observer	$V(1, 0)$	Opposition	Observer	$V(1, 0)$
1864–1865. . .	Zöllner	−1.59	1916.	King, Pv	−1.52
			1918.	King, Pv	−1.51
1877–1878. . .	Müller	−1.44	1920–1921. . .	King, Pv	−1.45
1879–1880. . .	Müller	−1.48	1922.	King, Pv	−1.56
1881–1882. . .	Müller	−1.57			
1883–1884. . .	Müller	−1.55	1952.	Kuiper, Harris	−1.56
1886.	Müller	−1.51			
1888–1889. . .	Müller	−1.52	1954.	Lowell Obs.	−1.51

4.25. *Jupiter.*—Accurate photometric data for Jupiter are quite limited, and it is not possible to give a completely satisfactory description of its brightness variations. Guthnick and Prager (1918) found no evidence for a variation with rotational phase during the opposition of 1917–1918, but during the 1920 opposition Guthnick (1920) found a variation of 0.14 mag.

A similar situation obtains with regard to variation with solar phase angle. Müller's extensive visual observations indicate a small value for the phase coefficient, of the order of 0.005 mag/degree, and a similar small variation is indicated by Güssow's (1929) observations of 1928. On the other hand, the observations at the 1917 opposition by Guthnick and Prager gave a well-determined variation amounting to 0.015 mag/degree, while the opposition of 1918–1919 gave a value about half as large (reported by Schönberg, 1921). Figure 3, *a*, *b*, and *c*, illustrates the phase differences found by Müller, by Güssow, and by Guthnick and Prager in 1917. It is seen that the slope fitting the 1917 results is not supported by the other observations. The phase coefficient may be variable; but from theoretical considerations the smaller value of 0.005 mag/degree appears more representative than the larger value (see Sec. 8.6).

Becker (1933) found the mean opposition magnitudes of Jupiter to vary with an amplitude of 0.34 mag. in a period of 11.6 ± 0.4 years. We have reduced the older observation to the V system in Table 5 and plotted them in Figure 4. The mean for 23 oppositions is $V(1, 0) = -9.25$, with the individual oppositions varying from -9.03 to -9.48, a range of 0.45. As the same observers' values for Mars and Saturn show, at most, very small variations, it is concluded that the larger changes found for Jupiter are real. The data in Table 5 are not sufficient to establish a period, but the variation may well be correlated with changes on the surface features, as

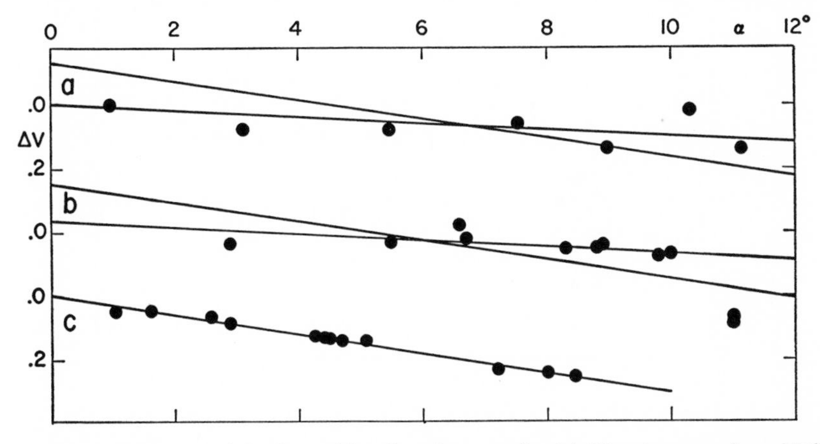

FIG. 3.—Variation of Jupiter with solar phase angle. (*a*) Visual means from Müller; (*b*) photoelectric observations by Güssow (1928); (*c*) photoelectric observations by Guthnick and Prager (1917).

TABLE 5

VALUES OF $V(1, 0)$ FOR JUPITER

Opposition	Observer	$V(1, 0)$	Opposition	Observer	$V(1, 0)$
1862........	Zöllner	−9.22	1889........	Müller	−9.09
1863........	Zöllner	−9.22	1890........	Müller	−9.07
1864........	Zöllner	−9.34			
			1915........	King, Pv	−9.39
1878........	Müller	−9.03	1916........	King, Pv	−9.18
1879–1880...	Müller	−9.14	1918........	King, Pv	−9.16
1880–1881...	Müller	−9.19	1920........	King, Pv	−9.24
1881–1882...	Müller	−9.28	1921........	King, Pv	−9.41
1883........	Müller	−9.24	1922........	King, Pv	−9.30
1883–1884...	Müller	−9.28			
1885........	Müller	−9.25	1951........	Kuiper	−9.46
1886........	Müller	−9.18	1952........	and	−9.48
1887........	Müller	−9.11	1954........	Harris	−9.39

Becker (1933) suspects. The reduction to mean opposition is −6.70 mag.; therefore, $\bar{V}_0 = -2.55$ (var.).

4.26. *Saturn.*—The presence of the Rings makes the study of Saturn's disk very difficult, but the brightness of the total light is surprisingly well known. In regard to a variation with rotational phase, the results of Guthnick and Prager (1918) indicated that the system did not vary more than ±0.01 mag. at the oppositions of 1914–1915 and 1917.

The variation in the magnitude of the Saturn system with solar phase angle is nearly linear over the observable range of 6°, and the phase coefficient, according to Müller, is 0.044 mag/degree. Guthnick and Prager's

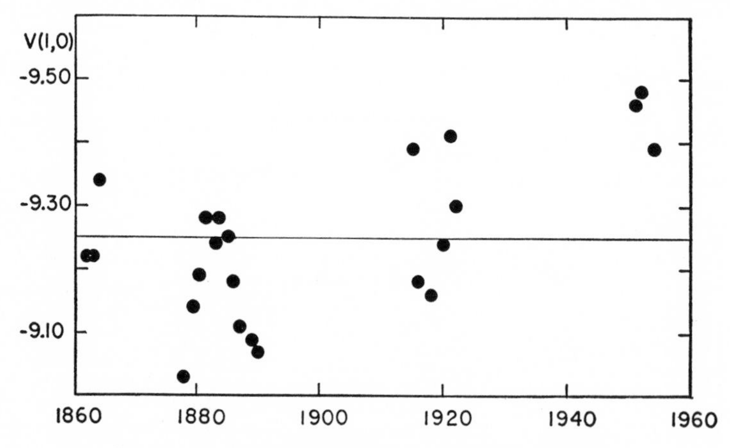

FIG. 4.—Long-period variation of Jupiter (see Table 5)

observations in 1914–1915 and 1917 give a somewhat smaller value, about 0.033 mag/degree, while Guthnick's observations in 1918 and 1920 (quoted by Schönberg, 1921) give 0.043 mag. and 0.049 mag/degree. The phase coefficient, therefore, seems to vary slightly with the inclination of the Rings and, hence, the fraction of the total light they contribute. It should be mentioned that Guthnick's (1918) observations, made very close to opposition, are nearly 0.15 mag. brighter than would be expected from the linear formula; however, the observation on January 8, 1918, with $\alpha = 2\overset{\circ}{.}60$, is also discordant, so that this brightening near opposition must be confirmed before the linear phase function is abandoned.

The foregoing discussion applies to the combined light of the system; the phase correction for the ball of the planet alone is undoubtedly much smaller (Schönberg, 1921) and is probably of the order of 0.01 mag/degree or less.

The magnitude of the system varies greatly with changing aspect of the Rings. Müller's empirical formula to reduce the brightness to "no-ring" is

$$\Delta V = +2.60 \sin B - 1.25 (\sin B)^2 \quad (6)$$

where B is the saturnicentric latitude of the earth. The quantity ΔV was determined from observations made between 1877 and 1891, when the earth was south of the planet's equatorial plane. Becker (1948) suggests that the correction depends on whether the earth is north or south of the plane, a suggestion that should be tested by observations during the coming oppositions.

Becker (1948) finds that the mean opposition magnitude of the planet, reduced to "no-ring," varies irregularly by 0.33 mag. The results of our reduction of the older measures are given in Table 6. It is seen that the

TABLE 6

VALUES OF $V(1, 0)$ FOR SATURN (NO-RING)

Opposition	Observer	$V(1, 0)$	Opposition	Observer	$V(1, 0)$
1862........	Zöllner	−8.84	1883–1884...	Müller	−8.87
1863........	Zöllner	−8.82	1884–1885...	Müller	−8.86
			1885–1886...	Müller	−8.92
1877–1878...	Müller	−8.86	1886–1887	Müller	−8.88
1878........	Müller	−8.88	1887–1888...	Müller	−8.87
1879–1880...	Müller	−8.87	1891........	Müller	−8.94
1880–1881...	Müller	−8.90			
1881–1882...	Müller	−8.89	1951........	Kuiper,	−8.87
1882–1883...	Müller	−8.89	1952........	Harris	−8.86

values of $V(1, 0)$ show very little scatter, from −8.94 to −8.82 mag. only, while Müller's observations from 1877 to 1891 show a total range of only 0.08 mag. It appears that the brightness of the planet is remarkably constant, the mean value being $V(1, 0) = -8.88$. The reduction to mean opposition is +9.55 mag., so that $V_0 = +0.67$.

4.27. *Uranus*.—The question as to whether Uranus varies in light with its rotation period of 10.84 hours (Lowell and Slipher, 1912; Moore and Menzel, 1930) is difficult to decide. To show such a variation, the surface of the planet must develop temporary spots, and the planocentric latitude of the earth must be small. Such conditions may have been satisfied in 1917, when Campbell (1917, 1936) found Uranus to vary by 0.15 mag. in a period of 10^h49^m, and during the interval from 1920 to 1930, when several visual and photographic series (Becker, 1928; Slavinas, 1928; Goddard, 1930) indicated variations with the same period and amplitudes ranging from 0.03 to 0.15 mag.

On the other hand, photoelectric observations by Stebbins and Jacobsen (1928), Güssow (1929), Calder (1936), Hardie and Giclas (1955), and Kuiper and Harris, at the oppositions of 1927, 1928, 1934–1935, 1935–1936, and from 1950 to 1955, respectively, show no indications of a variation in brightness with rotational phase, in spite of the much higher accuracy of these observations. Until the rotational variation has been confirmed by high-precision photoelectric series, it is probably best to use the period of rotation as determined spectroscopically.

The solar phase angle of Uranus attains only 3°, but Stebbins and Jacobsen (1928) found a phase coefficient of 0.0028 mag/degree. The Lowell observers find a somewhat smaller value, of the order of 0.001 mag/degree.

TABLE 7

VALUES OF $V(1, 0)$ FOR URANUS

Opposition	Observer	$V(1, 0)$	Opposition	Observer	$V(1, 0)$
1864.........	Zöllner	−7.19	1928........	Güssow	−7.20
1878.........	Müller	−6.90	1935........	Calder	−7.26
1879.........	Müller	−7.06	1936	Calder	−7.27
1880.........	Müller	−7.12			
1881.........	Müller	−7.13	1949–1950...	Lowell Obs.	−7.19
1884.........	Müller	−7.14	1950–1951...	Lowell Obs.	−7.20
1885.........	Müller	−7.08	1951–1952...	Lowell Obs.	−7.19
1886.........	Müller	−7.00			
1888.........	Müller	−7.01	1952........	Kuiper, Harris	−7.18
1927.........	Stebbins and Jacobsen	−7.16	1953–1954...	Lowell Obs.	−7.17

Becker's investigation (1948) of the mean opposition magnitude of Uranus would indicate that Uranus varies 0.29 mag. with a period of the order of 8 years, superposed on a variation of 0.26 mag. with a period of 82 years. The latter variation, interpreted as the effect of the oblateness of the planet, gives too large a value of the polar compression.

As far as the 8-year variation is concerned (which seems to have found some confirmation, Ashbrook, 1948), the observations at the Lowell Observatory from 1949 to 1955, already mentioned, which cover more than half the suspected period, show no evidence for a variation larger than about 0.01 mag., which may well be due to unavoidable uncertainties.

The results of our reduction of the older series of measures is given in Table 7. It is seen that Müller's observations show considerable scatter. The mean of his 92 observations is 0.12 mag. fainter than our adopted value of $V(1, 0) = 7.19$, while Calder's observations are about 0.08 mag.

brighter than our adopted value. As Hardie and Giclas (1955) have emphasized, the strong absorption bands in the visual region make the reductions of one series to another rather uncertain. The reduction to mean opposition is +12.71 mag. or $\bar{V}_0 = +5.52$.

4.28. *Neptune.*—The rotation period of Neptune determined spectroscopically by Moore and Menzel (1928) is 15.8 ± 1 hours. A number of photographic and visual investigations of Neptune have indicated periodic variations in brightness which were attributed to rotation. However, as in the case of Uranus, the photoelectric observations fail to show any evidence of variation (Calder, 1938; Hardie and Giclas, 1955; and this chapter). To indicate the attainable accuracy of photoelectric observations as applied to the possible short-period variation of Neptune, we find from Hardie's (1953) observations on 15 nights from June 10 to July 3, 1953,

TABLE 8

VALUES OF $V(1, 0)$ FOR NEPTUNE

Opposition	Observer	$V(1, 0)$	Opposition	Observer	$V(1, 0)$
1864.........	Zöllner	−6.77	1886–1887...	Müller	−6.89
1878.........	Müller	−7.01	1949–1950...	Lowell Obs.	−6.79
1881–1882....	Müller	−6.96	1950–1951...	Lowell Obs.	−6.83
1883.........	Müller	−6.87	1951–1952...	Lowell Obs.	−6.83
1883.........	Müller	−6.97	1951–1952...	Kuiper	−6.87
1884–1885....	Müller	−6.89	1952–1953...	and	−6.84
1885–1886....	Müller	−6.82	1953–1954...	Harris	−6.85

that two nights deviated by 0.006 mag. from the mean, two nights show deviations of 0.005 and 0.004 mag., while the remaining eleven nights deviate by 0.002 mag. or less. It should be remembered that these deviations also include possible variations in the brightness of the comparison star. As, furthermore, the announced variations do not agree with the spectroscopically determined period or with each other, the reality of these results is in doubt.

The period of rotation of a planet determined spectroscopically is independent of the assumed value of the planet's diameter (Moore and Menzel, 1930); recent suggestions that the period found by Menzel and Moore should be shortened because of Kuiper's (1949) new value for the apparent diameter of the planet are, therefore, incorrect.

The observed range in phase angle is very small, and no variation due to it has been observed. Becker (1933) suggested that the mean opposition magnitudes of Neptune showed a variation of 0.36 mag. in a period of 21 ± 0.6 years. Our reduction of the older measures are given in Table 8.

We find that the mean of thirteen oppositions gives $V(1, 0) = -6.87$, with a range of individual values of 0.24 mag. Most of this scatter is probably due to observational error and difficulties of reduction to the standard system. The difference between the Lowell and McDonald results in 1951–1952, for example, can be attributed to the heavy methane bands of Neptune and the slight difference in filter-cell combination employed at the two observatories. The reduction to mean opposition is +14.71 mag., so that $\bar{V}_0 = +7.84$.

4.29. *Pluto.*—Observations of Pluto made by Walker and Hardie (1955) in 1954 and 1955 combined with earlier observations by Kuiper made in 1952 and 1953 show that Pluto varies in light by about 0.11 mag. in a period of 6.390 ± 0.003 days. A plot of the observations is given in Figure

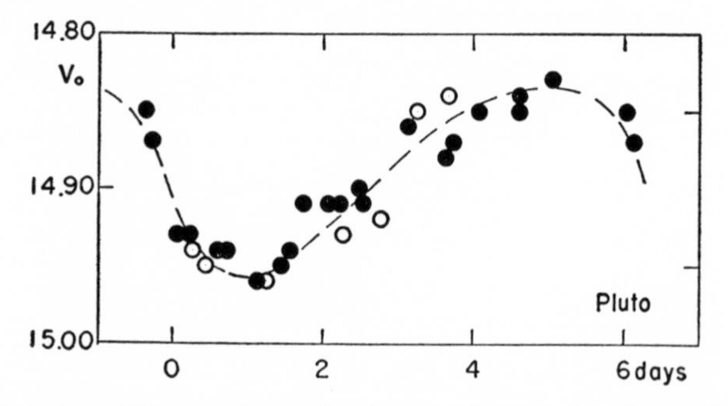

FIG. 5.—Variation of Pluto with rotational phase. *Open circles:* observations by Kuiper (1952 and 1953); *filled circles:* observations by Walker and Hardie (1954 and 1955).

5, where the open circles refer to the 1952–1953 observations by Kuiper. This variation is assumed to be due to the rotation of the planet; the significance of such a long period of rotation has been discussed by Kuiper (1953, 1956*a*).

Any variation over the small range of solar phase angle, $\sim 1\overset{\circ}{.}5$, would be difficult to establish because of the much larger rotational variation.

The mean opposition magnitude of $V_0 = 14.90$, and the reduction to mean opposition is +15.91 mag., so that $V(1, 0) = -1.01$.

4.3 Summary of the Photometric Data on the Planets

The values of the magnitudes at unit distance from the earth and sun, $V(1, 0)$, the corresponding magnitudes at mean opposition, and remarks concerning the variability of the brightness of the planets are summarized in Table 9. Observations of the solar constant by the Astrophysical Observatory of the Smithsonian Institution suggest that the solar radiation may vary by as much as ±1.5 per cent over long periods, with short-period variations of the order of ±0.4 per cent. Such variations are too small to be established by direct observation of the magnitude of the sun. However, as any change in the brightness in the sun would be noted in the reflected light from the planets, satellites, and asteroids, a number of investigations have been made to check the constancy of the sun's radiation by observations of the planets.

Uranus and Neptune appear to be most suitable for this purpose, and observations by the Lowell observers (Hardie and Giclas, 1955) indicate that for the 5-year period 1949–1954 the sun varied by less than ±0.4 per cent, the same upper limit obtained earlier by Stebbins and Jacobsen (1928). Such investigations should be continued to shed more light on the long-period variations suggested by the solar constant data.

The magnitude of the earth, as observed from the sun, may well show the largest variations of any of the planets, while variations with reduced

TABLE 9

PHOTOMETRIC DATA FOR PLANETS

	$V(1, 0)$	V_0	Variation Due to Rotation	Other Variations
Sun	−28.81	....	....	Constant within ±1.5 per cent
Mercury	− 0.36	....	....	....
Venus	− 4.29	....	Probably very small	....
Earth	− 3.87	....	Probable—irregular	Annual variation; short-period variations due to changing cloud coverage
Mars	− 1.52	− 2.01	Variable amplitude, about 0.15.	Probable annual variation; short-period variations due to changes in atmospheric transparency
Jupiter	− 9.25	− 2.55	Probable—irregular	Long-period variation correlated with change in surface features
Saturn	− 8.88	+ 0.67	Probably very small	No evidence of long-period variation
Uranus	− 7.19	+ 5.52	Suspected, but not confirmed	A small, long-period (84 years) variation expected due to oblateness and high obliquity of equator
Neptune	− 6.87	+ 7.84	Suspected, but not confirmed	No evidence of long-period variation
Pluto	− 1.01	+14.90	Amplitude 0.11	No evidence, but data limited

amplitude have been observed and are expected for Mars. In addition, Jupiter shows irregular variations over long periods, and Pluto a rotational variation of some 10 per cent. With these exceptions, after correction for distance and solar phase variation, the planets are remarkably constant in brightness; the variations suspected for the other planets have not been confirmed.

It is probable that accurate photoelectric observations of the planets, particularly in the blue region of the spectrum, will reveal small variations that are not apparent from the present data. It would appear that the amplitudes of such variations are less than the accuracy of visual or photographic photometric methods which can be profitably employed in other astronomical investigations.

5. MAGNITUDES AND COLORS OF THE SATELLITES

5.1. Introduction

The photometric data for the satellites in the solar system are still quite incomplete; in several cases only rough estimates of the mean opposition magnitudes, V_0, are available. Usually this scarcity of data is due to the satellite's proximity to a very much brighter planet or to its faintness, which makes observation difficult except with the largest telescopes.

For the satellites closest to their planets the scattered light makes it practically impossible to obtain accurate photoelectric or even photographic measures. For such objects, visual estimates of the magnitude difference between the inner and an outer satellite when nearly coincident will be the most reliable. In a few cases, where no new data were available, we have adopted the values given by Russell, Dugan, and Stewart (1945), denoted by the abbreviation RDS.

For the faintest satellites, only photographic magnitudes have been derived. These are listed in the text but have been converted to V magnitudes in the final compilation (Table 14, below) with the assumed mean color index of 0.8 mag. This conversion introduces an uncertainty in addition to that of the photographic magnitude scale.

5.2. The Moon

The detailed photometry of the moon is covered in chapter 6 of this volume. The most comprehensive study of the integrated brightness of the moon is that of Rougier (1933, 1937), who determined the variation with phase photoelectrically. The variation is not quite symmetrical, but the maximum difference between waxing and waning phases is only about

0.08 mag. near the quadratures. The mean variation can be expressed in the form

$$\Delta m(a) = 3.05\left(\frac{a}{100°}\right) - 1.02\left(\frac{a}{100°}\right)^2 + 1.05\left(\frac{a}{100°}\right)^3. \qquad (7)$$

There is considerable uncertainty in the magnitude of the full moon as defined by Rougier's phase-curve. Zöllner's (1865) accordant observations, made by comparing the moon with Capella and the moon with the sun, give $V_0 = -12.54$. Rougier's observations give $V_0 = -12.83$, and this is in reasonable agreement with that derived from Nikonova's (1949) photoelectric observations, which give $V_0 = -12.76$. The latter's observations also confirm the phase variation obtained by Rougier. Giving half-weight to Zöllner's visual observations, we obtain $V_0 = -12.74$, corresponding to $V(1, 0) = +0.21$.

Observations of a number of areas on the moon give a mean $B - V$ color of $+0.92$ with little difference between maria and bright regions. The mean $U - B$ color is $+0.46$ (Hardie, 1956).

5.3. The Satellites of Mars

The magnitudes and colors of the Martian satellites were observed by Kuiper at the 1956 opposition, when they were most favorably situated because of the close approach of the planet. The scattered light from Mars was still very large, and the results can be considered as only approximate ($\pm 0^{m}.1$). His observations give $V(1, 0) = 12.1$ for Phobos and $V(1, 0) = 13.3$ for Deimos, with $B - V = 0.6$ for both satellites. Reduced to mean opposition, these correspond to $V_0 = 11.6$ and 12.8, which are in good agreement with the earlier visual estimates.

5.4. The Satellites of Jupiter

Our knowledge of the variation in brightness of the Galilean satellites is quite complete, largely because of the extensive observational programs by Stebbins (1927) and by Stebbins and Jacobsen (1928). These observations were made with a quartz-potassium cell (maximum sensitivity for constant energy at 4600 A) and comparison stars of approximately the same color as the satellites. The magnitudes and colors of these stars have been observed at the McDonald Observatory, so that the Lick observations could be reduced to the V system. The 1926 series reduced with the linear relation, $V = Pe + 0.015 - 0.96(B - V)$, gives the mean opposition magnitudes shown in Table 10, while the 1927 series reduced with the relation, $V = Pe + 0.032 - 0.96(B - V)$, gives the mean opposition magnitudes shown in Table 11.

Magnitudes and colors of the satellites observed at the McDonald Observatory between 1950 and 1954 serve as a check on the reductions, and no difficulties were encountered for Jupiter II, III, and IV. It was not possible to effect a satisfactory transformation in the case of Jupiter I, for which the color is considerably redder than the comparison stars used and, moreover, varies with rotational phase.

Stebbins separated the variation in brightness with rotational phase from the solar phase variation by a method of successive approximations, and his results have been adopted in our discussion, except in the case of Jupiter IV, where we have redetermined the mean solar phase variation, using his individual observations.

5.41. *Jupiter I.*—The solar phase function (Stebbins and Jacobsen, 1928) is

$$\Delta m(\alpha) = 0.046\alpha - 0.0010\alpha^2 . \qquad (8)$$

From their observations in 1926 and 1927 the satellite was found to vary by about 0.21 mag., the leading side being the brighter. The McDonald observations made in 1951, 1952, and 1954 show considerable scatter but

TABLE 10

MEAN OPPOSITION MAGNITUDES OF JUPITER'S SATELLITES (1926)

Object	P_e	V	$B-V$	V (Stebbins)
ι Cap	5.114	4.28	0.88	4.28
42 Cap	5.796	5.19	0.65	5.19
μ Cap	5.451	5.11	0.37	5.11
Jupiter I	5.802	....	1.17 var.	4.69
J II	5.988	....	0.87	5.17
J III	5.303	....	0.83	4.53
J IV	6.263	....	0.86	5.45

TABLE 11

MEAN OPPOSITION MAGNITUDES OF JUPITER'S SATELLITES (1927)

Object	P_e	V	$B-V$	V (Stebbins)
20 Psc	6.334	5.44	0.96	5.44
27 Psc	5.720	4.84	0.94	4.85
44 Psc	6.544	5.72	0.90	5.71
Jupiter I	5.798	....	1.17 var.	4.71
J II	5.976	....	0.87	5.17
J III	5.293	....	0.83	4.54
J IV	6.261	....	0.86	5.47

indicate the same form of variation, with possibly somewhat smaller amplitude, as shown in Figure 6, *a*.

Slight color variations have been noted for other satellites, but the large variations in color with orbital phase for Jupiter I, shown in Figure 6, *b* and *c*, and amounting to 0.18 mag. in $B - V$ and 0.5 mag. in $U - B$, are outstanding.

The mean opposition magnitude from the McDonald observations is $\bar{V}_0 = 4.80$, which is 0.10 mag. fainter than given in Tables 10 and 11 but is to be preferred because of the uncertainty arising from the reduction of Stebbin's observations of this satellite to the V system.

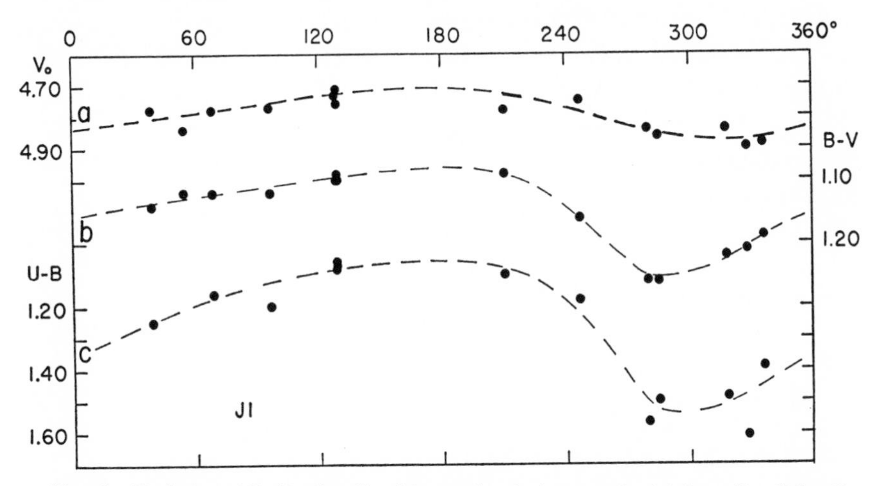

FIG. 6.—Variation of Io (Jupiter I) with rotational phase angle. (*a*) In yellow light, V_0; (*b*) in the $B - V$ color; (*c*) in the $U - B$ color. McDonald observations.

5.42 *Jupiter II*.—The solar phase function (Stebbins and Jacobsen, 1928) is

$$\Delta m(\alpha) = 0.0312\alpha - .00125\alpha^2 . \quad (9)$$

The variation with orbital phase obtained in 1927 is shown in Figure 7 as full-drawn lines, while the McDonald observations are plotted as circles and connected by dashes. The mean opposition magnitude is $\bar{V}_0 = 5.17$, and the $B - V$ color is 0.87 mag., showing no variation with orbital phase, θ. $U - B$ observations are available on 4 nights when θ was less than 180° and on 6 nights when θ was greater than 180°. Surprisingly, the mean $U - B$ color of the leading side is 0.43 mag., and of the trailing side 0.62 mag.; i.e., the trailing side is fainter in the ultraviolet by about 20 per cent.

5.43. *Jupiter III*.—The solar phase function (Stebbins and Jacobsen, 1928) is

$$\Delta m(a) = 0.323a - 0.00066a^2 . \quad (10)$$

The variation with orbital phase obtained in 1927 is shown in Figure 8 as a line, while the McDonald observations are plotted as circles. The mean opposition magnitude is $\bar{V}_0 = 4.54$, the $B - V$ color is 0.83 mag. and shows no variation with orbital phase. The mean $U - B$ color is 0.51 mag., with evidence of only a slight variability. The average for the leading side is 0.48 mag. (4 nights) and the trailing side 0.52 mag. (10 nights).

5.44. *Jupiter IV*.—Stebbins and Jacobsen (1928) found the brightness variation of Jupiter IV to be exceptional, in that near opposition the satellite brightened up more on the leading side than on the trailing side. To demonstrate this phenomenon, we have plotted the three light-curves shown in Figure 9. Curve *a* is based on observations with $a < 1.^\circ 5$, *b* from observations with $2.^\circ 7 < a < 8^\circ$, and *c* from observations made with $a > 8^\circ$. The phase function,

$$\Delta m(a) = 0.078a - 0.00274a^2 \quad (11)$$

makes the average of all the light-curves reduced to mean opposition $\bar{V}_0 = 5.50$, which is 0.04 mag. fainter than given in Tables 10 and 11.

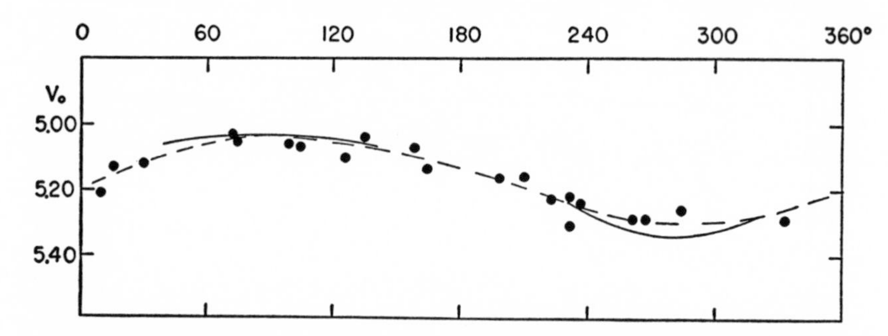

FIG. 7.—Variation of Europa (Jupiter II) in yellow light, V_0, with rotational phase angle. *Full-drawn curves:* Stebbins and Jacobsen (1927); *dashes and heavy dots:* McDonald observations.

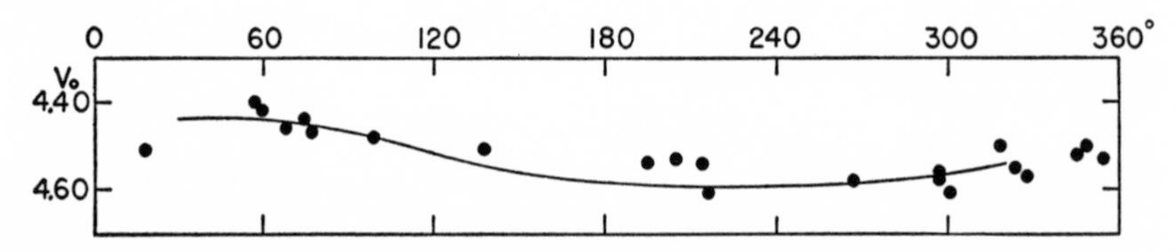

FIG. 8.—Variation of Ganymede (Jupiter III) in yellow light, V_0, with rotational phase angle. *Curve:* Stebbins and Jacobsen (1927); *dots:* McDonald observations.

Near opposition, the surface of Jupiter IV shows little variation with rotational phase, but, as the solar phase angle increases, the leading side decreases in brightness considerably more than does the trailing side. The McDonald color observations show no variation in either $B - V$ or $U - B$ with orbital phase or with difference in solar phase angle; the values of the colors are $B - V = +0.86$ and $U - B = +0.55$.

Comparing the decrease in brightness from opposition to $\alpha = 10°$, we find that for $\theta = 140°$ the decrease is 0.59 mag., while at $\theta = 250°$ it is only 0.41 mag. As Stebbins has noted, these changes in brightness are very large compared with those observed for other objects in the solar system. The moon shows a decrease of only 0.26–0.30 mag. over the same range.

As for the moon, the rapid decrease in brightness near opposition in Juipter IV is assumed to be due to small irregularities of the surface, casting shadows. It is therefore interesting that the leading side shows this effect more than the trailing side, possibly indicating a sweeping action through a particle cloud having a lesser angular velocity around Jupiter.

5.45. *The other satellites.*—The magnitude of Jupiter V is given as +13.0 in RDS. Van Biesbroeck and Kuiper (1946) have observed the satellite under moderately favorable conditions near western elongation and sug-

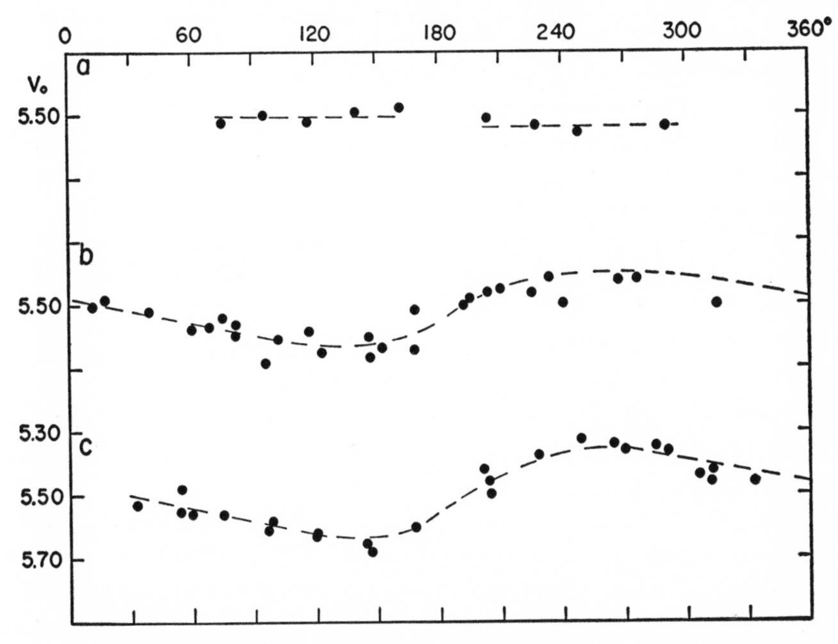

FIG. 9.—Variation of Callisto (Jupiter IV) in yellow light, V_0, with rotational phase angle. (*a*) Solar phase angle less than 1°.5; (*b*) solar phase angle between 2°.7 and 8°; (*c*) solar phase angle greater than 8°. Data by Stebbins and Jacobsen (1927).

gest that the satellite may be fainter than when it is near eastern elongation.

Jupiter VI is a moderately bright object, for which $V_0 = 13.7$ (RDS), and should be investigated for variation of brightness.

Jupiter VII is given as $V = 16$ in RDS, and no other determinations have been published.

The magnitudes of the last five satellites of Jupiter are given below and are photographic measures and estimates kindly made available by Dr. S. B. Nicholson for inclusion in this section. They are revisions of Nicholson's earlier values (published in the *Pub. A.S.P.*), based on corrections to the photographic magnitude scale indicated by recent photoelectric observations.

	P_g		P_g
Jupiter VIII......	19.6	Jupiter XI.......	18.9
Jupiter IX.......	19.1	Jupiter XII......	19.6
Jupiter X........	19.4		

We have assumed a mean color index of 0.8 mag. for these objects to reduce them to V magnitudes.

5.5. The Satellites of Saturn

5.51. *Satellite I, Mimas.*—This satellite has not been measured photoelectrically because of its close proximity to the ring. The magnitude derived from earlier visual measures is about $\bar{V}_0 = 12.1$ (RDS).

5.52. *II, Enceladus.*—Two observations at McDonald (made in 1951, 1956) give $V_0 = 11.77$ and $B - V = 0.62$. No measures of the ultraviolet were obtained because of the proximity to the ring.

5.53. *III, Tethys.*—The magnitudes and colors based on four nights at McDonald (in 1951, 1953, 1956) are $V_0 = 10.27$, $B - V = 0.73$, $U - B = 0.34$.

5.54. *IV, Dione.*—The magnitudes and colors based on five nights at McDonald (in 1951, 1953, 1956) are $V_0 = 10.44$, $B - V = 0.71$, $U - B = 0.30$.

5.55. *V, Rhea.*—The mean magnitude and colors based on 11 nights at McDonald (in 1951, 1952, 1953, 1956) are $V_0 = 9.76$, $B - V = 0.76$, $U - B = 0.35$. Although the observations are not well enough distributed with regard to orbital phase (Fig. 10, *a*) to give a well-determined light-curve, it appears that the magnitude varies about 0.20; maximum (9.68) occurs near $\theta = 30°$ and minimum (9.88) near $\theta = 240°$. The colors show little or no variation.

5.56. *VI, Titan.*—The magnitudes and colors observed at McDonald on

19 nights from 1951 to 1956 give $V_0 = 8.39$, $B - V = 1.30$, $U - B = 0.75$. Although this satellite has been considered to be variable in light with a period equal to that of its revolution about Saturn and an amplitude of 0.24 mag. (Pickering, 1913), there is no definite evidence of a variation in our measures (Fig. 10, *b*). Examination of the observations by Wendell (1913) used by Pickering indicates that the assumed variation is probably due to the errors in the magnitudes of the comparison stars and not to a variation in the brightness of Titan. For example, Wendell measured

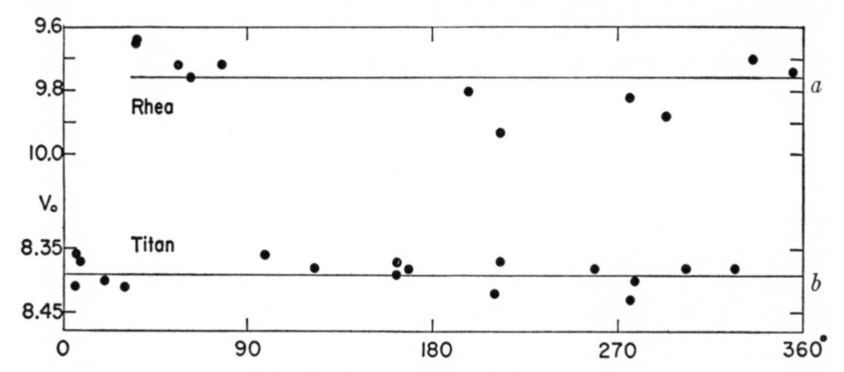

FIG. 10.—The yellow magnitudes, V_0, of Rhea and Titan as a function of rotational phase angle.

TABLE 12

WENDELL'S OBSERVATIONS, TITAN −(BD−21°4540)

Phase	Δm	Phase	Δm
23°..........	−0.535	201°........	−0.535
67..........	− .535	246..........	− .535
90..........	− .49	269..........	− .555
113..........	− .51	337..........	−0.58
134..........	−0.52		

Titan with respect to BD −21°4540 on 9 nights when Saturn was near its stationary point. The magnitude differences listed in Table 12, on the assumption that Titan is constant in light, give a mean error of only ±0.024 mag. for one observation.

5.57. *VII, Hyperion.*—The mean magnitude of Hyperion as determined at McDonald in 1953 is $V_0 = 14.16$ (5*n*), while the colors are $B - V = 0.69$, $U - B = 0.42$:. The magnitude is considerably fainter than the earlier estimates.

5.58. *VIII, Iapetus.*—The variation in brightness of Iapetus is exceptional because of its large range of over 2 mag. It was discovered by J. D.

Cassini in 1671, who could see the satellite only at western elongation. Herschel could follow the satellite around the planet and obtained an approximate light-curve. A visual photometric light-curve was obtained by Pickering (1879). A more recent investigation by Widorn (1950) indicates that the brightness (not the magnitude) of the satellite can be represented quite well by the simple formula

$$H = 0.571 - 0.429 \sin \theta \tag{12}$$

with the maximum occurring at western elongation.

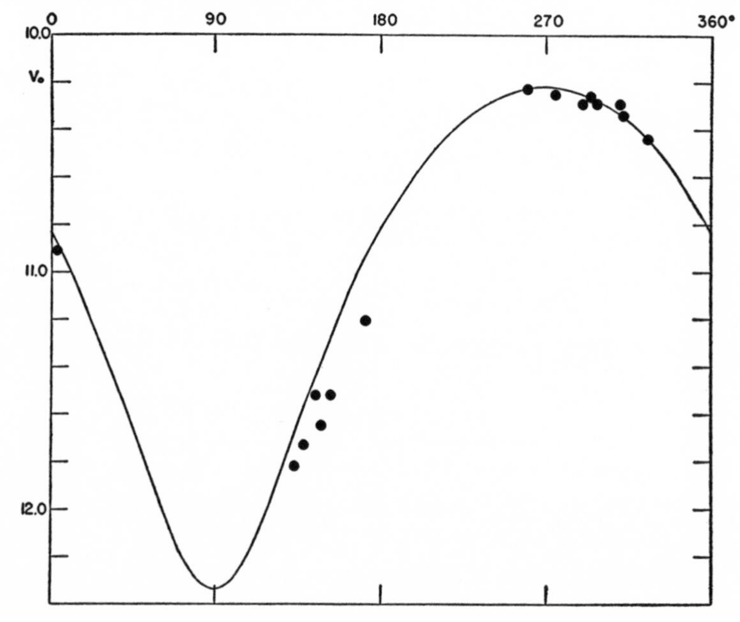

FIG. 11.—Variation of Iapetus in yellow light, V_0, with rotational phase angle. *Solid line:* Widorn's mean light-curve; *filled circles:* McDonald observations.

The satellite has been observed at McDonald on 17 nights from 1951 to 1953. These observations indicate a phase coefficient of about 0.03 mag/degree. The individual observations have been plotted in Figure 11, where Widorn's mean light-curve is shown as a solid line adjusted so that the maximum is 10.22 mag. In spite of the large range in magnitude, the colors are nearly the same at all phases; the mean colors for $0° < \theta < 180°$ are $B - V = 0.73$, $U - B = 0.30$, while for $180° < \theta < 360°$ the mean colors are $B - V = 0.69$, $U - B = 0.26$.

Widorn, from a consideration of his and earlier series of visual observations, has discussed the possibility that the light-curve of Iapetus is not constant in shape or mean magnitude. This must be further investigated, as the early observations appear to be rather unreliable.

5.6 The Satellites of Uranus

The magnitudes and colors of the two outer satellites of Uranus have been measured photoelectrically on 3 nights at McDonald. For the inner three satellites, Mr. T. Gehrels has determined the relative magnitudes from photographs taken with the 82-inch reflector. The results are given in Table 13.

5.7. The Satellites of Neptune

5.71. *Satellite I, Triton.*—The close satellite of Neptune is difficult to observe photometrically because of its proximity to the much brighter

TABLE 13

Magnitudes and Colors of Uranus Satellites

Satellite	V_0	$B-V$	$U-B$
V Miranda	16.5	....	....
I Ariel	14.4	....	....
II Umbriel	15.3	....	....
III Titania	14.01	0.62	0.25
IV Oberon	14.20	0.65	0.24

planet. Five observations made at McDonald (in 1950, 1951, 1953, 1956) give $V_0 = +13.55, B - V = 0.77, U - B = 0.40$. There is an indication that Triton's leading side is brighter than the following side by approximately 0.25 mag. Because of the limited number of observations and the difficulty in making them, this apparent variation is considered with reservation.

5.72. *II, Nereid.*—The distant satellite of Neptune has an apparent photographic magnitude of $Pg = 19.5$, determined by Kuiper. If its color is $B - V = 0.8$, the value of V_0 will be about 18.7.

5.8. Summary of Satellite Magnitudes

The magnitudes of the satellites reduced to unit distance, $V(1, 0)$, their mean opposition magnitudes, and remarks concerning their variability due to their rotation are summarized in Table 14. It appears that the periods of rotation that have been found are identical with the periods of revolution of the satellites around their primaries, indicating that the satellites keep the same face toward the planet as does the moon.

Photometrically it is seen that the moon ranks seventh in brightness; the four Galilean satellites, Titan, and Triton are considerably brighter, while Rhea is equal to the moon.

The two satellites of Mars and the outer satellites of Jupiter are intrinsically the faintest objects, and it is obvious that their discovery is due to the comparative proximity to the earth; Deimos at a distance of Neptune would be an object of the twenty-eighth magnitude.

TABLE 14

SUMMARY OF SATELLITE MAGNITUDES

	Satellite	$V(1, 0)$	$\bar{V}_0$	Rotational Variation (Mag.)
	Moon.........	+ 0.21	−12.74	±0.08*
Mars	I Phobos......	+12.1	+11.6	....
	II Deimos.....	+13.3	+12.8	....
Jupiter	V............	+ 6.3	+13.0	....
	I Io..........	− 1.90	+ 4.80	0.21
	II Europa.....	− 1.53	+ 5.17	0.34
	III Ganymede..	− 2.16	+ 4.54	0.16
	IV Callisto.....	− 1.20	+ 5.50	0.16†
	VI...........	+ 7.0	+13.7	....
	VII..........	+ 9.3:	+16	....
	VIII.........	+12.1‡	+18.8‡	....
	IX...........	+11.6‡	+18.3‡	....
	X............	+11.9‡	+18.6‡	....
	XI...........	+11.4‡	+18.1‡	....
	XII..........	+12.1‡	+18.8‡	....
Saturn	I Mimas.......	+ 2.6	+12.1	....
	II Enceladus...	+ 2.22	+11.77	....
	III Tethys.....	+ 0.72	+10.27	....
	IV Dione......	+ 0.89	+10.44	....
	V Rhea........	+ 0.21	+ 9.76	0.20
	VI Titan......	− 1.16	+ 8.39	....
	VII Hyperion..	+ 4.61	+14.16	....
	VIII Iapetus...	+ 1.48	+11.03	2.12
Uranus	V Miranda.....	+ 3.8	+16.5	....
	I Ariel.........	+ 1.7	+14.4	....
	II Umbriel.....	+ 2.6	+15.3	....
	III Titania.....	+ 1.30	+14.01	....
	IV Oberon.....	+ 1.49	+14.20	....
Neptune	I Triton.......	− 1.16	+13.55	0.25:
	II Nereid......	+ 4.0‡	+18.7‡	....

* Slight difference between waxing and waning phases.

† For $\alpha > 8°$; small near opposition.

‡ *Pg* magnitudes reduced with a mean color index $+0^{m}8$.

6. COLOR MEASUREMENTS

6.1. The Sun

Because of the difficulties encountered in determining the color of the sun directly, we generally assume that its colors are the same as those of stars with the same spectral type and luminosity. The spectral type of the sun has been determined by Keenan and Morgan (1952) to be G2 V; it refers to the solar spectrum as reflected from one of Jupiter's satellites and is directly comparable with spectra of stars.

As the relationship between spectral type and color is well defined for MK types (e.g., Morgan, Harris, and Johnson, 1953), the mean values of G2 V, $B - V = +0.63$ and $U - B = +0.14$, should closely approximate the colors of the sun. The values of $V - R = 0.45$ and $R - I = 0.29$ were derived by Hardie (1956).

Stebbins and Kron (1956) have determined the $B - V$ color index of the sun directly in conjunction with their determination of the magnitude of the sun (Sec. 2). Their adopted value of 0.636 ± 0.006 (m.e.) for $B - V$ agrees with the value based on the spectral type.

6.2 The Planets

The colors used in this chapter are based on two unpublished sources: the McDonald measures of U, B, V and the R and I values derived by

TABLE 15

Mean Colors of the Planets

Planet	$U-B$	$B-V$	$V-R$	$R-I$
Sun..........	0.14	0.63	+0.45	+0.29
Mercury......	....	0.93	+0.85	+ .52
Venus........	.50	0.82	....	....
Mars.........	.58	1.36	+1.12	+ .38
Jupiter.......	.48	0.83	+0.50	− .03
Saturn.......	.58	1.04	....	....
Uranus.......	.28	0.56	−0.15	− .80
Neptune......	.21	0.41	−0.33	− .80
Pluto.........	0.27	0.80	+0.63	+0.28

Hardie at the McDonald and Lowell Observatories, kindly made available in advance of publication. They are collected in Table 15.

Table 16 gives the corresponding difference, planet − sun, for each wave-length region, the values of ΔV being taken equal to zero. The data of Table 16 are plotted in Figure 12, which therefore represents the reflectivity as a function of wave length on a logarithmic scale.

The observed colors for Mercury are rather uncertain, as the observa-

TABLE 16

COLOR DIFFERENCES BETWEEN PLANETS AND SUN

Planet	U	B	V	R	I
Neptune.......	−0.15	−0.15	0.00	+0.78	+1.87
Uranus........	+0.07	− .07	.00	+ .60	+1.69
Pluto..........	+0.30	+ .17	.00	− .18	−0.17
Venus.........	+0.55	+ .19	.00	....	....
Jupiter.........	+0.54	+ .20	.00	− .05	+0.27
Mercury.......	....	+ .30	.00	− .40	−0.63
Saturn.........	+0.85	+ .41	.00	....	....
Mars..........	+1.17	+0.71	0.00	−0.67	−0.76

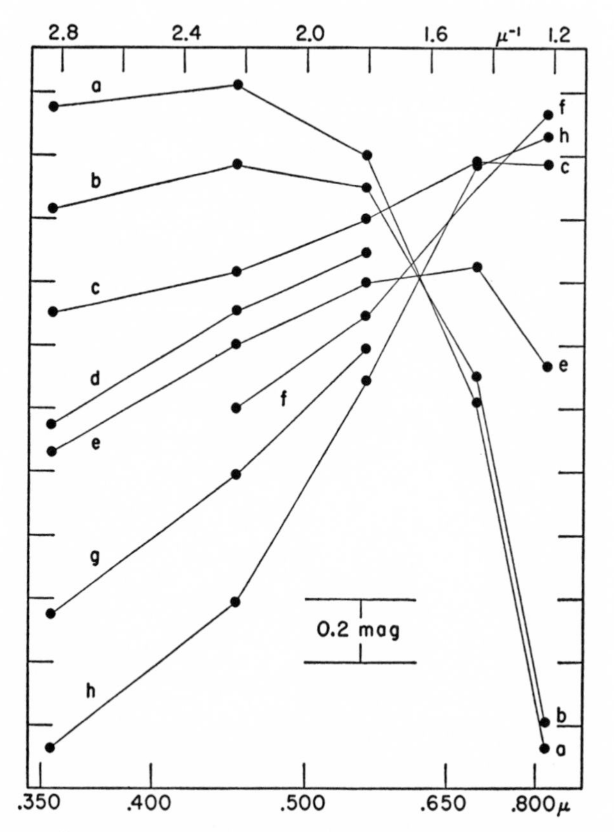

FIG. 12.—Wave-length dependence of the reflectivity of the planets. (*a*) Neptune; (*b*) Uranus; (*c*) Pluto; (*d*) Venus; (*e*) Jupiter; (*f*) Mercury; (*g*) Saturn; (*h*) Mars.

tions have to be made against a bright sky or at twilight when the zenith distance is large. The values for Mars are based on the results at two oppositions and have been freed from the effects due to rotation. For Saturn the color of the combined ball and ring depends on the appearance of the ring, as the latter is considerably bluer than the planet alone. The values given in the table are extrapolations obtained from measures when the inclination of the ring to the line of sight was small to "no-ring."

It must be remembered that our colors are based on broad-band filters and are affected by differences in absorption bands. The first two planets plotted in Figure 12 show that the effect of the absorption bands is very prominent in the *R* and *I* measures and is also present to a lesser extent in the *V* measures.

In the case of Mars we have additional data concerning the variation of color with wave length from the monochromatic magnitudes by Woolley and collaborators (1953, 1955), made during the 1952 and 1954 oppositions. The results obtained in 1954 are plotted in Figure 13, *a*, as circles

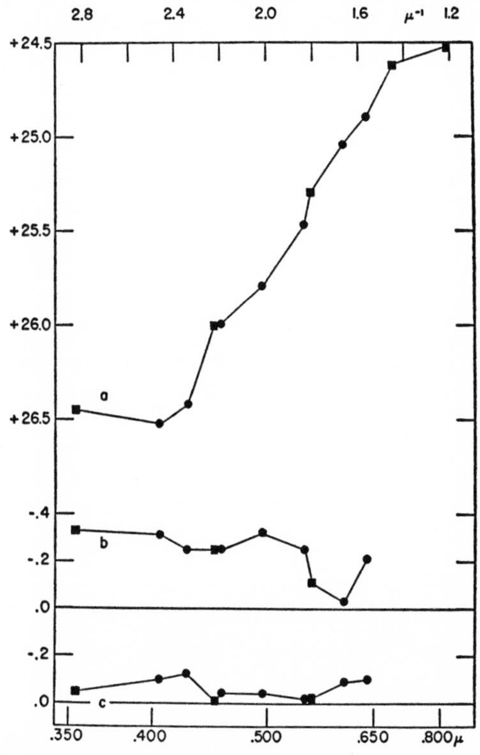

FIG. 13.—(*a*) Wave-length dependence of the mean reflectivity of Mars. (*b*) Deviations for abnormal night, May 3–4, 1952, from mean curve. (*c*) Deviations for nearly normal nights, March 8 and 12, 1952, from mean curve. Squares are Woolley's (1954) observations of monochromatic magnitude differences, Mars − sun; circles are photoelectric means.

and show good agreement with the photoelectric values plotted as squares. It is of interest to note that the observed U point is brighter than the observation at 0.405 μ, indicating the presence of Rayleigh scattering in the Martian atmosphere.

The Mount Stromlo opposition magnitudes and phase coefficients derived for 1952 appear adversely affected by the observations of May 3–4, 1952, when the planet was close to opposition ($\alpha = 2^\circ.5$). Fortunately, McDonald observations were secured on that same night, and the two sets are accordant. They show that the brightness and the color of Mars on that night were abnormal. We have plotted the differences, May 3–4, 1952, *minus* normal, in Figure 13, *b*, from which it is seen that both the Mount Stromlo and the McDonald data differ systematically from the remainder of the observations. On May 3–4, 1952, the planet was both brighter and bluer than normal. A similar comparison was made for more typical nights, March 8 and 12, 1952, as shown in Figure 13, *c*. The abnormal nature of the night May 3–4 is apparent. These conditions may have been extreme, but variations of a smaller amount may be more common; such variations have been noted by Johnson and Gardiner (1955).

6.3 The Satellites

The adopted values of the colors of the satellites are given in Table 17, while the differences, satellites − sun, are given in Table 18. These latter results are plotted in Figure 14, in a manner similar to that for the planets.

The results for Oberon and Titania, though not of high weight because of the proximity to Uranus and their apparent faintness, show that the surfaces of these satellites must be nearly neutral reflectors of solar radiation. In general, the surface reflectivity is higher in the red part of the spectrum than in the blue and ultraviolet, the extreme example being that of Io.

The bottom two curves—those of the moon and Titan—are of particular interest in that they represent spectral distributions for a satellite without an atmosphere and one with an atmosphere. The effect of the atmosphere is apparent in the ultraviolet but is very obvious in the infrared, where the methane absorption bands are prominent.

For the faintest satellites, only photographic magnitudes are available at present; they have been reduced to visual with the median value of +0.75 for $B - V$ obtained from the brighter satellites without atmospheres. Because of the exceptional behavior of Io, some uncertainty is attached to this reduction.

TABLE 17

COLORS OF THE SATELLITES

Satellite	$U-B$	$B-V$	$V-R$	$R-I$	Remarks
Earth:					
Moon	0.46	0.92	0.80	0.46	Variable over surface
Mars:					
I Phobos	....	0.6	....	....	
II Deimos	....	0.6	....	....	
Jupiter:					
I Io	1.30	1.17	.66	.32	Variable colors
II Europa	0.52	0.87	.57	.31	$U-B$ variable
III Ganymede	0.50	0.83	.59	.31	Nearly constant
IV Callisto	0.55	0.86	.61	.32	Constant
Saturn:					
II Enceladus	....	0.62	....	....	
III Tethys	0.34	0.73	....	....	
IV Dione	0.30	0.71	.48	.32	
V Rhea	0.35	0.76	.61	.26	Constant
VI Titan	0.75	1.30	.88	.11	Constant
VIII Iapetus	0.28	0.71	....	....	Nearly constant
Uranus:					
III Titania	0.25	0.62	.52	.41	
IV Oberon	0.24	0.65	.49	.33	
Neptune:					
I Triton	0.40	0.77	0.58	0.44	

TABLE 18

DIFFERENCE IN COLORS: SATELLITE – SUN

Satellite	U	B	V	R	I	n
Earth:						
Moon	0.61	+0.29	0.00	−0.35	−0.52	9
Jupiter:						
I Io	1.70	+ .54	.00	− .21	− .24	10
II Europa	0.62	+ .24	.00	− .12	− .14	8
III Ganymede	0.56	+ .20	.00	− .14	− .16	6
IV Callisto	0.64	+ .23	.00	− .16	− .19	7
Saturn:						
IV Dione	0.24	+ .08	.00	− .03	− .06	3
V Rhea	0.34	+ .13	.00	− .16	− .13	4
VI Titan	1.28	+ .67	.00	− .43	− .25	11
Uranus:						
III Titania	0.10	− .01	.00	− .07	− .19	1
IV Oberon	0.12	+ .02	.00	− .04	− .08	2
Neptune:						
I Triton	0.40	+0.14	0.00	−0.13	−0.28	5

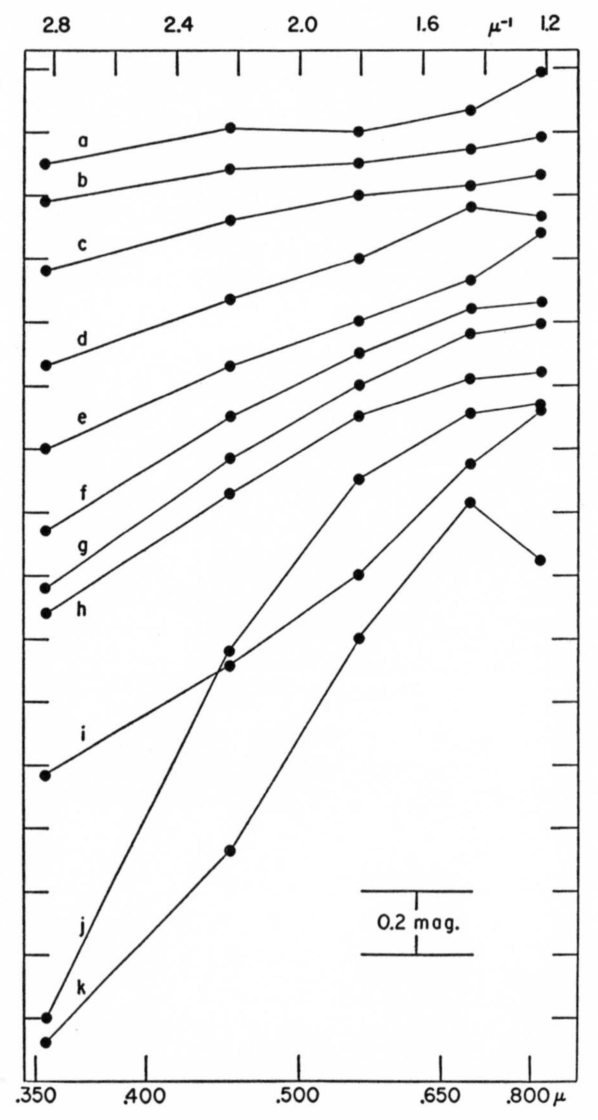

FIG. 14.—Wave-length dependence of the reflectivity of the satellites. (*a*) Titania; (*b*) Oberon; (*e*) Dione; (*d*) Rhea; (*c*) Triton; (*f*) Ganymede; (*g*) Callisto; (*h*) Europa; (*i*) moon; (*j*) Io; (*k*) Titan.

6.4. Observations of the Planets and Satellites in the Infrared

The observed energy distribution of the planets between 1.0 and 2.5 μ has been studied by Kuiper (1952, pp. 351 ff.) with a lead sulfide photoconductive cell and a spectrometer. From the published curve (Kuiper, 1952, Fig. 85) for the moon, the effects of the strong terrestrial water-vapor bands illustrate the difficulty of interpretation of broad-band photometry in this region, especially as these bands are variable in strength.

From the published curves for Mercury the effect of planetary radiation temperature is apparent in the region near 2.1 μ. For Mars the comparison with the moon shows that, with the relative intensities adjusted so that they are equal at 1.3 μ, Mars is about 1.5 times as bright as the moon at 0.8 μ and only 0.5 times as bright at 2.1 μ; i.e., Mars is much "bluer" than the moon in the infrared, contrary to the difference in the blue visual region.

For Venus, and particularly Jupiter and Saturn, respectively, there are additional infrared absorptions by CO_2 and by CH_4 and NH_3. The relative energy distributions of Jupiter and Saturn, however, are quite different in the infrared, Jupiter being much the "bluer" of the two.

Kuiper (1956*b*) has measured the intensities of several of the planets and satellites at 1 and 2 μ, using a lead sulfide cell and broad-band filters. This method does not give the detailed information that is obtained from the spectrometer observations, but the limiting magnitude is much fainter.

The results for the planets and satellites are given in Table 19, where the data are presented in the form $I(2\ \mu)/I(1\ \mu)$. This ratio has been adjusted so that the mean for a number of stars with average spectral type close to that of the sun is 1.

TABLE 19

Infrared Observations of Planets and Satellites

Planet or Satellite	$I(2\mu)/I(1\mu)$	Planet or Satellite	$I(2\mu)/I(1\mu)$
Sun	1.00	Mars	1.00
Venus	1.61	Io	1.06
Saturn	0.47	Europa	0.66:
Saturn's Ring	0.45	Ganymede	0.63
Jupiter	0.21	Callisto	0.95
Uranus	0.06:	Titan	0.20:
Mercury	3.5		

The large value of the ratio, $I(2\ \mu)/I(1\ \mu)$, observed for Mercury is due to planetary radiation (see above). Venus is also relatively bright at 2 μ, but the absorption bands in Saturn, Jupiter, and especially Uranus reduce the intensity at 2 μ considerably below that of the incident radiation from the sun.

For the satellites the methane atmosphere of Titan (Kuiper, 1944, 1952) reduces the intensity at 2 μ below that of Jupiter's satellites. Whether the lower values of Jupiter II and III are to be attributed to similar absorption must await further observations. At the present time no atmospheres have been definitely found for them, but Kuiper (1952, p. 309) does not regard the matter as closed.

7. ALBEDOS OF THE PLANETS AND SATELLITES

7.1 Introduction

Of the several possible definitions of *albedo* (Schönberg, 1929), the ratio of the total flux reflected in all directions to the total incident flux, suggested by Bond (1861) and adopted by Russell (1916*b*), appears to be most suitable for use in the discussion of the planets and satellites.

Consider the planet to be at unit distance from the sun and the earth. Then the apparent diameter at unit distance, σ_1 radians, and the incident solar flux per unit area, πF, give the total flux incident on the planet,

$$\pi F \sigma_1^2 .$$

Let $j(\alpha)\sigma_1^2$ be the flux reflected by the planet in the direction of the earth in which α is the phase angle; then the total flux reflected in all directions is

$$2\pi \int_0^{\pi} j(\alpha)\, \sigma_1^2 \sin \alpha d\alpha .$$

The *Bond albedo, A*, is given by

$$A = \frac{2}{\pi F} \int_0^{\pi} j(\alpha) \sin \alpha d\alpha = \frac{j(0)}{\pi F} \times 2 \int_0^{\pi} \Phi(\alpha) \sin \alpha d\alpha = pq . \quad (13)$$

The factor p, which is called the *geometric albedo*, can be easily evaluated, since

$$m(1, 0) = \text{Constant} - 2.5 \log j(0)\, \sigma_1^2 \quad (14)$$

and

$$m(\text{sun}) = \text{Constant} - 2.5 \log \pi F , \quad (15)$$

so that

$$\log p = 0.4 [m(\text{sun}) - m(1, 0)] - 2 \log \sigma_1 . \quad (16)$$

If the mean radius of a planet is expressed in terms of the earth's mean radius, R/R_E,

$$\sigma_1 = \frac{R}{R_E} \sin 8''.79 \quad (17)$$

and

$$\log p = 0.4 [m(\text{sun}) - m(1, 0)] - 2 \log \frac{R}{R_E} + 8.741 . \quad (18)$$

The factor q, which is called the ***phase integral***, can be computed from the observed variation in brightness with solar phase angle for Mercury, Venus, earth, and the moon but must be estimated from theoretical considerations for the remaining objects.

7.2 The Geometric Albedos of the Planets

Provisional values of the mean radii of the planets as given by Kuiper are listed in the second column of Table 20. Using the values of $V(1, 0)$ given in Table 9 with the adopted value of V (sun) $= -26.81$, the computed values of $p(V)$ listed in Table 20 follow from the definition of p. These values have been computed to three figures although an uncertainty of ± 0.05 mag. in V (sun) introduces an uncertainty of nearly 5 per

TABLE 20

Albedos of the Planets

	R/R_E	$p(U)$	$p(B)$	$p(V)$	$p(R)$	$p(I)$	q_V	$A(v)$
Mercury..	0.38	....	0.076	0.100	0.145	0.179	0.563	0.056
Venus....	0.961	0.353	.492	.586	....	....	1.296	.76
Earth....	1.000	....	....	.367	....	....	1.095	.36*
Mars.....	0.523	.052	.080	.154	.286	.310	1.04	.16
Jupiter...	(11.20, 10.46)	.270	.370	.445	.466	.347	1.65	.73
Saturn...	(9.48, 8.48)	.211	.316	.461	....	....	1.65	.76
Uranus...	3.72	.530	.603	.565	.325	.119	1.65	.93
Neptune..	3.38	.585	.624	.509	.248	.091	1.65	.84
Pluto.....	0.45	0.099	0.111	0.130	0.154	0.152	1.04	0.14

* See text.

cent in p, while additional uncertainty is introduced in the case of Mercury and Pluto, whose diameters are not well determined.

For the other wave-length regions the differences in color, planet − sun (Table 16), give the values of $p(U), \ldots, p(I)$ listed in the table. For Mars, the monochromatic measures by Woolley and his collaborators (1955) provide additional information regarding the variation of p with wave length for this planet. The results derived from their data, using Kuiper's diameter for Mars and our adopted value of V(sun), are listed in Table 21.

The values of p have been plotted against reciprocal wave length in Figure 15. The lines connecting the observed points for the planets Mercury, Mars, and Pluto are probably rather reliable indications of the variation of p with wave length. However, for the remaining planets the presence of strong atmospheric absorption bands makes the variation of p with wave length irregular.

7.3 THE BOND ALBEDOS OF THE PLANETS

The visual Bond albedo for each planet is listed in the last column of Table 20, based on the values of the phase integral, q, given in the penultimate column. For the first three planets the values of q have been derived from the observed phase-curves. For Mars (and the asteroids) we have followed a suggestion by Russell (1916*b*), who noted that the ratio of $q/\Phi(50°)$ was nearly constant. From the modern phase-curves for Mercury, Venus, earth, and moon we find that $q = 2.17\Phi(50°)$ represents the observed values very well; and the value derived for Mars (and the asteroids) from the slight extrapolation of the observed phase-curve should be reliable. The value of q adopted for Pluto is that derived for Mars, although it is possible that a value nearer to that found for Mercury would be more appropriate.

TABLE 21

MONOCHROMATIC ALBEDOS OF MARS

(After Woolley)

	λ(A)						
	4050	4250	4550	4945	5430	5980	6360
$p(\lambda)$	0.049	0.054	0.081	0.097	0.131	0.194	0.227
$q(\lambda)$	.95	.99	1.04	1.09	1.19	1.25	1.31
$A(\lambda)$	0.047	0.053	0.084	0.106	0.16	0.24	0.30

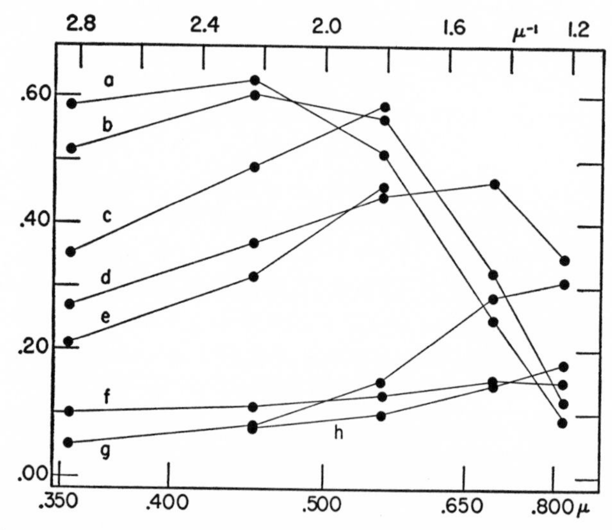

FIG. 15.—Wave-length dependence of the geometrical albedos of planets. (*a*) Neptune; (*b*) Uranus; (*c*) Venus; (*d*) Jupiter; (*e*) Saturn; (*f*) Pluto; (*g*) Mars; (*h*) Mercury.

From the major planets, Russell (1916*b*) assumed $q = 1.50$, which obtains for Lambert's law of diffuse reflection. Based on the discussion of theoretical phase-curves in Section 8, we have adopted $q = 1.65$ for the phase integral in the visual region.

The Bond albedos have not been tabulated for wave-length regions other than the visual because of the uncertainty in the phase integral that should be used. Only in the case of Mars are there data concerning the variation of q with wave length, based on Woolley's (1955) observations and tabulated in Table 21. For the major planets the theoretical discussion in Section 8 indicates that q probably decreases slightly with increasing wave length, while for Venus the low value found for the visual region is probably applicable also in the blue and ultraviolet. It would be desirable to observe the phase-curves for Mercury, Venus, earth, moon, and Mars in wave-length regions other than the visual region and to obtain accurate limb-darkening-curves for Jupiter and Saturn in different colors, so that the Bond albedos could be made more reliable.

For the earth, the visual albedo determined from p and q is 0.40, but the value given in Table 20 is 0.36, based on the discussion by Danjon in Volume **2**, chapter 15, Section 3; the former value is based on observations made in France only, while the latter is a mean value for the entire earth.

It has been pointed out by several investigators (e.g., Sharonov, 1939) that the evaluation of the Bond albedo in the manner outlined in Section 7.1 is only approximate, in that it assumes that the brightness of the planet depends only on the phase angle, α, e.g., that the surface has uniform properties of diffuse reflection. This is obviously not true, for example, in the case of Mars with regard to its north or south poles; but it is difficult to see how such effects of non-uniformity could be quantitatively allowed for.

7.4 Albedos of the Satellites

The values of the geometric albedos, p, for the satellites tabulated in Table 22 are derived from the mean radii given by Kuiper (see Vol. **5** of this series), the values of $V(1, 0)$ given in Table 14, and the color differences given in Table 18. These values are also plotted in Figure 16.

The value of the phase integral required in the determination of the Bond albedo is known only for the moon, for which $q = 0.585$; this has been adopted in the table for all the satellites.

The high values of the geometric albedo for the Saturn satellites are well known, but the extremely high values for Jupiter I and II have not been pointed out before. The magnitudes of Jupiter II, III, and IV as derived from Stebbins' observations in 1926 and 1927 and the magnitudes

TABLE 22

ALBEDOS OF THE SATELLITES

	R/R_E	$p(U)$	$p(B)$	$p(V)$	$p(R)$	$p(I)$	q	A_V
Moon......	0.273	0.066	0.088	0.115	0.16	0.17	0.585	0.067
Jupiter I....	.255	.19	.56	.92	1.12	1.15	.585	.54
Jupiter II...	.226	.47	.67	.83	0.93	0.95	.585	.49
Jupiter III..	.394	.29	.41	.49	0.56	0.57	.585	.29
Jupiter IV..	.350	.14	.21	.26	0.30	0.31	.585	.15
Mimas.....	.04:	....	....	.49	......	......	.585	.29
Enceladus..	.05:	....	.53	.54	......	......	.585	.32
Tethys.....	.08	.64	.76	.84	......	......	.585	.49
Dione......	.07:	.75	.87	.94	0.96	0.99	.585	.55
Rhea.......	.120	.60	.73	.82	0.96	0.93	.585	.48
Titan......	.377	.06	.12	.21	0.32	0.27	.585	.12
Triton......	0.29	0.25	0.32	0.36	0.41	0.47	0.585	0.21

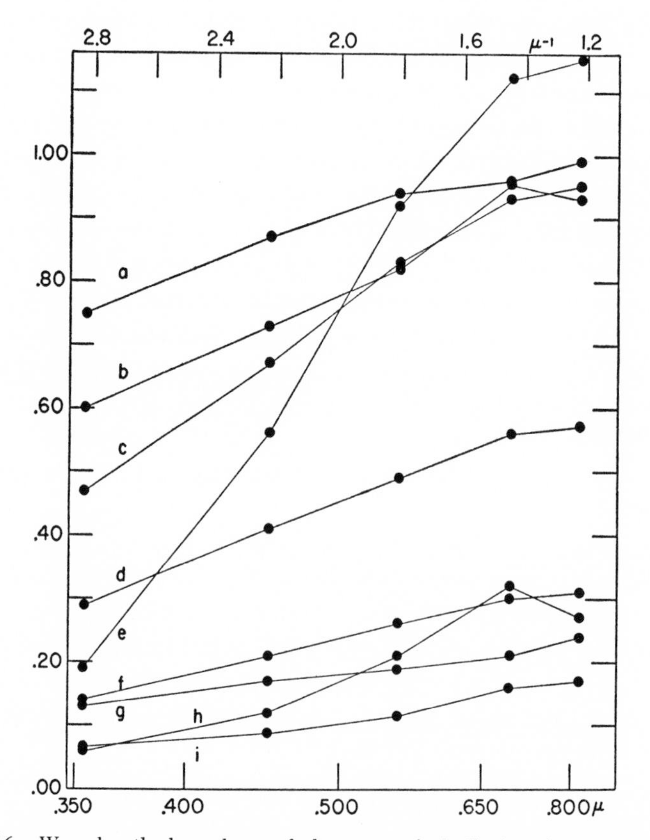

FIG. 16.—Wave-length dependence of the geometrical albedos of satellites. (*a*) Dione; (*b*) Rhea; (*c*) Europa; (*d*) Ganymede; (*e*) Io; (*f*) Callisto; (*g*) Triton; (*h*) Titan; (*i*) moon.

Note added in proof: To be consistent with the value now accepted for the radius of Triton, the albedos shown above (*g*) should be multiplied by 1.9 (see Table 22).

observed at the McDonald Observatory agree in making these satellites considerably brighter than the earlier measures by visual observers. In the case of Jupiter I the reduction of Stebbins' observations to the *V* system was uncertain and would make this satellite even brighter than the adopted value in Table 22, which is based on the McDonald measures.

8. THEORETICAL LIMB DARKENING, PHASE VARIATIONS, AND ALBEDOS

8.1 Introduction

The observations of planets' and satellites' brightness at various phase angles, α, enable one to determine the Bond albedo, A, as summarized in Section 7. It was shown that the albedo could be expressed as a product of two factors, p and q, where the factor p depends only on observations made at full phase and can be derived for all the planets and satellites whose diameters are known. The factor q, on the other hand, can be derived from observations only in the cases of Mercury, Venus, earth, and the moon; for the major planets the factor must be estimated from theoretical considerations, guided in part by observations of limb darkening.

In surveying the theoretical aspects of planetary photometry, we consider three types of "model planets": (*a*) planets with optically thick atmospheres (Secs. 8.4–8.6); (*b*) planets with no atmosphere (Sec. 8.7); and (*c*) planets with optically thin atmospheres (Sec. 8.8). The "models" are assumed to be spherical and to possess homogeneous atmospheres and surfaces. As the real planets are not strictly spherical, their atmospheres not homogeneous, and their surfaces not without definite features, we must consider the results as applying to the average or smoothed-out characteristics of the real objects. This is obviously true for the reflective properties of the lunar surface and the limb darkening of Jupiter with its belts; but it is true also in the case of a rotating planet where convection currents from the bright to the dark side may cause differences of particle size, which affect the brightness systematically over the surface (e.g., van de Hulst, 1952, p. 91).

8.2 Definitions

The co-ordinates of a point on the surface of a planet can be described in terms of a longitude, ω, and a latitude, ψ (Fig. 17). In photometric investigations the natural great circle to employ is that one which passes through the subsolar point, S, and the subterrestrial point, E; longitude is measured along this intensity equator from the subterrestrial point, while latitude is measured along meridians perpendicular to this equator. The longitude of the subsolar point is the solar phase angle, α, already referred to in our discussions of the phase variations of the planets.

For any point on the surface, $P(\omega, \psi)$, we have

$$\begin{aligned} \mu &= \cos \epsilon = \cos \psi \cos \omega \,, \\ \mu_0 &= \cos i = \cos \psi \cos (\omega - \alpha) \,, \end{aligned} \tag{19}$$

where i and ϵ are the angles of the incidence and reflection. Let the angle EPS equal $\pi - \phi$, where ϕ is the azimuthal angle of the reflected ray, given by

$$\cos \phi = \frac{\cos \epsilon \cos i - \cos \alpha}{\sin \epsilon \sin i} \,, \qquad \sin \phi = \frac{\sin \psi \sin \alpha}{\sin \epsilon \sin i} \,. \tag{20}$$

At full phase, $\alpha = 0°$, and all points have $\mu = \mu_0$ and $\phi = 180°$.

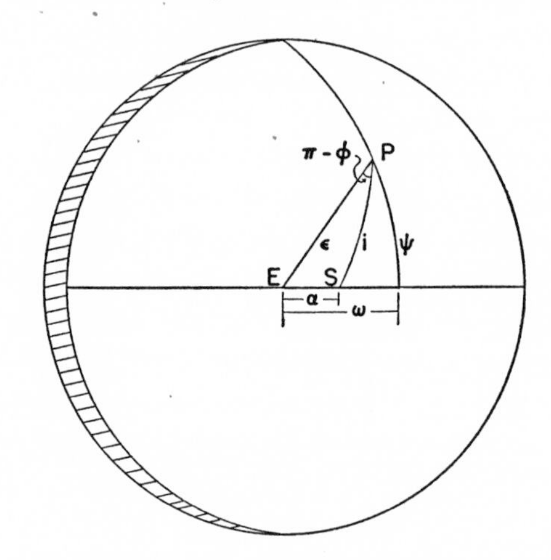

FIG. 17.—Co-ordinate system for photometric investigations of planets

Because of the great distances of the planets from the earth, the apparent disk of the planet is an orthographic projection on the plane of the sky. If R is the apparent radius of the planetary disk, the distance of a point from the center, E, of the disk may be expressed in terms of $r/R = \sin \epsilon$.

8.3 THE PRINCIPLE OF RECIPROCITY

The principle of reciprocity, first formulated by Helmholtz, has been discussed with regard to planetary photometry by Minnaert (1941, 1946). If we consider a point, $P(\omega, \psi)$, on the planet when the solar phase angle is α, we may express the intensity at this point either as $I(\mu, \phi; \mu_0)$ or as $I(\mu, \mu_0, \alpha)$. The first form is generally used in theoretical discussions, but Minnaert has commented on the fact that the latter notation is convenient in photographic photometry of planets, since, for a given photograph, α is constant.

According to the principle of reciprocity, a planet having a surface with homogeneous photometric properties will satisfy the relation

$$\frac{I(\mu, \mu_0, \alpha)}{\mu_0} = \frac{I(\mu_0, \mu, \alpha)}{\mu}, \tag{21}$$

where we have merely interchanged the angles of incidence and reflection.

It is seen that this principle of reciprocity permits one to check the homogeneity of a planetary atmosphere and/or surface, irrespective of the specific form of the law of reflection. If the reciprocity principle is not satisfied, one must look for an explanation outside the scope of the present remarks (see also chap. 6).

8.4 Planets with Optically Thick Atmospheres

As van de Hulst (1952) remarks, a deductive theory of scattering in a planetary atmosphere consists of three distinct parts:

(i) The properties of single scattering have to be computed from the assumed composition and particle size. For those small compared with the wave length of light, the Rayleigh scattering formulae are applicable, while for larger particles Mie's formulae must be used. As the exact formulae are complex and depend on the unknown particle size, it is convenient to assume simple analytical forms for the particle phase function, $f(\cos\Theta)$, where $\Theta = \pi - \alpha$ is the angle between the incident ray and the reflected ray. We shall have occasion to refer to the following phase functions:

Rayleigh scattering with allowance for polarization:	$f_l(\cos\Theta) = \frac{3}{4}\cos^2\Theta,\ f_r(\cos\Theta) = \frac{3}{4}$;
Rayleigh phase function:	$f(\cos\Theta) = \frac{3}{4}\varpi_0(1+\cos^2\Theta)$;
Isotropic scattering:	$f(\cos\Theta) = \varpi_0$;
Anisotropic scattering:	$f(\cos\Theta) = \varpi_0(1+a\cos\Theta)$.

The last three are special cases of

$$f(\cos\Theta) = \varpi_0 + \varpi_1 P_1(\cos\Theta) + \varpi_2 P_2(\cos\Theta), \tag{22}$$

where the $\tilde{\omega}_i$'s are constants and the $P_i(\cos\Theta)$'s are Legendre polynomials.

(ii) For a given particle phase function it is necessary to solve the equation of transfer for the problem of diffuse reflection giving $I(\mu, \phi; \mu_0)$, allowing for multiple scattering. This problem has been the subject of extensive investigations by Chandrasekhar (1950), whose methods are used in our discussion (see van de Hulst, 1952, for a discussion of the inadequacies of earlier investigations). The comparison of the theoretical values of $I(\mu, \phi; \mu_0)$ with observed *limb-darkening* results can be made at this point.

(iii) From the computed values of $I(\mu, \phi; \mu_0)$ the properties of the *total light* reflected by the planet may be determined by integration over the visible portion of the planetary disk. The necessary cubature formulae and coefficients are tabulated by Horak (1950). Letting the incident flux density be πF, the integrations give $j(\alpha)/F$, from which the geometric albedo, $p = j(0)/\pi F$, and the phase variation, $\Phi(\alpha) = j(\alpha)/j(0)$, follow. As $\Delta m(\alpha) = -2.5 \log \Phi(\alpha)$, the theoretical results can be compared with the observed phase variation, $\Delta m(\alpha)$.

(iv) The final step is the determination of the *Bond albedo*. Since $A = p \times q$, A is found from p and the integral,

$$q = 2\int_0^{\pi} \Phi(\alpha) \sin \alpha d\alpha \,.$$

8.5 Limb Darkening, Optically Thick Atmospheres

8.51. *Theory.*—Though the principal interest in the theory of diffuse reflection relates to the total light of the planet and not to local studies of planetary detail, the determination of $I(\mu, \phi; \mu_0)$ is a necessary step in the computations, and comparison of the theoretical discussion of limb-darkening observations may give the only clue to the actual form of the particle phase function.

The limb-darkening-curves for different particle phase functions at full phase can be approximated by the simple formula

$$I(\mu) = I(1.0)\,[1 - b(1-\mu) + c(1-\mu)^2]\,, \text{ valid for } 0 \leqslant \mu \leqslant 1\,. \quad (23)$$

Observationally, it is very difficult to obtain reliable observations beyond $r/R = 0.9$, which includes the range of $0.4 < \mu < 1.0$. Therefore, it is very difficult to determine both b and c from observations, and normally only the coefficient b is obtained, on the assumption that c is zero; this is often called the "limb-darkening coefficient." It appears that, instead of comparing limb-darkening coefficients, it is preferable to consider *the intensity at the characteristic point*, $\mu = 0.5$ or $r/R = 0.866$, and define *the ratio* $x = 2\,I(0.5)/I(1.0)$. This ratio has the important property that, to the extent that $I(\mu)$ can be represented by the second-degree expression given above, $p = j(0)/\pi F = I(1.0)(1 + 2x)/3$. As the quantity p is more easily determined from observations than is $I(1, 0)$, we have used p and x to classify the results for a number of theoretical limb-darkening-curves listed in Table 23. For Lambert's law, $p = 2\lambda_0/3$ and $x = 1.00$.

The results tabulated in Table 23 show the following characteristics.

a) For a given form of the particle phase function, $f(\cos \Theta)$, the values of p decrease rapidly with the *particle albedo*, $\tilde{\omega}_0$. At the same time, x increases, indicating that the limb darkening becomes less pronounced.

b) For a given value of $\tilde{\omega}_0$, the values of p and x increase as the anisotropy of the particle phase functions changes from forward to backward scattering.

c) The values of p and x found for Rayleigh scattering, which properly allows for polarization, show small but significant differences from the results obtained with the Rayleigh phase function. Although the latter has often been employed in the past, only the former has any physical significance.

d) The values of p and x found for Rayleigh scattering (and also for the Rayleigh phase function approximation) appear to be essentially the result of the backward-scattering component.

8.52. *Limb darkening of the major planets.*—The values of the geometric

TABLE 23

THEORETICAL LIMB-DARKENING PARAMETERS

	p	x
Isotropic scattering and ϖ_0 = 1.000	0.690	0.96
0.999	.636	1.01
0.995	.575	1.07
0.990	.534	1.12
0.985	.505	1.15
0.980	.482	1.18
0.975	.462	1.20
0.950	.392	1.30
0.925	.345	1.36
0.900	.310	1.41
0.850	.257	1.50
Anisotropic scattering, with $1+\cos\Theta$	.611	0.73
$1+0.6\cos\Theta$	.647	0.84
$1+0.2\cos\Theta$	.677	0.92
$1-0.2\cos\Theta$	.702	0.99
$1-0.6\cos\Theta$	.723	1.05
$1-\cos\Theta$	.742	1.11
$0.975\,(1+\cos\Theta)$	.343	0.89
$0.975\,(1+0.6\cos\Theta)$	.398	1.05
$0.975\,(1+0.2\cos\Theta)$	.442	1.16
$0.975\,(1-0.2\cos\Theta)$	.480	1.24
$0.975\,(1-0.6\cos\Theta)$	.512	1.31
$0.975\,(1-\cos\Theta)$	.539	1.37
Rayleigh scattering	.798	1.09
Rayleigh phase function or $3\,(1+\cos^2\Theta)/4$	0.752	1.05

albedos, p, of the major planets are given in Table 20, and it is seen that none approaches the value of 0.80 given by Rayleigh scattering or the value of 0.69 given by isotropic scattering with $\tilde{\omega}_0 = 1.0$, indicating that we must consider the cases of non-conservative scattering, i.e., $\tilde{\omega}_0$ less than 1.

The values of x determined from the observed limb-darkening-curves are uncertain for the following reasons:

a) The photometric problem of measuring the variation of intensity over the small disk of a planet photograph only a few millimeters in diameter is complicated by atmospheric seeing effects and photographic effects that are difficult to allow for.

b) The apparent surface of the planet shows bands and/or spots. Therefore, it is necessary to combine many limb-darkening-curves in order to obtain a representative mean that can be compared with theory.

For Jupiter and Saturn the measures by Barabaschev and Semejkin (1933, 1934) and Sharonov (1939, 1940) indicate that x is less than 1—say 0.8—in the red region and increases to about 1.2 in the blue region of the spectrum. Kuiper's measures of Jupiter at 3500 A (quoted by van de Hulst, 1952, p. 100) indicate that x increases to 1.52, while in the infrared, at 13000 and 16000 A, x is a little less than 1.

The difficulties of observation increase for the more distant planets, Uranus and Neptune. However, Richardson (1955) has observed Uranus with the 100-inch reflector and finds that the intensity variation across the disk is similar to that of Jupiter.

The ultraviolet results for Jupiter, $p(u) = 0.25, x = 1.52$, can be interpreted as being due to isotropic scattering with albedo $\tilde{\omega}_0 = 0.85$. However, isotropic scattering cannot explain the results in the blue, visual, and red regions, as the computed values of x are too large for the corresponding values of p. In order to obtain agreement with the observed measures, it appears that one must postulate a forward-scattering particle phase function, a simple form of which might be $f(\cos\Theta) = \tilde{\omega}_0(1 + a\cos\Theta)$.

This case was first examined by Horak (1950), at which time the necessary auxiliary functions had not yet been computed (Appendix I), and approximations were obtained by interpolation between $f(\cos\Theta) = \tilde{\omega}_0$ and $f(\cos\Theta) = \tilde{\omega}_0(1 + \cos\Theta)$. Using this method, he suggests that, for the visual region, Saturn's limb-darkening-curves can be explained by $f(\cos\Theta) = 0.985\,(1 + 0.9\cos\Theta)$, for which $p_V = 0.40, x = 0.91$; similarly in the blue region, $f(\cos\Theta) = 0.925\,(1 + 0.65\cos\Theta)$ and $p_V = 0.27$, $x = 1.18$.

Using the same method but giving less weight to the measures near the limb, we obtain a less extreme form of scattering function, $f(\cos \Theta) = 0.98\,(1 + 0.6 \cos \Theta)$, for which $p = 0.43$, $x = 1.01$, to explain the observations of Jupiter and Saturn in the visual region.

It is recognized that the simple adopted form of anisotropic scattering is only a mathematical simplification; but the limited accuracy of the observed limb-darkening-curves does not at present appear to justify an extension of the theoretical investigation to more complicated forms for $f(\cos \Theta)$.

8.6. Variation with Solar Phase Angles and the Bond Albedo

The integrated or total light, diffusely reflected toward the earth by the planet at a given phase angle α, is

$$j(\alpha) = \int I(\mu, \phi; \mu_0)\, \mu dS \,. \tag{24}$$

TABLE 24

Theoretical Phase Variations, $\Phi(\alpha)$

α	Rayleigh Scattering	Isotropic Scattering		Forward Scattering		Lambert's Law
		$\varpi_0 = 1.00$	$\varpi_0 = 0.95$	$0.975\,(1+0.5 \cos \Theta)$	$0.985\,(1+0.9 \cos \Theta)$	
0°	1.00	1.00	1.00	1.00	1.00	1.000
20	0.91	0.94	0.93	0.95	0.96	0.944
40	0.71	0.78	0.79	0.82	0.84	0.800
50	0.59	0.69	0.70	0.74	0.77	0.708
60	0.48	0.58	0.60	0.65	0.68	0.609
80	0.31	0.39	0.42	0.46	0.51	0.410
100	0.18	0.22	0.25	0.29	0.33	0.236
120	0.091	0.106	0.13	0.16	0.18	0.109
140	0.037	0.037	0.049	0.062	0.070	0.034
160	0.009	0.006	0.010	0.012	0.016	0.006
180°	0.000	0.000	0.000	0.000	0.000	0.000
q	1.25	1.45	1.52	1.65	1.77	1.50
$q/\Phi(50°)$	2.11	2.12	2.19	2.24	2.30	2.12

At full phase, $\alpha = 0°$, and we obtain $j(0)/\pi F = p$, already considered. For other phase angles the brightness is expressed in terms of the brightness at full phase, or $\Phi(\alpha) = j(\alpha)/j(0)$. From $\Phi(\alpha)$ we compute the integral

$$q = 2 \int_0^\pi \Phi(\alpha) \sin \alpha d\alpha$$

and obtain the Bond albedo, $A = p \times q$.

The values of $j(\alpha)/F$ for several different particle phase functions have been computed by Horak, from which we have obtained the values of $\Phi(\alpha)$ tabulated in Table 24; the results for Lambert's law have been taken from Schönberg (1929). The last two lines of the table give the integral q and the ratio q/Φ (50°). The latter ratio was found to be about 2.17 from the observed phase-curves of the moon and terrestrial planets; it appears that the ratio increases slightly with the brightness at phase angle 50°.

From the examples listed in Table 24 it is seen that Rayleigh scattering gives the most rapid brightness decrease with phase angle, while the forward-scattering phase functions show the slowest decrease. For all the phase functions considered, the decrease in brightness over the first few degrees is very small; for example, Talley and Horak's (1956) accurate integration for isotropic scattering ($\tilde{\omega}_0 = 1.0$) gives Δm (10°) $= -2.5 \log \Phi$ (10°) $= 0.020$ mag., so that a linear phase coefficient of the order of only 0.002 mag/degree would be expected. This is the order of magnitude obtained empirically for Uranus; it suggests that the higher values found for Jupiter are due to irregularities, such as spots.

TABLE 25

THEORETICAL ALBEDO PARAMETERS, p, q, AND A

	p	q	A
Isotropic scattering and $\varpi_0 = 1.000$	0.690	1.45	1.00
0.975	.462	1.52	0.70
0.950	.392	1.52	0.60
Anisotropic scattering with $1+\cos\Theta$	.611	1.64	1.00
$1+0.6\cos\Theta$	.647	1.55	1.00
$1+0.2\cos\Theta$	.677	1.48	1.00
$1-0.2\cos\Theta$	.702	1.42	1.00
$1-0.6\cos\Theta$	.723	1.38	1.00
$1-\cos\Theta$	.742	1.35	1.00
$0.975\,(1+0.5\cos\Theta)$	.41	1.65	0.67
$0.985\,(1+0.9\cos\Theta)$	.40	1.77	0.71
Rayleigh scattering	.798	1.25	1.00
Rayleigh phase function, $\varpi_0=1.000$	.752	1.33	1.00
$3\varpi_0(1+\cos^2\Theta)/4$, $\varpi_0=0.975$	.52	1.32	0.69
$3\varpi_0(1+\cos^2\Theta)/4$, $\varpi_0=0.950$	0.45	1.33	0.60

Using q as a parameter to characterize the variation with solar phase angle, we may derive values of $q = 1/p$ for the cases of conservative scattering discussed earlier. The results are collected in Table 25. From these we see that q increases slightly with decreasing particle albedo, while the asymmetry of the particle phase function introduces considerable changes, in the sense that forward-scattering functions have the larger q's. The low values of q for Rayleigh scattering (and for the Rayleigh phase-function approximation) can be attributed to the backward-scattering component. The values listed for $f(\cos\Theta) = \tilde{\omega}_0[\frac{3}{4}(1 + \cos^2\Theta)]$ are based on approximations discussed in the next section.

8.61. *The phase variation of Venus.*—The planet Venus can be observed over a large range in phase angle, and the variation in brightness is well determined by Danjon's (1949) measures. From the observed variation one finds that the integral q is only 1.30, which, combined with $p = 0.55$, gives a Bond albedo in the visual region of 0.71. The low value of q indicates that the variation with phase angle must be similar to that found for Rayleigh scattering, implying an anisotropic particle phase function with

TABLE 26

OBSERVED AND THEORETICAL PHASE VARIATION OF VENUS

α	Observed (Danjon)	Rayleigh Scattering	$0.975\,[\frac{3}{4}(1+\cos^2\Theta)]$
0°	1.00	1.00	1.00
20	0.91	0.91	0.92
40	0.71	0.71	0.73
60	0.49	0.48	0.52
80	0.31	0.31	0.33
100	0.18	0.18	0.20
120	0.107	0.091	0.102
140	0.062	0.037	0.042
160	0.036	0.009	0.009
180°	0.023	0.000	0.000
p	0.548	0.798	0.52
q	1.296	1.33	1.32
A	0.71	1.00	0.69

a strong backward component. We have tabulated the values of $\Phi(\alpha)$ as defined by Danjon's results in the second column of Table 26, along with the values given by Rayleigh scattering. It is seen that the agreement is quite good up to about phase angle 120° but that there is an excess observed brightness for larger phase angles. The value given for 180° results from an extrapolation of Danjon's measures from his upper limit of 171°, and indicates that the excess is about 2.3 per cent. It is well known that Venus can be observed as an annulus near inferior conjunction, the ring being considerably brighter than the sky in its neighborhood. This excess brightness is attributed to light diffused around the edge of the planet by the atmosphere—it is probable that the excess brightness at large phase angles over the Rayleigh values can be accounted for in the same manner.

As the Bond albedo is less than unity, we must introduce a particle albedo $\tilde{\omega}_0$, less than unity, and we are led to investigate the possibility that the particle phase function has the form $f(\cos\Theta) = \tilde{\omega}_0[\frac{3}{4}(1 + \cos^2\Theta)]$ which would reduce p and at the same time presumably have a phase variation similar to that of Rayleigh scattering. The accurate solution for a phase function of this form has not been attempted. However, Chandrasekhar (1950, p. 146) has pointed out that the effects of higher-order scattering for the Rayleigh phase function and for isotropic scattering are very similar. Accordingly, we have made the approximation that first-order scattering obeys our modified Rayleigh phase function, while the higher-order scattering is that given by isotropic scattering with the same albedo. This seems to be a rather good approximation for cases where the deviation from isotropy is small, but it becomes increasingly poorer as the phase function becomes more and more asymmetrical.

The solutions for this approximation are summarized in the last section of Table 25, and the computed phase variation for $\tilde{\omega}_0 = 0.975$ is given in the last column of Table 26. Although this is only an approximate solution and must be confirmed by accurate computations, it appears that an exact solution will probably explain the general features of the photometric behavior of Venus.

Horak (1950), who investigated the problem of Venus, suggested that isotropic scattering with $\tilde{\omega}_0 = 0.95$ would represent the phase variation for angles less than 130° as well as the limb-darkening measures. In regard to the phase variations, the difference between his computed q of 1.52 and the observed value is considerable, while Minnaert (1946) has shown that the limb-darkening measures do not satisfy the reciprocity principle and therefore cannot be represented by any of the simple theoretical cases considered here.

8.62. *The adopted values of q for the major planets.*—It appears that three courses are open in the selection of an appropriate value of q for use in albedo determinations of the major planets. First, one might adopt the value of $q = 1.30$ found for Venus, on the premise that both Venus and the major planets possess optically thick atmospheres. Second, one might assume that the major planets are covered by a cloud surface that reflects light according to Lambert's law, which would be in rough agreement with the limb-darkening results, for which x varies from, say, 0.8 to 1.2. This is the course adopted by Russell (1916*b*), who takes $q = 1.50$.

The third alternative is to assume that the major planets have optically thick atmospheres and to obtain from the limb-darkening measures the result that the particle phase function must have a forward component which decreases with wave length, becoming nearly isotropic in the ultraviolet. This is the course adopted by Horak (Sec. 8.5); his particle phase functions lead to $q = 1.77$ in the visual region and about 1.65 in the blue region. We are inclined to adopt a somewhat less extreme particle phase function, which gives less weight to the observed darkening results near the limb. We have adopted $q = 1.65$ for the visual, $q = 1.60$ for the blue, and $q = 1.55$ for the ultraviolet measures.

For a definitive estimate of the albedos of the outer planets, it is imperative that mean limb-darkening observations of high precision be obtained in the various wave-length regions. These, combined with more extensive theoretical investigations involving different forms of the particle phase function, should lead to considerably better values of the albedos than those given here.

One aspect that must be kept in mind is that the observations of the major planets can be made only near full phase, and it is possible that two radically different particle phase functions may lead to the same law of darkening. A simple example is that afforded by

$$f(\cos \Theta) = 0.925\,(1 + \cos \Theta),$$

which leads to a law of darkening that agrees with Lambert's law over the whole disk to within a fraction of a per cent. It is probable that q for this phase function is of the order of 1.7 instead of the Lambert value of 1.5.

8.7. Planet with No Atmosphere

The planet Mercury and the moon show very similar variations with solar phase angle, and this variation deviates markedly from any known law of diffuse reflection—for example, Lambert's law or the Lommel-Seeliger law. As the photometry of the moon has been reviewed by Minnaert in chapter 6, we limit our remarks to an attempt to discover an empirical law of reflection from the lunar observations.

We have examined Bennett's (1938) observations of the variation in brightness of a number of lunar positions with phase angle in an effort to find an interpolation formula that would represent his observations of relative brightness, h/h_0, as a function of phase angle. An early investigation by Öpik (1924) led him to adopt the empirical law,

$$\frac{h}{h_0} = C\,(\cos i)^{k}, \qquad (25)$$

where h_0 is the brightness of an area at full phase and C and k are constants that depend on α only. Minnaert (1941, p. 409) pointed out that this relation did not satisfy the reciprocity principle and suggested that it might be modified to

$$\frac{h}{h_0} = C\,(\cos i)^{k}\,(\cos \epsilon)^{k-1}, \qquad (26)$$

where again C and k depend on α only. The value of k is found from measures made on the same photographic plate, but the constant C depends on the adopted zero point of the plate, which in Bennett's observations was adjusted to fit Russell's (1916*a*) lunar phase-curve.

From Bennett's data it appears that, for phase angles greater than 90°, k is nearly equal to 1.0, so that Minnaert's law reduces to Lambert's law in this case. At full moon we take $k = 0.5$, to agree with the observed fact that there is no limb darkening at this phase. For intermediate phase angles we have obtained the value of C and k given in Table 27. A check of a number of individual areas shows that our interpolation formula is quite satisfactory for both maria and terrae as was to be expected from Bennett's own discussion (Fig. 18).

TABLE 27

CONSTANTS DERIVED FOR MINNAERT'S LAW OF REFLECTION

(Bennett's Data)

α	C	k	α	C	k
0°...	1.00	0.50	50°...	0.54	0.82
10...	0.88	0.58	60....	0.51	0.87
20...	0.78	0.64	70....	0.50	0.92
30...	0.69	0.70	80....	0.50	0.96
40°...	0.61	0.76	90°...	0.50	1.00

The integration of our interpolation formula over the disk at any given phase angle should reproduce Russell's mean light-curve for the moon, provided that we assume that the distribution of light and dark areas is uniform over the disk. The fact that Rougier's curve shows little difference between increasing and decreasing phase angles indicates that this is a quite good approximation, in spite of the fact that the fraction of the illuminated surface covered by maria varies considerably (Danjon, 1933). For phase angles greater than 90°, Rougier's curve would give $C = 0.363$ instead of 0.50, and it is a simple matter to compare with Rougier's curve in this region as well. The comparison of observed and theoretical variations given in Table 28 appears to be satisfactory, although slight adjustment of the constants might bring them into better agreement.

Figure 19, showing the form of the law of reflection for different angles of incidence, is comparable to that given by Minnaert in chapter 6, Figure 9.

8.8. PLANET WITH AN OPTICALLY THIN ATMOSPHERE

The case of an optically thin atmosphere overlying the surface of a planet will, in the absence of clouds, apply to the earth, where in the visual

TABLE 28

OBSERVED AND COMPUTED PHASE VARIATION OF THE MOON

α	Russell	Computed	Rougier	Computed
0°	1.00	1.00	1.00	1.00
20	0.65	0.65	....	....
40	0.41	0.39	....	....
60	0.26	0.23	....	....
80	0.15	0.14	....	....
100	0.075	0.079	0.057	0.057
120	0.031	0.036	0.026	0.026
140	0.010	0.012	0.009	0.008
150°	0.004	0.005	0.005	0.004

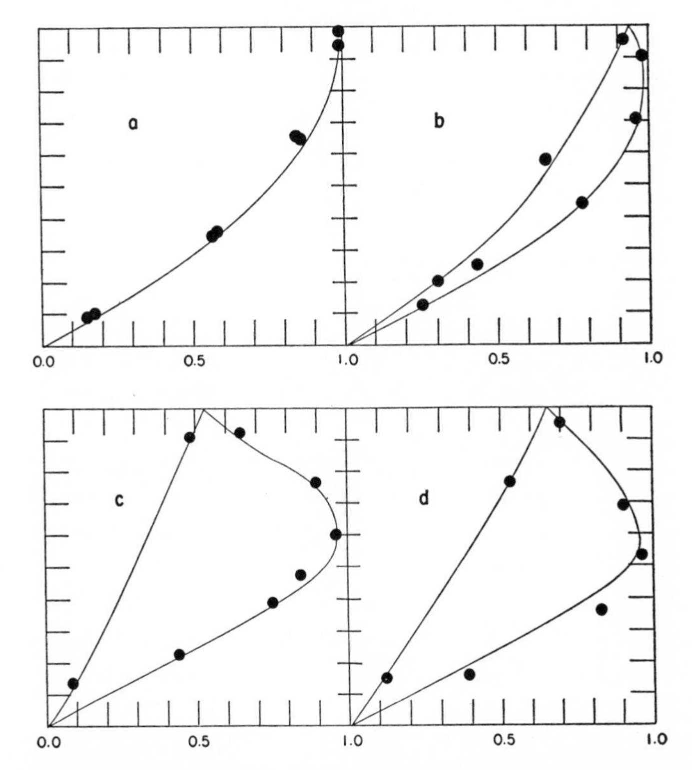

FIG. 18.—Comparison of Bennett's lunar observations with Minnaert's empirical law of diffuse reflection; *abscissae:* cos i; *ordinates:* h/h_0. (*a*) Centrally located maria; (*b*) terrae; (*c*) maria near limb; (*d*) dark crater floor near limb.

region $\tau = 0.15$ and increases to about 0.5 at 3540 A, and also to Mars, where the optical thickness is about one-third that of the earth. The case will also apply to a thin atmosphere overlying a cloud surface, such as might be encountered for Venus or the major planets.

The exact solutions for the emergent intensity are given by Chandrasekhar (1950, p. 270) for the special case of a ground reflecting according to Lambert's law with an albedo λ_0. The necessary functions for numerical application of the formulae have been computed for isotropic scattering and Rayleigh scattering (see Appendix I, this chapter).

The possible application to Venus and the major planets is suggested by the fact that small particles and molecules will not be thoroughly mixed; instead, the particles will tend to settle in the atmosphere. The "atmosphere" will then consist of molecules, which obey the Rayleigh scattering law, and will largely account for the observed polarization. It is assumed that below this "atmosphere" a cloud surface exists which reflects light according to Lambert's law.

Assuming $\lambda_0 = 0.60$ and a thin isotropic atmosphere with $\tilde{\omega}_0 = 1.0$, we find the values of p and x given in Table 29. The addition of an optically

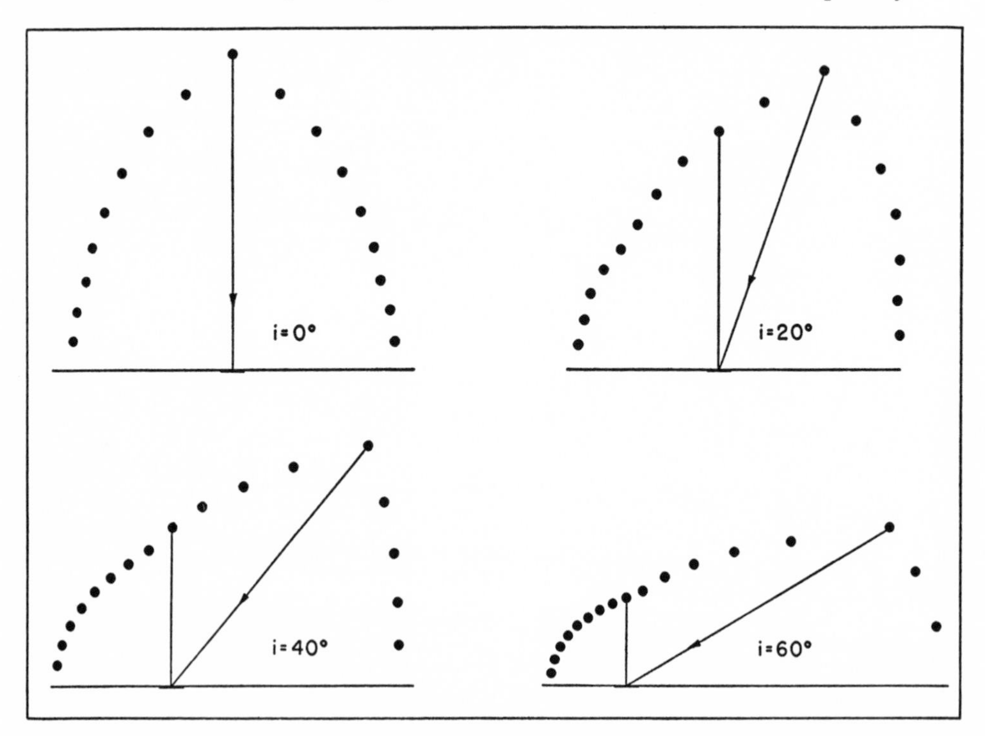

FIG. 19.—Indicatrices for Minnaert's law of diffuse reflection for various angles of incidence

thin atmosphere to a Lambert sphere appears to increase p only slightly; but the limb darkening decreases, because of the increased brightness near the limb, where the effective optical thickness becomes large because of the secant factor.

From an examination of a number of combinations of λ_0, $\tilde{\omega}_0$, and τ it does not appear that the Lambert "cloud" model can give the small values of x observed for the major planets in the visual and red regions. It appears that a model consisting of an optically thin layer of molecules obeying Rayleigh scattering, overlying a semi-infinite atmosphere of particles with a backward-scattering phase function, would be required to satisfy both the observed limb darkening and Lyot's polarization-curve.

For the case of a thin atmosphere overlying the solid surface of the planet, which is assumed to obey Lambert's law with a low albedo, the brightening at the limb by the atmospheric contribution modifies the results considerably. The limb darkening computed for a Lambert sphere with $\lambda_0 = 0.10$ and a thin atmosphere, scattering isotropically with $\tilde{\omega}_0 = 0.95$, is shown in Table 30. The influence of the atmosphere is most impor-

TABLE 29

LIMB DARKENING FOR OPTICALLY THIN ATMOSPHERE AND LAMBERT GROUND, $\varpi_0 = 1.0$, $\lambda_0 = 0.6$

τ	p	x	τ	p	x
0.00...	0.40	1.00	0.50....	0.44	1.16
.10...	.41	1.03	1.00....	0.46	1.22
0.25...	0.42	1.09			

TABLE 30

LIMB DARKENING FOR OPTICALLY THIN ATMOSPHERE AND LAMBERT GROUND, $\lambda_0 = 0.10$, $\varpi_0 = 0.95$

μ	r/R	$I(\tau=0.00)$	$I(\tau=0.05)$	$I(\tau=0.25)$	$I(\tau=1.00)$
1.0...	0.00	0.100	0.108	0.144	0.277
0.8...	0.60	0.080	0.094	0.136	0.274
0.6...	0.80	0.060	0.078	0.134	0.266
0.4...	0.92	0.040	0.065	0.140	0.245
0.3...	0.95	0.030	0.064	0.147	0.226
0.2...	0.98	0.020	0.068	0.156	0.199
0.1...	0.995	0.010	0.089	0.154	0.165
0.0...	1.00	0.000	0.119	0.119	0.119
p....	....	0.067	0.085	0.139	0.261
x....	....	1.00	1.31	1.89	1.86

tant very close to the limb but is still appreciable in the observable region, $\mu = 1.0$–0.4.

If an arbitrary law of diffuse reflection for the ground is substituted for Lambert's law, an exact solution appears to be extremely difficult to compute; but for small optical thickness and ground albedo the emergent intensity can be estimated with considerable accuracy. We have used the empirical law of diffuse reflection derived from the lunar observations (Sec. 8.7), which we shall denote as the "lunar law," and an isotropic scattering atmosphere with $\tilde{\omega}_0 = 1.00$ and $\lambda_0 = 0.10$. The albedo of the "ground" was adjusted to give a total albedo of 0.14. This result is com-

TABLE 31

PHASE VARIATION OF PLANET WITH OPTICALLY THIN ATMOSPHERE

α	Lambert Ground	"Lunar" Ground	Earth Observed	Mars Observed
0°	1.00	1.00	1.00	1.00
20	0.93	0.76	0.78	0.74
40	0.78	0.55	0.59	0.55
60	0.63	0.41	0.42	....
80	0.46	0.29	0.27	....
100	0.33	0.22	0.16	....
120	0.21	0.13	0.09	....
140	0.10	0.06	0.04	....
160°	0.03	0.02	0.02	....
p	0.083	0.117	0.366	0.144
q	1.70	1.16	1.095	1.04
A	0.14	0.14	0.40	0.15

pared in Table 31 with the solution for Lambert's law with $\lambda_0 = 0.067$ and the same atmosphere, which leads to the same Bond albedo. The phase variations of the earth as derived by Danjon (Vol. **2**, chap. 15) and of Mars are also tabulated for comparison.

It is seen that the observed phase variation of the earth is more nearly like that computed with the lunar law of reflection than that computed with Lambert's law. Although it is not reasonable to assume that the surface of the earth reflects like the surface of the moon, it is possible that the Martian surface with no open water reflects light in a manner quite similar to the moon and Mercury; it should be remembered that the photometric properties of the moon are determined by its small-scale irregularities and not by the visible mountains. It is seen that the Mars observations are well represented by the model.

9. SATELLITE ECLIPSES AND OCCULTATIONS OF STARS

9.1. Observations of Eclipses of Jupiter's Satellites

Photometric observations of the eclipses of Jupiter's satellites were initiated at the Harvard Observatory in 1878 by Pickering (1907), and the results for the times of mid-eclipse were made the basis of the *Tables of the Four Great Satellites of Jupiter* by Sampson (1910). Photographic photometry of eclipse light-curves was carried out by King (1917); he found that the scattered light from Jupiter produced an appreciable sky background which was difficult to correct for in the reductions.

In 1950 Kuiper made observations of the disappearances of Jupiter I and II with the photoelectric photometer at the Cassegrain focus of the 82-inch McDonald reflector, and in 1954 Kuiper and Harris obtained at least one light-curve of each of the satellites as they went into eclipse. The observing procedure was to start observations of the satellite approximately 10 minutes before predicted time of first contact and then adjust the rate of the telescope drive so that it followed the satellite. Although this adjustment could be made with considerable accuracy, the observations were made through a diaphragm of diameter 12 seconds of arc to insure that guiding errors would be negligible. The bright sky background was allowed for by making measurements on regions close to the satellite and interpolating for the position of the object. At the end of the run when the satellite had diminished by about 9.0 mag. (or a factor of 4000 in brightness), the sky deflection was given directly.

The measurements were made through the yellow filter, and a continuous eclipse-curve was traced out on the recorder, the record being interrupted only to check the centering or to change the amplification of the signal. The traces were then reduced by taking readings at equal time intervals sufficiently far apart that they would be essentially independent—the time constant of the amplifier being 1 second.

Measurements made before first contact indicate that the accidental error is about ± 0.3 per cent for one point. During the intervals before the start of the eclipses all four satellites remained constant in brightness within the error of measurement.

Table 32 gives a summary of the eclipses observed at the McDonald Observatory. The observed time of mid-eclipse is taken to be the time when the brightness, B, drops from the adopted value of 100.0 outside eclipse to 50.0. The times should be accurate to better than ± 2 seconds. The predicted times of mid-eclipse, taken from the *American Ephemeris*, yield the residuals, $O - C$, given in the last column.

TABLE 32

OBSERVED U.T. OF MID-ECLIPSE

Ec. D. I.	1950	Aug. 11^d 8^{h}26^{m}22^s	−1^{m}2
	1954	Nov. 19 9 56 38	−1.2
	1954	Nov. 26 11 50 16	−1.2
Ec. D. II.	1950	July 28 9 52 51	−0.9
	1954	Nov. 16 10 34 14	−0.9
Ec. D. III.	1954	Nov. 19 8 21 02	+0.2
	1954	Nov. 26 12 19 12	+0.3
Ec. D. IV.	1954	Dec. 31 8 54 17	−2.0

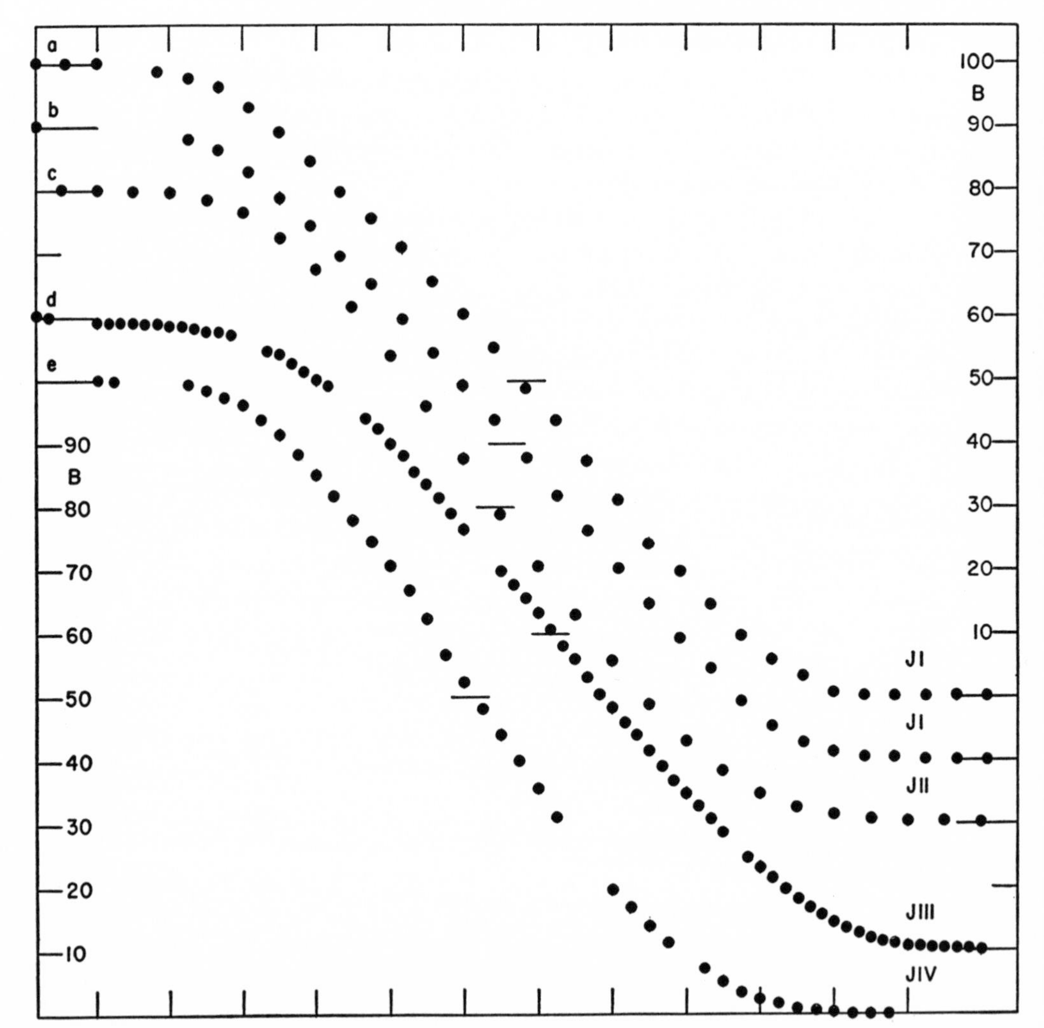

FIG. 20.—Eclipses of Jupiter's satellites on an intensity scale. (*a* and *b*) Jupiter I; (*c*) Jupiter II; (*d*) Jupiter III; (*e*) Jupiter IV. The measures are listed in Tables 33–37. Short horizontal lines near center indicate half-intensity points.

The observed light-curves for the eclipses which were considered the most reliable are given in detail in Tables 33–37, and the observations are plotted in Figure 20, *a–e*, on an intensity scale. The light-curves of I and III observed on November 26 are also plotted on a magnitude scale, in Figure 21, *a–b*.

9.2. Theoretical Eclipse Light-Curves

Sampson (1909) has computed theoretical light-curves which were used in the reduction of the Harvard series of eclipses. As his calculated curves are not quite accurate enough for use in comparisons with the McDonald data, we have redetermined theoretical curves based on the following assumptions. For any point on the satellite's disk, during the progress of the eclipse, the limb of Jupiter will cut off a portion of the sun's disk, and the

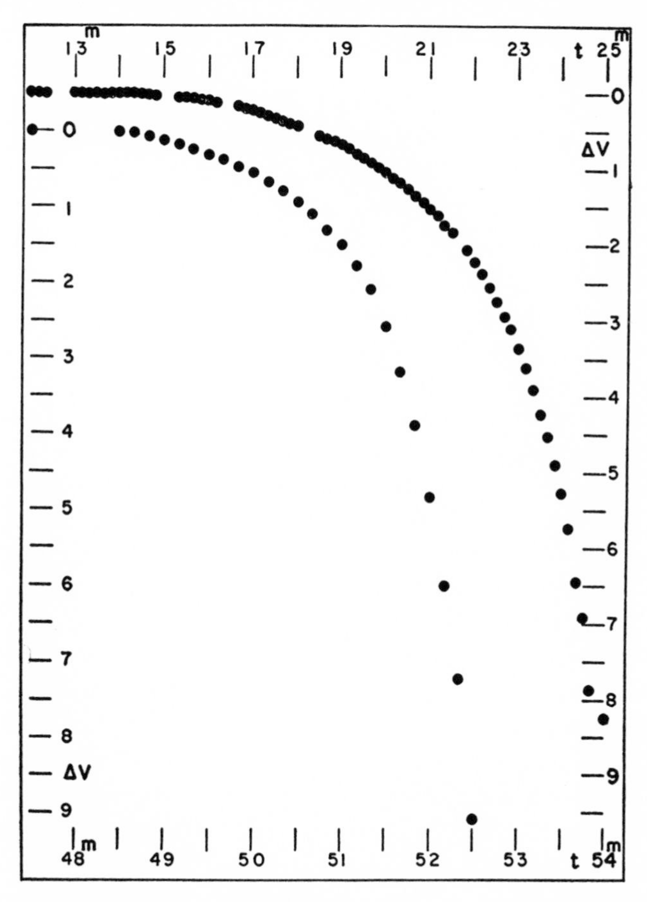

Fig. 21.—Observations of the eclipse of Jupiter III (*upper curve*) and of Jupiter I made on November 26, 1954, plotted on a magnitude scale.

illumination of this point is given by the brightness of the unobscured portion of the sun. If it be assumed that the brightness of the point is proportional to the illumination, integration of the illumination over all points on the disk will give for each epoch the integrated brightness of the satellite. In our calculations we have assumed a limb-darkening coefficient for the sun of 0.6 but assumed that there is no limb darkening on the satellite. In the case of Jupiter IV the deviation of the projected limb of Jupiter on the sun from a straight line necessitates a small correction, while for the other satellites the deviation is negligible. Also, in the cases of Jupiter III and IV, the variable motion of the satellite during the eclipse has been allowed for by using the results given by Sampson (1910).

The diameters of the satellites used in the calculations (expressed in terms of the earth's diameter) are as follows: I, 0.255; II, 0.226; III, 0.394; and IV, 0.350; these values are essentially the values listed by Kuiper in chapter 4.

The residuals, $O - C$, for each of the observed points are given in Tables 33–37, and these residuals are plotted against phase in Figure 22, *a–e*. The diameter of the circles corresponds to 0.4 units in B, slightly larger than the estimated accuracy of the measures. It is apparent that all the curves of residuals show systematic trends, the residual at mid-eclipse being arbitrarily adjusted to zero.

Sampson in his discussion enumerated several reasons why the observed and theoretical curves might not agree; some of these will be examined here.

(i) Refraction in Jupiter's atmosphere. The effect of refraction in Jupiter's atmosphere is discussed in some detail in Section 9.3. However, it appears that, for a reasonable assumption concerning the refraction and over the range in brightness which our observations cover, the effect is similar to a very small change in the satellite radius. For Jupiter IV, where refraction plays the most important role, the correction never amounts to more than 0.3 unit.

(ii) The adopted satellite diameter may not be appropriate. The measured diameters of the satellites used in our computations are about 4 per cent smaller than the "photometric" diameters adopted by Sampson, which gave the best fit for the Harvard eclipse observations. The effect of increasing the adopted diameter of I by 5 and 14 per cent reduces the residuals, $O - C$, for the November 26 curve to those shown in Figure 23, *a* and *b*. It appears that the McDonald observations of all the satellites are also better represented by "photometric" diameters that are larger than the measured diameters.

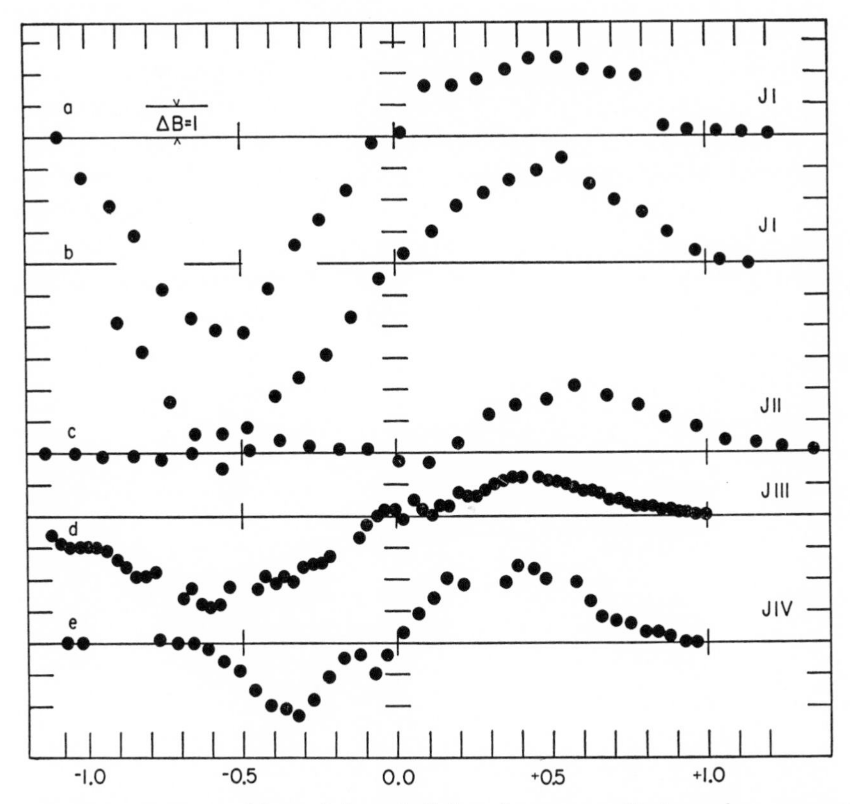

FIG. 22.—Differences (observed-theoretical) for eclipse-curves. *Abscissae:* phase measured from mid-eclipse; *ordinates:* marked at intervals of $B = 1.0$. (*a* and *b*) Jupiter I; (*c*) Jupiter II; (*d*) Jupiter III; (*e*) Jupiter IV.

TABLE 33

ECLIPSE DISAPPEARANCE OF JUPITER I ON NOVEMBER 19, 1954

U.T.	B	O−C	U.T.	B	O−C
09:54:00	100.0	0.0	09:56:40	48.8	+0.1
10	100.0	0.0	50	43.7	+1.6
20	100.0	0.0	57:00	37.2	+1.6
30	....	....	10	31.1	+1.8
40	98.7	−1.3	20	25.3	+2.1
50	97.7	−2.2	30	19.9	+2.5
55:00	96.2	−3.1	40	14.6	+2.5
10	93.0	−4.8	50	9.6	+2.1
20	89.2	−5.7	58:00	5.9	+2.0
30	84.7	−6.1	10	3.4	+1.9
40	79.8	−6.2	20	0.6	+0.3
50	75.6	−4.8	30	0.2	+0.2
56:00	71.1	−3.4	40	0.15	+0.15
10	65.8	−2.6	50	0.1	+0.1
20	60.3	−1.7	59:00	0.1	+0.1
30	55.2	−0.2	10	0.0	0.0

(iii) Non-uniform surface brightness of the satellites. In the simple model for the satellites considered above, we assume that the brightness of a surface element is proportional to the illumination incident upon it. If, instead, we adopt a brightness distribution over the satellite disk corresponding to the empirical law of diffuse reflection found for the moon at phase angle 10° (Sec. 8.7), then the differences between the computed curve and that for a uniform surface are positive before mid-eclipse, as shown in Figure 23, *d*. A better representation, as shown in Figure 23, *e*,

TABLE 34

ECLIPSE DISAPPEARANCE OF JUPITER I ON NOVEMBER 26, 1954

U.T.	B	$O-C$	U.T.	B	$O-C$
11:47:00	100.0	0.0	11:49:50	64.2	−2.9
10	100.0	0.0	50:00	59.2	−1.5
20	100.0	0.0	10	53.6	−0.5
30	100.0	0.0	20	47.7	+0.3
40	....	....	30	41.8	+1.0
50	....	....	40	36.1	+1.8
48:00	....	....	50	30.2	+2.2
10	....	....	51:00	24.8	+2.8
20	....	....	10	19.2	+2.9
30	98.0	−1.9	20	14.4	+3.3
40	96.3	−2.8	30	9.2	+2.5
50	92.9	−4.4	40	5.3	+2.0
49:00	88.8	−5.4	50	2.8	+1.6
10	84.5	−5.4	52:00	1.2	+1.0
20	79.7	−5.2	10	0.4	+0.4
30	75.1	−4.2	20	0.1	+0.1
40	69.7	−3.6	30	0.0	0.0

TABLE 35

ECLIPSE DISAPPEARANCE OF JUPITER II ON NOVEMBER 16, 1954

U.T.	B	$O-C$	U.T.	B	$O-C$
10:31:00	100.0	0.0	10:34:45	32.9	+0.3
15	100.0	.0	35:00	25.8	+1.2
30	100.0	.0	15	18.8	+1.5
45	99.9	− .1	30	12.9	+1.7
32:00	99.6	− .1	45	8.3	+2.1
15	98.5	− .2	36:00	4.7	+1.8
30	96.5	.0	15	2.5	+1.5
45	92.4	− .5	30	1.3	+1.1
33:00	87.7	+ .1	45	0.8	+0.8
15	81.6	+ .4	37:00	0.4	+0.4
30	73.9	+ .2	15	0.3	+0.3
45	66.0	+ .1	30	0.2	+0.2
34:00	57.6	+ .1	45	0.15	+0.15
15	48.9	− .3	38:00	0.1	+0.1
30	40.6	−0.1	15	0.0	0.0

may be obtained with the same law of diffuse reflection, provided that the satellite diameter is increased by 6 per cent. If ordinary limb darkening is added, a still larger satellite diameter needs to be postulated to explain the observed eclipse-curve. Such a large increase over the micrometric diameters appears ruled out.

Thus one is led to consider the effects produced by surfaces which are no longer of uniform brightness at full phase but which have large spots, somewhat like the moon. The actual presence of such spots on the Jupiter satellites follows from the fact that all of them show variations in brightness with rotational phase and that, moreover, several observers during

TABLE 36

ECLIPSE DISAPPEARANCE OF JUPITER III ON NOVEMBER 26, 1954

U.T.	*B*	*O−C*	U.T.	*B*	*O−C*
12:12:00	100.0	0.0	12:18:40	57.8	−0.3
10	100.0	0.0	50	55.6	0.0
20	100.0	0.0	19:00	53.2	+0.2
30	...	...	10	50.7	+0.2
40	...	...	20	47.9	−0.1
50	...	...	30	45.9	+0.5
13:00	99.4	−0.6	40	43.1	+0.2
10	99.1	−0.9	50	40.4	0.0
20	99.0	−1.0	20:00	38.2	+0.3
30	99.0	−1.0	10	35.8	+0.3
40	99.0	−1.0	20	33.8	+0.7
50	99.0	−1.0	30	31.3	+0.6
14:00	98.7	−1.3	40	28.9	+0.6
10	98.5	−1.4	50	26.8	+0.8
20	98.2	−1.6	21:00	24.8	+1.0
30	97.8	−1.9	10	22.7	+1.1
40	97.5	−1.9	20	20.6	+1.2
50	97.2	−1.8	30	18.6	+1.2
15:00	...	...	40	...	...
10	...	...	50	14.7	+1.2
20	94.7	−2.6	22:00	12.9	+1.1
30	94.1	−2.3	10	11.2	+1.1
40	92.6	−2.8	20	9.5	+1.0
50	91.3	−2.9	30	8.0	+0.9
16:00	90.1	−2.8	40	6.7	+0.8
10	89.2	−2.2	50	5.6	+0.8
20	...	...	23:00	4.5	+0.7
30	...	...	10	3.5	+0.5
40	83.8	−2.3	20	2.8	+0.5
50	82.2	−1.9	30	2.0	+0.4
17:00	79.9	−2.1	40	1.5	+0.3
10	77.9	−1.9	50	1.1	+0.3
20	75.5	−2.1	24:00	0.8	+0.3
30	73.7	−1.6	10	0.5	+0.2
40	71.4	−1.5	20	0.3	+0.2
50	69.0	−1.5	30	0.2	+0.1
18:00	66.8	−1.3	40	0.1	+0.1
10	...	...	50	0.05	+0.04
20	...	...	25:00	0.05	+0.05
30	59.9	−0.7	10	0.0	0.0

TABLE 37

ECLIPSE DISAPPEARANCE OF JUPITER IV ON DECEMBER 31, 1954

U.T.	B	$O-C$	U.T.	B	$O-C$
08:44:00	100.0	0.0	08:54:30	48.2	+0.3
30	100.0	0.0	55:00	43.9	+0.9
45:00	....	....	30	39.6	+1.4
30	....	....	56:00	35.5	+2.0
46:00	....	....	30	30.9	+1.8
30	99.2	+0.1	57:00	....	....
47:00	98.4	0.0	30	....	....
30	97.4	0.0	58:00	19.5	+1.9
48:00	95.9	−0.2	30	16.8	+2.4
30	93.8	−0.6	59:00	13.9	+2.3
49:00	91.4	−0.9	30	11.1	+2.0
30	88.4	−1.5	09:00:00	....	....
50:00	85.0	−2.0	30	7.1	+1.9
30	81.6	−2.1	01:00	5.1	+1.3
51:00	77.8	−2.3	30	3.4	+0.8
30	74.3	−1.8	02:00	2.4	+0.7
52:00	70.8	−1.1	30	1.6	+0.6
30	66.8	−0.5	03:00	0.9	+0.3
53:00	62.2	−0.4	30	0.6	+0.3
30	56.7	−1.0	04:00	0.3	+0.2
54:00	52.4	−0.4	30	0.0	0.0

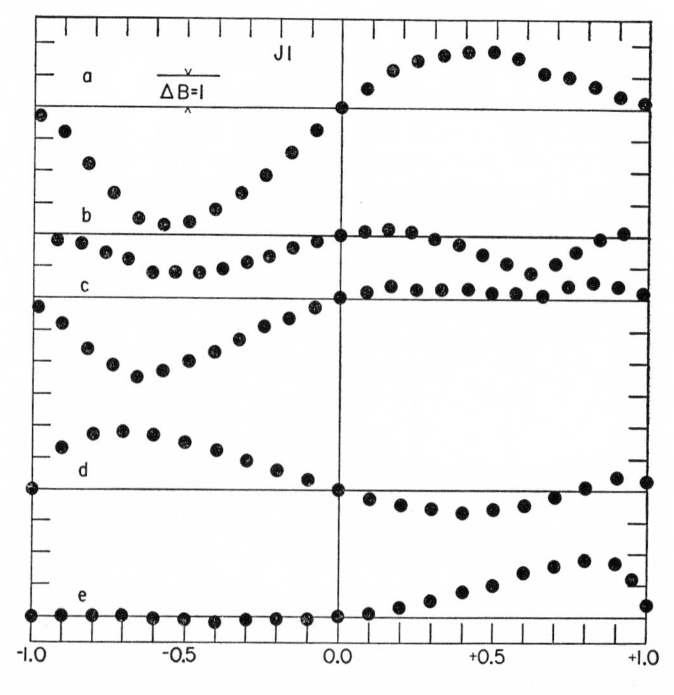

FIG. 23.—Same as Fig. 22, *b* (Jupiter I, November 26, 1954) with different radii and spots (see text for details).

the past half-century or more have noted, by direct inspection, uneven distributions of brightness over the disks (see chap. 15, Pl. 40). The presence of spots makes the computation of theoretical light-curves indeterminate. However, one may derive the effect produced by a bright equatorial zone (latitudes $< 30°$), assumed twice as bright as the polar regions; this has some resemblence to the brightness distribution found for I. With the same diameter as that used in computing Figure 23, *a*, the residuals for I are those shown in Figure 23, *c*. It is seen that the run of the residuals simulates one valid for a larger diameter but a uniform surface brightness which, of course, is not surprising.

9.3. Refraction and Absorption in Jupiter's Atmosphere

The effects of refraction and absorption on the observed light-curves of satellites as they go into eclipse have been considered by Link (1933) and others. In the simplest problem of this type the illumination is a point

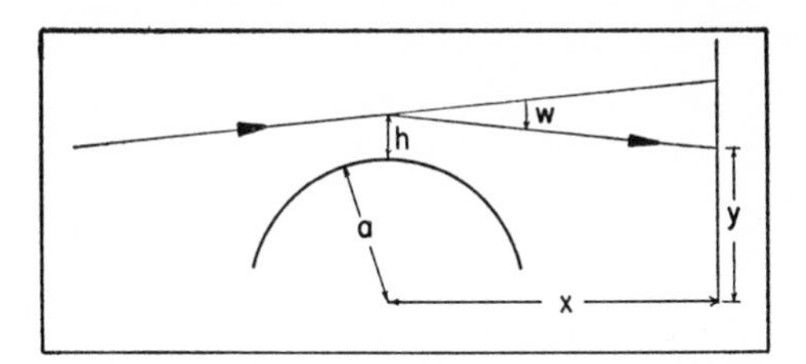

Fig. 24.—Schematic diagram of an occultation of a star by Jupiter

source at infinite distance, which applies to a stellar occultation by Jupiter. This case was considered earlier by Pannekoek (1903) and Fabry (1929). If the brightness of the star outside the occultation is taken as 100.0, the light-ray passing through the atmosphere will be refracted through an angle, w, and reduced to $B = 100/s$, where $s = 1 + wx/H$, if x is the distance of the observer from Jupiter and H is the scale height in Jupiter's atmosphere (Link, 1933, p. 245). This attenuation is caused by the light contained in a small solid angle being refracted into a larger solid angle; the attenuation due to absorption or scattering in the atmosphere is practically negligible by comparison.

If one imagines a screen set up at a distance x from Jupiter (Fig. 24), the intensity of illumination on the screen at a point y will be given by $B = 100/s$ and $y/H = \text{constant} - \ln (s - 1) - (s - 1)$ (cf. Link, 1933, p. 245). As the observer's motion on earth carries him across the screen, he will observe a light-curve, from which he can obtain the unknown, $1/H$.

The occultation of σ Arietis by Jupiter in 1952 was successfully observed by Baum and Code (1953) in the light of the K line of Ca II. From their observed light-curve they concluded that $1/H$ lies between 0.08 and 0.16 km^{-1}; their adopted value of 0.12 leads to a mean molecular weight of 3.3, which is direct evidence for the predominance of H and He in the Jupiter stratosphere and is in agreement with Kuiper's (1952) model of the Jupiter atmosphere.

The formulae given for the attenuation of light assume that the angle of refraction is small. In the case of the occultation, this is an excellent approximation; for example, when $B = 1.0$ and hence $s = 1 + wx/H = 100$, $w = 0''.28$, x and H being known (see above). However, if one moves in to the distance of Jupiter IV, the same amount of refraction gives a brightness as high as $B = 77$. Similarly, the refraction which causes a star's brightness to decrease to 50 units reduces the brightness of Jupiter IV almost imperceptably to $B = 99.7$.

The effect of refraction on the satellites as they go into eclipse can be evaluated accurately if one assumes a model atmosphere for Jupiter; however, to a good approximation the results depend only on the scale height, H. It is necessary in the case of the eclipses to take the finite size of the sun into account, with allowance for limb darkening, and to perform the integration over the satellite's disk. This was done for the model atmosphere considered by Kuiper, which gives a scale height of 8.44 km, and applied to the case of Jupiter IV. Between $B = 100$ and $B = 0.2$ the resulting light-curve differs very little from the one computed by ignoring refraction; the small differences could easily be confused with residuals arising from a small error in the adopted satellite diameter. However, when refraction is allowed for, the light-curve of the eclipse does not terminate but decreases asymptotically to zero as $1/t$. In the case of Jupiter IV we find that, with no refraction, the brightness will be zero about 8 minutes after mid-eclipse; with allowance for refraction, the brightness at this time will be about $B = 0.2$ and 6 minutes later will be $B = 0.05$, corresponding to a total decrease of about 8.3 mag.

Because of the bright background, the limiting value of B that has been observed photoelectrically is approximately 0.2. However, visually the eye can follow the satellite to fainter limits: Kuiper (1947) has observed the satellites as they were eclipsed and noted a "halt" in the light-curves near fourteenth magnitude which lasts 1–2 minutes. It appears that his observations can be explained by the effects of refraction, as he had suggested. A summary of Dr. Kuiper's visual eclipse observations is given in Appendix III of this chapter.

Eropkin (1931), using a short-focus telescope, made photographic observations of the eclipses of Jupiter's satellites. These showed decreases in brightness, amounting to 30 or 40 per cent, several minutes before the predicted time of first contact, which led him and later Link (1933) to suggest the presence of an absorbing layer very high in Jupiter's atmosphere. This large pre-eclipse decrease is not present in the Harvard visual observations, in King's photographic observations, or in any of our photoelectric light-curves, though we made a special effort to discover any such peculiarity. Any such decrease could not exceed 0.2 per cent.

APPENDIX I

The exact solutions of the equations of radiative transfer for the problem of diffuse reflection are given by Chandrasekhar in his monograph, *Radiative Transfer* (Oxford, 1950), here denoted by "*R. T.*" These solutions are expressed in terms of numerical functions which have been tabulated for various values of the parameters. The following brief outline is an index to the formulae and tables of functions, used in the preparation of this chapter.

A. Planets with Optically Thick Atmospheres (Secs. 8.4–8.6 of This Chapter)

1. Isotropic scattering: Section 47 and Table XI in *R. T.;* D. L. Harris, *Ap. J.*, **136,** 408, 1957. More accurate values given by D. W. N. Stibbs and R. E. Weir, in *M.N.*, **119,** 512, 1959.
2. Anisotropic scattering: Section 47 and Tables XIX and XX in *R. T.;* D. L. Harris, *Ap. J.*, **136,** 408, 1957.
3. Rayleigh phase function: Section 47 and Table XXI in *R. T.*
4. Rayleigh scattering function: Section 70.3 and Table XXV in *R. T.*

B. Planets with Optically Thin Atmospheres (Sec. 8.8 of This Chapter)

1. Isotropic scattering: Section 72.2 in *R. T.* The allied functions are given by Chandrasekhar, Elbert, and Franklin, *Ap. J.*, **115,** 244, 1952, and Chandrasekhar and Elbert, *Ap. J.*, **115,** 269, 1952.
2. Anisotropic scattering: Section 72.2 in *R. T.* The necessary functions have not been evaluated.
3. Rayleigh phase function: Section 72.2 in *R. T.* The necessary functions have not been evaluated.
4. Rayleigh scattering function: Sections 72.3 and 72.4 in *R. T.* The allied functions are given by Chandrasekhar and Elbert, *Trans. Amer. Phil. Soc.*, Vol. **44,** Part 6, 1954.

APPENDIX II

Refraction Effects

The formulae quoted in Section 9.3 giving the effect of the Jovian atmospheric refraction on occultations of a star or on the light-curves of a satellite are based on Link's (1933) discussion. Consider a simplified model with the following characteristics:

1. Jupiter is a spherical planet with radius a, surrounded by an isothermal stratosphere with a scale height H.
2. The source of illumination, the sun or star, may be considered to be a point source located at a distance of $-X$ from Jupiter.
3. A screen is set up at a distance, $+x$, from Jupiter perpendicular to the line between the source and Jupiter.

In the absence of refraction a light-ray passing Jupiter at a radial distance of $a(1 + h)$ will strike the screen at a radial distance given by

$$y_0 = a\,(1+h)\left(1+\frac{x}{X}\right),$$

so that light-rays passing Jupiter at distances between $a(1 + h)$ and $a(1 + h + dh)$ will illuminate an annulus on the screen whose area is $dA_0 = 2\pi y_0 dy_0$.

Taking a total refraction, w, into consideration, the refracted ray will strike the screen at a radial distance given by

$$y = y_0 - wx\,,$$

and the area of the illuminated annulus is $dA = 2\pi y dy$. It is convenient to transform out the effect of the distance of the source by letting $y_0' = y_0(1 + x/X)^{-1}$, and $x' = x(1 + x/X)^{-1}$. In the rest of the discussion we assume that this has been done and drop the superscript.

The brightness of the illuminated portion of the screen will be attenuated by $1/s$, where

$$s=\frac{y\,(dy/dh)}{y_0\,(dy_0/dh)}=\left(1-\frac{wx}{y_0}\right)\left[1-\frac{x\,(dw/dh)}{dy_0/dh}\right].$$

For the case where w = constant $\times\, e^{-ah/H}$ this reduces to

$$s=\left[1-\frac{wx}{a\,(1+h)}\right]\left(1+\frac{wx}{H}\right)\approx 1+\frac{wx}{H}.$$

Therefore, $(s - 1)H = wx =$ constant $\times\, x \times \exp(-ah/H)$, which gives $ah/H =$ constant $- \ln(s - 1)$. Substituting in the equation for y, we obtain

$$\frac{y}{H}=\text{Constant}-\ln\,(s-1)-(s-1)$$

and

$$s = 1 + \frac{wx}{H}.$$

As noted in Section 9.3, the above formulae obtain in the case of an occultation to a high degree of approximation. For eclipses of Satellite IV the distance of the screen is only 0.0126 a.u. compared with 4.2 a.u. in the case of an occultation observed near opposition. For eclipses it is necessary to consider much larger refraction angles, as well as allow for the finite solar diameter, the solar limb darkening, and the finite disk of the satellite.

APPENDIX III

Visual Observations of Jupiter Satellite Eclipses

(Prepared by G. P. Kuiper)

Prior to the precision measures made photoelectrically, begun in 1950 and recorded in Dr. Harris' chapter, some visual observations of eclipses were made with the 82-inch telescope. These observations first called attention to the existence of a "tail" to each eclipse-curve near the fourteenth magnitude, which was tentatively attributed to refraction in the Jupiter atmosphere (Kuiper, 1947). Since the photoelectric data obtained so far do not go faint enough to bring out this tail clearly, the visual estimates are added here. The complete records are given as made at the telescope, although the estimates for the brighter portions have, of course, been entirely superseded by our photoelectric measures.

Jupiter I was observed on March 2, 1946. Mid-eclipse was predicted in the *Nautical Almanac* at $9^h49^m.0$ U.T.; corrected for the apparently constant ephemeris error of $-1^m.2$ (Table 32, above), the expected time was $9^h47^m.8$ U.T. The following are the observing records, made with power 660×:

9:47:47	about ½ disk left	9:50:01	about thirteenth magnitude
9:48:42	about ¼ disk left	9:50:07–22	near fourteenth magnitude
9:49:51	about eleventh magnitude	9:50:27	lost

The satellite stayed near the fourteenth magnitude for 10–15 seconds; only a crescent, not the entire disk, was visible (end of record). On the basis of the extrapolated light-curve of Figure 19, *a*, the satellite should have been lost about 9:50:05 U.T.; or, if 135 seconds were added to the predicted time of mid-eclipse (Sampson, 1907, p. 16), complete disappearance would have occurred at 9:50:03 U.T., 24 seconds before the satellite was actually lost.

Jupiter II was observed on March 2, 1946, also. Mid-eclipse was predicted for $11^h01^m.6$ U.T.; if the correction of Table 32 is applicable, the expected time is $11^h00^m.7$ U.T. The observing records are:

10:59:28	about ½ disk left	11:02:33	about eleventh magnitude
11:00:33	about ⅓ disk left	11:02:53	about twelfth magnitude
11:01:01	about ¼ disk left	11:03:03	about thirteenth magnitude
11:02:02	about ninth magnitude	11:04:18	lost

The limiting magnitude was estimated to be 14.5 mag. For a whole minute the satellite was around the fourteenth magnitude (end of record). With the half-duration of the partial phase 149 seconds (Sampson, 1907, p. 16), the predicted time of zero intensity is 11:03:11 U.T., 67 seconds before the satellite was actually lost.

Note added in proof.—DeVaucouleurs (1960) has reported some multicolor photometry of Mars, in which the stellar magnitude and integral albedo for Mars at five wave lengths (from 3300 to 6900 A) are derived on the basis of observations made at the Lowell Observatory in 1958.

REFERENCES

Ashbrook, J.	1948	*Pub. A.S.P.*, **60**, 116.
Barabashev, N. P., and Semejkin, B. E.	1933	*Zs. f. Ap.*, **7**, 290.
	1934	*Ibid.*, **8**, 179.
Baum, W. A., and Code, A. D.	1953	*A.J.*, **58**, 108.
Biesbroeck, G. van	1946	*A.J.*, **52**, 114.
Becker, W.	1928	*A.N.*, **235**, 85.
	1930	*Ibid.*, **239**, 19.
	1933	*Sitz. preuss. Akad. Wiss., phys.-math. Kl.*, **28**, 839.
	1948	*A.N.*, **277**, 65.
Bennett, A. L.	1938	*Ap. J.*, **88**, 1.
Bond, W. C.	1861	*Proc. Amer. Acad. Arts and Sci.*, N.S., **8**, 232.
Calder, W. A.	1936	*Harvard Bull.*, No. 903, p. 11.
	1938	*Ibid.*, No. 907, p. 26.
Campbell, L.	1917	*Harvard Circ.*, No. 200.
	1936	*Harvard Bull.*, No. 904, p. 32.
Chandrasekhar, S.	1950	*Radiative Transfer* (Oxford: Clarendon Press).
Danjon, A.	1933	*Ann. Obs. Strasbourg*, **2**, 170.
	1949	*Bull. Astr.*, **14**, 315.
	1954	*Ibid.*, **17**, 363.
Eropkin, D. J.	1931	*Zs. f. Ap.*, **3**, 163.
Fabry, C.	1929	*J. Observateurs*, **12**, 1.
Goddard, R.	1930	*A.J.*, **40**, 98.
Güssow, M.	1929	*A.N.*, **237**, 229.

GUTHNICK, P. 1918 *A.N.*, **206**, 157.
1920 *Ibid.*, **212**, 39.

GUTHNICK, P., and PRAGER, R. 1918 *Veröff. Berlin-Babelsberg*, Vol. **2**, Part 3.

HARDIE, R. H. 1953 *Lowell Obs. Progress Repts.*, No. 4.
1956 Unpublished results.

HARDIE, R. H., and GICLAS, H. L. 1955 *Ap. J.*, **122**, 460.

HORAK, H. G. 1950 *Ap. J.*, **112**, 445.

HULST, H. C. VAN DE 1952 *The Atmospheres of the Earth and Planets* (rev. ed; Chicago: University of Chicago Press), chap. 2.

JOHNSON, H. L., and GARDINER, A. J. 1955 *Pub. A.S.P.*, **67**, 74.

JOHNSON, H. L., and MORGAN, W. W. 1951 *Ap. J.*, **114**, 522.
1953 *Ibid.*, **117**, 313.

KEENAN, P. C., and MORGAN, W. W. 1952 *Astrophysics*, ed. J. A. HYNEK (New York: McGraw-Hill Book Co.), chap. I.

KING, E. S. 1917 *Harvard Ann.*, Vol. **80**, No. 10.
1919 *Ibid.*, Vol. **81**, No. 4.
1923 *Ibid.*, Vol. **85**, No. 4.

KUIPER, G. P. 1938 *Ap. J.*, **88**, 429.
1944 *Ibid.*, **100**, 378.
1947 *A.J.*, **52**, 147.
1949 *Ap. J.*, **110**, 93.
1952 *Atmospheres of the Earth and Planets* (rev. ed.; Chicago: University of Chicago Press), chap. 12.
1953 *Proc. N.A.S.*, **39**, 1156.
1956*a* *J.R.A.S. Canada*, **50**, 170 ff.
1956*b* Unpublished.

LINK, M. F. 1933 *Bull. Astr.*, **9**, 227.

LOWELL, P., and SLIPHER, V. M. 1912 *Lowell Obs. Bull.*, No. 53.

MILFORD, N. 1950 *Ann. d'ap.*, **13**, 249, Table 2.

MINNAERT, M. 1941 *Ap. J.*, **93**, 403.
1946 *B.A.N.*, No. 367.

MOORE, J. H., and MENZEL, D. H. 1928 *Pub. A.S.P.*, **40**, 234.
1930 *Ibid.*, **42**, 330.

MORGAN, W. W., HARRIS, D. L., and JOHNSON, H. L. 1953 *Ap. J.*, **118**, 92.

MÜLLER, G. 1893 *Potsdam Pub.*, Vol. **8**, Part 4.

NIKONOVA, E. K. 1949 *Izvest. Crimean Astr. Obs.*, **4**, 114.

ÖPIK, E 1924 *Pub. Tartu (Dorpat)*, Vol. **26**, No. 1.

PANNEKOEK, A. 1903 *A.N.*, **164**, 5.

PICKERING, E. C.	1879	*Harvard Ann* , **11**, 266.
	1907	*Ibid.*, Vol. **52**, Part 1.
	1913	*Harvard Bull.*, No. 538.
RICHARDSON, R. S.	1955	*Pub. A.S.P.*, **67**, 355.
ROUGIER, G.	1933	*Ann. Obs. Strasbourg*, Vol. **2**, Part 3.
	1937	*Ibid.*, Vol. **3**, Part 5.
RUSSELL, H. N.	1916*a*	*Ap. J.*, **43**, 103.
	1916*b*	*Ibid.*, p. 173.
RUSSELL, H. N., DUGAN, R. S., and STEWART, J. Q.	1945	*Astronomy* (rev. ed.; Boston: Ginn & Co.), Appendix Table V.
SAMPSON, R. A.	1907	*Harvard Ann.*, **52**, Part 1.
	1909	*Ibid.*, Part 2.
	1910	*Tables of the Four Great Satellites of Jupiter* (London: W. Wesley & Son).
SCHÖNBERG, E.	1921	*Photometrische Untersuchungen über Jupiter und das Saturnsystem* (Helsinki: Finnish Academy of Science).
	1929	*Hdb. d. Ap.*, Vol. **2**, chap. 1.
SLAVINAS, P.	1928	*A.N.*, **233**, 125.
SHARONOV, V. V.	1939	*Pulkovo Circ.*, No. 27, p. 37.
	1940	*Ibid.*, No. 30, p. 48.
STEBBINS, J.	1927	*Lick Obs. Bull.*, No. 385.
STEBBINS, J., and JACOBSEN, T. S.	1928	*Lick Obs. Bull.*, No. 401.
STEBBINS, J., and KRON, G. E.	1956	*Ap. J.*, **123**, 440.
TALLEY, R. L., and HORAK, H. G.	1956	*Ap. J.*, **123**, 176.
VAUCOULEURS, G. DE	1960	*Multicolor Photometry of Mars in 1958*. (Scientific Rept. No. 3, ARDC Contract AF 19[604]-3074).
WALKER, M. F., and HARDIE, R. H.	1955	*Pub. A.S.P.*, **67**, 224.
WENDELL, O. C.	1913	*Harvard Ann.*, **69**, Part 2, 223.
WIDORN, TH.	1950	*Sitz. Öster. Akad. Wiss*, Ser. IIa, **159**, 189.
WOOLLEY, R. V.D.R.	1953	*M.N.*, **113**, 521.
WOOLLEY, R. V.D.R., and GASCOIGNE, S. C. B.	1948	*M.N.*, **108**, 491.
WOOLLEY, R. V.D.R., GOTTLIEB, R. K., and DE VAUCOULEURS, G.	1955	*M.N.*, **115**, 57.
ZÖLLNER, J. C. F.	1865	*Photometrische Untersuchungen* (Leipzig: W. Engelmann).

CHAPTER 9

Polarization Studies of Planets

By AUDOUIN DOLLFUS
Observatoire de Paris, Meudon

1. INTRODUCTION

LIGHT-RAYS striking an object may be reflected, refracted, diffracted, scattered, or absorbed. Differently oriented vibrations may be affected differently, in which case the light becomes partially polarized.

The *plane of vision* is defined by the directions of illumination and observation; the angle between these directions is the *angle of vision*, V, or the phase angle. The observed ray has two component vibrations, the intensity I_1 being normal to the plane of vision and the intensity I_2 being in the plane. The *degree of polarization* is defined as $P = (I_1 - I_2)/(I_1 + I_2)$, a quantity that may be either positive or negative.[1] Sunlight illuminating the planet may be regarded as natural light. The polarization is therefore due to the optical properties of the planetary surface and atmosphere, and the *polarization-curve*, which shows P plotted against V, will be an important source of information on the planet.

A polarization-curve may be obtained for a planet as a whole or for various parts of its disk. In this manner and with the aid of laboratory and field studies of terrestrial phenomena, much detailed information may be obtained. The power of the method is greatly increased by an extension of the observations over a wide range of wave-length bands.

[1] The more general definition is with I_1 and I_2 as the intensities of electric vector maximum and minimum, respectively. A third parameter is then needed, namely, the position angle of electric vector maximum, either in the equatorial frame of reference, θ, or, in planetary studies, with respect to the normal to the plane of vision, θ_r. In this chapter, θ_r is almost always equal to 0° or 90°, corresponding to the above definitions of positive and negative polarization, respectively.

2. MEASURING DEVICES

2.1. Lyot's Visual Polarimeter

A narrow-fringe Savart polariscope, consisting of two calcite disks 1.4 mm in thickness, is followed by a birefringent prism of small angle, which acts as an analyzer. The prism produces two images of the fringe systems of opposite polarization which are separated by an angle equal to half the spacing between two fringes. The two systems thus superpose their maxima and minima, and Lyot's polariscope is therefore twice as luminous as the classical Savart type. The polariscope, placed behind the eyepiece of a telescope set on a planet, shows fringes wherever the planetary image is polarized in a direction either parallel or normal to the fringes. In yellow light the spacing of the fringes is 10′ in the field of view, so that, with a high-power eyepiece, polarization can be detected on the planet in areas of only a few seconds of arc.

In making an observation the observer compensates the polarization by making the fringes vanish by means of an inclined celluloid film, L_1 (Fig. 1). To increase the precision, Lyot introduced a further polarization, by means of a second inclined film, L_2 (turned by the knob N), which produces faint fringes over the entire field. The rod K permits the axis of rotation of L_2 to be oriented either parallel or normal to the axis of L_1. In either position the fringes have the same contrast and are complementary. If L_1 does not exactly compensate the planetary polarization, the residual fringes are added to those of the auxiliary polarization in one position of L_2, while they overlap in the other, so that the two sets are no longer equal in contrast. The inclination of L_1 is adjusted until equality is established. By this device one can detect polarizations as small as 0.001 on small areas of the planets.

2.2. Visual Polarimeter for Faint Sources

The sensitivity of the Lyot polarimeter may become insufficient when planets are observed under high magnification, when narrow-band filters are used, or when the earth-light on the moon is studied. In 1949 Dollfus made a polarimeter which is especially adapted for low light intensities (Fig. 2).

The planetary image is projected, with adjustable magnification, onto a half-wave cellophane film (G in Fig. 2). This film is cut into 25 strips, each 0.4 mm wide and separated by 0.4-mm spaces. The strips are placed parallel to the expected direction of planetary polarization. The optical axis of the film is at 45° to the direction of the strips. In contact with film G, and following it, is placed an analyzer, consisting of a rhomb of calcite of such thickness (about 4 mm) that the images are separated by the same spacings as the strips. Thus dark bars and clear interspaces are produced in the image. The analyzer may be turned 90°, and the pattern changes to that of clear bars and dark interspaces.

The grid polariscope, placed in the plane of the enlarged image, is covered by a field lens projecting an image of the objective onto the

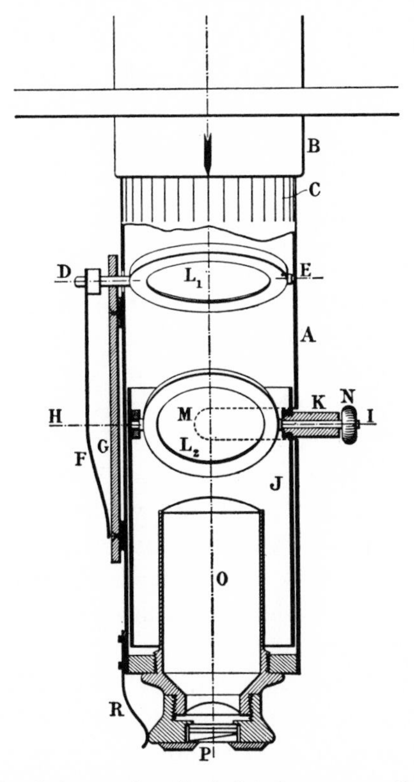

FIG. 1.—Lyot's visual polarimeter. L_1=tilted plate for compensation of polarization; L_2= tilted plate for further polarization; P=Savart polariscope with birefringent analyzer.

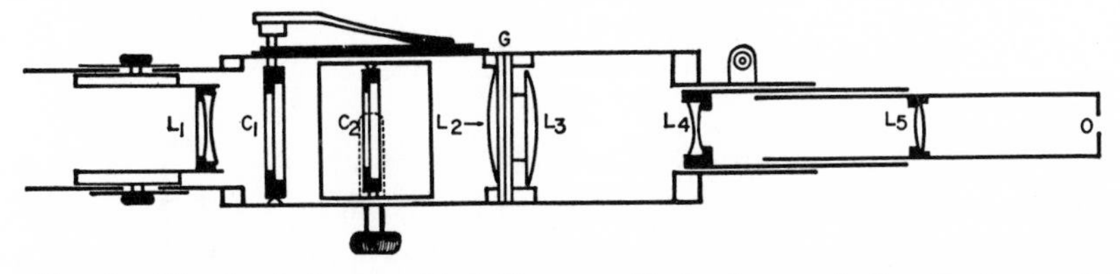

FIG. 2.—Visual polarimeter with half-wave grid. L_1=lens projecting the image of the planet on the grid G (this lens can be shifted along the axis to vary the size of the image); C_1= tilted plate for compensation of polarization; C_2=tilted plate for further polarization; G= half-wave grid; L_2 and L_3=field lenses; L_4=negative lens giving with L_5 a Galilean telescope, reducing the angular size of the planetary image; O=exit pupil.

observer's eye. The grid is observed with a magnifier or small telescope. The polariscope is preceded by the compensating and auxiliary polarization films of a Lyot polariscope.

Figure 2 shows the mechanical and optical details of an instrument in which the apparent spacing of the grid, as well as the magnification, is adjustable. In this instrument the sinusoidal photometric profile of the Savart polariscope is replaced by a system of bands of rectangular profile of greater contrast. It is more sensitive for low luminosities than Lyot's instrument. On areas having only one-millionth of a stilb in brightness it is still possible to detect a polarization of 1–2 per cent. The precision increases proportionally to the square root of the brightness to 0.001, when the brightness reaches 0.0001 stilb. For sources brighter than 0.0001 stilb the instrument has no advantage over the Lyot polarimeter.

2.3. Device for Measurements in Different Colors

To measure the polarization in narrow spectral bands, the observer adds a birefringent filter to the polarimeter (Fig. 3). P is a polaroid, L_1 and L_3 are quartz plates, and B_1 and B_2 are birefringent prisms of calcite. If this filter is placed in front of the eyepiece, four images of the planet are produced, two of which have complex spectral compositions and are therefore not used. The third and fourth images are transmitted in narrow bands, with half-widths of about 190 A, at 5770, 4940, and 4320 A for the one image and at 6290, 5320, and 4610 A for the other; the desired bands are isolated with normal color filters. The transparency can be increased if the polaroid is replaced by a calcite rhomb which produces two images whose polarizations are made parallel by half-wave plates. Thus both components of the polarized light are utilized, and the transmission attains 80 per cent. Figure 4 shows this filter F, with its two-component polarizer R, mounted on an adjustable holder behind a Lyot polariscope. The filter is of special interest when used in connection with the half-wave grid polarimeter. For that purpose the grid is cut into two superposed half-wave plates, the directions of whose axes differ by 40°. Then the contrast of the bars is practically independent of the wave length (Dollfus, 1955).

2.4. Sky-Light Compensators

When observations are made in daylight or twilight, spurious polarization is introduced. To correct for this, Lyot placed over part of the objective a stack of inclined glass plates which produced an offset image superposed on the sky background. By changing the orientation of the glass plates or the amount of light through the glass plates, Lyot succeeded in

PLATE 1.—Vest-pocket polarimeter for measures in the field

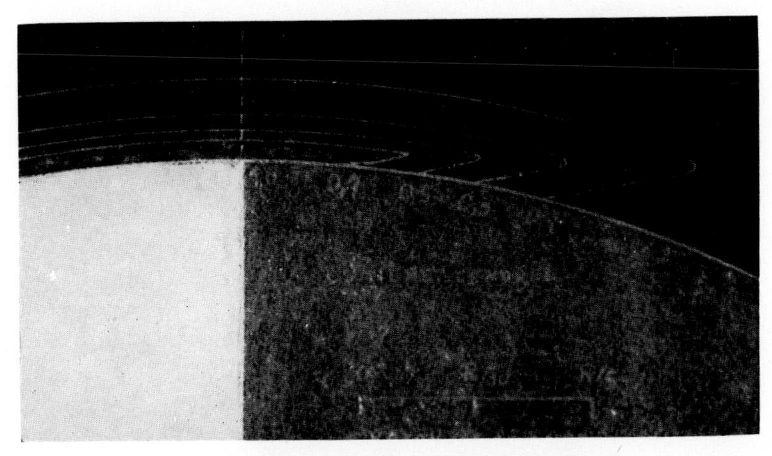

PLATE 2.—Brightness of a hypothetical lunar atmosphere extending from the cusps at phase 90°.

PLATE 3.—Appearance of Venus through the polarimeter on June 23 and 24, 1924. (After Lyot.)

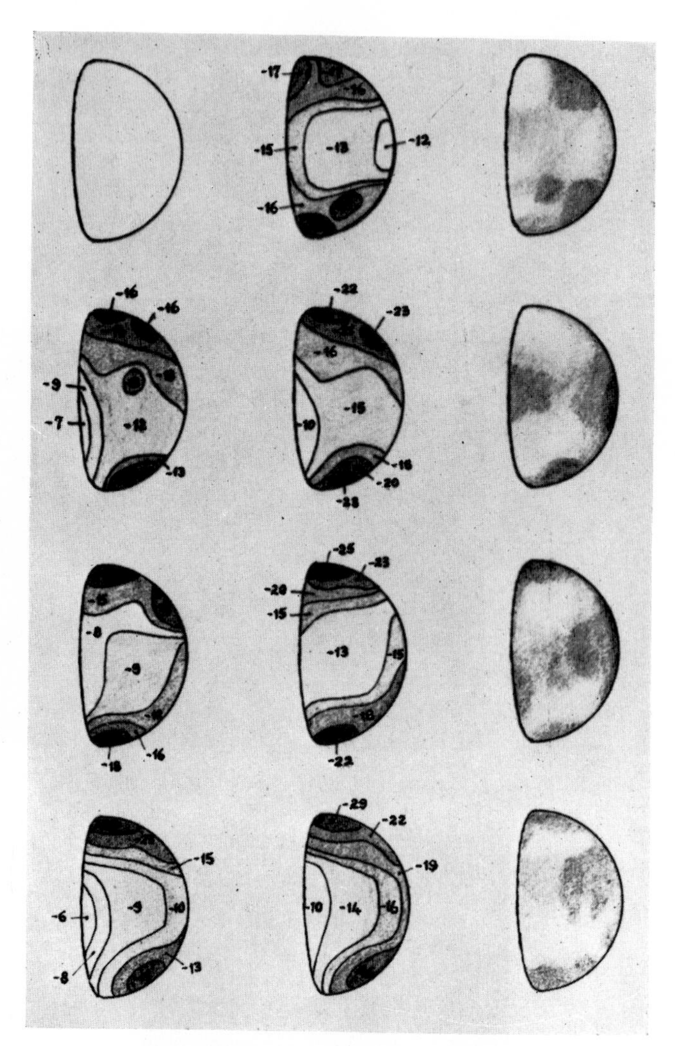

PLATE 4.—Comparison of polarimetric and visual observations of Venus. *Left:* polarization in green light; *center:* polarization in red light; *right:* position of clouds deduced from visual observations. Dates: August 29, 1953, $V = 56\overset{\circ}{.}6$; August 30, 1953, $V = 56\overset{\circ}{.}3$; August 31 1953, $V = 55\overset{\circ}{.}8$; and September 1, 1953, $V = 55\overset{\circ}{.}6$.

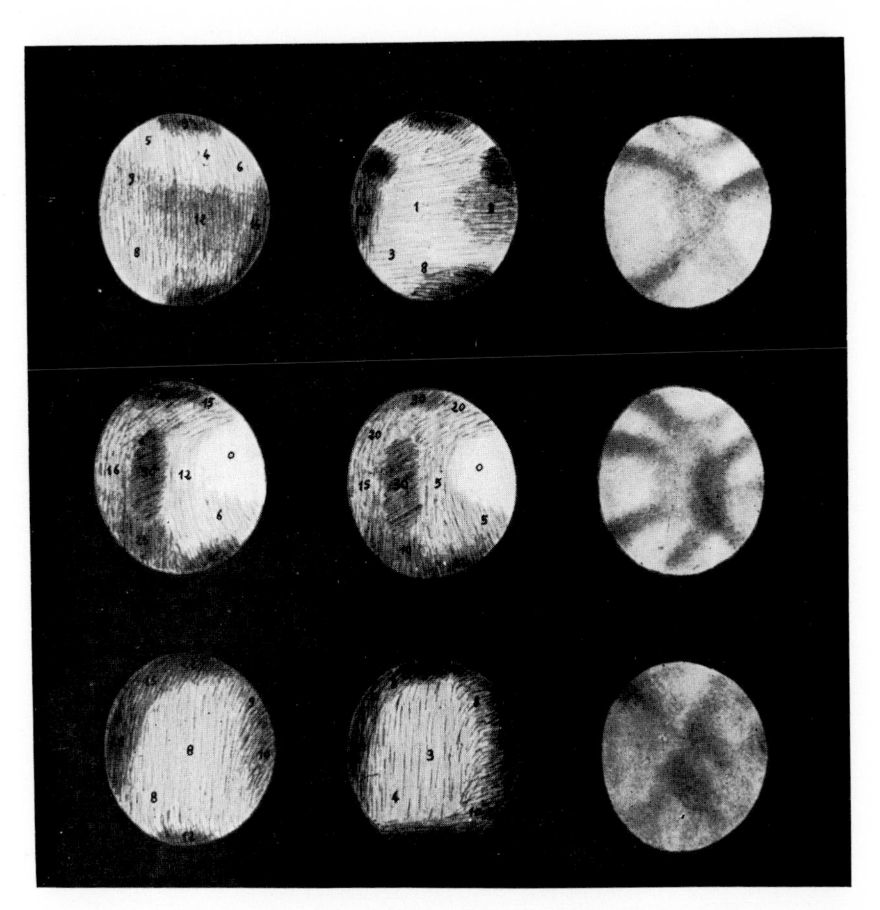

PLATE 5.—Comparison of polarimetric and visual observations of Venus. The direction of shading shows direction of polarization. *Left:* polarization in green light; *center:* polarization in red light; *right:* visual observations of faint markings. Dates: October 3, 1950, $V = 15\overset{\circ}{.}2$; October 12, 1950, $V = 11\overset{\circ}{.}1$; and October 20, 1950, $V = 9\overset{\circ}{.}0$.

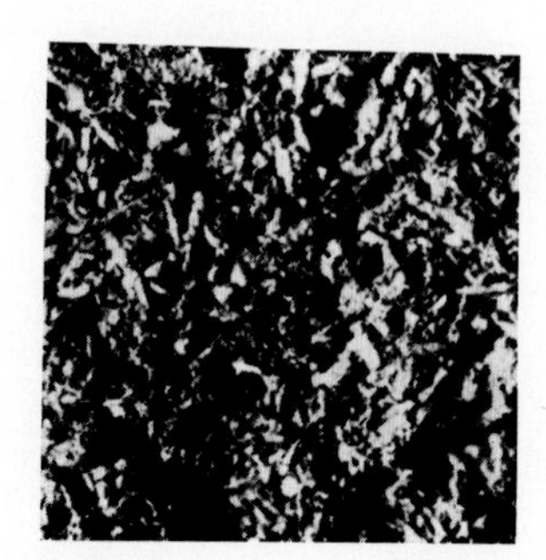

PLATE 6.—Microscopic structure of sample, 1 mm², of pulverized limonite

PLATE 7.—Jupiter, modulated spectrum of south polar cap of Jupiter, red region, January 24, 1940. (After Öhman.)

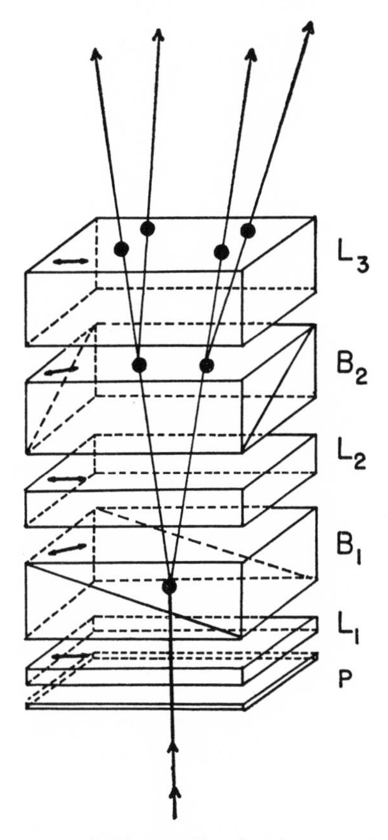

FIG. 3.—Birefringent filter for planetary studies

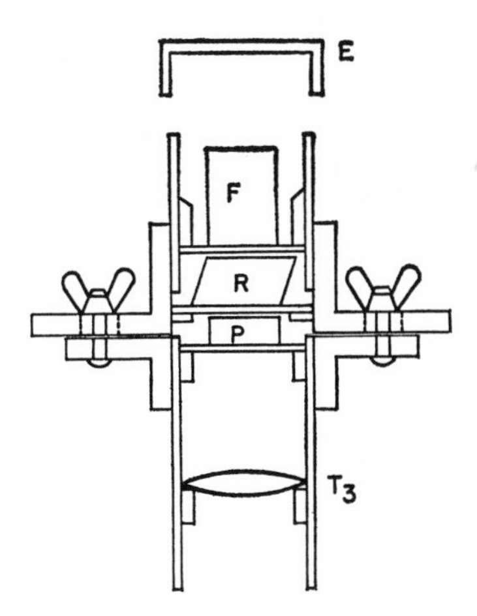

FIG. 4.—Mounting of a birefringent filter, *F*, with a two-component entrance polarizer, *R*, on a Savart polariscope, *P*.

equalizing the polarizations of sky and planet. Thus the polarization of the planet can be corrected. Later Lyot devised a birefringent device which automatically compensated for atmospheric polarization (Lyot, 1929*a*).

Since 1950, Dollfus has used a simpler arrangement consisting of a birefringent prism of calcite and glass behind the polariscope. This prism forms two separate images of the planet while the sky backgrounds are superposed. The prism is oriented in such a way that the fringes of the two images overlap and disappear for the sky (Dollfus, 1957).

2.5. Photographic Polarimeters

Lyot (1934) has described a photographic polarimeter used when the light is too faint for visual observations or beyond the spectral sensitivity of the eye. This instrument was used for asteroids. The telescopic image is centered in a diaphragm which is followed by a birefringent prism and a short-focus lens producing two contiguous images of the objective. A series of exposures is then made, with the plate shifted normal to the line joining the two images. Further, between successive exposures, the plane of polarization is turned 90° by means of a half-wave plate in front of the focus. Thus the plate will show a series of circular spots aligned in two rows, with the magnitude differences alternating between the pairs.

Öhman (1947) added a quartz-plate depolarizer to a birefringent prism, the depolarizer being followed by a second similar prism, finally producing four images. The instrument is in use at Stockholm Observatory for the study of interstellar polarization; similar instruments can be used for planetary studies.

Dollfus (1955) described a polarimeter in which the light is photographed after passing through a compensating plate that can be tilted, a grid cut in two half-wave plates, the directions of whose axes differ by 40°, and a calcite rhomb. For photographic polarimeters the precision increases with the square root of the total exposure time and of the quantum efficiency of the emulsion. With the Dollfus instrument, the precision reaches 0.001 in P on a sixth-magnitude object with a 24-inch telescope, for an exposure time of 20 minutes. The method is applicable to a planet and makes it possible to measure different parts of its surface. A birefringent filter and a compensator for atmospheric polarization may also be added to these polarimeters.

2.6. Photoelectric Polarimeters

With a multiplier phototube it becomes possible to investigate accurately very faint light-fluxes. Öhman first used a photoelectric polarimeter in 1944 for the study of the moon; he has given a polarization-curve for a spot in Mare Serenitatis close to the crater Bessel. In 1949 he improved the instrument and its sensitivity. The photoelectric polarimeter made by Hall for the measurement of stellar polarization has also been used for asteroids. In 1948, Lyot made an instrument of very high sensitivity, by improving the efficiency of each part.

In the more elaborate of Lyot's polarimeters, the light passes through an inclined compensating plate, then through a modulator in the form of a glass disk turned by a motor at a speed of 20 cps. On the disk are fastened pairs of half-wave plates oriented in such a way that the plane of polarization turns suddenly through 90° at each half-revolution. A double-image prism, used as analyzer, splits the light into two rays polarized at right angles, which are received by two photomultiplier cells. The unpolarized part of the light illuminates the two cells simultaneously. The polarized parts received by each of the two cells modulate the current with opposite phases. After phase changing, these currents are added, amplified, and rectified by a mechanical contact synchronized with the modulator. The resulting d.c. current, which is proportional to the amount of polarized light, is fed into a microammeter in which the deviation can be integrated over more than a minute of time.

Two successive measures with different inclinations of the compensator give the degree of polarization by interpolation, with a precision proportional to the square root of the flux and the integration time. For a flux of 1 microlumen a polarization of 10^{-4} can be detected in 2 minutes of integration with Lyot's instrument. With the new seven-stage multipliers by Lallemand, it is now possible to increase the photoelectric efficiency ten times.

Gehrels and Teska (1960) converted a photoelectric photometer at the McDonald Observatory into a polarimeter by using two multiplier phototubes and six filters for the range 3250–9900 A.

In 1959 Dollfus designed a new photoelectric infrared polarimeter with lead sulfide photocells capable of detecting 0.1 per cent polarization between 1 and 1.3 μ, 1.65 and 1.7 μ, and 1.9 and 2.6 μ, using a 40-inch telescope and a few seconds' integration time.

2.7. Measurements in the Laboratory

The polarizing properties of terrestrial samples can be studied in the laboratory with a goniometer. Lyot depolarized the light illuminating the laboratory material by means of a plate formed by two quartz blocks cut parallel to their optical axes, of thicknesses in the ratio 2:1, cemented together with their axes inclined at 45°. Light passing through this device comes out natural at all wave lengths.

With the sample illuminated with natural light, the measures themselves are made with a Savart polariscope having adjustable fringe widths to allow measurement on objects with different angular diameters (Dollfus, 1957). The fringe width is adjusted by varying the relative orientation of the two rhombs in the polariscope; one rhomb is followed and the other preceded by a quarter-wave plate whose axis is inclined at 45° to the principal section of the associated rhomb. When strong polarization is to be measured, two additional compensating plates, mechanically connected to form a symmetric pair at an adjustable angle, are provided.

2.8. Measurements in the Field

Measurement in the field requires an instrument giving simultaneously the phase angle, the angle between the line of sight and the nadir, and the degree of polarization. Plate 1 shows a sturdy vest-pocket instrument. The vertical angle is set by changing the point of suspension. The phase angle is obtained by sighting the sun's image in an unsilvered small convex mirror, supported at a pivot (bottom of Plate 1). The polarization is then measured as in Lyot's instrument. Natural polarization phenomena can be examined with this instrument under conditions not easily reproduced in the laboratory.

2.9. Measurements at the Telescope

When a refractor is used, it is necessary to determine beforehand the partial depolarization that an objective may produce, as a result of birefringence due to excessive annealing.

Reflectors coated with aluminum in a vacuum chamber sometimes produce partial polarization, which is added to the polarization to be measured. The origin of this polarization is not clearly understood. It varies from one coat to another and from point to point on a given coat. This effect is never shown by silver coats deposited chemically (cf. Dollfus, 1957).

3. LABORATORY MEASUREMENTS

Light reflected by illuminated substances shows polarization whose properties depend on the nature of the surface and on the angles of incidence and of observation. This is caused by reflections, scattering, diffraction, single and double refraction, and absorption. All these processes are taking place simultaneously in the surface layers to which the light penetrates.

Arago was the first to investigate this field. Other early studies of completely polarized incident light by Gouy (1884) and Lafay (1894) expressed results by means of curves using spherical or stereographic coordinates. In later experimental work, done chiefly in France, the samples were characterized by polarization-curves as a function of the phase angle V for varying orientations of the surface. Alternatively, the polarization was represented as a function of the inclination of the surface relative to the line of sight.

Materials may be divided into several groups according to the nature of the polarization. The polarization-curve of a certain substance is dependent on its structure, transparency, and index of refraction and may in certain cases suffice to identify the material.

A fraction I_s of the light with polarization P_s is reflected by the surface. The remainder, I_i, enters the substance, where it is diffused and partially

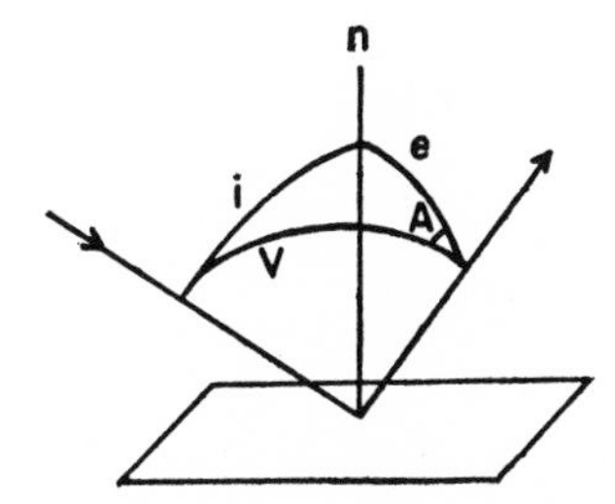

FIG. 5.—Angles defining the polarization

absorbed. A small fraction τ of I_i subsequently emerges from the substance, refracted with a polarization P_i. The total polarization P of the combined emergent flux in the simple case of linear polarization is

$$P = \frac{P_s I_s + P_i \tau I_i}{I_s + \tau I_i}.$$

Figure 5 shows the relations between the plane of vision, the normal to the illuminated surface, and the angles V, the phase angle; A, the azimuth of the plane of vision with respect to the plane of emergence; i, the angle of incidence; and e, the angle of emergence. If $A = 0°$, the planes of vision and emergence coincide with the incident plane; then e is represented by I in subsequent figures. If $A = 90°$, the angle e is called L in subsequent figures.

3.1. Vitreous Materials

The quantities P_s, P_i, I_s, and I_i are known for all azimuths of illumination and observation, from Fresnel's laws, as functions of the index of refraction; τ depends on the absorption, the scattering, and the surface structure of the substance. When the surface is plane, the flux I_s is concentrated in the direction of the specular reflection and shows strong polarization (Fig. 6). In other directions the refraction of the ray penetrating the material produces negative polarization, with the absolute value increasing with the inclination i, the angle between the incident beam and the normal to the surface.

A granular surface reflects light in many directions, so that the sharp peaks of Figure 6 are not encountered here. Figure 7 shows the polarization-curve for a transparent granular material; the amplitudes are greatly reduced. Figure 8 corresponds to a more absorbent material. The contribution τI_i is smaller, and the result is an increase in positive polarization with V.

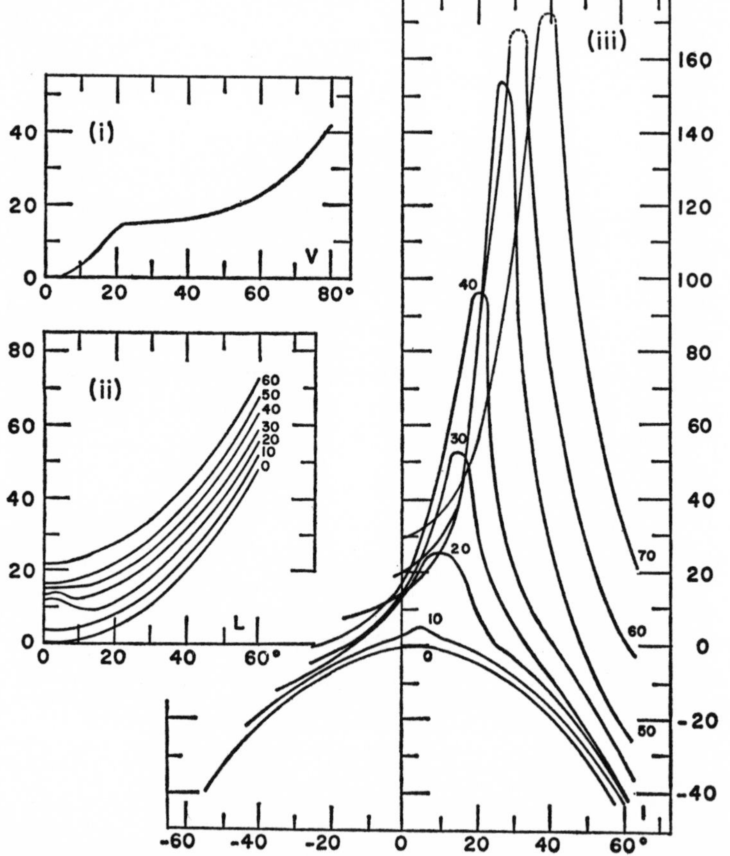

Fig. 6.—Polarization of light reflected from homogeneous diffusing material with an almost plane surface (milky quartz). *Ordinates:* polarization (unit 0.001). (*i*) $L=0$; (*ii*) and (*iii*), values of the phase angle V are given for each curve.

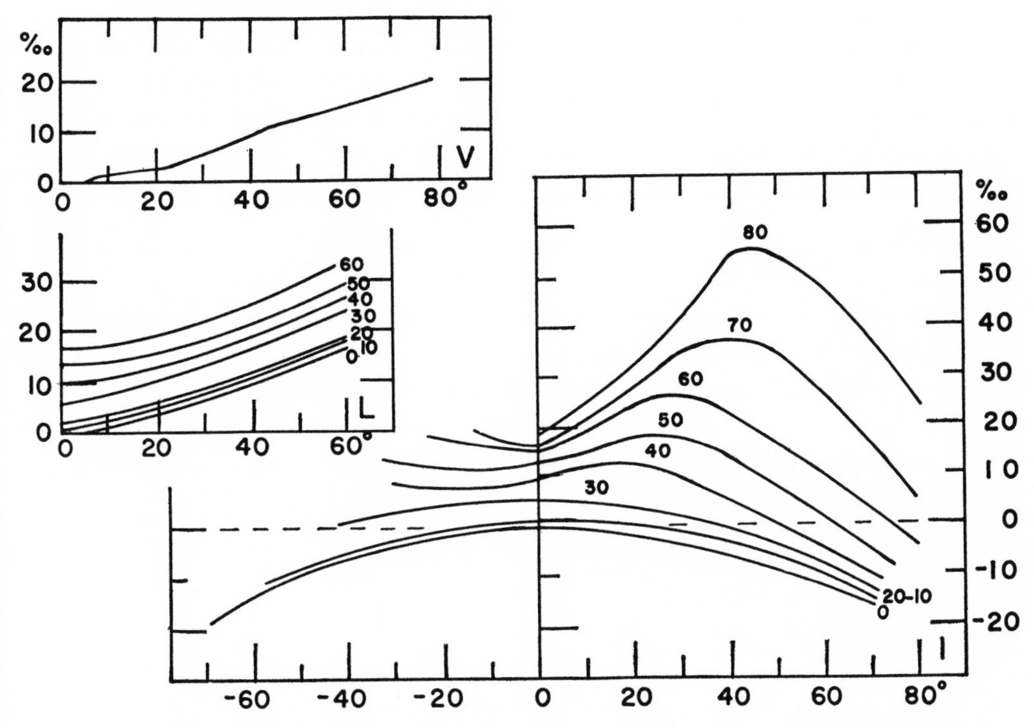

Fig. 7.—Polarization of light reflected from a piece of white sugar (inhomogeneous material with granular surface). Units and symbols as in Fig. 6.

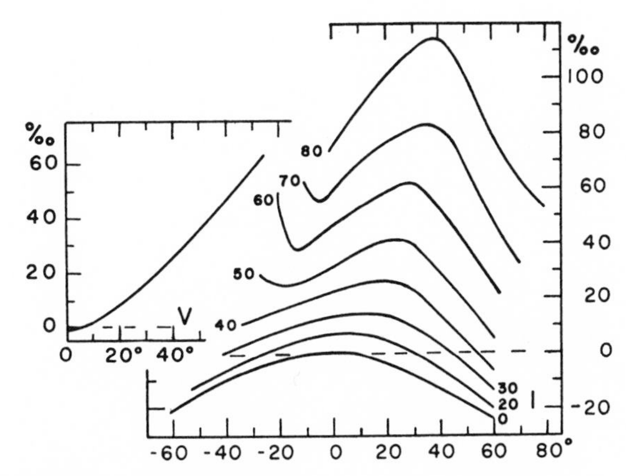

Fig. 8.—Polarization (unit 0.001) of light reflected from sandstone containing quartz (partially absorbing with rough surface).

If vitreous materials existed on the surface of a planet, a bright spot should appear in the direction of specular reflection. Even if this spot were not bright enough to be visible to the eye, the polarimeter would reveal strong polarization. No planet shows such effects, and planetary surfaces can therefore not be vitreous to any significant extent.

3.2. Nearly Transparent Powdered Materials

Figure 9 shows the measures obtained for pulverized glass. The facets in the mass of grains reflect light in all directions, which causes the polarization to be reduced by the spread in the inclinations. The *P-L* curves are seen to show a minimum.

An increase in the opacity of the material spreads the curves apart (see Fig. 10, which relates to sand). As early as 1929, Lyot noted that light coming from the interior of a powder shows greatly diminished polarization, with the maximum of polarization shifted toward larger phase angles. For larger grains the polarization follows Fresnel's laws and reaches a maximum for tan $V/2 = n$, although the general shape of the

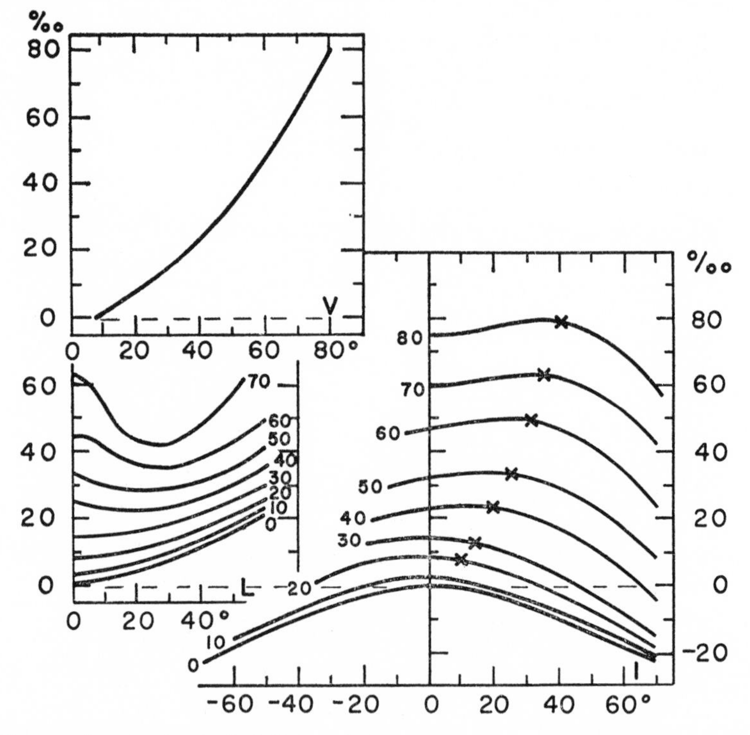

Fig. 9.—Polarization (unit 0.001) of light reflected from pulverized glass (powder with transparent grains).

polarization-curve seems to depend more on the shape of the grains than on the refractive index n.

Sands and clays have been studied by Cailleux and Dollfus (1950), who have shown the influence of absorption, diffusion, condition of surface, and size of grain. The curves are related to the microscopic structure of the grain surface. The clear difference between these curves and those obtained for the moon, Mercury, and the dark and bright areas of Mars is evidence of the absence of quartz in powdered form in the surface layers of these bodies. From that point of view, they differ from the earth. These polarization-curves are not compatible with the assumption of a large contribution of the reflected light coming from the interior of the surface layer; hence the presence of ordinary lime sediments is improbable.

3.3. Opaque Powdered Materials

The polarization of the light reflected from an opaque powder must be due to surface phenomena. It is found empirically that such powders produce a negative branch of the polarization-curve for V less than 10°–25°, depending on the substance. An example is shown in Figure 11. This

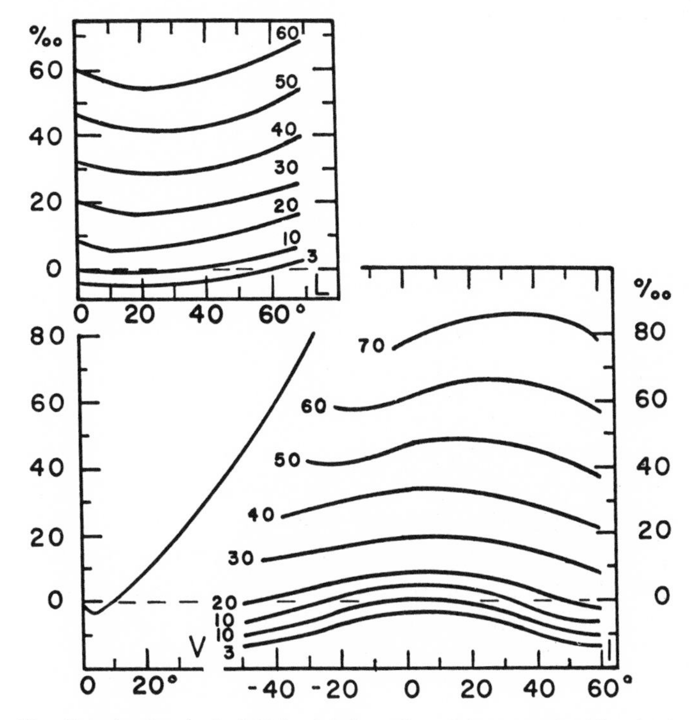

Fig. 10.—Polarization (unit 0.001) of light reflected from sand with mixed rough and polished spherical grains.

negative branch is found to depend on multiple scattering between the surface grains. If the grains are separated, as in a free-falling stream of particles, the negative branch disappears. This negative branch was discovered by Lyot (1929*a*), partially explained theoretically by Öhman (1955), and examined in great detail empirically by Dollfus (1956).

Figure 11 also shows that the *P-L* and *P-I* curves are very flat. This is attributed to the multiple scattering by the grains of the powder which tend to make the reflected light isotropic.

The greater the absorption by the powder, the deeper the negative polarization branch becomes. There is also a dependence on particle size. Figure 12 shows the curves for coarse, medium, and fine iron filings, respectively, which show the finest filings to have the deepest negative branch. The branch is especially deep for opaque materials ground to very fine grains which themselves are combined into larger grains. This type of curve is observed for the moon, Mercury, both the bright and the darker areas on Mars, and the asteroids.

3.4. Igneous Rocks

Rough cuts of basalt, diabase, periodotite, serpentine, etc., were studied by F. E. Wright (1927). Lyot measured several igneous rocks and lavas by observing them at the angle of specular reflection (i and e equal; see Fig. 5). Figures 13 and 14 give some of Lyot's results. Lower albedos are seen to be accompanied by higher maximum polarizations.

Negative branches of the polarization-curves are present but not pronounced; the inversion points occur at about 15° or less. The negative branch is the more developed, as to both the depth of the minimum and the position of the point of inversion, the more porous the lava is.

3.5. Volcanic Ash

In his comprehensive program, Lyot (1929*a*) included the study of volcanic ash from different eruptions of Vesuvius. Figure 15 shows his polarization-curves for ashes from the eruptions of 1872, 1906, and 1908. Lyot found the angle of inversion to vary from 17° to 23°, with the higher values associated with the higher albedos. The depth of the minima appeared, on the whole, the largest for the lowest albedos, varying between 0.008 and 0.014. The positive maxima are highest for the lowest albedos. Lyot found the same correlations with albedo on the moon.

3.6. Water Droplets and Fogs

Water droplets and artificial fogs were studied by Lyot (1929*a*) in considerable detail.

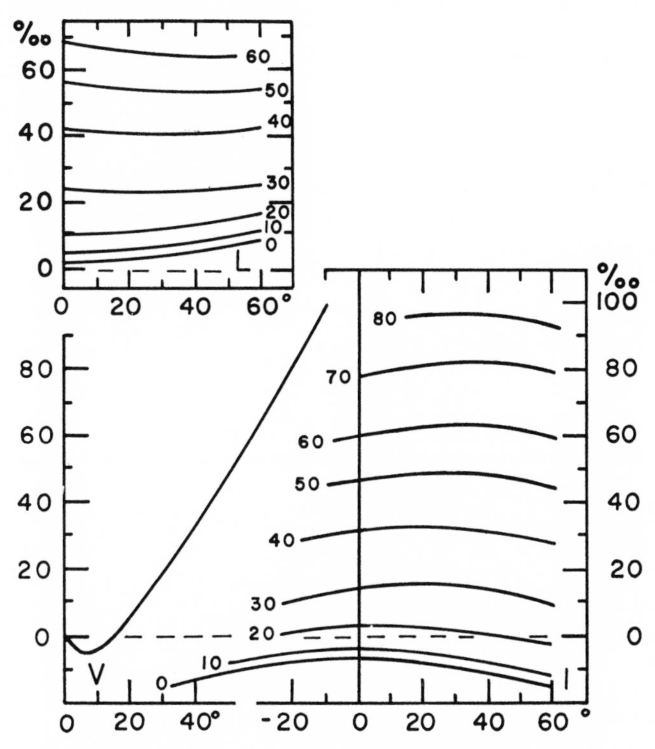

FIG. 11.—Polarization (unit 0.001) of light reflected from emery powder (small opaque grains).

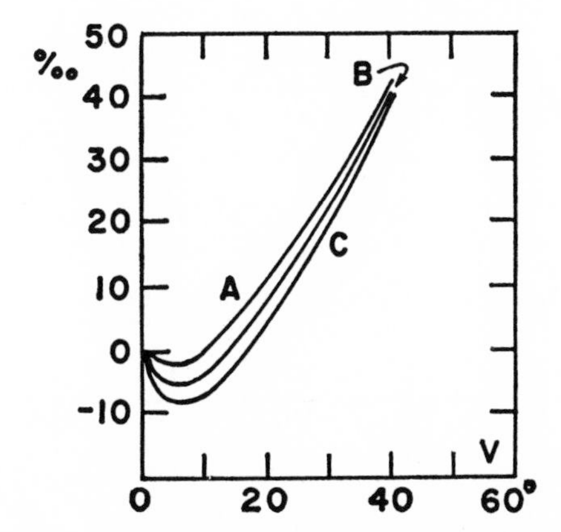

FIG. 12.—Polarization (unit 0.001) of light reflected from iron filings; *A:* coarse, *B:* medium, *C:* fine.

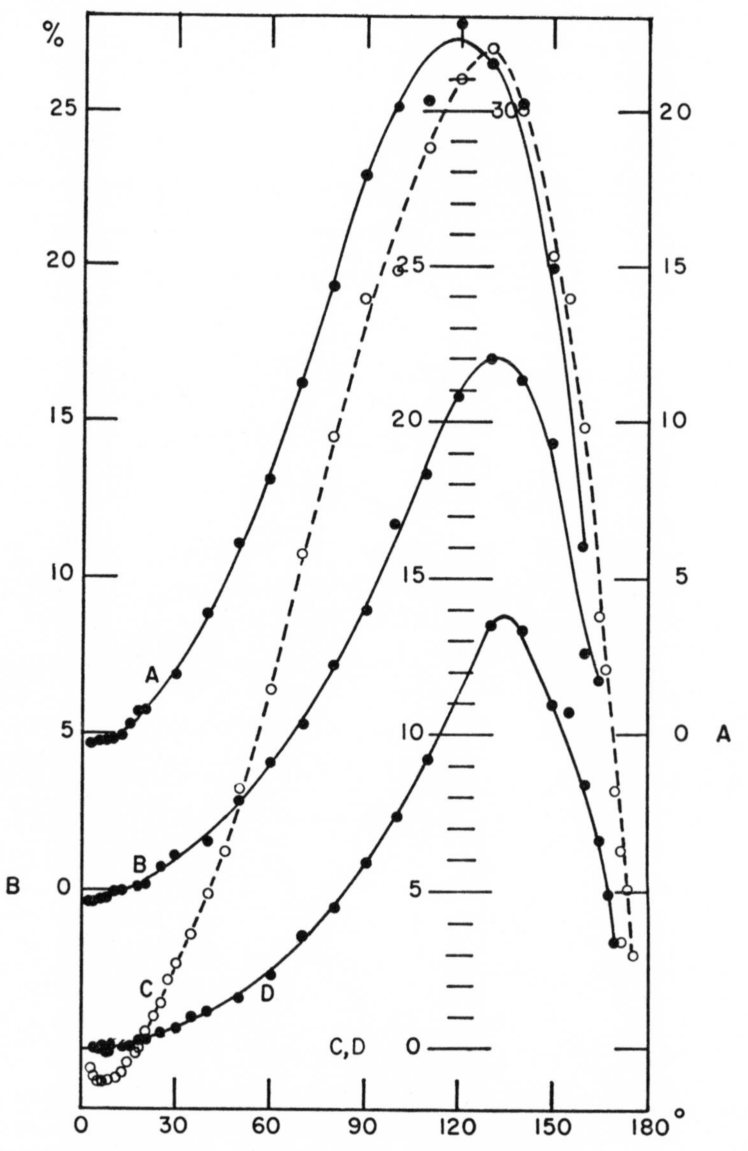

FIG. 13.—Polarization (unit 0.01) of light reflected from igneous rocks. *A:* basalt (albedo 0.08), *B:* granite (albedo 0.14); *C,* red sandstone (albedo 0.115); *D:* quartz sandstone (albedo 0.34). Scales have different zero points as indicated. (After Lyot.)

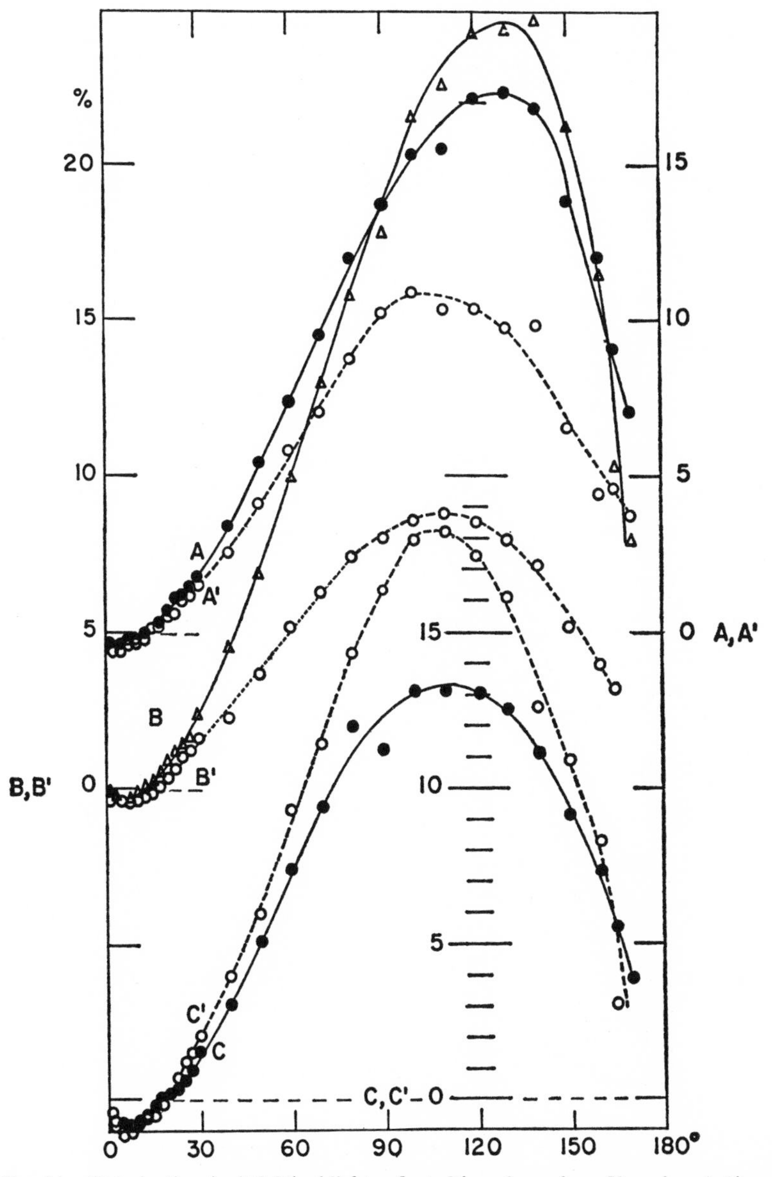

Fig. 14.—Polarization (unit 0.01) of light reflected from lavas from Vesuvius. *A*,*A'*: porous coarse-grained lava (albedos 0.23 and 0.17); *B*,*B'*: gray lava with crystals (albedos 0.15 and 0.17); *C*,*C'*: porous brown lava (albedos 0.08 and 0.15). *Full-drawn lines*, fractured lavas; *dashed lines*, crushed lavas. Scales have different zero points as indicated. (After Lyot.)

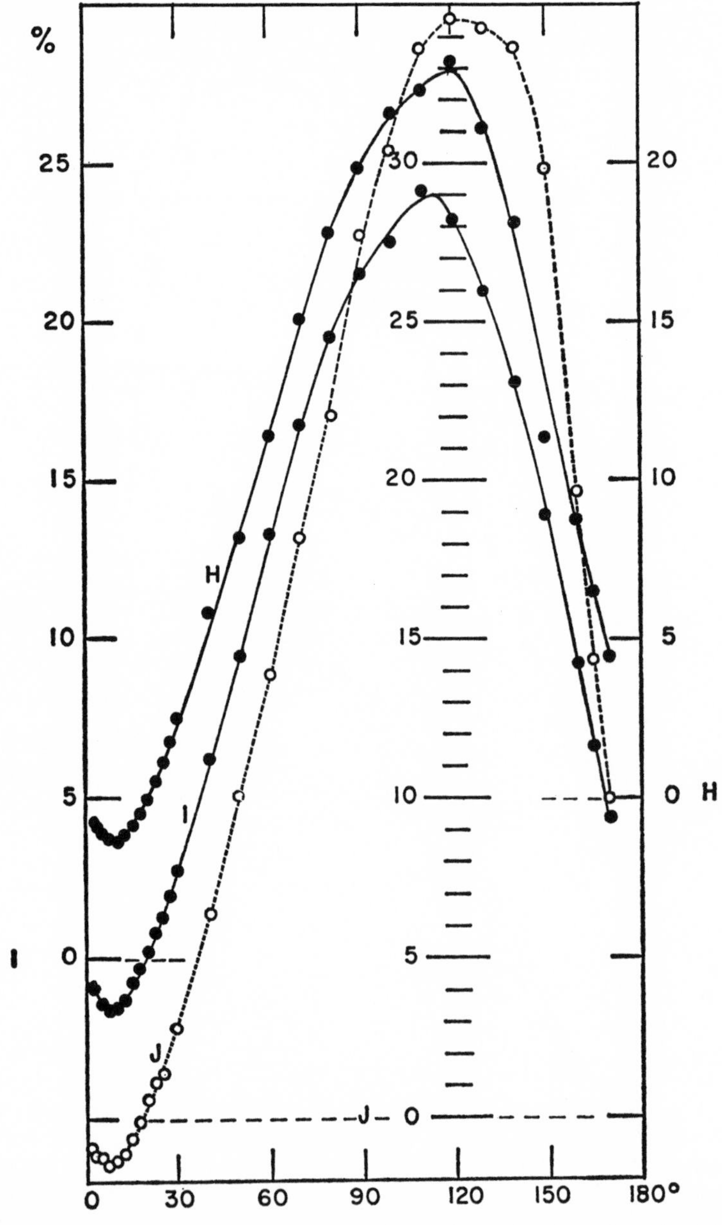

Fig. 15.—Polarization (unit 0.01) of light reflected from ashes from Vesuvius. *H:* grains with diameter between 0.3 and 1.0 mm (albedo 0.063); *I:* grains with diameter between 1 and 4 mm (albedo 0.051); *J:* grains with diameters between 0.5 and 3 mm (albedo 0.048). Scales have different zero points as indicated. (After Lyot.)

3.61. *Single droplets, 1 mm in diameter.*—The principal types of reflected and refracted rays that play a role in the light scattered by a single droplet are shown in Figure 16, upper-right corner insert. There is a reflected ray A, a doubly refracted and singly reflected ray B, a doubly refracted and reflected ray C, and a transmitted, doubly refracted ray D. Each of these rays has its particular intensity and degree of polarization,

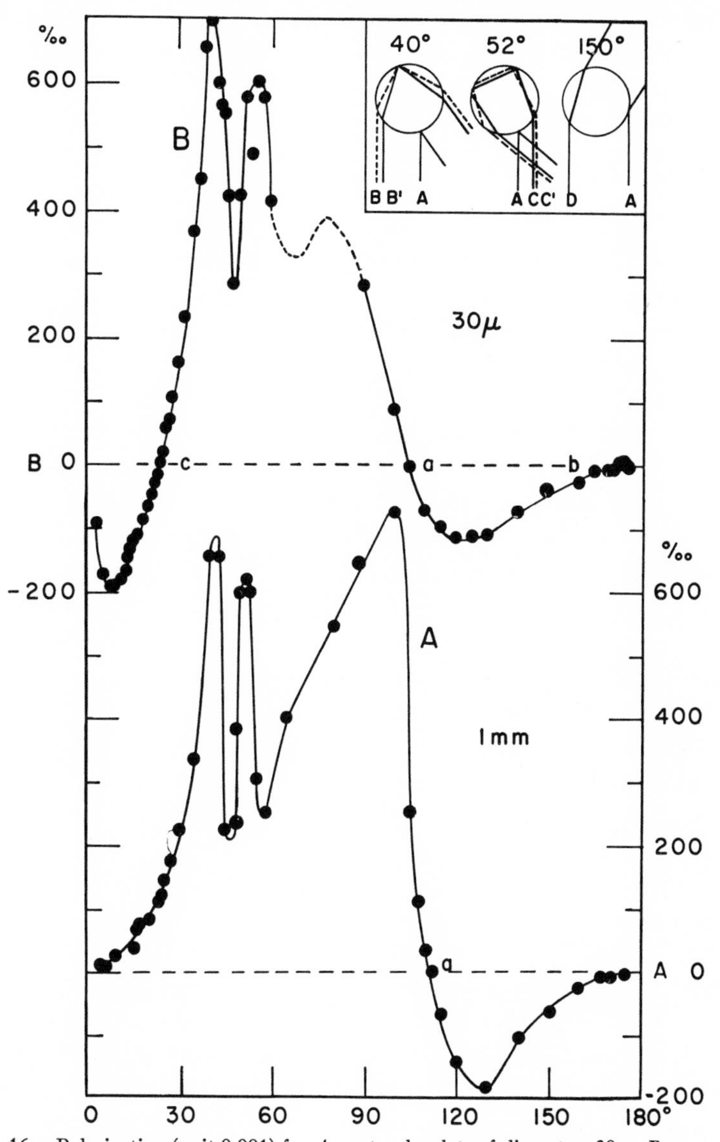

FIG. 16.—Polarization (unit 0.001) for A: water droplets of diameters 30 μ; B: water droplets of diameters 1 mm. (After Lyot.)

depending on the phase angle *V;* the rays *B* and *C* are present only in a very limited range of *V*, since they depend on total reflection. These rays are responsible for the first and second rainbows at 40° and 52°. The relative intensities and polarizations are responsible for the maxima and minima of the polarization curve *A*, shown in Figure 16. The two rainbows are highly polarized positively, because of the great intensities of the total reflections. For small phase angles, ray *A* is the most intense, which contributes positive polarization, increasing with *V*, beyond 100°. Thereafter, ray *D*, which is negatively polarized (since it has lost intensity by two reflections which are positively polarized), becomes of increasing intensity; rays *A* and *D* together determine the shape of the polarization-curve beyond 90°. Beyond 120° ray *D* is dominant, and the net polarization is negative. The inversion point, *a*, is at 112°. The polarization can be computed by Fresnel's formulae for all phase angles, taking into account absorption in the water.

It is clear from this description that the drop will not appear uniformly illuminated but that the scattered light received appears to come from discrete images on the drop, whose position can be found from the inset parts of Figure 16; the images are situated on the diameter in the plane of vision. The drop itself will appear dark. The images have low intensity except at the rainbow angles of 40° and 52°.

3.62. *Droplets of about 35 μ.*—The curve for an artificial cloud of droplets of about 35 μ is similar to that for 1-mm single droplets, but there are some deformations due to diffraction. The angle of inversion, *a*, is reduced from 112° to 105°. A new point of inversion, *c*, has appeared near 24° and is preceded by an important negative branch (Fig. 16, curve *B*).

3.63. *Cloud of droplets of about 5 μ.*—The polarization-curve is shown in Figure 17, *C*. Here we can see evidence for increasingly important diffraction phenomena. The principal angle of inversion, *a*, has decreased to 76°; the angle *c* has also decreased, while a new inversion has been formed at *d*, near 5°. The diffraction also greatly lowers and displaces the maximum of the second rainbow, while the first rainbow is displaced and widened (Fig. 17, curve *C*).

3.64. *Cloud of droplets of about 2.3 μ.*—The principal angle of inversion, *a*, is now only 45°. The inversions *c* and *d* have disappeared, and *e* and *f* have separated. The positive branch is considerably diminished, and the two rainbows have vanished (cf. Fig. 17, curve *D*).

3.65. *Natural clouds.*—Natural clouds may be observed from a mountain, close to a cloud layer with little intervening atmosphere, as was done by Dollfus (1956). A dense stratocumulus cloud, at 3000-meter altitude, was found to be composed of droplets 14 μ in diameter. The observed polarization-curve is found in Figure 18. It is seen to be intermediate between the laboratory curves for 35 μ and 5 μ droplets shown above, but the polarization is reduced fourfold, because multiple scattering occurs as described in Section 3.67.

A thin cloud with droplets of 18 μ was also observed, with the result shown in Figure 19. The observations were made in two colors, green and red, with similar results but with a displacement of the maxima and

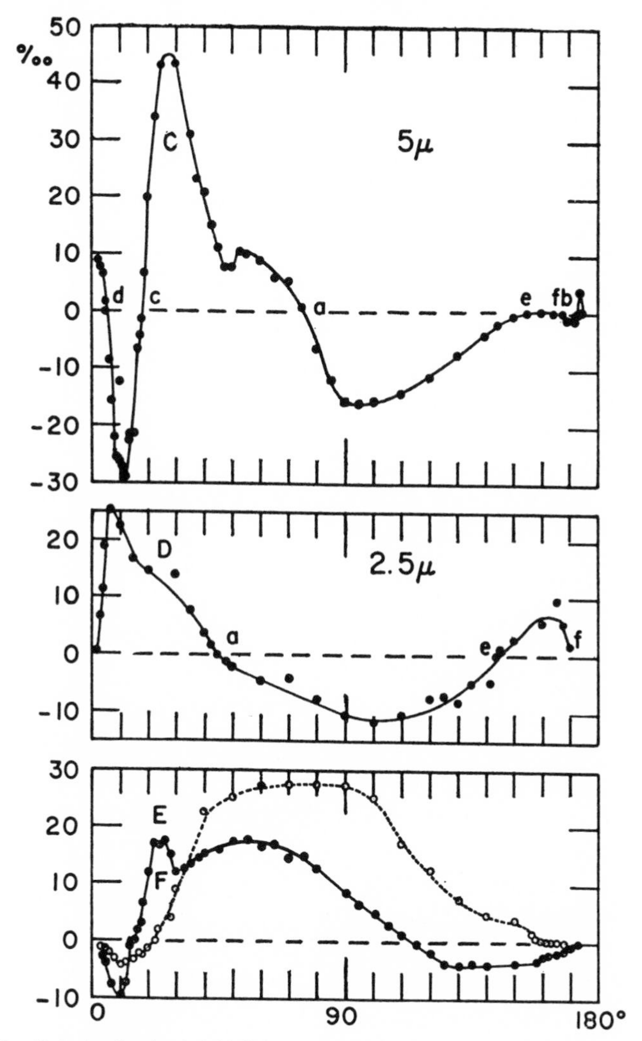

FIG. 17.—Polarization (unit 0.001) for *C:* water droplets of diameters 5 μ; *D:* water droplets of diameters 2.5 μ; *E:* fog consisting of ice crystals (partly melted); *F:* fog consisting of ice crystals (dry). (After Lyot.)

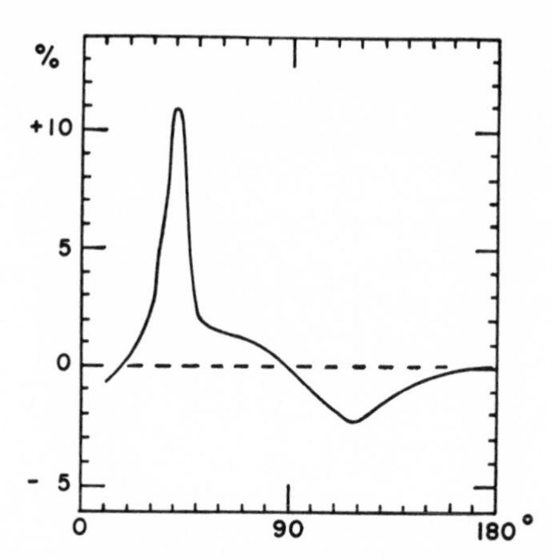

Fig. 18.—Polarization (unit 0.01) of stratocumulus cloud at 3000 meters. Droplet diameter 14 μ. (After Dollfus.)

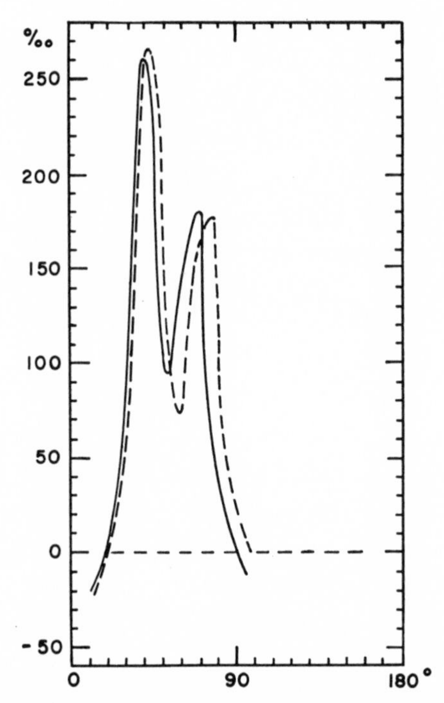

Fig. 19.—Polarization (unit 0.001) of light fog at 2800 meters. Droplet diameter 18 μ. (After Dollfus.)

minima. The displacement of the green curve toward the larger phase angles (constant drop size but decreasing wave lengths) is consistent with the shift of curve *B*, Figure 16, with respect to curve *C*, Figure 17 (constant wave length but increasing drop size).

3.66. *Cloud of ice crystals.*—Figure 17, *E* and *F*, shows the influence of partial melting on ice fogs produced in the laboratory. Diffraction phenomena are very important. The polarization is strongly affected by small variations in the dimensions of the ice crystals (assumed to be roughly 5 μ from the shape of the polarization-curve).

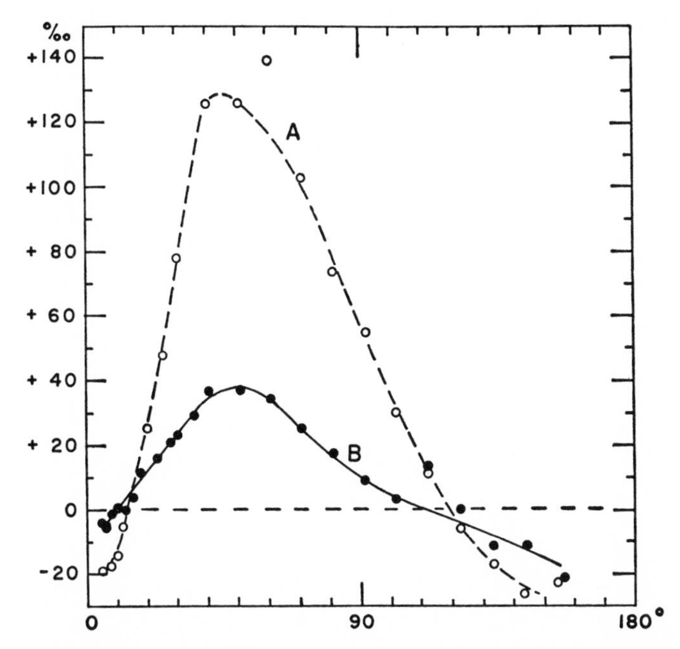

FIG. 20.—Polarization (unit 0.001) of colloidal solution of bromo-naphthalene in water. *A:* dilute; *B:* opaque. (After Lyot.)

3.67. *Cloud of opaque droplets.*—Lyot (1929*a*) has made experiments with suspensions which had refractive indices, with respect to the liquids used, equal to 1.33, which also is the refractive index of water with respect to air. By using opaque suspensions, a cloud layer of indefinite thickness could be simulated, illuminated, and observed from above. Figure 20 shows curves for bromonaphthalene in water, with particle sizes of about 2 μ, for two conditions: (*A*) when the suspension is highly diluted and (*B*) when it is concentrated and opaque.

It is noted that the polarization-curves are similar except for scale, the curve *B* showing an amplitude of one-third to one-fourth that of curve *A*.

Lyot explains this difference by multiple scattering (having practically no polarization) diluting the polarization present in the primary scattering process. The reduction of the polarization in a dense cloud is less for large grains than for small ones, partly because large grains have a stronger forward scattering.

Particles of soot over a smoky flame show polarization nearly identical with that of molecular scattering (Dollfus, 1956, Fig. 10).

3.68. *Very fine particles.*—The laws of diffusion and polarization by particles of sizes comparable to the wave length of light may be derived theoretically. Mie's theory (1908) has been applied to transparent or opaque particles by Schirman (1919), Born (1933), van de Hulst (1946, 1949), Sinclair (1947), Bouwkamp (1954), and others.

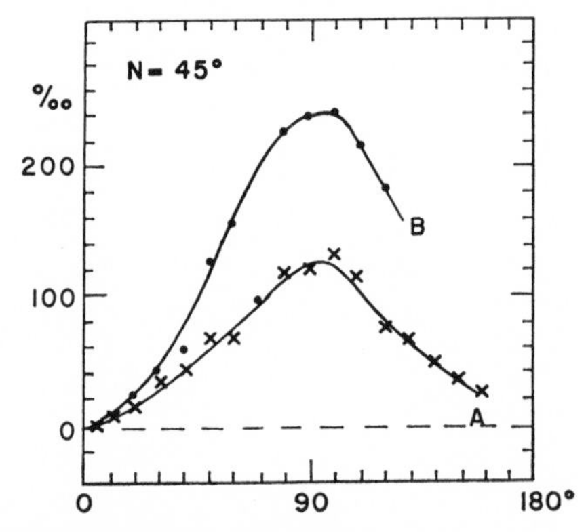

Fig. 21.—Polarization (unit 0.001) of grass field in summer, seen from the air. *A:* ground level; *B:* altitude 1000 meters. (After Dollfus.)

The *Tables of Scattering Functions for Spherical Particles* published by the United States National Bureau of Standards (1949) give some of these properties. More detailed results are found in Penndorf (1956).

3.7. Atmospheric Diffusion

The properties of molecular scattering are given by Rayleigh's formulae. Under normal conditions the intensity B for a thickness Z is

$$B = KZ\left(\frac{\lambda_0}{\lambda}\right)^4 (1+\cos^2 V)\,, \tag{1}$$

while the polarization P is

$$P = \frac{\sin^2 V}{1+\cos^2 V}\,. \tag{2}$$

The polarimetric properties of the earth's atmosphere between the ground and an observer at high altitude may be found from measures made from a balloon. Figure 21 shows the polarization-curves obtained for a green field from the ground (A) and from an elevation of 1000 meters (B). P varies inversely with the reflectivity of the ground and increases with the amount of overlying atmosphere. It is also wave-length-dependent. Such measures may be interpreted by considering the additive contributions of surface and atmosphere, as follows:

$$P=\frac{P_s I_s + P_a I_a}{I_s + I_a}, \tag{3}$$

with I_a and P_a given by equations (1) and (2). In this manner also the polarization of the earth as a whole as seen from outer space may be estimated (about 33 per cent at 90° from the sun).

4. POLARIZATION OF MOON AND PLANETS

Lyot's pioneering studies at the Meudon Observatory on the polarization of the moon and the planets, begun in 1922, were until recently the principal source of information on this subject. He obtained polarization-curves for the integrated light both of these bodies and of specific surface detail of different albedos, such as regions on the moon, the polar regions of Jupiter, and the Rings of Saturn.

Since 1948 Dollfus and collaborators, with the aid of special equipment attached to the 24-inch refractor at the Pic du Midi Observatory, have added new information. Their equipment enabled them to locate with precision the area on the planetary disk under study and to make measures at different wave lengths. The following discussions are based on the combined data now at hand.

4.1. The Moon

The data discussed in Section 4.11–4.14 were obtained by Lyot (1929*a*).

4.11. *Integrated moonlight.*—Near the quadratures, the lunar polarization reaches a maximum (Fig. 22): this maximum is 0.066 near first quarter and 0.088 near last quarter. This difference is attributed to the distribution of the maria, which have unusually large polarization and occupy about twice as much area at last quarter as at first quarter. The rapid decrease in polarization following last quarter coincides with the disappearance at the terminator of the darkest (and most highly polarized) regions of Mare Imbrium and Mare Nubium. As full moon is approached and the phase angle passes 23°30′, the polarization goes through zero and reappears a few hours later in the perpendicular plane. The minimum $P = -0.012$ is reached for phase angle 11°; the polarization vanishes near zero phase angle.

4.12. *Regional studies of the moon.*—The plane of polarization for the

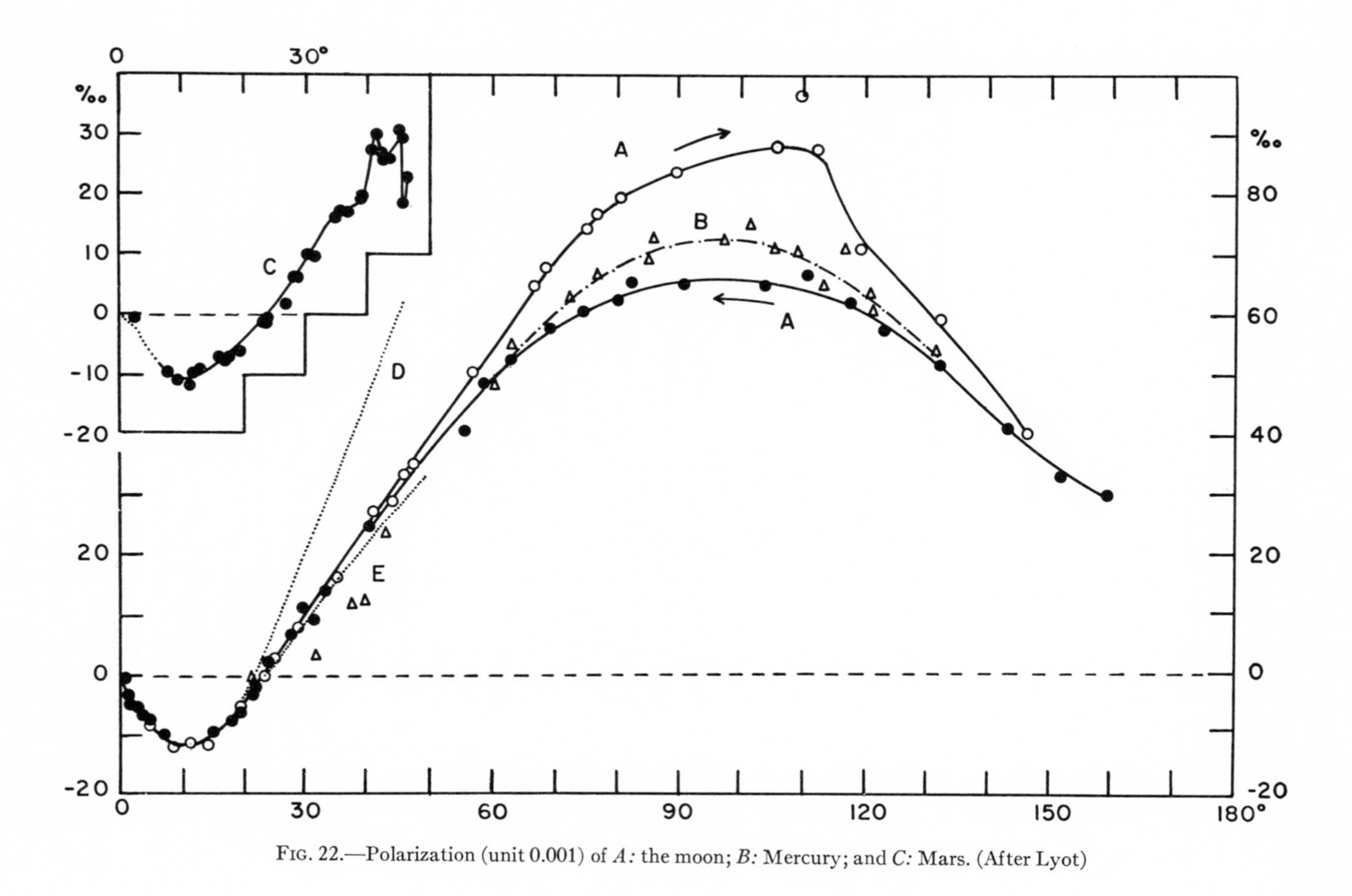

FIG. 22.—Polarization (unit 0.001) of *A:* the moon; *B:* Mercury; and *C:* Mars. (After Lyot)

uplands or terrae and the maria is the same within half a degree for all parts of the disk and is therefore independent of the orientation of the lunar surface relative to earth or sun. The polarization changes roughly inversely with the albedo. The polarization of the maria changes very irregularly from point to point, while there is less variation on the terrae. Only near the terminator does the polarization depend on the small angle of incidence of the solar rays. Close to the terminator the polarization sometimes increases by 50 per cent. From first quarter to full moon the polarization of all regions decreases in the same proportion as that of integrated moonlight, regardless of the albedo. This is indicated in Figure 22; the full curve A gives the general polarization of the moon, while the dotted curves D and E correspond to regions with the strongest and weakest polarization.

Lyot discovered a curious phenomenon on the lunar limb very near full moon. A weak polarization of 0.001–0.002 was found to be present in the direction parallel to the limb. This appears to be analogous to a phenomenon observed by Salet in the laboratory on vitreous materials.

Since 1927 Wright (1935) has measured 26 small lunar regions with a 6-inch refractor and fringe-polarimeter, and has given polarization-curves for these spots. These researches are summarized by this author in Volume **4**, chapter 1), together with a more complete review of earlier lunar polarization work.

4.13. *Lyot's interpretation of the polarization of moonlight.*—After studying many substances in the laboratory (cf. Sec. 3), Lyot concluded that the moon is covered almost entirely by powdery material closely resembling volcanic ashes found on the earth, but the layer could be thin.

4.14. *Polarization of the eclipsed moon.*—According to the measures made by Öhman (private communication), Dollfus, and Focas (1953), there remains in the shadow of the earth a weak polarization of 0.001–0.002 varying from point to point on the disk. This small residual polarization must be attributed in part to the fact that the phase angle is not zero everywhere but varies from point to point with the plane of vision. Part may be due to light scattered in the earth's atmosphere, which forms a small part of the refracted and scattered light that illuminates the eclipsed moon.

4.15. *Polarization of the earth-light.*—With a coronograph the light scattered by the sunlit part of the moon is reduced, and it becomes possible to observe the earth-light on the moon to within 1 day from full moon. With the half-wave grid polariscope one can measure the polarization for all phase angles larger than 35°. Figure 23 shows the polarization-curve of the earth-light measured on the dark areas. The polarization reaches as much as 0.100 for $V = 80°$. It is independent of the slope of the surface and has the same value for regions of equal brightness at the center or the limb. It is higher in the dark regions than in the bright ones by a factor 1.2. The polarization of the earth-light is strongly dependent on wave length. Dollfus (1957, p. 66) measured the polarization at $V = 70°$ for $\lambda = 6300$ A and $\lambda = 5500$ A, and found 0.035 and 0.054, respectively.

4.16. *Polarization of the earth.*—The polarization of the planet Earth could be derived from that of the earth-light (Fig. 23) if the depolarization by the lunar surface were known. This factor may be estimated by computing the polarization P of the earth from terrestrial data, including the

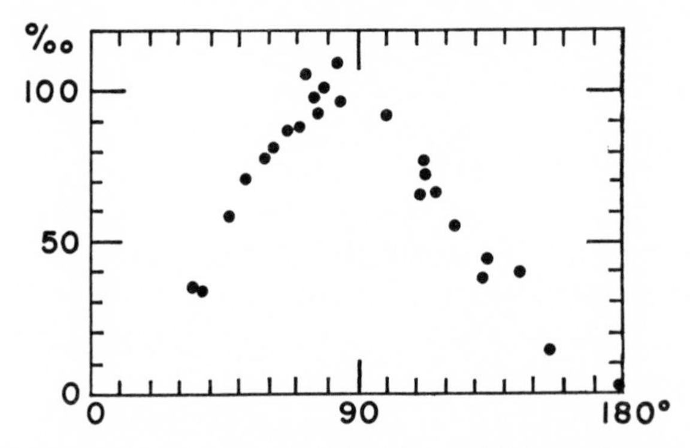

Fig. 23.—Polarization (unit 0.001) of the earth-light on the moon

polarization measures made during balloon flights. The result is $P = 0.33$ for $V = 90°$ (Dollfus, 1957, p. 67). Since the observed polarization of the earth-light is 0.100, it may be assumed that the polarization of the earth as a planet for other phase angles may be found from Figure 23 by increasing measures by the factor 3.3.

The residual polarization of lunar surface materials may now be derived from the observed polarizations of the earth light for areas of different albedo (Dollfus, 1957, p. 65, Fig. 26). Since the source (the earth) is common to the measured values for different albedo A and since the polarization P decreases with increasing A, it follows that the depolarization is greater, the greater the value of A. This result is consistent with laboratory measures of residual polarization R, made by illuminating samples resembling the lunar surface by polarized light (see Fig. 24). It is seen that the decrease in R with increasing A is a general result for opaque powdered material. The lunar data fit in with this relation by using the average scale factor 3.3 derived above, as is seen from the circles in Figure 24. This result strengthens the identification of the lunar surface materials as consisting of a powdery opaque substance.

4.17. *Search for a lunar atmosphere.*—A lunar atmosphere with a density one-millionth that of the earth would scatter a twilight glow at quadrature, extending the cusps of the crescent, whose intensity would be represented by Plate 2; the brightness at the base would be comparable with the earth-light, and observation would show it at once.

Much fainter glows can be revealed by the coronograph, which reduces the scattering of lunar light, this light being almost completely polarized. Under those circumstances, a polariscope would show at the tip of the

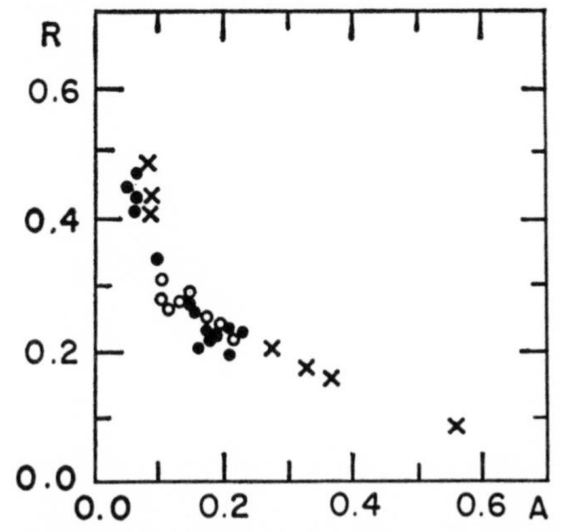

FIG. 24.—Residual polarization in direction of incident beam with fully polarized incident light. *Abscissae:* albedo; *crosses* and *dots:* volcanic ashes; *circles:* earth-light on the moon.

crescent a region crossed by fringes detectable even if the density were a thousand times smaller. The negative result obtained (Dollfus, 1952, 1956*a*) proves that the gaseous envelope of the moon must be at least 10^9 times less dense than that of the earth at sea level. Hence it contains less than 10^{10} mols/cc. Recently (Elsmore, 1957) a still lower limit has been set from radio observations of the Crab Nebula occulted by the moon; but this value is somewhat hypothetical.

4.2. MERCURY

4.21. *Integrated light.*—Lyot's polarization-curve of the integrated light of Mercury based on observations made in 1922 and 1926 is shown in Figure 22. Around quadrature, the polarization is intermediate between that of the waxing and waning phases of the moon. This agreement is remarkable when one considers the enormous differences between the polarizations of the lunar maria and the terrae. Closer to opposition, the polarization of Mercury agrees perfectly with the moon's. It is therefore probable that the surfaces of these two bodies are closely similar. It is noted that the albedos are essentially equal also.

4.22. *Regional studies.*—In 1950 Dollfus (1957, p. 38) was able to examine the distribution of polarized light on different parts of the planet's disk. This was possible because of the occasionally excellent seeing on the Pic du Midi, the use of a compensator for atmospheric polarization, and the use of an 8-meter-long sunshade ("parasoleil") shading the objective from direct sunlight. Figure 25 shows the polarization-curve obtained for the center of the disk, which is one of the brighter regions. It differs very little from the curve obtained by Lyot for the moon. Measures in green

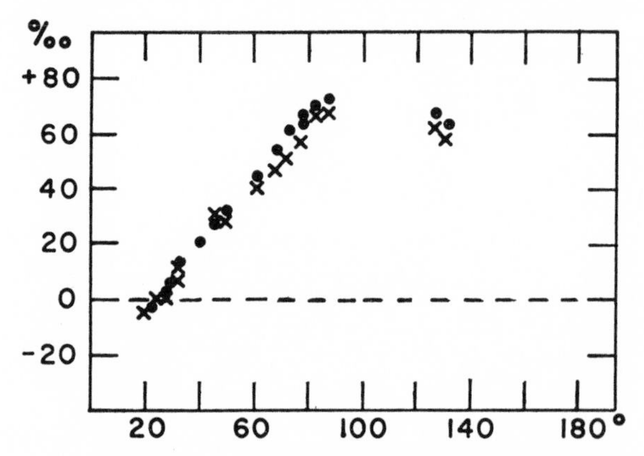

FIG. 25.—Polarization (unit 0.001) of Mercury. *Crosses:* red light; *dots:* green light. (After Dollfus.)

and red light do not differ for small phase angles, but when the phase increases, the polarization in green light becomes slightly the greater. For small phase angles there appears to be no difference for bright and dark regions, but around quadrature the polarization is definitely stronger for the dark regions, changing roughly inversely as the albedo. This property, also found on the moon, further confirms the similarity of the surfaces of the two bodies.

4.23. *Polarization of Mercury's atmosphere.*—For small phase angles the polarization in the polar regions is practically the same as that in the bright central parts. When the phase angle increases, the polarization becomes stronger, increasing more rapidly at the cusps of the crescent than at the center and also faster in the green than in the red. At quadrature the difference between the cusps and the center is 0.006 in the green and 0.003 in the red.

If Mercury has the same polarizing properties as the moon and if the polarization does not depend on the inclination of the surface (as is true for the moon), then the excess polarization at the cusps must be due to a weak atmosphere.

The surface brightness of a tenuous planetary atmosphere seen from the earth changes approximately with sec θ, provided that $\theta < 60°$ (θ measured from the center of the disk); with the phase angle V as $(1 + \cos^2 V)/2$; and with the wave length as λ^{-4}. If B_a' is the brightness at the center of the disk for zero phase angle and $\lambda = \lambda_0$, then the brightness of the atmosphere is generally

$$B_a = B_a' \left(\frac{\lambda_0}{\lambda}\right)^4 \frac{1 + \cos^2 V}{2 \cos \theta} . \quad (4)$$

The light scattered by the atmosphere is polarized according to equation (2) above. This polarization combines with that produced by the planetary surface at the distance θ from the center. If we call the brightness at that point of the disk $B_s = B_s' \phi(V) \psi(\theta)$ and the polarization $P_s(V)$, the observed polarization will be

$$P = \frac{P_s B_s + P_a B_a}{B_s + B_a} . \quad (5)$$

Since $B_a \ll B_s$, we have, approximately, $P = P_s + P_a B_a / B_s$. Then

$$P(V, \theta, \lambda) = P_s(V) + \frac{B_a'}{B_s'} \frac{\sin^2 V}{\phi(V)} \left(\frac{\lambda_0}{\lambda}\right)^4 \frac{1}{2\psi(\theta) \cos \theta} , \quad (6)$$

from which we derive the polarization difference between green and red and between the center and a point θ from the center. Figure 26 shows the measures plotted against the function of V found in equation (6).

If Mercury's atmosphere consists of air, its equivalent thickness is found to be about 25 meters, or 0.003 of the earth's atmosphere. The pressure at the ground would be about 1 millibar.

4.24. *Abnormal polarization.*—Sometimes local regions of the planet seem to show departures in polarization reaching 0.005–0.008. These results, if real, may indicate temporary veils in the atmosphere. For instance, a dust cloud could weaken the local polarization.

4.3. Venus

4.31. *Integrated light.*—The observational data, given in Figure 27 (*full-drawn curve*), are due to Lyot (1929*a*). Within the accuracy of measurement he always found the plane of polarization to be either normal or parallel to the plane of vision.

The polarization-curve of Figure 27 differs completely from those for

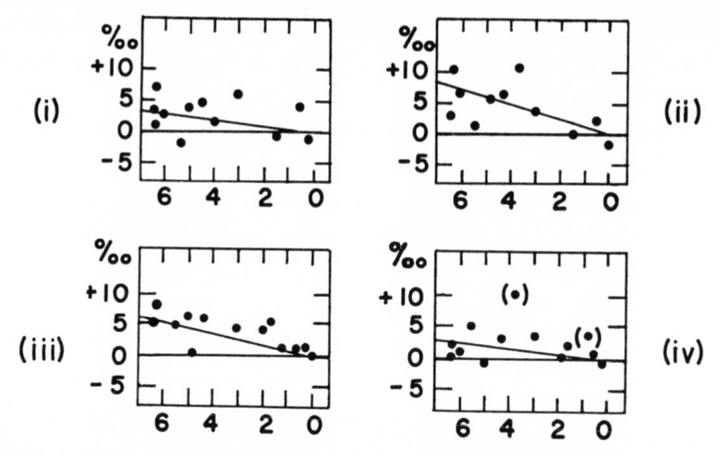

FIG. 26.—Differential polarization, ΔP, by the atmosphere of Mercury (unit 0.001), based on observations of center of disk and cusps, in both red and green light. *Abscissae:* $10 \sin^2 V/(1 + \cos^2 V)$. (*i*) $\Delta P = P_{\text{green}} - P_{\text{red}}$ (center of disk); (*ii*) $\Delta P = P_{\text{green}} - P_{\text{red}}$ (near cusps); (*iii*) $\Delta P = P_{\text{cusp}} - P_{\text{center}}$ (green); (*iv*) $\Delta P = P_{\text{cusp}} - P_{\text{center}}$ (red).

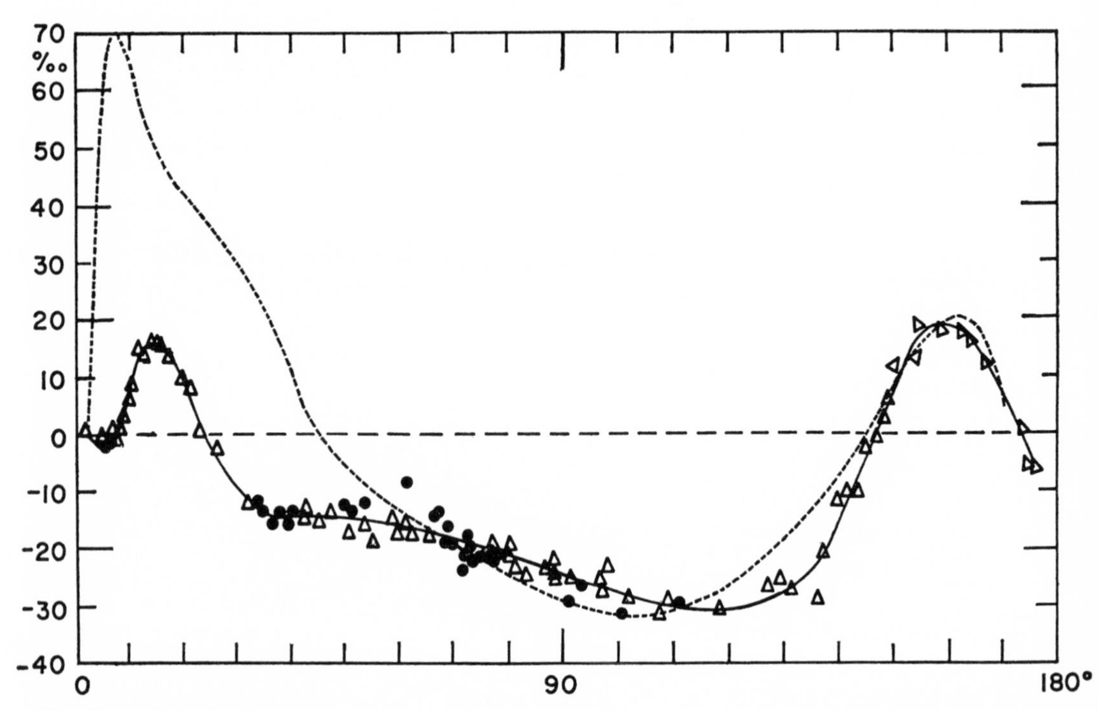

FIG. 27.—Polarization (unit 0.001) of Venus. The full curve represents observations during different cycles. The dotted curve is the experimental curve for water droplets with a diameter of 2.5 μ, the ordinates being divided by a factor of 3.5. (After Lyot.)

solid materials, while its several points of inversion make it analogous to curves found for clouds of droplets. This result, together with the very high albedo, led to the assumption that Venus is covered entirely by clouds. From his laboratory experiments Lyot concluded that it was not possible to account for the polarization-curve by assuming droplets as large as those of terrestrial clouds. The similarity increases when the droplet size decreases. Figure 17 shows the polarization-curve for the smallest droplets observed, 2.5 μ, to have two maxima and a minimum, located at the same abscissae as in the Venus-curve, and three points of inversion, two of which are also present in Figure 27. If we could increase the thickness or the density of the cloud, the polarization-curve would keep the same shape, but, according to the discussion accompanying Figure 20, the ordinates would be reduced by a factor of 3–4. In Figure 27 the curve of Figure 17, *D*, after reduction of the ordinates by a factor of 3.5, is shown as a dotted curve. The similarity between the two curves in Figure 27 is quite remarkable, considering the differences among the other curves studied.

The polarization of Venus therefore corresponds closely to that of opaque clouds whose upper parts consist of fine droplets a little over 2 μ in diameter, with the refractive index of water. However, very small transparent crystals could give similar polarization, and these could well exist and be stable.

4.32. *The atmosphere of Venus.*—The permanent clouds on Venus may obstruct observation of the lower layers of the atmosphere. According to Lyot, there is a small difference in polarization between the center and the limb which he interpreted as due to a medium having a scattering power equal to that of a layer of air 1500 meters thick. Dollfus has examined the polarization through red and green filters and found that when the phase angle is near 60°, polarization is -0.012 in the green and -0.015 in the red, remaining systematically the same from day to day. The polarization-curve of dense clouds of small particles should be nearly independent of the wave length at this phase angle. If the observed difference is due to molecular scattering in the clear atmosphere above the clouds, the corresponding height of this layer is 2 km. This value is found from the integrated light and corresponds to the apparent thickness of the atmosphere averaged over the disk. The vertical thickness is less by an amount depending on the limb darkening.

The observations just reviewed may be interpreted as resulting either from a clear atmosphere over a cloud layer or from a mixture, with the denser layers below the clouds strongly contributing also. The effective height of the atmosphere above the clouds may be about 800 meters, one-tenth the earth's atmosphere. The value of gravity is $0.88g$, which results in a barometric pressure of 90 millibars at the cloud level. In the earth's atmosphere this pressure is found at a height of 17 km.

4.33. *Regional studies.*—Although the Venus disk shows hardly any details, the polarimeter detects regions showing exceptional polarization. Unequally distributed, but sometimes well-defined, patches are indicated. Some of these patches occupy preferred positions, such as the poles. Others appear in any part, change in shape in a matter of hours, and usually last less than a day. These regions are observable in both red and green, but sometimes with different polarizations. The direction of polarization can differ by several degrees from that of surrounding regions. There is no indication that these deviations show a preferred direction. There are three series of data on these patches.

a) In 1924 Lyot observed the narrow Venus crescent as it passed phase angle 170° (where the polarization of the integrated light passes through zero) and noted that the polarization of the terminator was less algebraically than that of the limb. On May 26 the difference between the two borders of the image attained 0.014. The polarization became negative first on the limb, about June 15, and then at the terminator; for several days the two borders had opposite polarization, and the image in the Savart polarimeter displayed the curious appearance shown in Plate 3. It disappeared when the width of the crescent fell below the resolving power of the telescope.

b) In 1953 Dollfus observed the planet at phase angle about 60°, when the disk was gibbous (Pl. 4). The average polarization was -0.015; generally the regions near the cusps showed higher negative polarization, often reaching -0.020 and sometimes beyond -0.030. These regions were not always adjacent to the cusps, and their outlines changed capriciously. The directions of predominant vibration deviated sometimes by as much as 18° from the plane of vision, the divergence being in either direction.

The polarization often changed across the disk, as shown in Plate 4. This figure makes it possible to compare the above-mentioned polarized patches, measured in green and red, with the location of clouds deduced from visual observations. There does not seem to be a close relation. The changes in both sets of images are varied and rapid and seem independent of their shape.

c) In October, 1950, the phase angle was around 10° (where the polarization changes sign), and Dollfus (1957, p. 48) observed the distribution of polarization over the Venus disk with all precautions indicated by the proximity of the planet to the sun. The polarization observed in both red and green was found to be positive in some areas of the disk and negative elsewhere, but not necessarily of the same sign in the two colors (cf. Pl. 5). The pattern of polarization seemed to have a lifetime of less than a day and did not correspond to visible surface detail, simultaneously recorded and shown in the right column of Plate 5. On two of the dates shown in Plate 5 the patterns in the two colors are similar, but on the third there is an appreciable difference. The directions of polarization were found to deviate from the normal direction much more at this phase than at quadrature; the deviations were usually, but not always, in the same sense in the two colors. The polar regions are particularly irregular, in that the plane of polarization may deviate up to 45° from the plane of vision. The cause of this deviation can therefore not be due to multiple scattering in a molecular atmosphere, as is true for the Jupiter polar caps (cf. Sec. 4.64), but may be a cover of permanent clouds.

4.34. *Interpretation of the polarized patches.*—Lyot explained the difference in polarization between the limb and the terminator in this manner: near the terminator the sun illuminates only the outermost layers of the clouds; these may differ from the deeper layers.

The observed deviations in the orientation of polarization require the clouds to be non-isotropic. The lack of isotropy could be on the scale of the small particles forming the veils or, as is the case in cirrostratus or stratocumulus clouds, on the larger scale of their formation. The light is returned from the clouds by direct reflection from each particle and by multiple scattering between particles. The multiply reflected part can be strongly polarized when the clouds are striated, with the direction of polarization either parallel or perpendicular to the striae. Near the phase angle of 10°—corresponding to near-zero polarization—the multiply scattered light prevails, and its plane of polarization may be oriented in any direction.

The distribution of the polarized patches is not the same as that of the visible clouds. An algebraically larger polarization could exist for regions with less cloud coverage, since the contribution through molecular scattering would be larger. The observed result does not agree with this hypothesis and leads to the conclusion that the atmosphere cannot be completely clear between the veils.

The polarimeter measures further indicate that the polar regions are always covered with veils. Visual and photographic observations indicate the presence of clouds there. If, as seems possible (Dollfus, chap. 15), the period of rotation is equal to the period of revolution around the sun, all points on the terminator would have the same climatic conditions. One would then be tempted to attribute a polar cloud formation to the effect of relief.

4.4. Mars

The surface of Mars is known to show both bright and dark regions, treated separately in the following sections.

4.41. *Bright regions.*—The first investigations are those of Lyot in 1922. However, most of the following results were obtained with the 60-cm telescope of the Pic du Midi since 1948. Figure 28 shows the polarization measures made on bright spots near the center of the disk. When corrected for the Martian atmosphere (see below), the averages give the dashed polarization-curve in Figure 29. The angle of inversion is 29°.

When a bright region near the equator is followed crossing the disk, one finds a variation in the polarization (cf. Fig. 30). Apart from disturbances

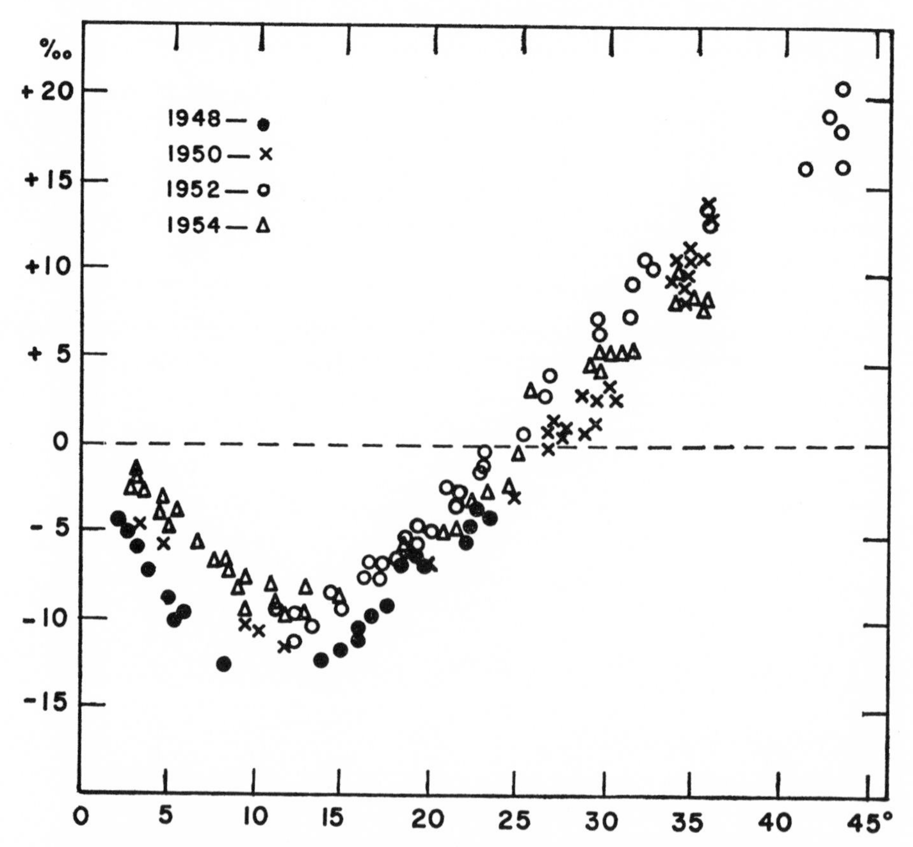

Fig. 28.—Mars, polarization (unit 0.001) for bright orange regions at center of disk during oppositions of 1948, 1950, 1952, and 1954.

due to clouds, this effect can be attributed to the presence of the Martian atmosphere combined with the change in the inclination of the Martian surface. For an inclination of 60° the effect due to the surface is found to be about −0.004.

Very few materials show the characteristics of Figure 29. This makes possible precise identification of the surface materials. From many comparisons it appears that the polarization of the desert regions on Mars is especially well reproduced by limonite, a hydrated iron oxide, $Fe_2O_3 \cdot 3H_2O$, in a finely pulverized condition. The curve for this yellow powder

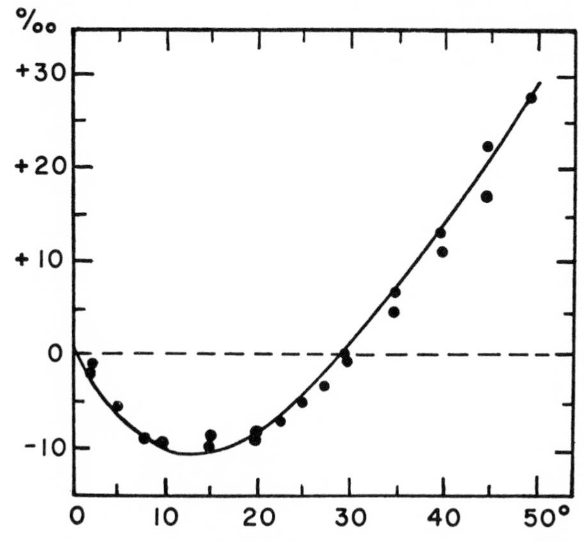

Fig. 29.—*Curve:* polarization (unit 0.001) for orange regions of Mars; *filled circles:* polarization measurements of pulverized limonite (ferric oxide).

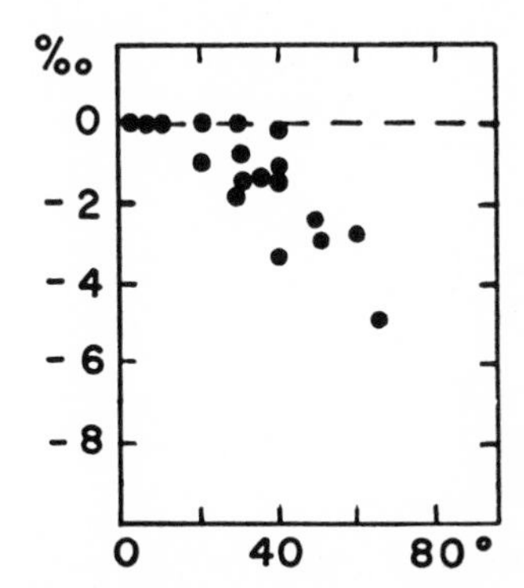

Fig. 30.—Mars, polarization of orange regions, depending on their position on the disk, during opposition, reduced to no atmosphere. *Abscissae:* longitudes measured with respect to center of disk; *ordinates:* differential polarization (unit 0.001) with respect to center of disk.

(Fig. 29), of albedo 0.20, is seen to be identical with that of the Martian surface. Plate 6 shows the appearance of this sample through the microscope. All other geologically possible materials lacked agreement with Figure 29.

4.42. *Dark markings.*—The polarization of the dark markings varies with the Martian seasons and the latitude. The seasonal variations appear related to the seasonal variations in contrast and color of these regions. In Figure 31 the open circles (A) refer to the regions north of latitude $+45°$ at the end of the Martian spring. The dots (B) correspond to the equatorial and southern regions during the northern spring and summer.

Figure 32 shows the differences between the dark and bright regions plotted against heliocentric longitudes (or seasons) for different zones of latitude. The curves indicate that seasonal changes occur in the surface texture. Comparative studies of this seasonal variation of the surface and of the planetary meteorology indicate that the changes in the surface texture are related to the changes in water-vapor content of the atmosphere

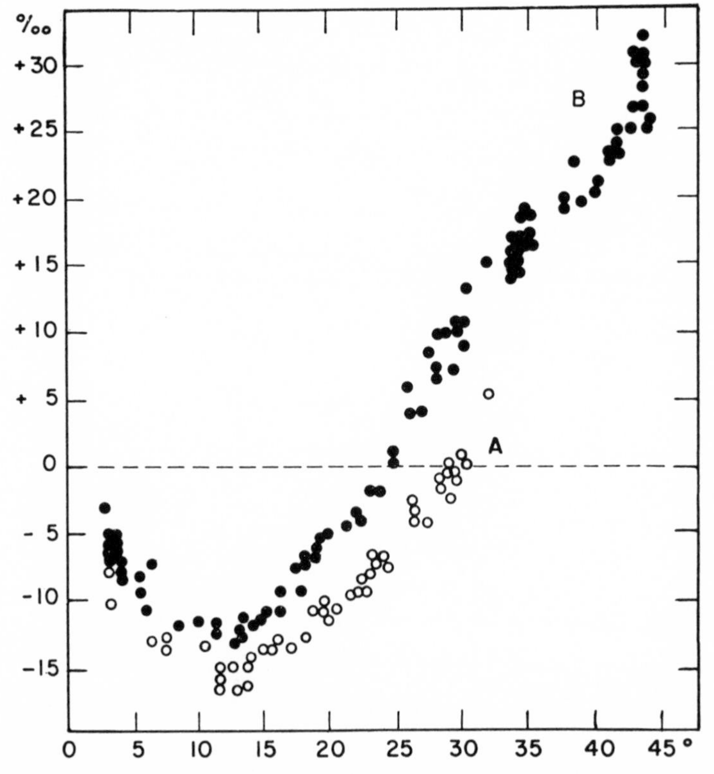

Fig. 31.—Mars, polarization (unit 0.001) for dark markings. *Filled circles:* equatorial markings at Martian spring; *open circles:* markings in Northern Hemisphere at Martian spring.

during the spring and summer, after evaporation of the polar caps (cf. Secs. 4.43 and 4.49, below).

It is tempting to assume that the changes in texture are due to vegetation. Many phanerogams (seed plants) and cryptogams with thalli have been measured; they show polarizations always quite different from the Martian surface. This is not true for microcryptogams. Scattered over the ground like a powder, these plants modify the polarization of the surface only slightly but in a varying manner. On the other hand, these micro-organisms have great powers of adaptation. Some protect themselves against extremes of temperature by means of a brightly colored superficial pigment. Monocellular algae, such as diatomic *Pheophyces* or

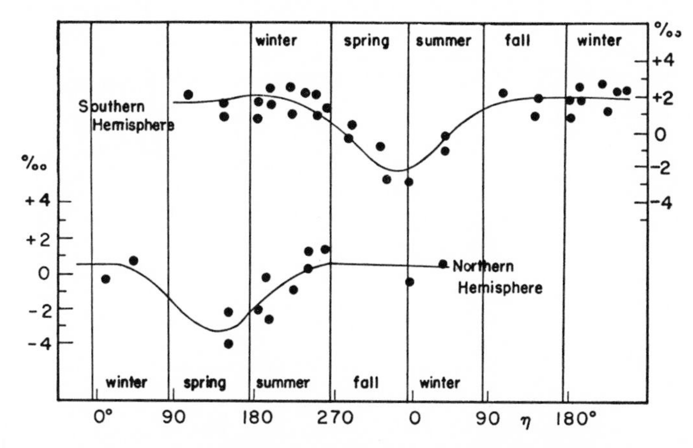

FIG. 32.—Mars, seasonal variation of polarization. Polarization differences between dark markings and orange deserts, for phase angle $V = 25°$, plotted against heliocentric longitude.

Clamidomonas chlorophyce, behave that way. The variety *Nivallis* flourishes in the severe climate of the high mountain snows and colors the mountains seasonally over large areas.

The chromogenic bacteria produce deep colors such as the red *Micrococcus prodigiosus* or the blue-violet *Bacillus pyocyaneus*, whose pigment scatters outside the organism and vividly colors the surroundings. These comparisons suggest a possible interpretation of the changing tints on Mars.

4.43. *Polar caps.*—The polar caps are quite small and difficult to measure. They are often covered by clouds or light veils and should be observed only with the Martian atmosphere clear. Nevertheless, it has been possible to make a number of measures during the spring regression, shown by circles in Figure 33. The polarization of the polar material remains very small and variable during that season for all phase angles.

Ice, hoarfrost, and snow deposited on mountains show, when melting and refreezing, strong positive polarization, if seen at the phase angles used to observe the Martian poles. This polarization comes from the light reflected in the interior of the ice and refracted with a plane of vibration normal to the plane of vision.

Fresh snow also shows positive polarization, though less than melting snow. Polarization in natural or artificial hoarfrost, deposited in small crystals in the laboratory at normal atmospheric pressure, is also higher than in the Martian deposits. However, the atmospheric pressure on Mars is only one-tenth of that on the earth. Hoarfrost formed under these conditions is found to consist of smaller grains. When heated by an electric

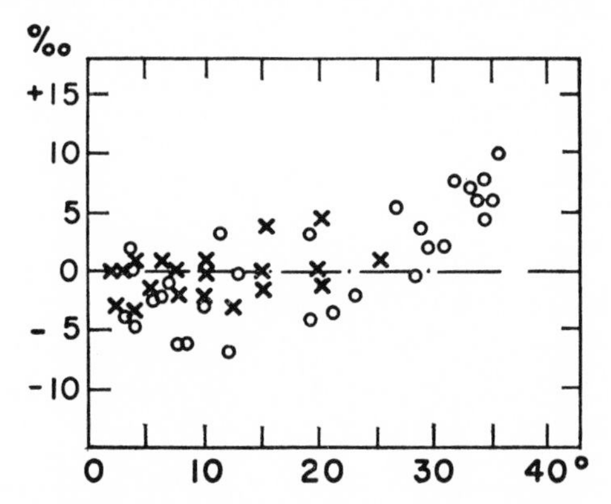

FIG. 33.—Mars, polarization (unit 0.001) of polar cap in spring. *Open circles:* polar cap; *crosses:* laboratory deposit of hoarfrost, sublimating under 80-millibar pressure.

arc, the hoarfrost sublimates without melting, and the remainder takes on the appearance of opal glass, full of small pores and cavities. In the process the albedo is reduced, and the polarization becomes very similar to that of the Martian polar caps. In Figure 33 the crosses refer to sublimating hoarfrost and are seen to agree remarkably well with the circles representing the Martian polar caps. It thus seems probable that the white spots at the poles of Mars are a deposit of hoarfrost which sublimates under the sun's radiation.

The edges of the polar caps show bright local spots. The light from these is polarized differently. The minimum at $V = 14°$ has the exceptional negative value of -0.022. No laboratory source has shown such values, but clouds of ice crystals show exactly this phenomenon. Visual observations show that these spots are located where the polar deposits last longest, presumably elevated areas. These spots would then be due to material evaporated from the polar cap, recondensed locally because of higher elevation of the ground.

4.44. *Different types of clouds.*—Visual, photographic, and polarimetric observations indicate several distinct types of atmospheric formation. We designate them by the color of the filter best adapted to bring them out: white clouds, blue clouds, yellow veils, and violet hazes.

The latter obliterate the surface details on photographs in the blue and violet; they could be tenuous because the natural contrast of the spots is already weak in blue light. To be transparent in yellow light, the "violet haze" must consist of very small particles. These are likely to polarize the visual light according to a law differing little from that for molecular scattering. The polarization of the haze is then similar to that of the Martian atmosphere, although much smaller.

The polarizing properties observed on the other types of atmospheric formation are as follows:

4.45. *White clouds.*—These extend all the way from bright formations, developing vertically and projecting from the terminator like prominences, to diffuse fogs and hardly perceptible veils. The polarimeter can detect even more tenuous haze, not apparent visually. In spite of this great diversity in appearance, all these formations have common polarizing properties. They are therefore of the same nature and differ only in the degree of opacity.

Figure 34 shows the measures made on dense clouds in the equatorial and temperate zones (Dollfus, 1957, p. 81). The shape of the curve is radically different from that of a cloud of droplets, but similar to that for a fog of ice particles. The white clouds on Mars must therefore consist of veils of crystals like our cirrus clouds.

The polarization of tenuous veils increases at first with the particle concentration but then remains nearly stationary for denser veils, as shown in Figure 34: for the brightest formations some decrease may even occur. This behavior can be accounted for by the onset of multiple scattering between crystals and the absorption of the light from the ground.

During the Martian winter the polar cap is covered by an opaque, cloudy veil which is permanent and uniform. Mr. Focas obtained the polarization-curve during the 1954 opposition (Fig. 35). The minimum occurs at $V = 16°$, with a value of -0.016 and the inversion at 25°, which is somewhat different from Figure 34, describing the veils in the temperate zone.

Slight variations in the plane of polarization of the clouds are probably due to striations or rolls, as found in terrestrial cirrus clouds.

The veils that are too faint to appear visually occur mostly in the Martian winter or spring. They are often connected with well-defined clouds which at times cover large areas. The regions of Syrtis Major, Mare Acidalium, and Mare Erythraeum seem to favor their formation.

The regions of Nix Olympia and Candor often show the above-mentioned abnormal polarization due to high clouds. Topographic relief may influence the haze structure.

The polarimeter detects the presence of faint morning haze at the sunrise limb; it usually disappears after a few hours of sunshine. Other properties of the white clouds are described in chapter 15, Sec. 3.310.

4.46. *Blue clouds.*—These clouds, brought out in dense blue filters, are almost always present at the equator at the rising and setting limbs, but they disappear in the middle of the day. These clouds may be regarded as morning and evening tropical fogs. They are sometimes connected with formations of white clouds. In such cases the polarization measures show that blue clouds are situated higher than the white clouds (see also chap. 15, Sec. 3.12).

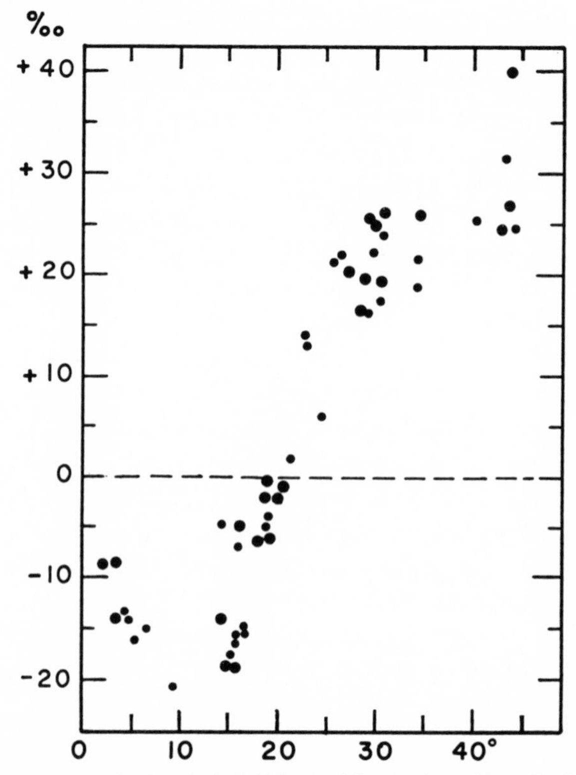

FIG. 34.—Mars, polarization (unit 0.001) of white clouds and haze in the equatorial and temperate zones. Size of dots is proportional to cloud opacity.

Although invisible in orange light, these veils show an appreciable differential polarization at this wave length (Fig. 36). The polarization of the cloud is nearly proportional to its brightness. The polarization is zero for $V = 0$, becomes positive with a maximum around $V = 14°$, drops to zero at 25°, and remains negative thereafter.

The fogs that produce such polarization consist of droplets about 3 μ in diameter, intermediate between curves *C* and *D* of Figure 17, above. A very thin veil suffices to produce such polarization effects. The atmosphere of Venus was found to contain similar veils.

The structure of these veils may also be compared with that of the mother-of-pearl clouds that can be seen or photographed in blue light in the terrestrial polar atmosphere at altitudes of 20–30 km. On the earth they are formed under conditions of temperature and pressure similar to those of the upper Martian atmosphere. According to Strömer, they consist of particles with diameters between 1.5 and 3 μ.

The large Martian desert region of Amazonis is often free of such blue veils and seems unfavorable for their formation.

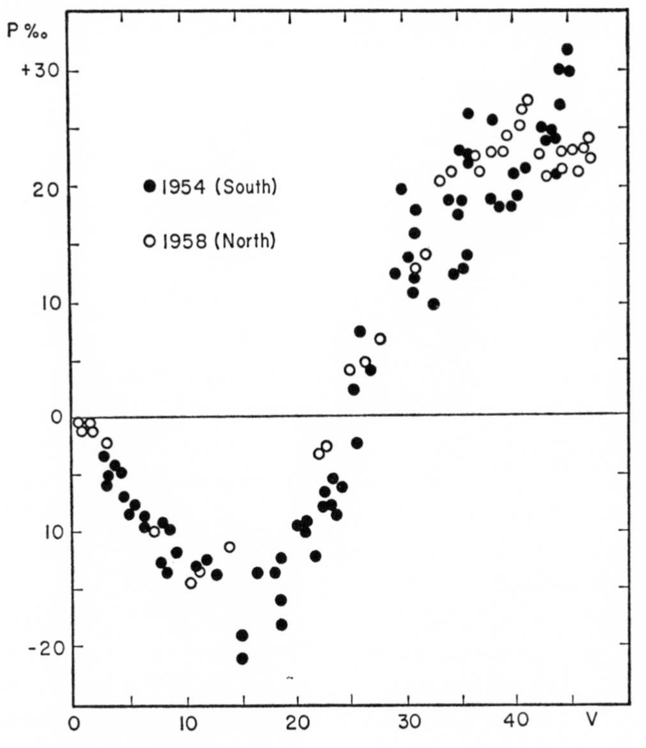

FIG. 35.—Mars, polarization (unit 0.001) of white clouds covering the polar caps in winter, plotted separately for two hemispheres.

4.47. *Yellow clouds.*—In December, 1924, a large part of the Martian surface remained covered for weeks by a yellow veil of exceptional density. The phase angle was 42°, and Lyot discovered a considerable reduction in the polarization of the total light of the planet (Fig. 37). In May, 1952, Dollfus found unusual polarization over Margaritifer Sinus and Mare Erythraeum which spread for several days as far as Sabaeus Sinus, Hellespontus, and Hellas. During that time the phase angle was around 17°. The white veils showed hardly any polarization, and no nebulosity could be detected through a blue filter. The variability of the phenomenon, its rapid increase toward the limbs, and its motion showed that it was atmospheric. It was probably a yellow veil, very tenuous and completely undetectable by any means other than polarization.

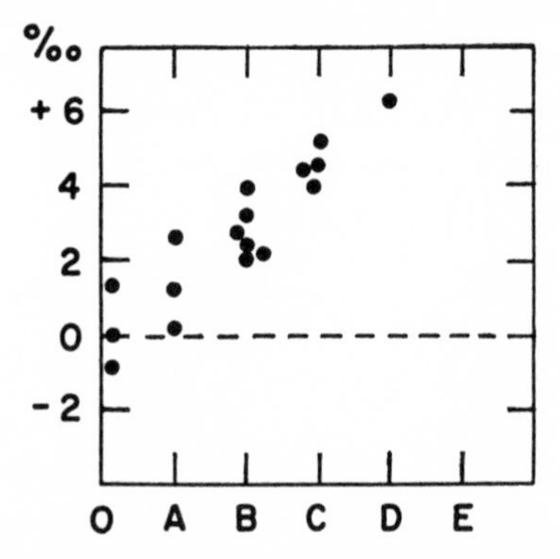

FIG. 36.—Mars, excess polarization (unit 0.001) caused by blue clouds as observed in yellow light. *Abscissae:* estimated brightness of cloud in blue; *ordinates:* differential polarization attributed to cloud. Phase angle 14°.

Yellow clouds have been attributed to dust storms. Polarimetric observations do not contradict this interpretation.

At the 1956 opposition of Mars, several very opaque yellow clouds blotted out the Martian features and the polar cap, and the atmosphere remained dusty for some time. Polarimetric observations at Pic du Midi showed not only a general decrease in polarization at all phase angles but also a variable polarization, correlated more or less with the opacity of the cloud. These are the combined effects of multiple scattering reducing the polarization and the average size of the particles decreasing with time because of settling. The polarization-curve is very sensitive to the particle size, and laboratory research has not yet succeeded in giving a critical identification. More detailed descriptions of the morphology and development of the dust storms are given in chapter 15, Section 3.311.

4.48. *Atmospheric pressure.*—The polarization of the Martian atmosphere varies as $\lambda^{-4} \sin^2 V \sec \theta$ (cf. Sec. 4.23). A computation similar to that made for Mercury allows us to combine this polarization with that of the light reflected by the surface. One may then compare the polarization at the center of the disk with that on the limb for various phase angles and colors. It was, however, necessary to limit the discussion to times when the Martian atmosphere was transparent, which is none too frequent. The measures as a whole indicate that at opposition and in orange light the brightness of the atmosphere at the center of the disk is nearly 0.028 times that of the cloudless planet.

Another determination is based on the spectral variation in the polarization of the integrated light of the planet. The visual and photo-

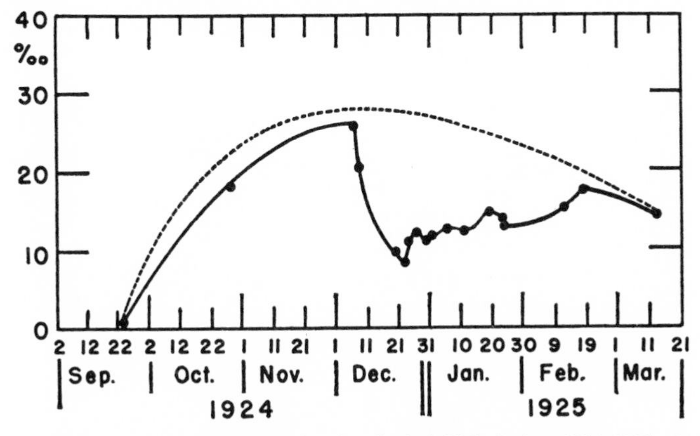

Fig. 37.—Mars, disturbance in polarization (unit 0.001) during 1924–1925, due to yellow cloud. *Dashed curve:* normal polarization; *dots* and *full line:* observed polarization. (After Lyot.)

graphic measures are plotted in Figure 38 for the phase angles near 12°, 25°, and 35°, which correspond to the minimum, the inversion, and the maximum of the polarization in orange light. The theoretical values are shown as curves, computed from the brightness determined above. The agreement of the two determinations is fair.

Accordingly, the thickness of the Martian atmosphere under normal conditions is 1.9 km or one-quarter of the terrestrial atmosphere. An aneroid barometer on the surface would show a pressure of 90 millibars. However, the presence of particles in the Martian atmosphere smaller than a wave length could alter this result somewhat.

4.49. *Water vapor in the Martian atmosphere.*—The polarimetric observations indicate the presence of water vapor in the Martian atmosphere, in agreement with Kuiper's (1952) infrared studies of the polar cap.

The vertical thermal gradient of the Martian atmosphere is estimated to nearly 3°.5 C/km. If the air is completely saturated at all altitudes, the amount of precipitable water would be, as a function of the poorly known temperature at the base of the atmosphere, 3 mm for −20° C, 1.0 mm for −30° C, and 0.3 mm for −40° C. Since the atmosphere must be very dry, the amount of precipitable water must be smaller yet. It is probably only a few tenths of a millimeter and perhaps much less. Such an amount of water exists in the earth's atmosphere at an altitude of 6–8 km and suffices to explain the clouds and the hoarfrost identified by the polarimeter. This amount is more than a hundred times smaller than that

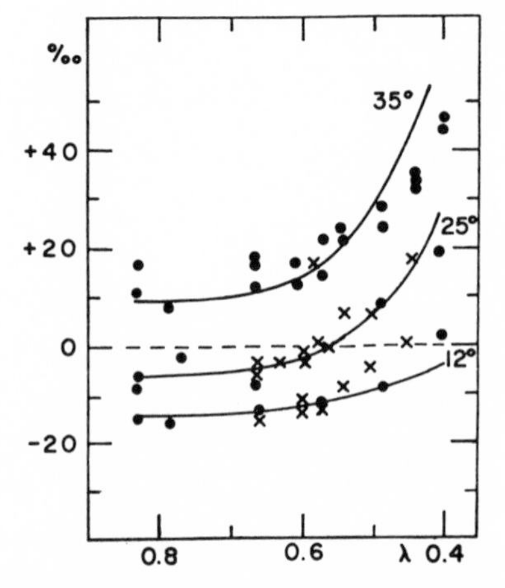

Fig. 38.—Mars, wave-length dependence of polarization (unit 0.001) for light of entire disk for phase angles 12°, 25°, and 35°. *Filled circles:* photographic measurements; *crosses:* visual measurements; *curves:* theoretical values for atmospheric pressure of 80 millibars. (Observations at 12° consist of 6 filled circles and 7 crosses.)

present in the earth's atmosphere and is therefore undetectable by ordinary spectroscopic investigations.

Under the small atmospheric pressure on Mars the rate of evaporation of the frost on the polar caps is at least ten times greater than on the earth. The molecular diffusion of water vapor in the atmosphere is eight times faster. The vapor transfer by turbulent transport, which is inversely proportional to the difference between the adiabatic vertical thermal gradient and the actual gradient, is also extremely fast. Water is never present in the liquid state. The total amount of water is smaller on Mars than on the earth; the diameter is only 0.54 of the earth's, and the year is almost twice as long. For all these reasons, the annual circulation of water must be much simpler on Mars than on the earth. The water which evaporates from one polar cap diffuses toward the equator and is next swept toward the opposite pole, where it freezes on the ground. Then the process reverses for the next half-year. This mechanism accounts for the behavior of the polar caps, of the ice clouds in winter or over high regions, and also for the color and polarization changes in the dark regions, as the seasonal wave of vapor arrives. Hess and de Vaucouleurs arrived independently at similar conclusions from their meteorologic and photometric investigations, and the result agrees with the information obtained by the polarimeter.

4.5. The Asteroids

With a specially built photographic polarimeter, Lyot (1934) determined the polarization of some asteroids in 1934 and 1936. Figure 39 shows the curve obtained for Vesta. The polarization is smaller than on the moon, but the angle of inversion is larger. It corresponds to a curve intermediate between the moon and chalk.

There is, further, an unpublished curve by Lyot for Ceres (Fig. 40).

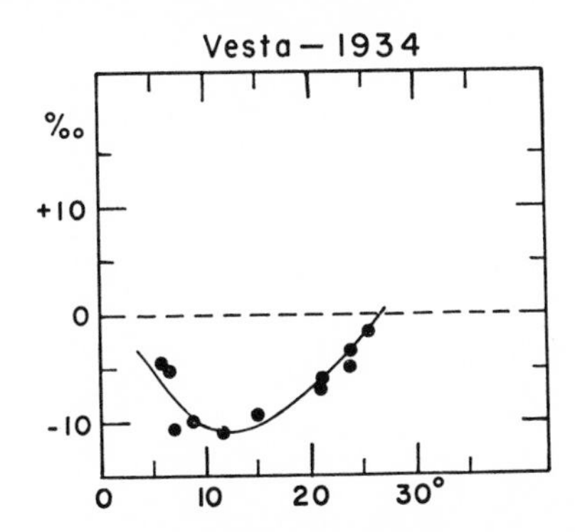

Fig. 39.—Vesta, polarization (unit 0.001) versus phase angle, 1934 opposition. (After Lyot.)

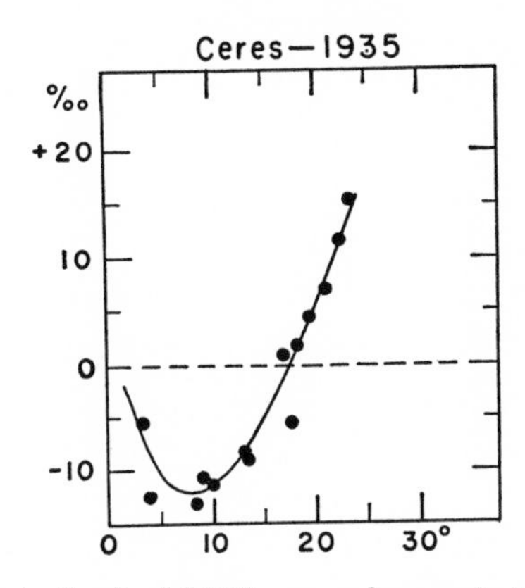

Fig. 40.—Ceres, polarization (unit 0.001) versus phase angle, 1935 opposition. (After Lyot.)

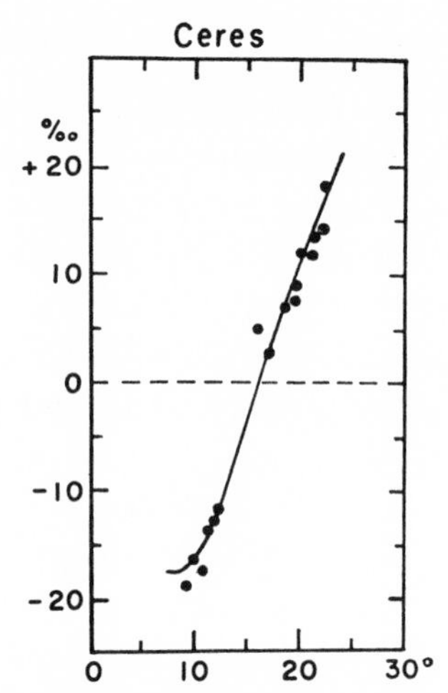

FIG. 41.—Ceres, polarization (unit 0.001) versus phase angle, 1954 opposition. (After Provin.)

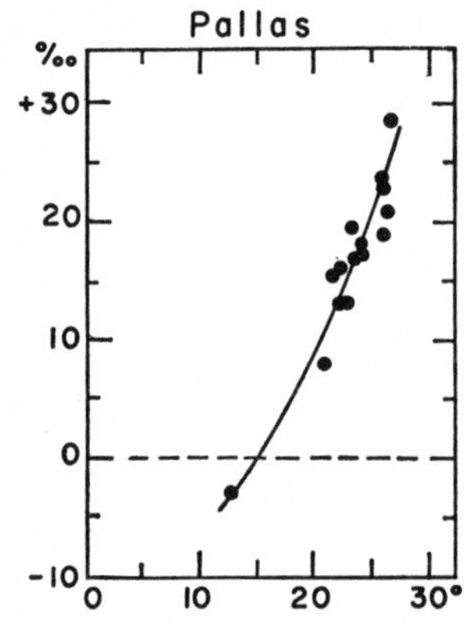

FIG. 42.—Pallas, polarization (unit 0.001) versus phase angle. (After Provin.)

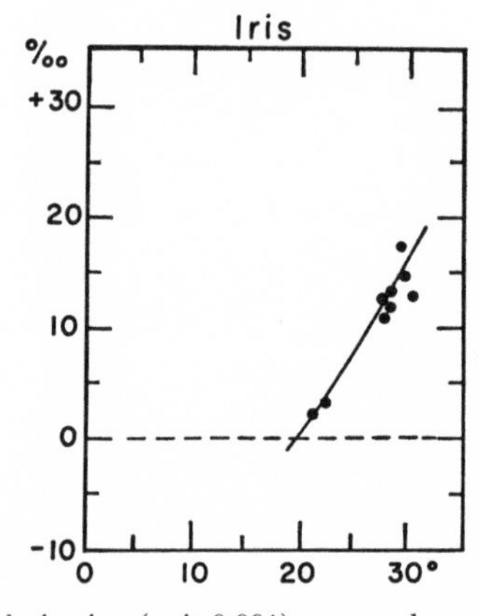

FIG. 43.—Iris, polarization (unit 0.001) versus phase angle. (After Provin.)

At the 1935 opposition the curve came out quite regular, with an angle of inversion of 17°.5. The next year the polarization was stronger and variable. This difference might indicate a change in the aspect presented by the asteroid to the earth. Fissured lavas and sandstone polarize light in the same manner.

New photoelectric measures have been made recently at the Naval Observatory, Washington, D.C. Provin (1955, and personal communication) obtained a new curve for Ceres and measures on Pallas and Iris (Figs. 41, 42, 43; shown here for the first time, as is Fig. 44). The curve for Iris is similar to that of the moon. This asteroid shows a marked periodic variation in light of 0.2 mag. Figure 44 shows the polarization

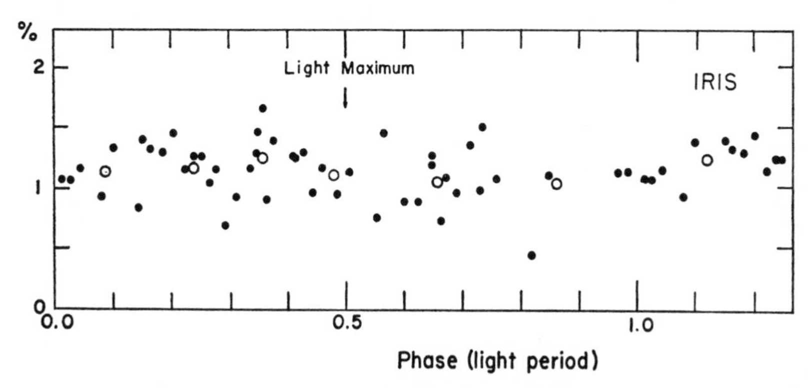

FIG. 44.—Iris, variation of polarization (unit 0.01) with phase of light-curve. Open circles are normals of 8 points each.

measures of Iris plotted against the phase of the light-curve. The change in polarization with rotational phase is small, about 0.002.

4.6. JUPITER

4.61. *Center of disk.*—Over a large central area the polarization on the Jupiter disk remains uniform. Lyot (1929*a*) studied the variation of this polarization with phase during the oppositions of 1923, 1924, and 1926, for which he derived the curves in Figure 45. The run in 1925 appeared intermediate between those of 1923 and those of 1924. These observations suggest rather uniform conditions along the equatorial zone of the planet, with variations occurring only from year to year, roughly simultaneously at all longitudes.

Figure 46, based on Dollfus' observations in 1952, shows that when the phase angle is 11°.7, the largest accessible, the polarization at the center is not strongly wave-length-dependent.

4.62. *Polar regions.*—Lyot (1929*a*) found an unexpected phenomenon in the gray-colored polar caps. The polarization was found to be independent of the phase angle. The predominant vibration is always normal to the limb. From a maximum at the pole the polarization decreases steadily toward the equator and is equal to that of the center of the disk at a latitude which may be as low as 35°. Figure 47 shows some of the measures (Lyot, 1929*a*, p. 75).

The regions of abnormal polarization cover about the same areas as do

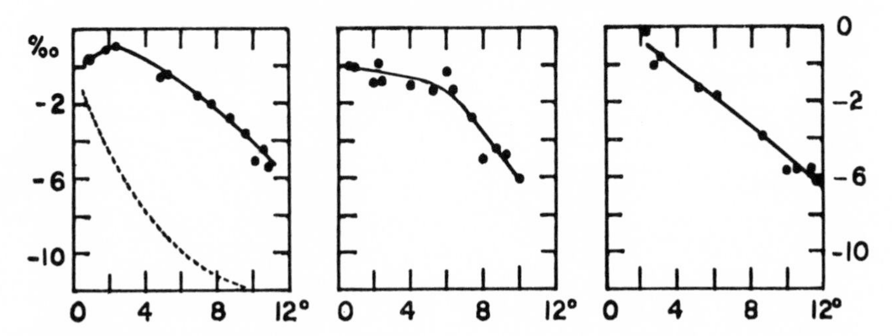

FIG. 45.—Jupiter, polarization (unit 0.001) for central part of disk during 1923, 1924, and 1926 oppositions. (After Lyot.)

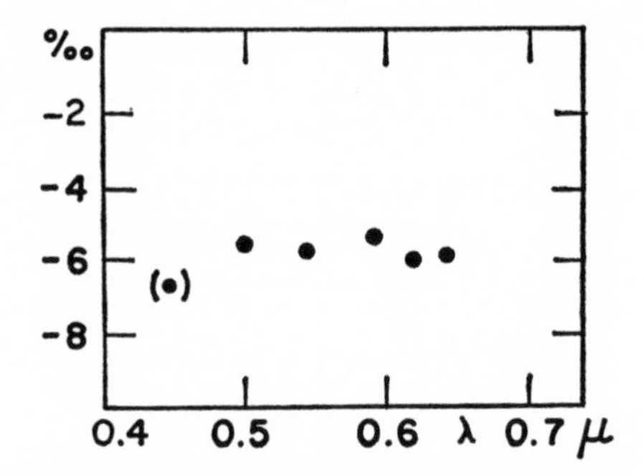

FIG. 46.—Jupiter, wave-length dependence of polarization for $V = 11\overset{\circ}{.}7$

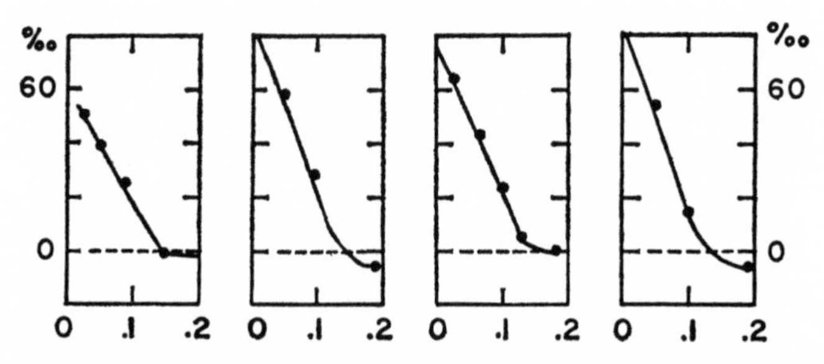

FIG. 47.—Jupiter, polarization near poles. *Abscissae:* distance from limb in terms of polar diameter. (After Lyot.)

the gray polar caps. On July 30, 1924, a gray cloud showing negative polarization, similar to the center of the disk, covered part of the north polar cap. The extent of the typical cap polarization was then less than usual and was limited by the edge of the cloud.

Dollfus found the two polar caps to be of equal size in 1952. In 1955 the area of abnormal polarization was unusually large at the north pole and unusually small at the south pole. No appreciable wave-length dependence was observed with blue, green, and red filters on several nights.

Öhman (1944), using a spectrograph whose slit was preceded by a quartz plate, cut parallel to its axis, and an analyzer, also found the polarization at Jupiter's poles to be independent of wave length (Pl. 7).

Lyot has interpreted this abnormal polarization at the poles as due to multiple scattering in a clear, dense atmosphere. The mechanism would then be analogous to that which produces the neutral points and the predominant vertical vibration found in the terrestrial atmosphere in a

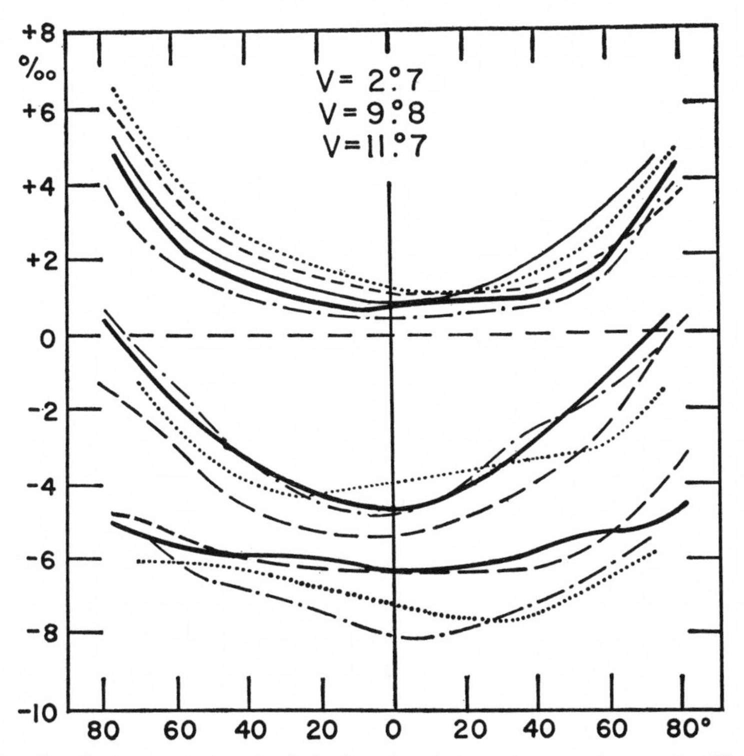

FIG. 48.—Jupiter, variation of polarization along belts and zones, phase angles 2°.7 (*upper set*), 9°.8 (*middle set*), and 11°.7 (*lower set*) in green and red light. *Full-drawn lines:* zone observed in red light; *dashes:* zone observed in green light; *dot-dash:* belt observed in red light; and *dots:* belt observed in green light. *Abscissae:* longitudes; *ordinates:* polarization (unit 0.001), with a common zero point.

direction opposite to the sun. The atmosphere, clear at the poles, is progressively veiled toward the equator, causing the polarization to decrease.

4.63. *Belts, zones, and limb.*—When one compares the polarization of the bright equatorial zone, the two dark tropical belts, and the two adjacent bright zones in blue, green, and red light, no permanent systematic differences are found, regardless of the phase angle (Fig. 48). The Red Spot measured on August 7, 1952, at different distances from the central meridian, also fails to show any marked polarization difference compared with its surroundings.

Except for the polar regions, polarization on Jupiter's disk is dependent only on the phase angle and the distance to the center. The curves in Figure 48 show the variation between center and limb at phases 2°.7, 9°.8, and 11°.7 for both belts and zones in red and green light.

4.64. *Interpretation of the observations.*—The atmosphere of Jupiter is often assumed to contain an optically opaque cloud layer, overlain by transparent gas. On this model, the polarization would increase rapidly with the phase angle and would vary as λ^{-4}. The observations contradict this. The atmosphere is therefore assumed to be contaminated by a thin fog, whose polarization is of a sign opposite to that of a pure gas (Dollfus, 1957). According to tables of scattering functions, particles of $d \simeq 1\ \mu$ would behave this way.

This conclusion is consistent with photographic records obtained in different spectral regions. The fog must have a smaller concentration over the dark belts than over the zones, and must disappear above the polar caps.

4.7. Saturn

Since Saturn is placed at nearly double the distance of Jupiter, its maximum phase angle attains only about 5°50′, half that of Jupiter, which renders only one-thirtieth of the polarization-curve accessible to observation. Nevertheless, some results of interest may be obtained. The presence of the Rings makes the Saturn system distinctive and of special importance. While the small range in phase angle hampers the study of the Rings also, there is a partial compensation due to the variable opening of the Rings, from 0° to about 28°. The variable tilt also assists in the study of the polar caps of the planet.

4.71. *The planet.*—The measures and discussion are due to Lyot (1929). He found the central part of the disk (about one-fourth of the area) to have very uniform polarization and obtained its variation with phase angle during the years 1922–1927. Three independent phase-curves were derived for 1923, 1924, and 1926 and are reproduced in Figure 49. The small portions covered are compatible with the assumption of a transparent gas seen over a dark background. On this model, however, multiple scattering would always cause a strong polarization along the limb, with predominant vibration perpendicular to the limb. Such an effect appears only near the poles. The atmosphere of Saturn must therefore be laden with particles of a somewhat different nature than Jupiter's.

The polar regions of Saturn behave like those of Jupiter except that they are much less constant. Their extent and intensities undergo large fluctuations. The atmosphere must therefore be more transparent than in the tropical regions; yet veils that vary in opacity must at times be present. Only the polarization shows their presence.

Lyot repeatedly compared the polarization of the bright equatorial zone with the darker north-tropical belt and found them nearly always identical. However, on June 9, 1924, he discovered that the bright zone

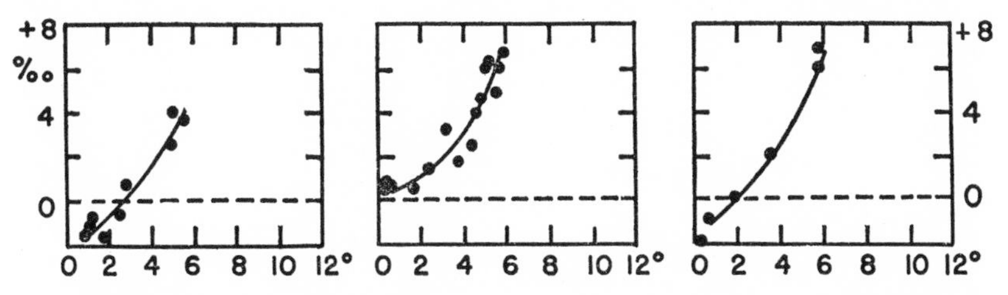

FIG. 49.—Saturn, polarization (unit 0.001) of center of disk versus phase angle and oppositions of 1923, 1924, and 1926. (After Lyot.)

between latitudes 45° and 65° N. had a negative polarization, opposite to that of the other parts of the disk. The polarization went through a minimum of −0.003 on June 15, became positive between June 16 and 25, but remained distinctly less than the surroundings until the end of the observations on July 7. This zone was brighter and less yellow than the equatorial zone. No other exceptional region was noted in the course of the observations.

Curiously, the two principal Rings have different polarization properties, as discovered by Lyot (1929). As long as they were measured together (Fig. 50, *upper left diagram*), only discordant results were obtained. Measures of the rings separately led to better correlations (Fig. 50).

4.72. *The Inner Ring (B).*—The direction of preferred vibration was found to be at right angles to the plane of vision (Fig. 50, *lower left*), and the polarization is thus negative (Fig. 50, *upper right*). The measures for Ring B are very consistent, although they refer to both ansae and to three different years. The polarization-curve near the tips of the ansae of Ring B was interpreted to be similar to that of non-pulverized minerals.

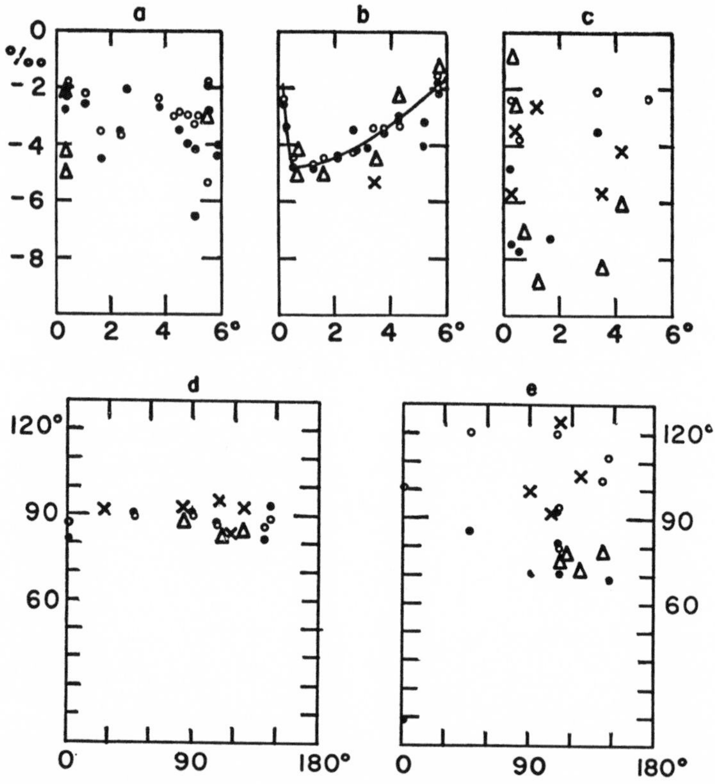

FIG. 50.—Rings of Saturn, polarization (unit 0.001) versus phase angle for (*a*) Rings A and B together; (*b*) Ring B; and (*c*) Ring A. Orientation of plane of polarization for Ring B (*d*) and Ring A (*e*). (After Lyot.)

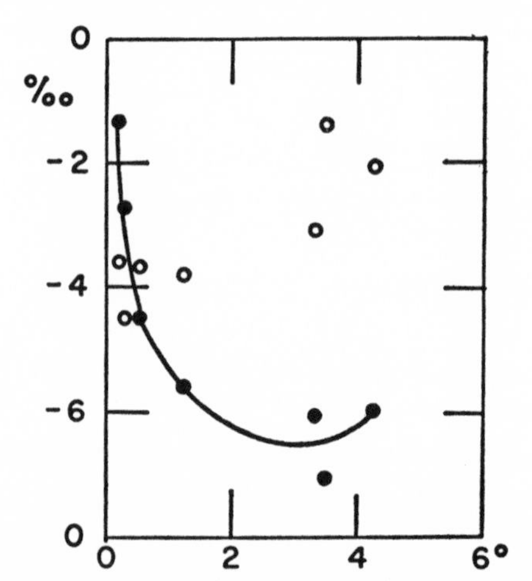

FIG. 51.—Rings of Saturn, separation of polarization of Ring A into two components. *Filled circles:* component due to direct sunlight; *open circles:* second component. (After Lyot.)

New measurements secured by Dollfus since 1958 all around the Ring revealed more details. The direction of polarization changes along the Ring. For each value of the phase angle the polarization may be assumed to be comprised of two contributions, one constant along the Ring in amount and direction and the other rotating along the Ring. The direction of this component is found to be either parallel or perpendicular to the tangent of the Ring, depending on the phase angle; the results of three different oppositions appear internally consistent. The second component appears due to the orientation of the Ring particles or scratches on the particles, parallel to their trajectories.

4.73. *The Outer Ring* (*A*).—The plane of polarization shows more scatter, and the degree of polarization shows no close relation with the phase angle (Fig. 50). The angle between the predominant vibration and the plane of vision is around 170° at the east tip and between 5° and 30° at the west tip. The amount varies irregularly from 0.001 to 0.005 for the west side and from 0.004 to 0.009 for the east.

Lyot has tried to explain this curious behavior by assuming that there are two causes acting simultaneously: (1) direct scattering of sunlight, (2*a*) scattering of planetary light, received from the illuminated hemisphere of the planet or (2*b*) multiple scattering of sunlight between aligned particles of the Ring. Cause 1 will give identical polarization on the two ansae, in the plane passing through the sun or normal to it. Cause 2 could produce different polarizations on the two ansae, but the plane of polarization would be symmetrical to the axis of Saturn. Lyot made the graph in Figure 51, where the dots represent the polarization caused by direct sunlight and the circles light of unknown cause. The curve given by scattered sunlight is similar to the curve for the Inner Ring and for many solid materials. The other component of unknown origin should have remained constant, but it seems to have changed from −0.004 to −0.002 for increasing phase angle. Lyot's hypothesis therefore does not explain all the facts. New observations are needed.

The author is greatly indebted to the editors for the translation, revision, and final drafting of the manuscript of this chapter. In addition, nearly all the figures in this chapter were revised and redrawn in the editorial office. The author would like to express his thanks also to Dr. G. P. Kuiper personally for making available the facilities of the Yerkes and McDonald Observatories, where some of the observations reported here were made.

REFERENCES

General

Arago, F. 1854 *Œuvres complètes* (Paris: Gide & Baudry).
Dollfus, A. 1957 *Ann. d'ap.*, Suppl. 4; also thesis, Paris, 1955.
Lyot, B. 1929*a* *Ann. Obs. Meudon*, Vol. **8**, Part 1.
Öhman, Y. 1949 *Polarization Measurements in Astronomy* (Boulder, Colo.: High Altitude Observatory).

Instruments

visual polarimeters

Dollfus, A. 1957 *Ann. d'ap.*, Suppl. 4; also thesis, Paris, 1955.
Lyot, B. 1929*b* *Rev. Opt.*, **5**, 108.
Wright, F. E. 1934 *J. Opt. Soc. America*, **24**, 206.

photographic polarimeters

Dollfus, A. 1957 *Ann. d'ap.*, Suppl. 4; also thesis, Paris, 1955.
Lyot, B. 1934 *C.R.*, **199**, 774.
Öhman, Y. 1939 *M.N.*, **99**, 624.
1942*a* *Stockholm Obs. Ann.*, Vol. **13**, No. 11.
1942*b* *Ibid.*, Vol. **14**, No. 4.
1942*c* *Nordisk astr. tidsskr.*, **23**, No. 77, 113.
1947 *Stockholm Obs. Ann.*, Vol. **15**, No. 2.

photoelectric polarimeters

Gehrels, T., and Teska, T. M. 1960 *Pub. A.S.P.*, **72**, 115.
Lyot, B. 1948*a* *C.R.*, **226**, 25.
1948*b* *Ibid.*, **226**, 137.
Öhman, Y. 1944 *Stockholm Obs. Medd.*, Vol. **55** (*Ark. f. mat., astr., fys.*, Vol. **328**, No. 1).
1949 *Stockholm Obs. Ann.*, Vol. **15**, No. 8.
Wille, H. 1952 *Optik*, **9**, No. 2, 84.

Polarization by Terrestrial Substances

Born, M. 1933 *Optik* (Berlin: J. Springer).
Bouwkamp, C. J. 1954 *Report on Progress in Physics*, *XVII*, p. 35: "Diffraction Theory."
Cailleux, A., and Dollfus, A. 1950 *C.R.*, **230**, 1411.
Dollfus, A. 1956 *Ann. d'ap.*, **19**, 83.
1957 *Ibid.*, Suppl. 4; also thesis, Paris, 1955.
Gouy, M. 1884 *C.R.*, **98**, 978.
Hulst, H. van de 1946 *Rech. astr. Obs. Utrecht*, Vol. **11**, Part 1.
1949 *Ibid.*, Vol. **11**, Part 2.
Lafay, A. 1894 *C.R.*, **119**, 154.
Lyot, B. 1929*a* *Ann. Obs. Meudon*, Vol. **8**.
Mie, G. 1908 *Ann. d. Phys.* (4), **25**, 377.

Öhman, Y. 1955 *Stockholm Obs. Ann.*, Vol. **18**, No. 8.

Sinclair, D. 1947 *J. Opt. Soc. America*, **37**, 475.

Schirman, M. A. 1919 *Ann. a. Phys.* (4), **59**, 493.

Umov, N. 1912 *Phys. Zs.*, **13**, 962.

Woronkoff, G., and Pokrowki, G. 1924 *Zs. f. Phys.*, **30**, 139.

Wright, F. E. 1927 *Proc. Nat. Acad. Sci.*, **13**, 535.

Sinclair, D. 1949 *Table of Scattering for Spherical Particles* (Nat. Bureau of Standards Appl. Math. Ser. No. 4).

Penndorf, R. B. 1956 *Geophys. Res. Papers*, No. 45, Parts 1, 4, and 5.

1957 *J. Opt. Soc. America*, **47**, No. 11, 1010.

Planetary Observations and Their Interpretation

Dollfus, A. 1948*a* *C.R.*, **227**, 331.

1948*b* *Ibid.*, **227**, 383.

1950 *Ibid.*, **231**, 1430.

1951 *Ibid.*, **232**, 1066.

1952*a* *Ibid.*, **233**, 467.

1952*b* *Ibid.*, **234**, 2046.

1952*c* *Ibid.*, **235**, 1013.

1956 *Ann. d'ap.*, **19**, 83.

1957 *Ibid.*, Suppl. 4; also thesis, Paris, 1955.

1960 *C.R.*, **250**, 463.

Dollfus, A., and Lyot, B. 1949 *C.R.*, **228**, 1773.

Elsmore, B. 1957 *Phil. Mag.*, **2**, 1040.

Focas, J. H. 1953 *C.R.*, **237**, 296.

1958 *Ibid.*, **246**, 1665.

Kuiper, G. P. 1952 *The Atmospheres of the Earth and Planets*, ed. G. P. Kuiper (2d ed.; Chicago: University of Chicago Press), p. 304.

1960 *Trans. I.A.U.*, **10**, 249.

Lyot, B. 1929*a* *Ann. Obs. Meudon*, **8**.

1929*c* *C.R.*, **189**, 425.

1930*a* *Ibid.*, **191**, 703.

1930*b* *Ibid.*, p. 834.

1934 *Ibid.*, **199**, 774.

Provin, S. 1955 *Pub. A.S.P.*, **67**, 115.

Salet, P. 1910 *Bull. Astr.*, Vol. **4**, Ser. 2, Part 1. (Mém. et Variétés, Vol. **27**, Chap. 1); thesis, Paris, 1910.

Wright, F. E. 1935 *Ann. Rept. Smithsonian Inst.*, p. 168.

CHAPTER 10

Planetary Temperature Measurements

By EDISON PETTIT
Mount Wilson and Palomar Observatories

1. INTRODUCTION

EARLY attempts in the nineteenth century to measure the radiation from stars and planets largely failed for lack of appreciation of the small masses required for the radiation receivers, which were then made thousands of times too heavy. The first successful measures were made on the whole lunar image by the Earl of Rosse (1869) with the 3-foot reflector, four-element, fine-wire thermopiles and a Thompson galvanometer in 1869. Deflections on the full moon reached 115 scale divisions.

S. P. Langley developed the bolometer for solar spectrum studies and, with F. W. Very (Langley and Very, 1889), applied it to a study of lunar radiation in the period 1885–1890. Nichols, using the radiometer, observed in 1898 and 1900 with considerable difficulty deflections on Vega, 0.5 mm; Arcturus, 1.1 mm; Jupiter, 2.4 mm; and Saturn, 0.4 mm (Nichols, 1901).

The primary principle on which modern measures of celestial thermal radiation depend was discovered in 1895 by Lebedev (1895), who found that a fine-wire thermocouple of iron-constantan was more sensitive in a vacuum than at atmospheric pressure. Pfund (1916) in 1913 and Coblentz (1915) in 1915 applied this principle to fine-wire thermocouples of high thermoelectric power and were able to observe considerable lists of stars and several planets. Between 1922 and 1932 the writer, in collaboration with Dr. S. B. Nicholson (1924, 1925), expanded the program and developed the methods and procedures which follow. During this period Coblentz and Lampland (1923, 1924; Coblentz, 1925*a*, *b*) made measurements of planetary radiation with the 40-inch reflector of the Lowell Observatory, and the temperature estimates made from their work are not greatly different from those described here. However, they are given in descriptive form only, and the student will find it difficult to derive them. The analytic form used in our work and detailed here will be found easily reproduced.

2. VACUUM THERMOCOUPLES

Thermocouples or bolometers can be pushed to high sensitivity in a vacuum, and the operating properties are about the same; but the shape and self-generating feature of the thermocouple make it better adapted to celestial measurements. Some properties of thermocouples are considered here.

The current, I, generated by a thermocouple depends on the thermoelectric power, p, resistance, R, and difference of temperature, ΔT, between the hot and cold junctions; ΔT in turn depends on the quantity, Q, of energy incident upon the junction receiver; the specific heat of the receiver and junction, S; the mass, M; the quantity of heat, C, conducted away along the wires; the quantity C_1 conducted through the air, the quantity C_2 convected away, and the quantity r radiated away. Then

$$I = \frac{kp}{SMR}[Q - (C_1 + C_2 + C + r)] , \tag{1}$$

where k is a proportionality constant.

Obviously, C_1 and C_2 may be eliminated by placing the couple in a vacuum. Then r will be a minimum if the back surface of the receiver is polished and will depend only on the absolute temperature. While C should be made as small as possible, C and R are ordinarily reciprocally associated, and we must find substances in which C is small when R is small. If C is very small and R is considered too large, the latter can be reduced by making the wires very short, 1 or 2 mm.

The optimum conditions are found in a couple made of bismuth against bismuth plus 5 per cent tin. Pure bismuth is necessary, since slight impurities greatly reduce its negative thermoelectric power. Since bismuth has a large R, the couples are made of wires about 2 mm long, and, with diameters of about 2 μ, the resistance of a compensated couple with connecting wires is about 50 ohms.

The design of a thermocouple for planetary studies involves its fabrication from parts of the smallest possible dimensions. Two wires of pure Bi of 2 μ in diameter are cut 2 mm long and attached to a middle section of like dimensions of Bi +5 per cent Sn in the form of an **N**; this is done with specks of solder under a microscope of 50 power. Heat is applied by radiation from an electrically heated needle of nichrome. $ZnCl_2$ is used as flux.

Short platinum wires of the same dimensions are attached to the free ends of the **N**, and these in turn are soldered to the free ends of the platinum lead wires of the vacuum cell.

Receivers 0.5 mm in diameter, round or square, are cut from gold leaf or copper 0.5 μ thick and attached to the junctions with white lead. Receivers so thin and much larger than 1 mm in diameter show a falling-off in response near the edge. They are blackened on one side with lampblack plus a 25 per cent volume of platinum black in a weak alcoholic solution of turpentine or by deposition in a partial vacuum of a mixture of bismuth and zinc. At a pressure of $\frac{1}{2}$ mm of mercury, this mixture, when heated to red-heat, evaporates in a black cloud which produces a thin coat on the receivers; this makes a thermocouple much faster in reaction to radiation than does lampblack, and it absorbs radiation of all wave lengths equally

TABLE 1

EFFECT OF AIR PRESSURE ON SENSITIVITY OF THERMOCOUPLE

P (Mm Hg)	Defl.	*P* (Mm Hg)	Defl.	*P* (Mm Hg)	Defl.
10	12	0.1	76	0.001	140
5	15	.05	105	.0005	145
1	25	.01	125	0.0001	146
0.5	38	0.005	136		

well. The fact that the temperature 613° K of the planet Mercury derived with our thermocouples is nearly the same as that from the solar constant and albedo (616° K, equation [35] below) is an argument for the impartial absorptivity of these blackened receivers.

The mass of the couple heated by the incoming radiation is of the order of 0.01 mg, and it is difficult to reduce this mass much further. The window of the cell is made of rock salt, about 2 mm thick, attached with picine wax, which melts at a temperature low enough to prevent cracking of the salt crystal.

The effect of a vacuum on the sensitivity of a thermocouple depends on its mass and specific heat. The writer has made them small enough to secure a factor of 30, and, while these were useful in the laboratory, they were too fragile to be used at the Newtonian focus of a telescope. Usually a factor of about 10 was employed in stellar and planetary investigations. Table 1 shows the effect of air pressure on a couple used for much of the stellar and planetary work on Mount Wilson. This couple has a vacuum amplification of 12, and Table 1 shows the necessity of maintaining a high vacuum to keep the sensitivity near its maximum.

Since it is impracticable to heat the whole cell, the vacuum is controlled by a calcium-filled quartz tube, the cell pumped out now and again while the tube is heated.

If the receivers are alike, such a thermocouple is compensated and can be exposed to the daylight sky without indicating much deflection. The radiation from the planet Mercury was measured as it crossed the meridian, 11° from the sun.

3. THERMOCOUPLE MEASUREMENTS

If the junctions of a compensated thermocouple are exposed alternately to the image of a radiating celestial object, the galvanometer records the total energy received. In order to compare radiation from different objects, the galvanometer deflections are reduced to the zenith by observations on a standard object at various air masses. The reduction takes the form

$$2.5 \log \frac{E_0}{E} = a (\sec Z - 1) , \tag{2}$$

where E_0 and E are the zenith values and observed values of the radiation and a is the extinction coefficient, which, on Mount Wilson, averages about 0.16 mag.

It is necessary to establish, first, stellar standards for easy calibration of the thermocouple. Once a magnitude scale is established with the thermocouple and comparisons with laboratory standards are made to calibrate the zero of the scale, series of stars can be intercompared and thereafter serve as direct sources of standardization in the telescope without making further reference to laboratory sources. This takes the form of a determination of the radiometric magnitudes of the stars.

4. RADIOMETRIC MAGNITUDES

The radiometric magnitude of an object is defined as the magnitude of an A0 star which gives the same galvanometer deflection as the object in question; all measures are first reduced to the zenith, with equation (2). To a considerable number of A0 stars were added stars of other types, to establish a representative group (Pettit and Nicholson, 1928). Table 2 lists, in order of right ascension, all the brighter stars of well-determined radiometric magnitude. After reduction to the zenith, the radiometric magnitude m_{r_0} of any object is determined by

$$m_{r_0} = m_r + 2.5 \log b - 2.5 \log b_0 , \tag{3a}$$

where m_r is the radiometric magnitude of a comparison star and b and b_0 are the deflections on the comparison star and the object, respectively, reduced to the zenith.

In practice, we measure about four standard stars and reduce them to the zenith and to zero radiometric magnitude. This is done by adding to the logarithm of the deflection of each star four-tenths of its radiometric magnitude and averaging these logarithms. This gives the logarithm of the average deflection b_s on a standard star of radiometric magnitude 0, and equation (3*a*) becomes

$$m_{r_0} = 2.5 \log b_s - 2.5 \log b_0 \,. \quad (3b)$$

TABLE 2

COMPARISON STARS FOR THERMOCOUPLE CALIBRATION

Star	m_r	Star	m_r	Star	m_r
β Ceti	+1.48	ι Orionis	+2.79	Arcturus	−0.98
α Arietis	+1.24	Betelgeuse	−1.67	Antares	−1.32
α Ceti	+0.72	Sirius	−1.27	α Herculis	−0.77
α Persei	+1.62	Castor	+1.74	Vega	+0.10
η Tauri	+2.98	Procyon	+0.22	Altair	+0.74
Aldebaran	−0.60	Pollux	+0.53	α Cygni	+1.24
Rigel	+0.23	β Cancri	+2.30	ζ Cephei	+2.24
Capella	−0.38	Regulus	+1.47	β Pegasi	+0.27
γ Orionis	+1.66	Spica	+1.00		

5. ENERGY CALIBRATION

Before we proceed to temperature determinations, the radiometric magnitudes must be reduced to energy units, e.g., cal cm^{-2} min^{-1}, as used in the solar constant. There are three sources easily available for comparison to convert radiometric magnitudes into energy units: (1) an amyl acetate or Hefner lamp radiates in a horizontal direction at 1 meter at the rate 156×10^{-5} cal cm^{-2} min^{-1} (Coblentz, 1915); (2) comparisons with the Hefner lamp show that a 100-watt Mazda floodlight lamp operated at 98 watts radiates 98 watts in the direction away from the lead wires; (3) the sun at mean solar distance radiates in the zenith on Mount Wilson 1.52 cal cm^{-2} min^{-1}.

Calibrations were made by removing the thermocouple from the telescope after the stellar observations were completed. Readings were made on the standard sources with two plane mirrors simulating the telescope's reflection losses. The result of many comparisons was that the radiation E of a star of radiometric magnitude m_r is

$$E = 17.3 \times 10^{-12} \times 2.512^{-m_r} \text{ cal cm}^{-2} \text{ min}^{-1} \,. \quad (4)$$

6. THE ISOLATION AND MEASUREMENT OF PLANETARY HEAT

Since the planetary radiation comes from extended objects, it is necessary to reduce the measurements to a standard solid angle. For convenience, we use 1 square second of arc. If the planetary image is smaller than the receiver, the angular area of the planet gives the number of square seconds of planetary radiation; and corrections must then be made to reduce the measures to an image which radiates uniformly at the intensity of the subsolar point. If the image of the planetary radiating surface is larger than the receiver, the radiating surface is equal to the angular area of the receiver in the field of the telescope. In either case we

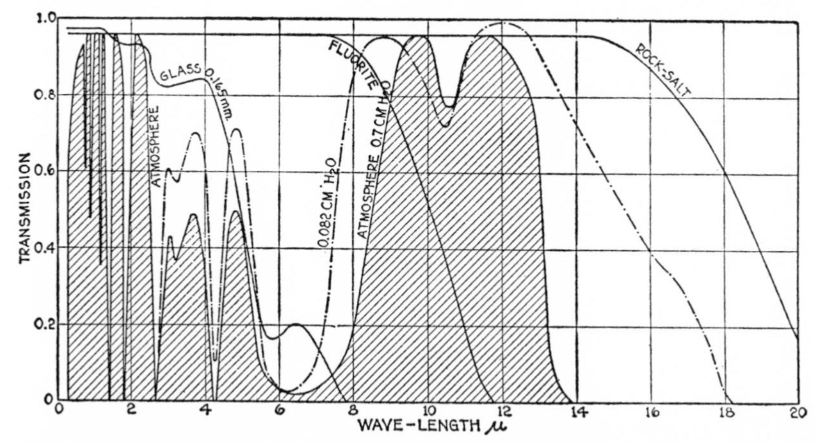

FIG. 1.—Transmission-curves of (*a*) the atmosphere above Mount Wilson (*shaded curve*) corrected to fit the theoretical lunar temperature; (*b*) water vapor, 0.082 cm precipitable, observed by Fowle; (*c*) microscope cover glass, 0.165 mm thick; (*d*) fluorite, 4 mm; and (*e*) rock salt, 2 mm.

establish the radiometric magnitude of 1 square second of planetary heat at a specified point, usually the subsolar point, on the disk.

Planetary radiation consists of reflected light plus radiation from its warmed surface. Both extend over a broad band of wave lengths, but the reflected sunlight is essentially confined to the region 0.3–3 μ, and its spectral distribution is approximately that of the sun. The effective range of planetary heat depends on the temperature. For all planets except Mercury the range is from 3 to 30 μ, while for Mercury it is 2–20 μ. The spectral distribution of planetary heat is assumed to be that of a black body. Clearly, a filter that transmits efficiently up to 3 μ can be used to separate reflected light from planetary heat. Such a filter is a piece of thin glass.

The water-vapor, oxygen, and carbon dioxide contents of the atmosphere greatly modify the spectral distribution of incoming radiation. Figure 1 shows the transmission-curves of (*a*) the atmosphere above Mount Wilson (*shaded curve*); (*b*) water vapor, 0.082 cm precipitable water; (*c*) microscope cover glass, 0.165 mm thick; (*d*) fluorite, 4 mm; and (*e*) rock salt, 2 mm thick. A water cell, 1 cm thick, transmits from 0.3 to 1.35 μ.

To separate the planetary heat, we have the choice of using a water

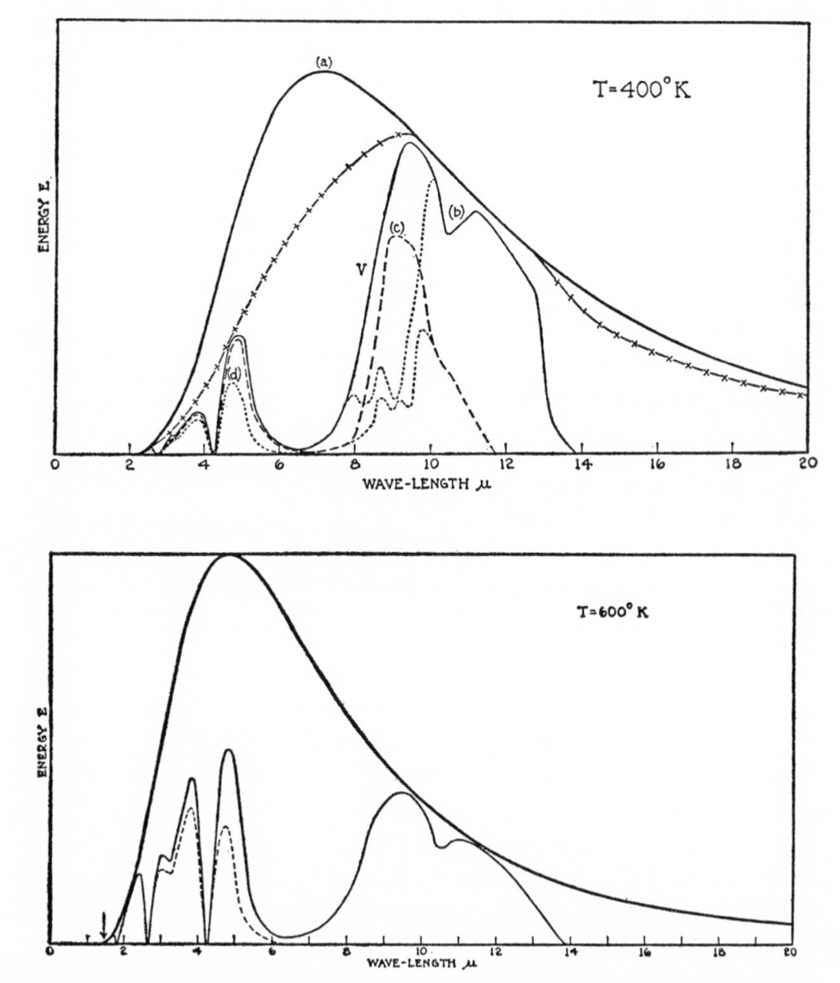

FIG. 2.—*Upper:* energy-curve (*a*) black body at 400° K; (*b*) after passage through atmosphere (*upper dotted line*, deformation by silica); (*c*) after passage through fluorite screen (*lower dotted line*, effect of hypothetical lunar silica); (*d*) after passage through microscope cover glass (*dashed line*). The dash-cross line shows deformation of *a* necessary to bring observed and theoretical lunar spectral curve together. *Lower:* energy-curve (*heavy line*), black body at 600° K; *light line*, same after passage through atmosphere; *dotted line*, same after passage through microscope cover glass. Arrow shows red limit of water cell.

cell, which cuts off some of the reflected light, or the microscope cover glass, which transmits some planetary heat. Two considerations are unfavorable to the water cell: (1) it is very difficult to clean internally and to keep free from air bubbles; (2) the spectral distribution of the reflected light from a planet makes the correction to the water-cell measures difficult. It also has a tendency to fog over or freeze, and the telescope must be refocused when it is used.

The microscope cover glass was therefore adopted as the principal screen to separate planetary heat from reflected light. For the moon and all planets except Mercury the planetary-heat transmission for the cover glass is less than 6 per cent, but for Mercury it rises to 35 per cent. In the case of Mercury, therefore, the water cell was preferable, particularly since the work was done in the daylight sky, which made frequent examination of the cell simple.

Figure 2 shows the effects of atmospheric transmission upon the spectral distribution of planetary heat passing through it. The upper diagram simulates the lunar conditions, and the lower diagram those of Mercury. They show the effect of temperature on the relative importance of the 8–14 μ and the 2–6 μ transmission bands.

7. CORRECTIONS FOR VARIABLE ATMOSPHERIC WATER VAPOR

Since the atmospheric transmission in Figure 1 is based on the mean water-vapor content 0.7 cm precipitable, this curve must be raised or lowered to suit the actual conditions at the time of observation. An estimate of the water-vapor content is given by Hann's formula,

$$p = 1.7\, e_w \sec Z\,, \tag{5a}$$

where p is precipitable water in cm, e_w is the vapor pressure determined with the sling psychrometer, and Z is the zenith distance. A variation of 50 per cent in water vapor changes the over-all transmission 11 per cent. The correction to a measure of planetary heat, B, due to variation of the water vapor from the standard 0.7 cm precipitable, is given by

$$\Delta B = \frac{p - 0.7}{3} B \qquad (0.3 < p < 1.5 \text{ cm})\,. \tag{5b}$$

Now ΔB is an additive term, and, to combine it in an equation of reduction, it must be expressed as a coefficient, namely,

$$\frac{B + \Delta B}{B} = \frac{p + 2.3}{3}\,, \tag{5c}$$

which will appear in the final equations (12) and (13) of reduction of the measures of planetary heat B.

8. EQUATIONS RELATING PLANETARY TEMPERATURE AND RADIATION

If we consider a uniformly illuminated planet or star, the familiar formula

$$T = \sqrt[4]{\left(\frac{D^2E}{r^2\sigma}\right)} \tag{6}$$

applies, where D is the distance of the object, r its radius, E the radiation received from it, and σ the constant in Stefan's formula in the units of E. Since, for astronomical objects, D/r is the reciprocal angular radius expressed in radians and E is the radiation constant in equation (4), equation (6) can be expressed by

$$\log T = 2.638 - 0.1\,(m_r - \Delta m_r) - \tfrac{1}{2}\log d\,, \tag{7}$$

where T is the absolute temperature in ° K, m_r the radiometric magnitude, Δm_r the loss in magnitudes due to atmosphere and telescope, and d the diameter of the planet in seconds of arc. Thus the linear diameter and distance of the object disappear from the calculation of temperature.

Since the radiation emitted by a planet or a star varies somewhat over the apparent disk and especially if the planet has sensible phase, E in equation (6) must be corrected for this effect. Let us take the sun, for example. The solar constant,[1] generally regarded as E, is 1.95 cal cm^{-2} min^{-1}. The derived solar temperature is 5750° K. However, it must be pointed out that this is an *effective temperature*, corresponding to integrated sunlight, and is lower than the photospheric temperature, on account of darkening at the limb. Now drift-curves in total radiation show that the radiation at the center of the disk is 16 per cent higher than the average over the disk; hence E should be 2.26 cal cm^{-2} min^{-1} for the undarkened photosphere, which gives 5973° K, a value nearly in line with that derived by other methods using radiation from the center of the disk.

Instead of correcting E in computing equation (7) for the planets, it would be better to use a small thermocouple receiver and measure the radiation from the subsolar point on the disk. For practical purposes, we shall reduce the measures to 1 square second of arc. Then $\frac{1}{2}\log d$ becomes 0.0262, and equation (7) reduces to

$$\log T = 2.612 - 0.1\,(\bar{m}_r - \Delta m_r)\,, \tag{8}$$

where $\bar{m}_r$ is the radiometric magnitude of 1 square second of arc of planetary heat.

[1] See note on p. 428.

The actual calculation of $\bar{m}_r$ is contained in the expression

$$\bar{m}_r = m_r + 2.5 \log b - 2.5 \log B - (a' \sec Z - a \sec z) - P + 2.5 \log A s^2 , \quad (9)$$

where m_r is the radiometric magnitude of the comparison star, b the deflection on the star, B that part of the deflection on the planet due to planetary heat, Z and z the zenith distances of the planet and star, respectively, a' and a the coefficients of atmospheric extinction for planetary heat and starlight, respectively, P the correction in magnitudes from equation (5c) due to the deviation of atmospheric water vapor from the average, A the area of the thermocouple receiver in square millimeters, and s the scale of the telescope in seconds of arc per millimeter.

The coefficients of extinction for star and planet are the same on Mount Wilson and equal to 0.16. Formerly a term $+(C - c)$ was used to correct for tarnished silver, but with aluminized surfaces this is neglected.

To obtain B, the part of the deflection due to planetary heat, the free deflection F is compared with F_g the deflection observed with the cover-glass screen. Let R be the part of the radiation reflected from the planet, k its transmission coefficient through the cover glass, and K the transmission coefficient of the planetary heat through the cover glass. Then

$$F = B + R , \quad (10a)$$

$$F_g = BK + Rk , \quad (10b)$$

and

$$B = \frac{F - F_g/k}{1 - K/k} . \quad (10c)$$

For the planets we substitute B from equation (10c) in equation (9), add the term $-2.5 \log \frac{1}{3}(p - 2.3)$ from equation (5c) to correct for variance of atmospheric water vapor, substitute the resulting $\bar{m}_r$ in equation (8), and collect the constants and those variables which are functions of the planetary temperature T. This gives

$$m_r + 2.5 \log b - 2.5 \log \left(F - \frac{F_g}{k}\right) - 0.16 (\sec Z - \sec z) - 2.5 \log \frac{p + 2.3}{3} + 2.5 \log A s^2 = 26.12 - 10 \log T + \Delta m_r - 2.5 \log \left(1 - \frac{K}{k}\right) = m_r' . \quad (11)$$

which applies to any planet for which the optical distribution of reflected light is known.

The color of the reflected light from the moon, Mercury, and Mars approximates that from a K0 star for which the absorption by the cover-glass screen is 0.03 mag., and the reflection on the surfaces is 0.08 mag., together making the value of k equal to 0.90. For the planet Venus a slightly different value obtains. If we introduce this value of k into equation (11), we obtain

$$m_r + 2.5 \log b - 2.5 \log (F - 1.11F_g) - 0.16 (\sec Z - \sec z)$$
$$- 2.5 \log \frac{p+2.3}{3} + 2.5 \log A s^2 = 26.12 - 10 \log T \quad (12)$$
$$+ \Delta m_r - 2.5 \log (1 - 1.11K) = m_r' .$$

In equation (12) the second and third terms are fixed by the conditions of observation. The fourth term seldom exceeds 0.16 mag. The fifth term seldom exceeds 0.1 mag. If it is much greater than this, a repetition of the measures on another night is advisable. The last term was usually about 5 mag. for the thermocouple commonly employed at the Newtonian focus of the 100-inch.

Now m_r' differs from $\bar{m}_r$ only in the term $-2.5 \log (1 - K/k)$. Since k is about 0.90 and K is a small fraction, this term is usually unimportant. At 400° K its omission changes the computed temperature only 6°, and below 350° K the change is negligible. Hence, except for Mercury, m_r' is essentially the radiometric magnitude of 1 square second of planetary heat, but the above qualifications must be kept in mind.

9. RADIATION FROM THE MOON AND THE CALCULATION OF m_r'

From equation (12) it is seen that the solution for temperature involves a computation of the function m_r' from the second member; the unknowns Δm_r and K must be computed from Fowle's measures of water-vapor transmission and the transmission of the cover glass shown in Figure 1. These, multiplied into the black-body curves for various temperatures, give Δm_r and K, and these in turn, applied to the second member of equation (12), give m_r' shown in Table 3*A*. The corrections $\Delta m_r'$ result from a study of lunar radiation made particularly for the purpose.

The very extended extrapolation which had to be made to the laboratory measures of H_2O and CO_2 transmissions necessitated the study of a celestial object of known radiative properties to test their accuracy. The moon satisfies these requirements. The problem then resolved into an empirical determination of the temperature of the subsolar point on the moon at mean solar distance, by measuring the emitted planetary heat with application of atmospheric corrections; and a theoretical temperature from the solar constant after the losses by reflection and by conduction into the lunar surface without re-radiation have been allowed for.

On account of the roughness of the lunar surface, it was necessary to determine the mean spherical distribution of planetary heat and reflected light from the subsolar point as this point swept across the lunar surface during the lunation.

Figure 3 (*upper*) shows the radiation diagram obtained in this way, from which the average radiation of planetary heat in all directions, using Fowle's transmission, is 1.93 cal cm^{-2} min^{-1}, while the average reflected light is 0.24 cal cm^{-2} min^{-1}, not involving Fowle's measures. The fact that

TABLE 3.4*

PRELIMINARY TABLE FOR m'_r

T (° K)	Δm_r	K	m'_r	$\Delta m'_r$
100	6.35	0.000	+12.47	−0.49
200	2.12	.000	+ 5.23	− .28
300	1.39	.013	+ 2.76	− .20
400	1.28	.063	+ 1.46	− .18
500	1.28	.147	+ 0.60	− .18
600	1.30	.249	− 0.01	− .17
700	1.31	.364	− 0.46	− .17
800	1.25	0.468	− 0.86	−0.16

* Computed from eq. (12) with the extrapolated water-vapor transmissions of Fowle applied to atmospheric transmissions Δm_r and cover-glass transmission K; $\Delta m'_r$ is obtained from a study of the moon.

the total emission by the moon appears to exceed the solar constant is considered below.

The heat conducted into the moon cannot be observed while the temperature is slowly changing during the long lunar day but is readily observed as re-radiation during the rapid changes of insolation that occur during a lunar eclipse. Figure 3 (*lower*) shows the graphs of radiation and temperature at the lunar subsolar point and at the limb, as these points swept through the earth's shadow. Note the gradual decrease in temperature during total phase. This is accounted for by a conductivity into the surface in full sunlight of 0.1 cal cm^{-2} min^{-1} (note that the unit of time is the *minute*).

We are now in position to compute the theoretical temperature of the moon and compare it with the temperature derived from the lunar heat measured through our atmosphere.

The solar energy, E, converted into planetary heat is the solar constant

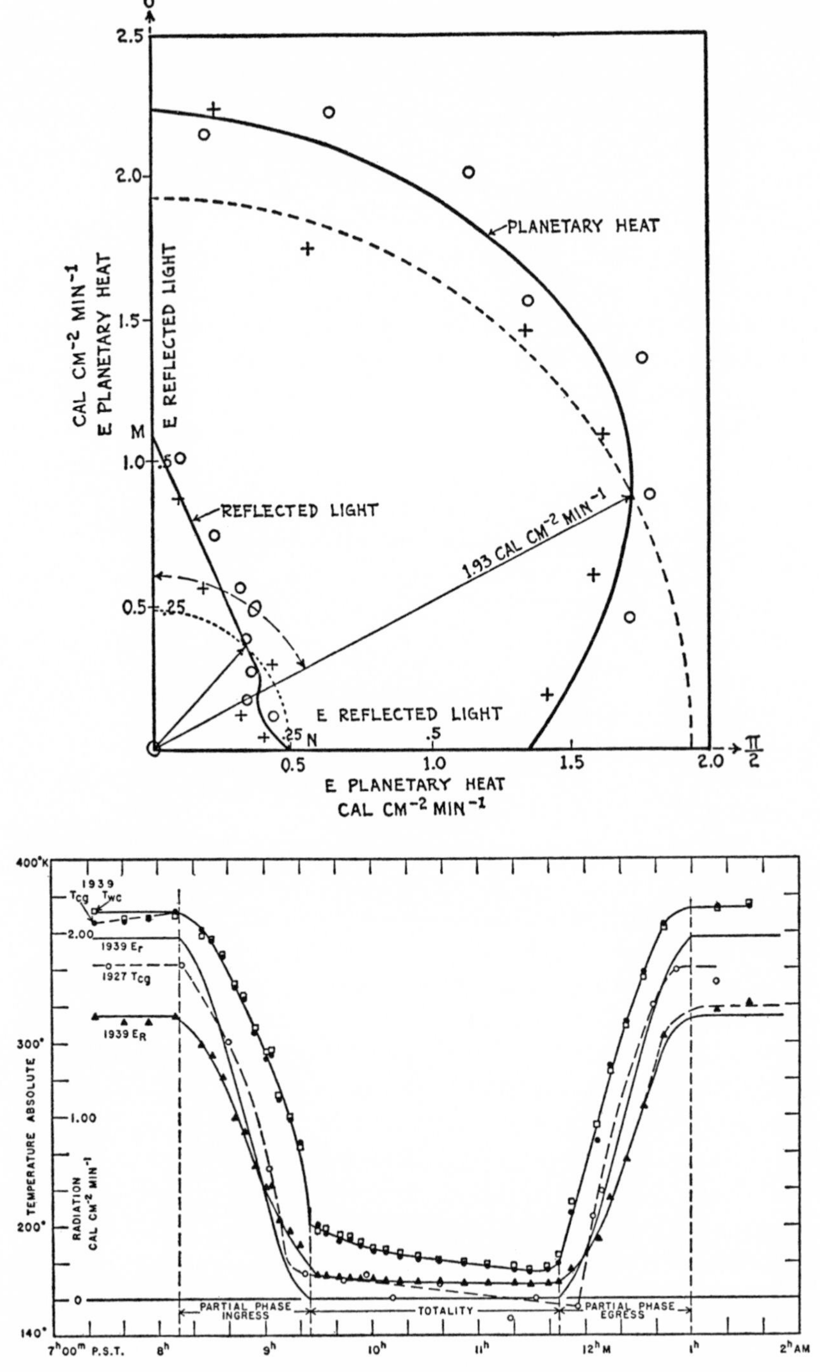

FIG. 3.—*Upper:* distribution of planetary heat and reflected sunlight about lunar subsolar point. Corrected for atmosphere by Fowle's extrapolated transmissions of water vapor. *Lower:* march of temperature (from water-cell and cover-glass measures), energy received from sun E_r and radiant energy, E_R during the lunar eclipses of 1927 (point near limb) and 1939 (subsolar point).

minus the sum of the reflected light and conducted heat. The temperature $T\,^\circ$ K then results from Stefan's law,

$$\begin{aligned} E &= 1.95 - (0.24 + 0.1) = 1.61 \text{ cal cm}^{-2} \text{ min}^{-1} , \\ T &= 374^\circ \text{K} . \end{aligned} \tag{13}$$

If we apply the simple formula (34), below, with albedo 0.07, the temperature comes out as 385° K.

Equation (13) is based on the atmospheric transmission of sunlight, which is well known, used in the value of the solar constant and the reflected-light measurements. The term 0.1 cal cm^{-2} min^{-1} is planetary heat and, while small, suffers the uncertainties of atmospheric transmission at long wave lengths. The value used here, however, is based on revised atmospheric transmissions, and the temperature given in equations (13) may be accepted as effectively free from such uncertainties.

From the distribution-curve of planetary heat in Figure 3, *A*, we find that the mean spherical intensity which is to be compared with the theoretical value in equations (13) is

$$E = 1.93 \text{ cal cm}^{-2} \text{ min}^{-1} , \qquad T = 391^\circ \text{K} , \qquad m_r' = 1.54 \text{ mag} , \tag{14}$$

where T is calculated from Stefan's law and m_r' is taken from Table 3*A*.

The measured planetary heat from the lunar surface, when corrected for atmospheric extinction in equations (14), is therefore 0.32 cal cm^{-2} min^{-1}, or 20 per cent in excess of that which can be accounted for by the theory in equations (13), and corresponds to a temperature 17° higher than the solar constant permits, after correcting for the losses.

Since the solar constant is well established, we are left with two alternatives to explain this discrepancy: (1) The emissivity is not unity. The dash-and-cross line in Figure 2 shows the deformations of the black-body-curve required, which observations with the fluorite filter show are actually not present. (2) The atmospheric transmission of water vapor determined by Fowle is too low, and the 391° K obtained in equations (14) resulted from overcorrection of the black-body radiation. This requires a correction of Table 3*A* amounting to -0.19 mag. in m_r' at the lunar temperature. Therefore, we go back and everywhere raise the ordinates in Fowle's atmospheric curve (which resulted in the shaded curve of Fig. 1) to account for this defect and thus formally compute corrections $\Delta m_r'$ which must be applied to Table 3*A* to allow for it.

These corrections $\Delta m_r'$ also apply to Δm_r, since it is this function which is fundamentally at fault. It should be understood, however, that K, which is the ratio of the planetary heat transmitted by the cover glass to the whole amount transmitted to it by the telescope, is a ratio of energies —the whole amount transmitted by the atmospheric curve in Figure 1 to that to the violet of 6 μ. Since we are multiplying the atmospheric transmissions by a constant to make our corrections, this ratio is little affected. In computing planetary temperatures from equation (11) or equation (12), we shall now use Table 3*B*.

The effect of $\Delta m_r'$ is to lower the temperature computed from the radiation measurements of Mercury by about 45° C, Venus 9°, Mars 10°, and the moon 17°, compared with the values obtained by use of the Fowle water-vapor transmissions in their original extrapolated form. This correction to the infrared atmospheric absorption is the principal reason for

TABLE 3*B**

FINAL TABLE FOR m_r'

T (° K)	Δm_r	*K*	m_r' (Corr.)
100	5.86	0.000	+11.98
200	1.84	.000	+ 4.95
300	1.20	.013	+ 2.56
400	1.11	.063	+ 1.28
500	1.12	.147	+ 0.42
600	1.16	.249	− 0.19
700	1.21	.364	− 0.63
800	1.20	0.468	− 1.02

* Obtained by correcting m_r' (prov.) by $\Delta m_r'$ of Table 3*A*; corrected values of Δm_r are also given.

the disparity between the planetary temperatures obtained elsewhere and those obtained at Mount Wilson with Table 3*B*.

In a more recent study (1955) of the temperatures of Venus, the writer in collaboration with S. B. Nicholson has recomputed m_r' in Table 3*B* from the atmospheric transmissions given by Adel and Lampland (1940). The resulting changes are −0.09 mag. for 100° K and +0.06 mag. for 200° K; the values for other temperatures are unaffected.

10. TEMPERATURE OF THE DARK SIDE OF THE MOON

Measurements of the radiation from the dark side of the moon were made during a crescent phase on July 5, 1927, at a point on the earth-lit side nearly opposite the sun. The sensitivity was such that a star of radiometric magnitude 0 would produce 19 mm deflection. The observed deflection was 0.09 ± 0.05 mm. This corresponds to radiometric magnitude 10.6 per square second of arc (essentially m_r'), for which Table 3*B* gives a temperature of about 120° K. The deflection plus its probable error would give 125° K, which may be regarded as its upper limit. It is seen that the temperature on the moon opposite the sun is near the lower limit of measurement.

11. DISTRIBUTION OF RADIATION OVER THE LUNAR DISK

Drift-curves of radiation over a lunar diameter at full phase show that the reflected light is greatest at the center, falls to about 60 per cent near the limb, then climbs to more than 90 per cent at the limb itself. The curve is irregular because of the surface features.

The planetary heat is not so abruptly changed by lunar features and falls smoothly toward the limb according to the formula

$$E = a \cos^{2/3}\theta \ , \qquad (15a)$$

where θ is the angular distance of a point from the subsolar point. A smooth, slowly rotating sphere receives solar energy and re-radiates it according to the formula

$$E = a \cos\theta \ . \qquad (15b)$$

The roughness of the lunar surface qualitatively explains the difference between equations (15*a*) and (15*b*).

12. POROUS NATURE OF THE LUNAR SURFACE

Silicates, of which the lunar surface may contain a high percentage, have low emissivity between 8 and 10 μ, a region which can be isolated by the fluorite and cover-glass screens (Fig. 1). We shall now compare the ratio H of the planetary heat from the moon transmitted by the fluorite screen to that without the screen and H computed by assuming (*a*) unit emissivity and (*b*) emissivity of silica.

If F is the deflection with free aperture, F_f with fluorite screen, and F_g with cover glass, our measures for the subsolar point on the moon are

$$F = 145 \text{ mm} \ , \quad F_f = 79 \text{ mm} \ , \quad F_g = 41 \text{ mm} \ , \quad H = \frac{F_f - F_g}{F - F_g} = 0.37 \ . \qquad (16a)$$

From these data, F_g, which represents essentially the reflected light, R, is 0.39 of the planetary heat, the latter being essentially $F - F_g$.

Now, if we multiply the black-body curve for 400° K (Fig. 2, *a* in upper diagram), essentially the temperature of the lunar subsolar point, into the atmospheric transmission-curve (Fig. 1), we obtain *b* in the upper diagram of Figure 2. The area which represents F is 2054 units, F_f is 773. Before we can compute H, we must correct the deflections represented by these areas for the fraction of R reflected by the cover glass, $0.10R$, and by the fluorite, $0.05R$. We then have

$$F_f - F_g = 0.05R + 773\,, \qquad F - F_g = 0.10R + 2054\,,$$

$$H = \frac{F_f - F_g}{F - F_g} = 0.38\,. \tag{16b}$$

Here we have assumed the emissivity in Figure 2, *a*, to be unity.

Next we compute H on the assumption of a lunar surface of silica. The energy-curves shown in the upper part of Figure 2 must be multiplied by the emissivity coefficients e of silica before the ratio H is computed. The coefficients of e are shown in Table 4, as given by Wood (1911) and Sosman (1927).

TABLE 4

EMISSIVITY *e* OF SILICA

λ	e	λ	e	λ	e	λ	e
7.75.......	1.00	8.5.......	0.30	9.0.......	0.20	9.75.....	0.68
8.0........	0.76	8.62......	0.40	9.25......	0.24	10.0......	0.98
8.25.......	0.38	8.75......	0.32	9.5.......	0.42	10.25.....	1.00

From Table 3*B*, Figure 2, *a*, and equation (16*a*), we find, for a silica surface,

$$H = \frac{41 + 505}{82 + 1691} = 0.30\,. \tag{16c}$$

This result is so radically different from equations (16*a*) and (16*b*), which latter are in fair agreement, that we must assume that the lunar surface is not silica in the form giving the data of Table 4, but is presumably very porous or covered with a powdered layer.

From the data in Figure 3 (*lower*, 1927 curve), Epstein (1929) has shown that $(\kappa\rho C)^{-1/2}$, where κ is the conductivity, ρ the density, and C the specific heat of the lunar crust, is about 120, corresponding to pumice or volcanic ash, for which the conductivity is about 0.001.

On the basis of polarization measurements (cf. chap. 9), it has been assumed that the moon is covered with a powdered layer of volcanic ash or its equivalent. A test of this proposition can be made by comparing the radiation from a steep lunar slope (e.g., the face of Pico) during an eclipse, with a mare. The temperature of the cliff ought to fall more rapidly because of the absence of ash cover. It might be difficult to set the thermocouple upon the image.

13. THE PHASE RELATION

The planet Mercury offers so small an image and is observed in daylight so near the sun that the seeing conditions preclude the study of any but the entire image. Since the apparent diameter of the planet never exceeds 0.7 mm at the Newtonian focus of the 100-inch telescope, it was easy to provide a thermocouple with receivers of these dimensions. This created the problem of interpreting such measures in terms of temperature at the subsolar point. On account of the undesirable effects of exposure of a great mirror to sky so near the sun, the measures of Mercury were limited to 26 days between 1923 and 1925, covering phase angles 30°–125°. It was then decided to observe the moon in a like manner with a short-focus mirror to assist in the reduction of the Mercury measures.

For this purpose a parabolic mirror 6.4 cm in diameter and 8.8 cm in focal length was arranged with special vacuum thermocouple at the focus, which was small enough to exclude only 20 per cent of the radiation from the moon. The image of the whole moon was formed on the receivers of the thermocouple, 1 mm square and 2.5 mm apart.

The instrument was mounted on a telescope with electrical arrangements to throw the image of the moon from one junction to the other, and the galvanometer deflections were read on a scale. Calibration was by a 100-watt Mazda floodlight lamp on the 60-foot tower telescope, 71 meters away. A microscope cover glass was inserted electrically to eliminate the planetary heat. A simple weight-controlled gauge read the air mass. Corrections were applied to reduce the measures to mean distance from the center of the earth and from the sun.

If we consider a smooth, slowly rotating, non-conducting black sphere exposed to solar radiation and viewed at any phase angle i, the whole planetary heat emitted, E_B, can be computed from the radiation E_0, at zero phase angle, by the formula

$$E_B = \frac{E_0}{\pi} [(\pi - i) \cos i + \sin i] . \tag{17}$$

In order to compare the observed measures of the planetary heat from the whole lunar image with equation (17), it was necessary to reduce them to outside the atmosphere. Since the temperature varies over the apparent surface, the atmospheric transmission K does also. The lunar surface was therefore divided into ten temperature zones, assuming a temperature of 374° K at the subsolar point and an energy distribution proportional to $\cos^{2/3} \theta$, where θ is the angular distance from the subsolar

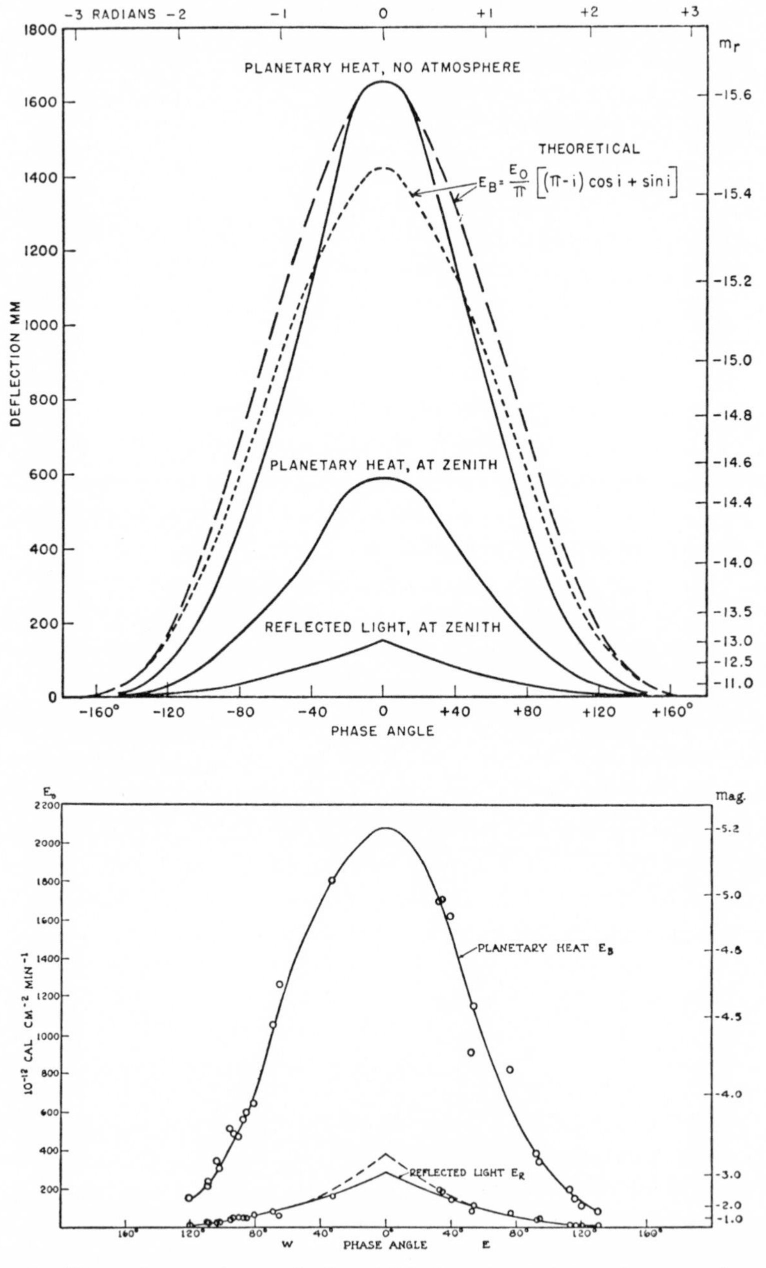

FIG. 4.—*Upper:* planetary heat and reflected light from the whole lunar image as a function of phase angle; 532 mm equals 10^{-5} cal cm^{-2} min^{-1}. *Lower:* planetary heat and reflected light from Mercury corrected to no atmosphere. *Broken line:* visual curve of Müller.

point. The factor to transfer the measures to no atmosphere was found to be practically constant at 2.82 (or -1.13) magnitudes for all phase angles.

Figure 4 shows the radiometric magnitude of the whole lunar image in reflected light and planetary heat at all observable phase angles. The curve for no atmosphere (*full line*) is shown with the theoretical curve (*dashed line*) coincident at phase angle 0°. It will be noted that the theoretical curve is above the observed curve everywhere else. If we make the area of the two curves the same, the dotted line marks the theoretical curve. It is probable that the shadowing of the rough mountain surfaces explains this discrepancy.

14. THE RADIATION OF MERCURY TRANSFERRED THROUGH OUR ATMOSPHERE

Since in the case of Mercury the whole image of the planet must be observed, we can derive observationally only the temperature at the subsolar point. The free deflection and the radiation transmitted by the water cell, cover glass, and fluorite window were measured, and from them the corresponding radiometric magnitudes of the planet were deduced at unit distance and mean radius vector.

The water-cell absorption of the reflected light of the planet was obtained from the visual minus water-cell magnitudes, using Müller's (1897) formula for m_v corrected to the Harvard scale. This gave a water-cell index, *WCI*, for Mercury. When this value was read from a plot of *WCI* against water-cell absorption for stellar radiation, it gave the water-cell absorption *WC* for Mercury, -0.66 mag. From this the radiometric magnitudes of the whole planetary heat, m_B, and that transmitted by cover glass and fluorite, m_g and m_f, were determined. Specifically,

$$m_R = m_{WC} - 0.66 \,, \tag{18}$$

$$m_B = -2.5 \log (2.512^{-m_B} - 2.512^{-m_R}) \,, \tag{19}$$

$$m_g = -2.5 \log (2.512^{-m_g} - 2.512^{-(m_R + 0.11)}) \,, \tag{20}$$

$$m_f = -2.5 \log (2.512^{-m_f} - 2.512^{-(m_R + 0.08)}) \,, \tag{21}$$

where, as before, the constant 0.11 corrects the light reflected by the planet for reflection from and absorption by the cover glass and $+0.08$ is the correction for reflection by the fluorite screen.

To reduce m_B, m_g, and m_f to no atmosphere, the method used for the moon was not applicable, since the mean spherical distribution of energy was not determinable. An indirect method was therefore employed.

The mean temperature on a planet, $\bar{T}$, is obtained from Stefan's law by

$$\log \bar{T} = 2.713 - 0.1\,(m_r - \Delta m_r) - 0.25 \log [d^2 (1 + \cos i)]\,, \quad (22)$$

where d is the diameter of the planet in seconds of arc and m_r is the radiometric magnitude of that part of the planetary heat observed. Either m_B or m_g may be used in equation (22) with the corresponding Δm_r.

As a first approximation, both Δm_B and Δm_g were computed from black-body-curves and atmospheric absorption coefficients for various temperatures. Curves were then drawn showing the relation between mean temperature and $\bar{m}_r$ (cf. eq. [8]). Since Δm_B and Δm_g are functions of T, so is their difference, and we have

$$m_g - m_B = m_K = \Delta m_g - \Delta m_B\,. \quad (23)$$

While the first member is observed and the last computed as a function of temperature, this temperature is $\bar{T}$ and is independent of area or the deflection in energy units. Since the temperature of the radiating surface is non-uniform, the atmospheric absorption corresponding to the mean temperature is not the mean atmospheric absorption, and the first approximations of $\bar{T}$ obtained with that assumption need a correction, depending on i and the temperature of the subsolar point, the latter being a function of r. For different values of the subsolar temperature, the planetary radiation from each isothermal zone transmitted by the atmosphere was summed, and these sums, divided by the corresponding totals without atmospheric absorption, gave theoretical atmospheric transmissions; from these a table of corrections, $\delta\bar{T}$, was constructed to be applied to the first approximations of $\bar{T}$. As expected, the corrections increased with phase angle, owing to the larger temperature variations over the disk. For a subsolar temperature of 600° K and phase angle 0°, $\delta\bar{T}$ corresponding to m_B is +1° C; for phase angle 120° it is −2° C. For m_g the figures are −5° and −14°, respectively, and for m_K they are −13° and −26° C.

15. PLANETARY HEAT AND REFLECTED LIGHT

Since we now have the atmospheric transmission for all values of i and r, we are in a position to compute M_B, the magnitude of the planetary heat outside our atmosphere at unit distance from the earth and mean distance, r_0, from the sun. If the distance planet to earth be ρ, then

$$M_B = m_r - \Delta m_r - 5 \log \rho - 5 \log \frac{r}{r_0}\,, \quad (24)$$

and the radiation in cal cm^{-2} min^{-1} is

$$E_B = 17.3 \times 10^{-12} \times 2.512^{-m_B}\,. \quad (25)$$

There are two similar equations for the reflected light. The planets have a color resembling that of a gG8 star, the atmospheric absorption for which is 0.41 mag. Therefore, we replace $m_r - \Delta m_r$ in equation (24) by $m_R - 0.41$, in order to find M_R and so E_R from equation (25).

16. MERCURY AND THE MOON COMPARED WITH A BLACK BODY

Figure 4 shows a plot of the reflected light E_R and planetary heat E_B observed on Mercury at different phase angles (lower diagram). Scales of energy and radiometric magnitude are given.

Figure 5 shows the phase radiation-curves for Mercury and the moon compared with that of a black body. For east phase angles the two objects

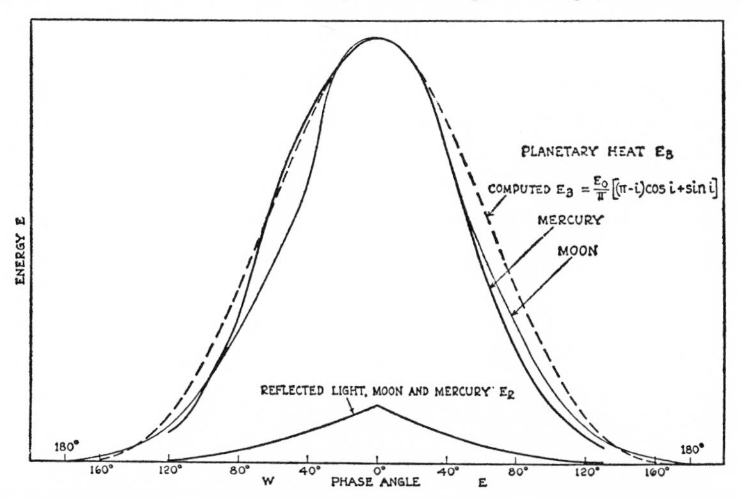

FIG. 5.—Radiation from Mercury and the moon as a function of phase angle compared with a smooth, slowly rotating black sphere.

behave alike, but for west phase angles Mercury closely follows the black-body condition. These observations were compatible with Mercury keeping the same face to the sun and the side presented at western elongation being less mountainous.

Figure 5 shows that the planetary heat received from the whole disk at zero phase angle, mean distance from the sun, and unit distance from the earth is

$$E_0 = 2080 \times 10^{-12} \text{ cal cm}^{-2} \text{ min}^{-1} . \qquad (26)$$

If the whole disk radiated at the same rate as the subsolar point and the actual distribution over the disk were the same as that of the moon (eq. [15*a*]), the total radiation received by an observer at 1 astronomical unit would be

$$E_S = \tfrac{4}{3} \times 2080 \times 10^{-12} = 2770 \times 10^{-12} \text{ cal cm}^{-2} \text{ min}^{-1} . \qquad (27)$$

However, it can be estimated from Figure 5 that the western side of the planet radiates 20 per cent in excess of the eastern side; the latter seems to follow formula (15*a*). This means that the right member of equation (27) must be multiplied by 1.10, since it applies to the whole illuminated hemisphere. We therefore have

$$E_S = 1.10 \times 2770 \times 10^{-12} = 3047 \times 10^{-12}\ \text{cal cm}^{-2}\ \text{min}^{-1}\ . \quad (28)$$

Applying the inverse-square law, we find that the radiation from the subsolar point on Mercury is

$$E_m = 3047 \times 10^{-12} \times 3.815 \times 10^{9} = 11.62\ \text{cal cm}^{-2}\ \text{min}^{-1}\ ,$$

which corresponds to the temperature

$$T = 613^{\circ}\ \text{K}\ . \quad (29)$$

If we had used formula (15*a*) without correction for excess radiation, the temperature would have been 600° K.

17. TEMPERATURE OF MERCURY FROM THE SOLAR CONSTANT

With the solar constant at 1.95 cal cm^{-2} min^{-1}, Mercury receives, at mean distance,

$$E_0 = 13.02\ \text{cal cm}^{-2}\ \text{min}^{-1}\ . \quad (30)$$

The reflected light at zero phase angle in Figure 4, *B*, is

$$E_R = 291 \times 10^{-12}\ \text{cal cm}^{-2}\ \text{min}^{-1}\ , \quad (31)$$

which, at the surface of the planet, becomes

$$E_R = 291 \times 10^{-12} \times 3.81 \times 10^{9} = 1.11\ \text{cal cm}^{-2}\ \text{min}^{-1}\ . \quad (32)$$

Hence the net energy radiated as planetary heat is

$$E_0 = 13.02 - 1.11 = 11.91\ \text{cal cm}^{-2}\ \text{min}^{-1}\ ,$$

and the corresponding temperature is

$$T = 617^{\circ}\ \text{K}\ . \quad (33)$$

We may compute the temperature from the solar constant corrected for albedo A and the distance in astronomical units r_0 from Stefan's law. If the constants are collected, this relation becomes

$$T = 392 \sqrt[4]{\left(\frac{1 - A}{r_0^2}\right)}. \quad (34)$$

If we put $A = 0.07$, we have

$$T = 616^{\circ}\ \text{K}\ , \quad (35)$$

which is nearly the same as equation (33).

If we assume the temperature at mean distance to be 613° K, the temperature of the subsolar point at perihelion will be 688° and at aphelion 558° K.

18. VENUS

While for Mercury, as for the moon, the radiation properties are comparatively simple, its small size makes temperature determinations dependent on a study of the phase relation. Venus, on the other hand, is complicated by an ever clouded atmosphere; but it presents an image so large that detailed study with the thermocouple may be made. As with Mercury, the measures were made in daylight, when the planet crossed the meridian.

The upper limit we can expect on the temperature of Venus is given by equation (34), where A is 0.59 and r_0 is 0.72. This comes out at 370° K; hence the cover glass alone can be used to separate the planetary heat.

Near superior conjunction the whole image was measured, but near inferior conjunction discrete points on the surface were observed. Perhaps the best approach is to study drift-curves, made by moving the image of the planet slowly across the thermocouple receiver with the telescope slow motion in right ascension. Such curves are shown in Figure 6.

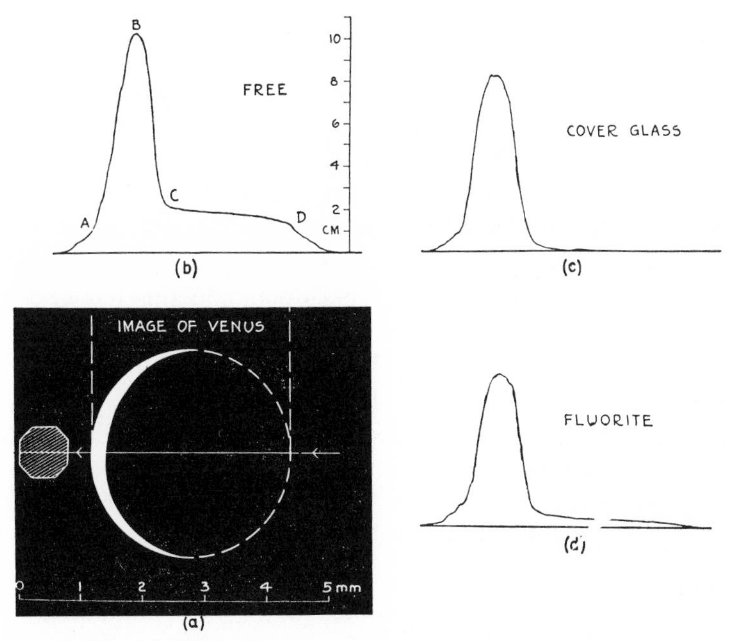

FIG. 6.—Drift-curves across Venus: (a) image drifts across thermocouple junction while motor drives recorder; (b) galvanometer trace without screens (note energy CD across dark side); (c) with cover glass, no energy across dark side transmitted; (d) fluorite transmits only a little energy from dark side. Mount Wilson observations, 1924.

The observational arrangement is shown in Figure 6, *a*, with the octagonal thermocouple receiver at the left of the planetary image. In *b* is seen a drift-curve taken without screens (free); *ABC* shows the radiation from the visible crescent, but *CD* indicates a considerable radiation from the dark side of the planet. In *c* the cover glass is seen to cut out this radiation, and even the fluorite filter (*d*) transmits only a little, indicating a low temperature for this part of the planet. From sets of deflections made in the standard manner and the application of equation (11), we find m_r' on the dark side to be 3.70, which, by Table 3*B*, yields

$$T = 241^\circ \text{K} . \tag{36}$$

The earlier reduction (Pettit and Nicholson, 1924*a*), using Table 3*A without* $\Delta m_r'$, gave 250° K. In a later study (1955), the author in collaboration with S. B. Nicholson found from 17 observations a temperature of 240° K on the dark side of Venus and 235° K on the bright side from 14 observations. Actually, because of its greater area, the dark side of Venus at the phase shown in Figure 6, *a*, sends us three times as much planetary radiation as the bright side, and at inferior conjunction its radiometric magnitude is −4.8 (essentially the dark hemisphere), which is 0.8 mag. brighter than the radiometric magnitude at superior conjunction.

The drift-curves over the dark side of the planet are nearly uniform; again the planetary heat radiated at superior conjunction is nearly the same as that radiated at inferior conjunction in terms of m_r'. These observations indicate a nearly uniform temperature over the whole planet. Actually, the measures show the radiation from the dark hemisphere to be 8 per cent higher after inferior conjunction than before, indicating a difference of 5° C in temperature. Considerable observing would be required to establish this important conclusion, and this has never been done.

19. MARS

Mars is a planet with a rare atmosphere without any considerable water-vapor interference; the only planet presenting a solid radiating surface large enough for detailed radiation studies. According to equation (34), the temperature of the subsolar point at mean distance can be as high as 305° K or 32° C or 90° F. We can therefore use the cover glass to measure planetary heat.

The dryness of the Martian atmosphere is amply demonstrated by drift-curves of planetary heat across the image. At favorable opposition, the image is nearly 1.6 mm in diameter at the Newtonian focus of the 100-

inch; hence a receiver 0.2 mm in diameter gives sufficient resolution to test the position of the point of maximum radiation.

Figure 7 shows drift-curves across the image made by slowly moving the thermocouple with a motor-driven screw, (*A*) near opposition and (*B*) when the planet showed sensible phase. The drift-curves (*1*) are free, without screen, and (*2*) with cover glass. Below is plotted the difference, or planetary heat, above a diagram of the planet with local Martian time indicated. It will be seen that, in both phases, maximum radiation and therefore temperature come at noon, and therefore there is little water vapor or other blanketing effect. Coblentz (1925*b*) at the Lowell Observatory, using the same thermocouple resolution (one-eighth of the planetary image diameter), but spotting deflections, reported a postmeridian maxi-

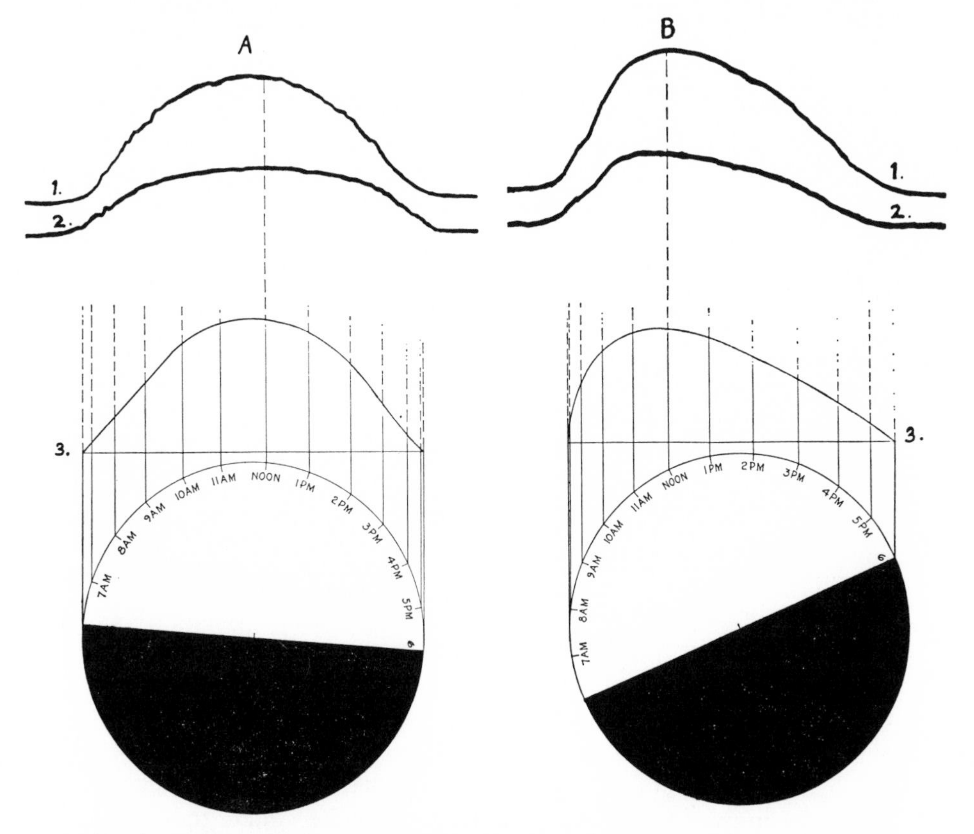

FIG. 7.—Drift-curves across Mars: (*A*) near opposition and (*B*) showing phase; (*1*) free and (*2*) with cover glass. Below is the planetary heat. Note that maximum occurs at local noon in both figures. Thermocouple $\frac{1}{8}$ image diameter, same ratio as that used at Flagstaff. Mount Wilson observations, 1924.

mum. We do not know why this result was obtained, for we made many drift-curves, all like those of Figure 7.

The temperatures at the subsolar point obtained from equation (11) and Table 3*B*, resulting from twelve sets of measures near perihelion between July 6 and November 21, 1924, all reduced to perihelion, varied from 269° to 309° K with a mean of 293° K. Five measures made while the planet was at perihelion on August 28–29 gave a mean of 297° K.

The temperature of the subsolar point was measured by m_r' (Table 3*B*), at every opposition from perihelion 1924 to aphelion 1933. A smooth curve drawn through the measures gave for perihelion 300° K and for aphelion 273° K (Adams, 1933).

Since the temperature varies with the reciprocal square root of the radius vector, its theoretical range is 10 per cent or 27° between perihelion

TABLE 5

TEMPERATURES ON MARS FROM MEASURES AT OPPOSITIONS 1924–1932

	Subsolar Point	Limb	South Polar Cap
Perihelion.	300° K	279	221
Aphelion.	273	254	201
Mean distance.	286	266	211

and aphelion, which agrees with the observed measures cited above. If we normalize the temperatures of 1924 to agree with the trend of the measures of succeeding oppositions, we obtain Table 5.

20. REMARKS ON MARTIAN TEMPERATURES

The subsolar temperature at mean distance is found to be 19° below the theoretical value. By comparison, the terrestrial desert temperatures are at least 50° below theoretical values.

The temperatures in Table 5 are compatible with the idea of *some* planetary moisture to produce the green-colored areas and canals at perihelion. The white areas seen near the limb, not identifiable as clouds, might also be frost. However, the green areas are seen at aphelion, though of a lighter, frosty-green, color. The writer has seen part of the canal system with the 60-inch telescope near aphelion when the subsolar temperature according to Table 5 should have been 275° K.

It is difficult to reconcile these low temperatures and the noontime maximum (Fig. 7) with the presence of sensible surface water. The nature of the lighter markings on the planet, including the polar caps, must still be in the category of some conjecture, but Kuiper (1948) detected infrared bands of ice in the spectra of the polar caps, which seems to fix their identity with water. The march of the green areas and canal system with Martian season must then be reasonably associated with the idea of a definite, but minute, water supply. How it can function at such low temperatures is not understandable.

21. JUPITER AND SATURN

The planets Jupiter and Saturn have a very small amount of planetary heat mixed with the reflected light; but the cover glass is sufficient to make the separation. Some measures showed the cover-glass deflections corrected for reflection to be nearly as large as the whole energy. This means that the temperatures are near the limit of determination.

If we take the albedos of Jupiter and Saturn at 0.44 and 0.42, respectively, the theoretical temperatures according to equation (34) of the subsolar points come out 149° and 111° K, respectively. The dense atmospheres of these planets make it difficult to surmise what departure the measured temperatures will have from those given by equation (34).

Only 8 per cent of the radiation from Jupiter, Saturn, and the bright side of Venus is planetary heat (Pettit and Nicholson, 1924*b*). For Jupiter the magnitude of 1 square second of integrated radiation at the subsolar point measured on 15 occasions averaged 4.94, with a range from 4.83 to 5.14. Since the planetary heat is only 8 per cent of the whole radiation, we must add 2.74 mag. to this, which gives 7.68 mag. for m_r'. The temperature interpolated from Table 3*B* is near 150° K, nearly the theoretical value.

In the case of Saturn the magnitude of 1 square second of integrated radiation from the subsolar point is 1.58 mag. fainter than from Jupiter, which makes m_r' 9.46. Interpolation of Table 3*B* gives a temperature of 125° K. It seems, then, that these two planets sustain temperatures nearly like those given by equation (34).

That the measured temperatures of Jupiter and Saturn are slightly above the theoretical may be due to imperfect measures of m_r' or faulty estimates of albedo.

REFERENCES

Adams, W. S. 1933 *Pub. A.S.P.*, **45**, 273 (extract from *Ann. Rept. Mt. W. Obs.*, 1932–1933).

Adel, A., and Lampland, C. O. 1940 *Ap. J.*, **91**, 481.

COBLENTZ, W. W. 1915 *Bull. Bur. Stand.*, **11**, 185.
1925*a* *Sci. Papers Bur. Stand.*, No. 512.
1925*b* *A.N.*, **224**, 361.

COBLENTZ, W. W., and LAMPLAND, C. O. 1923 *Lowell Obs. Bull.*, **3**, No. 10 (Bull. No. 85), 91.
1924 *Pop. Astr.*, **32**, 546 and 570.

EPSTEIN, P. S. 1929 *Phys. Rev.*, **33**, 269.

GOLDBERG, L., and PIERCE, A. K. 1959 *Handbuch der Physik*, ed. S. FLUGGE (Berlin: Springer-Verlag), Vol. **52**, chap. 1.

JOHNSON, F. S. 1954 *J. Meteorology*, **11**, 431.

KUIPER, G. P. 1948 *A.J.*, **54**, 72.

LANGLEY, S. P., and VERY, F. W. 1889 *Temperature of the Moon* ("Memoirs of the National Academy of Sciences, Washington," Vol. **4**, Part 2), pp. 107–212. There are 10 papers between 1885 and 1889 inclusive which deal with lunar radiation and temperature, published as "Memoirs of the National Academy of Sciences" or in the *Amer. J. Sci.*

LEBEDEV, P. N. 1895 *Ann. d. Phys.*, **56**, 12.

MÜLLER, G. 1897 *Photometrie der Gestirne* (Leipzig: W. Engelmann), p. 353.

NICHOLS, E. F. 1901 "On the Heat Radiation of Arcturus, Vega, Jupiter, and Saturn," *Ap. J.*, **13**, 101.

PETTIT, E., and NICHOLSON, S. B. 1924*a* *Pop. Astr.*, **32**, 614.
1924*b* *Ibid.*, p. 601.
1925 *A.N.*, **225**, 331, contains accounts of Mars observations (see also *Mt. W. Contr.*, Nos. 246–705, for a number of papers).
1928 *Ap. J.*, **68**, 279; *Mt. W. Contr.*, No. 369.
1955 *Pub. A.S.P.*, **67**, 293.

PFUND, A. H. 1916 *Allegheny Obs. Pub.*, **3**, 43.

QUIRK, A. L. 1955 *Final Report*—Contract AF 19 (122)-249 of Geophys. Res. Directorate, University of Rhode Island, Dept. of Physics.

ROSSE, EARL OF 1869 *Proc. R. Soc. London*, **17**, 436.

SOSMAN, R. B. 1927 *The Properties of Silica* (New York: New York Chem. Cat. Co.), p. 738.

WOOD, R. W. 1911 *Physical Optics* (New York: Macmillan Co.), p. 602.

Note added in proof.—The value for the solar constant given by Johnson (1954) is (2.00 ± 0.04) cal cm^{-2}min^{-1}. Rocket observations in 1950 (Quirk, 1955) suggested a value between 1.99 and 2.02 cal cm^{-2}min^{-1}. Later determinations from balloon flights indicated 1.96 ± 0.03 cal cm^{-2}min^{-1} (Goldberg and Pierce, 1959). Further work on the determination of the solar constant is desirable.

CHAPTER 11

Recent Radiometric Studies of the Planets and the Moon

By WILLIAM M. SINTON
Lowell Observatory

1. INTRODUCTION

In THE last two decades the procedures of making infrared measurements have been greatly improved. As a result of their employment by industry, the techniques have also been made dependable, and the equipment has been made commercially available.

The procedure of chopping the radiation to be measured provides a "tag" in the electrical output of the thermocouple or other detector and identifies the desired signal. A transformer or coupling condenser separates the a.c. component from the slowly changing d.c. signal that is produced by variation in the ambient temperature of the detector.

Filters that transmit only the long-wave infrared are now available. The use of these avoids the necessity of subtracting large deflections to obtain the net infrared emission, as is the case when water-cell or cover-glass filters are used. Dr. Strong and the writer (Sinton and Strong, 1960*a*, *b*) at various times have employed a double monochromator, the residual rays from quartz (giving a narrow band at 8.8 μ), and an Eastman Kodak silver sulfide filter (transmitting beyond 5 μ) in infrared work on the planets with the 100-inch and 200-inch telescopes.

Black bodies that consisted of cavities within temperature-regulated aluminum blocks were used to calibrate the planetary deflections for the infrared wave lengths at which the planets were measured. The calibration with visible light from stars or from planets, as used by earlier investigators, may introduce errors due to imperfect blackening of the thermocouple or from inaccurate knowledge of the atmospheric transmission in the visible.

It was found possible to obtain spectra of Mars and Venus in the 8–13 μ atmospheric window in addition to the temperature studies. Both prism and grating spectrometers were employed in this work, and a resolution of 0.08 μ was obtained with a grating having approximately 100 lines/mm. The resolution with the prism spectrometer was about 0.36 μ.

During this work we investigated the transmission of the atmosphere, using the spectrometers and observing the moon at different zenith distances. It was found that a plot of the deflection against the square root of the air mass is demonstrably better than the customary log deflection versus air-mass diagram (Strong, 1941; Sinton and Strong, 1960*a*). Accordingly, the square-root dependence is used in deducing the temperatures presented here.

2. VENUS

Venus was observed with the 100-inch telescope at Mount Wilson in 1952, with the 200-inch at Palomar in 1953 and 1954, and by the writer with the 42-inch at Lowell Observatory in 1959. In the preliminary work with the 100-inch the radiation of the whole disk of the planet was measured, and a temperature of −48° C was obtained. In subsequent work, scans across the disk were made with small apertures. At Palomar, scans were made through the center of the disk parallel to the terminator (referred to as "meridional scans"), and others were made perpendicular to the terminator ("equatorial scans"). The writer made east-west scans at the time of inferior conjunction in 1959. Figure 1 presents means of the scans obtained at the different times and shows in insets the size and shape of the aperture and the appearance of the disk. Temperatures obtained in the centers of these curves are all within a few degrees of −38° C, the temperature found by Pettit and Nicholson (1955).

A striking feature of the Venus heat radiation, which was noted by Pettit and Nicholson, is the near-equality of the temperatures of the bright and dark hemispheres. Indeed, 1959 measurements of the effective temperature of the whole disk manifest, at most, a few degrees' change between superior and inferior conjunctions, and they agree, on the average, with the 1952 temperatures.

One may use the equatorial drift-curves to study the limb darkening. The similarity of the three curves in Figure 1 at various aspects of the planet and the already mentioned constancy of temperature with the variation of phase show that the temperature change is certainly small from the day to the night hemisphere and that the change of temperature across the disk must be small. The east and west halves of the curves are averaged, in order to reduce any effect of a variation in temperature across the disk. In Figure 2 most of the smoothing, produced by the finite size of the aperture and seeing, has been removed, and the average data are plotted separately for 1953 and 1954. Limb darkening is quite evident, and it is nearly identical for the two years.

King (1956) studied theoretically the radiation properties of planetary atmospheres, and in Figure 2 several of his computed curves are compared with the observations. The lower two curves were computed for semi-

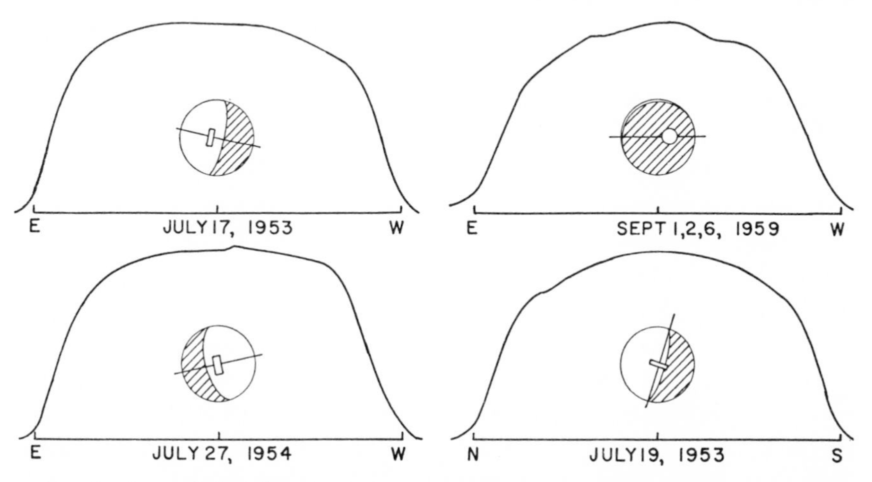

FIG. 1.—Scans of the infrared emission across the disk of Venus. The size of the scanning aperture and the direction of its traverse relative to the appearance of Venus are shown in the insets. The upper right curve is the mean of 13 scans made with the 42-inch telescope in a 1.5 μ band at 11.9 μ. The others are each means of 4 scans made with the 200-inch telescope in the 8–13 μ band.

infinite atmospheres in radiative equilibrium. The lowest curve is for an atmosphere with absorption by lines and a grayness parameter of $\beta =$ 0.1, where β represents the ratio of line width to line spacing. The middle curve is for a gray atmosphere (Chandrasekhar, 1950). The upper curve is derived for a gray atmosphere of unit optical depth, $\tau = 1$. It is impossible to judge the exact edge of the planet from the observed curves, and therefore the abscissa scale is somewhat arbitrary. However, with any reasonable adjustment, the observations suggest that the part of the atmosphere in radiative equilibrium is not optically thick but has a finite thickness greater than 1. Unfortunately, it is not possible to say whether the absorption is gray or produced by lines. The scans obtained at Lowell have greater smoothing because of the larger size of the aperture, but they definitely exhibit limb darkening similar to the 1953 and 1954 observations.

It can be noticed in Figure 1 that the polar regions in the meridional scans of July 19, 1953, are colder than similar regions near the limbs of the equatorial curves. The difference in temperature is about 8°–10° C. Further, we notice that there is a cold notch in the curve near to the north limb. On photographs taken in ultraviolet light on the same day, a bright cloud is found in this area. The north pole, determined by Kuiper (1954), pointed 28° away from the earth on this date, and therefore this cloud was at fairly low latitudes. In measurements at Lowell at inferior conjunction, it was again found that the poles were colder than the east and west limbs, but this time the south pole was colder than the north. No ultraviolet photographs were taken of the very thin crescent.

By assuming that the observed limb darkening applies along any diameter, the effective temperature for the whole disk may be calculated. It is found that a disk with the observed limb darkening will have 0.80 of the emission of a uniformly emitting disk with the intensity of the center. From the central temperatures measured at Palomar an effective whole-disk temperature of −47° C (226° K) is derived, which is in good agreement with the measurements at Mount Wilson and Lowell.

In 1953 spectra in the 8–13 μ window were obtained with a prism spectrometer. Absorptions by the terrestrial atmosphere have been removed from these spectra, and the resulting spectrum is shown in Figure 3. Prior to obtaining these observations, it was thought that the excited-state bands of CO_2 at 9.4, 10.4, and 12.6 μ would produce very prominent absorptions in the emission of Venus. These bands are not found in the 1953 spectra, and perhaps only slight indications of them are found in grating spectra made in 1954. Their absence signifies either that the radiating atmosphere is gray or that the wave lengths between these CO_2 bands are absorbed by unknown molecules. An indication of such occurrence is shown by the band at 11.2 μ. This band was found to be very weak and diffuse in the grating spectra. One possible molecule for this band is carbon suboxide, C_3O_2. This molecule has already been suggested by Sinton (1953) and Kuiper (1959) as a constituent of the Venus atmosphere. Spectra of this gas obtained by Lord and Wright (1937) showed bands at 7.2, 8.2, 8.9, 9.8, 11.0, 11.2, and 12.8 μ. The strongest bands were the ones at 11.0 and at 11.2 μ. Carbon suboxide is an unstable molecule and readily polymerizes to a reddish or whitish mass. Recent spectra by Rix (1954) do not show the 11.0 and the 11.2 μ bands, and it now appears that the 11 μ bands found by Lord and Wright were due to an impurity. It may well be that the strongest bands were due to the polymerized molecule.

The spectra seem to agree with the drift-curves, in that the observed emission arises within a region in radiative equilibrium. From the rotational spectrum of CO_2 near λ 8000 A, and therefore for a region near the top of the cloud layer, Chamberlain and Kuiper (1956) have found a temperature of 12° C, which is significantly higher than the temperature of the emitting layer. This presumably means that the emitting layer is well above the cloud layer.

3. MARS

In 1954, scans were made approximately parallel or perpendicular to the equator across the disk of Mars. Most of these were made with an aperture of 1.5 seconds of arc in diameter. Several scans are reproduced in Figure 4, along with charts giving their location on the disk. Numbers on one of the charts and the associated scans indicate positions of the scanning aperture determined from photographs taken during the scans.

On one day a yellow cloud on Mars was well shown in Lick Observatory photographs by H. M. Jeffers. Scans *2* and *11* crossed the cloud, and in the region of the cloud they differ quite appreciably from scans in adjacent unclouded regions. They give a temperature of −25° C for the cloud.

Coblentz and Lampland (1925) showed that the albedo of the surface affects the temperature reached. This is seen in Figure 4, where scan *10* crosses Sinus Meridiani. In scan *13*, Sinus Margaritifer is 8° warmer than the adjacent bright area. This difference is about what one would expect from the albedos of the two regions.

Table 1 gives temperatures of the disk at positions corresponding to various local times on Mars. In all the east-west scans, except *2* on July 20, which was affected by the yellow cloud, the maximum temperature occurred between $\frac{1}{2}$ and $\frac{3}{4}$ hour after noon. We may compare these temperatures with the theoretical diurnal variation, assuming that the ground at each location on Mars has the same characteristics and follows the same variation. Scans *8* on July 20 and *9* on July 21 are near the equator and remain almost entirely on bright areas. The different regions covered by these scans should therefore follow the same variations.

The theory of heat conduction of a planet with no atmosphere that receives energy from the sun and re-radiates to space has been investigated by Wesselink (1948) and by Jaeger (1953) and was applied by them to the moon. Using their methods, the writer has applied the theory to Mars,

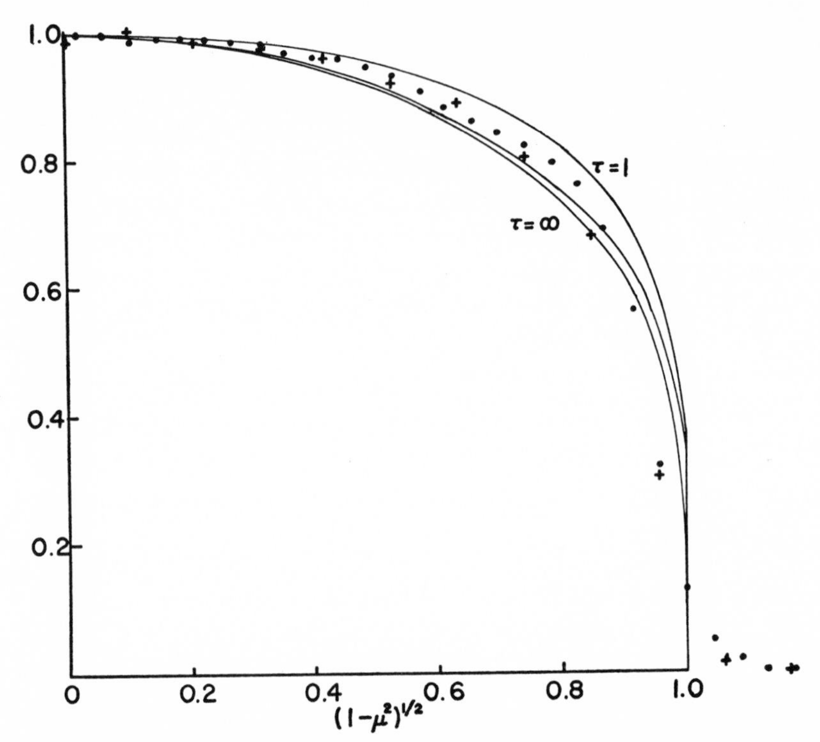

FIG. 2.—Comparison of the Venus limb darkening determined in 1953 (*dots*) and 1954 (*crosses*) with theoretically determined curves. The middle curve applies to a gray semi-infinite atmosphere. The lower curve is for a semi-infinite atmosphere with line absorption and a grayness parameter, $\beta = 0.1$. The upper curve is for a gray atmosphere with optical depth $\tau = 1$.

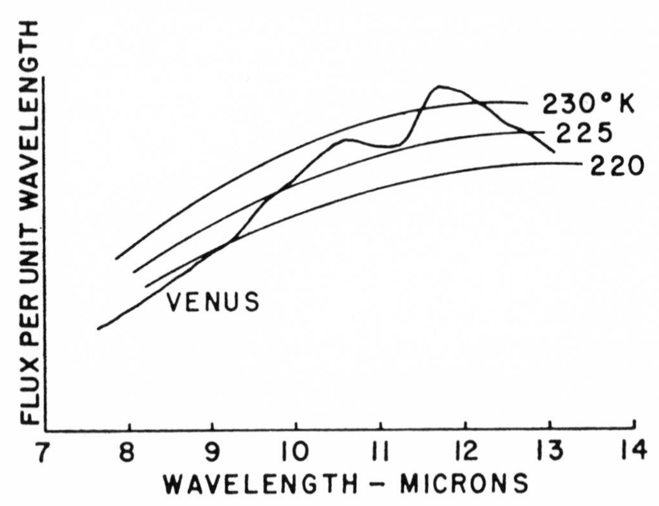

FIG. 3.—Spectrum of Venus reduced to outside the earth's atmosphere. The slit width corresponded to 0.36 μ at 10 μ.

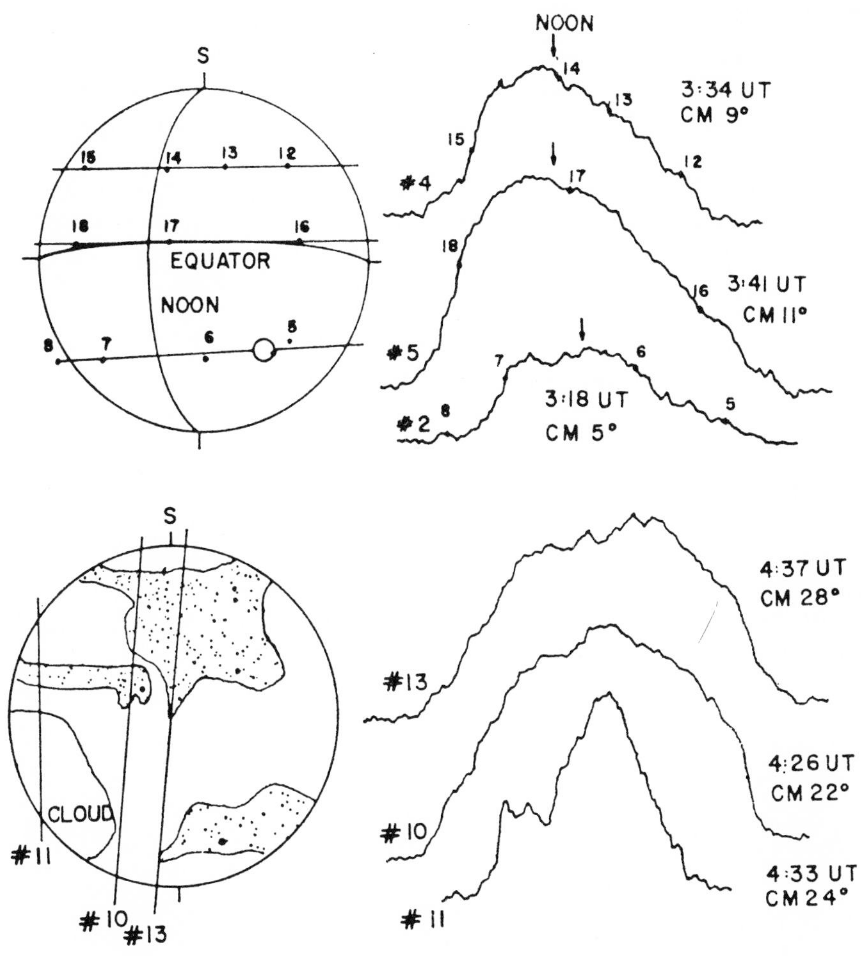

FIG. 4.—Scans across Mars on July 20, 1954. The numbers on the scans correspond to those on the chart and indicate times at which photographs were made.

TABLE 1

TEMPERATURE DEPENDENCE ON LOCAL TIME ON MARS

Date July	Scan No.	Lat.	0700	0800	0900	1000	1100	1200	1300	1400
20.	5	− 2°	−64° C	−32° C	−16° C	−1° C	16° C	22° C	22° C	7° C
20.	8	+10	−64	−41	−19	−2	13	26	28	15
21.	9	+ 8	−54	−35	−11	+2	18	26	23	20
23.	2	+14	−78	−46	−21	0	10	16	14	8
23.	3	−12	−42	−32	− 6	+8	18	18	18	8
23.	4	− 8	−55	−39	−10	+9	19	20	20	14

with the recognition that differences between the observed and theoretical diurnal variations are most probably caused by the atmosphere.

The theory assumes the planet's surface to be a plane, semi-infinite, homogeneous solid. The well-known heat-conduction equation applies,

$$\frac{\partial T}{\partial t} = \frac{k}{\rho c}\frac{\partial^2 T}{\partial x^2}, \tag{1}$$

where T is the absolute temperature; t is time; k, ρ, and c are the thermal conductivity, density, and specific heat of the surface material; and x is a linear co-ordinate measured downward from the surface. In the following discussion, c.g.s. units will be employed for these quantities.

The heat flux at any level is given by

$$F = \frac{k\partial T}{\partial x}. \tag{2}$$

At the surface, this flux must obey the condition

$$\sigma T^4 = I + F, \tag{3}$$

where σT^4 is the heat emitted, I is the heat received from the sun, and F is the heat conducted from below the surface. It is assumed that during the daytime the heat received from the sun is given by

$$I = \frac{1 - A}{r^2} G \cos\frac{2\pi t}{P}, \tag{4}$$

where A is the albedo, r is the distance from the sun in astronomical units, G is the solar constant at the earth, and P is the planet's rotation period. At night, $I = 0$. For known values of A, r, G, and P the theory yields a series of curves that depend on the unknown properties of the surface k, ρ, and c through their product, which is customarily expressed as $(k\rho c)^{1/2}$ and called the *thermal inertia*.

In Figure 5 the data from the most suitable scans in Table 1 are compared with curves for three values of $(k\rho c)^{1/2}$. The data appear to agree in phase with the curve for $(k\rho c)^{1/2} = 0.004$, while the amplitude is more nearly that of the curve with 0.01 for the inertia. Undoubtedly, the amplitude is reduced by the moderating effect of the atmosphere, and it is believed that a thermal inertia of 0.004 is nearly correct.

The maximum temperature at a favorable opposition for a desert area near the equator appears to be close to 25° C, and for a dark area it is about 8° C hotter. It is emphasized that these temperatures are surface temperatures and are certainly higher than the air temperature near the ground. The minimum surface temperature is uncertain, but for a bright area on the equator at perihelion it is probably near −70° C.

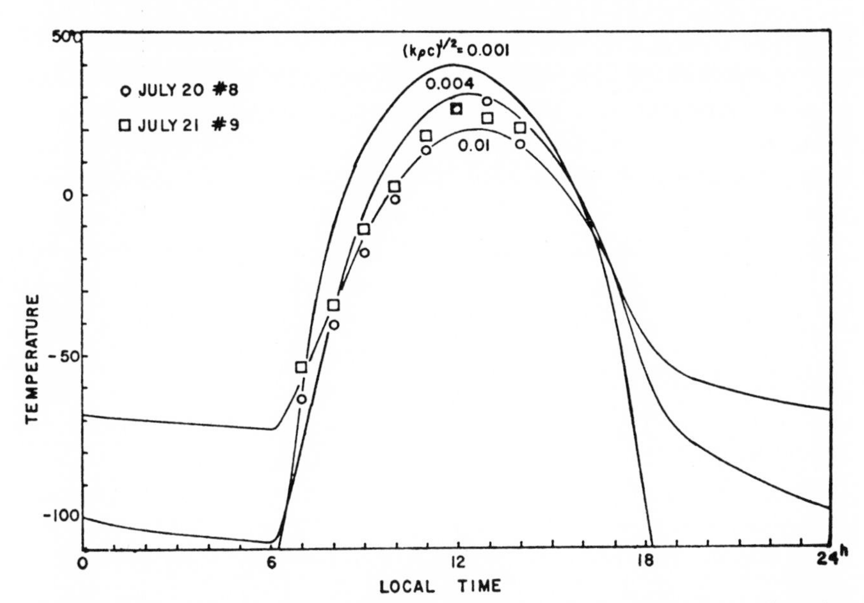

FIG. 5.—Theoretical and observed diurnal temperature variation of Martian bright regions on the equator.

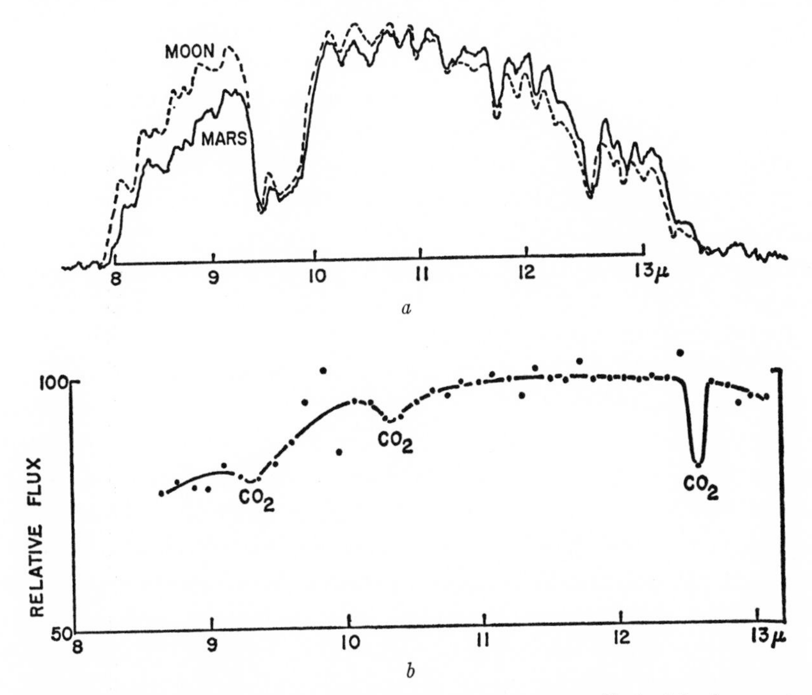

FIG. 6.—(a) The observed spectrum of Mars and of the moon; (b) the Martian spectrum corrected to outside the earth's atmosphere. The ordinate scale is the relative emission within a fixed wave-length interval.

Terrestrial rocks have inertias near 0.05, while sandy or dusty soils range between 0.01 and 0.02. The presence of moisture appears to raise the inertia. A dark volcanic ash at the Grand Falls of the Little Colorado River, near Flagstaff, Arizona, has $(k\rho c)^{1/2} = 0.01$ and a measured surface temperature change from 3° to 72° C through a day near the summer solstice. The observed lag of the maximum for this ash was a little over an hour, and this is the lag expected of a surface with an inertia of 0.01. On the other hand, the air 1 meter above the ground did not attain its maximum temperature of 30.5° C until 3 hours after the local noon. From these observations we may conclude that the lag is not sensibly affected and that the amplitude is greatly moderated by the presence of an atmosphere. For the generally dry, dusty surface of Mars an inertia of 0.004, a lag of about $\frac{1}{2}$ hour, and a total variation near 100° C appear quite reasonable.

Grating spectra of Mars were obtained in 1954, and one is reproduced in Figure 6, *a*. Most of the absorptions appearing here are of terrestrial origin and appear also in the spectrum of the moon (*dashed curve*). A few, however, are strengthened by the Mars atmosphere. These may be seen in Figure 6, *b*, where the terrestrial absorptions have been removed through the use of transmission measurements made with the moon as a source. Figure 6, *b*, exhibits the carbon dioxide bands at 9.4, 10.4, and 12.6 μ. The irregularity near 9.6 μ is caused by the very strong ozone band, whose removal was imperfect. In the moon spectrum the 10.4 μ band appears as a doublet because of the two rotational branches. In Figure 6, *b*, it appears only as a singlet and may signify that the Martian CO_2 is very cold and that only lines of low quantum number near the band origin are excited.

4. THE MOON

A total eclipse of the moon is very exciting to observe with a pyrometer, for the moon's temperature drops 150° C within an hour, as was found by Pettit and Nicholson (1930) and Pettit (1940). Epstein (1929) and later Wesselink (1948) showed how from this datum the thermal inertia may be derived. From the eclipse observations they have derived an inertia of 0.001. The thermal inertia can be derived also from the temperature variation throughout a lunation just as the inertia of Mars was derived. The variation through a lunation has not been adequately observed, but it is in fair agreement with the thermal inertia derived from the eclipse observations. From the theory and the inertia determined from the eclipse observations the temperature of lunar "midnight" should be −175° C. At the Lowell Observatory the author has obtained

-151° C $\pm$ 3° (p.e.) for the point opposite the sun, which is in good agreement with -148° C $\pm$ 5° obtained by Pettit and Nicholson (1930).

The theories developed by Wesselink and Jaeger assume that the thermal conductivity and the specific heat of the material are constant. Actually, there is some evidence that powdered materials have a conductivity which is proportional to absolute temperature, and from data in the *International Critical Tables* it appears that the specific heats of materials are also proportional to the temperature. Muncey (1958) has revised the theory with the assumption that these two parameters are indeed proportional to the temperature. With this modification, equation (1) becomes

$$\frac{\partial (T^2)}{\partial t} = \frac{k}{\rho c} \frac{\partial^2 (T^2)}{\partial x^2}, \tag{5}$$

which is the same as equation (1) with T^2 as the variable. Using this analysis, Muncey finds a higher inertia; at a temperature of 300° K this parameter is near 0.003.

Dr. Strong and the writer observed the July 25, 1953, eclipse at Palomar with a heliostat. We made drift-curves across the moon during the initial partial phases. The calibration was poor, but we determined the temperature scale by assuming that the temperature of the center of the moon before the eclipse was 100° C. Points near the limbs were observed, as well as the center of the disk. We considered that we might find the limb regions to have the larger thermal inertia because steep rocks might be devoid of dust and such rocks would be largely observed at the limbs. As the data in Figure 7 show, an opposite effect was present. Regions *I* and *V* were about 2′ from the west and east limbs, *III* was at the center, and *II* and *IV* were midway between the center and *I* and *V*. Figure 7 shows that at entrance into the umbra the temperature of the limb regions was only -80° C, while the central areas were -60° C. Though the limb areas were colder to begin with, they had lost a much larger fraction of their total heat than had the central regions; they appear to have a lower effective thermal inertia. Pettit and Nicholson made measurements in 1927 of a point only 48 seconds of arc from the south limb. At 17 minutes after entering the umbra, this point was -99° C, much colder than a point near to the center of the moon at a similar time in their 1939 measurements. The explanation of this phenomenon probably lies in the roughness of the lunar surface. At the center of the moon, valleys, slopes, and peaks of the lunar surface contribute to the measured emission. However, the valleys cannot radiate to a hemisphere of space, as assumed in the theory or as the material on the peaks does. The valleys receive some radiation back from their neighboring slopes and consequently cannot cool as fast as predicted by a theory based on a semi-infinite plane surface. Near the limbs of the moon the valleys cannot be observed, only peaks and slopes contribute to the emission, and the observed cooling is more rapid. Consequently, the present theory should be applied to observations of the limbs, or the theory must be modified to include the effect of roughness.

The correction for the reduced cooling of the center of the moon will produce a thermal inertia lower than the present value. The correction will

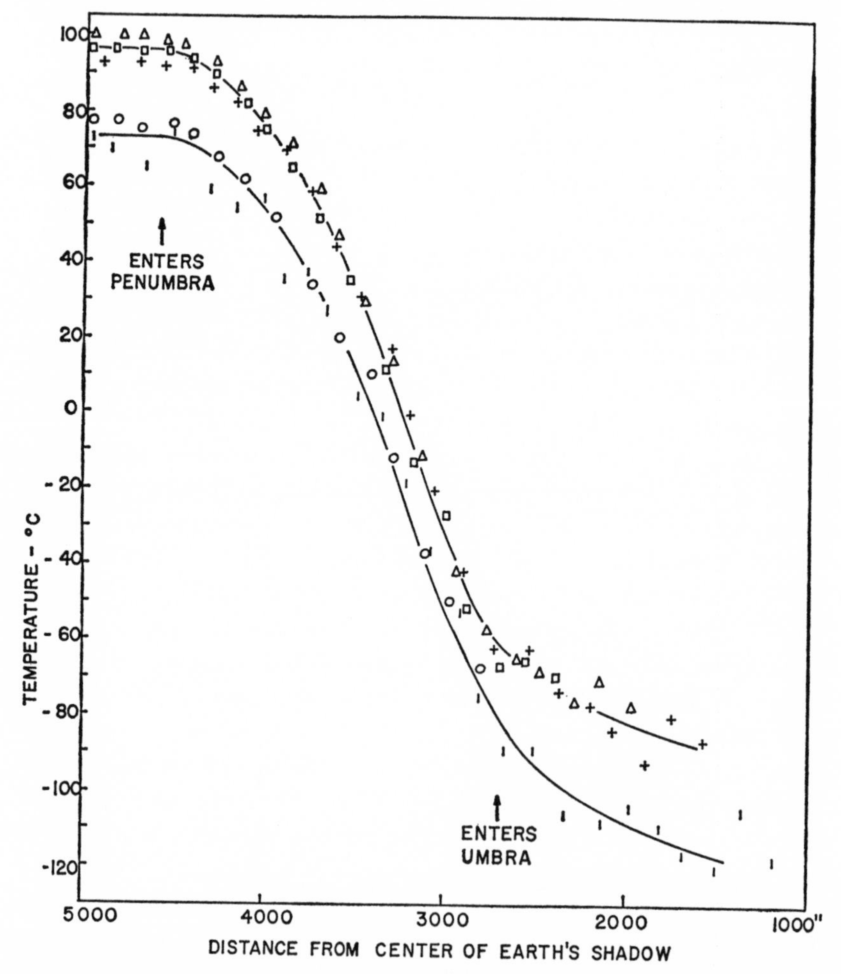

FIG. 7.—The temperature-curve of five areas of the lunar surface during the first part of the total eclipse of July 26, 1953. Vertical bars refer to area *I*, crosses refer to area *II*, triangles refer to area *III*, squares refer to area *IV*, and the circles refer to area *V*. The relative velocity of the moon's center through the earth's shadow was 33″.81 per minute.

no doubt offset much of Muncey's higher value that results from the variation of k and c with temperature. The discrepancy between the observed and theoretical midnight temperature will also have to be resolved, but perhaps this will be accounted for by the above corrections.

REFERENCES

CHAMBERLAIN, J. W., and KUIPER, G. P.	1956	*Ap. J.*, **124,** 399.
CHANDRASEKHAR, S.	1950	*Radiative Transfer* (Oxford: Clarendon Press), p. 125.
COBLENTZ, W. W., and LAMPLAND, C. O.	1925	*J. Franklin Inst.*, **200,** 103.
EPSTEIN, P. S.	1929	*Phys. Rev.*, **33,** 269.
JAEGER, J. C.	1953	*Cambridge Phil. Soc.*, **49,** 355.
KING, J. I. F.	1956	*Ap. J.*, **124,** 272.
KUIPER, G. P.	1954	*Ap. J.*, **120,** 603.
	1957	*The Threshold of Space*, ed. M. ZELIKOFF (New York: Pergamon Press), p. 85.
LORD, R. C., and WRIGHT, N.	1937	*J. Chem. Phys.*, **5,** 642.
MUNCEY, R. W.	1958	*Nature*, **181,** 1459.
PETTIT, E.	1940	*Ap. J.*, **91,** 408.
PETTIT, E., and NICHOLSON, S. B.	1930	*Ap. J.*, **71,** 102.
	1955	*Pub. A.S.P.*, **67,** 293.
RIX, H. D.	1954	*J. Chem. Phys.*, **22,** 429.
SINTON, W. M.	1953	Dissertation, Johns Hopkins University.
SINTON, W. M., and STRONG, J.	1960*a*	*Ap. J.*, **131,** 470.
	1960*b*	*Ibid.*, p. 459.
STRONG, J.	1941	*J. Franklin Inst.*, **232,** 1.
WESSELINK, A. J.	1948	*B.A.N.*, **10,** 356.

CHAPTER 12

Radio Emission of the Moon and Planets

By CORNELL H. MAYER
Radio Astronomy Branch, U.S. Naval Research Laboratory, Washington, D.C.

1. INTRODUCTION

It has recently become practical to use the radio emission of the moon and planets as a new source of information about these bodies and their atmospheres. The results of present observations of the thermal radio emission of the moon are consistent with the very low thermal conductivity of the surface layer which was derived from the variation in the infrared emission during eclipses (e.g., Garstang, 1958). When sufficiently accurate and complete measurements are available, it will be possible to set limits on the thermal and electrical characteristics of the surface and subsurface materials of the moon.

Observations of the radio emission of a planet which has an extensive atmosphere will probe the atmosphere to a greater extent than those using shorter wave lengths and should in some cases give otherwise unobtainable information about the characteristics of the solid surface. Radio observations of Venus and Jupiter have already supplied unexpected experimental data on the physical conditions of these planets. The observed intensity of the radio emission of Venus is much higher than the expected thermal intensity, although the spectrum indicated by measurements at wave lengths near 3 cm and 10 cm is like that of a black body at about 600° K. This result suggests a very high temperature at the solid surface of the planet, although there is the possibility that the observed radiation may be a combination of both thermal and non-thermal components and that the observed spectrum is that of a black body merely by coincidence. For the case of Jupiter, the radio emission spectrum is definitely not like the spectrum of a black-body radiator, and it seems very likely that the radiation reaching the earth is a combination of thermal radiation from the atmosphere and non-thermal components.

Of the remaining planets, only Mars and Saturn have been observed as radio sources, and not very much information is available. Mars has been observed twice at about 3-cm wave length, and the intensity of the observed radiation is in reasonable agreement with the thermal radiation which might be predicted on the basis of the known temperature of Mars. The low intensity of the radiation from Saturn has limited observations, but again the measured radiation seems to be consistent with a thermal origin. No attempts to measure the radio emission of the remaining planets have been reported, and, because of their distances, small diameters, or low temperatures, the thermal radiation at radio wave lengths reaching the earth from these sources is expected to be of very low intensity. In spite of this, the very large radio reflectors and improved amplifying techniques which are now becoming available should make it possible to observe the radio emission of most of the planets in a few years.

The study of the radio emission of the moon and planets began with the detection of the thermal radiation of the moon at 1.25-cm wave length by Dicke and Beringer (1946). This was followed by a comprehensive series of observations of the 1.25-cm emission of the moon over three lunar cycles by Piddington and Minnett (1949). They deduced from their measurements that the radio emission from the whole disk of the moon varied during a lunation in a roughly sinusoidal fashion; that the amplitude of the variation was considerably less than the amplitude of the variation in the infrared emission as measured by Pettit and Nicholson (1930) and Pettit (1935); and that the maximum of the radio emission came about $3\frac{1}{2}$ days after Full Moon, which is again in contrast to the infrared emission, which reaches its maximum at Full Moon. Piddington and Minnett explained their observations by pointing out that rocklike materials which are likely to make up the surface of the moon would be partially transparent to radio waves, although opaque to infrared radiation. The infrared emission could then be assumed to originate at the surface of the moon, while the radio emission originates at some depth beneath the surface, where the temperature variation due to solar radiation is reduced in amplitude and shifted in phase. Since the absorption of radio waves in rocklike material varies with wave length, it should be possible to sample the temperature variation at different depths beneath the surface and possibly detect changes in the structure or composition of the lunar surface material.

The radio emission of a planet was first detected in 1955, when Burke and Franklin (1955) identified the origin of interference-like radio noise on their records at about 15 meters wave length as emission from Jupiter. This sporadic type of planetary radiation is discussed by Burke (chap. 13) and Gallet (chap. 14). Steady radiation which was presumably of thermal origin was observed from Venus at 3.15 and 9.4 cm, and from Mars and Jupiter at 3.15 cm in 1956 (Mayer, McCullough, and Sloanaker, 1958*a*, *b*, *c*), and from Saturn at 3.75 cm in 1957 (Drake and Ewen, 1958). In the relatively short time since these early observations, Venus has been observed at additional wave lengths in the range from 0.8 to 10.2 cm, and Jupiter has been observed over the wave-length range from 3.03 to 68 cm.

The observable characteristics of planetary radio radiation are the intensity, the polarization, and the direction of arrival of the waves. The maximum angular diameter of any planetary disk as observed from the earth is about 1 minute of arc. This is much smaller than the highest resolution of even the very large reflectors now under construction, and consequently the radio emission of different regions of the disk cannot be resolved. It should be possible, however, to put useful limits on the diameters of the radio sources by observing with large reflectors or with interferometers. Measurements of polarization are presently limited by apparatus sensitivity and will remain difficult because of the low intensity of the planetary radiation at the earth. There have been few measurements specifically for the determination of the polarization of planetary radiation. The measurements made with the NRL 50-foot reflector, which is altitude-azimuth-mounted, would have shown a systematic change with local hour angle in the measured intensities of Venus and Jupiter if a substantial part of the radiation had been linearly polarized. Recent interferometer measurements (Radhakrishnan and Roberts, 1960) have shown the 960-Mc emission of Jupiter to be partially polarized and to originate in a region of larger diameter than the visible disk. Other than this very significant result, most of the information now available about the radio emission of the planets is restricted to the intensity of the radiation.

The concept of apparent black-body temperature is used to describe the radiation received from the moon and the planets. The received radiation is compared with the radiation from a hypothetical black body which subtends the same solid angle as the visible disk of the planet. The ***apparent black-body disk temperature*** is the temperature which must be assumed for the black body in order that the intensity of its radiation should equal that of the observed radiation. The use of this concept does not specify the origin of the radiation, and only if the planet really radiates as a black body, will the apparent black-body temperature correspond to the physical temperature of the emitting material.

The radio radiation of the sun which is reflected from the moon and planets should be negligible compared with their thermal emission at centimeter wave lengths, except possibly at times of exceptional outbursts of solar radio noise. The quiescent level of centimeter wave-length solar radiation would increase the average disk brightness temperature by less than 1° K. At meter wave lengths an increase of the order of 10° K in the average disk temperatures of the nearer planets would be expected. Therefore, neglecting the extreme outbursts, reflected solar radiation is not expected to cause sizable errors in the measurements of planetary radiation in the centimeter- and decimeter-wave-length range.

2. THE MOON

2.1 Observations

Radio observations of the moon have been made over the range of wave lengths from 4.3 mm to 75 cm, and the results are summarized in Table 1. Observations have also been made at 1.5 mm, using optical techniques (Sinton, 1955, 1956; see also chap. 11). Not all the observers have used the same procedures or made the same assumptions about the lunar brightness distribution when reducing the data, and this, together with differences in the methods of calibrating the antennae and receivers, must account for much of the disagreement in the measured radio brightness temperatures.

In the observations at 4.3 mm (Coates, 1959*a*), the diameter of the antenna beam, 6.′7, was small enough to allow resolution of some of the larger features of the lunar surface, and contour diagrams have been made of the lunar brightness distribution at three lunar phases. These observations indicate that the lunar maria heat up more rapidly and also cool off more rapidly than do the mountainous regions. Mare Imbrium seems to be an exception and remains cooler than the regions which surround it. These contour diagrams also suggest a rather rapid falloff in the radio brightness with latitude.

Very recently, observations have been made at 8-mm wave length with a reflector 22 meters in diameter with a resultant beam width of only about 2′ (Amenitskii, Noskova, and Salomonovich, 1960). The constant-temperature contours are much smoother than those observed at 4.3 mm by Coates (1959*a*), and apparently the emission at 8 mm is not nearly so sensitive to differences in surface features. Such high-resolution observations as these are needed at several wave lengths in order that the radio emission of the moon can be properly interpreted.

The observations of Mayer, McCullough, and Sloanaker at 3.15 cm and of Sloanaker at 10.3 cm have not previously been published and will be briefly described. Measurements at 3.15 cm were obtained on 11 days spread over the interval May 3 to June 19, 1956, using the 50-foot reflector at the U.S. Naval Research Laboratory in Washington. The half-intensity diameter of the antenna beam was about 9′, and the angle subtended by the moon included the entire main beam and part of the first side lobes. The antenna patterns and the power gain at the peak of the beam were both measured (Mayer, McCullough, and Sloanaker, 1958*b*), so that the absolute power sensitivity of the antenna beam over the solid angle of the moon was known. The ratio of the measured antenna temperature change during a drift scan across the moon to the average brightness temperature of the moon over the antenna beam (assuming that the brightness temperature of the sky is negligible) was found, by graphical integration of the antenna directivity diagram, to be 0.85. The measured brightness temperature is a good approximation to the brightness temperature at the *center of the lunar disk* because of the narrow antenna beam and because the temperature distribution over the central portion of the moon's disk is nearly uniform. The result of the observations is (in ° K)

$$T = (195 \pm 25) - (12 \pm 5) \cos(\omega t - 44 \pm 15°) , \qquad (1)$$

where the phase angle, ωt, is measured in degrees from new moon and the probable errors include absolute as well as relative errors. This result is plotted along with the 8.6-mm observations of Gibson (1958) in Figure 1, *a*. The variation in the 3-cm emission of the moon during a lunation is very much less than the variation in the 8.6-mm emission, as would be expected from the explanation of Piddington and Minnett (1949). In the discussion which follows, the time average of the radio emission will be referred to as the *constant component*, and the superimposed periodic variation will be called the *variable component*.

The 10.3-cm observation of Sloanaker was made on May 20, 1958, using the 84-foot reflector at the Maryland Point Observatory of the U.S. Naval Research Laboratory. The age of the moon was about 2 days. The half-intensity diameter of the main lobe of the antenna was about 18′.5, and the brightness temperature was reduced by assuming a Gaussian shape for the antenna beam and a uniformly bright disk for the moon. An estimate for the power gain of the antenna was made by measuring the intensities of discrete radio sources, which have also been measured with other apparatus over a range of wave lengths. The measured antenna temperature of 105.6° K was found to correspond to a radio brightness temperature averaged over the disk of the moon of 207° K ± 27° (p.e.).

The radio emission of the moon (for $\lambda \geq 7.5$ mm) has been observed during a number of *lunar eclipses*, but no change in emission during an eclipse has been found. This result is consistent with the other observations, since the duration of the temperature change during an eclipse is much shorter than the period of a lunation and for this reason the change in radio emission to be expected during an eclipse is less than one-tenth that during a lunation. Such small changes would be difficult to measure.

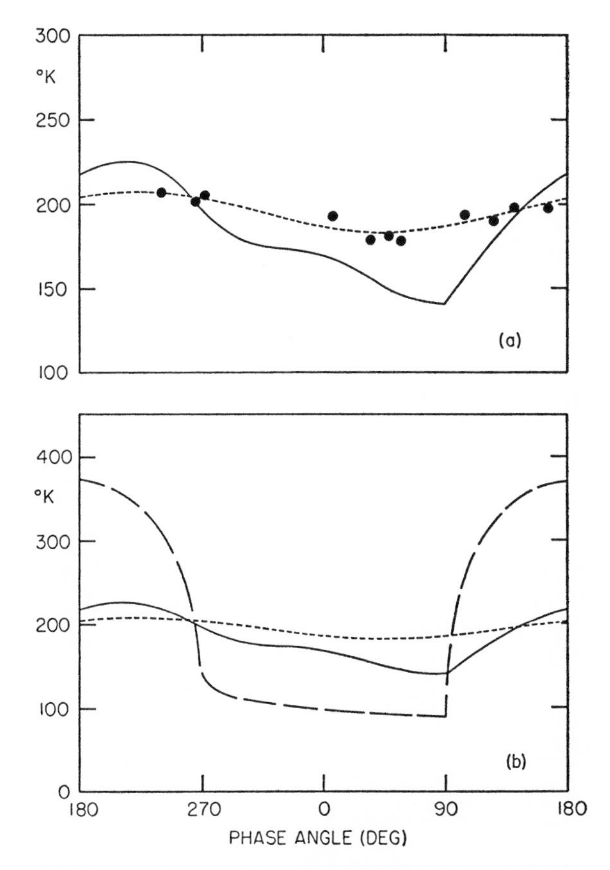

FIG. 1.—(*a*) Brightness temperature of the central portion of the lunar disk during a lunation. *Circles and dashed sine curve:* 3.15-cm measurements (Mayer, McCullough, and Sloanaker, unpublished); *solid curve:* 0.86-cm measurements from Gibson (1958). (*b*) Data of Fig. 1, *a*, compared with the surface temperature according to Wesselink (1948). Phase angle counted from new moon.

2.2. The Constant Component of Radio Emission

An inspection of Table 1 shows that there is a large spread in the values of the constant part of the radio brightness temperatures of the moon measured by different observers, even when allowance is made for the differences between the average temperature over the disk quoted by some observers and the equatorial temperature quoted by others. The values from Table 1 have been plotted in Figure 2, *a*, as a function of wave

TABLE 1

Radio Observations of the Moon

Wave Length (Cm)	Apparent Black-Body Temperature* (° K)	Uncertainty	Observer
0.43...	T_c=170 to 290	25 per cent (p.e.)	Coates (1959*a*, *b*)
0.75...	T_d=125 to 175	....	Mitchell and Whitehurst (1958)
0.80...	T_c=197−32 cos [ωt−40 (±5)°]	10 per cent	Salomonovich (1958)
0.86...	T_c=150 (single observation)	40 per cent	Hagen (1949)
0.86...	T_c=145 to 225	....	Gibson (1958)
1.25...	T_d=270 (single observation)†	....	Dicke and Beringer (1946)
1.25...	T_c=249−52 cos (ωt−45°)‡	5 per cent	Piddington and Minnett (1949)
1.63...	T_c=224−36 cos (ωt−40°)	±10–15 per cent	Zelinskaya, Troitskii, and Fedoseev (1959)
2.20...	T_c=200 (constant to ±10° K)	5 per cent	Grebenkemper (1958)
3.15...	T_c=195−12 (±5) cos [ωt−44 (±15)°]	±25° K (p.e.)	Mayer, McCullough, and Sloanaker (unpublished)
3.20...	T_d=183 (constant to ±9° K)	±13° K	Zelinskaya and Troitskii (1956)
3.20...	T_d=170 (constant to ±12° K)	±20 per cent	Troitskii and Zelinskaya (1955)
3.20...	T_d=133 (constant to ±10° K)	±20° K	Kaidenovsky, Turusbekov, and Khaikin (1956)
10.0....	T_d=130 (single observation)	20 per cent	Kaidenovsky, Turusbekov, and Khaikin (1956)
10.0....	T_c=315 (constant to ±50° K)	±50° K	Akabane (1955, private communication)
10.3....	T_d=207 (single observation)	±27° K	Sloanaker (unpublished)
21.0....	T_c=250 (constant to ±5° K)	12 per cent	Mezger and Strassl (1959)
22.0....	T_d=270 (single observation)	±60° K	Westerhout (1958)
32......	T_d=246 (constant to ±5 per cent)	±40° K	Ko (private communication)
33......	T_d=220 (constant to ±9° K mean dev.)	±33° K	Denisse and LeRoux (private communication)
75......	T_d=160 (single observation)	....	Seeger, Westerhout, and van de Hulst (1956)
75......	T_d=185 (constant to 10 per cent)	....	Seeger, Westerhout, and Conway (1957)

* T_c is the radio brightness temperature at the center of the disk; T_d is the radio brightness temperature averaged over the disk.

† Corrected by Piddington and Minnett (1949).

‡ Corrected for an emissivity of 0.9 by the observers.

length; however, the single observations are omitted, since they give no indication of phase dependence. A correction (shown by arrows in Fig. 2, *a*) has been applied to the results averaged over the disk, in order to raise them to a value more comparable to that at the center.[1]

There is no strong wave-length dependence in the constant component of equatorial brightness temperature, although in the 10–33-cm range the radio emission seems to be higher than at shorter wave lengths. The mean of the measured temperatures at all wave lengths which are plotted in Figure 2 is 211° K after the correction of the average disk temperatures.

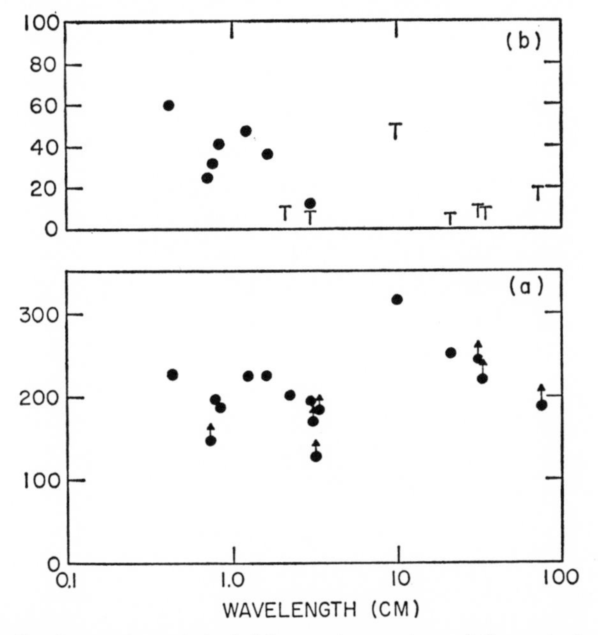

FIG. 2.—Radio observations of the brightness temperature of the central portion of the lunar disk from the data of Table 1. *Abscissae:* wave length in cm (logarithmic scale); *ordinates:* apparent temperature (° K). (*a*) Constant component; (*b*) variable component. Average (disk) to center of disk corrections shown as arrows (*a*).

The radio observations then indicate that the time average of the radio brightness temperature at the center of the moon's disk is about 210° K. This value may be compared with the time average of the surface temperature at the equator of the moon. For this purpose the surface tem-

[1] The formula of Troitskii (1954), which includes the effect of the variation in emissivity over the disk, has been used to estimate the correction. For an assumed latitude temperature distribution of $\cos^{1/4} \varphi$, the ratio of the observed average disk temperature to the observed temperature at the center of the disk is estimated as about 0.9. This correction was applied for illustrative purposes only, and its effectiveness will depend on the details of the original data reduction.

perature on the illuminated face of the moon will be assumed to vary as $\cos^{1/4} \theta$, where θ is the angle between the normal to the surface and the direction of the sun, and the surface temperature on the dark side will be assumed to be constant and equal to the temperature at the midnight point. Using the values of 374° K for the surface temperature at the subsolar point and 120° K for the midnight point (Pettit and Nicholson, 1930), the time average of the surface temperature at the equator is about 230° K. This value and the time average of the equatorial radio brightness temperature are in agreement if the radio emissivity is about 0.9. If Wesselink's (1948) time average of 212° K is used, the radio emissivity would be 1.0.

2.3. The Variable Component of Radio Emission

The reported amplitudes of the variable component of lunar radio emission measured at different wave lengths are plotted in Figure 2, *b*. There is a great deal of scatter in the measurements, and about all that can be said is that the amplitude of the variable component probably decreases from about 40° K at 8-mm wave length to about 12° K at 3-cm wave length. Corrections which take into account the fraction of the moon's disk from which radiation was accepted by the antenna and the wave form of the measured variation in the emission are needed to make the observations comparable but would require a detailed analysis.

The variable component measured by Piddington and Minnett (1949) was very nearly sinusoidal, while that measured by Gibson (1958), which is shown in Figure 1, is rich in harmonics and has somewhat the same shape as the variation in radio brightness temperature at the equatorial point predicted by Jaeger (1953). The main lobe of Gibson's antenna was only about 12′ in diameter and would be expected to sample more nearly the temperature at the center of the disk, while, with broader antenna beams which accept radiation from all parts of the disk at one time, the harmonics would tend to average out.

2.4. Comparison of the Observations with Theory

Piddington and Minnett (1949) developed the theory to relate their radio observations to the physical characteristics of the moon, using the variation in surface temperature obtained from the infrared observations. They related the change in temperature beneath the surface of the moon to the surface temperature variation and summed up the thermal emission of electromagnetic radiation at different depths beneath the surface. Because their measurement accepted radiation from the whole disk of the moon, they neglected the higher harmonics in the temperature variation.

The result for the observed temperature at the center of the disk can be written as follows:

$$T = (1 - R)\left[0.44\,(T_s - T_m) + T_m + \frac{0.6\,(T_s - T_m)\cos(\omega t - \Phi)}{(1 + 2\,\delta + 2\,\delta^2)^{1/2}}\right], \quad (2)$$

$$\delta = \frac{\beta}{\alpha} = \frac{(\omega \rho s / 2K)^{1/2}}{4\pi\sigma / c\sqrt{\epsilon}}, \quad (3)$$

$$\tan\Phi = \frac{\delta}{1 + \delta}, \quad (4)$$

where R = power reflection coefficient; T_s = surface temperature at the subsolar point; T_m = surface temperature at the midnight point; ω = angular velocity of solar illumination change; t = time; Φ = phase angle of brightness temperature variation; ρ = density; s = specific heat; K = thermal conductivity; σ = electrical conductivity; c = velocity of light; and ϵ = dielectric constant.

This formula is based on a $\cos^{1/4}$ variation of temperature with lunar latitude, and the temperature of the unilluminated side is assumed to be constant and equal to the temperature opposite the subsolar point. A similar development has been made by Troitskii (1954), who included an analysis of the dependence of the emissivity of the moon on the angle of incidence and the polarization of the wave and used different temperature distribution functions. This type of treatment has been put into graphical form by Pawsey and Bracewell (1955). Since the observations give a result of the form

$$T = A + B\cos(\omega t - \Phi), \quad (5)$$

a direction solution is possible for

a) δ,

b) $\{(1 - R)[0.44\,(T_s - T_m) + T_m]\}$,

and

c) $\left[\frac{(1 - R)(T_s - T_m)}{(1 + 2\,\delta + 2\,\delta^2)^{1/2}}\right]$.

In principle, δ, which is proportional to the depth of emission of the electromagnetic wave and inversely proportional to the depth of penetration of the thermal wave (3), can be found from equation (4), since Φ is measured. However, since the observed values of Φ are near 45° where δ approaches infinity, the phase lag, Φ, of the variation in the radio emission of the moon must be measured very accurately to be of any use in determining δ and hence the properties of the lunar surface from this simple theory.

For example, if $\Phi = 40° \pm 5°$, δ will be between about 2 and infinity, and the amplitude of the variable component will be between 0° and about 40° K.

We are then left with b and c, from which may be found δ and $(1 - R)$, the emissivity of the moon, if the results are used of the infrared measurements of the midnight temperature, $T_m = 120° \pm 15°$ K, and the theoretical value for the temperature of the subsolar point, $T_s = 374°$ K, both taken from Pettit and Nicholson (1930). We have found that the average of the measurements of the constant component of the equatorial temperature was about 210° K, which, together with b above, gives about 0.91 for $(1 - R)$. For a poorly conducting material with $\mu \simeq 1$, this would correspond to a dielectric constant of about 3.5.

Even though the observations do not accurately specify the amplitude of the variable component of the radio emission of the moon, it is of interest to use the previously estimated values to make illustrative calculations. The estimated variable component at near 8-mm wave length is about 40° K, for which c above gives $\delta \simeq 2$. For the value at 3-cm wave length of about 12° K, $\delta \simeq 8$. We can now follow some of the individual investigators and derive the most consistent values for electrical conductivity from the values of δ. For this purpose we shall use the values for the thermal constants which appear to be consistent with the infrared observations (Garstang, 1958) and which are given in Table 2 under the column headed "Powder B." In addition, we shall assume that the dielectric constant is 3. We then calculate electrical conductivities at 3-cm wave length of $\sigma \simeq 2 \times 10^8$ e.s.u., and at 8-mm wave length of $\sigma \simeq 9 \times 10^8$ e.s.u. The value at 3 cm is about 3.5 times higher than the measured electrical conductivity of dry sandy soil reported by von Hippel (1954) for this wave length. However, the thermal constants of the "Powder A" column would halve the electrical conductivities derived from the measurements, and the thermal properties of pumice would reduce them by a factor of $\frac{1}{20}$. These estimates would also indicate average depths of emission of about 20 cm for the 3-cm wave length and about 10 cm for the 8-mm wave length. Troitskii and Khaikin (1957) derive electrical conductivities which are about twice as high and suggest that such high conductivities may indicate considerable quantities of calcium, sodium, and iron oxides in the lunar surface material.

It is also of interest to estimate the values for δ which might be expected for terrestrial materials under conditions approximating those on the moon. These estimates will necessarily be crude, as some of the required thermal and electrical constants are not known and estimates must be used. We shall use the values for the thermal constants that are quoted by Garstang (1958) for rock, pumice, and dust and values for the electrical constants that are estimated from values quoted by Kerr *et al.* (1951), Pawsey and Bracewell (1955), Strutt (1947), and von Hippel (1954). The constants used and the resulting calculated values for δ and the amplitude of the variable component at wave lengths of 8 mm and 3 cm are listed in Table 2 for materials similar to rock, pumice, and dust in vacuum. These estimates assume that the electrical conductivity is roughly proportional to frequency and that the value for dust is similar to the values for dry sand or dry sandy soil quoted by Kerr *et al.* (1951), Strutt (1947), and von Hippel (1954). A comparison of these illustrative calculations with the

TABLE 2

ILLUSTRATIVE CALCULATIONS OF THE VARIABLE COMPONENT OF LUNAR THERMAL RADIATION, USING THERMAL AND ELECTRICAL CONSTANTS SIMILAR TO THOSE OF TERRESTRIAL MATERIALS

	λ (CM.)	ROCK	PUMICE	POWDER IN VACUUM A	POWDER IN VACUUM B
Thermal constants from Garstang (1958):					
K = Thermal conductivity, cal /cm²/sec/(° K/cm)	......	5×10^{-3}	3×10^{-4}	1×10^{-5}	3×10^{-6}
ρ = Density, gm/cm³	......	3	0.6	2	2
s = Specific heat, cal/gm	......	0.2	0.2	0.2	0.2
$\beta = (\omega\rho s/2K)^{1/2}$	......	1.3×10^{-2}	2.3×10^{-2}	2.2×10^{-1}	4.1×10^{-1}
Estimated electrical constants:					
σ = Electrical conductivity, e.s.u.	0.8	4×10^{8}	2×10^{8}	2×10^{8}	2×10^{8}
	3	1×10^{8}	5×10^{7}	5×10^{7}	5×10^{7}
ϵ = Relative dielectric constant	......	5	2	3	3
$\alpha = 4\pi\sigma/c\sqrt{\epsilon}$	0.8	7.6×10^{-2}	6×10^{-2}	4.8×10^{-2}	4.8×10^{-2}
	3	1.9×10^{-2}	1.5×10^{-2}	1.2×10^{-2}	1.2×10^{-2}
$\delta = \beta/\alpha$	0.8	0.17	0.38	4.6	8.5
	3	0.7	1.5	18	34
$T_{\text{variable component}} = 138/(1+2\delta+2\delta^2)^{1/2}$, ° K	0.8	118	97	19	11
	3	74	46	5	3
Phase lag, degrees	0.8	8	16	39	40
	3	22	31	42	44

results of the observations suggest that the radio observations are consistent with a lunar surface material whose electrical and thermal characteristics are similar to those of pumice or dust.

A different theoretical approach was used by Jaeger (1953), who took into account the heat loss by radiation during the lunar night. He presents his solution in the form of graphs of the derived variations of the radio brightness temperature during a lunation for different combinations of thermal and electrical characteristics—in particular, those corresponding to rock, pumice, and dust. Gibson (1958) has compared his radio measurements at 8.6 mm with the calculated curves of Jaeger (1953) and concludes that the observations are consistent with a homogeneous surface of dust-like or pumice-like material. The calculated curve of Jaeger (1953) which Gibson considers the best fit to his data corresponds to the thermal properties of dust (Gibson, 1958).

The theoretical treatments described have assumed that the characteristics of the lunar material do not depend on temperature, although Wesselink (1948) and others pointed out early that the thermal characteristics may very well be temperature-dependent, and this is also true for the electrical characteristics. Recently, Muncey (1958) has suggested that thermal conductivity and specific heat may be proportional to absolute temperature and that the mean square temperature should be constant rather than the average value of the temperature. The present radio data do not seem to give this suggestion much support, although there may be compensating effects.

An objection can be raised to the direct application of the results of the present radio observations averaged over very large areas of the moon to the determination of the composition of the lunar surface. The visible surface is not uniform, and no determination of surface characteristics from the radio emission will be satisfactory until angular resolution at least comparable to the gross features of the lunar surface is available. Recently Fremlin (1959) has pointed out that in localized areas considerable deviation from the observed average thermal conductivity may be permitted. He calculated that a spotty moon made up of two types of surface material—(1) bare rock with relatively high thermal conductivity and (2) dust with about one-fifth the mean thermal conductivity—fits the infrared eclipse observations of Pettit (1940) better than the uniform model having the mean conductivity, if the bare rock makes up about 5 per cent of the surface. The first approach to detailed observations of the radio emission of different areas of the surface of the moon is given in the work of Coates (1959*a*) (although his beam width of about 6′.7 still averages over a considerable portion of the surface), and in the observations of Amenitskii, Noskova, and Salomonovich (1960) with a beam width of 2′.

To summarize, the radio observations of the moon have shown an average radio brightness temperature at the center of the disk that is comparable with the average surface temperature found from infrared observations and a variation in radio emission during a lunation that is much less than that measured in the infrared and lags behind the solar illumination in phase. These results are consistent with emission of the radio radiation beneath the surface of the moon, with an effective depth of emission which increases with wave length. The results of the radio observations are generally consistent with predicted results for a lunar subsurface material having average thermal and electrical properties similar to those expected of dust, dry sand, or porous rock, such as pumice, although either a higher thermal conductivity than is indicated by the infrared measurements or a higher electrical conductivity than is expected for materials like dry sand is suggested. Further radio observations, particularly at the very short wave lengths, with improved accuracy and higher resolution, will be of great interest.

3. VENUS

3.1. Observations

Venus approaches the earth closer than any other planet, and during these approaches the radio emission at centimeter wave lengths is relatively easy to observe. At other times the distance to Venus is so great that the observations are very difficult. For this reason, the present observations were all made near inferior conjunction, the farthest removed being those made near maximum elongation, using a maser radiometer of very high sensitivity (Alsop, Giordmaine, Mayer, and Townes, 1958, 1959). The maser radiometer is described by Giordmaine, Alsop, Mayer, and Townes (1959) (see also Drake, 1960). On the other hand, observations very near inferior conjunction are also difficult because of interference from solar radio noise. The intensity of the centimeter radiation of the sun is so much higher than that of Venus that the signal from the sun when observed in one of the near side lobes is comparable to that of Venus seen in the main lobe. The solar interference is less serious at the short wave lengths, where the flux from Venus is relatively high; thus at 3.15 cm it was possible to make meaningful observations throughout the inferior conjunction of 1956. At longer wave lengths it has been found difficult to make observations for a period of several weeks centered on inferior conjunction.

A summary of the circumstances and results of the radio observations of

Venus is given in Table 3. The radio emission has been observed extensively at wave lengths near 3 cm and near 10 cm, and single sets of observations have been made at 8.6 and 8 mm. The estimated probable errors for the measurements listed, except those of Drake (1959) at 3.75 cm, include not only the accidental errors of measurement but also the systematic errors from the determination of the effective aperture area of the antenna and other calibrations. The apparent black-body disk temperatures have been calculated, using the diameter of the optical disk, since there is no measurement of the diameter of the radio source and the optical

TABLE 3

RADIO OBSERVATIONS OF VENUS

Wave Length (Cm)	Black-Body Disk Temperature (° K)	Dates of Observations	Phase Angle (Deg.)	Observer
0.80....	315±70*	Sept. 18–Nov. 13, 1959	216–271	Kuzmin and Salomonovich (1960)
0.86....	410±160	Jan. 29, 1958	180	Gibson and McEwan (1959
3.15....	595± 55	May 5–June 23, 1956	106–186	Mayer, McCullough, and Sloanaker (1958*a*, *b*)
3.37....	575± 58	Apr. 18–19, 1958	275	Alsop, Giordmaine, Mayer, and Townes (1958, 1959)
3.4.....	575± 60	Feb. 12–Mar. 5, 1958	212–241	Mayer, McCullough, and Sloanaker (1960*b*)
3.75....	585± 7†	July 7–Oct. 4, 1959	101–238	Drake (private communication)
9.4.....	580±160	June 25 and July 27, 1956	214 av.	Mayer, McCullough, and Sloanaker (1958*a*, *b*)
10.2.....	600± 65	Sept. 17–Oct. 10, 1959	214–245	Mayer, McCullough, and Sloanaker (1960*a*, *b*)

* The value quoted is that for the beginning of the period rather than the middle, as quoted by the other observers.

† The probable error quoted by Drake includes only the experimental scatter, while the other errors listed include, in addition, the errors in the calibrations of the antenna and receiver.

disk is likely to be a good approximation. The observations at 8.6 mm and 9.4 cm were near the limit of sensitivity, and a number of individual observations were used to obtain a measurement.

A closer comparison of the Naval Research Laboratory observations at 3 to 10 cm can be made by using Figure 3, where the daily means are plotted against the angle between the sun and the earth as seen from Venus. The average black-body disk temperatures derived from the different series are nearly the same, although they were spread over a period of $3\frac{1}{2}$ years and were made at different wave lengths with different apparatus. Average disk temperatures of 550° K near inferior conjunction and 600° K about a month on either side of conjunction suggest a minimum in radio emission, but more accurate observations are needed to define the details. In particular, additional series of measurements are needed extending through all phases of Venus and made with a single well-calibrated apparatus. Although the intensity of the radiation from Venus near superior conjunction is only about one-thirtieth that at inferior conjunction, such measurements are now technically possible. More complete observations will be of great interest, as they may lead to the detection of a phase or a rotation effect. Since the visible surface of Venus does not show permanent markings, this would give important basic information.

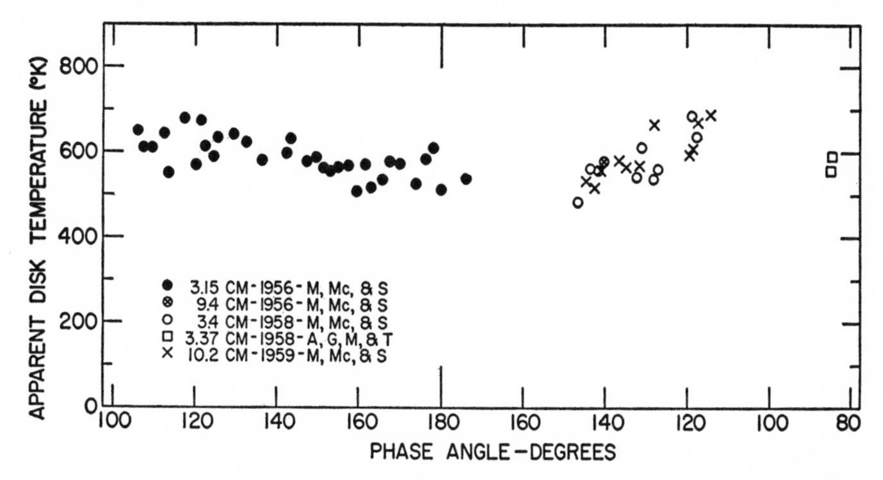

Fig. 3.—Radio observations of Venus as a function of phase angle from observations made at the United States Naval Research Laboratory in the wave-length range 3–11 cm. *A, G, M & T* = Alsop, Giordmaine, Mayer, and Townes (1958, 1959). *M, Mc & S* = Mayer, McCullough, and Sloanaker (1958*a, b;* 1960*a, b*).

3.2 Discussion

The observational evidence summarized in Table 3 indicates that Venus emits radiation at centimeter wave lengths with a black-body spectrum corresponding to about 580° K. The observations at 8–8.6 mm give disk temperatures which are much lower, in the range of 300°–400° K. The tentative interpretation of these results is that the centimeter radiation is emitted thermally at some level deep in the atmosphere, perhaps at the solid surface, where the temperature is much higher than that inferred from the infrared observations, and that a part of the radiation at 8–8.6 mm is produced at a higher level in the atmosphere, where the temperature is lower. However, a temperature as high as 580° K at the surface of Venus requires explanation.

It seems reasonable to expect the atmosphere of Venus to be relatively transparent to centimeter-wave-length radiation. Although the composition of the atmosphere is not completely known, the abundances of oxygen and water vapor, which are the principal absorbers at these wave lengths in the earth's atmosphere, are small (Dunham, 1952; Strong, 1960). CO_2 is known to exist in large quantities in the atmosphere of Venus, but its radio wave-length absorption is small (Birnbaum, Maryott, and Wacker, 1954). No other likely atmospheric constituents with significant radio wave-length absorption are known. A very extensive ionosphere would be required to provide appreciable absorption at centimeter wave lengths, and electron densities or path lengths orders of magnitude greater than those in the earth's ionosphere would be necessary. The cloud or haze which obscures the surface of Venus may be made up of particles which are a few microns in diameter, according to polarization measurements (Lyot, 1929; van de Hulst, 1952; Kuiper, 1952, 1957; see also chap. 9). Kuiper (1952) suggests that the observed variations in the CO_2 absorption might indicate particles as large as 10 μ. Clouds of such small particles are nearly transparent at radio wave lengths.

At millimeter wave lengths, the absorption of CO_2 and H_2O in the atmosphere of Venus may become important (Barrett, 1960), with the result that an increasing proportion of the emission would occur in the cooler atmosphere as the wave length is decreased. Barrett (1960) estimates that, for a model atmosphere composed of 75 per cent CO_2, 3 per cent H_2O, and the remainder N_2 and for a surface temperature of 580° K, a surface pressure of about 10 atm. would be required to produce the observed apparent disk temperature of about 410° K at 8.6 mm. If the atmosphere were to contain no water vapor, a pressure of 30 atm. would be required.

Observations of the emission at infrared wave lengths between 8 and 13 μ indicate that the apparent black-body temperature is nearly the same for the day and night hemispheres of Venus and is equal to about 235° K (Pettit and Nicholson, 1924, 1955; Coblentz and Lampland, 1925; Sinton and Strong, 1960). As a number of investigators have pointed out, the infrared emission is expected to be a complex combination of the emission of the CO_2 above the clouds and the emission of the clouds or other materials, and the emission has been observed to change, possibly as the cloud level changes (see, for example, Adel, 1941; Lampland, 1941; Kuiper, 1952, 1957; Sinton and Strong, 1960). The measurements of Sinton and Strong (1960), which give the distribution of the infrared emission over the disk of Venus, show limb darkening, indicating that the emission takes place in a region where the temperature decreases with height.

The temperature of the observed CO_2 has been derived from the intensities of the rotational transitions at about 0.8 μ (Adel, 1937; Chamberlain and Kuiper, 1956). Kuiper (1957) points out that if the levels are determined by CO_2 absorption rather than by particle scattering, the more accurate temperature of 285° K from the recent determination (Chamberlain and Kuiper, 1956) will refer to a lower level in the atmosphere than will the temperature derived from the emission in the 8–13-μ band. The 285° K rotational temperature is the average for the CO_2 along the path of the reflected sunlight, and Kuiper (1957) estimates that the temperature at the bottom of this path, which is still presumably above the cloud, is about 320° K.

These temperatures derived from the infrared measurements refer to the upper atmosphere of Venus and are much lower than the temperatures derived from the radio measurements. Wildt (1940) predicted that the temperature at the surface of Venus might be as high as 408° K because of the greenhouse effect due to the large amount of CO_2 in the atmosphere. This prediction was based on an albedo to incident solar radiation of 0.6 and an estimate of the partial trapping of the long-wave-length re-radiation extrapolated from a laboratory measurement of the infrared emissivity of CO_2 made by E. Eckert. Kuiper (1952) has re-evaluated this estimate, using a higher albedo of 0.76 and other more recent data, and finds 350° K for the temperature at the tropical midday surface. Thus the greenhouse effect of CO_2 will not alone account for a surface temperature of 600° K. In order to raise the temperature at the solid surface of Venus to values near 600° K when the only source of heat is solar radiation, the absorption or reflection of the surface radiation by the atmosphere must be nearly complete. The effect of the clouds or haze at infrared wave lengths is not known, but Öpik (1956) has suggested that the albedo may be very high in the infrared, possibly higher than 0.75.

Sagan (1960*a*, *b*) has shown that, for a surface temperature of 600° K on a rapidly rotating Venus, the transmissivity of the atmosphere to the surface radiation must be only about $\frac{1}{2}$ per cent, which requires that the atmosphere be opaque from the near infrared to about 40 μ. Sagan also points out that the necessary opacity at wave lengths longer than about 20 μ cannot be accounted for by CO_2 absorption, and he concludes that it is not likely to be accounted for by particle scattering in the cloud layer. Sagan finds the most reasonable explanation for a 600° K surface temperature to be an enhanced greenhouse effect by water-vapor absorption and that the necessary quantity of water vapor corresponds to a partial pressure of about 9 gm cm^{-2} at the surface if Venus rotates on its axis with a period of less than 30 days. If Venus always keeps one side toward the sun, an abundance corresponding to a partial pressure of only 1 gm cm^{-2} is required. Sagan points out that the total quantity of water required is greater than that in the atmosphere of the earth but that the relative humidity is very low throughout most of the atmosphere of Venus. In his model, saturation is reached at a level where the temperature is about 220°–233° K, where an ice-crystal cloud layer is predicted. He shows that amount of the water vapor which would exist above the cloud layer on the basis of this model is in reasonable agreement with the amount derived from the balloon observations of Moore and Ross (Strong, 1960). The work of Sagan shows that the results of the different types of observations are consistent with a reasonable physical model for the atmosphere of Venus.

A possible source of radio emission is in thermal radiation of an extensive ionized atmosphere. However, it is difficult to see how this mechanism could account for an important part of the observed centimeter-wave-length emission of Venus, as the measured spectrum would require that $\int N_e^2\, dl \sim 10^{26}\ cm^{-5}$. If the electrons were uniformly distributed in a layer 1000 km thick, the required electron density would be about $10^9\ cm^{-3}$, which is about 10^3 times the density in the ionosphere of the earth.

Another possibility is that all or part of the radio emission of Venus may be of non-thermal origin. At the present time it does not seem likely that any significant part of the observed radiation is of non-thermal origin, since a mechanism of origin has not been found that will satisfactorily explain the observations. The mechanism of synchroton radiation of relativistic electrons trapped in a magnetic field of the planet, which has been proposed to explain the radio observations of Jupiter (Drake, 1959), will not provide the observed spectrum (see chap. 13, Burke), as has been pointed out by Drake (1959) and by Sagan (1960*b*). In addition, no polarized component of the observed radiation has been detected.

To summarize, the observations of the radio emission of Venus indicate that the apparent disk temperature is very nearly the same at wave lengths of 3 cm and 10 cm and that it does not change greatly with time, while the disk temperature near 8 mm is appreciably lower. The centimeter-wave-length spectrum is that of a black body at about 600° K, which suggests that the emission occurs deep in the atmosphere or at the solid surface. In order to account for such a high surface temperature, it is necessary that almost all the absorbed solar radiation be trapped or else that there be some other source of heat. It seems that the atmospheric opacity at infrared wave lengths, which is required to provide the necessary trapping of absorbed solar radiation, may be provided by the absorptions of CO_2 and H_2O in the atmosphere of Venus. The present observations are not sufficiently accurate or complete to define a dependence of the radio emission on the phase or rotation of Venus. It is important to make further observations to establish whether such a dependence exists, both to obtain this basic information and to allow a better understanding of the origin of the radio emission.

4. MARS

The radio emission of Mars has been observed at two different times. Observations made at the favorable opposition of September, 1956, at 3.15-cm wave length (Mayer, McCullough, and Sloanaker, 1958*b*, *c*) were sensitivity-limited, and it was necessary to average about 70 observations to obtain a measurement of reasonable accuracy. The result of this measurement corresponds to an apparent black-body disk temperature for Mars of 218° ± 50° K.

Mars was again observed at about 6 weeks past the opposition of November, 1958, with a very sensitive maser radiometer at 3.14-cm wave length (Giordmaine, Alsop, Townes, and Mayer, 1959). This measurement yielded an apparent black-body disk temperature for Mars of 211° ± 20° K. The agreement between the two observations is good, in view of the uncertainties in the measurements.

The two measurements made near 3-cm wave length indicate an apparent black-body disk temperature for Mars of about 212° K. This value may be compared with the apparent black-body disk temperature derived from measurements of the infrared emission of Mars. The average disk temperature is given by Pettit and Nicholson (1924) as 260° K and by Menzel, Coblentz, and Lampland (1926) as between 237° and 254° K, where in both cases the quoted temperatures were derived from the measured infrared radiation using the fourth-power law of total radiation, apparently the more reliable method (Coblentz, 1942). The apparent black-body disk temperature derived from the radio observations is seen to be about 40° C less than that derived from the infrared observations.

If, as seems to be the case for the moon, the radio radiation is emitted beneath the surface of Mars, while the infrared is emitted at the surface, it can be expected that the infrared temperature will approximate the average temperature over the sunlit surface, while the radio temperature will be that beneath the surface and may be expected to approximate more nearly the average temperature over the whole surface. An estimate of the average temperature over the entire surface of Mars is given by the average radiation temperature of 217° K computed by Kuiper (1952), using the visual albedo of 0.148. Considering the uncertainties in the observations and in the emissivity of the surface of Mars, the observed radio emission is consistent with the thermal radiation which is expected on the basis of previous knowledge.

5. JUPITER

5.1. Observations

The early observations of Jupiter at about 3-cm wave length gave black-body disk temperatures similar to those expected on the basis of the infrared radiometric observations. However, more recently Sloanaker (1959; McClain and Sloanaker, 1959) discovered that the emission at 10.3 cm was unexpectedly intense. The steady radio emission of Jupiter has now been observed over a relatively wide range of wave lengths, from about 3 to about 70 cm. The measured values for the flux density of radiation and the corresponding black-body disk temperatures are listed in Table 4, and the results of all but the most recent observations are plotted in Figure 4. They show that the emission of Jupiter does not fall off rapidly with in-

TABLE 4

Radio Observations of Jupiter

Wave Length (Cm)	Black-Body Disk Temperature (° K)	Flux Density for Disk Diameter=45.4″ watts/m²/(c/s)	Dates of Observations	Observer
3.15...	140± 38	1.5×10^{-25}	May 13 and 31, 1956	Mayer, McCullough, and Sloanaker (1958*b*, *c*)
3.15...	145± 18	1.5	Mar. 23–Apr. 1, 1957	Mayer, McCullough, and Sloanaker (1958*b*, *c*)
3.03...	171± 20	2.0	Aug. 22–Sept. 4, 1958	Giordmaine, Alsop, Townes, and Mayer (1959)
3.17...	173± 20	1.8	May 24–July 29, 1958 Jan. 31–Feb. 7, 1959	Giordmaine, Alsop, Townes, and Mayer (1959)
3.36...	189± 20	1.8	Apr. 16–May 8, 1958	Giordmaine, Alsop, Townes, and Mayer (1959)
3.75...	200	1.5	July 24, 1957	Drake and Ewen (1958)
10.3....	640± 57	0.63	June 10–Aug. 20, 1958	Sloanaker (1959); McClain and Sloanaker (1959)
10.3....	315± 45	0.31	Oct. 16–30, 1959	Sloanaker and Boland (1960)
21......	2,500±450	0.60	May 15–July 31, 1959	McClain (1959); McClain, Nichols, and Waak (1960)
21......	3,000	0.71	May 16–June 2, 1959	Epstein (1959)
22......	3,000	0.65	May, 1959	Drake and Hvatum (1959)
31......	5,500	0.60	April 15–June 17, 1959	Roberts and Stanley (1959)
68......	50,000	1.1	May 26–27, 1959; July 20–30, 1959	Drake and Hvatum (1959)
31......	Polarization and source size		Apr. 2–21, 1960	Radhakrishnan and Roberts (1960)

creasing wave length, as would be expected for radiation from a black body at a fixed temperature. As a result, it has been possible to observe Jupiter at the longer wave lengths with existing apparatus.

In addition to the rapid increase in the apparent black-body disk temperature with increasing wave length, the observations suggest that the radio emission may change with time. The apparent black-body temperature at 3.15 cm was measured in 1957 and found to be 145° ± 18° K (Mayer, McCullough, and Sloanaker, 1958*b*, *c*), but measurements made in 1958, using the same antenna and calibration procedures, gave a value of 173° ± 20° K (Giordmaine, Alsop, Townes, and Mayer, 1959). The apparent increase is one and a half times the probable error and indicates a possible change in the emission of Jupiter. Also an anomalously high disk temperature of about 268° K may have been observed on April 30–May 1, 1958 (Giordmaine, Alsop, Townes, and Mayer, 1959), but no changes in emission at this wave length were noted which could be correlated with the planet's rotation.

On the other hand, Sloanaker (1959) reports that his observations suggest a periodic variation of about 30 per cent in the 10.3-cm emission with a period which is between 40 seconds and 2 minutes longer than the rotation

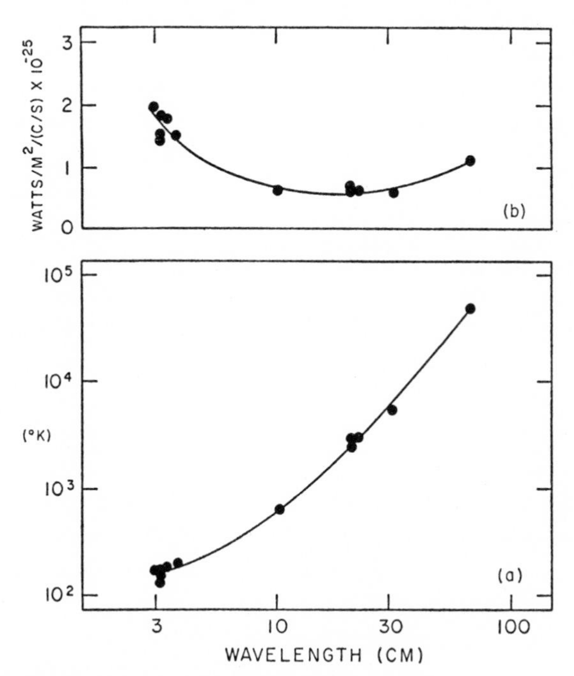

FIG. 4.—Radio observations of Jupiter plotted from the data of Table 4. (*a*) Apparent disk temperature; (*b*) flux density for a 45″.4 disk diameter.

period of the higher latitudes on Jupiter (System II), although the amplitude of the variation is only slightly higher than the expected standard deviation of the measurement and no definite conclusion is possible. Similarly, McClain (1959) states that the measurements at 21 cm are suggestive of a possible variation of about 30 per cent in the 21-cm emission of Jupiter with the System II rotation period and that there was also a possible increase in emission following a solar flare of importance 3+, but again the measurement uncertainty is nearly as large as the observed variations, and the associations are only tentative. Epstein (1959) reported variations in his observations at 21 cm of as much as a factor of 2 in a few hours' time. Drake and Hvatum (1959) observed variations of about 30 per cent over a few days' time in the 22-cm emission of Jupiter but found no significant correlation with the rotation of Jupiter. Roberts and Stanley (1959) found no correlation of the 31-cm emission with the rotation of Jupiter at either the System I or the System II periods or with solar activity as indicated by the Sacramento Peak Flare Index. They conclude from comparisons with the observed variations in the intensities of other weak sources that their observations did not show a significant variation in the emission of Jupiter.

More recent observations at 10.3 cm made in October, 1959 (Sloanaker and Boland, 1960), gave a black-body disk temperature of $315° \pm 45°$ K, which is only about half the value of $640° \pm 57°$ K measured the previous year at this same wave length (Sloanaker, 1959). Since the same apparatus was used for the two series of measurements, Sloanaker and Boland (1960) conclude that the difference between the radiation measured in 1958 and that in 1959 is probably real.

Still more recently, Radhakrishnan and Roberts (1960) have observed Jupiter at 31 cm, using the interferometer at the radio observatory of the California Institute of Technology. The elements of the interferometer are two 90-foot-diameter reflectors which were used for these observations at two different spacings, 400 and 800 feet. By comparing the interference patterns obtained with the two antennae linearly polarized in the same plane with those obtained with the two antennae cross-polarized, they found values for the degree of polarization of 25–43 per cent, with a mean value of 33 per cent. A comparison of the intensities measured with the antennae polarized parallel to each other but with the plane of polarization rotated by 90° on successive days gave about 22 per cent for the degree of polarization. Their measurements indicate essentially linear polarization, with the electric vector aligned with the equator of Jupiter to within 12°. Radhakrishnan and Roberts (1960) also report that the use of the wide antenna spacing of 800 feet (fringe spacing 4.4 minutes of arc) made it possible to distinguish an angular extent for the source of the polarized radiation which is consistent with an equatorial ring of diameter about 2 minutes of arc, that is, about three times the diameter of Jupiter.

These important observations show that it is necessary to consider the polarization of the antenna when evaluating the radio radiation received from Jupiter. The measurements of Radhakrishnan and Roberts gave a measured intensity at 31 cm, which was about 1.7 times greater for a linearly polarized antenna with the electric vector parallel to the equator of Jupiter than for the orthogonal polarization. Sloanaker and Boland (1960) point out that the 10.3-cm observations of 1958 were made with the electric vector of the antenna at an angle of 67° from the equator of Jupiter, while the inclination in 1959 was about 79°, and that this could partly account for the lower measured intensity in 1959. However, a very large polarized component of about ten times the measured intensity would be required to account for the whole change in this way.

5.2. Discussion

The results of the radio observations of Jupiter can be summarized as follows: (*a*) the spectrum is not at all similar to that of simple black-body radiation at a constant temperature—in fact, the flux density of the radiation is nearly constant and the apparent black-body disk temperature rises from about 150° K at 3-cm wave length to about 50,000° K at 68 cm; (*b*) there are indications of long time changes in the measured radio emission of Jupiter; (*c*) some observers report suggestions of changes in the measured radiation at nearly the period of rotation of the planet's higher latitudes, but other series of observations do not show such a correlation and cast doubt on these results; (*d*) the 31-cm radiation is about 30 per cent linearly polarized, with the electric vector approximately parallel to the equator of Jupiter; and (*e*) the diameter of the source of 31-cm radiation is larger than the diameter of Jupiter—in particular, the source of the polarized radiation is consistent with an equatorial ring of about three times the diameter of Jupiter.

The measured spectrum of Jupiter leading to the very high apparent black-body disk temperatures at decimeter wave lengths led F. D. Drake (1959; Drake and Hvatum, 1959) to propose that the decimeter radiation was of non-thermal origin and could be explained by the radiation of relativistic particles trapped in a magnetic field of Jupiter, although he estimated that for a magnetic field strength of 5 gauss the number of particles would have to be about 10^6 times that estimated for the Van Allen belts of the earth. The recent discovery by Radhakrishnan and Roberts (1960) of linear polarization of the radiation and a diameter for the radio source which is several times greater than the diameter of Jupiter give experimental support to this hypothesis.

Prior to these recent observations a number of possible explanations for the high intensity of the decimeter radiation had been explored theoretically. It was generally considered that the early observations at about 3-cm wave length were consistent with thermal radiation. The atmosphere of Jupiter contains an appreciable quantity of ammonia, which absorbs strongly at 1.28-cm wave length. Consequently, the atmospheric opacity will vary greatly with wave length from being nearly complete at about 1.28 cm to nearly zero at decimeter wave lengths, and the depth of emission and intensity of radiation will be a function of wave length. Both Field (1959) and Giordmaine (1960) have calculated the probable wave-length variation in the intensity of Jupiter's thermal radiation, assuming model atmospheres similar to those of Kuiper (1952) and an adiabatic temperature gradient. Although their results do not agree in detail, it seems possible to account for the 3-cm radiation as thermal emission, but it seems that only part of the radiation at 10 cm and longer wave lengths can be accounted for in this way.

The possibility that free-free transitions in an ionized atmosphere might be responsible for the observed emission was investigated by Field (1959) and Roberts and Stanley (1959). Both concluded that a very extensive ionized region is necessary with $\int N_e^2\, dl > 10^{25}$ cm^{-5} for an electron temperature of about 10^5 ° K and that it is difficult to find an adequate source for such an ionized region. Based on the suggestion of Chapman (1957) that the solar corona extends throughout the solar system, Roberts and Stanley (1959) investigated the possibility that Jupiter might have collected an extensive ionized atmosphere with a temperature of about 10^5 ° K from this material. They concluded that gravitational attraction is not capable of concentrating the required density of particles but that they could not adequately evaluate the possibility of magnetic trapping.

The possibility of synchrotron emission was also investigated by Roberts and Stanley (1959). They found that, for a magnetic field of 7 gauss, a density of electrons with energies greater than 1 Mev of about 10^{-2} cm^{-3} is necessary. This is 3×10^4 times greater than their estimate of the density in the Van Allen belt of the earth, but they observe that densities which exceed their estimate by a factor of 10^2 have been reported. They also point out, however, that the observed radio spectrum of Jupiter is not consistent with the energy distribution observed for the relativistic electrons in the Van Allen belt of the earth and that some different energy distribution would be necessary for the particles associated with Jupiter.

Field (1959, 1960) has also investigated radiation by particles trapped in a magnetic field of Jupiter, with the conclusion that synchrotron radiation, by high-speed cosmic-ray electrons or relativistic electrons from the sun, is an unlikely source because there are not enough of these particles to supply the energy radiated at radio wave lengths. He finds, however, that there is enough energy in nonrelativistic particles from the sun to account for the radio emission and that the cyclotron radiation of these particles may explain the observed spectrum, although a very large magnetic field, greater than 1200 gauss at the poles, would be required.

These investigations suggest that cyclotron and synchrotron radiation theories are the best possibilities for explaining the decimeter radiation of Jupiter. In addition, these are the only mechanisms which predict the linear polarization that has now been observed (Radhakrishnan and Roberts, 1960). Radhakrishnan and Roberts (1960) point out that polarization and source size measurements made as a function of frequency may make it possible to distinguish between these two mechanisms, as, for example, Field (1960) has predicted a rapid change in the polarization with frequency for the cyclotron radiation case. Although cyclotron and synchrotron radiations seem to be promising possibilities for the explanation of the observed decimeter radiation of Jupiter, there are still difficulties. The problem of the theoretical explanation of the radio emission of Jupiter is considered in detail by Burke in chapter 13.

It now seems that three distinct types of radio emission from Jupiter have been observed. At the very short centimeter wave lengths, the observed radiation is predominantly the thermal emission of the atmosphere. At decimeter wave lengths, the predominant source of radio emission is non-thermal and may be the radiation of charged particles trapped in the magnetic field of Jupiter. The third type of Jupiter emission is that observed at very long wave lengths near 15 meters (see chaps. 13 and 14). Thus the study of the radio emission of Jupiter has emerged as a powerful tool for the study of the physical conditions surrounding this planet.

6. SATURN

The 3.75-cm emission of Saturn was observed by Drake and Ewen (1958) about 2 months after opposition in 1957. The intensity of the radiation was very weak, and the antenna temperature due to Saturn was only 0.04° K. Observations of Saturn and Jupiter were made in quick succession, and a ratio of 4.3 was measured between the antenna temperature of

Jupiter and that of Saturn. Drake and Ewen (1958) suggest that this gives evidence that the rings of Saturn do not emit 3.75-cm radiation as efficiently as the body of the planet, since they estimate from predicted temperatures for Jupiter and Saturn (Allen, 1955) that the ratio for transparent rings would be about 4.7, while the ratio for opaque rings would be about 3. However, this conclusion is uncertain in view of present knowledge of the relatively high apparent temperature of Jupiter observed at radio wave lengths.

More recently, measurements of the emission of Saturn at 3.4 cm were made at the University of Michigan with the 85-foot reflector and a ruby maser (Cook, Cross, Bair, and Arnold, 1960). The average of 14 measurements gave an apparent black-body disk temperature of 106° ± 21° K, neglecting any possible radiation by the rings. This apparent black-body disk temperature is in reasonable agreement with the radiation temperatures calculated by Kuiper (1952) and with the temperature derived from infrared radiometric observations of 123° K (Menzel, Coblentz, and Lampland, 1926).

REFERENCES

ADEL, A. 1937 *Ap. J.*, **86**, 337.
1941 *Ibid.*, **93**, 397.
AKABANE, K. 1955 *Proc. Japan Acad.*, **31**, 161.
1960 Private communication.
ALLEN, C. W. 1955 *Astrophysical Quantities* (London: University of London; Athlone Press), p. 163.
ALSOP, L. E., GIORDMAINE, J. A., MAYER, C. H., and TOWNES, C. H. 1958 *A.J.*, **63**, 301.
1959 *Proc. I.A.U. Symp., No. 9—U.R.S.I. Symp., No. 1*, ed. R. N. BRACEWELL (Stanford: Stanford University Press), pp. 69–74.
AMENITSKII, N. A., NOSKOVA, R. I., and SALOMONOVICH, A. E. 1960 *A.J. (U.S.S.R.)*, **37**, 185.
BARRETT, A. H. 1960 *J. Geophys. Res.*, **65**, 1835.
BIRNBAUM, G., MARYOTT, A. A., and WACKER, P. F. 1954 *J. Chem. Phys.*, **22**, 1782.
BURKE, B. F., and FRANKLIN, K. L. 1955 *J. Geophys. Res.*, **60**, 213.
CHAMBERLAIN, J. W., and KUIPER, G. P. 1956 *Ap. J.*, **124**, 399.
CHAPMAN, S. 1957 *Smithsonian Contr.*, **2**, 1.

COATES, R. J. 1959*a* *A.J.*, **64**, 326.
1959*b* Private communication.

COBLENTZ, W. W. 1942 *J. Res. Bur. Stand.*, **28**, 297.

COBLENTZ, W. W., and LAMPLAND, C. O. 1925 *J. Franklin Inst.*, **199**, 785; **200**, 103.

COOK, J. J., CROSS, L. G., BAIR, M. E., and ARNOLD, C. B. 1960 *Nature*, **188**, 393.

DENISSE, J. F., and LEROUX, E. 1960 Private communication.

DICKE, R. H., and BERINGER, R. 1946 *Ap. J.*, **103**, 375.

DRAKE, F. D. 1959 Private communication.
1960 *Stars and Stellar Systems*, ed. G. P. KUIPER and B. M. MIDDLEHURST (Chicago: University of Chicago Press), Vol. **1**, chap. 12.

DRAKE, F. D., and EWEN, H. I. 1958 *Proc. Inst. Radio Eng.*, **46**, 53.

DRAKE, F. D., and HVATUM, H. 1959 *A.J.*, **64**, 329.

DUNHAM, T., JR. 1952 *The Atmospheres of the Earth and Planets*, ed. G. P. KUIPER (2d ed.; Chicago: University of Chicago Press), chap. 11.

EPSTEIN, E. 1959 *Nature*, **184**, 52.

FIELD, G. 1959 *J. Geophys. Res.*, **64**, 1169.
1960 *Ibid.*, **65**, 1661.

FREMLIN, J. H. 1959 *Nature*, **183**, 1317.

GARSTANG, R. H. 1958 *J. British Astr. Assoc.*, **68**, 155.

GIBSON, J. E. 1958 *Proc. Inst. Radio Eng.*, **46**, 280.

GIBSON, J. E., and MCEWAN, R. J. 1959 *Proc. I.A.U. Symp., No. 9—U.R.S.I. Symp., No. 1*, ed. R. N. BRACEWELL (Stanford: Stanford University Press), pp. 50–52.

GIORDMAINE, J. A. 1960 *Proc. Nat. Acad. Sci.*, **46**, 267–276.

GIORDMAINE, J. A., ALSOP, L. E., MAYER, C. H., and TOWNES, C. H. 1959 *Proc. Inst. Radio Eng.*, **47**, 1062.

GIORDMAINE, J. A., ALSOP, L. E., TOWNES, C. H., and MAYER, C. H. 1959 *A.J.*, **64**, 332.

GREBENKEMPER, C. J. 1958 *U.S. Naval Res. Lab. Rept.*, No. 5151.

HAGEN, J. P. 1949 *U.S. Naval Res. Lab. Rept.*, No. 3504.

HIPPEL, A. VON 1954 *Dielectric Materials and Applications* (New York: Technology Press of M.I.T. and John Wiley & Sons, Inc.), p. 314.

HULST, H. C. VAN DE 1952 *The Atmospheres of the Earth and Planets*, ed. G. P. KUIPER (2d ed.; Chicago: University of Chicago Press), pp. 102–106.

JAEGER, J. C. 1953 *Australian J. Phys.*, **6**, 10.

KAIDANOVSKII, N. L., TURUSBEKOV, M. T., and KHAIKIN, S. E. 1956 *Trans. Fifth Conf. on Problems of Cosmogony* (in Russian) (Moscow: Academy of Sciences, U.S.S.R.), pp. 347–354.

KERR, D. E., FISHBACK, W. T., and GOLDSTEIN, H. 1951 *Propagation of Short Radio Waves*, ed. D. E. KERR (New York: McGraw-Hill Book Co., Inc.), p. 398.

KO, H. C. 1960 Private communication.

KUIPER, G. P. 1952 *The Atmospheres of the Earth and Planets*, ed. G. P. KUIPER (2d ed.; Chicago: University of Chicago Press), chap. 12.

1957 *The Threshold of Space*, ed. M. ZELIKOFF (New York: Pergamon Press), pp. 78–86.

KUZMIN, A. D., and SALOMONOVICH, A. E. 1960 *A.J. (U.S.S.R.)*, **37**, 297.

LAMPLAND, C. O. 1941 *Ap. J.*, **93**, 401.

LYOT, B. 1929 *Ann. Obs. Paris, Meudon*, **8**, 66, 151.

MCCLAIN, E. F. 1959 *A.J.*, **64**, 339.

MCCLAIN, E. F., NICHOLS, J. H., and WAAK, J. A. 1960 Paper read at the XIIIth General Assembly, U.R.S.I., September 5–15, 1960.

MCCLAIN, E. F., and SLOANAKER, R. M. 1959 *Proc. I.A.U. Symp. No. 9—U.R.S.I. Symp. No. 1*, ed. R. N. BRACEWELL (Stanford: Stanford University Press), pp. 61–68.

MAYER, C. H. 1959 *A.J.*, **64**, 43.

MAYER, C. H., MCCULLOUGH, T. P., and SLOANAKER, R. M. 1958*a* *Ap. J.*, **127**, 1.

1958*b* *Proc. Inst. Radio Eng.*, **46**, 260.

1958*c* *Ap. J.*, **127**, 11.

1960*a* *A.J.*, **65**, 349.

1960*b* Paper read at the XIIIth General Assembly, U.R.S.I., September 5–15, 1960.

MENZEL, D. H., COBLENTZ, W. W., and LAMPLAND, C. O. 1926 *Ap. J.*, **63**, 177.

MEZGER, P. G., and STRASSL, H. 1959 *Planet. Space Sci.*, **1**, 213.

MITCHELL, F. H., and WHITEHURST, R. N. 1958 *A Radio Study of the Sun and Moon at Millimeter Wavelengths* (University of Alabama Radio Astronomy Laboratory Rept.), pp. 5–10.

MUNCEY, R. W. 1958 *Nature*, **181**, 1458.

ÖPIK, E. J. 1956 *Irish A.J.*, **4**, 37.

PAWSEY, J. L., and BRACEWELL, R. N. 1955 *Radio Astronomy* (Oxford: Clarendon Press), pp. 280–292.

PETTIT, E. 1935 *Ap. J.*, **81**, 17.

1940 *Ibid.*, **91**, 408.

PETTIT, E., and NICHOLSON, S. B. 1924 *Pub. A.S.P.*, **36**, 269.

1930 *Ap. J.*, **71**, 102.

1955 *Pub. A.S.P.*, **67**, 293.

PIDDINGTON, J. H., and MINNETT, H. C. 1949 *Australian J. Sci. Res., A*, **2**, 63.

RADHAKRISHNAN, V., and ROBERTS, J. A. 1960 *Phys. Rev. Letters*, **4**, 493.

ROBERTS, J. A., and STANLEY, G. J. 1959 *Pub. A.S.P.*, **71**, 485.

SAGAN, C. 1960*a* *A.J.*, **65**, 352.

1960*b* Private communication.

SALOMONOVICH, A. E. 1958 *A.J. (U.S.S.R.)*, **35**, 129.

SEEGER, C. L.; WESTERHOUT, G.; and HULST, H. C. VAN DE 1956 *B.A.N.*, **13**, 89.

SEEGER, C. L., WESTERHOUT, G., and CONWAY, R. G. 1957 *Ap. J.*, **126**, 585.

SLOANAKER, R. M. 1959 *A.J.*, **64**, 346.

SLOANAKER, R. M., and BOLAND, J. W. 1960 Paper read at the XIIIth General Assembly, U.R.S.I., September 5–15, 1960.

SINTON, W. M. 1955 *J. Opt. Soc. America*, **45**, 975.

1956 *Ap. J.*, **123**, 325.

SINTON, W. M., and STRONG, J. 1960 *Ap. J.*, **131**, 470.

STRONG, J. 1960 Private communication.

STRUTT, M. J. O. 1947 *Ultra- and Extreme Short Wave Reception* (New York: D. Van Nostrand Co., Inc.), p. 6.

TROITSKII, V. S. 1954 *A.J. (U.S.S.R.)*, **31**, 511.

TROITSKII, V. S., and KHAIKIN, S. E. 1957 *Proc. I.A.U. Symp., No. 4: Radio Astronomy*, ed. H. C. VAN DE HULST (Cambridge: Cambridge University Press), pp. 406–407.

TROITSKII, V. S., and ZELINSKAYA, M. R. 1955 *A.J. (U.S.S.R.)*, **32**, 550.
WESSELINK, A. J. 1948 *B.A.N.*, **10**, 351.
WESTERHOUT, G. 1958 *B.A.N.*, **14**, 215.
WILDT, R. 1940 *Ap. J.*, **91**, 266.
ZELINSKAYA, M. R., and TROITSKII, V. S. 1956 *Trans. Fifth Conf. on Problems of Cosmogony* (in Russian) (Moscow: Academy of Sciences, U.S.S.R.), pp. 99–105.
ZELINSKAYA, M. R., TROITSKII, V. S., and FEDOSEEV, L. I. 1959 *A.J. (U.S.S.R.)*, **36**, 643.

CHAPTER 13

Radio Observations of Jupiter. I

By B. F. BURKE
Carnegie Institution of Washington, Department of Terrestrial Magnetism

1. INTRODUCTION

IN THE few years since radio techniques were first applied to astronomy, various celestial objects have been shown to be sources of radio emission. In some cases this emission is simply the low-frequency end of the thermal spectrum of a hot body and hence is called "thermal." From the observed radio fluxes, temperatures can be inferred which can be useful in astrophysical studies, since the radiation mechanisms are well understood. At the present time (September, 1960) such thermal emissions have been detected with certainty from the planets Venus, Mars, Jupiter, and Saturn (see chap. 12).

For the so-called "non-thermal" sources, however, equilibrium processes appear inadequate to explain either the observed spectrum or the intensity of the radiation. The noise storms and outbursts of the active sun have been well known for a number of years as examples of the latter class, and recently it has also been established that the planet Jupiter is a source of non-thermal radio emission. The mechanism for producing non-thermal radio noise in these and other objects is a problem of first importance, but all hypotheses proposed so far exhibit weaknesses. Plasma oscillations, possibly excited by shock waves, offer promise in some cases (notably the active sun), but the difficulties associated with coupling the longitudinal plasma oscillations to the transverse radiation field have as yet not been satisfactorily resolved. The Crab Nebula is a well-known source of radio emission in which a radically different mechanism—that of "synchrotron" radiation by relativistic electrons—appears promising. The problem then becomes that of generating the necessary magnetic fields and supplying energy for electron acceleration.

The high internal velocities observed in most radio sources form one unifying feature, the degree of velocity dispersion being related, qualitatively at least, to the surface brightness of the source. Whether more than one mechanism is acting among the wide variety of known radio sources or whether only one basic mechanism will suffice for all cannot be stated in the absence of further observations.

Physical conditions on the planets of our solar system would not seem to favor the production of intense radio noise, and it is somewhat surprising that at least one, the planet Jupiter, is at times the most prominent object in the sky at frequencies below about 30 Mc/s.

2. IDENTIFICATION

Although bursts of radio noise from Jupiter had been recorded at least as early as 1950 by C. A. Shain in Australia and later, in 1954, by F. G. Smith and B. F. Burke at Washington, the intermittent character of the noise prevented identification. In early 1955 the 22 Mc/s Mills Cross at the Carnegie Institution of Washington (1955), in the course of a survey in the vicinity of the Crab Nebula, detected a discrete source of intermittent radio noise which changed its right ascension slowly over a period of several months. Both the position and the change of position with time agreed with that of Jupiter, and, since no other object could fit the data, this identification was the only possibility (Burke and Franklin, 1955). The planet was in retrograde motion when first detected, but the observations were continued through May, 1955, when the planet had resumed its direct motion, thus confirming the identification. Several of the early records are shown in Figure 1, aligned for constant sidereal time. The Mills Cross, being a narrow-beam transit instrument, could observe a given source for only a few minutes a day, but from the first observations several characteristics of the radiation were clear. The bursts of radio noise were not observed every day, but only on an average of 1 day out of 3. Furthermore, since there was no definite periodicity related to the rotational period of either the equatorial belt or the non-equatorial regions, there must have been several active regions on the planet at that time. The bursts were usually intense, at times exceeding all other radio sources in peak intensity, except for the active sun. The duration of individual bursts was usually of the order of $\frac{1}{2}$ or 1 second, apparently superimposed on a weaker, more continuous noise background, but it was not clear whether the background was real or simply a superposition of weak bursts. An interesting implication, in view of the failure of intensive surveys made at higher frequencies to notice such radiation, was the possibility that the emission was strongest at relatively low frequencies and extremely weak at frequencies of 40 Mc/s and higher, where most radio observations had been made. This suspicion received support from the 22.2 Mc/s prediscovery Washington observations of Smith and Burke in 1954, in which a simultaneous recording at 22.2 and 38.7 Mc/s showed no detectable radiation at the higher frequency, although an unusually strong event was observed for

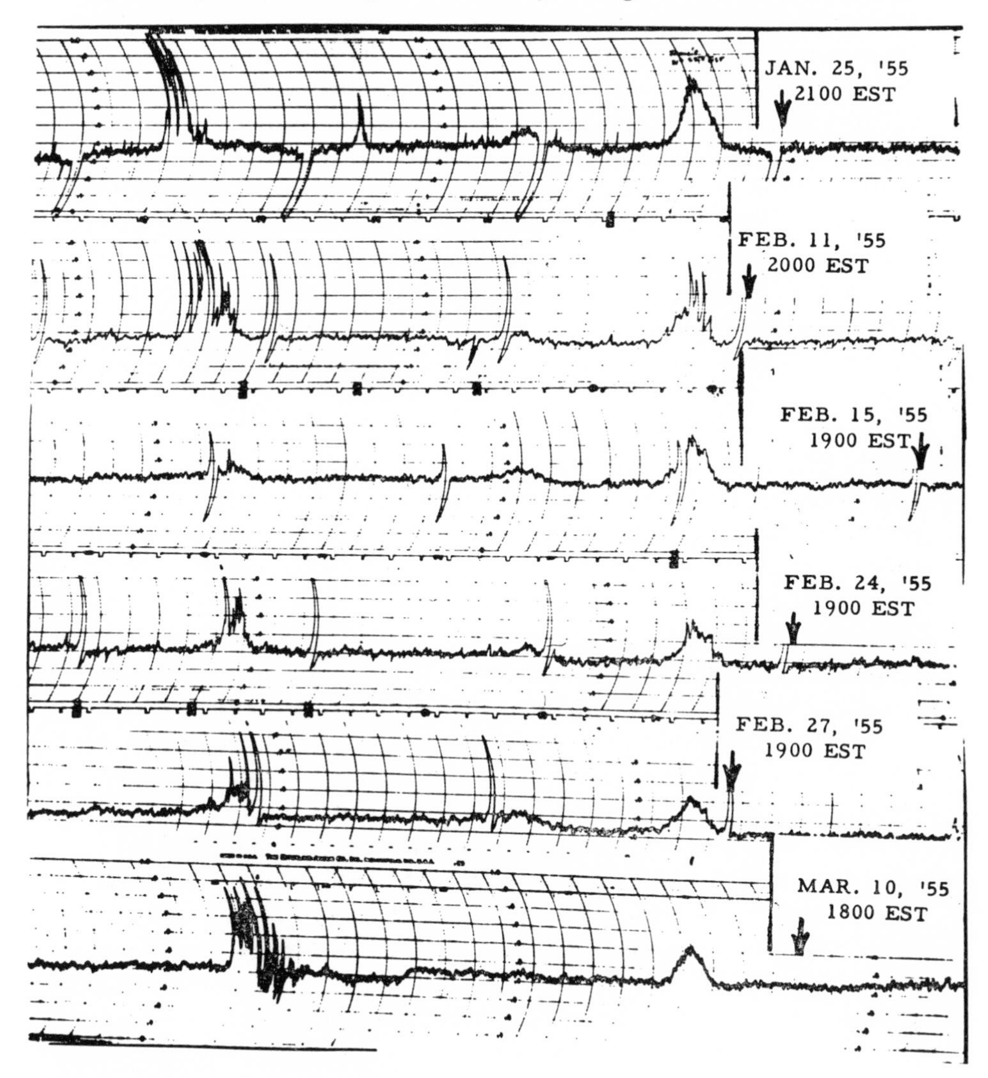

FIG. 1.—Mills Cross records of Jupiter, time increasing from right to left. Crab Nebula large source at far right, IC 443 the weaker source 45 minutes later, and Jupiter burst appears farthest left. Calibration marks interrupt the records once an hour (Burke and Franklin, 1955).

nearly 2 hours at the lower frequency. The signal-to-noise ratio indicated a lower limit of 200 to 1 for the intensity ratio at the two frequencies (Burke and Franklin, 1955).

At first, it was thought that the bursts might be due to electrical discharges similar to lightning, but comparison of the peak intensities (at times greater than 10^{-21}watt/m^{2}sec) with the power received at the same frequency from a lightning stroke of known distance required the absolute intensity of the Jovian lightning bolt to be at least 10^9 times larger than its terrestrial counterpart. Smith (1955) has pointed out that the failure to observe noise simultaneously at a frequency only a factor of 2 higher implied that the spectrum fell off more sharply than the spectrum of terrestrial lightning, which would decrease by no more than a factor of 16 or so, as compared with the lower limit of 200 observed.

In Section 6 it will be seen that more recent spectral investigations raise still more serious objections to the lightning hypothesis. It would appear that if electrical discharges are responsible, they must be of entirely different character from those observed in lightning storms.

3. LOCALIZATION OF SOURCES

Observations of galactic radio noise at 18.3 Mc/s had been made in 1950–1951 by Shain and Higgins (1953), using an antenna of relatively wide beam width. A review of these records enabled Shain (1955, 1956) to confirm that Jupiter had been a strong source of radio noise at that time. A particularly interesting series was available in which only a part of the original antenna had been used, covering such a wide range of hour angle that Jupiter was observable for nearly 8 hours each day, or almost one complete revolution. From the 7 weeks' records available (August 15–October 2, 1951) Shain showed that a definite periodicity could be established, as can be seen from Figure 2. The occurrence of radio noise plotted as a function of the longitude of central meridian showed an obvious systematic drift when referred to System I (the equatorial belt, assumed rotational period $9^h50^m30\overset{s}{.}004$) but was noticeably concentrated with respect to the System II co-ordinates. It was clear, therefore, that the active region or regions had a mean rotational period close to that of the non-equatorial regions (assumed rotational period $9^h55^m40\overset{s}{.}632$). The possibility could not be ignored that the radio emission was stimulated by solar radiation and consequently was most probable when the source was near the limb, but the simplest hypothesis, in the absence of further evidence, was to assume that the radiation of radio noise was most likely to be observed when the source was near the central meridian. Shain deduced that the most probable longitude was 67° in System II on August 14, 1951, with a period of $9^h55^m13^s$ fitting the observations best. At first, it appeared that the most active radio region was well correlated with one of several white spots in the South Temperate Belt. Optical observations by amateurs of the British Astronomical Association, summarized by Fox (1952), showed good agreement in both position and rotational period during August and September, 1951. Other observations have made it abundantly clear, however, that no long-term correlation with this optical feature exists. A major purpose of subsequent observations has been to establish whether the radio emission originates from long-lived active regions or is a transient phenomenon associated with particular regions of Jupiter for only a few months at a time.

The 1950–1951 Australian observations demonstrated the importance of observing the planet for as many hours per day as possible, with apparatus that has some direction-finding capability, in order to distinguish between Jovian radio bursts and terrestrial interference, both natural and man-made. Steerable antennae are so cumbersome at the longer wave lengths that the Michelson interferometer, using relatively low-gain antennae is clearly the most suitable (for a survey of these instruments, see Wild, 1953; Vol. **1**, chap. 9, Sec. 10, p. 676; also Bolton, 1960). Over the period September, 1955, to March, 1956, radio observations of Jupiter were made at several places, some form of interferometer being used in all cases. These included the Radio Physics Laboratory in Sydney, the Central Radio Propagation Laboratory at

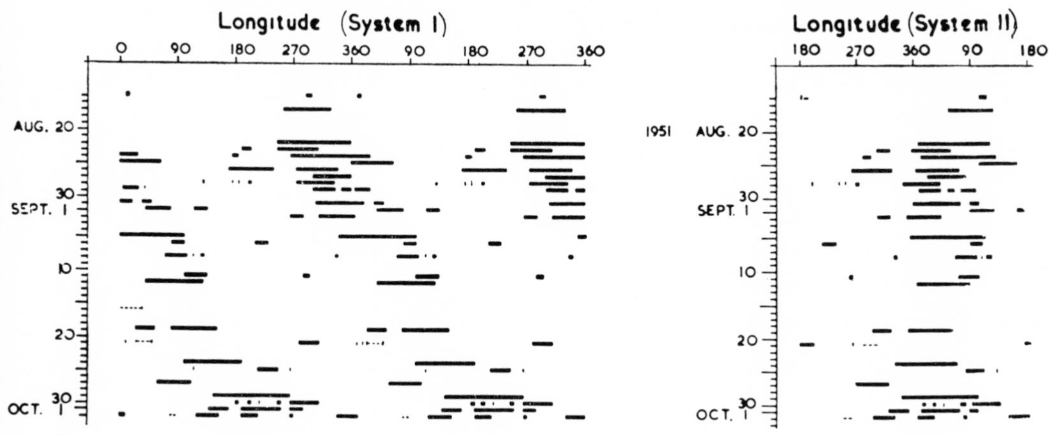

FIG. 2.—Central meridian of co-ordinate systems on Jupiter while radio noise was being observed at 18.3 Mc/s in Sydney during August–September, 1951 (Shain, 1956).

Boulder, Colorado, Ohio State University, and the Carnegie Institution of Washington. The CRPL continued observations over the 1956–1957 apparition. Since 1957, observations have been made at the University of Florida, Gainesville, and at Yale University.

As an example of the activity observed during this period, the Carnegie Institution and Sydney data are shown in Figures 3 and 4, with the same representation as in Figure 2 but with qualitative indication of intensity. Only the System II representation is used, since no obvious periodicity with respect to System I has been detectable in this series. The positions of several visual features during this period, as reported by Alexander (1956) in his summary of the BAA observations, are also

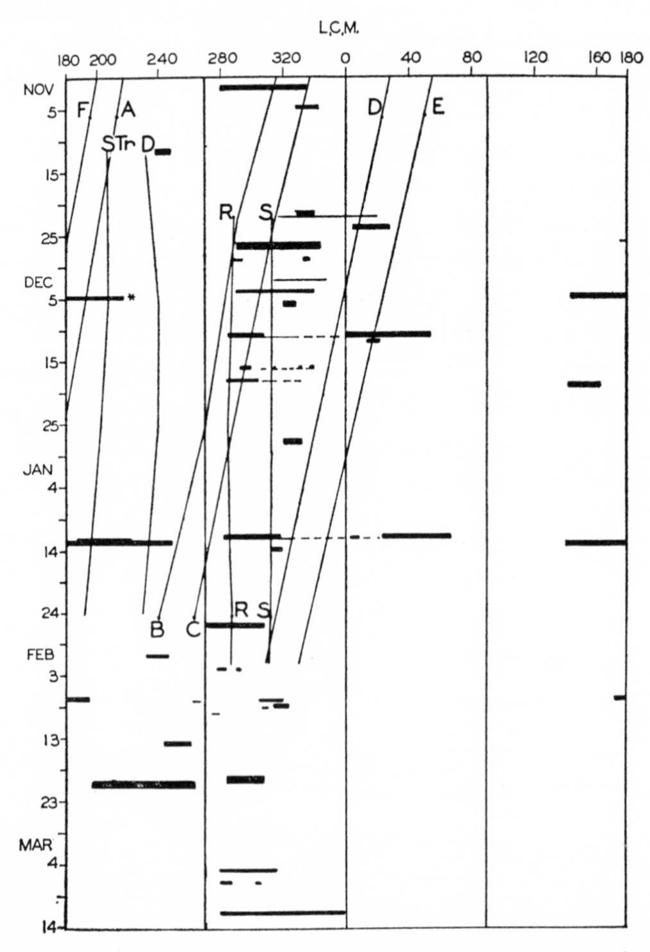

FIG. 3.—Central meridian, System II, for noise occurrences at 22.2 Mc/s observed at Washington, 1955–1956 (Franklin and Burke, 1958).

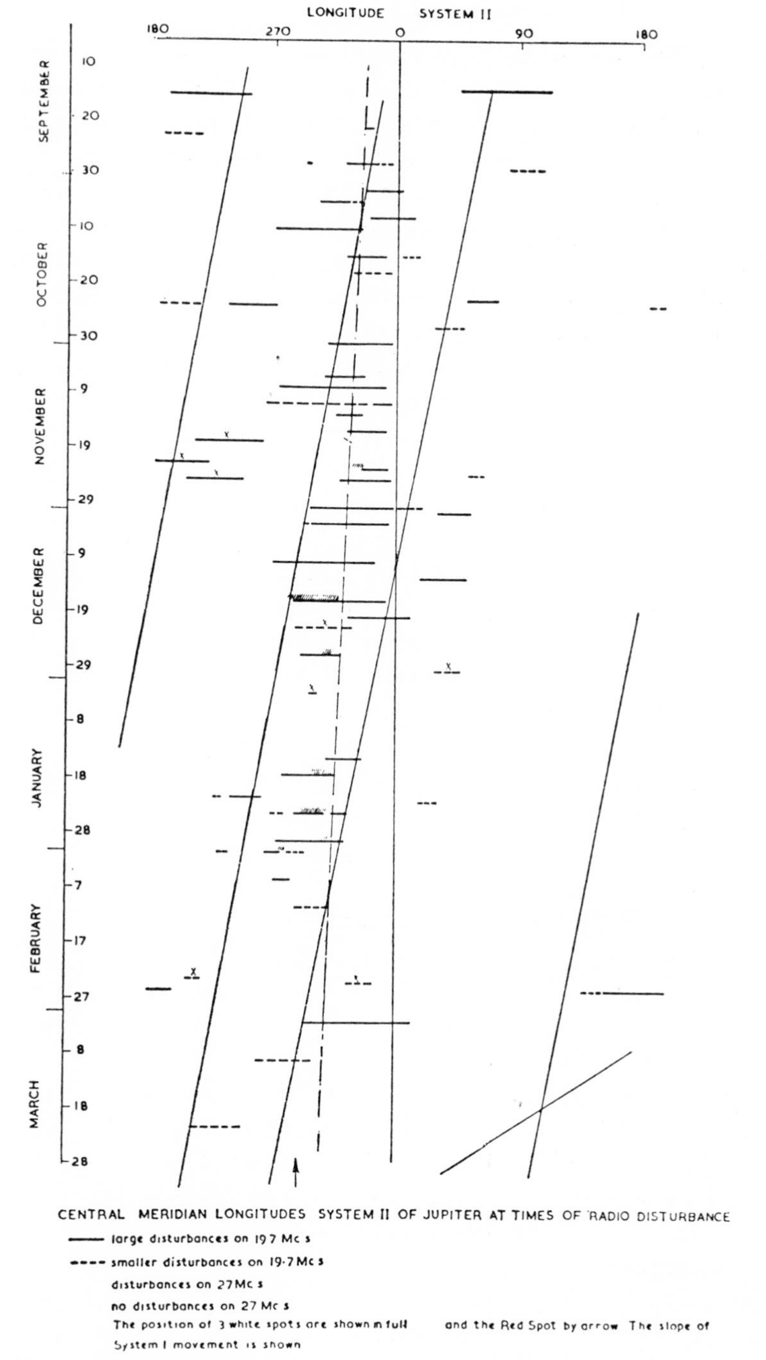

FIG. 4.—Central meridian, System II, for noise occurrences at 19.7 Mc/s observed at Sydney, 1955–1956 (Shain and Gardner, 1958).

shown. The slightly faster rotational period for the three white spots FA, BC, and DE in the South Temperate Belt can be easily seen. The spot DE was clearly not identifiable in this series with the majority of the radio-noise storms observed. The most marked active region had a longitude not far from that of the Red Spot, RS, if the central meridian marks the most probable position when radio noise is observed. Mean rotational periods were fitted to the observations, Gardner and Shain (1958) quoting $9^h55^m30^s$ for the Australian observations and Franklin and Burke (1958) giving $9^h55^m33^s$. Both estimates of the short-term periodicity favor a slow drift in System II longitude, with a period about 10^s faster than that of the Red Spot.

The probability of long-term persistence was recognized independently by Gallet (1957), Burke (1957), and Smith *et al.* (1958), who, in combining the 1950–1951 Australian series with later observations, deduced rotational periods of $9^h55^m29^s.5$, $9^h55^m28^s.5$, and $9^h55^m28^s.6$, respectively. The most active radio region has continued to persist and shows in later observations as well. Figure 5 is a representation covering ten years of observations, showing frequency of occurrence of noise storms as a function of longitude of central meridian in System II, a representation first used by Shain (1955). The base line for each histogram is drawn for the mean epoch of observation, the duration of observations being indicated by the shading at the right. Only one set of observations for each opposition is given, the selection being made solely to give as homogeneous a set of data as possible. The sources are as follows:

1. January, 1951: 18.3 Mc/s, by Shain (1956), who gave LCM of principal active region but no occurrence histogram.
2. September, 1951: 18.3 Mc/s, by Shain (1956).
3. June, 1954: 22.2 Mc/s, by Smith and Burke, quoted by Franklin and Burke (1958).
4. March, 1955: 22.2 Mc/s, by Burke and Franklin (1955), point observations only, each point extended over 40° to approximate other results.
5. January, 1956: 22.2 Mc/s, by Franklin and Burke (1958).
6. January, 1957: 18 Mc/s, by Carr *et al.* (1958).
7. February, 1958: 22.2 Mc/s, by Smith and Carr (1959).
8. May, 1960: 22.2 Mc/s, by Smith *et al.* (unpublished).

A quantitative statistical analysis of the periodicities present in all data available through mid-1960 has been performed by Douglas (1960), who considered two possible models. The first model is the simplest one suggested by the histograms and consists of a randomly active region on Jupiter, emitting radiation outward in a cone whose size may vary but

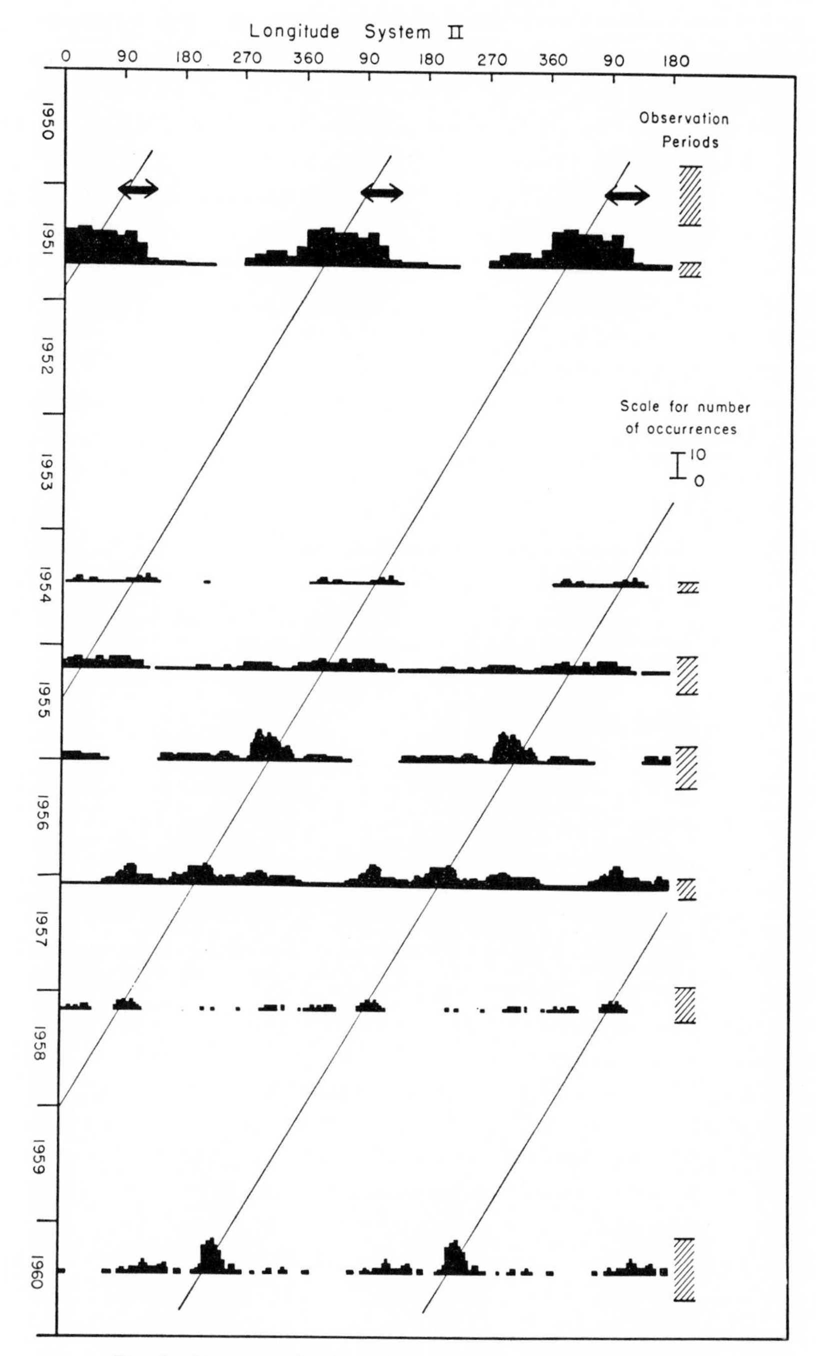

Fig. 5.—Summary of radio noise occurrence on Jupiter, 1950–1960

is symmetrical about the zenith. The histograms then give direct indication of the longitude of the source. The second model is of a randomly active region on the planet, but stimulated by solar radiation. The simple source model should show periodicity only with respect to the longitude of central meridian (LCM) when the assumed rotation period of the longitude system is close to the true period. Douglas used the variance from the mean as a criterion for the period, maximum variance indicating the best guess at the true period. Table 1 summarizes his results, region 2 referring to the main peak, with regions 1 and 3 referring to the active regions on either side of the main peak at smaller and larger longitudes, respectively. His periodograms for regions 1 and 2 had well-defined maxima in all cases, but region 3 for some samples did not exhibit as clear a maximum. His mean period, combined with the assigned weights, is $0^{d}.4135344 \pm 0^{d}.0000015$ or $9^{h}55^{m}29^{s}.37 \pm 0^{s}.13$ (s.d.)

TABLE 1

DATA FROM THE WORK OF DOUGLAS (1960)

	Period	Weight
18 Mc/s Region 1	$0^{d}.4135332$	69
Region 2	337	134
Region 3	344	107
20 Mc/s Region 1	368	31
Region 2	372	74
Region 3	279	33
22.2 Mc/s Region 1	352	56
Region 2	351	86
Region 3	358	37

Periodograms for individual years indicated that no significant fluctuations could be detected within the uncertainties, which were necessarily greater. The mean rotational periods for 1950–1955 and 1956–1960 were apparently within $0^{s}.5$ of the mean period for the full ten years.

The histograms of Figure 5 have been plotted against the System II longitude system and hence do not represent the true profile, which shows somewhat sharper peaks when the best radio period is used.

The simple source hypothesis is not the only possibility, although it is certainly the most straightforward interpretation of the results. Douglas, in his second model, sought a periodic effect which would indicate solar stimulation. If the bursts arose from solar stimulation, there should be a periodicity in the data corresponding to the variation in longitude of the sun as seen from a source on Jupiter. No statistically significant effect could be found, but neither could the possibility of solar stimulation be completely ruled out.

4. DIRECTIONAL PROPERTIES OF THE RADIATION

If an isotropic point source of radiation were located on the surface of the planet, one would expect radio noise from a given active region to be observable over a 180° range of longitude. Shain (1956) pointed out that such was not the case for the August–September, 1951, observations, which showed that the most active region was visible over only 135° of longitude, indicating that some mechanism was limiting the range of angles within which radiation could escape from the planet. The effect was even more striking at 22.2 Mc/s in 1955–1956, as can be seen from the histogram in Figure 5. The peak is even sharper when plotted with respect to the best radio rotation period, exhibiting a width of about 50°. Shain and Gardner's observations for the same period give a similar width, 60°, for an observing frequency of 19.7 Mc/s, although the edges of the histogram, when drawn for their data, are not so sharp. Bowles

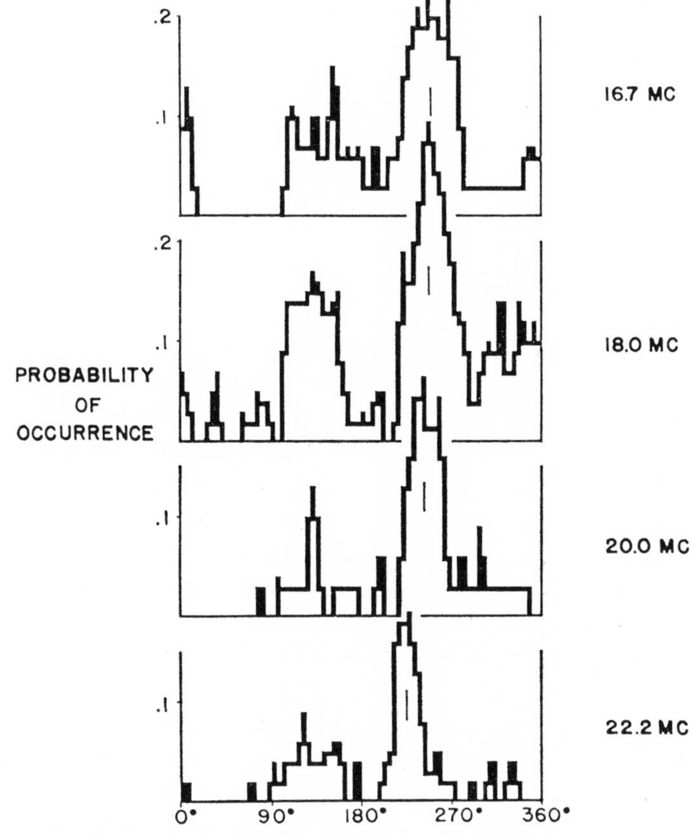

Fig. 6.—Histograms of radio noise occurrence at various frequencies obtained at Maipu, Chile, during 1960 (Smith and Carr, unpublished).

and Gallet (1956) also called attention to the narrow polar diagrams exhibited at their two observing frequencies of 18 and 20 Mc/s.

Franklin and Burke (1958), in comparing observations during 1955–1956 at a number of frequencies covering the range 18.5–27 Mc/s, noted that the angular width of the main peak seemed to decrease with increasing frequency, a tendency which was demonstrated more convincingly by Smith and Carr (unpublished) during 1960 at the Florida State University station at Maipu, Chile. The histograms are shown in Figure 6. The peak noticeably decreases in width with increasing frequency, accompanied by a progressive shift in LCM.

Shain (1956) has briefly discussed a few of the possibilities which could account for such a restricted angle of emission. The restriction would be an apparent one only if the source were stimulated by solar radiation. The threshold amount of radiation required would appear to have changed, however, between 1951 and 1955, if such a mechanism were acting. Atmospheric refraction does not act in the right direction, although ionospheric refraction can. A source situated below an ionosphere would exhibit such a narrowing of the polar diagram, due to total internal reflection at the critical angle (cf. also chap. 14, Sec. 4.1). Since the angle of total reflection is frequency-dependent, the observed range in LCM should also be frequency-dependent. Franklin and Burke (1958) showed that the approximate variation with frequency should be given by

$$\cos \tfrac{1}{2}\lambda = \frac{f_0}{f} \sec \phi ,$$

where λ is the range of LCM, f_0 is the critical frequency at the ionization maximum, and ϕ is the latitude of the source on Jupiter. The quantity f_0 would vary, of course, and so the cone of emission should vary. Nevertheless, the observed decrease in angle with increasing frequency poses a direct conflict with the simple ionospheric hypothesis.

More complicated models could be considered, for it is not clear that the source need be located below the ionosphere, nor need the main peak be composed of a single source. The difficulty does not yet appear to have been adequately solved.

5. SCINTILLATION EFFECTS

It is well known that radio sources scintillate, primarily because of ionospheric density fluctuations, although the details are not well understood. The degree of scintillation is frequency-dependent, increasing as the frequency decreases. In the vicinity of 20 Mc/s, the intensity fluctuations are nearly 100 per cent at times, and Shain and Higgins (1953) demonstrated that the smaller the angular size of the source, the greater was the degree of scintillation, just as for optical objects. If dynamic details of the Jupiter bursts are to be studied, the effects of scintillation should be well understood, or one might be determining properties of our own ionosphere rather than that of Jupiter. This prompted Shain and Gardner (1958), to observe at two different locations with receivers tuned accurately to the same frequency. A comparison of records from the two sites, 20 km apart, revealed a disturbing dissimilarity so great that they concluded that scintillation effects on Jovian noise were far more severe than for any other known source. There was almost no burst-for-burst correlation, and frequently a group of bursts visible at one station was undetectable at the other. Shain and Gardner pointed out that the active regions, which are certainly smaller than the disk of the planet, must have an angular diameter smaller than that of any other known source of radio noise. Consequently, scintillation which is known to be severe at these lower frequencies for sources several minutes of arc in size could easily be even more violent for a source having a diameter of only a few seconds of arc or less.

These results have been confirmed by A. G. Smith *et al.* (1960), who made simultaneous comparisons between Maipu, Chile, and Gainesville, Florida. Experiments at Yale (unpublished) do not show such great dissimilarity, possibly because the spacing was not large enough to escape coherence effects. The question might be raised whether the burst-like structure is entirely a scintillation effect, the true burst length being many seconds long. This has not been demonstrated but appears unlikely from inspection of fast records of individual bursts.

A related phenomenon was noticed by Shain (1956) in the 1951 observations. A change in appearance of the bursts attributed to the planet Jupiter could be seen as the planet approached conjunction with the sun, the impulsive character of the bursts being smoothed out markedly. A similar smoothing was noticeable in the 1954 observations at Washington. Shain interpreted the smoothing as a multiple-scattering phenomenon analogous to that inferred by Hewish (1955) from measurements of the occultation of the Crab Nebula by the solar corona.

6. FREQUENCY AND TIME DEPENDENCE

The success which has been met in searching for regularities in the dynamic structure of bursts from the active sun raised hopes that similar effects might be observable in the noise occurrences on Jupiter as well (a comprehensive survey of solar noise has been given by Pawsey and Smerd, 1953). The most complete observations during 1955–1956 in this respect were those of Gallet and Bowles at Boulder, who observed simultaneously at 18 and 20 Mc/s with identical equipment. In Sydney, Gardner and Shain (1958), in addition to their principal equipment at 19.7 Mc/s, operated single antennae at 14 and 27 Mc/s, while, at Washington, Franklin and Burke (1958) had access to records taken by H. W. Wells at 18.5 and 27 Mc/s, in addition to their 22.2 Mc/s observations. In chapter 14 the Boulder results are discussed in detail. The remaining observations, since they were made with dissimilar equipment, are difficult to discuss quantitatively, but several qualitative features were common to all observations. There appeared no definite burst-for-burst correlation between the different frequencies, although, in view of the severity of scintillations, this need not mean that no such relation exists. A more striking difference between observations at different frequencies, noted by all observers, was the complete absence at times of noise at one frequency when at an adjoining frequency a strong noise storm was in progress. Douglas (1960), Lee and Warwick (1960*b*), and A. G. Smith *et al.* (unpublished) have made observations with equipment that was better suited to spectral studies, and a few facts seem clear.

The sweep-frequency equipment of Lee and Warwick (1960*a*) made observations over the band 16–35 Mc/s but integrated the output so that an average of noise activity as a function of frequency and time was obtained, and individual bursts could not be seen. Plate 1 shows one of their earliest records, with the Jovian noise showing as a banded structure through the noise. The bands are an instrumental effect, an interferometer being used in which the interference maxima and minima are functions of frequency. The first traces of noise show around 0425, increasing in intensity as time progresses, with a general tendency to progress from low to high frequency. The noise apparently continues into the range above 35 Mc/s, beyond the range of the equipment. Lee and Warwick have reported that the tendency to progress from low to high frequency with time is the most usual pattern, although reverse cases have also been seen. The records also show the tendency of the noise storms to last longer at the lower frequency, and the progression from low to high frequency would distort the LCM histograms in precisely the manner shown in Figure 6.

Douglas and H. J. Smith observed Jovian noise with a series of ten fixed-tuned receivers, spaced to cover a range of about 2 Mc/s. Since all channels were operating continuously, more sensitivity was obtained,

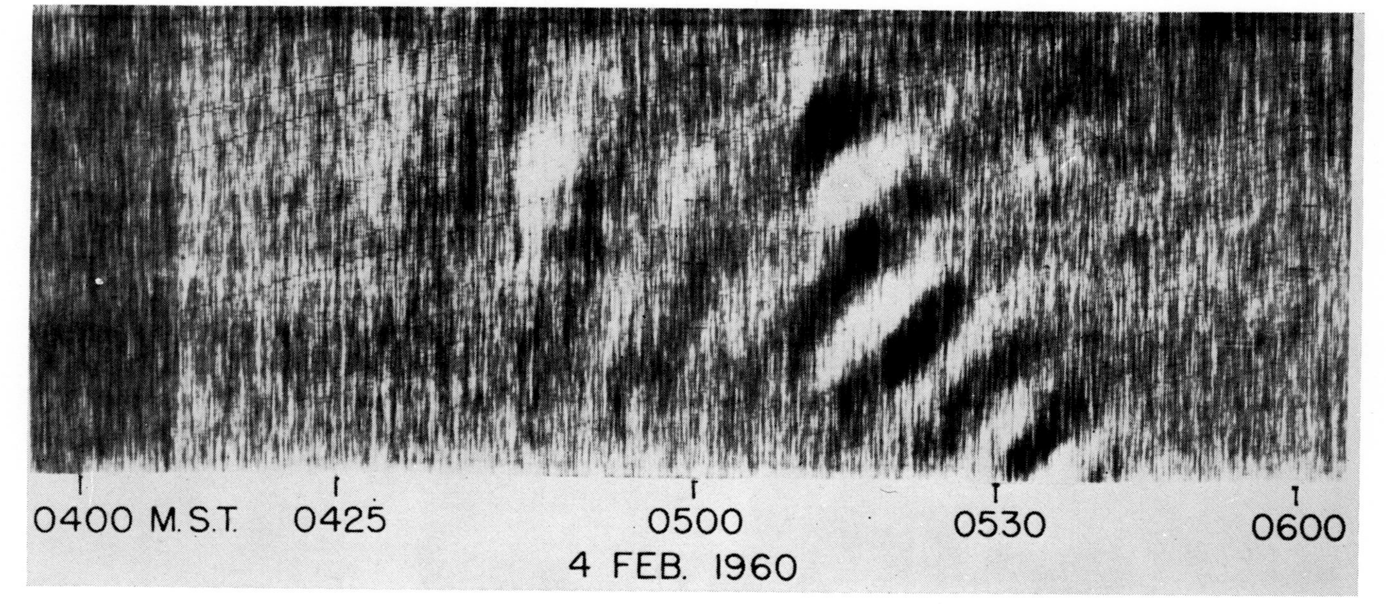

PLATE 1.—Sweep-frequency record of Jovian radio noise, covering the range 16–35 Mc/s, obtained at Boulder, February 4, 1960. Low frequency at the top, time increasing left to right. The banded structure is instrumental, since a phase-switching interferometer was used. (Lee and Warwick, unpublished.)

and individual bursts could be studied. The band width of each burst could be measured and is apparently of the order of 1 megacycle, although bursts could sometimes be seen extending across the entire band. The burst occurs almost simultaneously across the entire band, within less than a tenth of a second. A group of bursts tends to move in frequency with time, and Douglas reports that the movement of the center of gravity of such a burst group usually, but not always, progresses from high to low frequency. This is not necessarily in disagreement with the observations of Lee and Warwick but may represent fine structure within the total noise storm.

The duration of individual bursts has been studied by Douglas (1960), who finds that the duration of isolated, well-defined bursts varies from $0^{s}.2$ to over 1^{s}, occurring in a rough Rayleigh distribution peaked at about $0^{s}.6$. The records of isolated bursts often show a characteristic triangular shape, rather different in form from the noise bursts recorded from the active sun.

Both Kraus (1956) and Gallet and Bowles (1956) have reported noise bursts of much shorter period, lasting only a few milliseconds. On the other hand, Gardner and Shain (1958) report negative results in a search for such bursts. All noise events much shorter than 1 second were attributable to terrestrial ionospherics, their system of interferometer permitting direction finding on single bursts. Subsequently, no evidence for bursts lasting only a few milliseconds has been reported.

In discussing long-period variations of Jupiter activity, covering periods of many months or years, complications arise because different observations are not directly comparable if different antennae are used. Terrestrial interference is so variable that, even with the same instrument, the relative sensitivity may vary diurnally, seasonally, and with the solar cycle. As an example of this, Gardner and Shain (1958) compared the relative Jupiter activity for 1955–1956 as observed by their small interferometer, which was easily disturbed by interference, with the relative activity observed with a large fan-beam instrument which discriminated better against local interference. A definite maximum appeared in November, 1955, using the first group of observations, although the number of occurrences seen by the latter antenna in this month showed only a slight maximum, which could easily have been a chance fluctuation. Some estimate can be made, probably, by comparing the percentage of days on which activity was observed with the fan-beam instrument with the activity observed with the 22.2 Mc/s Mills Cross in Washington the previous year. For the interval

January–March, 1956, noise was observed at 19.7 Mc/s in Sydney on 1 day out of every 10, while for the same period in 1955 radio-noise bursts were detectable on 1 day out of 3, despite the slightly higher observing frequency. The marked change in activity over this period indicates that the degree of activity varies over a period as short as a year. The Australian observations in 1951 showed a degree of activity so great that one can certainly say that during that period the planet was more active than for 1955–1956.

From 1956 to 1957 activity fell markedly as the sunspot number increased, as described by Gallet in chapter 14. During the years 1958–1959, at sunspot maximum, noise storms were so rare that observations were a most discouraging and tedious task, but, in early 1960, observations of Jovian noise became more frequent. It is difficult to make a quantitative analysis of the effect, since many different equipments were involved, operating under a great variety of interference conditions, but it appears that the more active the sun is, the more infrequent Jovian noise storms become.

7. POLARIZATION

The radio noise from noise storms observed on the sun are usually nearly circularly polarized (see Pawsey and Smerd, 1953), and it was natural to extend measurements of polarization to the planet Jupiter. In 1956, both Shain and Gardner (1958) and Franklin and Burke (1958) demonstrated that circular polarization was present in the radio noise from Jupiter. Shain and Gardner obtained one good measurement, when the longitude of central meridian was nearly 300° (System II), the approximate apparent longitude of the most active region. The polarization was approximately circular with right-hand sense of rotation, in the radio convention (looking along the direction of propagation). Franklin and Burke measured the sense of polarization for seven events, of which five exhibited right-hand, one left-hand, and one a mixed character, with different bursts exhibiting different senses, and some bursts polarized little, if at all (radio convention). Whenever the most active region was near central meridian, the polarization was always right-hand. Both the Australian and the Washington measurements showed the phase difference of the two linear radiation components to be close to 90°, with amplitude ratios ranging from equal to 3:1. The observation that the most active region appeared to have right-hand polarization in both hemispheres suggests that the polarization mechanism is not in our own ionosphere. Simultaneous observations by A. G. Smith and associates (unpublished) at Maipu, Chile, and Gainesville, Florida, confirmed these conclusions.

8. THEORIES OF RADIO NOISE GENERATION

8.1. Thermal Radiation

At least three components of the radio noise spectrum of Jupiter can be easily recognized. At wave lengths of 3 cm and shorter, the thermal emission from the lower atmosphere of Jupiter appears to be dominant and does not concern us here, although its characteristics will undoubtedly prove useful in future astrophysical studies. At decimeter wave lengths, the unexpectedly intense, relatively steady, radio emission requires explanation, as does the third component, the low-frequency noise bursts. While no relation between the two latter components has been demonstrated, neither has it been shown that they are completely independent. In the present state of our knowledge, the same theoretical consideration can be applied to both.

TABLE 2

Radio Phenomena for Decimeter Wave-Length Range (from Field 1959)

Wave Length (Cm.)	Disk Temperature (° K)	Flux: Unit 10^{-26} $wm^{-2}(c/s)^{-1}$
3.15	145±18	....
3.37	165±17	....
10.3	650±250	5.5±2
21.4	3500±1700	6.8±3
31.0	5500±1500	5.2±1.5
68.0	70,000±30,000	14±6

It has been customary in radio astronomy to describe objects in terms of total received flux, S, surface brightness of the source, B, and brightness temperature, T_B, which is not necessarily a physically meaningful temperature but refers to the temperature that a black body of the same angular size would have to be, in order to give the flux observed. The quantities, for a uniformly bright source of angular size ω, are related by

$$B(f)\,df = \frac{2kTf^2}{C^2}\,df\,, \tag{1}$$

$$S(f) = B(f)\,\omega\,. \tag{2}$$

Equation (1) is simply the Rayleigh-Jeans approximation to the low-frequency end of the Planck distribution, but expressed per unit frequency interval rather than the more customary unit wave-length interval. The radio phenomena observed on Jupiter can be briefly summarized in these terms for the decimeter range by Table 2. Within experimental error, the flux, S, remains approximately constant at 6×10^{-26} wm^{-2} sec (a slight increase in flux with decreasing frequency also fits the data), while the brightness temperature, assuming that the emission comes uniformly from the visible disk of the planet, varies from 145° K at 3 cm to 50,000° K at 74 cm. The angular size observations of Radhakrishnan and Roberts (1960) at 31 cm have shown that the angular size at that wave length is considerably greater, requiring a downward revision of the brightness temperature. Such a revision is doubtless required for all the longer wave lengths.

In the case of the long-wave-length bursts, even higher temperatures are required. For example, the bursts at 15 m wave length often exceed 10^{-20} wm^{-2} sec flux at the earth, requiring a brightness temperature (again assuming emission from the entire visible disk) greater than 10^{12} ° K.

The radio radiation from the quiet sun is well explained by the bremsstrahlung arising from electron-ion collisions in a high-temperature plasma, the temperature of the solar corona being of the order of 10^{6} ° K. The essential points, for the solar case, have been summarized by Pawsey and Smerd (see Vol. 1, chap. 7, of this series). The same considerations also are relevant if Jupiter has a corona and have been applied to the problem of the decimeter radiation by Fields (1959) and Roberts and Stanley (1960). If numerical values are inserted in Pawsey and Smerd's expressions for the absorption coefficient, κ, and emissivity η, one finds that for a completely ionized hydrogen plasma of temperature T° K and density n electron/cm³, for wave length λ cm,

$$\kappa = 2.51 \times 10^{-23} (3.38 + \log_{10} T - \tfrac{1}{3} \log_{10} \eta) T^{-3/2} \eta^2 \lambda^2 \text{ cm}^{-1} , \quad (3)$$

$$\eta = 2KT\lambda^{-2} \text{ (per unit band width)} , \quad (4)$$

where it is assumed that the index of refraction is close to unity.

One notices immediately that the emissivity per unit band width, η, is independent of frequency, while k varies as f^{-2}. If the frequency is sufficiently high, absorption can become negligible, and the ionized gas will appear optically thin, with surface brightness that is independent of frequency. At the low-frequency limit, for a gas of uniform temperature, the surface brightness must appear as that of a black body (eq. [1]). Consequently, the thermal radiation from electron-ion collisions in an ionized gas can only exhibit a spectrum that increases or is constant with increasing frequency, a theorem proved in general form by Westfold (1959).

The brightness temperatures encountered in the low-frequency bursts are so high that even if the peculiar frequency effects could be explained by refraction and scintillation phenomena, the mechanism fails completely.

The mechanism of thermal emission from a Jovian corona cannot be dismissed so abruptly in the case of the decimeter radiation. If the intensities of Table 2 only were to be explained, an optically thin corona of temperature 10^5 ° K, as mentioned by Fields and by Radhakrishnan and Roberts, would suffice, although the 74-cm intensity would have to be in error by a factor of 2. The difficulty lies in the diameter, 10^{11} cm, required. The 960-Mc/s angular size measurements of Radhakrishnan and Roberts imply a somewhat smaller diameter. A more serious objection is the linear polarization, 20 to 30 per cent, also reported by the same workers. It is not clear how an ionized gas, even in the presence of a magnetic field, could produce such strong linear polarization, and so the mechanism must be set aside. The possibility cannot be eliminated, however, that a significant fraction of the decimeter radiation may arise from such a corona.

8.2. Radiation from Electrons in Magnetic Fields

The observed polarization of the radio noise received from Jupiter implies the existence of some ordering mechanism, acting either on the radio source or on the medium through which the radiation passes. A magnetic field offers particularly interesting possibilities and is, in fact, strongly indicated by the circular polarization of the low-frequency bursts. Of the possible radiation mechanisms, the radiation from electrons gyrating in such a field has shown particular promise and is especially attractive because quantitative calculations are possible which can be compared with experiment. There is little difference, in principle, between the radiation from non-relativistic and relativistic electrons, since the effects are completely classical. Both relativistic and non-relativistic cases can be accurately described as magnetic bremsstrahlung but are commonly referred to as "cyclotron" and "synchrotron" radiation, respectively. In the limit, however, the differences are superficially so great that it is often convenient to consider the limiting cases separately. The low-energy limit is well known, and its application to the decimeter radio emissions of Jupiter have been discussed in detail by Field (1960). Radiation from highly relativistic particles is more complicated analytically and has been discussed by many authors, the most complete study probably being that of Westfold (1959). Application of the theory has been made by Field (1959) and Roberts and Stanley (1959) to the case of the Jovian decimeter radiation.

A particle of any energy can be characterized by γ, its ratio of mass to rest-mass, where $\gamma = (1 - v^2/c^2)^{-1/2}$. If γ is nearly unity (the non-relativistic limit), the radiation field associated with a particle of charge q and mass m as it gyrates in a magnetic field is well known: the lines of E are instantaneously meridians of a sphere, with the axis along the direction of acceleration, $(q/m)\, \boldsymbol{v} \times \boldsymbol{B}$. The instantaneous field strength goes to zero at the poles and reaches a maximum at the equator of the sphere. Since the trajectory of the particle is, in general, a helix, with gyration frequency equal to the cyclotron's frequency $(q/mc)B$, it is clear that the resulting radiation in the plane defined by $\boldsymbol{v}$ and $\boldsymbol{v} \times \boldsymbol{B}$ will be linearly polarized but will appear circularly polarized to an observer looking along the direction of the magnetic field. The observed polarization will depend on the geometry of the region, while the energy radiated from an assemblage of particles can be calculated easily from the relation

$$P = \frac{32}{q} \frac{f^2 e^4}{m^2 c^3 E_0} \int_v n_e B^2 E \, d v , \tag{5}$$

where E_0 is the rest-energy and E is the kinetic energy of the electron. Numerically,

$$P = 3.88 \times 10^{-21} \int_v n_e B^2 E' \, d v \quad \text{ergs/sec} ,$$

where E' is the electron energy expressed in electron volts. Detailed calculations have been performed by Field (1960) for a specific model to explain the decimeter radiation. Field postulates a general dipole field for the planet, with a supply of low-energy electrons trapped in the outer regions, much in the same fashion as particles are trapped in the earth's outer Van Allen belts, where large quantities of non-relativistic electrons, whose energies range from 60 to 200 kev, have been noted by Van Allen (1959) and further studied by Arnoldy, Hoffman, and Winckler (1960).

The particles are constrained to motions along magnetic surfaces, spiraling along the field lines until they reach the mirror points appropriate to their energy and inclination at the equator, when they are reflected and travel in the reverse direction to the conjugate mirror point. As they bounce back and forth, the individual electrons precess in longitude, but diffusion from one magnetic surface to another is probably not rapid. The outer radiation belt of the earth is not confined to one surface, of course, but the spread is not great, being perhaps ± 10 per cent in radius at the equator. For convenience of calculation, Field idealizes the Jovian case, assuming that all the electrons lie in one magnetic surface, of radius R_0 at the equator. He further assumes that the axis of the dipole field is coincident with the axis of rotation and that the distribution of velocities is isotropic at the equator.

A particular electron, although instantaneously radiating at the cyclotron frequency appropriate to the magnetic field in which it lies, traverses a range of field strengths between its mirror points; and so the total spectrum can be obtained by suitably averaging over the time spent at any particular field point, for a particular pitch angle at the equator, and then averaging over all pitch angles. It is clear that at some locus of the magnetic surface, a level of the atmosphere is reached at which the electrons are lost, and thus a maximum frequency is defined. Similarly, at the equator one reaches a minimum field strength, and hence a minimum emission frequency is also specified. By choosing R_0 sufficiently large, however, a broad range of frequencies can be embraced. Ordinarily, this requires a magnetic surface which passes close to the dipole axis.

The exact expression for the spectrum is somewhat lengthy, but its principal characteristics are easily described. Near the minimum frequency f_0 (the cyclotron frequency at the equator, at radius R_0) there is a divergence, caused by the assumption of isotropic velocity distribution and hence not serious. As one goes to higher frequencies (electrons emitting at higher latitudes), the spectrum flattens out, and for $f \gg f_0$,

$$P_v = \frac{0.68\, e^3}{m^2\, c^4} H_0 \epsilon \left(\frac{f}{f_0}\right)^{-1/3}, \tag{6}$$

where

$$\epsilon = \int_0^\infty N(E)\, E dE$$

is the total electron energy. At the highest frequencies, the emission region is close to the pole and obviously must have a cutoff at the frequency corresponding to the polar field. In order to fit the observed decimeter radio flux, which exceeds the thermal flux at least below 3400 Mc/s, the polar field must exceed 1200 gauss, which seems remarkably high. A reasonable fit to the observed fluxes from 10.3 to 68 cm can be achieved if f_0 in formula (6) is taken to be 1.9 Mc/s. The total energy requirement can be calculated by fitting to the observations; if $f_0 = 1.9$ Mc/s, then $\epsilon \sim 10^{24}$ ergs. If the electrons all have 40-kev energy and are extended over a small range of R_0—say 0.2 R_0, as for the earth—only 0.1 electron per cm^3 is needed, a density considerably less than that found in the earth's outer belt. The rate of energy loss is not excessive and apparently can easily be supplied by the same source that supplies the earth's radiation belts.

The greatest difficulty with the model lies in its polarization predictions. For $f \approx f_0$ (electrons close to the equator) and for $f \gg f_0$ (electrons near the pole), there should be a net linear polarization, with E parallel to the equator, and for some intermediate frequency, when the electrons are circling about the magnetic field in planes perpendicular to the equator, there should be a net polarization of E at right angles to the equator. Consequently, there are two frequencies, above and below this intermediate frequency, at which the net polarization is zero. Unless there is a polar field orders of magnitude in excess of 1000 gauss, the radiation at decimeter wave lengths corresponds to the case $f \gg f_0$ and hence should be emitted largely from the polar regions, with polarization parallel to the equator. The observations of Radhakrishnan and Roberts (1960) showed that there is indeed 20–30 per cent linear polarization parallel to the equator, but the angular extent is so large, 4 or 5 planetary diameters, that the radiation must be coming at least in part from the equatorial regions. It does not seem possible to reconcile this observation with Field's model, even with a wide range of magnetic surfaces and velocity distribution.

No detailed treatments have been given for the case of the low-frequency noise bursts, although simple estimates can be made from formula (5). For a typical burst, of intensity 10^{-21} wm^{-2} sec at the earth, spread over a band width of 1 Mc/s at 20 Mc/s, P must be approximately 10^{17} ergs/sec. Assuming that the emitting region has dimensions of the order of 10^9 cm, the electron energy density would have to be 10^9 ev/cm^3, and even higher, if the dimensions of the region are smaller. The uniformity of the magnetic field would have to be better than 10 per cent over the entire region, of course.

8.3. Relativistic Electrons in a Magnetic Field

When the energy of an electron circulating in a magnetic field is very much greater than its rest-mass ($\gamma \gg 1$), the radiation is radically altered by Lorentz transformation effects. The radiation field seen by an observer no longer varies in sinusoidal fashion, for the poles of zero field strength, which lay along the axis of acceleration for $\gamma \approx 1$, are bent forward in the direction of motion and the forward field strength is greatly strengthened. Most of the radiated power is concentrated in a narrow cone whose angle is of the order of $1/\gamma$, and consequently the observer sees a pulse of strongly linearly polarized radiation which can be described in terms of a Fourier series or Fourier integral, depending on whether or not the magnetic field is sufficiently uniform to insure periodicity. The gyration frequency in a magnetic field B_0 is

$$\omega_0 = \frac{eB_0}{\gamma m_0 c} = \frac{1}{\gamma} \omega_c ,$$

where m_0 is the electron rest-mass and ω_c is the classical cyclotron frequency ($=2.80\ B_0$ Mc/s for B_0 in gauss). For periodic motion, the spectrum consists of harmonics of ω_0, reaching maximum intensity at a frequency ω_m given by

$$\omega_m = \gamma^3 \omega_0 = \gamma^2 \omega_c$$

and then dropping rapidly in intensity at higher harmonics. Arzimovitch and Pomeranchuk (1945) and Schwinger (1949) have given analyses of synchrotron radiation, which first found astrophysical application in the suggestion of Alfvén and Herlofson (1950) that the anomalous radio noise from the discrete sources could be generated by this mechanism. The suggestion was revived by Shklovsky (1953) for the Crab Nebula and has since been treated by a number of authors. A particularly useful, rigorous first-order theory has been given by Westfold (1959), who deals in detail with the polarization characteristics as well.

When γ is sufficiently large to allow replacement of discrete harmonics by a continuous spectrum, the mean power $\langle P(f)\rangle\, df$ radiated into the frequency interval $(f, f + df)$ by an electron whose velocity vector makes an angle α with respect to the magnetic field B_0 can be written

$$\langle P(f)\rangle = C\,(B_0 \sin\alpha)\, F\left(\frac{f}{f_c}\right), \tag{7}$$

with $C = 2.34 \times 10^{-25}$ for B_0 in gauss, $\langle P\rangle$ in watts $(\mathrm{c/s})^{-1}$, where

$$f_c = (\tfrac{3}{4}\pi)\, \omega_c \gamma^2 \sin\alpha$$

and

$$F(x) = x \int_x^\infty \kappa_{5/3}(\eta)\, d\eta\,.$$

The function F involves the modified Bessel function $K_{5/3}$ and has been tabulated by Westfold, who also gives the related function F_p, which describes the linearly polarized part of the field. For $f \ll f_c$, F increases as $x^{1/3}$, and for $f_c \ll f$, F decreases as $x^{1/2}\, e^{-x}$. The total power radiated is given by integrating over all frequencies, with the result

$$P = \int P(f)\, df = 1.59 \times 10^{-19} \gamma^2 B_0^2 \sin^2\alpha \text{ watts}\,. \tag{8}$$

The formula is useful for rough estimates, but in most cases it is more useful to know the emissivity of an ensemble of electrons distributed in energy and direction. A particularly simple distribution of interest is that in which the electron velocities are isotropically distributed and the energy distribution is given by a simple power law,

$$N(E)\, dE = A E^{-\beta} dE \quad (E_1 < E < E_2)\,,$$

where the cutoff energies E_1 and E_2 insure convergence. The general expression given by Westfold is not simple, but for a frequency f in the range $f_{c1} \ll f \ll f_{c2}$, where f_{c1} and f_{c2} are the values of f_c appropriate to E_1 and E_2, the emissivity at an angle θ to the magnetic field takes the approximate form

$$\eta(f) \cong (\text{const.})(B_0 \sin\theta)^{(\beta+1)/2} f^{-(\beta-1)/2} . \qquad (9)$$

Consequently, if an observed spectrum can be described by a simple power law, the power law describing the electron energy distribution can be inferred.

If the spectrum of the decameter noise bursts described in § 6 is a true emission spectrum, essentially unmodified by ionospheric cutoff effects, the narrow band width observed (of the order of 10 per cent or less) rules out the possibility of generation by relativistic electrons. The function $F(x)$ in equation (7) has a sufficiently broad peak to guarantee a much wider band width even for a monoenergetic electron distribution. Unless the difficulties associated with an ionospheric cutoff mentioned in § 4 can be overcome, the synchrotron mechanism appears to offer little promise for solving the low-frequency noise problem.

In the case of the decimeter radiation, a very different situation prevails, and it is here that the synchrotron mechanism appears to offer most promise. Drake made the original suggestion that Jupiter possesses a far more extensive Van Allen belt than does the earth, with a sufficiently large population of relativistic electrons to give the observed radio noise intensity. Both Field (1959) and Roberts and Stanley (1959) have compared the observation with the theoretical possibilities with encouraging results. The strong linear polarization and large angular extent of the noise at 31 cm observed by Radhakrishnan and Roberts (1960) provides particularly strong support. A density of the order of 10^{-2} electrons/cc of energy greater than 1 Mev in a magnetic field of a few gauss would suffice, although this density is several orders of magnitude greater than the density of the terrestrial Van Allen belt.

Several awkward points remain, for the terrestrial Van Allen belt exhibits an energy distribution of the form

$$n(E) = AE^{-6}$$

from satellite measures quoted by Roberts and Stanley. The radio spectrum, from equation (9), would then be

$$P(f) = Kf^{-2.5} ,$$

which is much steeper than the observed spectrum. The values of flux given in Table 2 require that the spectral index lies between 0 and 0.5, implying an energy index β between 1 and 2.

When β is less than 2, there are interesting implications. A simple particle injection mechanism, in which electrons of energy E are injected and allowed to slow down by radiation losses only, results, from equation (8), in an E^{-2} distribution, giving a spectral index 0.5. If a distribution of electrons, with number diminishing with increasing energy, is injected, the spectral index will be larger than 0.5. If the spectral index is smaller than 0.5, continuous acceleration processes and/or extra loss mechanisms, effective for low energies only, must be invoked. The length of time required for an electron to lose half its energy is

$$T_{1/2} = 23.5\gamma^{-1} (B_0 \sin\theta)^{-2} .$$

The radiation lifetime is, therefore, about ten years for 1-Mev electrons in a magnetic field of 1 gauss. Consequently, it is quite possible that processes other than radiative losses determine the energy distribution.

8.4. Plasma Oscillation

In addition to the various electron bremsstrahlung processes, magnetic and otherwise, that have been treated in §§ 8.1–8.3, it has been suggested that plasma phenomena such as plasma oscillations can serve as a means of generating radio noise. The narrow-band nature of the low-frequency noise bursts has served as an indication that such collective phenomena, close to the plasma frequency, are involved. Zhelezniakov (1958) has discussed the possibility of exciting such oscillations in a Jovian ionosphere, which would have to have a density of the order of 10^7 electrons/cm^3. The calculated relaxation time, a few tenths of a second, agrees well with the observations. The problem of converting the longitudinal plasma oscillations into the transverse radiation field, which causes such difficulty in the case of the sun (cf. Pawsey and Smerd, Vol. 1 of this series) may not be so serious in the case of Jupiter, since sharper density gradients can occur. The principal weakness with all plasma oscillation theories is still theoretical, for the relevance of the simple theories to physical reality is not yet accepted by everyone. The excitation of the plasma oscillations, either by turbulence or by shock waves from below, and the maintenance of an extremely dense Jovian ionosphere, with ion densities one or two orders of magnitude greater than in our terrestrial ionosphere, are points which need further detailed discussion.

REFERENCES

Alexander, A. F. O'D. 1954 *J. British Astr. Assoc.*, **64**, 281.
1956 *Ibid.*, **66**, 208.

Alfvén, H., and Herlofson, N. 1950 *Phys. Rev.*, **78**, 616.

Arnoldy, R. L., Hoffman, R. A., and Winckler, J. R. 1960 *J. Geophys. Res.*, **65**, 1361.

Arzimovitch, L., and Pomeranchuk, I. 1945 *J. Phys. U.S.S.R.*, **9**, 267.

Bolton, J. 1960 *Stars and Stellar Systems*, ed. G. P. Kuiper and B. M. Middlehurst (Chicago: University of Chicago Press), Vol. **1**, chap. 10.

Burke, B. F. 1957 *Carnegie Institution of Washington Yearbook*, **56**, 90.

Burke, B. F., and Franklin, K. L. 1955 *J. Geophys. Res.*, **60**, 213.

Carnegie Institution of Washington 1955 *Yearbook*, **54**, 43.

Carr, T. D., Smith, A. G., Pepple, R., and Barrow, C. H. 1958 *Ap. J.*, **127**, 274.

Douglas, J. N. 1960 *A Study of Non-Thermal Radio Emission from Jupiter* (Ph.D. dissertation, Yale University).

Field, G. B. 1959 *J. Geophys. Res.*, **64**, 1169.
1960 *Ibid.*, **65**, 1661.

Fox, W. E. 1952 *J. British Astr. Assoc.*, **62**, 280.

Franklin, K. L., and Burke, B. F. 1958 *J. Geophys. Res.*, **63**, 807.

Gallet, R. M. 1957 *Trans. I.R.E. AP-5*, p. 327.

Gallet, R. M., and Bowles, K. L. 1956 *A.J.*, **61**, 194.

Gardner, F. F., and Shain, C. A. 1958 *Australian J. Phys.*, **11**, 55.

Hewish, A. 1955 *Proc. R. Soc. London, A*, **228**, 238.

Kraus, J. D. 1956 *A.J.*, **61**, 182.

Lee, R. H., Boischot, A., and Warwick, J. W. 1960 *Ap. J.*, **131**, 61.

Pawsey, J. L., and Smerd, S. F. 1953 *The Sun*, ed. G. P. Kuiper (Chicago: University of Chicago Press), chap. 7.

Radhakrishnan, V., and Roberts, J. A. 1960 *Phys. Rev. Lett.*, **4**, 493.

Roberts, J. A., and Stanley, G. J. 1959 *Pub. A.S.P.*, **71**, 485.

SCHWINGER, J.	1949	*Phys. Rev.*, **75**, 1912.
SHAIN, C. A.	1955	*Nature*, **176**, 836.
	1956	*Australian J. Phys.*, **9**, 61.
SHAIN, C. A., and HIGGINS, C. S.	1954	*Australian J. Phys.*, **7**, 130.
SMITH, A. G., and CARR, T. D.	1959	*Ap. J.*, **130**, 641.
SMITH, A. G., CARR, T. D., BOLLHAGEN, H., CHATTERTON, N., and SIX, F.	1960	*Nature*, **187**, 568.
SMITH, F. G.	1955	*Observatory*, **75**, 252.
VAN ALLEN, J. A.	1959	*J. Geophys. Res.*, **64**, 1683.
WESTFOLD, K. C.	1959	*Ap. J.*, **130**, 241.
WILD, J.	1953	*The Sun*, ed. G. P. KUIPER (Chicago: University of Chicago Press), chap. 9, Sec. 10.
ZHELEZNIAKOV, V. V.	1958	*A. J. U.S.S.R.*, **35**, 230, trans. *Soviet Astronomy*, **2**, 206.

CHAPTER 14

Radio Observations of Jupiter. II

By ROGER M. GALLET
National Bureau of Standards, Boulder, Colorado

I. INTRODUCTION

SOON after the discovery of the radio emission from Jupiter by Burke and Franklin (1955), the author undertook a study of this unexpected behavior of the planet. This led to the adoption of an experimental program at the Laboratories of the National Bureau of Standards at Boulder, Colorado, since not only information of astronomical interest would be obtained but also practical problems were involved: the nature of an ionosphere other than our own and the associated problems of the propagation of radio waves.

It was clearly an advantage to use differential methods of observation rather than absolute measurements. The discovery data had already indicated that the emissions consisted of *pulses*, of a few seconds' duration, and that they were not received every day, but irregularly. On 22.2 Mc they occurred on about one out of four observations. On frequencies well above 22 Mc the emissions were absent; this was shown by simultaneous observations at the Carnegie Institution of Washington and by the absence of emissions from Jupiter in the radio astronomy surveys at frequencies above 40 Mc. But at 22 Mc, Jupiter, when radiating, is by far the strongest radio source in the sky. The problem of whether one or several localized sources are present on the planet requires continuous observation for at least half a rotation, or 5 hours. But if an ionosphere limits the emission to a cone of semiangle less than 90°, we shall receive the emissions from one source during intervals always *less* than 5 hours. Overlapping cones of two sources at different longitudes may cause confusion, but the sources will not always both be active. Simultaneous observation at two frequencies not too far apart seemed indicated because, for a given critical frequency in the ionosphere above a source, the aperture of the cone is the larger for the higher frequency, and simultaneous observation should show a longer period of reception at the higher frequency, at least if the emission were continuous on both frequencies. For intermittent radiation, *average* intervals may be compared. At the same time, indications on the spectrum of the emissions would thus be provided. The fine structure of the pulses which might throw light on the emission mechanism required investigation also.

The equipment, permitting simultaneous observation on 18.0 and 20.0 Mc, was designed by K. L. Bowles of the Boulder Laboratories, who also took a large share in obtaining the observational data. The equipment consists of two independent phase-switching interferometers (cf. Wild, 1953) erected as close together as possible. In order to achieve the longest possible run, each arm consisted of one collinear array of five half-wave dipoles, oriented parallel to the earth's axis. A similar array, parallel to the first and placed a quarter wave length beneath as a reflector, roughly doubled the gain. This arrangement permits reception of a source near the equator for practically its entire track above the horizon. Adjustment

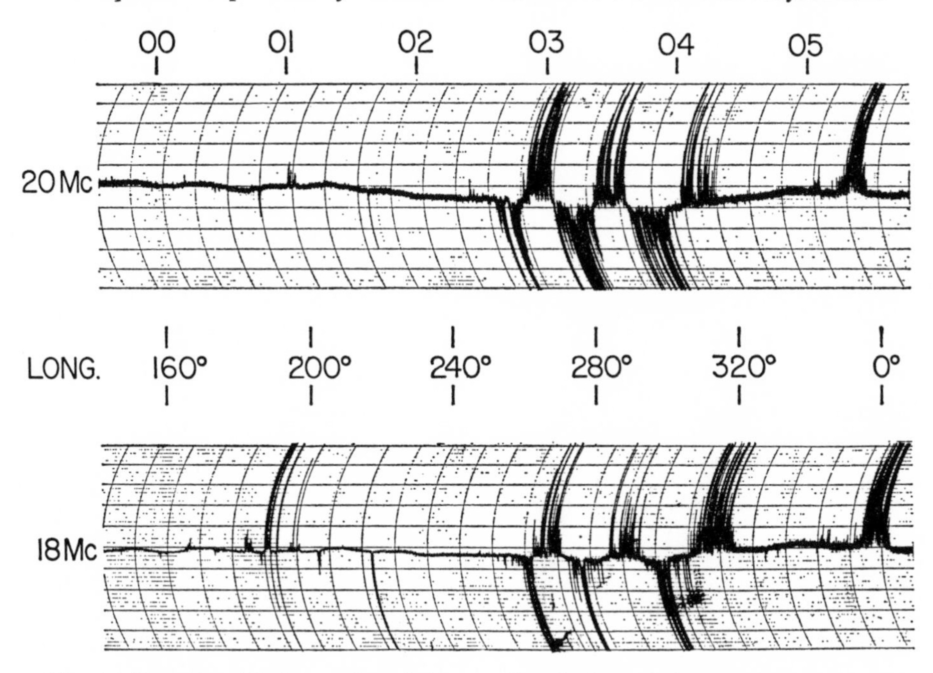

FIG. 1.—Example of Jupiter radio emissions at 20 and 18 Mc, February 10–11, 1956. Hours are M.S.T. (U.T. − 7^h), longitudes on System II. Note that the density of the emission is stronger at 20 Mc, a rather rare case.

of the directivity for other declinations was achieved by suitably phasing the dipoles.

The band width of the receivers was normally 100 kc but could be reduced when there was interference; then 16 or 5 kc were often used. Phase switching eliminates the general sky background due to the galactic emission; but a point source is recorded as alternate positive and negative deflections as it sweeps through the lobes of the interferometer pattern. The records show the signals alternatively as positive and negative. The duration of the passage through a lobe was about 13 minutes near the meridian, and the amplitude is positive during meridian passage. Near the horizon the duration increases progressively in a known manner. This permits easy identification of the Jupiter emissions.

Two types of observations were obtained, always simultaneously on both frequencies. First are the *synoptic records* made whenever Jupiter was above the horizon, at low chart speed, 3 inches per hour. The time constant of the mechanical recorder was about $\frac{1}{2}$ second, but the resolution of the record is much less because of the low speed, and many pulses forming

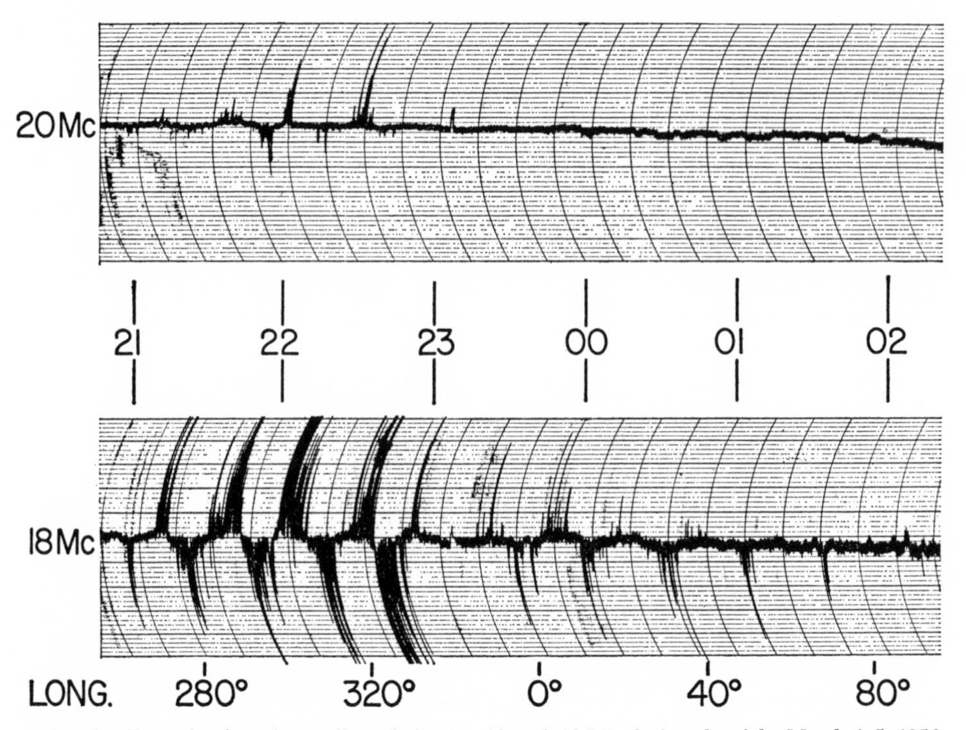

Fig. 2.—Example of Jupiter radio emissions at 20 and 18 Mc, during the night March 4–5, 1956. Hours are M.S.T. (U.T. − 7^h), longitudes on System II. Intensities much stronger at 18 Mc, a common situation.

a group are often blended together. These records give the times of the emissions; typical samples are shown in Figures 1 and 2. The central trace is the noise due to the fluctuating part of the galactic emission, the steady part being eliminated by the phase-switching system. Jupiter's signals were traced alternatively as positive and negative, following the interferometric pattern. Inevitably, the pattern reduced the sensitivity during the 1–2 minutes near each crossover, which may give a slight indeterminacy in the emission times. From January to March, 1956, it was often possible at night to record continuously on both frequencies for more than 10 hours, i.e., more than a complete Jupiter rotation.

Second, special series of high-speed records were made for studying the fine structure of the pulses and the degree of simultaneity of their arrival on both frequencies. For that purpose, different recording devices were available: a medium-speed pen recorder, resolving about $\frac{1}{2}$ second; a Brush Electronics Co. two-channel, high-speed pen recorder which records correctly 60 periods per second and will separate two sharp pulses separated by 0.01 second; an oscilloscope, synchronized with the switching frequency between aerials of 1000 cps and equipped with a special photographic camera for still shorter pulses; magnetic-tape recording; and an aural signal from earphones or a loudspeaker. The use of such equipment required the presence of an operator, and some or all of the special records were made whenever Jupiter was active.

2. SURVEY OF THE PRINCIPAL RESULTS

The program described started in the fall of 1955, and the most important results obtained are reviewed here. The discussion is based principally on the Boulder data, but, since meanwhile other data have been published, an effort is made to integrate all present information.

The synoptic records show the following characteristics:

a) The reception period was usually around $1\frac{1}{2}$ hours; it was sometimes shorter, but rarely as long as 3 hours consecutively. Never was the duration as long as would correspond to a source emitting continuously during one half-rotation of Jupiter. Usually the beginning and the end of the reception are defined within a few minutes.

b) There is usually a conspicuous difference between 18 and 20 Mc, in spite of their closeness. Normalized in percentage of time when Jupiter is emitting, the ratio (20 Mc/18 Mc) of the activities is about 0.5–0.6. From the data obtained by Burke and Franklin, with somewhat different equipment working on 22.2 Mc at the Carnegie Institution, it is seen that the decrease continues at higher frequencies. This is an important property of the *statistical spectrum*, defined by activity as a function of frequency, as distinct from the *dynamic spectrum* of each pulse, which gives the intensity versus frequency as a function of time.

c) There is a strong recurrence tendency for emissions at the same longitude, measured in System II. The histograms, giving the number of emissions versus the longitude of Jupiter's central meridian, show definite maxima and minima. The histograms we have used are normalized; they give the percentage of observations (not masked by interference) showing radio reception for each longitude of Jupiter's central meridian. These histograms are called *longitude profiles*. Their maxima are remarkably sharp and agree well on the two frequencies. They apparently define several local sources spaced in longitude, of different *specific activity*. Early in 1956 the most active of these was received on 50 per cent of the 18-Mc observations.

d) The sharpness of the longitude profiles indicates that a cone limits reception. The semiaperture of the cone seems to vary from day to day, but the average semiangle for the best-defined sources is only about 20° at 18 Mc. At 20 Mc the longitude profile is less sharp, so that the cones appear broader. These conclusions and those of Section 1 suggest the presence of a *Jovian ionosphere.*

e) The longitude profiles obtained for the periods January–March, 1956, and November, 1956—March, 1957, apart from an amplitude scale factor, have essentially the same shape. Thus the relative positions of the sources appear unchanged. The same seems to apply to profiles and histograms obtained from the other radio observations, particularly the 1951 records of Shain (1956). This indicates stability of the relative positions of the Jovian sources for at least 7 years, suggesting that they are related to the *solid body* of the planet.

f) The *rotation period* of this system of sources is found from the shifts in longitude in System II needed for obtaining good superposition of the longitude profiles for different years. All data of 1951, 1956, and 1957 agree closely and yield $9^h55^m30^s \pm 0^s.5$, some 10.5 seconds shorter than System II.

g) During the 1956–1957 opposition the total percentage of time when Jupiter was received was less than that during the preceding opposition, in the ratio about 0.65 on both frequencies. This could be because the sources maintained, on the average, the same true activity, with the critical frequency of the Jovian ionosphere increasing with solar activity, causing the cones of transmission through the Jovian ionosphere to decrease. The relation between solar activity and a Jovian ionosphere thus indicated is verified by subseries from the 1956–1957 data taken at intervals of about one solar rotation. A few years of additional data, properly normalized, should suffice to verify this relation through the declining branch of the present solar cycle.

Complementary results, mentioned under *h* and *i* below, have been obtained from the high-speed records.

h) Two kinds of pulses are received. The usual kind, called "long pulses" (*L*-pulses), have a duration of about 2 seconds, with a spread of from 1 to 3 seconds. One may follow the rise and fall of the *L*-pulses aurally or by watching the pen recorder. Occasionally, very short and sharp pulses are received, with a duration of the order of 10^{-2}–10^{-3} seconds. These *S*-pulses have a peak amplitude usually larger than the *L*-pulses and may occur at the rate of as high as 10 per second.

i) Simultaneity on 18 and 20 Mc was never observed, either for *L*- or for *S*-pulses. Also, a careful search for time displacements up to at least 20 seconds between recognizable groups of pulses on the two frequencies failed to show any similarity in the fine structure of the emissions. One may even receive *S*-pulses on one frequency and *L*-pulses on the other or have quiet on one frequency while the other is very active. While there is definitely a gross correlation between the two frequencies in synoptic activity, there is complete dissimilarity in the details during intervals of several minutes. Thus the *dynamic spectrum* of the pulses must be very narrow. Now the pulses obtained with band widths of 16 kc and separated by 200 kc are essentially identical. Thus the relative band width of a given pulse is bracketed,

$$0.01 < \frac{\Delta f}{f} < 0.1 .$$

The *S*-pulses are not shown on the synoptic records, and their relative contribution to the total emission is still unknown. It is estimated that they are not present more than a few per cent of the emission time.

j) The pulses exhibit a remarkable tendency toward *grouping*. In between groups practically no pulses are present. A group may last some minutes for the *L*-pulses and 10–20 seconds for the *S*-pulses, while the gaps are usually of longer duration. Within intervals of 30 minutes or more, the groups show a tendency to cluster, with larger gaps between them. Inside a cluster, usually one group stands out, with a larger density of pulses, a longer duration, and a somewhat larger amplitude of pulses. This grouping property resembles an irregular relaxation phenomenon.

No adequate mechanism has yet been found to account for all observations. Some theoretical discussion is given in Section 3, along with a more quantitative presentation of the data. A fuller account will be published later as a report of the National Bureau of Standards.

3. SYNOPSIS OF THE BOULDER DATA

The observations at Boulder were started January 6, 1956, and were interrupted late in March, 1956, when increasing interference masked the Jupiter emissions. The interference was due to the increased ionization of the terrestrial F2 layer following increased solar activity and to normal seasonal and daily variations in the ionosphere for the progressively shifting times that Jupiter was above the horizon. Attempts to observe Jupiter when it was behind the sun (September, 1956) failed. Observations were

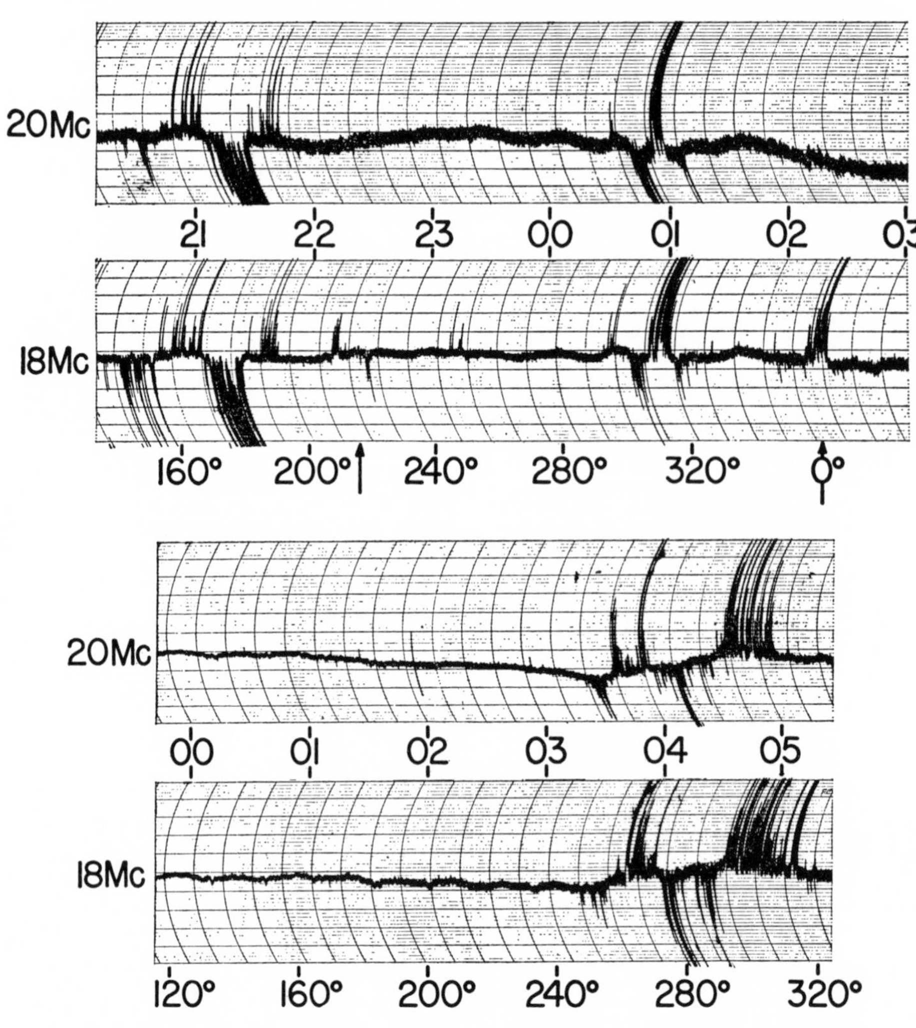

Fig. 3.—Further examples of Jupiter radio emission, with comparable intensities on both frequencies. Arrow indicates short bursts of interference out of phase on the interferometric pattern for the direction of Jupiter. Dates: January 6–7 and February 17–18, 1956.

resumed on November 14, 1956, and continued until the end of March, 1957; during this interval Jupiter was observable at night, and the F2 layer was less ionized.

Typical synoptic records are found in Figures 1–4. Figure 1 shows three emission periods in the course of 6 hours—one very weak, one very strong, and one very strong but of short duration. The corresponding longitudes of Jupiter's central meridian computed in System II are indicated. Besides the unusual intensity ratio, Figure 1 shows that the two strong emissions started and finished some minutes earlier on 20 Mc than on 18 Mc. This feature was often observed, though it was not a normal systematic effect. An earlier start on 20 Mc would be expected because of the larger ionospheric cone for a source which is continuously emitting on both frequencies; but then the 20 Mc emission should end later also. The situation of Figure 1 may indicate an asymmetry of unknown origin, perhaps related to the morning-afternoon asymmetry of the critical frequencies in Jupiter's ionosphere or to the distribution of magnetic field at the source. Figure 2 presents the more usual case where the emission is the stronger on 18 Mc, while Figure 3 shows no strong differences between 18 and 20 Mc. Figure 4 illustrates how interference will disturb observation. In Figure 5 emissions of short and long duration are grouped for showing the various gradations from a very dense record to a very sparse one. The sparser groups (classes I and II) are difficult to recognize, and their limits are

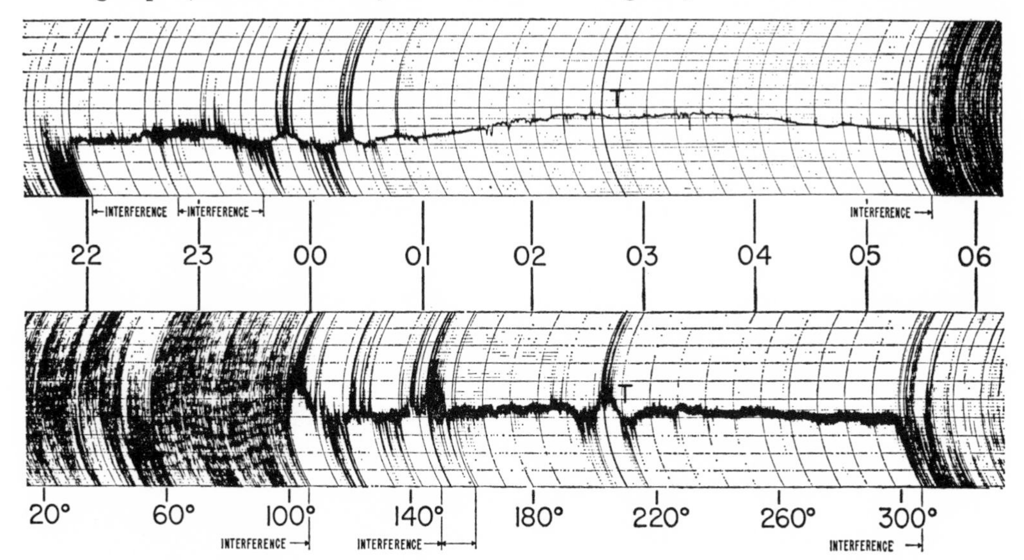

Fig. 4.—Typical limitation of Jupiter observation by interference, February 8–9, 1957. *Above:* 20 Mc; *below:* 18 Mc. Hours M.S.T. (U.T. − 7^h) given between records. Longitudes on System II at bottom. *T* marks meridian passage of planet.

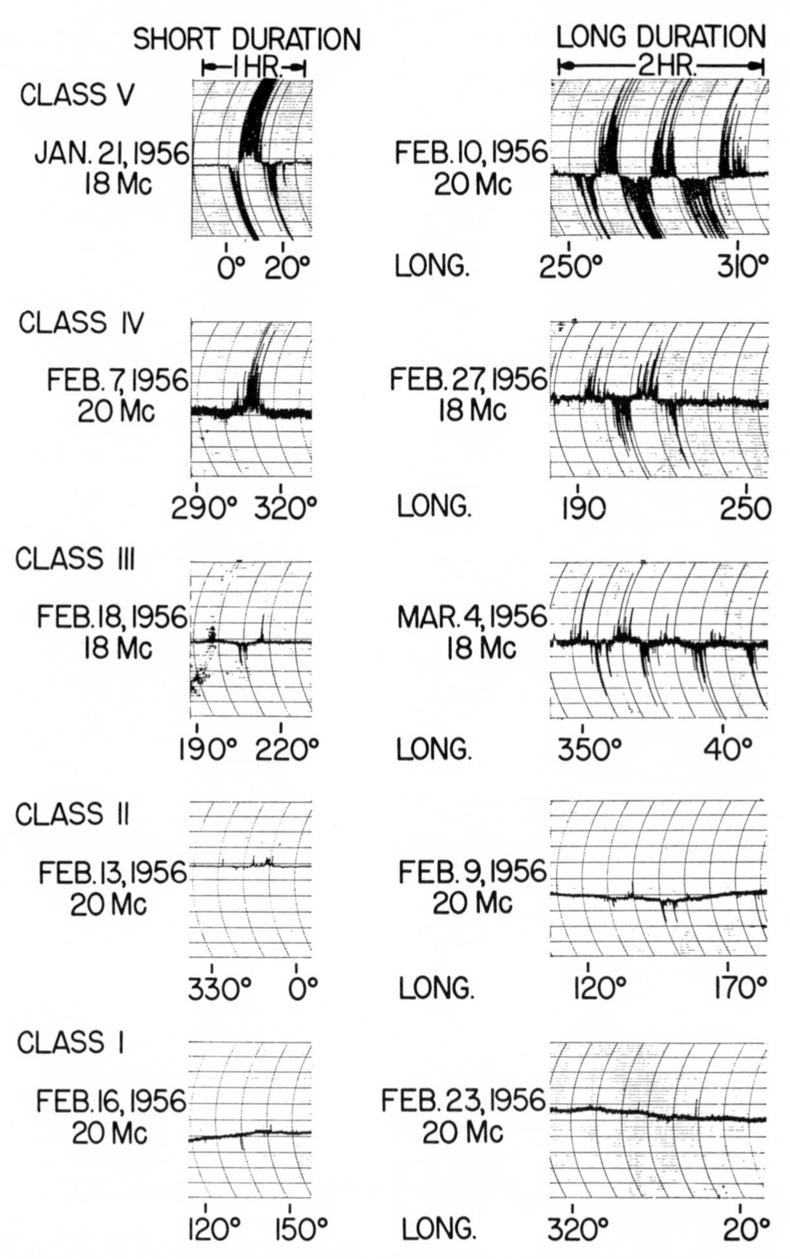

FIG. 5.—Samples of Jupiter radio emission, in order of pulse density during the emission. The two lowest classes give indefinite measurements of emission duration and contribute little to the total activity; they are not used in the statistics.

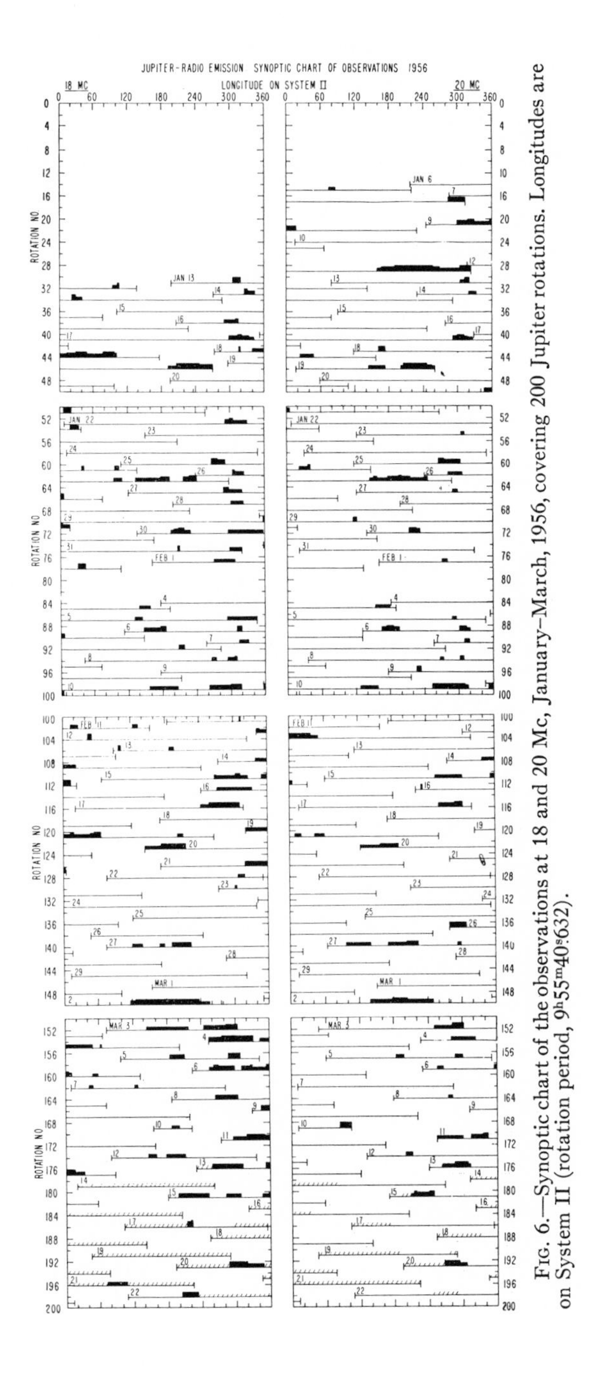

FIG. 6.—Synoptic chart of the observations at 18 and 20 Mc, January–March, 1956, covering 200 Jupiter rotations. Longitudes are on System II (rotation period, $9^h55^m40^s.632$).

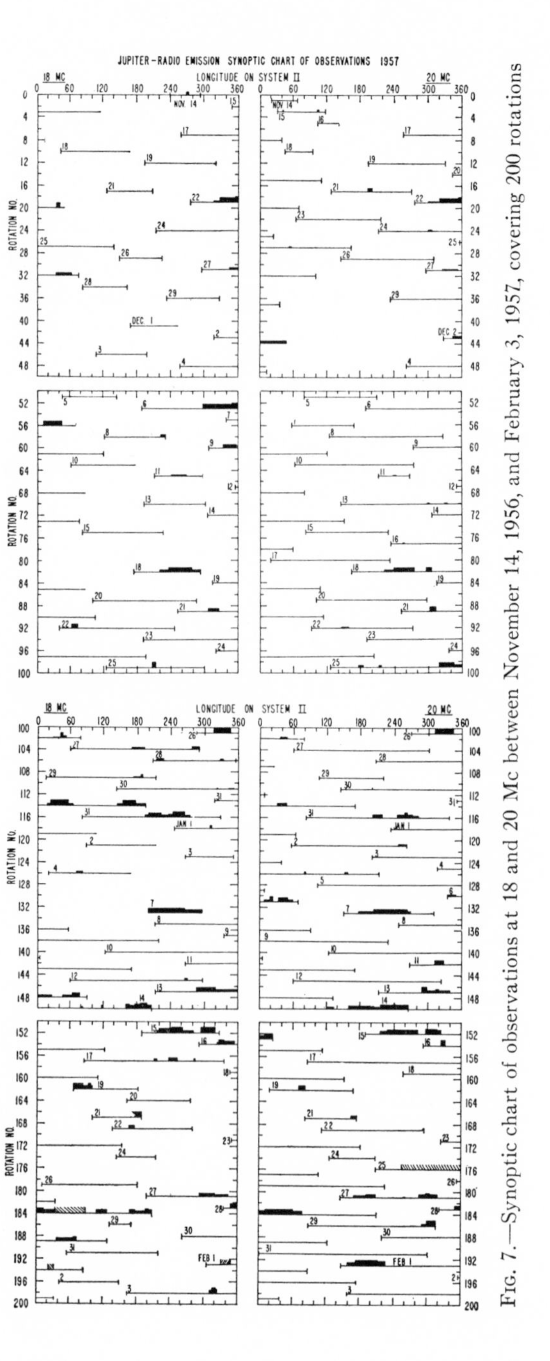

FIG. 7.—Synoptic chart of observations at 18 and 20 Mc between November 14, 1956, and February 3, 1957, covering 200 rotations

uncertain; they were omitted in the following statistics, which include only the well-defined classes III–V.

The entire series of observations is plotted in the synoptic charts of Figures 6 and 7. The January–March, 1956, data covered slightly less than 200 planetary rotations. The 1957 data covered about 300 rotations from November, 1956, to March, 1957, of which the first 200 are shown. These charts permit one to compare the degree of simultaneity in the emissions on the two frequencies. The longitudes are recorded with a precision of about 1° in System II. It is seen that in 1956 the observations often covered continuously more than a complete rotation. The gross statistics from the synoptic records are summarized in Table 1.

The longitude profiles obtained from the synoptic charts are presented in Figure 8. Also given is a histogram obtained from the synoptic chart

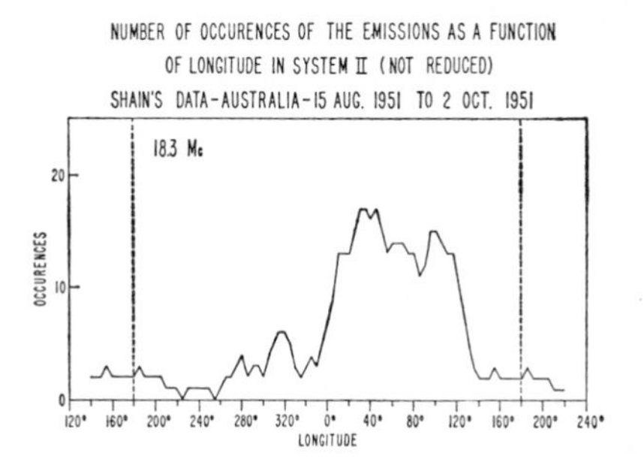

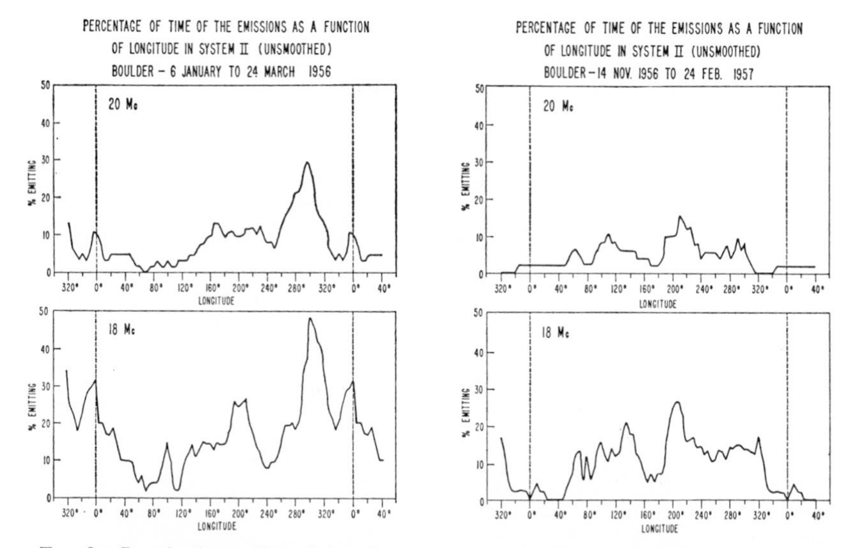

FIG. 8.—Longitude profiles, giving the percentage of radio reception time as a function of the longitude of Jupiter's central meridian on System II. *Above:* the histogram for 1951 was derived from Shain's data (Shain 1956) (cf. explanations in the text, Sec. 4.4); *below:* Boulder data, 1956 and 1957.

published by Shain (1956) for his 1951 data at 18.3 Mc (cf. Sec. 4.4). These profiles were computed for each 5° in longitude, giving the fraction in per cent of the number of positive cases of reception to the total number of observations available at this longitude. Cases where interference was present were not used. Because of interference and stroboscopic effects between the rotations of earth and Jupiter, all longitudes will not be observed equally, even for series extending over several months. Thus histograms which show merely the number of cases of reception as a function of longitude would not be significant. The normalization procedure used here is necessary.

4. DISCUSSION OF THE RESULTS

4.1. Comparison of Longitude Profiles

The profiles for different frequencies during the same period show a good similarity (cf. Fig. 8). At least four main sources can be distinguished, but a more complex pattern probably exists, made up of weaker sources that cannot be separated because of blending and statistical fluctuations. A weak source at 275° in the profile for 1956, on the preceding edge of the principal source at 300°, is probable, being separated on the 18-Mc profile and indicated by the asymmetry of the main peak at 20 Mc. It is shown below that the rotation period is about 10.5 seconds shorter than System II used for computing the longitudes. This difference produces a broadening of the profile peaks. By this effect a given point source is spread over 7°.5 of longitude per month. The 1957 data are distributed over more than 4 months, and thus the profiles should be even less sharp than those for 1956. In future work it will be necessary to adopt the radio rotation period in computing the longitudes. Even so, the width of the peaks is remarkably small, particularly in 1956, and seems larger on 20 Mc than on 18 Mc. At 18 Mc the half-amplitude width of the strongest peak is about 38°.

These small widths may be readily interpreted on the hypothesis of a Jovian ionosphere. For emissions occurring below the maximum of ionospheric ionization, the outward propagation of a wave of frequency f is restricted to a cone whose axis is the vertical of the emission point. The semiangle α of the cone is given by

$$\cos \alpha = \frac{f_c}{f}, \tag{1}$$

where f_c is the critical frequency. The critical frequency is related to the maximum electron density in the ionosphere, $N_{\max}$, according to

$$f_c^2 = \frac{N_{\max} e^2}{\pi m},$$

or, numerically, for $N_{\max}$ in electrons per cubic centimeter and f_c in cycles per second,

$$N_{\max} = 1.24 \times 10^{-8} f_c^2 , \quad (2)$$

where e and m are the charge and the mass of the electron, respectively. In order that the emission of a source situated at latitude ϕ be received at the earth, it is necessary that $\alpha > \phi$. When two frequencies, f_1 and f_2, are observed simultaneously, the cones are such that $\alpha_2 > \alpha_1$ if $f_2 > f_1$. This situation is illustrated in Figures 9 and 10. These diagrams show that the reception time can be very short, being limited by the latitude of the source as well as the semiangle of the cone. In the terrestrial ionosphere the

FIG. 9.—Reception cones of Jupiter's radio emissions for two different frequencies, limited by Jupiter's ionosphere.

critical frequency f_c fluctuates strongly from day to day, particularly in the F2 region. Part of these fluctuations is due to purely geophysical effects (circulation of air masses, tides, etc.), and another part is correlated with solar activity, such as day-to-day fluctuations in the ultraviolet radiation of the sun and magnetic storms produced by corpuscular radiation. By analogy, we can expect similar fluctuations in the ionosphere of Jupiter. These fluctuations probably account for part of the day-to-day variation in the duration of the reception of a given source, as illustrated in synoptic charts (Figs. 6 and 7). An increase in the ionization narrows the cone and may even supress the reception at the earth altogether. Thus the absence of reception for a given source does not necessarily mean an absence of emission. Therefore, the peak amplitudes observed on the longitude profiles are lower limits to the intrinsic statistical activities. The different possible cases are shown in Figure 10. For two frequencies separated by 10 per cent, the two cones are not very different, and thus the intermediate case of Figure 10 should occur only rarely, in good agreement with the observations.

4.2. Influence of Solar Activity and Latitude

In the course of the solar cycle the ionization of the ionosphere varies in phase with the solar activity. Therefore, the cone of transmission will, on the average, be smaller during years of high solar activity, and the average percentage of reception will be smaller. During 1956 the solar activity was already high, with a mean Zurich sunspot number of 110 for the period covered. But for the 1957 observing period it increased to a mean sunspot number of 160. The average percentages of observed emission at the two frequencies are given in Table 1 and are shown in Figure 11. These results suggest that the ionization in the ionosphere of Jupiter is indeed controlled by solar activity. This conclusion is strengthened by a subdivision of the material. Figure 7 shows a low percentage of time for reception during the first hundred rotations, followed by a significant increase. Now the solar activity was high in November and December, 1956 (sunspot number about 180), and fell sharply during the following months. The small dots and crosses in Figure 11 show the results obtained by subdivision of the

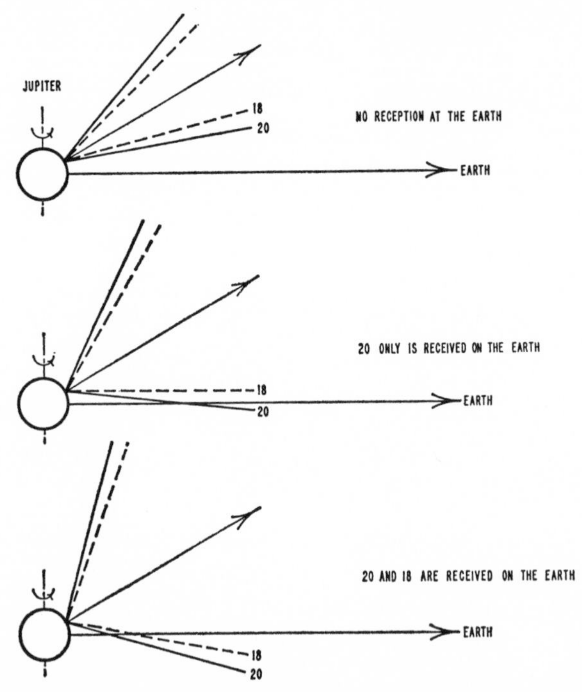

FIG. 10.—Geometry of radio reception for different critical frequencies determined by the emission cones through the ionosphere of Jupiter.

data into subseries of 50 planetary rotations each (about 21 days). The scatter is not large, and the regression lines agree well with the lines connecting the average values 1 year apart. However, good series over several years will be needed for establishing the suggested relation quantitatively.

The tentative relationship with solar activity has certain consequences. First consider the ratios of the total activities (line B 1, Table 1) at the two frequencies:

$$\frac{\text{Act. (20 Mc)}}{\text{Act. (18 Mc)}} = 0.50 \text{ in } 1956 \text{ and } 0.60 \text{ in } 1957 .$$

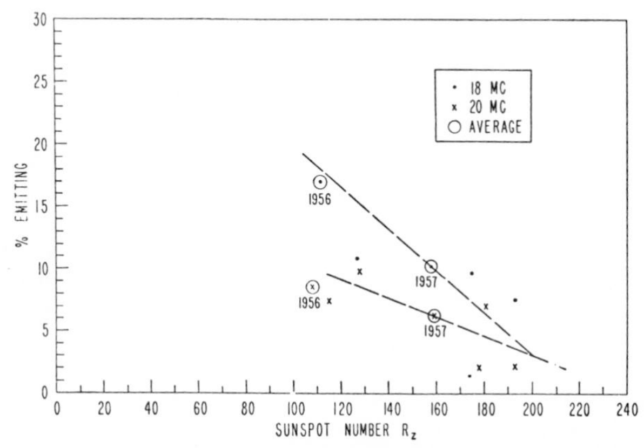

FIG. 11.—Dependence on solar activity of the total percentage of time when Jupiter was received on the two frequencies derived for two oppositions. The 1957 data go up to February 25, 1957.

These ratios indicate a strong reduction in the activity for an increase of only 10 per cent in frequency. Furthermore, they are actually upper limits to the intrinsic reduction factor given by

$$\text{Observed } \frac{(20)}{(18)} = \text{Intrinsic } \frac{(20)}{(18)} \times \frac{(\text{cone } 20)}{(\text{cone } 18)} , \tag{3}$$

where the last term represents the average ratio of the times that the earth occurs inside the respective cones at these two frequencies. Figure 9 shows that this second term is always greater than 1 and that the ratio can be large if α_{18} only slightly exceeds the latitude ϕ. On the other hand, the term is near unity if the cones are wide open, i.e., when the critical frequency is small. Thus the intrinsic factor is best approached during *minimum* solar activity. With increasing solar activity, the observed ratio, equation (3), should increase also. The two observed values quoted are in agreement with this expectation. The four quantities used in these two ratios are based on about 500 hours of observation each and therefore have considerable weight.

TABLE 1

GROSS STATISTICS* OF SYNOPTIC RECORDS AT NATIONAL BUREAU OF STANDARDS

		Frequency (Mc)	Series 1956	Series 1957	Notes
A	1. Total recording time available expressed in hours...........	20	605^{h}	590^{h}	Total: 2100^{h} of records
		18	494^{h}	412^{h}	
	2. Total recording time available expressed in degrees of longitude..............	20	21926	21248	
		18	17909	14948	
	3. Number of observational nights during period.............	20	70	114	
		18	62	103	
	4. Efficiency: mean length in longitude observable per night	20	313°	186°	
		18	289°	145°	
	5. Mean time duration.	20	$8^{h}.63$	$5^{h}.13$	
		18	$7^{h}.97$	$4^{h}.00$	
	6. Average number of opportunities to have each longitude in central meridian.......	20	61.8	59.1	
		18	50.0	41.9	
	7. Middle date of observational period.....	20	13 Feb.	14 Jan.	336 days apart } av. 334.25
		18	16–17 Feb.	14 Jan.	332.5 days apart }
	8. Weighted mean epoch (determined by emission)........	20	9 Feb.	24 Jan.	349 days apart } av. 347
		18	16–17 Feb.	26 Jan.	344.5 days apart }
B	1. Percentage of time Jupiter is emitting...	20	8.49	6.13	
		18	16.94	10.6	
	2. Average Zurich sunspot number during observational period.	20	108	159	
		18	111	158	

* These gross statistics do not include times during which radio observation of Jupiter was impossible, such as Jupiter rising or setting, terrestrial ionospheric limitation, interferences, or equipment failure, and can be used directly for normalization purposes. They do not represent the total observing time, which was considerably more.

A second consequence seems to be that Jupiter has maintained essentially the same intrinsic activity 1 year apart, the changes observed being explicable by the geometric factors related to ionospheric properties. This conclusion is confirmed by the study of the rotation.

A third consequence may permit a determination of the *latitudes* of the sources. The observed peak value of an equatorial source will not be much affected by changes in solar activity, provided always that $f > f_c$. Then only the width of the peak will be reduced at high solar activity, which will produce sharper longitude profiles and diminish the blending. For a high-latitude source, however, the peak amplitude for a given frequency will reduce with increasing solar activity, and the observed ratio 20 Mc/18 Mc for the peak amplitudes will increase—the more so, the higher the latitude. Year-to-year variations in intrinsic source activity will be a complicating factor, but this can be tested with several years of good data. With the few longitude profiles now available it is too early to propose a definite latitude classification. The main source at longitude 300° in 1956 suggests a rather high latitude, while the group of weak sources between 120° and 180° in 1956 fits the criteria for equatorial sources.

4.3. Intrinsic Properties of Sources

Properties other than the latitude may be obtained from intercomparisons of sources for a given time interval. First, we have the peak amplitudes of the various sources at a given frequency. To minimize the geometrical effects, these are best compared at minimum solar activity. A source at higher latitude has a lower probability of being received at the earth. The main source at 18 Mc emits at least 50 per cent of the time, which suggests quasi-permanent emission.

Second, the cut-off factor of the statistical spectrum is different for the different sources. The four most important sources observed in 1956 are listed in Table 2. The 100° source has the steepest spectral cutoff. Table 2 suggests a relation between the cut-off factor and the peak amplitude for a given frequency, with the most active sources having a statistical spectrum extending farthest toward high frequencies. This tentative conclusion is strengthened by noting that the histograms for the higher frequencies observed at other observatories show an apparently stronger concentration in longitudes corresponding to the main source at 300° in 1956. Continued study should lead to the determination of the statistical spectra of individual sources, provided that the data are properly normalized and the reduction is made to take account of solar activity (cf. Sec. 4.2).

4.4. Rotational Period

4.41. *Solid-body rotations.*—If it is correct that the sources represent properties of the solid body of the planet, a precise determination of the rotation period is important. No other means for doing this is presently available. Visual observations reveal complexities concerning the rotation periods of the atmospheric cloud systems (cf. chap. by Lenham, this series), with currents at different speeds, and time variations of the rotation period of a given current. Even the most permanent visual feature, the Red Spot, presents conspicuous long-term variations in its period. Thus observations from outside the upper cloud level have not revealed a "system of reference" to which the general atmospheric circulation can be compared; and a quantitative discussion of this circulation, including the

TABLE 2

Statistical Activity of Principal Sources, from 1956 Longitude Profile

	Longitude of the Sources			
	0°	100°	200°	300°
Peak activity in per cent at 20 Mc.	10.5	~ 2–3	~10	30
Peak activity in per cent at 18 Mc.	31	14.5	26	48
Ratio 20 Mc/18 Mc. . . .	0.34	~0.14 0.20	~ 0.38	0.62

nature of the sources of energy and the driving forces, becomes impossible. Real progress can be expected if the rotation period of the solid body can be established.

4.42. *Boulder data.*—Superposition of the longitude profiles obtained 1 year apart determines the longitude shift and the correction to the period. Since the profiles are not identical, owing to statistical fluctuations and systematic variations of peak intensities and widths, the adjustment must be made "on the average." It is found, however, that different persons agree remarkably well in their interpretation. For example, an estimate of the uncertainty was made from independent superposition by eight different persons, who used the original large-scale longitude profiles plotted on transparent paper. The determinations were made independently for 18 Mc and 20 Mc and were checked by the two cross-comparisons—18 Mc (1956) with 20 Mc (1957) and vice versa. The average displacement measured in System II obtained for 1956–1957 was $-77\overset{\circ}{.}7$ of longitude, for the two runs with mean epochs 324 days apart. The standard dispersion per determination was $\pm 2\overset{\circ}{.}6$, and for the average $\pm 0\overset{\circ}{.}7$.

Using the complete material available to the end of the 1957 series of the Boulder observations, the best superposition of the longitude profiles in System II gives an average longitude displacement of $-80°$, with the two mean epochs 334 days apart. This gives a correction of -9.99 seconds to the System II rotation period and a rotation period of $9^h55^m30\overset{s}{.}64$. The uncertainty of this average does not seem to be larger than $\pm 4°$, or 0.50 second on the period.

The correction of 10 seconds to System II spreads a point source over about $7\overset{\circ}{.}5$ per month in the longitude profiles. Since during an observing run lasting several months the emissions may not be uniformly scattered, the effective mean epoch may not coincide with the middle of the series. Also, the *shapes* of the peaks and valleys in the profile will be deformed and the centers of gravity shifted. It is therefore important to adopt the radio period as a new system of longitudes. This will permit a closer study of the properties discussed in the preceding sections and the derivation of a third approximation to the rotation speed and its possible fluctuations.

4.43. *Shain's 1951 data.*—Another way of improving the accuracy of a mean rotation period is to extend the interval between the two longitude profiles. Of very great importance in this respect, a set of data at 18.3 Mc has been obtained by C. A. Shain (1956) in a beautiful way from old records of galactic radiation taken in 1951. A simple histogram has to be used, because the data are not amenable to normalization, with the danger that all longitudes have not been observed evenly. Quite remarkably, however, there is a really good correspondence between this histogram and four longitude profiles at 18 Mc, particularly the one for the 1957 series. With plots made on comparable ordinates, the superposition is very easy; many independent determinations by different people, not aware of the nature of the adjustment requested from them, agree within limits of $\pm 5°$ from their average. Now the largest interval at our disposal is about $5\frac{1}{2}$ years. Using the System II histogram of Shain's data and the 1957 longitude profile at 18 Mc, the best correction to the System II period is -11.11 ± 0.06 seconds, giving an average rotation period of $9^h55^m29\overset{s}{.}70 \pm 0\overset{s}{.}06$ from 1951 to 1957.

This result immediately discloses two facts. The first is that the system of sources seems to be essentially the same over 6 years. The sources have maintained their gross characteristics in relative intrinsic activity and relative position in longitude, revealing a long-term stability. The second is that a difference of very nearly 1 second is indicated between the rotation periods for the intervals 1951–1957 and 1956–1957. If the rotation period for the last interval is accepted, the discrepancy in longitude will accumulate to 49° in the comparison between 1951 and 1957. Such an error seems to be out of the question. On the other hand, if we use the shorter 1951–1957 rotation period, the longitudes, which in System II were apparently decreasing by about 80° from 1956 to 1957, will now appear to increase by about 8°.4. Such a discrepancy seems to be larger than the possible uncertainty in the superposition of the four longitude profiles available, but it cannot be rejected without more refined treatment.

4.44. *A provisional System III.*—As outlined in Section 4.42, the small discrepancy between the two rotation periods may result from the use of longitude profiles obtained in the System II differing notably from the true rotation period. The need for a treatment of all the data measured with the best possible approximation of the true period is obvious. Such a reduction was undertaken before the series for 1957 was complete. This "Provisional System III" has a rotation period $9^h55^m29^s.4$. The total spread in the longitude profile for the longest series available (114 days) is now only 0°.8, far smaller than the statistical fluctuations. The new system has been made to coincide with the longitudes of System II on January 1, 1956, 00.00 U.T. The histogram for Shain's data has also been recomputed in the new system from the enlargement of the synoptic chart in System II.

It is apparent that the features of the profiles have become sharper. The writer has not found an adequate statistical test for the reality of peaks indicated by a progressive profile of several points but has tested the reality of the difference between several pairs of points on the profile. The standard deviation of the peaks is about 5 or 6 per cent.

The superposition of any two profiles is now more accurate also, and the residual longitude displacements are small. All intercomparisons confirm that Provisional System III makes a slight *overcorrection:* from several determinations the remaining longitude displacement for 1951–1957 amounts to $+7° \pm 1°$. The resulting correction to the period is +0.147 second (± 0.02 sec), and the final period is $9^h55^m29^s.55$ ($\pm 0^s.02$). This value is not changed by use of a slightly longer interval.

For the 1956–1957 interval a longitude displacement of $+12° \pm 2°$ is found from both 18- and 20-Mc profiles. This furnishes a correction to the period of System III of $+1.48 \pm 0.25$ seconds from 334 days, or 1.40 seconds from 349 days. This last number gives a rotation period of $9^h55^m30^s.80 \pm 0^s.25$.

Provisional System III appears close to the true average over the longest interval; the average period from 1956 to 1957 is again not equal to that from 1951 to 1957 and seems to be longer by an amount too large to be explained by the statistical errors. The results are summarized in Table 3.

If the present results are confirmed, of an essentially constant period of rotation (of about $9^h55^m29^s.7$) with fluctuations of the order of 1 second superposed, then one may have to consider an exchange of rotational momentum between Jupiter's core and mantle, as has been necessary for the earth (cf. **2**, 36). For the earth the core has only some 10 per cent of the total moment of inertia; but, for Jupiter, Ramsey's (1951, 1952) model allows a large ratio for core/mantle (83/17 for Jupiter versus 1/9 for the earth) if the boundary surface is taken to be the transition of metallic hydrogen to non-metallic hydrogen. Thus much larger fluctuations of the

TABLE 3

SUMMARY OF BEST DETERMINATIONS OF ROTATION PERIODS

I. Using Longitude Profiles in System II ($9^h55^m40^s.632$)	
From 1951 to 1957 (18 Mc).............	$9^h55^m29^s.70$ ($\pm 0^s.06$)
From 1956 to 1957 (18 and 20 Mc).......	$9^h55^m30^s.64$ ($\pm 0^s.50$)
Difference.........................	$0^s.94$
II. Using Longitude Profiles in "Provisional System III" ($9^h55^m29^s.40$)	
From 1951 to 1957 (18 Mc).............	$9^h55^m29^s.55$ ($\pm 0^s.02$)
From 1956 to 1957 (18 and 20 Mc).......	$9^h55^m30^s.80$ ($\pm 0^s.25$)
Difference.........................	$1^s.35$

period of the mantle may occur on Jupiter than on the earth. Recent work on the origin of the earth's magnetic field as a consequence of the convection inside the metallic core (Elsasser, 1955, 1956; Cowling, 1957) suggests that the magnetic field of Jupiter may be quite strong if the metallic core extends nearer to the surface. Also the irregular part of the magnetic field, which at the core's surface is of the same order of magnitude as the general dipolar field, may be much more pronounced and variable than at the surface of the earth.

5. THE FINE STRUCTURE OF THE RADIATION FROM JUPITER

5.1. Intensity and Radiated Power

It is a remarkable fact that the observed radiation consists of pulses. The observed *durations, spectral properties* of individual pulses, and the *grouping tendency* are mentioned in Section 2, under h, i, and j and are illustrated below.

The peak amplitudes of the emissions observed have a very wide range. Histograms have been constructed from the high-speed records giving the distribution of the peak amplitudes inside a group of pulses. The observable range, limited on the low side by the galactic noise, extends usually over two orders of magnitude in power. The contribution of numerous pulses of smaller amplitude to the general background noise is uncertain, but usually during times of very active emission the central trace of the noise on the synoptic records shows a sinuous wave in phase with the principal Jupiter emissions. On a phase-switching interferometer this indicates either a steady source or unresolved pulses. On the high-speed records this contribution is not apparent. The histograms of pulse amplitudes usually show a decreasing probability with increasing amplitude, but in a fairly large percentage of cases a secondary peak is present at high amplitudes.

A simple and convenient way to calibrate the amplitude of the pulses is comparison with the records of known strong radio sources observed in very nearly the same geometrical conditions, such as Taurus A (the Crab Nebula). The early measurements of Burke and Franklin (1955) indicated a peak power greater than 10^{-21} watt m^{-2} $(cps)^{-1}$. Recent figures from Shain show that the observed pulses reach values greater than 4×10^{-20} watt m^{-2} $(cps)^{-1}$. In this range of frequencies when Jupiter is emitting, it is by far the most intense radio source in the sky. At frequencies as low as 18 and 20 Mc the quiet radiation from the sun is undetectable, and the non-thermal emissions are very rare and inconspicuous.

5.11. *Comparison with the solar radio bursts.*—For comparison with the sun, the figure quoted above, 4×10^{-20} for the strong Jupiter pulses, must be multiplied by the square of the ratio of the distance of Jupiter from the earth to the distance of the sun from the earth, i.e., about $(4.5)^2 = 20$. Thus the stronger pulses from Jupiter are equivalent to solar events of magnitude 0.8×10^{-18}. The radio bursts from the sun are most intense at frequencies around 100 Mc. At these frequencies a burst of intensity 10^{-18} watt m^{-2} $(cps)^{-1}$ is a relatively rare and quite important event, being about one thousand times the radio intensity of the quiet sun. Such events occur at a rate of only one in several days during periods of intense solar activity, while they are occurring continually on Jupiter, apparently without much variation over a period of several years. If we note that the surface of Jupiter is only one-hundredth that of the sun and that the bulk of the planet is thought to be a cold body of solid hydrogen, we arrive at the surprising conclusion that in the domain of radio emission Jupiter is comparatively far more active than the sun with its "temperature" of a million degrees and all its phenomena.

If we make the approximation that the radiation is isotropic over one hemisphere and take 6.7×10^{11} meters as the typical distance between Jupiter and the earth, we find that the radiated power per pulse is of the order of

$$4 \times 10^{-20} \times 2\pi \times (6.7 \times 10^{11})^2 = 10^5 \text{ watts (cps)}^{-1} .$$

In order to evaluate the total radiated energy per pulse, we can accept as rough figures a duration of 2 seconds (L-pulses) and a band width of 0.5 Mc (cf. Sec. 2, i, and below). The total *radiated* energy per strong pulse is of the order of 10^{11} joules $= 10^{18}$ ergs.

5.12. *Comparison with some physical and geophysical quantities.*—While obviously the observed characteristics of the Jupiter pulses are very different from the radio emission of lightning discharges on the earth, it is of interest to compare the radiated energy. The lightning discharges are the most common natural source of radio pulses. But their dynamical spectrum extends from practically zero frequency to 10–20 Mc, with a strong concentration of the energy between about 5 and 20 or 30 kc, as opposed to the relatively narrow frequency band width of Jupiter pulses near 18 and 20 Mc. Recent measurements by the National Bureau of Standards give for the total *radiated* energy by very strong lightning discharges a value of 10^{12}–10^{13} ergs. Thus there is a ratio of at least 10^5 for the *total radiated* energies between the two phenomena. The ratio is much higher: 10^8–10^9 if the comparison is made of the spectral densities near 18 or 20 Mc.

Normally, the radiated electromagnetic energy in the radio domain is only a small fraction of the total energy available in the causal physical phenomena. The efficiency of the conversion is poor. The bulk of the energy is dissipated in heat, light, mechanical effects, ionization, etc. For example, the total electrical energy of the lightning discharge mentioned by Allen (1955, p. 123) is of the order of 10^{17} ergs. The efficiency for radio emission is thus of the order of 10^{-4}–10^{-5} only. Similar figures are obtained for other phenomena, such as in the theory of the solar radio outbursts in the solar corona produced by a shock wave (Sen, 1953, eq. [20]; 1955, eqs. [30] and [37]; Gould, 1956). Thus the causal phenomena of the Jupiter emissions are expected to involve energies of the order of 10^{22}–10^{23} ergs. It is interesting to compare the energies required with known effects in physics or in geophysics. If the source of the energy is *chemical* in nature, a possibility would be the formation, accumulation, and subsequent detonation of free radicals resulting from photochemical or electrical discharge activity in the atmosphere of Jupiter and stabilized at low temperatures. (This hypothesis, however, meets considerable difficulties of many types and does not look promising.) The energy released by the explosion of TNT is typical of the amount of energy available in chemical reactions. The explosion of 1 ton of TNT releases 4.2×10^{16} ergs. Thus the radio emission alone involves the energy of about 25 tons of TNT per pulse. For total causal energies of the order of 10^{22}–10^{23} ergs (efficiency of conversion 10^{-4}–10^{-5}), the equivalent amount of TNT is 2.5×10^5 to 2.5×10^6 tons. It is interesting to note also that the total energy involved in a lightning discharge is equivalent to about 2.5 tons of TNT.

Now comparing with some geophysical phenomena, the most interesting in connection with the processes at work on Jupiter are perhaps the earthquakes and the different forms of volcanism. Using the calibration of the intensity scale of earthquakes (Bullard, **2,** 87), 6×10^{22} ergs correspond to the energy radiated as seismic waves in the relatively minor earthquake in the Bay of San Francisco in February, 1957 (magnitude 5.5). The San Francisco earthquake of 1906 of magnitude 8.25 involved 10^{27} ergs. Another interesting phenomenon is the famous Krakatoa eruption of 1883. It has been possible (Pekeris, 1939) to estimate the total energy involved in the resulting gigantic atmospheric wave which was observed propagating several times around the earth at 10^{24} ergs. Here we have a phenomenon comparable in magnitude and perhaps in nature with the explosions involved in the Jupiter emissions. Of course, an earthquake or a volcanic eruption does not generate radio waves directly, but the resulting shock wave, in certain suitable circumstances, may.

5.13. *Average radiated power. Comparison with the sun's radio emissions.*—At 18 Mc, Jupiter's emissions, when present, have a pulse density greater than 6 pulses per minutes. The density decreases somewhat at 20 Mc, largely as a result of a decrease in the probability of the emission of groups. It is known that practically no detectable emission exists above about 25 Mc. Toward the low frequencies, observations are rendered impossible by our ionosphere. The effective band width of the statistical spectrum, with a density of 6 pulses per minute, does not seem to be larger than about 20 Mc. Let us adopt a conservative figure of 10 Mc. The individual pulses do not show any correlation on two frequencies 2 Mc apart, and the effective band width of their dynamic spectrum has been provisionally taken as 0.5 Mc. In 1956 at 18 Mc, the emissions were observed 15 per cent of the time. Thus, if we were able to observe all frequencies simultaneously, we could estimate that a rough average of $6 \times 10/0.5 \times 0.15 = 18$ pulses per minute, or 0.3 pulses per second, would have been observed. This figure may be taken as a *lower limit* for the intrinsic activity of Jupiter as a whole, since the observations are limited by the ionosphere of Jupiter and because the values used are very conservative. The intrinsic intensity of individual pulses does not seem to depend on the radio frequency. Thus the average integrated power received at the earth is: power per cps $\times$ dynamic band width $\times$ number of pulses per second $= 4 \times 10^{-20} \times 0.5 \times 10^{6} \times 0.3 = 0.6 \times 10^{-14}$ watt m^{-2}. At a distance of 1 astronomical unit, the power would be twenty times this amount, i.e., 1.2×10^{-13} watt m^{-2}.

The true figure can very well be from three to six times larger if 15 per cent does not reflect the true fraction of emission time on Jupiter because of geometrical and ionospheric limitations.

The average radio power received at the earth from the sun integrated from 0 to 300 Mc is 2×10^{-13} watt m^{-2} for the thermal radiation of the quiet sun (cf. van de Hulst, 1951), while the average *integrated* power, including the enhanced radio radiation and the bursts, is much more uncertain but about 1000×10^{-13} watt m^{-2}.

The integrated power radiated by Jupiter is thus about equal to the radio power of the quiet sun and is within a factor of 1000 of the total solar radio emission. These figures are still more impressive if we recall that the total surface of Jupiter is only a hundredth of the solar surface. Thus the radio power output of Jupiter per unit of surface is one hundred times that of the quiet sun. For more precise comparison it is necessary to take into account the localized nature of the sources of emission, which certainly do not cover more than a very small fraction of the total area of Jupiter.

5.14. *Radio power versus thermal radiation.*—In view of such powerful emission, it is necessary to ask whether the radio power represents a major fraction of the total radiation from Jupiter. The amount of thermal infrared radiation is given by Stefan's law: $S\sigma T^4$. Adopting the effective temperature $T = 150°$ K, the radiation is 2.9×10^{4} ergs sec^{-1} cm^{2}. With a mean diameter of 140,000 km, the total radiation power is 7.2×10^{25} ergs sec^{-1}, or 7.2×10^{18} watts.

The observed average radio power was estimated above to be at least 1.2×10^{-13} watt m^{-2} at the earth. If the average distance between earth and Jupiter during the observation period was 6.7×10^{11} meters, the total average radio power was

$$1.2 \times 10^{-13} \times 4\pi \times (6.7 \times 10^{11})^2 = 7 \times 10^{11} \text{ watts} .$$

This value would be somewhat reduced if the radiation was not isotropic (e.g., diminished toward the poles) but could quite easily have been larger if the intrinsic probability of emission was larger than 15 per cent.

Thus the ratio of the radio to the thermal emission is about 10^{-7}. On the average, therefore, the radio emission is an inconspicuous fraction of the thermal radiation.

But this result should be interpreted with some caution. On the one hand, it is likely that the causal mechanism involves 10^4–10^5 times more energy than the radio energy. Hence the real energy output detected by the radio emission is of the order 10^{-3}–10^{-2} of the total thermal radiation. Thus, if the source of energy is external to Jupiter, in some transformation from the solar radiation available equal to the total thermal energy radiated by the planet, a conversion of the order of 10^{-3}–10^{-2} at least of the total energy will be necessary. Considering further that this energy has to become concentrated and released in well-localized regions, solar radiation as the source of the energy is very unlikely. This rules out such possibilities as photochemical action with formation of unstable chemicals in the atmosphere or even electric energy from solar-driven meteorological motions of clouds.

On the other hand, if the total emissive area for radio energy is only 10^{-2}–10^{-3} of Jupiter's surface, the two energy outputs are comparable in these regions. The total effective area for the radio emissions may very well be much smaller. In such case the disturbance producing the radio emission is a major phenomenon dominating the thermal balance locally in the region of the sources.

5.2. Pulse Properties from High-Speed Records

In several parts of the preceding sections it has been necessary to indicate some of the observed properties of the pulses recorded at high speed. Little can be added to the condensation of results given in Section 2, paragraphs *h*, *i*, and *j*. Only a selection of different aspects from the records will be given here. Quantitative estimates for band width and average occurrence have been given in Section 5.13.

The more significant properties are the duration and spectral properties of the pulses. Both can be represented by dynamic spectra on a frequency-time diagram. Attempts to record the dynamic spectra with automatic instruments, sweep frequency recorder, and camera have not been successful, because of instrumental difficulties and lack of favorable time. The fact that the band width at any given moment is smaller than 2 Mc and greater than 0.2 Mc is, however, firmly established, as well as the absence of a relatively slow frequency drift over a frequency band extending from 18 to 20 Mc.

The existence of two types of pulses—long (L) and short (S)—is well established experimentally on both frequencies and during the two recording periods 1956 and 1957. Unfortunately, the S-pulses could not be dis-

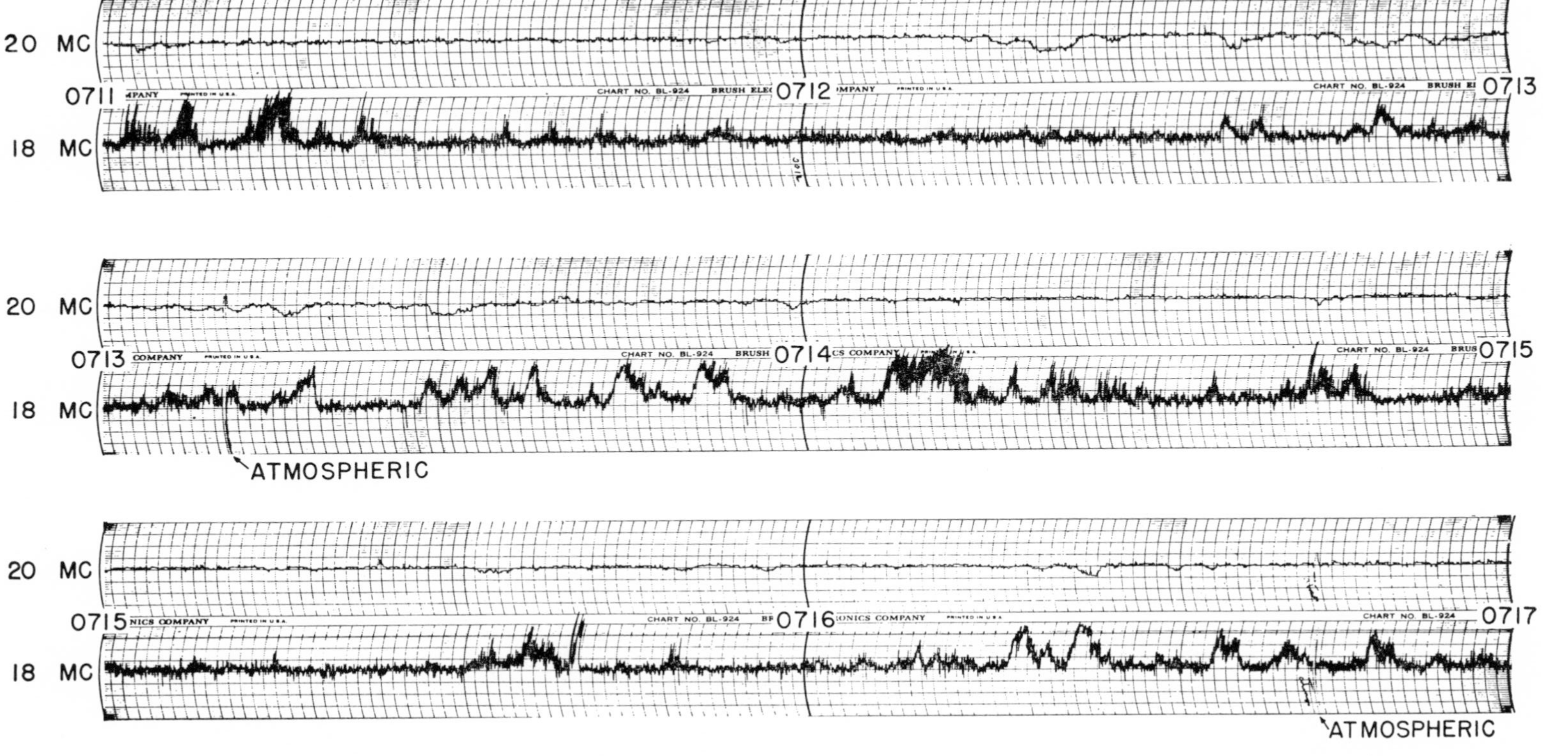

FIG. 12.—Fine structure of long pulses, observed only at 18 Mc, February 8, 1957, $7^{h}11^{m}$–17^{m} U.T. Scale: one division per second

tinguished on the synoptic records. The few occasions of their detection with special recording equipment indicated that, although they are rare, they may be observed for $\frac{1}{2}$ hour or more on occasion.

Sample records of L-pulses are presented in Figures 12 and 13 and of S-pulses in Figure 14. The sensitivities at the two frequencies were very nearly equal. The records exhibit the fine structure and duration of both types and show the conspicuous lack of similarity or correspondence in details at the two frequencies. On the records of S-pulses the sharpest features have a width of one- to two-hundredths of a second, being limited by

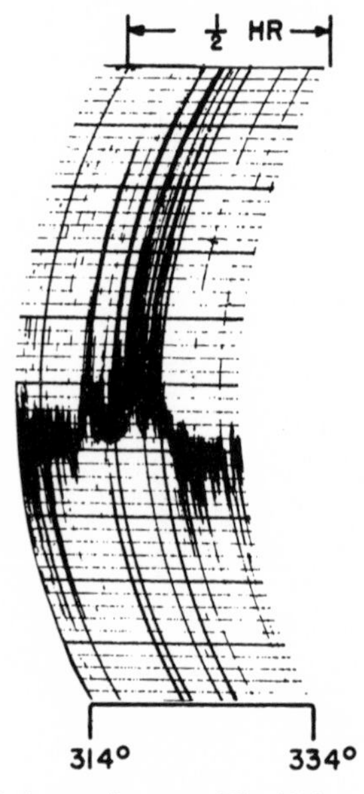

FIG. 13.—Same period of emission as shown on Fig. 12, but recorded on the 18-Mc synoptic records.

the instrumental time constant. With the help of photographic records made with a cathode-ray oscilloscope, pulses as short as a few thousandths of a second have been recorded.

It is not easy to reproduce records long enough to show the *grouping* of the pulses. Many statistical analyses were made which emphasize the grouping tendency without revealing simple properties other than those summarized in Section 2, paragraph j. Figure 15 is a block diagram showing such an analysis for 9 minutes of records of S-pulses. The bottom part presents the mean pulse density, in number of separate S-pulses per second, as well as the duration of each group; the area is proportional to the total number of pulses within the group. The upper part is an attempt to show, very roughly, the relative importance of groups in "intensity,"

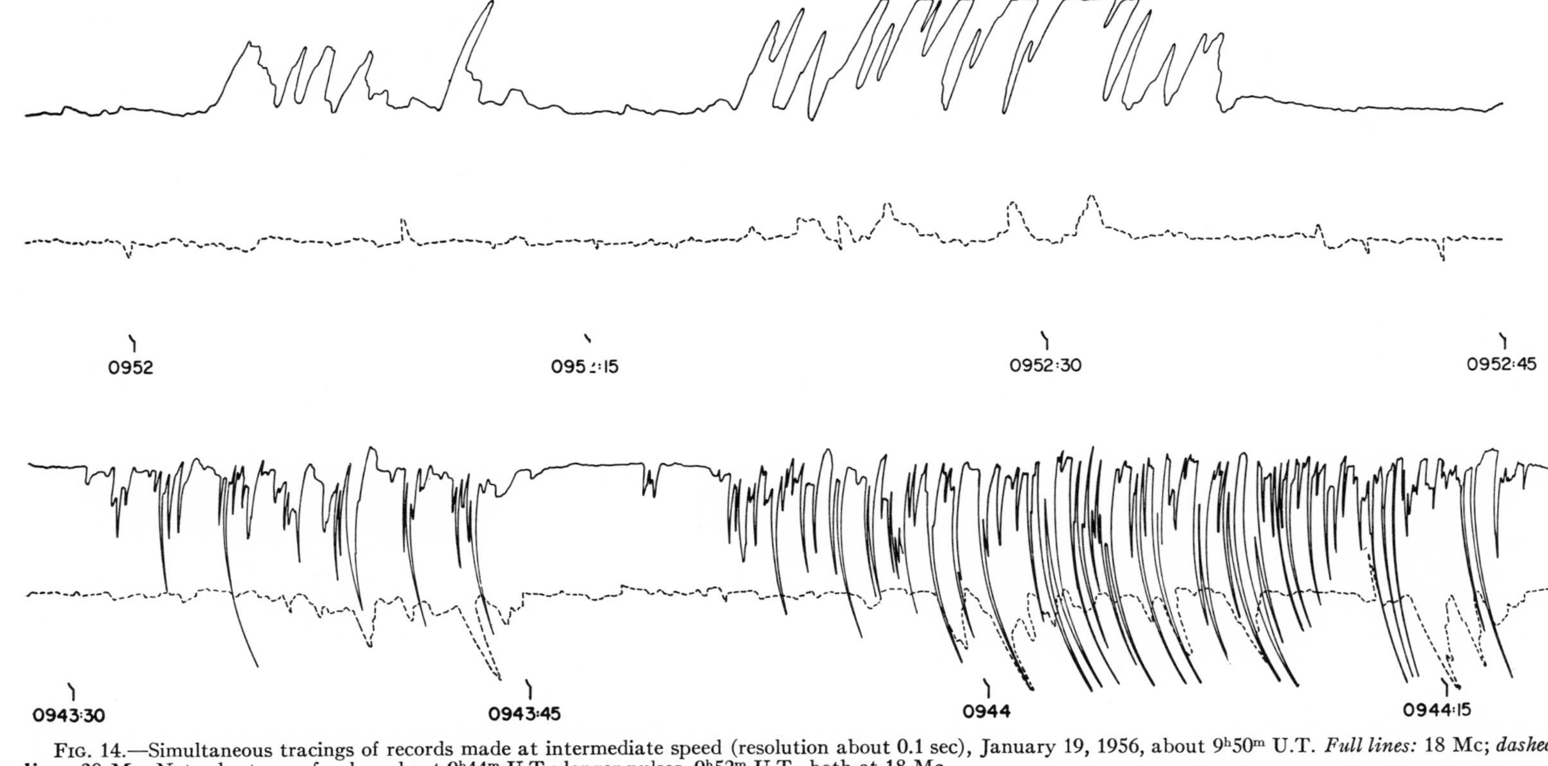

FIG. 14.—Simultaneous tracings of records made at intermediate speed (resolution about 0.1 sec), January 19, 1956, about 9^h50^m U.T. *Full lines:* 18 Mc; *dashed lines:* 20 Mc. Note shortness of pulses about 9^h44^m U.T.; longer pulses, 9^h52^m U.T., both at 18 Mc.

defined as the sum of the products of the number of pulses between two amplitude levels by the average amplitude level, contained inside a group; the amplitude levels are arbitrary and are given by the horizontal lines of the recording paper; ten to twelve levels are distinguishable. The area per group is proportional to this total intensity. It is seen that, as a rule, the densest groups also have a larger average amplitude. The grouping property suggests a relaxation phenomenon. Gardner and Shain (1958) have recently indicated that simultaneous records made 25 km apart as the same frequency do not have the same fine structure and that the pulses are not the same or present at the same time. This observation is extremely

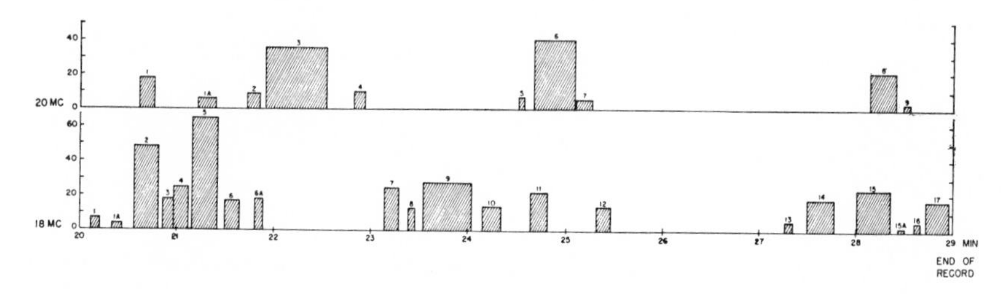

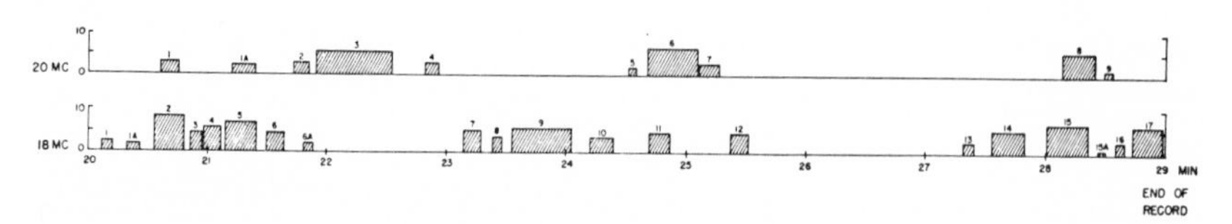

FIG. 15.—Block diagram showing grouping tendency of slow pulses at two frequencies

important because it implies, as these authors suggest, that the main features of the received signal are attributable to a strong scintillation effect caused by changing irregularities in the transmission medium. However, it is difficult to see how the sharp, well-separated pulses observed could be produced by an irregular diffraction process in the transmission medium or how this could account for the simultaneous observation of S-pulses at one frequency and L-pulses at another frequency only 2 Mc apart. Thus it seems very important to confirm this effect, taking great precautions to be sure that the frequencies, band widths, time constants, and sensitivities of the equipments are the same.

The question of the importance of scintillation affects only the interpretation of the fine structure of the received signal. Since scintillation involves only a redistribution of energy in space and time but does not create energy or change the frequency, the average power received and the statistical spectral properties remain essentially the same.

6. CONCLUSIONS

Jupiter's non-thermal radio emissions are important because they reveal entirely unexpected properties of the largest planet. In the past it has been very difficult to obtain quantitative physical data on the exterior parts. With the advent of radio astronomy, much new quantitative information is becoming available. But much precision work and new types of observation, such as polarization and dynamic spectra of bursts, remain to be made.

The great question is the nature of the mechanism producing Jupiter's emissions. The planet simulates the sun, in emitting these radio bursts; but its physical conditions are very different, and the characteristics of the planet's emissions make them more puzzling than those of the sun. The narrow frequency band width, the change in the central frequency from pulse to pulse, and the pulse duration suggest that a resonance phenomenon is involved in the production. The properties indicate excitation of an ionized gas with a well-defined plasma frequency. The source of the excitation may be a shock wave resulting from an explosion. But for Jupiter very stringent conditions have to be met. First, the ionized gas with plasma frequencies between, say, 10 and 25 Mc has to be explained. The indicated presence of an ionosphere suggested that the emission takes place in this region. But if a shock wave propagates nearly vertically through the ionosphere, a drift in frequency is expected, while careful examination has so far failed to reveal any correspondence between frequencies only 10 Mc apart. A simple mechanism is therefore ruled out. But if drifts were observed, the production of shock waves with the required energy at that level would be a difficult problem. Another possibility is that both the ionization and the excitation of the plasma are produced in the shock front itself. On this hypothesis one may visualize the production and localization of explosions with the necessary energy at great depths near the surface of the solid body.

Regardless of the mechanism, it seems that the solid body of the planet is involved. This conclusion calls for reconsideration of the model of the internal structure of the giant planets on such questions as internal temperature, convection in the metallic core, strength of the mantle, level of the solid surface, and extension of the atmosphere. The suggested existence of "volcanic activity" is puzzling, though this term is not meant too literally.

6.1. Desiderata

a) Rotation period. It is not likely that in a few years the period can be greatly improved. A long series of continuous synoptic records taken on at least two frequencies simultaneously will be needed. Besides furnishing a very precise rotation period of the solid body, such observations would test the constancy of the rotation, an important factor for the knowledge of the planet's internal structure. The adoption of the Provisional System III (period = $9^h55^m29\overset{s}{.}6$) for the reduction and the intercomparison of the data of different observatories is more necessary than for visual observations and is strongly recommended. In about 5 more years the precision could attain one-hundredth of a second.

b) Comparison of the longitude profiles on several frequencies for long-term stability in position and in activity of the sources. This would result in a better statistical separation of the sources in longitude and the determination of their intrinsic statistical spectra of activity.

c) Separation of the effect of solar activity. This effect will give from one side the sensitivity of the ionosphere to the variation of solar radiation and on another side some latitude distribution for the sources. The theory of the geometry of the ionospheric cones seen from the earth as a function of frequency, critical frequency, and latitude has to be worked out in detail.

d) Polarization measurements and determinations of the magnetic field of Jupiter, with its possible variations with position and time. Several methods are possible and can check one another.

e) Dynamic spectra of the pulses, simultaneously, at two separate stations, in order to isolate scintillation effects.

f) Studies of the pulses, when Jupiter is occulted by the solar corona. Alterations in the apparent length of the *S*-pulses and propagation differences for different frequencies are possible consequences of the ionization of the interplanetary medium of the solar corona.

g) Detailed study of visual features appearing near the source longitudes. Search for manifestations of "volcanic activity" at the upper cloud level (convection, dark or white spots, etc.).

h) Theory of the Jovian ionosphere. The formation and electronic density of the ionosphere in the atmosphere of hydrogen, with small proportions of methane and ammonia.

Perhaps the most important condition for further progress is the homogeneity of the observations and a proper normalization in percentage of time, when different equipments and different sensitivities are used. Continuity of the series is also essential.

REFERENCES*

ALLEN, C. W. 1955 *Astrophysical Quantities* (London: Athlone Press) (Sec. 5.1).

BURKE, B. F., and FRANKLIN, K. L. 1955 *J. Geophys. Res.*, **60**, 213 (Secs. 1, 5.1).

BULLARD, SIR EDWARD 1954 *The Earth as a Planet*, ed. G. P. KUIPER (Chicago: University of Chicago Press), pp. 87 and 128–129 (Secs. 4.4, 5.1).

COWLING, T. G. 1957 *Magnetohydrodynamics* (New York: Interscience Publishers), chap. 5 (Sec. 4.4).

ELSASSER, W. M. 1955 *Am. J. Phys.*, **23**, 590.

1956 *Ibid.*, **24**, 85 (Sec. 4.4).

GARDNER, F. F., and SHAIN, C. A. 1958 *Australian J. Phys.*, **11**, 55 (Sec. 5.2).

GOULD, R. W. 1956 *Bull. Amer. Phys. Soc.*, Ser. II, **1**, 35 (Sec. 5.1).

HULST, H. C. VAN DE 1951 *A Course in Radio Astronomy* (Leiden Observatory), p. 122 (Sec. 5.1).

JEAN, G. 1957 Private communication (Sec. 5.1).

JONES, SIR HAROLD SPENCER 1954 *The Earth as a Planet*, ed. G. P. KUIPER (Chicago: University of Chicago Press), chap. 1, particularly pp. 28–37 (Sec. 4.4).

MILES, B., and RAMSEY, W. H. 1952 *M.N.*, **112**, 234 (Sec. 4.4).

PEKERIS, C. L. 1939 *Proc. R. Soc. London, A*, **171**, 434 (Sec. 5.1).

RAMSEY, W. H. 1951 *M.N.*, **111**, 427 (Sec. 4.4).

SEN, H. K. 1953 *Australian J. Phys.*, **6**, 67 (Sec. 5.1).

1955 *Phys. Rev.*, **97**, 849 (Sec. 5.1).

SHAIN, C. A. 1956 *Australian J. Phys.*, **9**, 61 (Sec. 4.4).

WILD, J. P. 1953 *The Sun*, ed. G. P. KUIPER (Chicago: University of Chicago Press), chap. 9, Sec. 10 (Sec. 1).

1955 *Advances in Electronics*, **7**, 299 (Sec. 1).

* The section for which the paper cited is relevant has been given in parentheses at the end of each reference.

CHAPTER 15

Visual and Photographic Studies of Planets at the Pic du Midi

By A. DOLLFUS
Observatoire de Paris, Meudon

1. ASTRONOMICAL SEEING

1.1 INTRODUCTION

THE first telescopic observation of the planets was made by Galileo in 1609. For the first time the planets appeared as disks, like the earth, illuminated by the sun. This spectacle created a strong desire to know these neighboring worlds better. More powerful telescopes had to be constructed and erected in favorable locations. In 1610 Galileo observed with a resolving power of 10 seconds of arc. In 1630 Fontana, Grimaldi, and Riccioli reached 3″ and discovered the belts on Jupiter. In 1670 Huygens observed the Martian polar cap with 2″. In 1680, with 1″ resolving power, Cassini distinguished the dark division in Saturn's Ring, which bears his name.

In seventy years the power of the telescope had been increased tenfold, but subsequent progress was slow. No fundamentally new observations were made during the eighteenth century. However, by computation and experimentation, new ideas in pure optics were developed, and, by the work of Dollond, Clairaut, d'Alembert, Fresnel, Fraunhofer, and others, telescopic equipment with much greater resolution was available at the beginning of the nineteenth century. By 1830 Beer and Mädler could observe the planets with twice Cassini's resolution, and later Schiaparelli made further improvements. Meanwhile, the dependence of resolution on telescopic aperture had become understood as a natural consequence of the wave nature of light.

1.2 Diffraction Effects

Early this century there were great expectations for the large telescopes of Meudon, Lick, and Yerkes, with lenses up to 1 meter in diameter, and later for the still larger Mount Wilson reflectors. These large telescopes achieved only partially what had been expected. In spite of their optical quality, their usable resolving power generally remained around 0″4. Finer detail appeared only exceptionally. The definition was clearly no longer limited by the dimensions of the telescopic apertures but by the properties of the overlying atmosphere.

1.3 Atmospheric Effects

Before reviewing the planetary observations made at the Pic du Midi, we shall examine the optical disturbances in the atmosphere. The index of refraction of the air varies with the temperature. At 0° C the index is 1.000292 and, at 1° C, 1.000291, a difference of 10^{-6}. A parcel of air 1° warmer than its surroundings and only 15 cm thick will change the light-path by 0.15 μ, or $\lambda/4$ for red light, enough to affect the resolution. The resulting thermal disturbances can be traced to three sources: (1) the free atmosphere, (2) the ground, and (3) the instrument.

1.31. *Effects of the free atmosphere.*—These effects are of two kinds:

a) Waves and eddies between horizontal layers. Temperature discontinuities occur at all altitudes. A layer of warm air overlying a layer of cool air is dynamically stable. Since air is a poor heat conductor, the discontinuity will tend to persist. The boundary may undulate, and a light-ray entering the telescope will be deviated. As long as the telescopic aperture is smaller than the dimensions of the air waves, these disturbances are not too serious (cf. Sec. 1.6); but if the aperture is larger, they will spoil the definition. With strong relative motion of the layers, eddies will appear in the interface, and the images will deteriorate, even with small apertures.

The effect on the image is reduced with increasing altitude of the undulating layer, because of decreasing density of the air and the smoothing of the irregularities of the wave front with distance.

Danjon (1926) and more recently Keller (1955) and Protheroe (1955) have studied numerically the behavior of the wave front. The mean wave length usually varies between 10 and 25 cm but may be as large as 1 meter; the amplitudes are such that the image of a star may move in the focal plane by 1″–0″2 from the mean position and exceptionally by only 0″05. The undulating atmospheric layers also produce stellar scintillation, which increases toward the horizon.

One can study the structure of the waves at the telescope by removing the eyepiece and making the *Foucault test,* i.e., occulting the image of a bright star with a knife edge. The partially illuminated objective then shows undulations, moving with the wind.

With a large telescope, additional disturbances may be superposed, due to imperfections in the mirror or local atmospheric effects described in Section 1.4. In order to distinguish the upper-atmospheric components, one observes the exit pupil of the telescope with a second eyepiece. Then one sees the exit pupil crossed by dark and bright moving striations due to the convergence or divergence of the light produced by high-altitude undulations. Inside or outside the exit pupil, one sees shadows produced by the local atmospheric disturbances.

Temperature discontinuities and accompanying undulating layers occur near a cold front, with cold air masses wedging under pre-existing warmer layers. The writer has studied these thermal discontinuities during many balloon flights. The cooler layers give the balloon a greater lift than do the warmer layers. The balloon, rising slowly in the cooler air, may be arrested at the boundary layer, a circumstance used by the pilot to cover large distances under stable conditions. Crossing the boundary between layers in relative motion causes a swaying of the balloon.

b) Disturbances due to water vapor are caused largely by vertical motions of saturated air. If a dry mass of air rises in the atmosphere, it expands adiabatically. Its temperature drops very nearly 1° C for each 100 meters of vertical motion. But when the air is saturated with water vapor, this will partly condense. The air then cools with the wet-adiabatic gradient, which may be 0.4 to 0.6 times the dry lapse rate, depending on air temperature and altitude; furthermore, it will vary locally. Part of the heat budget of the atmosphere is latent in clouds. The latter move with the air, redistribute the heat, and may disturb the thermal equilibrium.

When the air is saturated, rising parcels of air will start condensation. The cloud rises with a reduced cooling rate as long as water condenses, while its lift increases as the surrounding air becomes increasingly cold. Convection domes arise in the unstable atmosphere and may produce storms. Filaments of cold air descend between the clouds. Even after the cumuli have dissolved, the atmospheric stirring continues, and telescopic images remain unsteady.

Precipitation in the form of rain may also destroy the thermal balance and impair the seeing. If the precipitation is snow and the flakes fall through air above 0° C, this will tend to cool the air to 0° C everywhere, which may temporarily lead to good seeing. Thereafter, the cool air will subside and disperse, and poor seeing may recur.

1.32. *Ground effects.*—Four typical examples are considered:

a) During the day, the sun heats up the ground by varying amounts, depending on the nature of the soil. Rock and sand become quite hot, and the rising eddies picked up by the wind spoil the day-seeing. At night, radiation cooling, if it is uneven, can also spoil the seeing. For example, a forest retains warm air, while the surroundings will cool. At night the warm air will rise in filaments and create seeing disturbances near the ground.

b) When the ground is level and uniform, the radiation to space will be uniform also. The air will stratify in stable layers, and there will be a temperature inversion, especially in deserts where dry air obstructs less radiation and the dry soil reduces conduction of heat by the ground. The author has often found steady images under such conditions.

When the ground is uneven, the cold air will flow into depressions, and the boundary layer will undulate. Consequently, the images are rarely good in deep valleys or on slopes.

c) When a very strong wind blows over terrain covered with obstacles, dynamical disturbances will result. The obstacles will cause compressions and expansions of the air. If the air is dry, the adiabatic changes in volume will induce changes in temperature, according to the adiabatic relation for air $TV^{0.4}$ = constant, in which T is the absolute temperature. A 1 per cent change in the volume will therefore alter the temperature by 1° C. Streamlets of air 15 cm in diameter, thus compressed, will suffice to affect the quality of the telescopic image. Such a volume change will require a high wind. When the air is nearly saturated, such turbulence becomes directly visible; then expanding parcels of air become saturated and show filaments of clouds that form and disappear. The effect may be complicated by the mixing of air parcels with initially different temperatures.

With the kind co-operation of Dr. H. Camichel, the dynamical turbulence has been studied in the following way. The refractor and the reflector of the Pic du Midi Observatory, both 60 cm in aperture and placed about 100 meters apart, were pointed at each other; the connecting path is parallel to and above the mountain crest. The refractor had an artificial star at its focus, which was examined with the reflector. During the experiments the seeing on the artificial star was good when no wind was present. With average wind, the image was rather poor, although a star at 45° altitude during the same nights would show good images, in spite of the fact that the light had traversed the entire overlying atmosphere! The poor seeing must therefore have resulted within a thin surface layer over the mountain top. On other days, the wind strength was rather similar, but a thin fog caused by the air rising along the slope, was present from time to time. Then, in spite of the wind, the seeing on the artificial image improved each time a layer of fog appeared. Apparently, the small local differences in temperature and refractive index in fog were conducive to obtaining better telescopic images.

d) Unsaturated water vapor has only a small effect on the refractive index of air, with humid air having the smaller index. The effect may be gauged from the thickness that will produce a differential path of one-quarter wave between saturated and dry air (which, of course, is extreme): 150 cm at $-10°$ C., 30 cm at $+10°$ C., and 16 cm at $+20°$ C. The actual air masses required will be much larger, and their dimensions will exceed telescopic apertures. Differences in water-vapor content, if unsaturated, will therefore affect the seeing only rarely.

1.4 Instrumental Effects

The telescope and the dome themselves will introduce seeing troubles.

1.41. *Thermal effects in the dome.*—In the evening, when the air temperature drops rapidly, the walls supporting the dome will cool more slowly. The inside walls cause the formation of parcels of warm air, which accumulate in the dome and gradually escape through the slit. Since the telescope points through the slit, the escaping warm air will tend to spoil the seeing; the larger the dome, the more pronounced will this effect be.

With the Foucault test, streamers and small whirls of air may be seen a few centimeters in diameter and moving slowly, though sometimes with sudden lateral motion due to gusts of wind. These phenomena depend on the wind velocity and the position of the telescope. They appear distinct from disturbances by the high atmosphere if one focuses an auxiliary eyepiece just inside or outside the exit pupil given by the eyepiece of the telescope itself. Because the warm air will escape principally through the upper part of the slit, the stellar images may be worse near the zenith than lower in the sky, as we have actually observed with the Meudon and the Yerkes refractors. In addition, there will be mixing of air within the dome itself. For this reason, reflectors that have the mirror in the lower part of the dome will give somewhat worse images than refractors that have their objectives close to the slit; and the images as seen in a finder may be worse than in the much higher refractor, as has also been observed. These effects will be most pronounced in large domes and during the beginning of the night. Opening and shutting the door to the heated observatory building enables one to experiment with the thermal dome effects.

1.42. *Dynamical effects of the dome.*—With the wind blowing, the dome acts as an obstruction, and the currents become turbulent in the leeward direction. Upwind the images may still be good, but downwind the seeing may be very poor. Under such conditions the Foucault test shows the wave front covered with small eddies, a few centimeters in diameter with intricate overlapping structures, moving with the wind. This dynamical dome effect will extend over a length related to the diameter of the dome.

1.43. *Convection in the telescope tube.*—Even within the telescope tube convection may develop. This may happen when the upper part cools by radiation to space; and while the temperature differential along the tube will be slight, the optical path may be considerable, so that the effects on the seeing are not negligible. The knife-edge test reveals long filaments, crossing the illuminated image of the objective, often moving or bending very slowly, a few centimeters per second, sometimes gently shaping themselves into temporary whirls. Rösch (1955) has studied these effects for the 60-cm refractor at the Pic du Midi and has given numerical results. For a reflector with a closed tube the turbulence is likely to be much more serious, because cool outside air will descend continually into the tube. Reflectors with open tubes (i.e., merely a framework) are less subject to these air drops, although some effects arising from the metallic structure itself remain.

Serious disturbances will result from solar heating if the sun itself or nearby objects are observed. At the Pic du Midi, during daytime observations of Mercury and Venus, the top of the tube was struck by sunlight. The heating caused convective motion, which, however, could be greatly reduced by painting the top of the tube white.

1.44. *Heat exchange at the mirror surface.*—Some troubles are caused by the mirror temperature being different from that of the ambient air. The mirror temperature lags behind the changing air temperature and is affected also by radiation, which will tend to make it lower than the air. As a result, an internal temperature gradient may be set up which will deform the mirror's surface, while temperature differences between mirror and air may give rise to cellular convection just above the mirror. The light-beam traverses this disturbed layer twice, with the addition of phases, which doubles the amplitude of the light scattered and therefore quadruples its intensity. For refractors, the glass disks are thinner than for mirrors of corresponding aperture, and, moreover, the light traverses the boundary layer only once. Also the wind destroys the cellular patterns at the surface of the objective. The disturbing effects are therefore less important for refractors. Couder (1953) has attempted to remove these effects by ventilation of dome, tube, and mirror. At the Cambridge Observatory, precooling of the mirror to the probable night temperature has been used. At Palomar and certain other observatories the entire dome is well insulated against the higher daytime temperatures. The problem of mirror distortions has been largely solved by the use of Pyrex and fused quartz disks.

1.5 Seeing Effects on Contrast and Resolving Power

The present section examines ways in which the telescopic image is affected by the atmospheric disturbances discussed above. One may class the atmospheric troubles into two categories: large-scale perturbations which affect principally the resolving power and small-scale perturbations which affect principally the contrast. For telescopic observations of the planets, the contrast and the resolving power are equally important.

1.51. *Large-scale perturbations.*—Typical examples of such perturbations are found in undulating, horizontal layers in the atmosphere (see Sec. 1.31). We first suppose that the telescopic aperture is well below the mean distance between two undulations. If one observes the star at the zenith, at right angles to the layer, and if a "low" of the undulating surface just passes the optical axis, then the wave front resembles a segment of a sphere on the optical axis. The resulting image will be a perfect diffraction pattern, focused at a point beyond the regular focal plane. On the other hand, if a "high" passes the optical axis, the diffraction pattern will be brought to focus inside the instrumental focus. Halfway between these points a boundary layer will simulate an inclined plane; the image will be a theoretical diffraction pattern laterally displaced. Intermediate cases will add astigmatism to these extra-focal effects and lateral displacements. Altogether, the telescope will show an image approximating the theoretical diffraction pattern, but slightly distorted and moving constantly in space, laterally and in depth. The relative variation in focal length is usually between about 10^{-3} with poor seeing and 10^{-4} with good conditions. The lateral displacements remain normally between 1″ and 0″.2, and total amplitudes smaller than 0″.1 may occur on good nights. Now and then, depending on conditions, the resolving power will attain its theoretical value, but visual observations will be strenuous. The motion in depth is followed most readily by a youthful observer, having a great range of accommodation, and may cause headaches to an older person. If the observation is made at low altitudes, the beam traverses the disturbed layer obliquely, and the amplitude and the astigmatism are increased.

If the telescope possesses an aperture much larger than the distance between "highs" and "lows" on the wave front, so that several of these are included within the aperture, we may suppose the objective to be divided into a number of equal adjacent areas of the type considered above. Each of these smaller areas will then give a moving diffraction pattern, and the different images will often have random phase relations. The resulting image will then be a spot with little motion but extended and diffuse. The increase in telescopic aperture will then cause an increase in intensity but not in resolving power.

The equivalent resolving power depends on the distance between the highs and lows in the wave front, as well as on the amplitude of the undulations. The first quantity determines the size of the independently moving diffraction patterns, while the second quantity determines the amplitude of their motion. This explains why an increase in telescopic aperture may decrease the resolving power when the amplitude of the undulations is large. It is clear that the above discussion represents a simplified picture, since in reality there will be intermittent coherence between adjacent elements.

1.52. *Small-scale perturbations.*—These perturbations result most often from local or instrumental effects and produce a wave front marked by fine striations, whirls, or eddies, temporary and in motion. We shall now consider one small striation or eddy. The light-beam from a star passing through the perturbed zone will experience a phase difference. According to the classical Fresnel construction, the vector representing the luminous vibrations undergoes a rotation. One can resolve the vector into two components, in the initial and perpendicular directions. The components in the initial direction will, together with the light passing non-perturbed regions, form the diffraction pattern with somewhat reduced amplitude, which will somewhat modify this pattern. The disturbance by the orthogonal component is more serious; this component gives a luminous spot with a diffraction pattern corresponding to the area covered by the disturbed region of the wave front. Because the perturbed area is small, the diffracted light is distributed over a wide area around the slightly disturbed stellar diffraction pattern. The different striations, eddies, or whirls which cover the surface of the wave front add their effects; as a result, the image of the star will be surrounded by a halo of light.

In the visual observation of double stars such a halo is not very disturbing. However, for the study of planetary surfaces it is very serious. The reason is that the different parts of the planetary surface produce halos which overlap, thereby reducing the contrast of surface detail to the point where it may disappear over areas far greater than the expected resolving power. Often the wave front is perturbed by fine ripples, with a fairly uniform distribution over the wave front. In this case the diameter and the relative brightness of the halo are independent of the aperture of the telescope; an increase in aperture may decrease the diameter of the central diffraction image but will not change the angular diameter of the halo. However, the brightness and the diameter of the disturbing halo may increase with increased telescopic aperture, because the larger instrument and the larger dome contribute to an increase in the local atmospheric disturbances. Reflectors are more sensitive to this effect than refractors for the reasons given in Sections 1.43 and 1.44.

1.6. Conditions for Good Images

There are several causes for local inhomogeneities in the refractive index of the atmosphere, and, in order that the images may be excellent, all these causes must be absent simultaneously. This will happen only rarely. During long periods of clear weather, such as occur during atmospheric highs, the disturbed horizontal layers in the atmosphere tend to disappear. The lower atmosphere will become steady when light haze or fog develops, because saturated air is more nearly isothermal. Or, when the air is dry, steadiness may result from radiation causing a temperature inversion. Seeing troubles caused by the telescope and dome are absent when there is neither wind nor a temperature variation. Most observatories have at least some nights of good seeing, so that, with patience and persistence, very good results may be obtained almost anywhere.

However, many planetary investigations need *frequent* observation under excellent conditions; this requires an atmosphere that is free from optical discontinuities during a large fraction of successive nights. Such locations may be found on completely level deserts, on seashores where the wind comes from the sea, on small islands, and on isolated mountain tops. Here we shall be particularly concerned with the advantages offered by mountain observatories.

1.7. High-Altitude Observatories

A high and isolated mountain offers advantages, not merely because of the rarified air at high altitude, but also because of the position of the telescope on the summit, as we shall see below.

The index of refraction, n, of the air varies with temperature and pressure. At sea level and at $+10°$ C, $n = 1.000288$. At 3000 meters altitude and $-5°$ C, $n = 1.000200$; $(n - 1)$ is therefore reduced by the factor 1.4. The phase effects and the resulting disturbances in the seeing produced by atmospheric waves will be reduced in the same proportion. Besides, one-third of the total atmosphere mass, the part having the highest refractive index in which most of the atmospheric turbulence occurs, is below the observatory. Now the advantage of a mountain observatory should not be permitted to be lost by the introduction of new local disturbances. A mountain pass, a mountain slope, a high plateau, or even a peak surrounded by other mountains has been found to suffer from disturbances of local atmospheric origin.

At night, in the absence of wind, the rocks on top of an *isolated* mountain will cool rapidly, and cool air formed by contact will slide down the slopes; a telescope on the summit will be free from such ground effects. If there is a moderately strong wind, coming from the direction of observation, the beam will traverse air in laminar flow, and the seeing will not be disturbed. A lateral wind is less favorable, because eddies form at the edge of the dome shutters. The worst case will be a strong wind from the opposite side of the shutters; this will spoil the seeing completely. In summary, one may say that the number of circumstances which can lead to bad seeing is minimized on an isolated mountain top but that one still has to cope with strong winds causing local effects; in addition, it will obviously be an advantage if the mountain is located in a geographic area having frequent anticyclones, which will reduce the number of general disturbances in the atmosphere above the observatory.

2. PLANETARY WORK AT THE PIC DU MIDI

The Pic du Midi Observatory is on top of a 2870-meter (9400-foot) mountain, favored with good atmospheric conditions. It is shown in Plate 1. The site was first used for planetary work by de la Baume Pluvinel and Baldet (1909), who obtained good photographs of Mars in yellow and blue light (see Pl. 10). In 1939, Lyot, with deep understanding of optics and atmospheric seeing conditions, began to use this site for planetary studies. New equipment was constructed, and Lyot with several collaborators began a visual and photographic survey of the planets with refractors of increasing aperture (23, 38, and 60 cm). Later were added optical studies of the properties of the light reflected by the different small parts of the planetary surfaces. These included measurements of brightness, polarization, and diameter and observations through narrow-pass-band filters. The present chapter is concerned with the visual and photographic observations of planetary and satellite surfaces. The polarization measures are discussed in chapter 9.

2.1 Telescopes Used

The first astronomical equipment on the Pic was installed by B. Baillaud at the close of the nineteenth century; it consisted of a moderately good Newtonian telescope of 50-cm aperture (which was used by Baldet in 1909 for observations of Mars) and an excellent 23-cm refractor. Around 1934, Lyot was struck by the remarkable steadiness of the images in the 23-cm refractor, and Camichel, after 1937, showed that observations were possible in winter also. For more effective study of the atmospheric conditions, Lyot replaced the 50-cm reflector by an excellent lens of 38-cm aperture in 1941, and planetary studies were begun jointly with Camichel and Gentili (Lyot, 1943). The images in that telescope were often perfect and frequently showed the theoretical resolving power of 0".32 on planetary disks. It was evident that the instrumental aperture by no means utilized the full possibilities offered by the site. Accordingly, Lyot decided to make further experiments with a more powerful instrument. The Paris Observatory put at his disposal the visual 60-cm objective of 18.2 meters focal length, which had been used on the large equatorial coudé. The installation was made by Dr. J. Baillaud. Since the dome had an internal diameter of only 8 meters, it was necessary to fold the beam by using two silvered plane mirrors, figured by Couder (1953). Lyot discovered that the thickness of the coating of the flats had to be extremely regular, in order not to disturb the wave front; currently, the coatings are made at Meudon by Mr. H. Grenat by evaporation. The lower flat mirror, which measures 50 cm in diameter, is held in place by 12 clasps which exercise a slight pressure on the rim of the polished surface. Each clasp is adjustable, which makes possible the correction of any slight astigmatism in the optical train. This installation has proved satisfactory; when the atmosphere is calm, the instrument gives stellar images close to the theoretical limit of 0".2. Actually, with perfect seeing, the diffraction pattern of a star is somewhat triangular, surrounded by a slightly fuzzy first diffraction ring. The faint halo around the theoretical diffraction pattern, on which the contrast on planetary disks so critically depends, in this telescope is usually produced by small residual atmospheric effects; i.e., if the mirrors have excellent coatings, the instrumental contribution to the halo is smaller than the atmospheric effect. A small variable aberration sometimes appears when the atmosphere cools suddenly; this effect may be corrected at the center of the field by use of a compensator placed at an adjustable distance inside the focus. Compared with the results obtained with the 38-cm refractor, the new installation showed finer details, the improvement being nearly in proportion to the ratio of the diameters. The contrast, however, is often somewhat reduced. It follows that the 60-cm objective still does not fully utilize the possibilities presented by the atmosphere at the Pic. This conclusion induced Lyot to consider a much more powerful telescope which at the same time would include all known devices for suppressing instrumental disturbances causing loss of contrast. With such an installation an effective resolving power of 0ʺ.1 was expected. This increase would, of course, be of the greatest value, because a comparatively small increase in resolving power greatly increases the number of observable details on the planet and makes new discoveries possible.

2.2 Methods of Observation

2.21. *Visual observation.*—The observation of planetary detail depends on contrast and resolving power (Sec. 1.5). Until now, visual observation has been the only method by which we have attained the theoretical resolving power of the 60-cm telescope for objects of sufficient contrast.

To observe visually, with optimum resolution and contrast, one must adjust the magnifying power P and the aperture A of the telescope. The exit pupil has the diameter $d = A/P$.

The aberrations of the eye and the discontinuous structure of the retina permit the eye to be diaphragmed to 1 mm without loss in resolution. Below 1 mm the product $ds(d)$ is constant, where d is the aperture of the pupil and $s(d)$ is the limiting resolution; this follows from the diffraction theory of light. The eye in combination with the telescope sees the image magnified by the ratio P, so that the resolution becomes $S = s(d)/P$, and hence $S = ds(d)/A$, in which the exit pupil of the telescope replaces the entrance pupil of the eye. Hence, to see the finest possible detail for a given telescopic aperture A, one must increase the magnification until the telescopic exit pupil decreases to 1 mm. Increasing the magnification further does not increase the resolving power because it correspondingly decreases the angular resolving power of the eye. This implies that, for maximum resolving power, *the magnification must be no less than the aperture of the telescope expressed in millimeters.*

However, the perception by the eye of small contrast differences sets further conditions, among which the surface brightness of the object under study is the most important. The sensitivity of the eye is maximum for a surface brightness of about 0.01 candle/cm^2 (0.01 stilb) and is nearly constant for the interval between 10^{-3} and 10^{-1}. Below 10^{-3} stilb, it drops steeply and becomes very small for values less than 10^{-6} stilb (Danjon and Couder, 1935, p. 51). The surface brightness of the moon is 0.25 stilb, of Venus 3, of Mars 0.2, of Jupiter 0.05. Therefore, if an exit pupil is used equal to one-tenth that of the pupil of the eye, i.e., 0.5 mm, the surface brightness of the moon, Venus, and Mars, as seen in the telescope, are reduced 100 times and are still sufficient to yield a maximum degree of visual contrast; but for Jupiter and particularly for Saturn, Uranus, and Neptune, smaller magnifications and larger exit pupils must be used. The above considerations give only an approximate picture; the planets are surrounded by a dark sky, so that the mean surface brightness of the field of vision is not well defined; also, the sensitivity of the eye to contrast is a function of the interval during which an observation is made and will be reduced for planetary images affected by poor seeing; but they show that, for maximum contrast, the outer planets—Jupiter to Neptune—must be observed with lower magnifications than corresponds to the condition of maximum resolution.

The difficulties of photographic observation also increase with decreasing brightness; one may therefore expect that, particularly for these planets, electronic image-recording techniques will be useful.

In the presence of an atmospheric layer having undulations of large amplitude the different parts of the objective will displace the light-beam without coherence, causing a continual blurring of the light in the focal plane (cf. Sec. 1.51). In such cases the observation will be improved by a reduction in the telescopic aperture. With large reflectors we have found it advantageous to put behind the eyepiece in the plane of the exit pupil a small screen with aperture of about 1 mm in diameter, the amount depending on the surface brightness of the planet. Then, if different magnifications are used, the apparent surface brightness of the image will not change. With increasing magnification, the resolving power will increase until a limit is reached set by the atmospheric seeing. Clearly, it is this maximum magnification that should be used under given conditions. This simple procedure automatically results in the most efficient use of the telescope.[1]

Sometimes the visual observer is uncertain about the reality of detail that he suspects on a planetary disk. Examples are bright or dark ripples in the Rings of Saturn and borders around the polar caps of Mars, increases in brightness of the cusps of Mercury and Venus, delicate shadings on the Venus disk, and spots on Neptune, the Jupiter satellites, and Titan. In such cases we have found the use of different magnifications helpful; the spurious features will not change their angular diameters proportionally. Alternatively, one may change the brightness, as well as the resolving power, by the use of diaphragms.

The fact that moments of good seeing are often of short duration causes further difficulties. One may then see on planetary disks, suddenly and

[1] E. E. Barnard and probably other observers were well aware of the advantages of a 1-mm diaphragm placed at the exit pupil of a large telescope.

briefly, straight dark lines, parallel lines, circular black dots, or a combination of these. The sensitivity of the eye adapts itself, within limits, to the brightness level of the object under study. As a result, contrast differences in a large image will tend to disappear if the eye does not scan across the image. As was well shown by Byram (1944), only the continuous movements of the eye will destroy this adaptation at the boundary of unequal areas and permit the observation of markings. If the atmospheric seeing blurs the telescopic image except for brief moments, the scanning remains incomplete and gives insufficient information on the complexity of the spots under study.

Special problems arise during the daytime observations of Mercury and Venus, when these planets are close to the sun. Three causes are mentioned: (1) multiple reflections between the optical surfaces, leading to diffuse light in the focal plane; (2) scattering by dust on the optical surfaces; and (3) diffraction by the edges of the lens (if a refractor is used). All three light-sources increase the sky background and may be studied by observing the objective without an eyepiece from the focal plane. Furthermore, the objective is heated by the sun, which may cause optical aberrations. It is therefore very desirable to protect it from direct sunlight. At the Pic du Midi we use a circular screen (*parasoleil*) whose shadow falls on the objective. The screen is held by an angle iron, 8 meters long, attached to the bottom of one of the halves of the dome slit and projecting outward in the direction of the sun. The angle iron is supported near the screen by a cable attached to a winch mounted at the top of the dome for adjustment in altitude; the point of support by the cable is so chosen as to minimize flexure of the angle iron. Rotation of the dome makes adjustment possible in azimuth. This device has proved efficient and on occasion has resisted very strong winds.

2.22. *Photographic observation.*—Photographic images are usually less detailed than drawings obtained visually; but they are more reliable, are obtained much more rapidly, and permit photometric studies and precise position measurements. Furthermore, reliable comparisons may be made between records obtained at different times, including minor changes in contrast. At the Pic du Midi the photographs have been taken chiefly by Dr. Camichel.

The photographic emulsion is chosen primarily for its ability to record delicate contrasts. This ability is measured by the *quantum efficiency*, which is the number of photons falling on the plate divided by the number of silver grains formed in the emulsion (this would be strictly true if the grains did not influence each other). The quantum efficiency depends on the nature of the emulsion, the wave length, the process of development, and the density obtained.

The quantum efficiency, θ, is proportional to C^2S/G^2, in which C is the contrast, S the sensitivity, and G the graininess. The plate sensitivity, S, is measured by the inverse exposure time required for obtaining the same density, using a constant laboratory source. The latter must not differ too much from the intensity of the planetary image; C is the contrast corresponding to that density. The graininess, G, is the root mean square of the fluctuations observed on a microphotometer record for the smallest areas recorded on the planetary image.

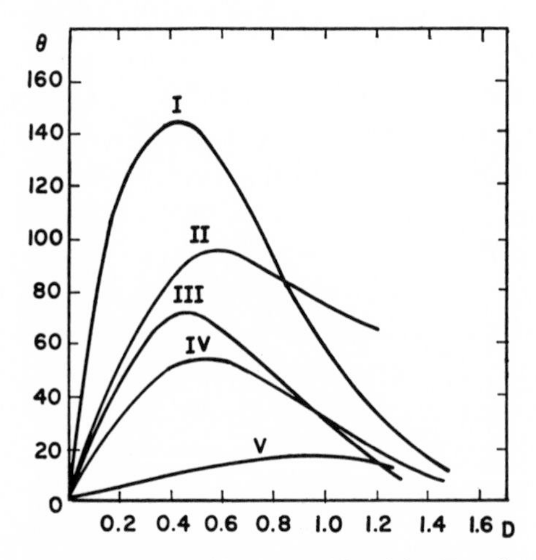

FIG. 1.—Relative quantum efficiency (θ) in blue light of five French orthochromatic emulsions as a function of density; *I*, Super-Fulgur, $S = 70$; *II*, Collodium, $S = 2.3$; *III*, Helioguil, $S = 6$; *IV*, Anecra, $S = 2.2$; *V*, Lactate, $S = 0.5$.

Figure 1 illustrates the relative quantum efficiency in blue light of five French orthochromatic emulsions as a function of the density. The relative sensitivities S are indicated. The shape of the curves is due to the action of the photons falling on the silver bromide crystals; several photons are required to sensitize one crystal. Under weak illumination the crystal will, on the average, not receive enough photons to become developable; some of the photons are not effective, and hence the quantum efficiency is low. With increasing illumination the quantum efficiency rises, reaching a maximum when each crystal receives a certain photon flux. Still stronger illumination will increase the density but with lower quantum efficiency.

We have found that the quantum efficiency of emulsions of either high or low sensitivity tends to be smaller than for plates of average sensitivity, 15–60 ASA. The plates used at the Pic du Midi for the yellow exposures were actually mostly of medium sensitivity, namely, Agfa Rot-Rapid, Kodak P-300, or Eastman IV-B.

Underdevelopment reduces the quantum efficiency; we found full development with metol or hydroquinone preferable to fine-grain developers, which often act incompletely. However, orthophenylene diamine developers or those with ammonium sulfo cyanide reduce, at the same time, the sensitivity, the contrast, and the graininess but do not alter the resulting quantum efficiency.

Figure 1 shows that the optimum quantum efficiency corresponds to a density of about 0.5. On account of the limited exposure time available, the size of the enlarged image on the plate has to be adjusted so as to obtain approximately such a density.

The detection of feeble contrasts improves with the image size and in proportion to the square root of the exposure time. Hence the latter should be made as long as possible. An opposite requirement is set by atmospheric tremors and guiding imperfections. To enhance the delicate halftones without increasing the exposure time, Lyot introduced the method of composite printing, which we have used on all our plates. (This technique was developed independently at the Lowell Observatory.) The observer at the telescope makes 20 or 30 planetary images in rapid succession on the same plate. The best of these are selected and successively enlarged on the same plate so as to obtain a single positive picture, in which each image contributes only fractionally. The resolution of the composite is then the same as that of the selected best exposures, while the graininess is reduced proportionally to the square root of the number N of images used. The markings appear as they would on a single exposure $\sqrt{N}$ times greater in diameter. Often we have superimposed up to 16 or 18 images.

The technique of composites makes it possible to reduce the size of the images and the time of exposure, thereby increasing the chances of finer resolution. The image cannot be reduced too much, however, because the resolution of the telescope must be retained. The image of a point source in the focal plane is the diffraction pattern. On the photographic plate such a source produces a diffusion disk in which light decreases radially according to the law $I_0 e^{-r/D}$, where I_0 and D are constants and r is the distance from the center. The diffusion disk should be smaller than the diffraction disk. The coefficient D is found equal to 26 μ on the fine-grain Panatomic Eastman plates and to 58 μ on the fast Superfulgur Guilleminot plates, which require minimum aperture ratios of $f/45$ and $f/100$.

The method of composite pictures can bring out very delicate shades. For that purpose the contrast and the number of images have to be increased together. The increase in contrast is possible only if the background is sufficiently uniform in brightness. For Jupiter and Saturn we have reduced the limb darkening by vignetting the edges during the enlargement. For Venus, faint markings are difficult to detect because of the strong brightness gradient between the limb and the terminator; markings merely produce faint waves in the isophotes. By compensating for the background gradient, the markings are given closed isophotes and become more readily visible (Sec. 3.22).

The photographic equipment used for 18 years by Dr. H. Camichel at Pic du Midi enlarges the image of the planet with a negative lens so as to obtain an adjustable equivalent focus on the 60-cm refractor between 38 and 50 meters. A double-slide plateholder can take 20–30 exposures in succession on one plate. The last exposure is shifted in right ascension as a reference for orientation. The exposure times are a fraction of a second on Mercury and Venus, 1 or 2 seconds on Mars and the moon, and a few seconds on Jupiter and Saturn. The observer watches the seeing in a guiding eyepiece or the 23-cm guiding refractor, so as to catch the most favorable moments. For the longer exposures he may correct the guiding by exerting a slight pressure on the telescope with a flexible rod. After exposure, each plate is photometrically calibrated by means of a tube sensitometer.

3. THE OBSERVATIONS

3.1. Mercury

Under the best conditions, the planet Mercury shows markings which may be likened to those on the moon as observed with the naked eye. In 1942 Lyot obtained 25 fine drawings with the 38-cm refractor and magnifications between 300 and 500 (Dollfus, 1953*a*). The theoretical resolving power of 0$''$.3 represents one-thirteenth of the minimum diameter of the planet and one-twentieth of the maximum diameter. The actual separation of details obtained in these drawings is close to this theoretical limit (Pl. 2, *upper half*). In 1950 the writer used the 60-cm refractor with powers of 750–900×. The observed features are more numerous, contrasty spots about 300 km apart being clearly resolved (Pl. 2, *lower half*).

From one day to the next the same regions appeared on the disk. They slowly moved under the effects of libration and the varying position of the line joining the cusps. The period of rotation of Mercury is thus found to be equal to the period of revolution, with a precision of better than one in ten thousand. The surface relief cannot be much higher than that on the moon; if it were twice as large, shadows near the terminator could have been detectable. The Pic du Midi observations gave no conclusive proof for the presence of atmospheric haze (Dollfus, 1953*a*).

Photographic observations with the 38-cm refractor were begun by Lyot and Camichel in 1942. A run from July 8 to August 13, 1942, covering about half the orbit, included the perihelion passage on July 25. Twelve of the composites obtained clearly show the markings; three of these are reproduced in Plate 3. The resolving power is about 0″.45. The positions of 23 selected markings were measured in a system of co-ordinates in which the Mercury orbit is the fundamental plane and the zero meridian on the planet is that which contains the subsolar point at perihelion. On the basis of these measures a chart was constructed from the photographic records alone. The chart is reproduced in Plate 4. Comparison with the visual observations of Plate 2 will be found of interest; because of libration, the terminator does not correspond precisely with the edge of the map in Plate 4.

A comparison between Plate 4 and charts obtained visually by earlier observers has been published elsewhere (Dollfus, 1953*a*). The maps of Schiaparelli and Antoniadi are in good accord with ours if they are rotated about 15° in a counterclockwise direction. These visual maps were based on evening observations, when the planet was close to aphelion, while Plate 4 is based on morning observations when the planet was close to perihelion. The observations are brought into coincidence by assuming an obliquity of about 7° for the axis of rotation.

Unpublished photographs obtained by Dr. Camichel with the 60-cm refractor confirm the 38-cm results described here.

3.2 Venus

3.21. *Visual observations.*—The brilliant surface of the planet Venus usually shows only markings of very low contrast. Their visibility is made even more difficult by the strong brightness gradient between the limb and the terminator. In order to see the markings, the observer must try to ignore this gradient by appropriate visual scanning across the disk. With moderate-sized telescopes, one perceives markings of two kinds: (1) slightly darker regions, which are regarded as real, and (2) shadings along the bright limb, which appear to be physiological effects resulting from the projection of the bright crescent on a dark background. These latter effects are reduced by increasing the apparent diameter of the planet, which is done by increasing the magnification or by using a larger telescope with correspondingly larger magnification. With the 38-cm refractor Lyot secured some 30 drawings of Venus, while the writer, with the 60-cm refractor, obtained about 100 more.

The observation of the shaded regions on Venus does not depend on the telescopic resolving power in the same critical manner as that of contrasty planetary or lunar detail; instead, telescopic and seeing conditions are desired that result in maximum contrast. Furthermore, the visibility of these regions is not constant, apparently depending on the state of the Venus atmosphere. Occasionally, no markings are visible at all. Usually, a few large areas are seen, while on rare occasions considerable detail is visible. Plate 5 shows two drawings by the writer, 35 days apart, made at times when finer detail could be seen than is usually the case. Exceptionally, a marking of considerable contrast is visible on the planet; such an observation was made by Dr. H. Camichel at the time he took the photograph shown in Plate 8, on November 6, 1943.

In comparisons of drawings made at consecutive days or within a week or two, the markings are found to recur more often in the same location than would be expected according to chance. On the other hand, sometimes markings visible on one day are definitely absent on the next. If all drawings made on separate days are copied on transparent paper and combined photographically by superposition, in groups having essentially the same phase angle (the intervals used are of the order of 2 or 3 weeks), then copies are obtained which will emphasize the common features. Plate 6, derived by this method, was based on approximately 60 drawings. (Approximately 40 additional drawings were made with smaller apertures or through filters, which were used, among others, to study the physiological effects.) For practical reasons, it was not possible to obtain the sequence of Plate 6 during a single synodic period.

The following conclusions are based on inspection of Plate 6:

a) The same general features can be traced from one composite to the next, allowing for the changing position of the terminator. This confirms an earlier result by Danjon (1943).

b) When one day's drawings are compared with the composites, one usually finds certain shaded features missing or entire areas blank.

Because of conclusion *a*, a general map of the planet may be composed; the result is shown in Plate 7. The horizontal diameter corresponds to the plane of the orbit; the center is the mean subsolar point, and the projections of the great circles passing through it are divided uniformly, with the 30°, 60°, and 90° distances marked. (The central part of the map is therefore essentially the heliocentric appearance of the planet, but the limb areas would be foreshortened; the quarter phases are represented by the right- and left-hand portions of the map, allowing for distortion.)

3.22. *Photographic observations.*—Before discussing possible interpretations of the visual observations, we shall review the results obtained from a study of a long series of photographs in yellow light obtained by Dr. Camichel since 1942.

Inspection of the original negatives shows very diffuse markings on only a few of the plates, while most plates show no detail at all. This lack of detail is due in part to the strong intensity gradient across the images. In order to increase the sensitivity of detection, the gradient had to be compensated for and the remaining brightness configuration copied with greatly increased contrast, while the grain was kept down by using the technique of composites, described in Section 2. As many as 15–20 images were combined into one frame.

A selection of the composites so obtained is given in Plate 8. With the increased contrast, all these images now show markings. In making comparisons between Plates 8 and 6, it should be remembered that the shading procedure used in Plate 8 tends to displace the terminator inward and may also deform it. With allowance for this, the positions of the dark markings are compatible with Plate 6; clouds may cause certain markings to be invisible at any one time.

3.23. *Interpretations.*—The observations lead to the following considerations. The quasi-fixed position of the markings with respect to the terminator is most readily explained on the assumption that the periods of rotation and revolution are equal. The dark markings may show either the surface of the planet or a low-level distribution of clouds connected with surface topography. However, the argument for synchronization is not conclusive; the atmospheric pattern may be diurnal in character. In either case the observations show that the dark markings are at times covered by clouds. These higher clouds appear to be transient, as their pattern may change completely from day to day; either they move very rapidly, or else they form and dissolve.

For a comparison of the distribution of ultraviolet markings with the clouds shown in yellow light, ultraviolet photographs were taken with a 60-cm reflector. Two series of ultraviolet pictures were obtained—about 20 plates by the writer in 1948 and a somewhat larger number by Dr. Camichel in 1953. The resolving power, 0″.6, was not so good as on the yellow plates because of the mirror figure. Plate 9 shows ultraviolet photographs for 4 days, as well as the cloud distributions deduced for these days from comparisons of the visual observations with the composites of Plate 6. The agreement between the position of the visual clouds and the bright markings in the ultraviolet is quite good. Between the clouds the planet is darker in the ultraviolet but does not show the discrete markings seen there in yellow light. The visibility of the clouds increases toward the shorter wave lengths; they become prominent for $\lambda < 3800$ A.

The observations presented suggest the following model for the Venus atmosphere. There is a lower layer, which may be the surface of the planet or a cloud layer associated with the surface topography and having a certain degree of permanence, as indicated by Plate 6. Above it are large, discrete clouds which either are temporary in nature or are moving, which obstruct our vision of the lower layer and become directly observable in the ultraviolet.

3.3 Mars

3.31. *Photographic observations.*—During the close opposition of 1909, plates were taken at the Pic by de la Baume Pluvinel and Baldet (1909), in yellow as well as blue light. Plate 10 shows a selection of the plates, four in yellow and two in blue, copied as composites by the writer. The originals were taken with the 50-cm reflector, and the image was enlarged at the telescope to 3.5 mm. In all, 80 plates were taken, each with about 20 images. The yellow exposures took 8–12 seconds, and the blue exposures (no filter) were 2 seconds. The latter clearly show the violet veil obscuring the surface markings, a fact to which the 1909 paper called attention.

Between 1941 and 1958, about 800 plates were obtained by Dr. Camichel, for the last three oppositions in collaboration with Mr. J. H. Focas of Athens. Only nights with good seeing were used. This period covers nine consecutive oppositions—1941, 1943, 1946, 1948, 1950, 1952, 1954, 1956, and 1958—and includes more than one cycle of synodic heliocentric longitudes, so that all aspects along the Martian orbit are included. The effective resolving power was usually between $0\overset{''}{.}3$ and $0\overset{''}{.}4$. Furthermore, Lyot in 1941 took a film showing the motion of the markings due to the rotation of the planet.

Plates 11 and 12 show the same Martian hemisphere for eight consecutive oppositions, using the same scale and the same orientation in equatorial co-ordinates. The variation of the apparent diameter and the different orientations of the Martian polar axis with respect to the north-south line are apparent (the position of the axis is characterized by the quantities P and $D_{\oplus}$ given in the *Nautical Almanac*). The 800 plates were copied by a group of astronomers at the Meudon Observatory by making composites on glass, each composed of 6–16 selected images from one original plate. The scale used in these glass copies is uniformly $1''$/mm. For purposes of examination, the diapositives so obtained were classified in a two-dimensional array, by opposition and by longitude of the central meridian. To this end, the longitudes were subdivided into blocks of 30°

PLATE 1.—Pic du Midi Observatory

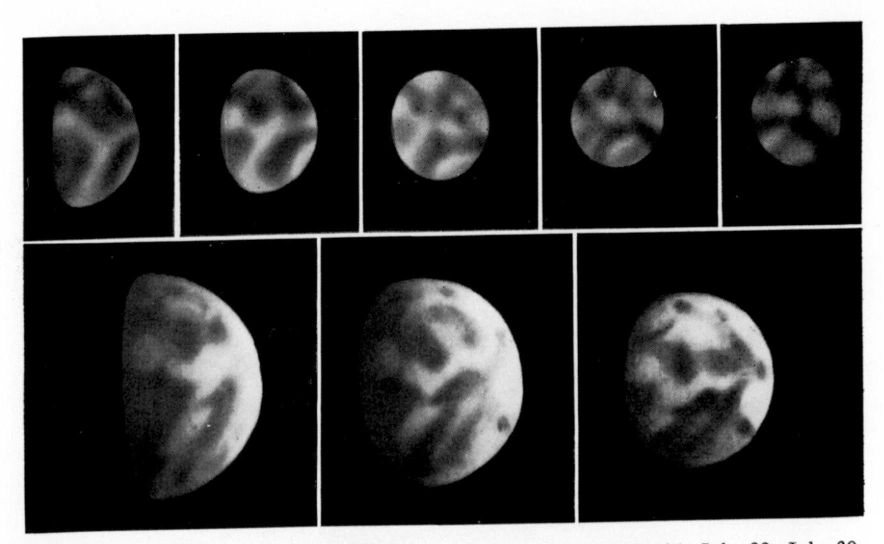

PLATE 2.—Drawings of Mercury. *Above:* 38-cm refractor; July 16, July 22, July 30, August 8, and August 13, 1942; each drawing based on 5 observations by Lyot. *Below:* 60-cm refractor; October 6, 12, and 19, 1950; each drawing based on 3–6 observations made by Dollfus on neighboring dates.

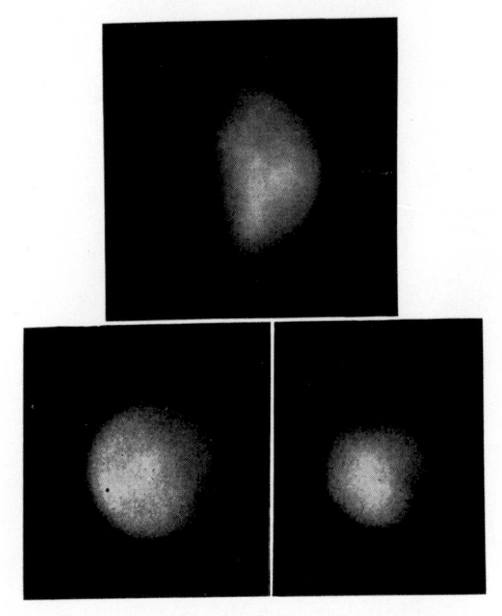

PLATE 3.—Photographs of Mercury by Lyot and Camichel, 38-cm refractor. *Above:* July 16, 1942, 6^h25^m U.T., phase angle $i = -73°$, composite of 10 images. *Left:* August 6, 1942, 7^h50^m U.T., $i = +11°$, composite of 16 images. *Right:* August 10, 1942, 9^h00^m U.T., $i = +22°$, composite of 12 images.

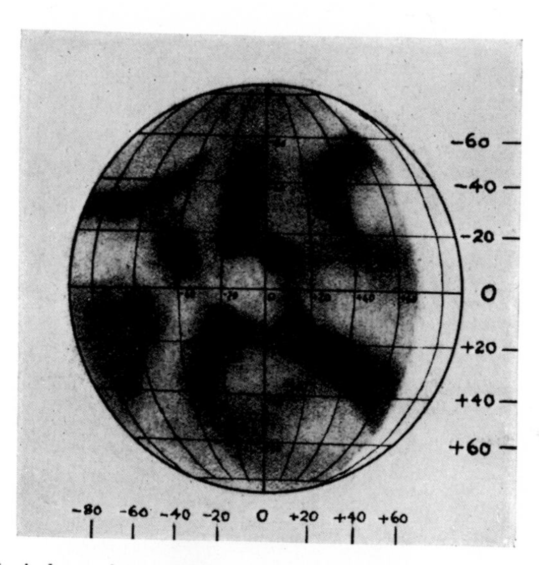

Plate 4.—Planisphere of Mercury, summarizing the spots observed on 10 composite photographs obtained in 1942 and 1944.

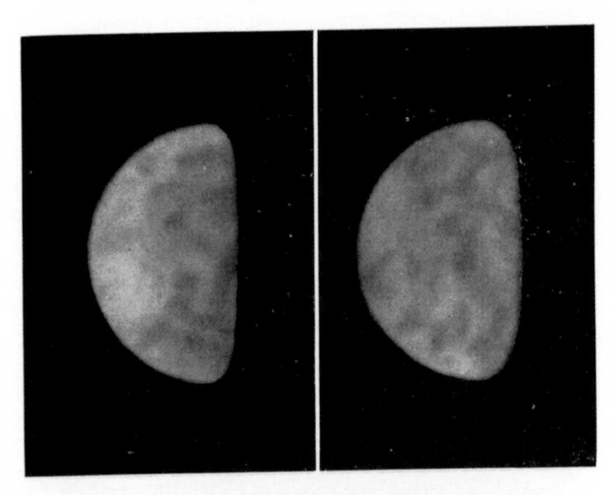

PLATE 5.—Drawings of Venus, obtained under especially transparent conditions of the Venus atmosphere. *Left:* March 17, 1948. *Right:* February 11, 1948; 60-cm refractor, power 700×. Large gray areas appear through veils of unusual transparency on both days. The intensity gradient, limb terminator, has been eliminated.

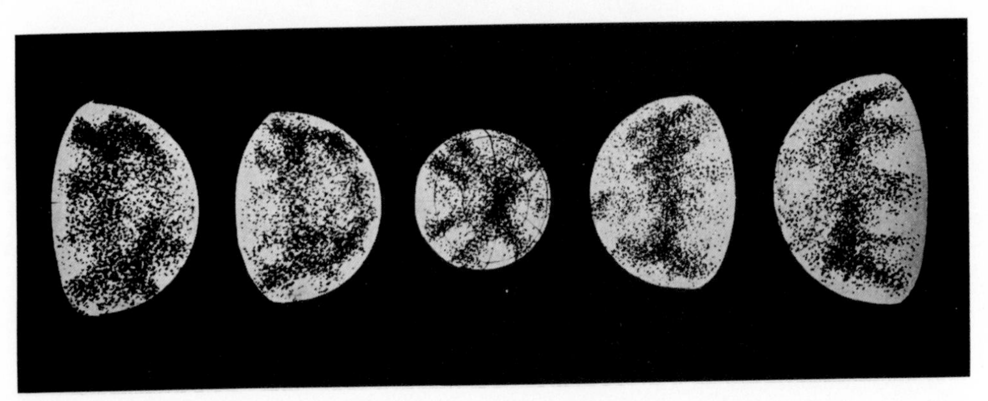

PLATE 6.—Composite images of the markings on Venus in yellow light, each obtained by superposition of several drawings to reduce effects of atmospheric veils. The data are, left to right:

Mean Date	No. of Drawings	Phase i	Latitude of Center $D\oplus$	Diameter d	Heliocentric Longitude, η
Aug. 10, 1953..............	8	−65°	+3°	15.″6	32°
Aug. 28, 1953..............	9	−57	+3	14.0	60
Oct. 7, 1950..............	10	−14	+1	10.0	170
Feb. 13, 1948..............	9	+58	−2	14.5	58
Mar. 11, 1948..............	8	+70	−3	17.2	103

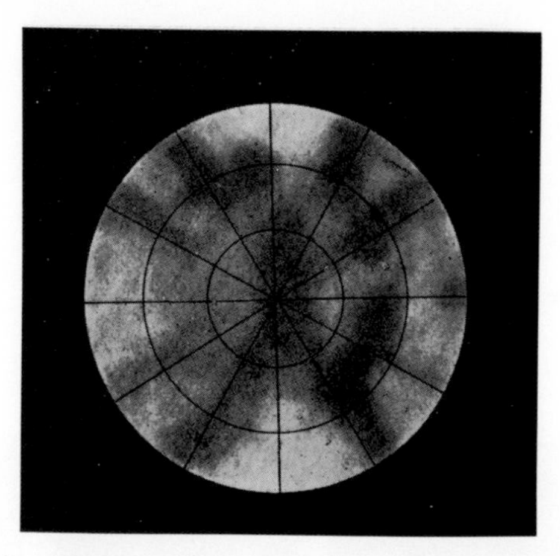

PLATE 7.—Planisphere of the permanent markings on Venus, based on the observations of Pl. 6. Molweida projection. The contrast is strongly increased.

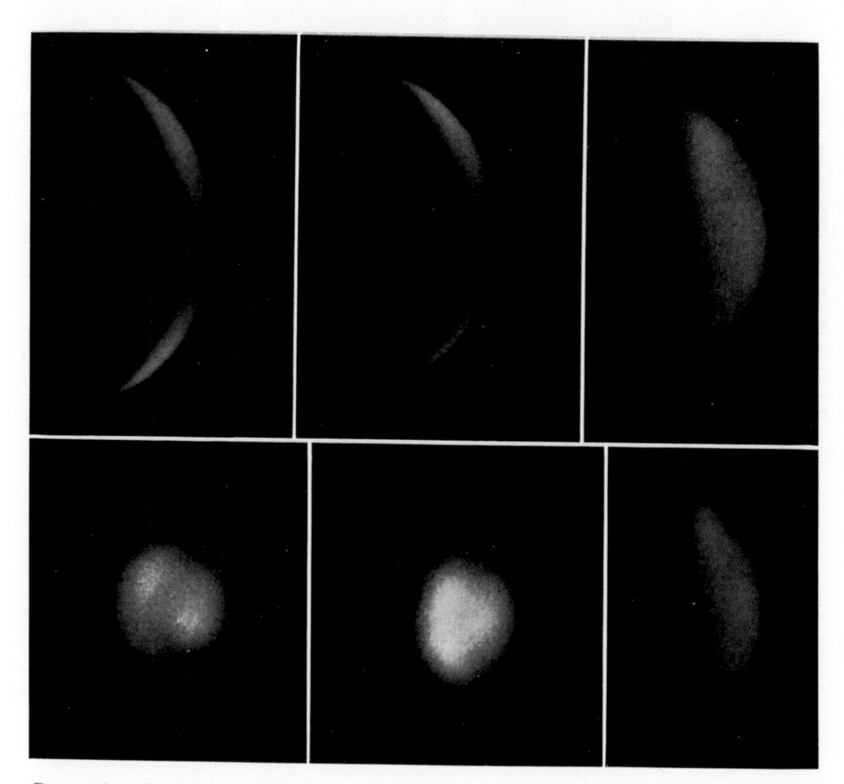

PLATE 8.—Composite photographs of Venus in yellow light showing markings, based on plates by Camichel with 38-cm refractor. Gradient between limb and terminator suppressed by shading and contrast increased by repeated copying. Dates:

Oct. 3, 1943 $k = 0.189$	Oct. 9, 1943 $k = 0.242$	Nov. 6, 1943 $k = 0.446$
Sept. 17, 1942 $k = 0.963$	July 26, 1942 $k = 0.866$	Nov. 22, 1943 $k = 0.534$

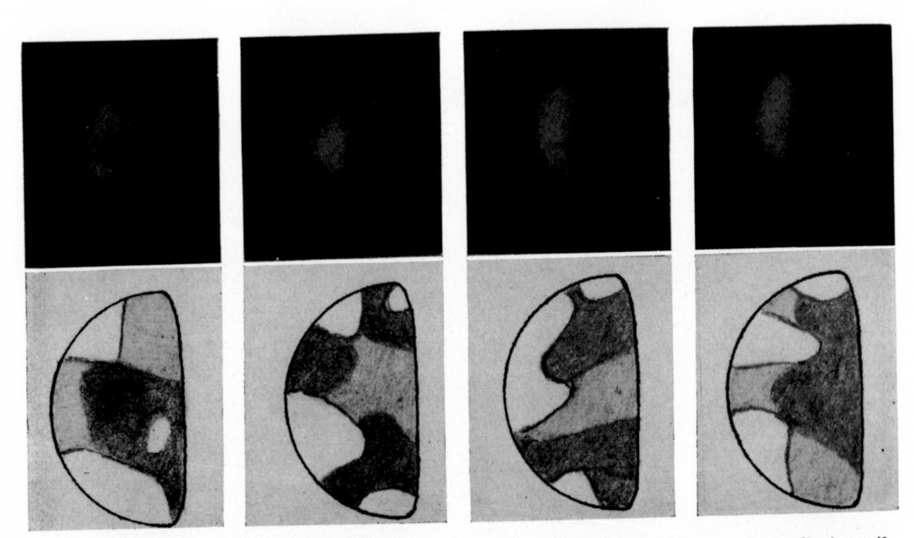

PLATE 9.—Comparison of ultraviolet photographs with distribution of atmospheric veils observed in yellow light. Photographs by Camichel obtained with reflector, diaphragmed to 50 cm. Distribution of veils seen in yellow light found by comparing 60-cm refractor drawings with composites of Pl. 6 by Dollfus. Bright areas are bright clouds; halftones, veiled regions in which the markings are invisible; and dark areas, regions showing gray markings. March 6, 10, 13, and 14, 1948.

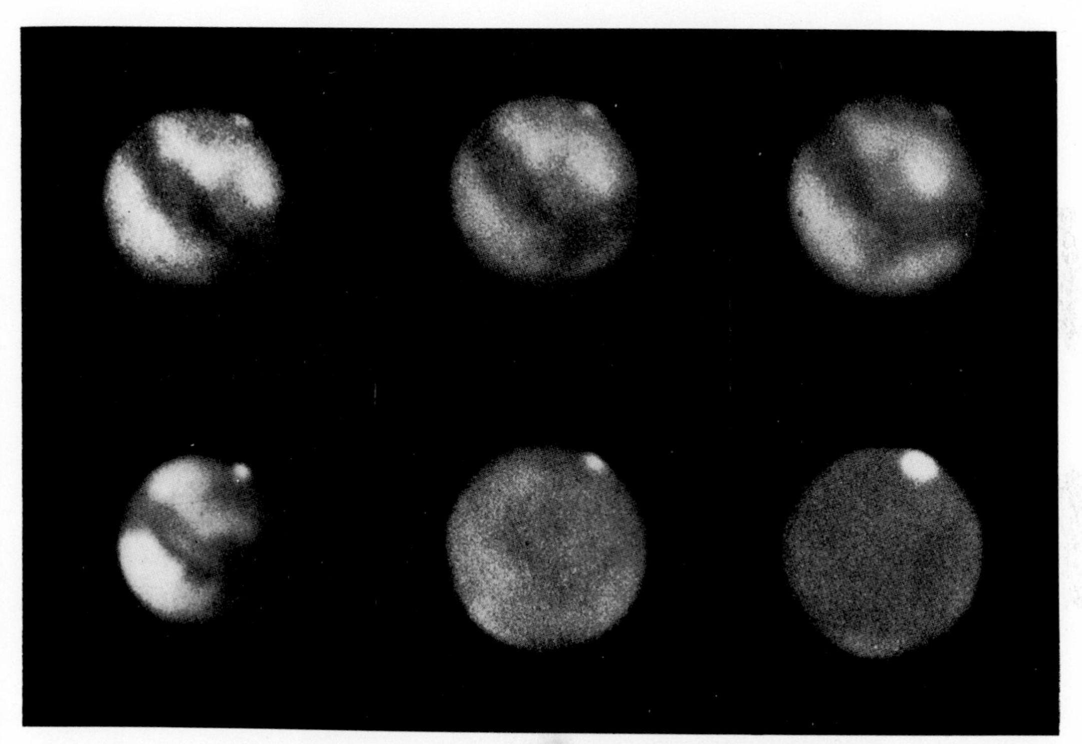

PLATE 10.—Photographs of Mars in 1909. Composites of originals taken by de la Baume Pluvinel and Baldet with 50-cm reflector. *Above:* September 27, yellow light. *Below:* October 20, yellow light; September 6 and 15, blue light.

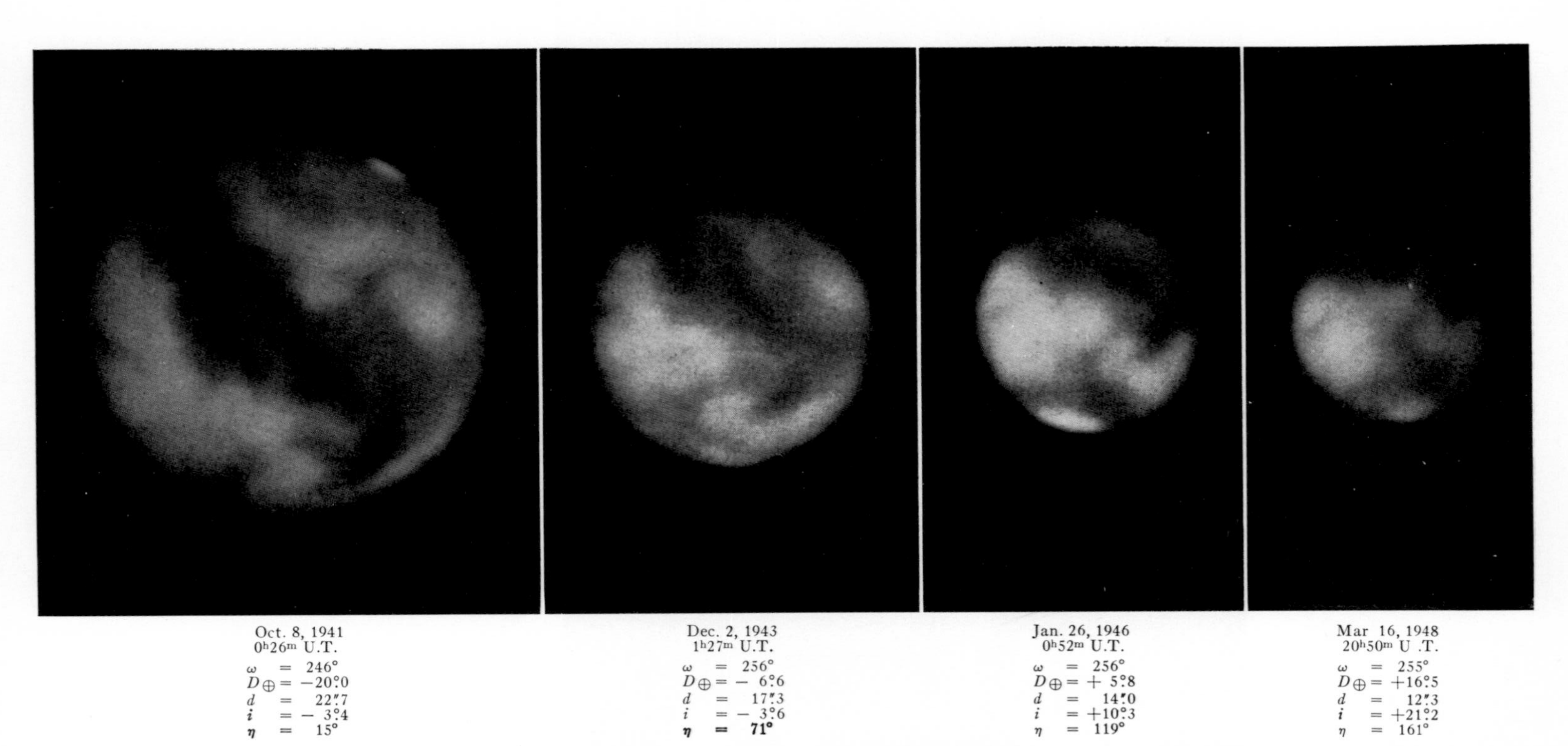

Oct. 8, 1941 0^h26^m U.T.	Dec. 2, 1943 1^h27^m U.T.	Jan. 26, 1946 0^h52^m U.T.	Mar 16, 1948 20^h50^m U .T.
$\omega = 246°$	$\omega = 256°$	$\omega = 256°$	$\omega = 255°$
$D_\oplus = -20\overset{\circ}{.}0$	$D_\oplus = -6\overset{\circ}{.}6$	$D_\oplus = +5\overset{\circ}{.}8$	$D_\oplus = +16\overset{\circ}{.}5$
$d = 22\overset{''}{.}7$	$d = 17\overset{''}{.}3$	$d = 14\overset{''}{.}0$	$d = 12\overset{''}{.}3$
$i = -3\overset{\circ}{.}4$	$i = -3\overset{\circ}{.}6$	$i = +10\overset{\circ}{.}3$	$i = +21\overset{\circ}{.}2$
$\eta = 15°$	$\boldsymbol{\eta = 71°}$	$\eta = 119°$	$\eta = 161°$

PLATE 11.—Photographs of Mars showing the same region at consecutive oppositions, with constant enlargement and common orientation with respect to the terrestrial equator.

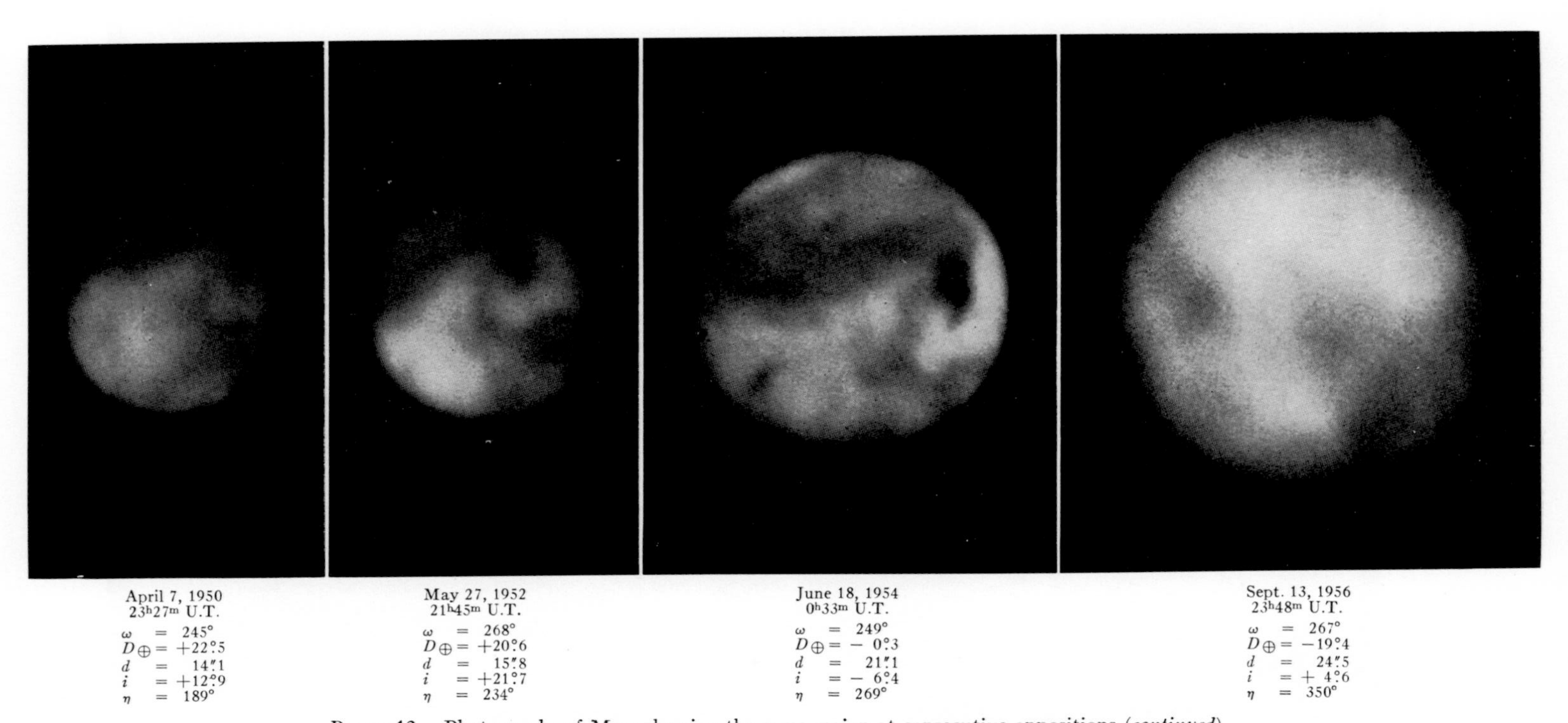

April 7, 1950
23^h27^m U.T.
$\omega = 245^\circ$
$D_\oplus = +22\overset{\circ}{.}5$
$d = 14\overset{''}{.}1$
$i = +12\overset{\circ}{.}9$
$\eta = 189^\circ$

May 27, 1952
21^h45^m U.T.
$\omega = 268^\circ$
$D_\oplus = +20\overset{\circ}{.}6$
$d = 15\overset{''}{.}8$
$i = +21\overset{\circ}{.}7$
$\eta = 234^\circ$

June 18, 1954
0^h33^m U.T.
$\omega = 249^\circ$
$D_\oplus = -0\overset{\circ}{.}3$
$d = 21\overset{''}{.}1$
$i = -6\overset{\circ}{.}4$
$\eta = 269^\circ$

Sept. 13, 1956
23^h48^m U.T.
$\omega = 267^\circ$
$D_\oplus = -19\overset{\circ}{.}4$
$d = 24\overset{''}{.}5$
$i = +4\overset{\circ}{.}6$
$\eta = 350^\circ$

PLATE 12.—Photographs of Mars showing the same region at consecutive oppositions (*continued*)

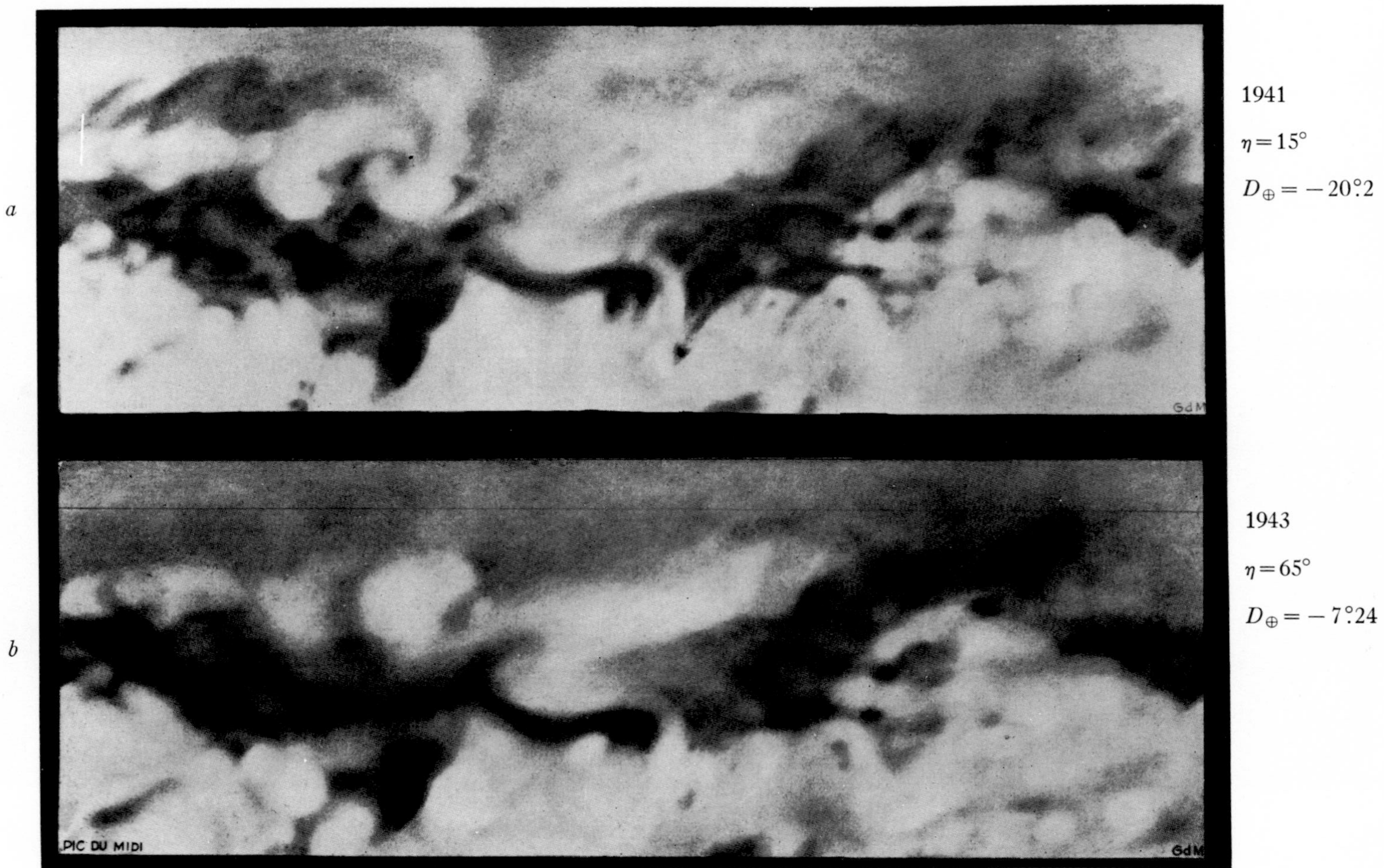

PLATE 13, *a*, *b*.—Planispheres of Mars drawn by G. de Mottoni from the Pic du Midi photographs. *a*, 1941, $\eta = 15°$; *b*, 1943, $\eta = 65°$

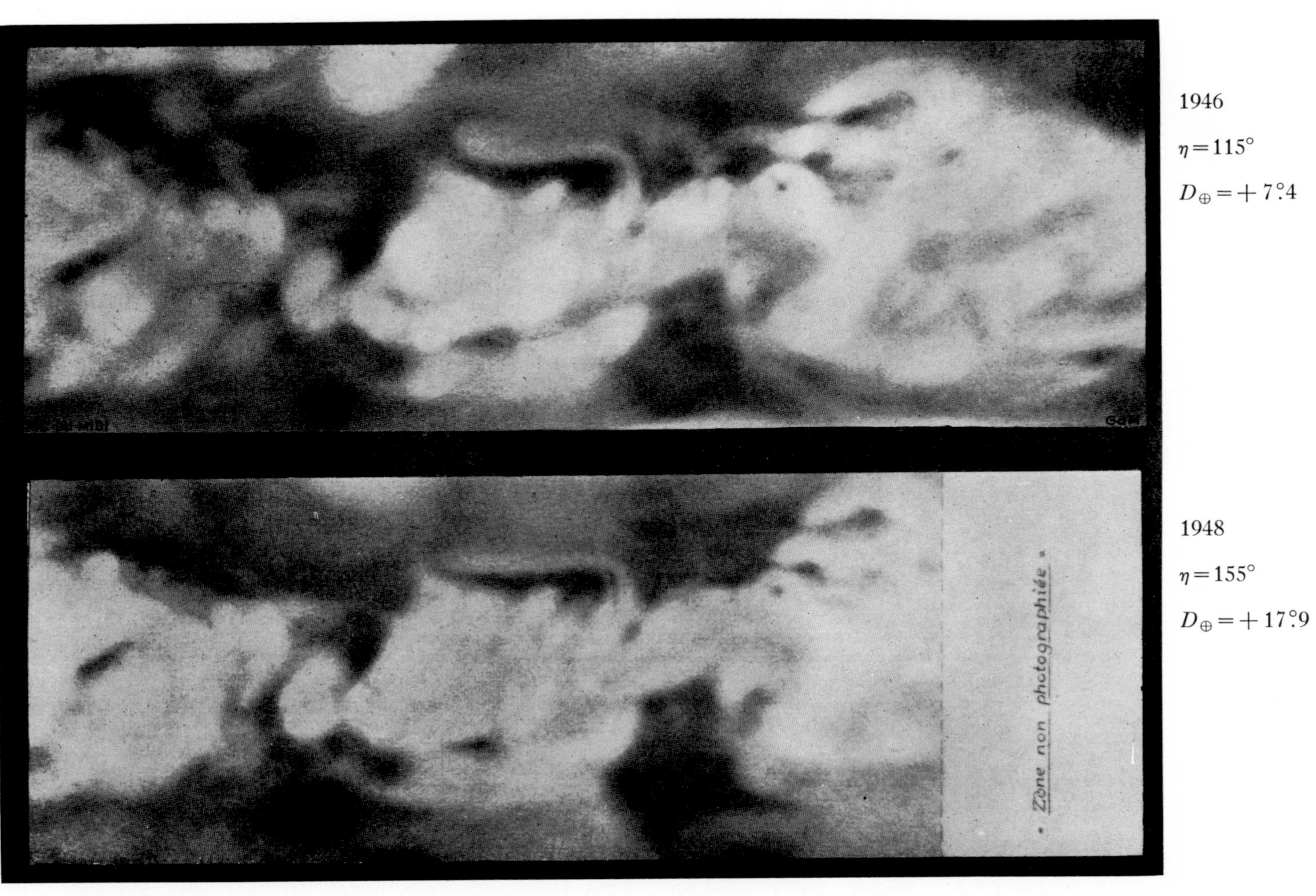

PLATE 13, *c*, *d*.—Planispheres of Mars drawn by G. de Mottoni from the Pic du Midi photographs. *c*, 1946, $\eta = 115°$; *d*, 1948, $\eta = 155°$

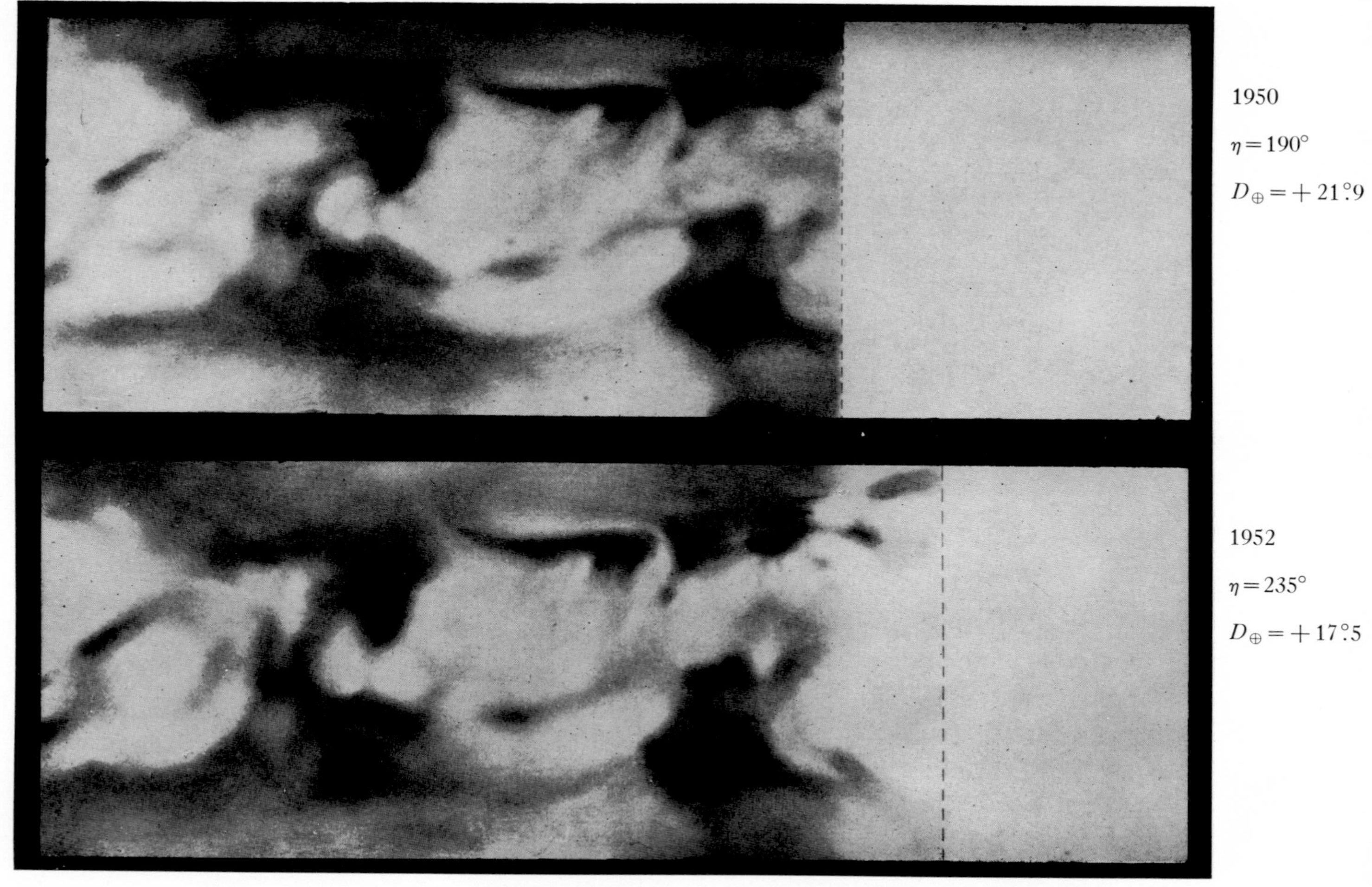

PLATE 13, *e*, *f*.—Planispheres of Mars drawn by G. de Mottoni from the Pic du Midi photographs. *e*, 1950, $\eta = 190°$; *f*, 1952, $\eta = 235°$

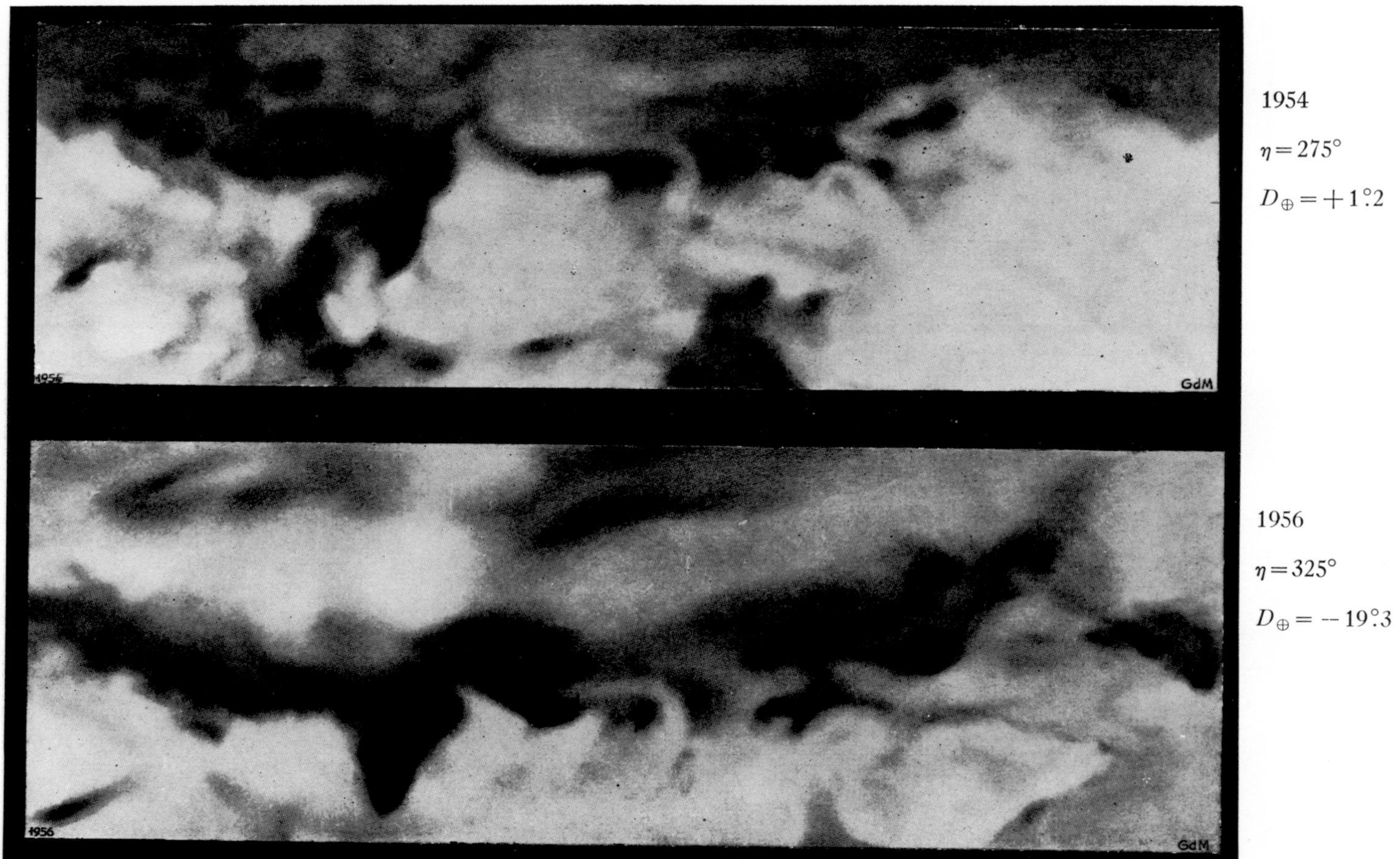

PLATE 13, *g*, *h*.—Planispheres of Mars drawn by G. de Mottoni from the Pic du Midi photographs. *g*, 1954, $\eta = 275°$; *h*, 1956, $\eta = 325°$

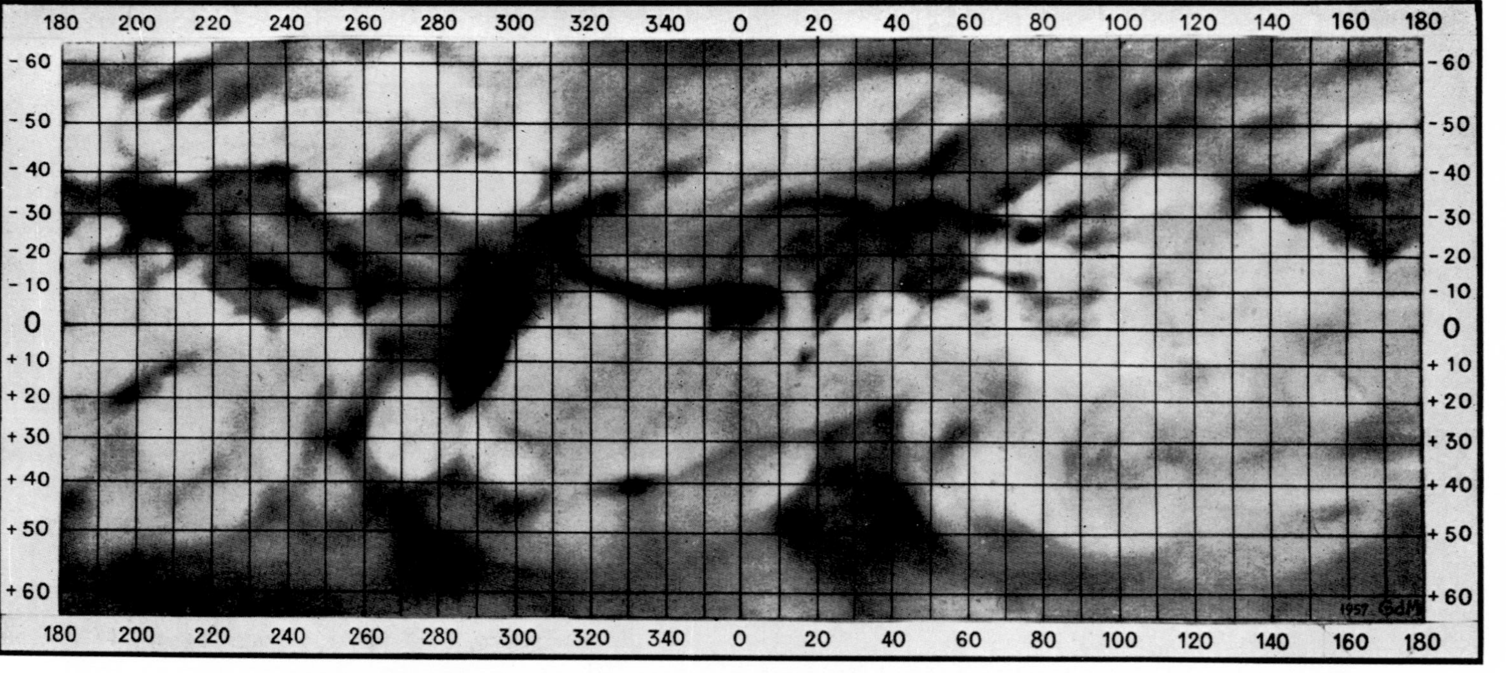

Longitude de l'équinoxe de printemps boréal	= 84°,0		= Beginning of northern spring.
Longitude du solstice d'été boréal	= 174°.0		= Beginning of northern summer.
Longitude de l'équinoxe d'automne boréal	= 264°,0		= Beginning of northern autumn.
Longitude du solstice d'hiver boréal	= 354°,0		= Beginning of northern winter.
Distance moyenne au Soleil	= 227,7.10⁶ km = 1,5237 U.A.		= Mean distance from the Sun.
Excentricité de l'orbite	= 0,0933		= Excentricity of the orbit.
Longitude du périhélie	= 334°35′		= Longitude of perihelion.
Inclinaison du plan de l'orbite	= 1°51′		= Inclination of the orbit.
Longitude du nœud ascendant	= 48°56′,5		= Longitude of ascending node.
Durée de révolution	= 686 j 23 h 30 m 41 s		= Martian year.
Coordonnées célestes de l'axe de rotation	= α = 316°,8 δ = + 53°,0		= Celestial coordinates of the axis of rotation.
Inclinaison de l'axe sur la normale à l'orbite	= 24°.8		= Inclination of the axis of rotation.
Durée de rotation du globe	= 24 h 37 m 22 s 6		= Martian day.
Diamètre équatorial du globe	= 0,530 = 6760 km		= Equatorial diameter.
Valeur de 1° planétocentrique sur le disque	= 60 km		= 1° Planetocentric at the surface.

Les petits détails sont désignés par leurs coordonnées planétographiques.
Small features are designated by their planetographic coordinates.

Les grandes régions sont désignées par un nom dont voici la liste et les coordonnées.
Main markings are designated by names according to the following record and coordinates.

Acidalium M. (30°, + 45°)
Aeolis (215°, — 5°)
Aeria (310°, + 10°)
Aetheria (230°, + 40°)
Aethiopis (230°, + 10°)
Amazonis (140°, 0°)
Amenthes (250°, + 5°)
Aonius S. (105°, — 45°)
Arabia (330°, + 20°)
Araxes (115°, — 25°)
Arcadia (100°, + 45°)
Argyre (25°, — 45°)
Arnon (335°, + 48°)
Aurorae S. (50°, — 15°)
Ausonia (250°, — 40°)
Australe M. (40°, — 60°)
Baltia (50°, + 60°)
Boreum M. (90°, + 50°)
Boreosyrtis (290°, + 55°)
Candor (75°, + 3°)
Casius (260°, + 40°)
Cebrenia (210°, + 50°)
Cecropia (320°, + 60°)
Ceraunius (95°, + 20°)
Cerberus (205°, + 15°)
Chalce (0°, — 50°)
Chersonesus (260°, — 50°)
Chronium M. (210°, — 58°)
Chryse (30°, + 10°)
Chrysokeras (110°, — 50°)
Cimmerium M. (220°, — 20°)
Claritas (110°, — 35°)
Copaïs Palus (280, + 55°)
Coprates (65°, — 15°)
Cyclopia (230°, — 5°)
Cydonia (0°, + 40°)
Deltoton S. (305°, — 4°)
Deucalionis R. (340°, — 15°)
Deuteronilus (0°, + 35°)
Diacria (180°, + 50°)
Dioscuria (320°, + 50°)
Edom (345°, 0°)
Electris (190°, — 45°)
Elysium (210°, + 25°)
Eridania (220°, — 45°)
Erythraeum M. (40°, — 25°)
Eunostos (220°, + 22°)
Euphrates (335°, + 20°)
Gehon (0°, + 15°)
Hadriacum M. (270°, — 40°)
Hellas (290°, — 40°)
Hellespontica Depressio (340° — 6°)
Hellespontus (325°, — 50°)
Hesperia (240°, — 20°)
Hiddekel (345°, + 15°)
Hyperboreus L. (60°, + 75°)
Iapygia (295°, — 20°)
Icaria (130°, — 40°)
Isidis R. (275°, + 20°)
Ismenius L. (330°, + 40°)
Jamuna (40°, + 10°)
Juventae Fons (63°, — 5°)
Laestrygon (200°, 0°)
Lemuria (200°, + 70°)
Libya (270°, 0°)
Lunae Palus (65°, + 15°)
Margaritifer S. (25°, — 10°)
Memnonia (150°, — 20°)
Meroe (285°, + 35°)
Meridianii S. (0°, — 5°)
Moab (350°, + 20°)
Moeris L. (270°, + 8°)
Nectar (72°, — 28°)
Neith R. (270°, + 35°)
Nepenthes (260°, + 20°)
Nereidum Fr. (55°, — 45°)
Niliacus L. (30°, + 30°)
Nilokeras (55°, + 30°)
Nilosyrtis (290°, + 42°)
Nix Olympica (130°, + 20°)
Noachis (330°, — 45°)
Ogygis R. (65°, — 45°)
Olympia (200°, + 80°)
Ophir (65°, — 10°)
Ortygia (0°, + 60°)
Oxia Palus (18°, + 8°)
Oxus (10°, + 20°)
Panchaia (200°, + 60°)
Pandorae Fretum (340°, — 25°)
Phaethontis (155°, — 50°)
Phison (320°, + 20°)
Phlegra (190°, + 30°)
Phoenicis L. (110°, — 12°)
Phrixi R. (70°, — 40°)
Promethei S. (280°, — 65°)
Propontis (185°, + 45°)
Protei R. (50°, — 23°)
Protonilus (315°, + 42°)
Pyrrhae R. (38°, — 15°)
Sabaeus S. (340°, — 8°)
Scandia (150°, + 60°)
Serpentis M. (320°, — 30°)
Sinaï (70°, — 20°)
Sirenum M. (155°, — 30°)
Sithonius L. (245°, + 45°)
Solis L. (90°, — 28°)
Styx (200°, + 30°)
Syria (100°, — 20°)
Syrtis Major (290, + 10)
Tanaïs (70°, + 50°)
Tempe (70°, + 40°)
Thaumasia (85°, — 35°)
Thoth (255°, + 30°)
Thyle I (180°, — 70°)
Thyle II (230°, — 70°)
Thymiamata (10°, + 10°)
Tithonius L. (85°, — 5°)
Tractus Albus (80°, + 30°)
Trinacria (268°, — 25°)
Trivium Charontis (198, + 20)
Tyrrhenum M. (255°, — 20°)
Uchronia (260°, + 70°)
Umbra (290°, + 50°)
Utopia (250°, + 50°)
Vulcani Pelagus (15, — 35)
Xanthe (50°, + 10°)
Yaonis R. (320°, — 40°)
Zephyria (195°, 0°)

PLATE 14.—General map of Mars for the years 1941–1952 drawn by G. de Mottoni, showing the principal formations with respect to a system of co-ordinates based on position measurements by Camichel, together with general data for the planet, its orbit and its rotation, as well as the list of named features and their approximate co-ordinates.

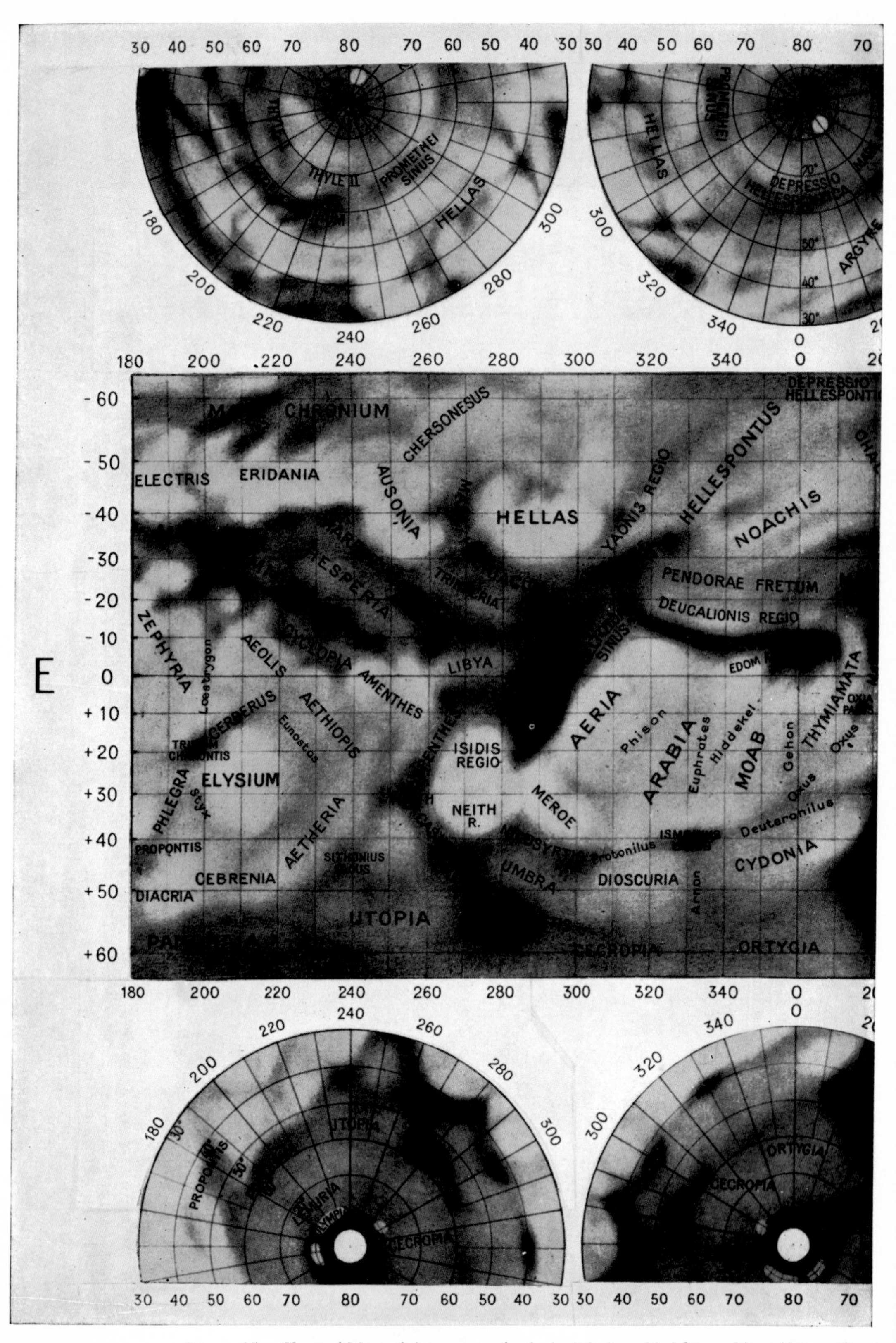

PLATE 15.—Chart of Mars giving names of principal dark and bright markings (*International*

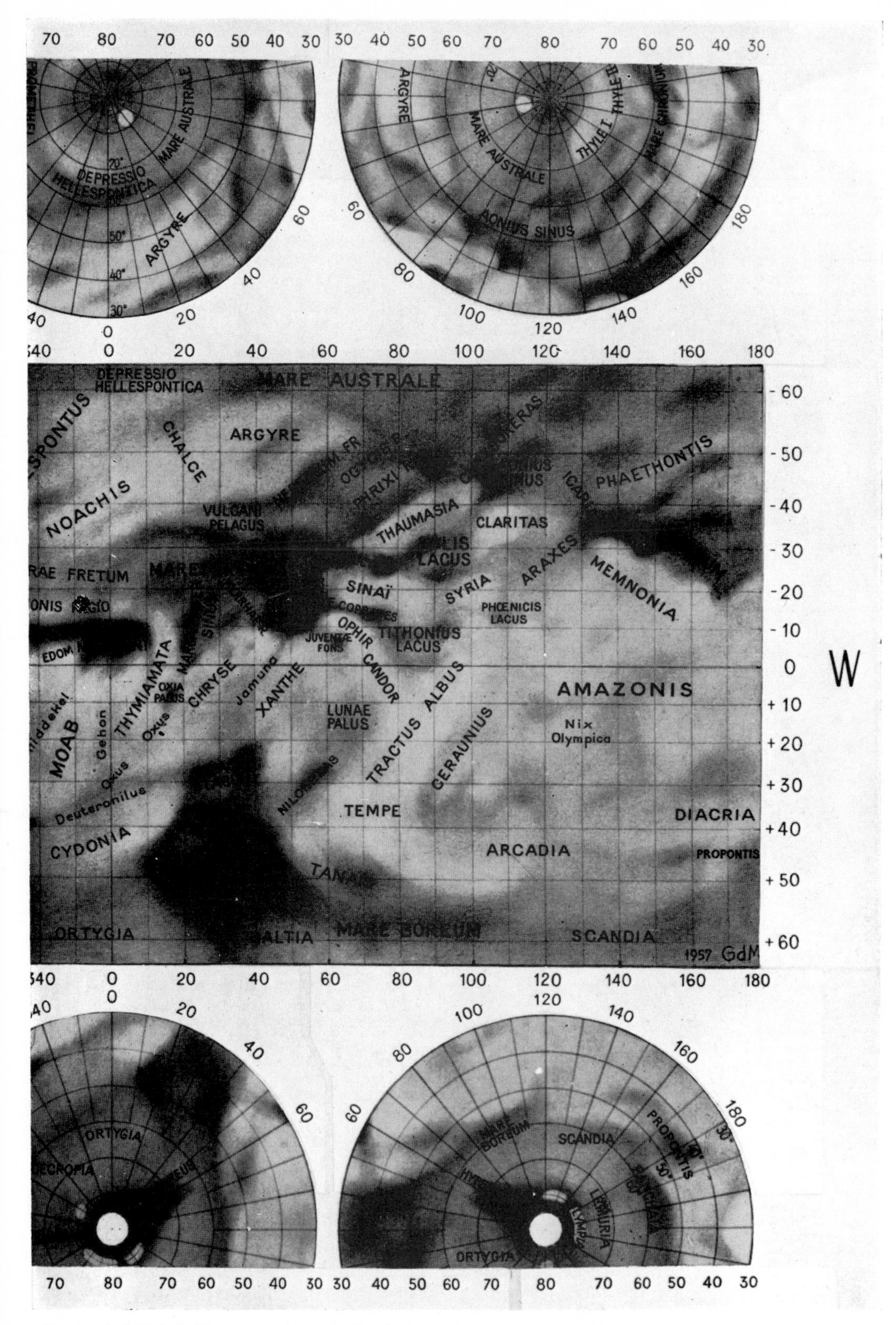

Astronomical Union). Two names (Iapygia, Pandorae Fretum) are misspelled on chart

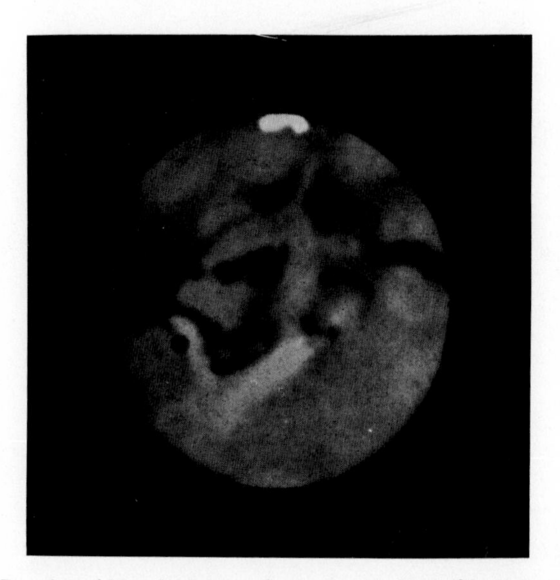

PLATE 16.—Drawing of Mars by Lyot on September 16, 1941, 1^h0^m U.T., 38-cm refractor

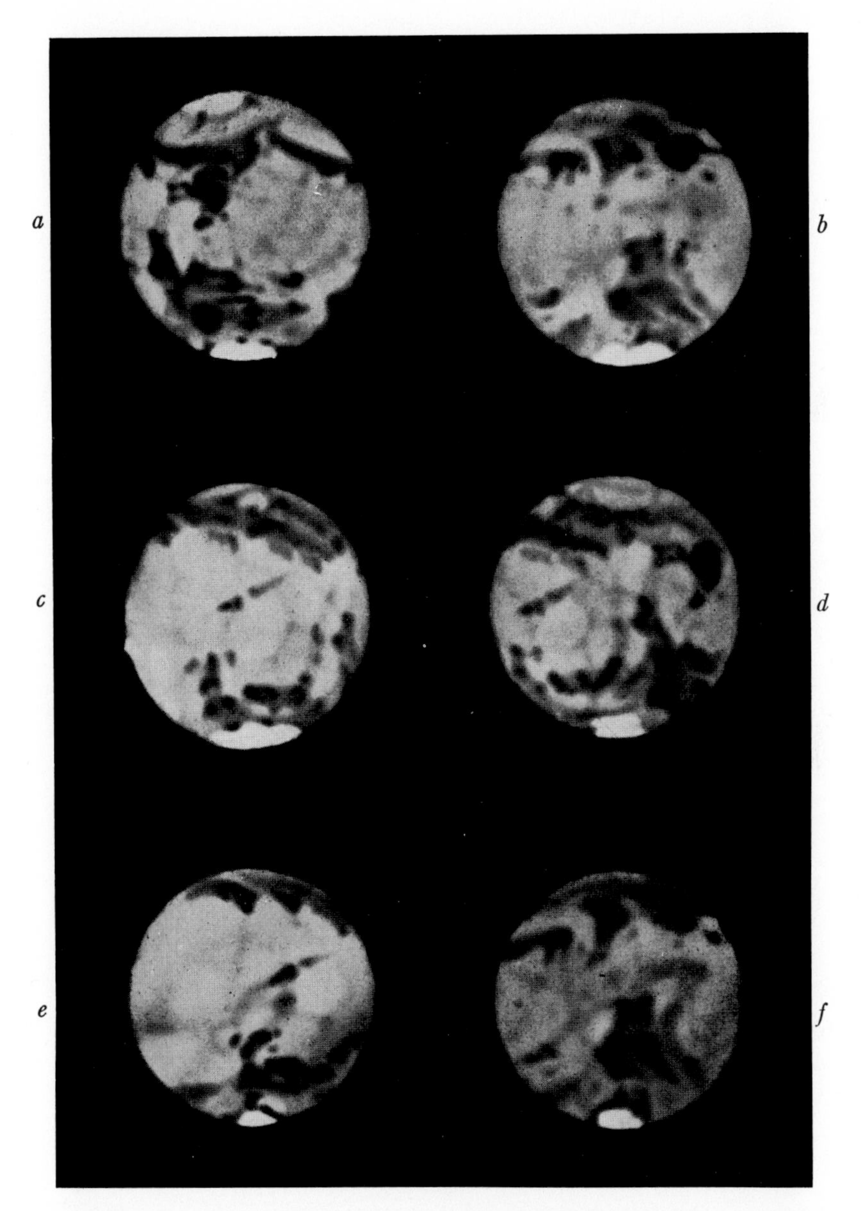

PLATE 17.—Drawings of Mars by author during the aphelion oppositions of 1948 and 1950, 60-cm refractor. Details in each drawing based on observations made during several nights. *a*, *b*, February, 1948; *c*, *d*, March, 1948; *e*, *f*, March–April, 1950.

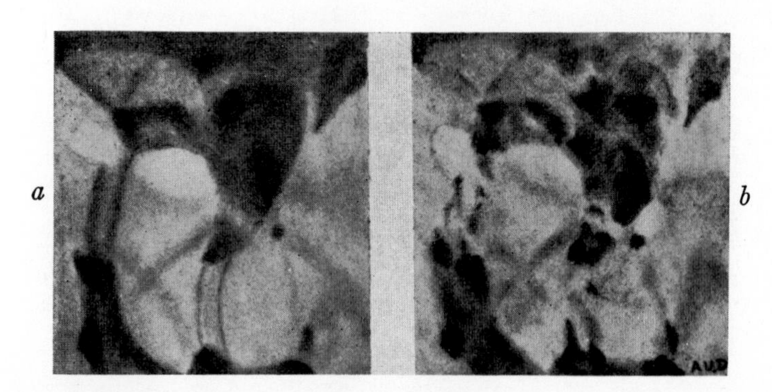

Plate 18.—Syrtis Major region and double streaks Thoth and Nilosyrtis, 60-cm refractor, 1948. *a*, Appearance with average seeing; *b*, appearance with perfect seeing. Mag. 900×.

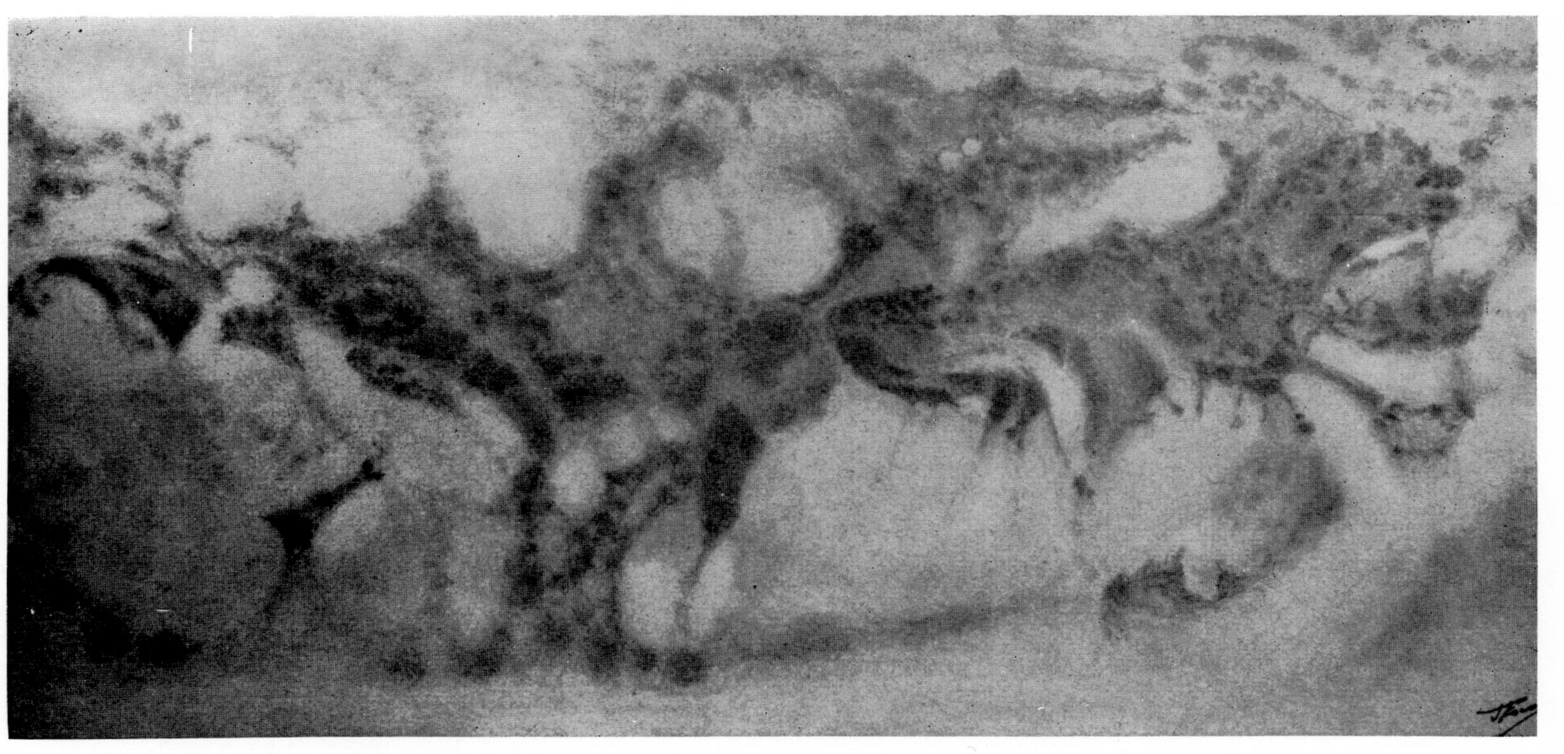

PLATE 19.—Chart of Mars by Dr. J. H. Focas, showing the fine structure of the dark areas, based on visual and photographic observations with the 60-cm refractor at the Pic du Midi Observatory in 1958.

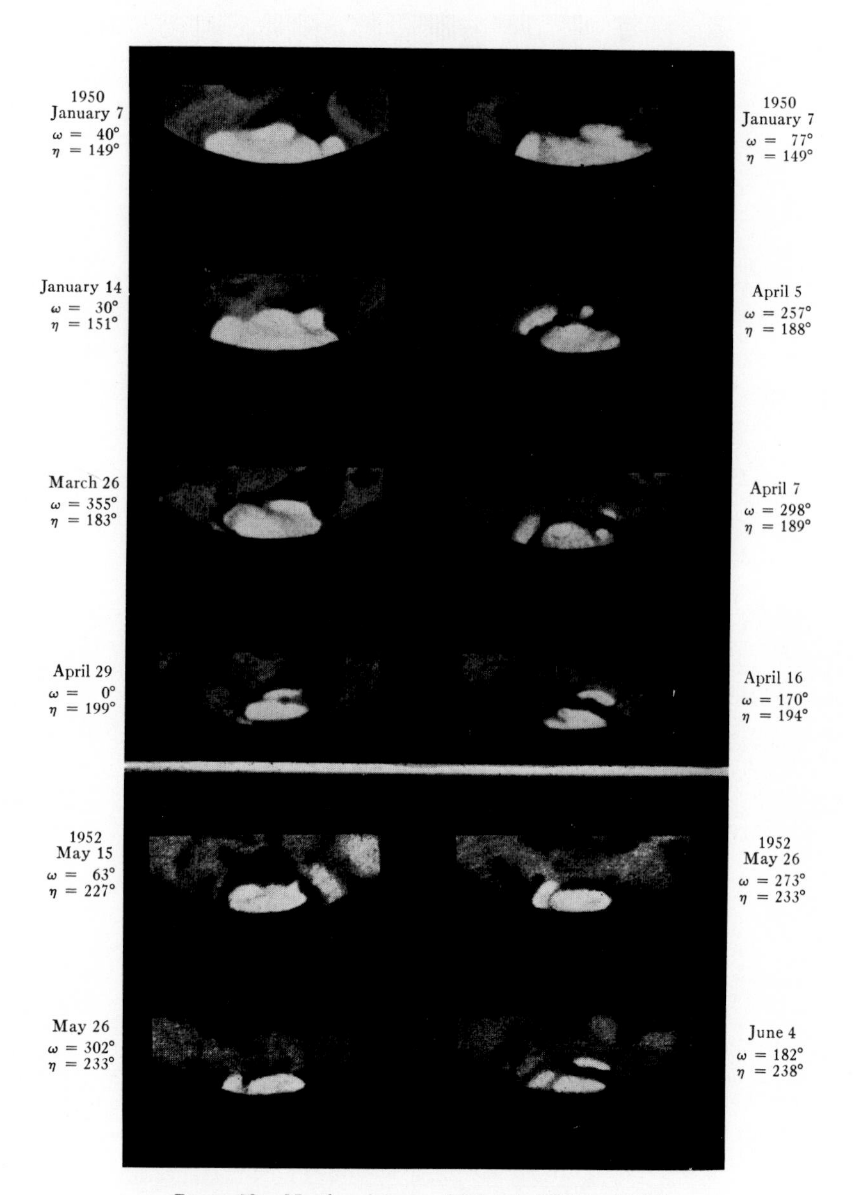

PLATE 20.—North polar cap of Mars in 1950 and 1952

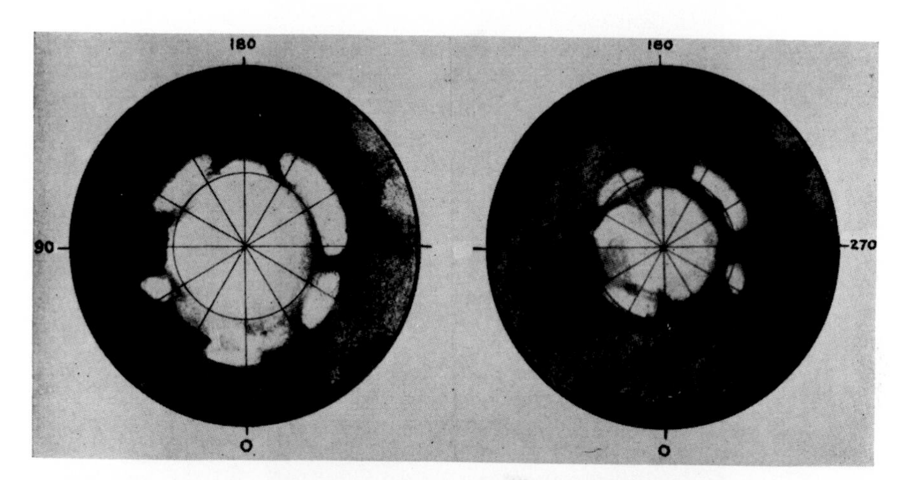

PLATE 21.—Two phases in average seasonal evolution of deposits near north pole, based on 1946, 1948, 1950, and 1952 oppositions. Latitudes 65°–90°. Representation shows north pole directed to observer. *Left:* heliocentric longitude $\eta = 155°$, *right:* $\eta = 195°$, corresponding to the seasonal dates June 2 and July 12.

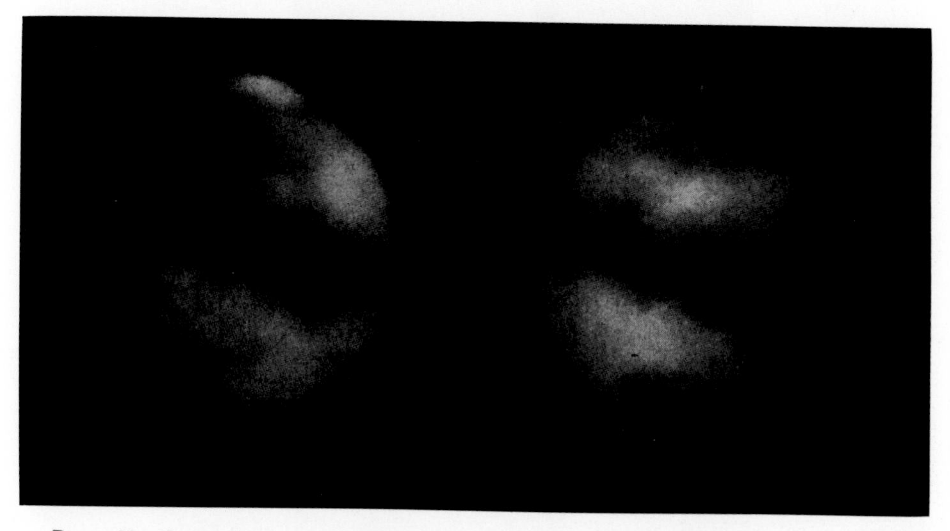

PLATE 22.—Seasonal variation of Mare Chronium. *Left:* August 15, 1956, 1^h34^m U.T. $\omega = 199°$; $D_\oplus = -19.^\circ4$; $\eta = 331°$. Late spring, Mare Chronium dark at top, left. *Right:* September 18, 1956, 23^h25^m U.T. $\omega = 218°$; $D_\oplus = -19.^\circ6$; $\eta = 352°$. Summer solstice in southern hemisphere; Mare Chronium at upper left is invisible.

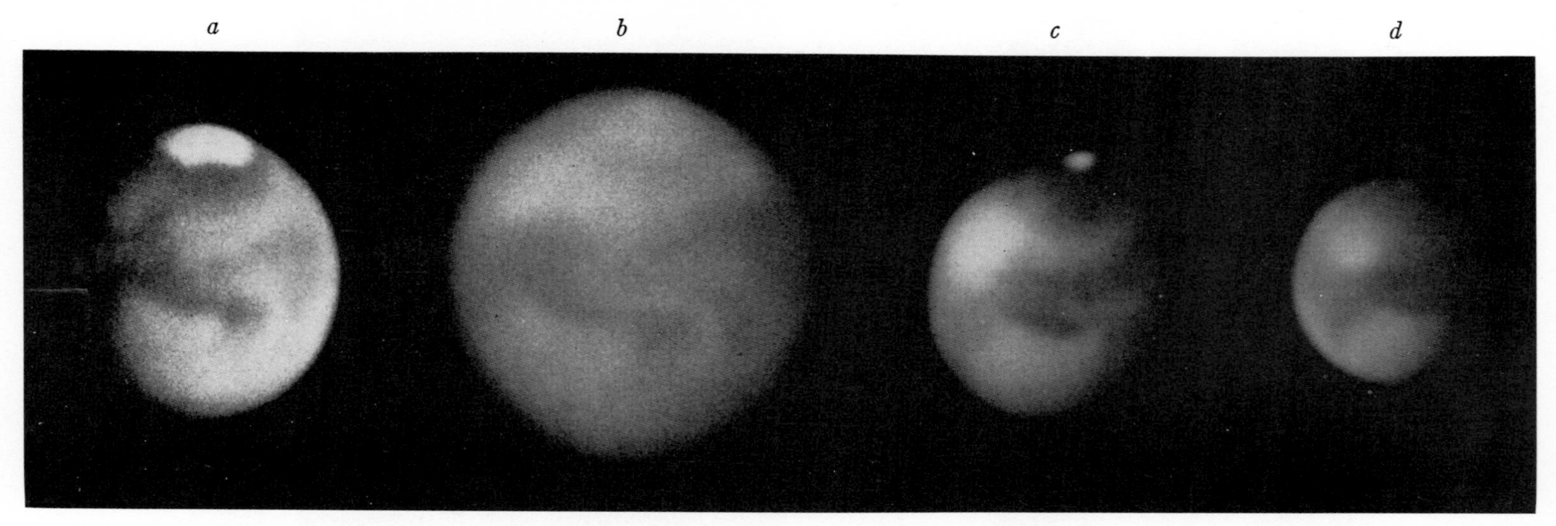

Plate 23.—Seasonal development of Depressio Hellespontica, Hellespontus, and Pandorae Fretum. *a*, July 31, 1956, $\omega = 355°$; $i = -25°$, $\eta = 329°$, mid-spring. Depressio Hellespontica, just below the polar cap, is dark; Hellespontus, the arm running downward through the center of the disk, is dark also; and Pandorae Fretum, the arm emerging from the center of the disk upward to the right, is pale. *b*, September 7, 1956, 1^h09^m U.T., $\omega = 348°$, $i = 5.°7$ (phase mostly NS., because of optical inclination), $\eta = 346°$, late spring. A dust storm masks part of the surface; Pandorae Fretum is darker than in *a*. *c*, October 8, 1956, 19^h52^m U.T., $\omega = 347°$, $i = +22.°6$, $\eta = 6°$, beginning of summer. Depressio Hellespontica is less dark; Hellespontus has disappeared; and Pandorae Fretum is very dark. *d*, November 17, 1956, 19^h40^m U.T., $\omega = 263°$, $i = 38.°2$, $\eta = 30°$, summer. Depressio Hellespontica is pale and reduced in area. Pandorae Fretum is still very dark and Mare Serpentis, at the center of the disk, very dark.

January 26, 1946, $1^{h}30^{m}$ U.T.

$\omega = 265°$
$D_{\oplus} = +5\overset{\circ}{.}8$
$\eta = 119°$

March 14, 1948, $20^{h}10^{m}$ U.T.

$\omega = 263°$
$D_{\oplus} = +16\overset{\circ}{.}2$
$\eta = 159°$

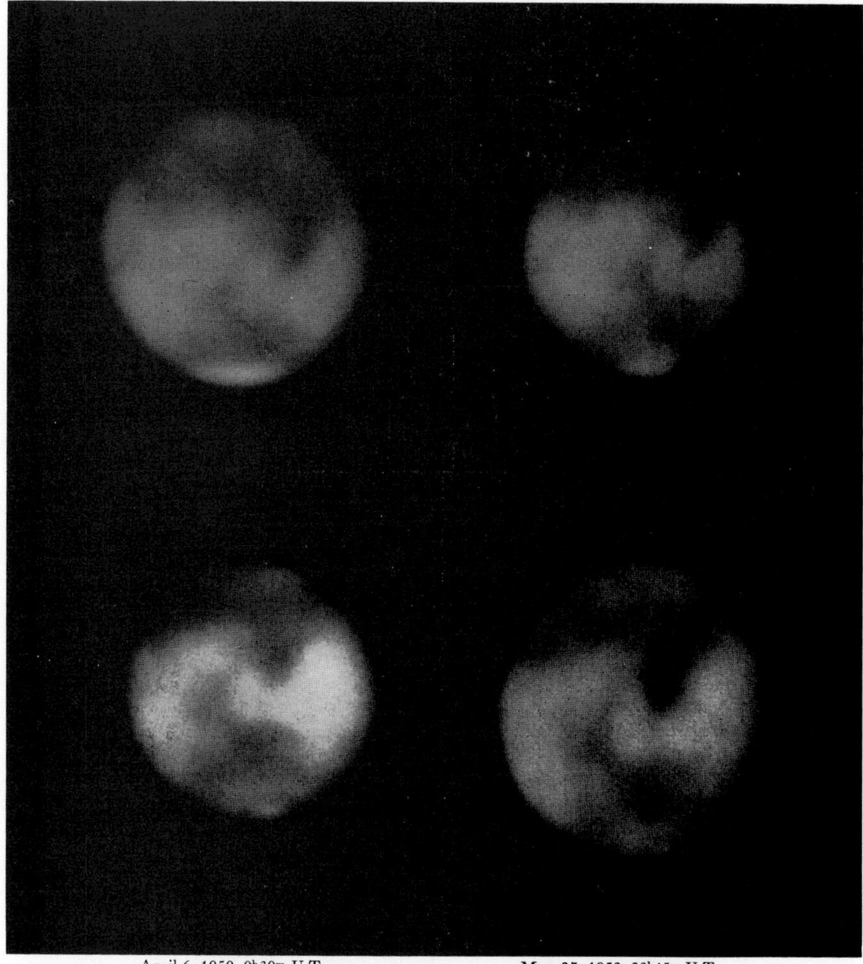

April 6, 1950, $0^{h}30^{m}$ U.T.

$\omega = 278°$
$D_{\oplus} = +22\overset{\circ}{.}6$
$\eta = 188°$

May 27, 1952, $22^{h}45^{m}$ U.T.

$\omega = 283°$
$D_{\oplus} = +20\overset{\circ}{.}6$
$\eta = 234°$

PLATE 24.—Seasonal variation of Syrtis Major, new aspect of Borecsyrtis Dioscuria, and changes in Nepentes, Thoth, and Utopia.

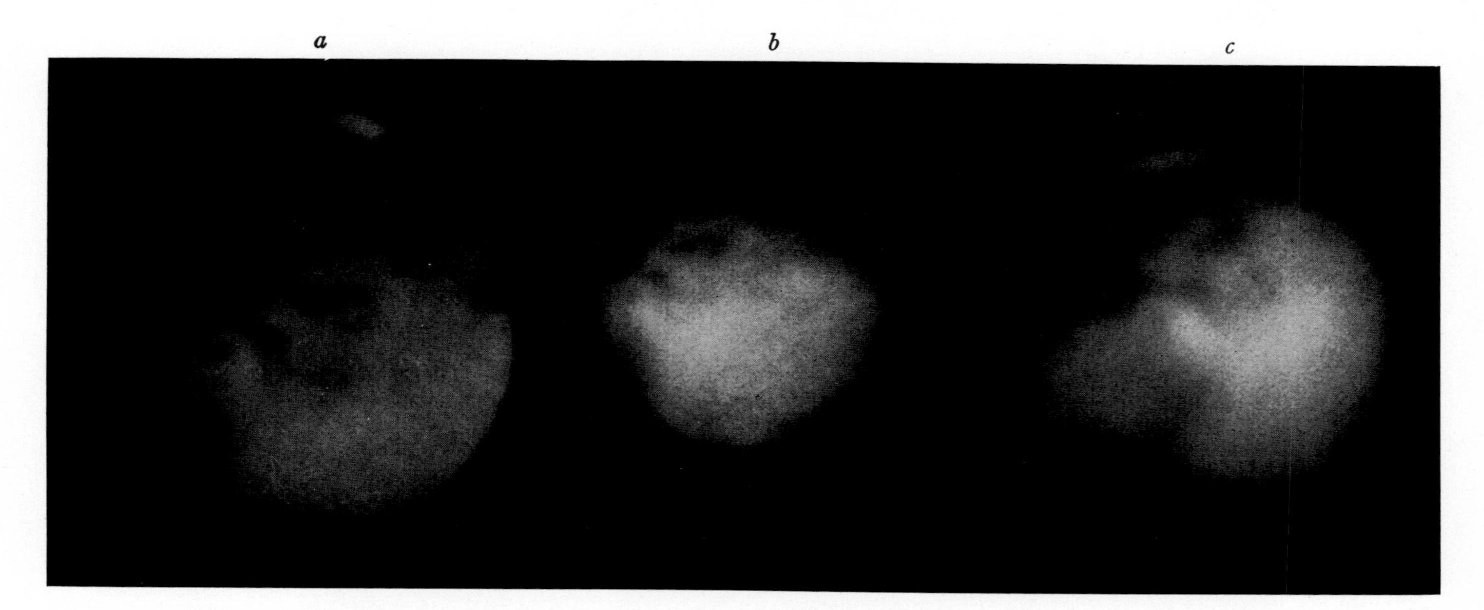

PLATE 25.—Different aspects of Solis Lacus and Thaumasia. *a*, September 19, 1941, 1^h45^m U.T. $\omega = 75°$, $D_\oplus = -18.°3$, $\eta = 4°$. *b*, December 17, 1943, 0^h25^m U.T. $\omega = 110°$, $D_\oplus = -9.°4$, $\eta = 78°$. *c*, July 8, 1954, 0^h41^m U.T. $\omega = 74°$, $D_\oplus = +3.°1$, $\eta = 280°$.

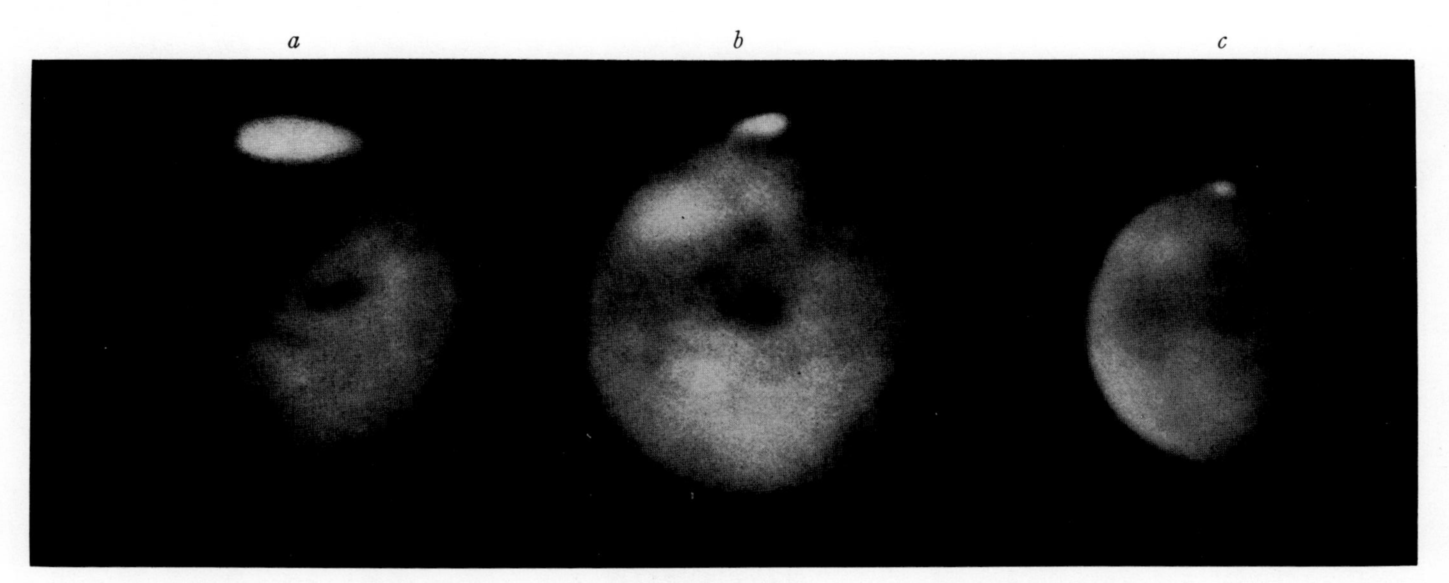

PLATE 26.—Seasonal changes in Solis Lacus, Thaumasia, and Phrixi Regio, mid-spring; Thaumasia bright, Phrixi Regio uniform. *a*, July 24, 1956, 4^h34^m U.T. $\omega = 61°$, $D_\oplus = -20°.2$, $i = +34°.4$, $\eta = 317°$. *b*, September 28, 1956, 21^h14^m U.T. Summer solstice; Thaumasia dark; Phrixi Regio spotted; Solis Lacus very dark. $\omega = 97°$, $D_\oplus = -20°.1$, $i = +15°.8$, $\eta = 359°$. *c*, November 6, 1956, 20^h37^m U.T. Early summer, dark spots are slightly reduced. $\omega = 86°$, $D_\oplus = -23°.3$, $i = +35°.6$, $\eta = 23°$.

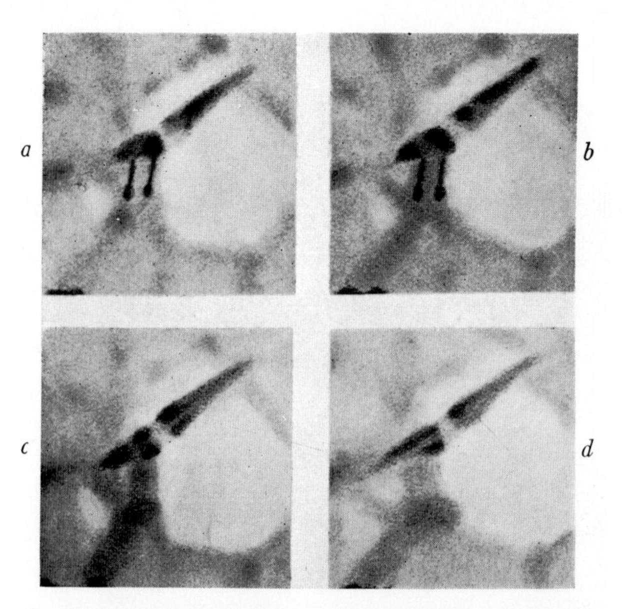

PLATE 27.—Drawings of region near Trivium Charontis at oppositions of 1943 (*a*), 1946 (*b*), 1948 (*c*), and 1950 (*d*). *a* by Gentili, 38-cm refractor; the others by author with 60-cm refractor.

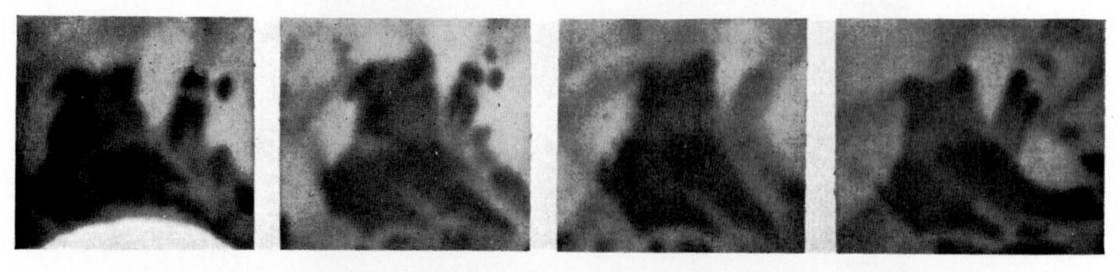

PLATE 28.—Changes in Nylokeras (*center right*), observed by the writer with 60-cm telescope during the oppositions of 1946, 1948, 1950, and 1952.

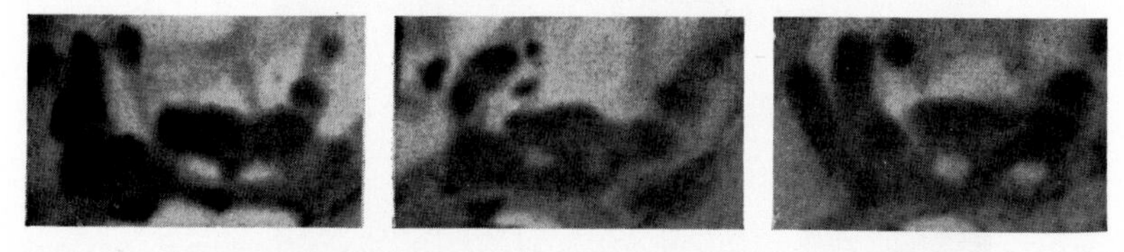

PLATE 29.—Changes in Propontis and Panchaia observed by the writer with 60-cm refractor in 1948, 1950, and 1952.

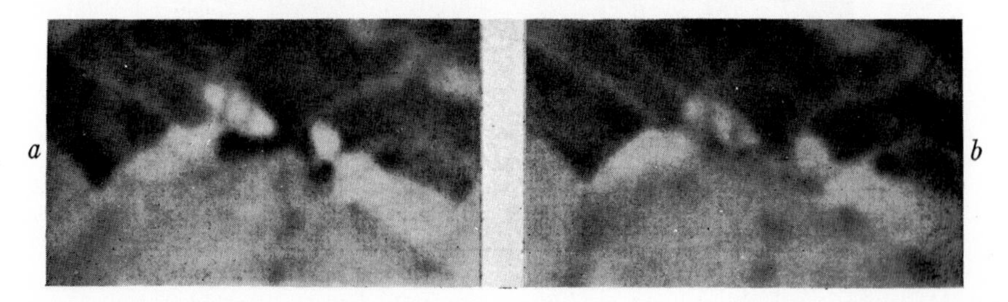

PLATE 30.—Temporary change south of Zephyria. *a*, altered appearance, which started during the opposition of 1939 and remained visible during the oppositions of 1941, 1943, and 1946. *b*, appearance during the oppositions of 1948 and 1950, being normal for this region.

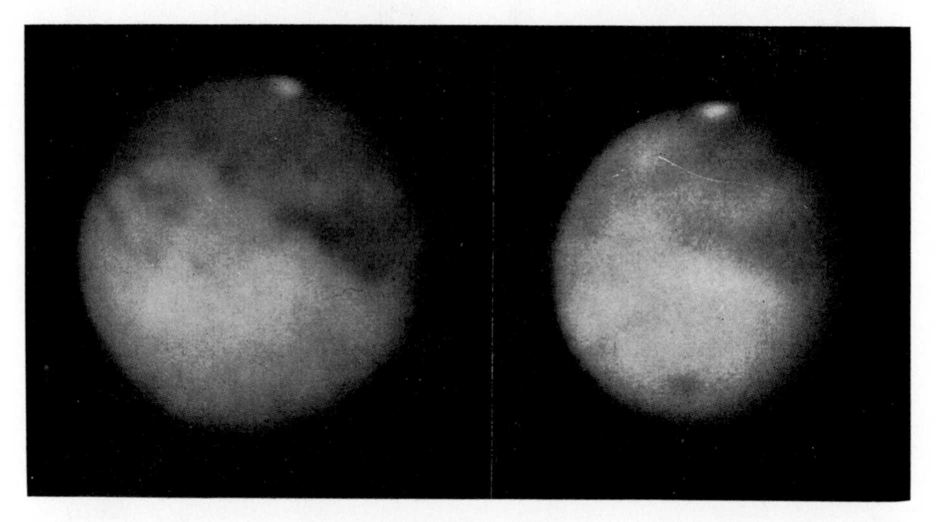

PLATE 31.—New spot in the region of Arcadia. *Left:* October 19, 1941, 23^h20^m U.T. $\omega = 125°$, $D_\oplus = -21°.4$, $\eta = 22°$. *Right:* September 29, 1956, 0^h30^m U.T. $\omega = 145°$, $D_\oplus = -20°.1$, $\eta = 359°$. Note spot toward bottom of second image.

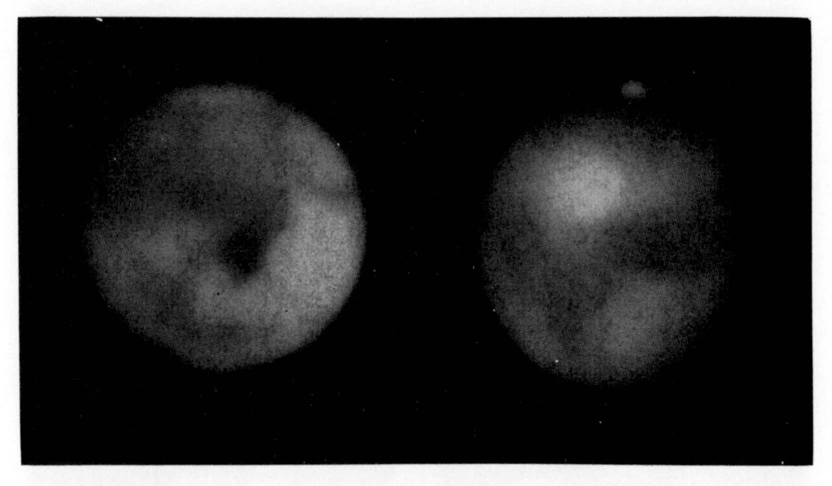

PLATE 32.—Remarkable intensity change in region of Hellas. *Left:* June 16, 1954, 1^h20^m U.T. $\omega = 279°$, $D_\oplus = -0°.6$, $\eta = 268°$. Hellas dark. *Right:* October 12, 1956, 20^h34^m U.T. $\omega = 320°$, $D_\oplus = -21°.2$, $\eta = 8°$. Hellas bright (normal appearance).

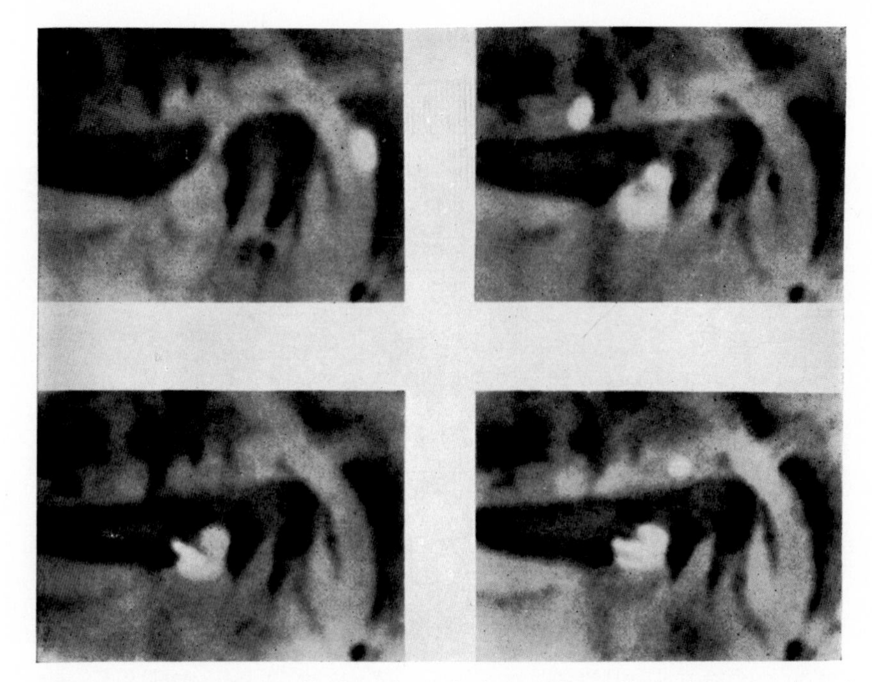

PLATE 33.—Persistent small cloud located above Edom. *Above:* September 7 and October 8, 1956. *Below:* October 12 and 13, 1956. First drawing by J. Focas; others by the writer.

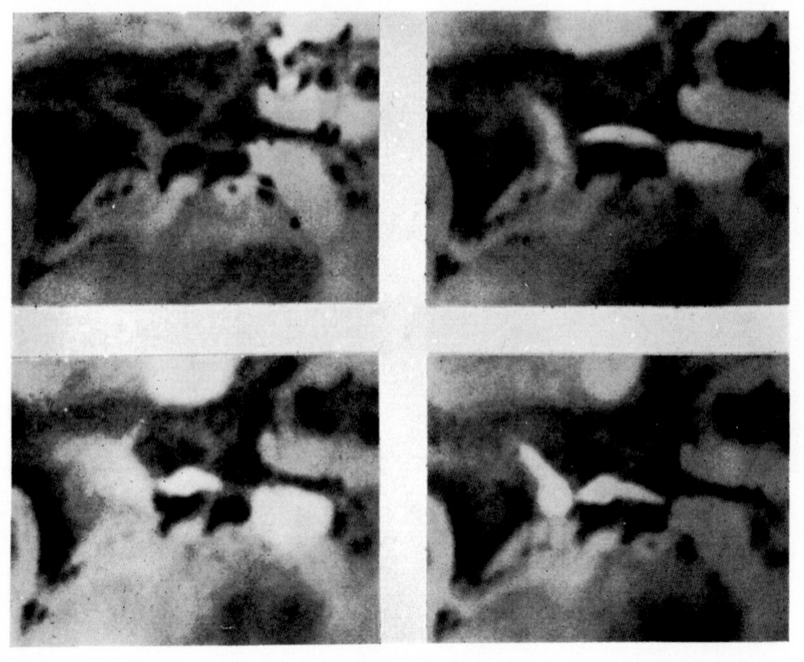

PLATE 34.—Small persistent cloud in region of Protei Regio, Pyrrhae Regio, and Ophir. *Above:* September 2 and 30, 1956. *Below:* October 4 and 8, 1956.

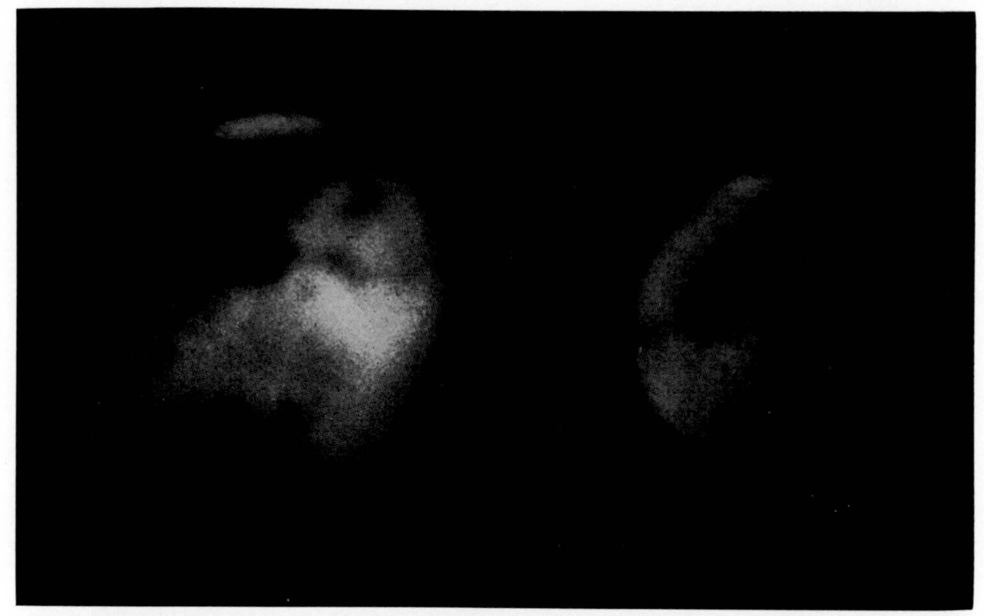

PLATE 35.—Evening haze over Mare Erythraeum and cloud over Mare Acidalium. *a*, July 8, 1954, 23^h36^m U.T. $\omega = 49°$, $D_\oplus = +3°.2$, $i = +13°.1$, and $\eta = 281°$, no evening haze, but light cloud over the lower part of Mare Acidalium. *b*, August 11, 1954, 20^h37^m U.T. $\omega = 55°$, $D_\oplus = +3°.8$, $i = +34°.7$, and $\eta = 301°$. Haze over left limb, dense cloud over entire Mare Acidalium.

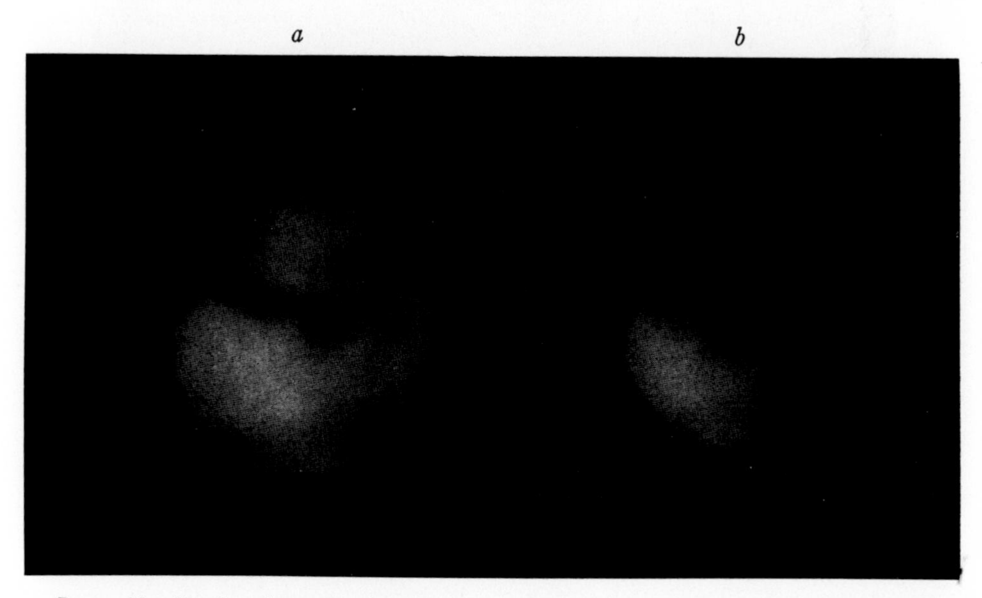

PLATE 36.—Cloud and morning haze over Margaritifer Sinus and haze cap over north pole. *a*, November 22, 1943, 1^h11^m U.T. $\omega = 338°$, $D_\oplus = -4°.7$, $i = -12°.4$, $\eta = 66°$; cloud and haze on right limb. *b*, December 22, 1943, 21^h42^m U.T. $\omega = 16°$, $D_\oplus = -10°.2$, $i = +14.5°$, $\eta = 81°$. No clouds.

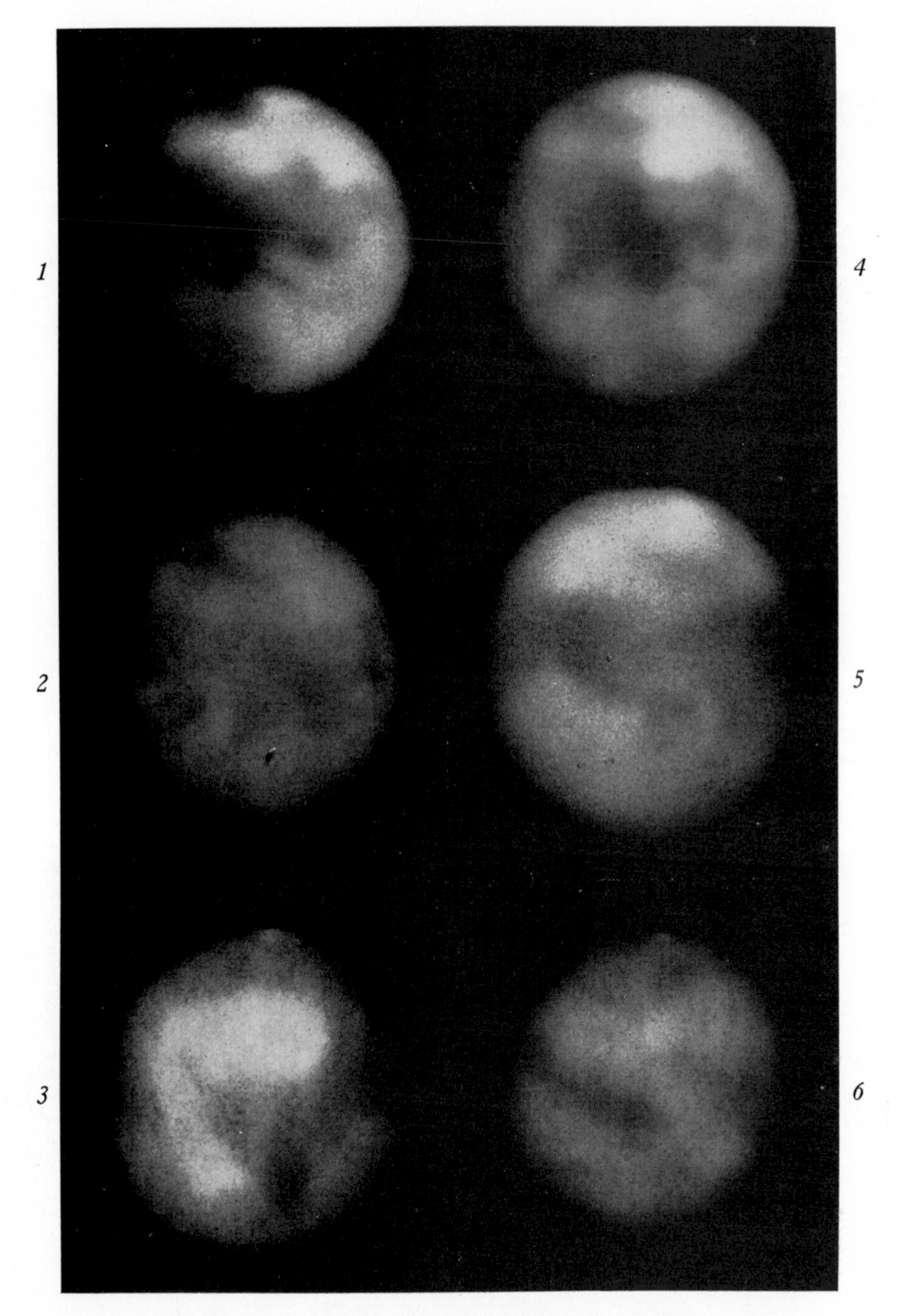

PLATE 37.—Development of yellow clouds during 1956

(1) August 30, 0^h50^m U.T.
$\omega = 57^\circ$
$D\oplus = -19^\circ.1$
$i = -11^\circ.6$
$\eta = 340^\circ$

(2) September 3, 2^h30^m U.T.
$\omega = 45^\circ$
$D\oplus = -19^\circ.1$
$i = -\ 8^\circ.4$
$\eta = 343^\circ$

(3) September 3, 0^h45^m U.T.
$\omega = 18^\circ$
$D\oplus = -19^\circ.1$
$i = -\ 8^\circ.4$
$\eta = 343^\circ$

(4) September 7, 1^h09^m U.T.
$\omega = 348^\circ$
$D\oplus = -19^\circ.2$
$i = -\ 5^\circ.7$
$\eta = 346^\circ$

(5) September 14, 0^h56^m U.T.
$\omega = 284^\circ$
$D\oplus = -19^\circ.4$
$i = +\ 5^\circ.1$
$\eta = 350^\circ$

(6) September 13, 22^h07^m U.T.
$\omega = 240^\circ$
$D\oplus = -19^\circ.4$
$i = +\ 5^\circ.1$
$\eta = 350^\circ$

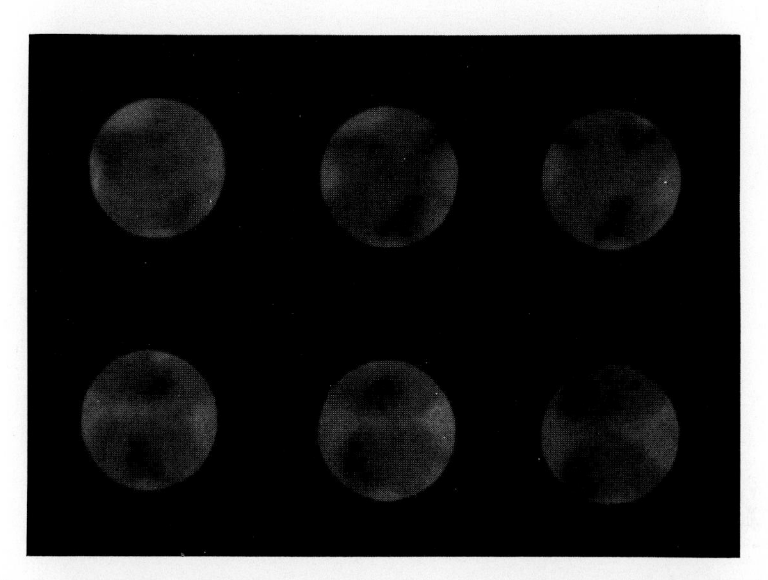

PLATE 38.—Mars observed visually through a dark-blue filter, showing the development of blue clouds, March 5 and 6, 1948. Times: 20^h40^m, 22^h00^m, 23^h35^m, 0^h45^m, 1^h35^m, and 2^h45^m. Note that morning and evening clouds retain approximately the same position in spite of rotating surface features also shown.

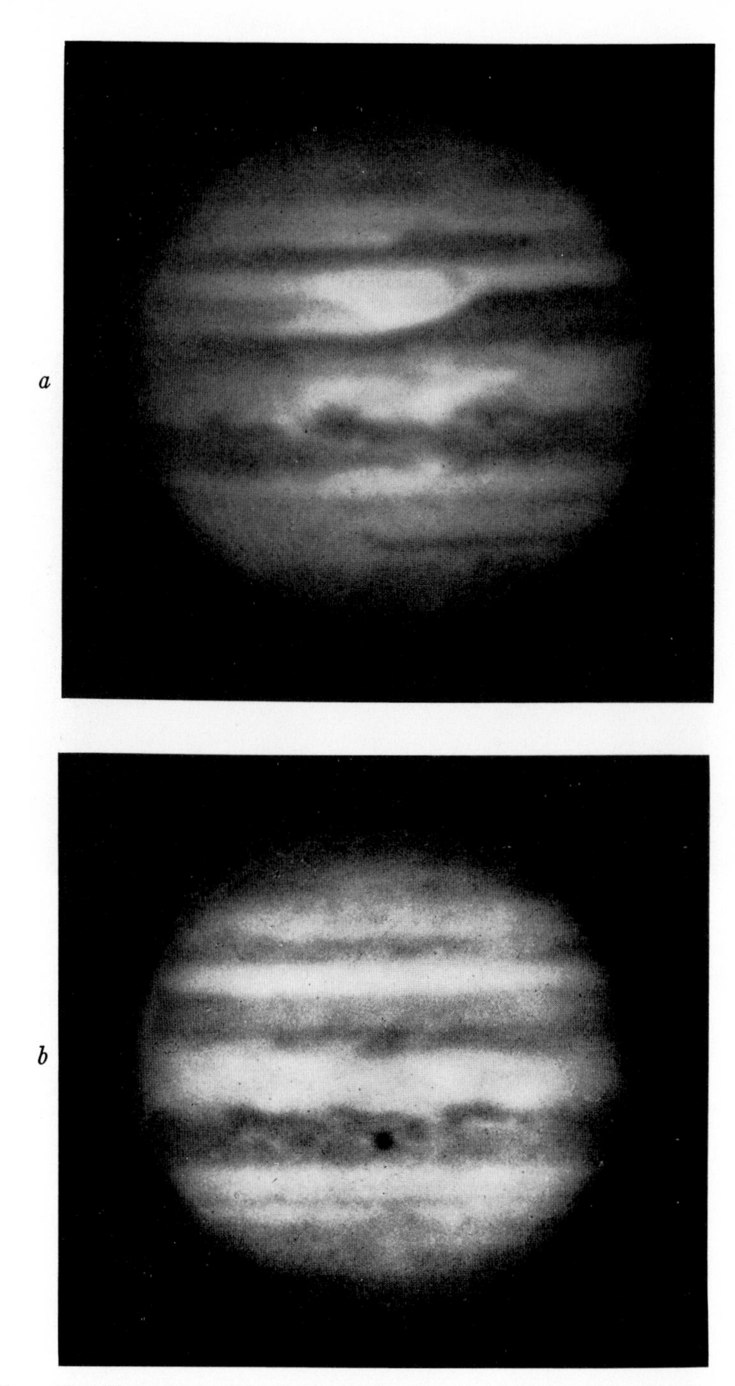

PLATE 39.—Photographs of Jupiter in 1945. *a,* March 16, $00^{h}05^{m}$ U.T.; Red Spot on meridian. *b,* April 14, $21^{h}03^{m}$ U.T., shadow of Satellite II.

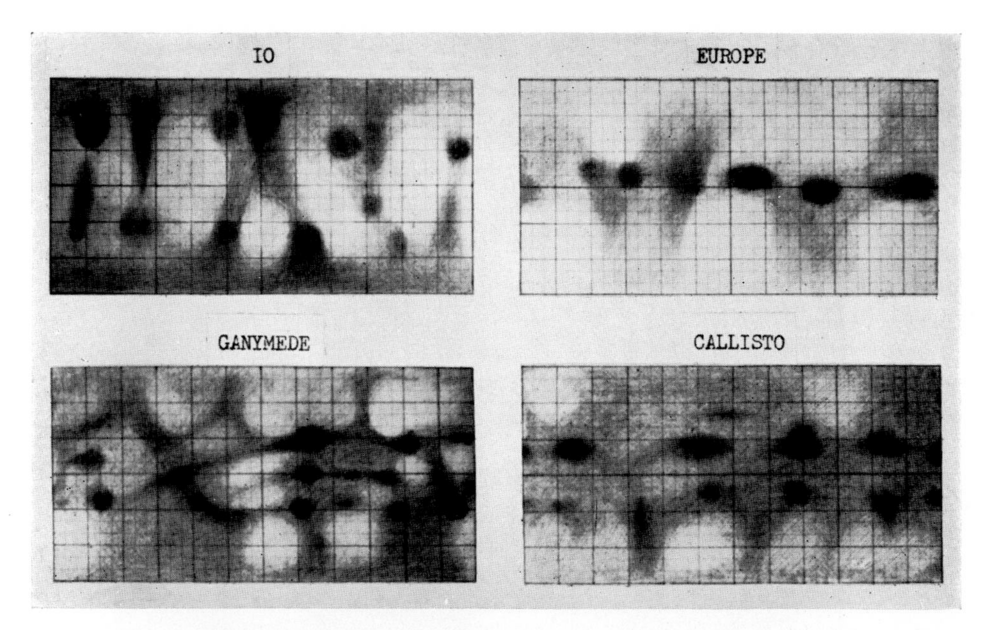

PLATE 40.—Planispheres of the Jupiter satellites according to Lyot. The center of each map corresponds to the center of the satellite when seen in projection on Jupiter. The contrast has been exaggerated. Note curious differences between appearances of Satellites I and II.

PLATE 41.—Drawing of the Saturn Rings according to Lyot, 60-cm refractor. Contrast somewhat exaggerated.

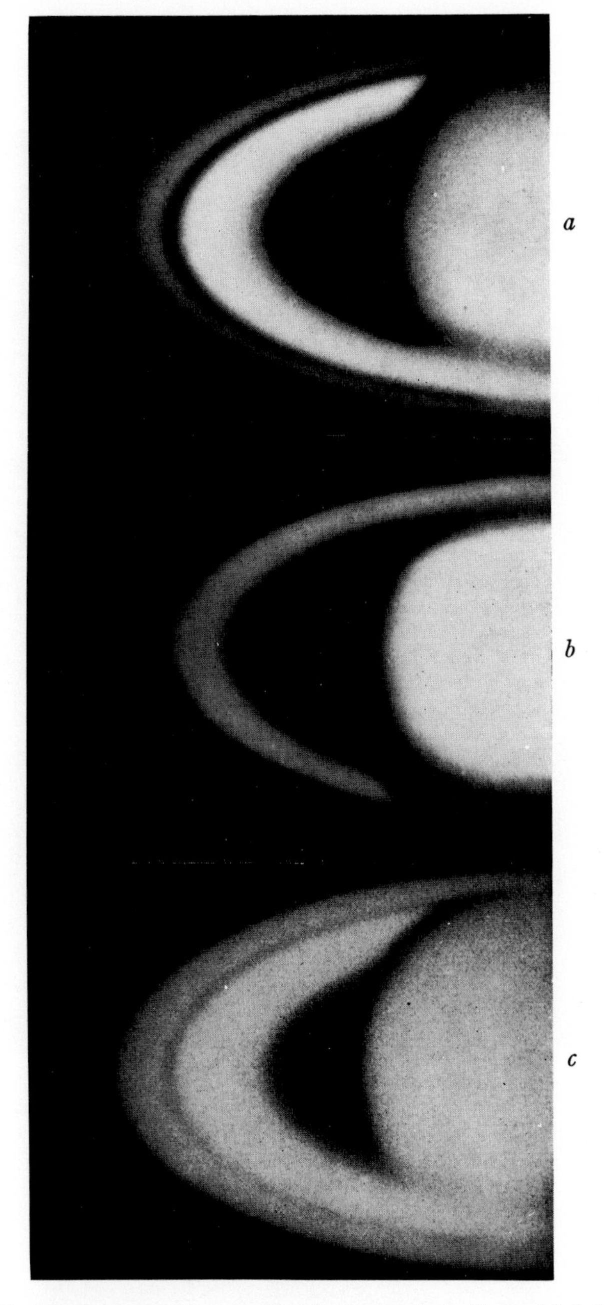

PLATE. 42.—Enlargements and composites of photographs by H. Camichel of the Rings of Saturn: *a*, contrast appropriate to show up details in the Outer Ring A; *b*, contrast appropriate to show the shading in Ring B; *c* shows the Crepe Ring C.

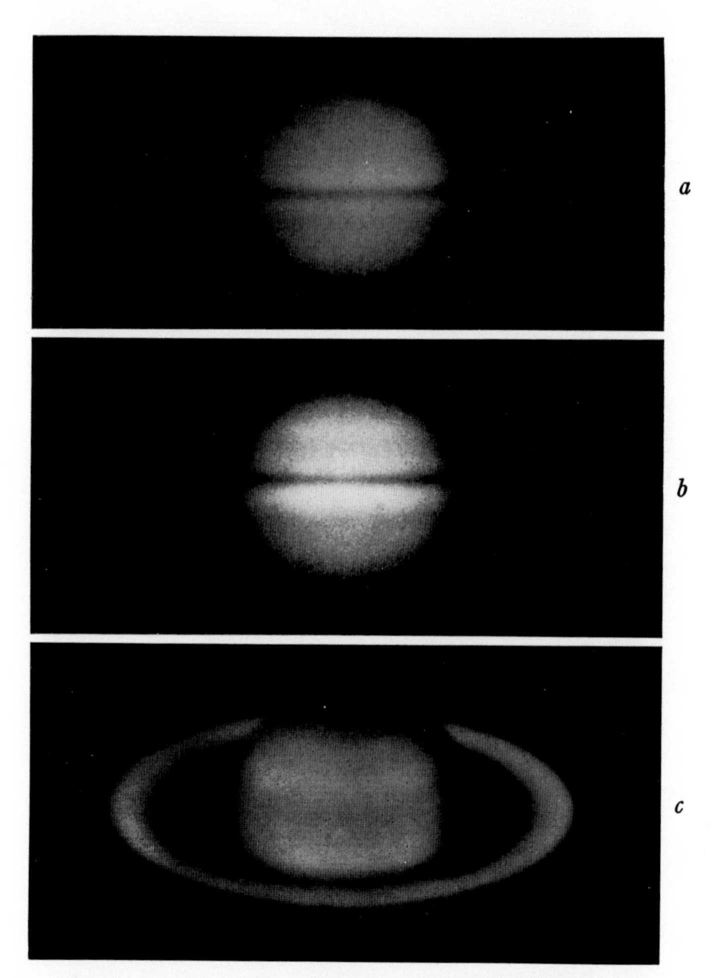

PLATE 43.—Photographs of Saturn by Camichel. *a*, April 7, 1950, planetocentric latitude of earth, $-4.3°$, and of sun, $-2.5°$. As a result of the low angle of illumination, the Ring is very dim. *b*, April 15, 1951, planetocentric latitude of earth $+1.6°$ and of sun $+3\overset{\circ}{.}1$. The shadow of the Ring is now projected above the center of the disk. *c*, February 11, 1956, $21^{h}30^{m}$ earth at $24\overset{\circ}{.}7$, sun at $+23\overset{\circ}{.}7$. Note white spot on the meridian.

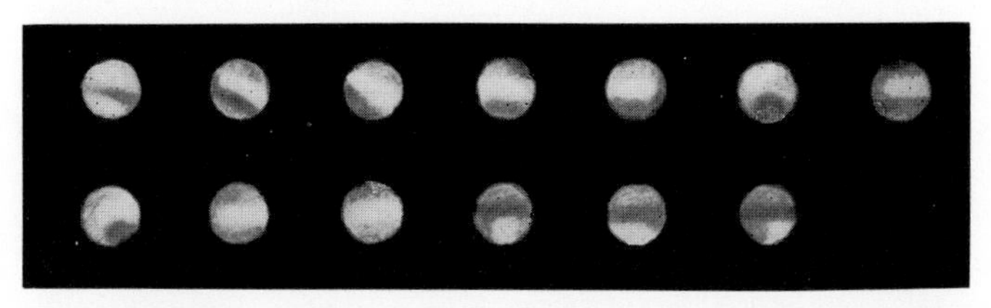

Plate 44.—Drawings of the satellite Titan by four observers, arranged in order of increasing planetocentric longitude, measured from superior conjunction: 21°, 62°, 83°, 107°, 164°, 173°, 203°, 211°, 245°, 265°, 291°, 312°, and 356°. The observations covered several revolutions.

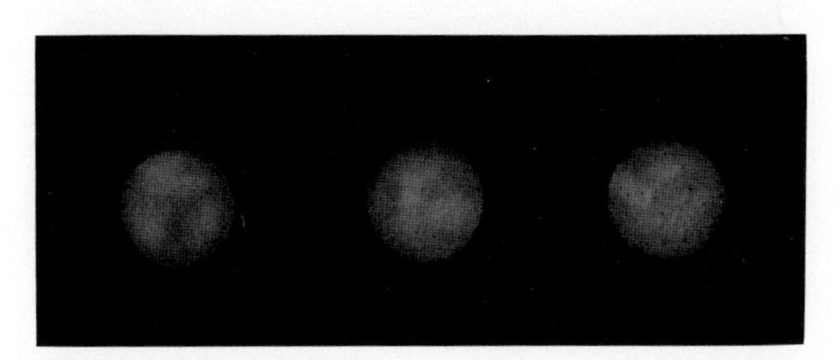

Plate 45.—Drawings of Neptune by three observers, March 15, 17, and 21, 1948. Contrast of the faint spots enhanced.

each—0°–30°, 30°–60°, etc.—and within each block the plates were arranged in chronological order. Inspection of the array in one dimension gives a comparison of the same area of the planet at different oppositions; examples of this are given in Plates 11, 12, and 24, 25. In the other dimension it gives the various aspects in longitude during the same opposition; an example of this is given in Plate 37. The diapositives within a given 30° cell show the seasonal variations and the development of the clouds above the same region within an interval of approximately four months; examples are Plates 23 and 26.

The arrangement of the diapositives by longitude has been used by Dr. G. de Mottoni as the basis for the construction of a set of maps on the Mercator projection, one for each Martian opposition (the map for 1958 was not completed at the time of writing). With his kind consent, these eight maps are included in this chapter and are reproduced as Plate 13 (4 pages). They give the mean aspect of the markings for each opposition and will greatly facilitate the discovery and study of seasonal and irregular variations on the planet.

3.32. *Topography and nomenclature.*—Camichel (1954, 1956*a*) has used his original negatives to make a critical study of the Martian diameter and of the orientation of its rotational axis. The method is based on measurements of the daily path of a small prominent spot (Juventae Fons) with respect to the center of the disk. The adopted diameter reduced to 1 a.u. is 9″.40, with some scatter (9″.25–9″.48) for the different years, depending on the image size used. The position of the pole of rotation was found to be $\alpha = 316^\circ.8$ and $\delta = +52^\circ.9$ (1950), with an estimated accuracy in each co-ordinate of perhaps 0°.2. With this information, a co-ordinate system could be constructed to which all other measurable points on the planet could be referred. Camichel measured 260 points for which he derived planetocentric co-ordinates. Some 40 of these were later selected for their suitability and used as the basis for a map on the Mercator projection. The map was constructed by Dr. de Mottoni and was presented to the meeting of the International Astronomical Union at Moscow in 1958, as part of the work of Commission 16 on Planets. It is reproduced in Plate 14. It was made with meticulous care and intends to represent the *average* position and appearance of the Martian surface detail during the oppositions of 1941–1952.

It is customary to designate the dark and the unusually bright markings on the planet by *names*, many of which were assigned by Schiaparelli, Fournier, and Antoniadi. For a summary of all earlier studies, reference is made to the monographs of Flammarion (1892, 1909) and Antoniadi

(1930). Since, however, inconsistencies occur among these designations and since altogether too many details, many transient, had been named, the I.A.U. appointed Subcommission 16*a* to deal with this problem. Its recommendations are embodied in Plate 14, which was officially adopted at the 1958 I.A.U. meetings. It is based on the co-ordinate system and the markings shown in Plate 14. The names of Plate 15 are used in the text below, while for small detail, not named in Plate 15, a designation by co-ordinates in degrees is used—e.g., (202°, −11°).

3.33. *Visual observations.*—Between 1941 and 1944, Lyot and Gentili made detailed drawings with the 38-cm refractor, in order to supplement the photographs. Plate 16 shows a sample drawing. After 1945, the writer examined the planet at each opposition with the 60-cm refractor and made drawings of successive regions. During the aphelion oppositions of 1946, 1948, and 1950, drawings of the entire disk were made, such as Plate 17, often from observations during several nights, to supplement the small photographs. During perihelion oppositions, on the other hand, smaller regions of special interest were drawn.

The resolving power of 0″.2 breaks up the dark regions into innumerable small spots; at perihelion the smallest is only 60 km in diameter. These regions change in shape, contrast, and color during the Martian seasons and years. The variations are due in part to changes in the individual small detail—some spots darkening, while others fade. In addition, the more extended markings seem to change on a larger scale. Examples can be found in Plates 13 and 27–30.

Canals may sometimes be noted across ochre-colored deserts; they appear only when the seeing is not very good, either as fleeting filaments or as soft bands. When the seeing becomes excellent, one sees at the same place only small spots or markings, more or less aligned, as in Plate 18. We have explained in Section 2.21 how fragmentary vision during imperfect seeing can explain the appearance of these lines.

During the 1958 opposition Dr. Focas observed the planet visually with the 60-cm refractor and, from a large number of detailed drawings and photographs, composed the planisphere shown in Plate 19. Attention is called to the high resolution obtained during this favorable opposition. These observations were a continuation of his 1954 and 1956 series.

3.34. *Polar caps.*—The dimensions of the polar cap were measured both visually with a double-image micrometer and photographically on the plate collection. For each opposition, spiral curves showing the regression of the border at increasing longitudes (owing to rotation) were derived, and curves were drawn showing the variation of the mean diameter with heliocentric longitude. Finally, the area covered by the white deposit was deduced and plotted against heliocentric longitude. The successive oppositions show consistent variations from one Martian year to another.

From 1945 to 1952 the north polar cap was visible, as may be seen in Plates 11 and 12. The development of the cap was followed during periods when it was free from overlying clouds. Sample drawings are given in Plate 20. The brightness of the cap is not uniform; one sees bright spots, dark regions, and streaks. These markings gradually change during the spring season, some brightening, others fading. The outer rim is ragged, broken up, often with bright spots. In the shrinking process, bright promontories may develop, which may later become detached from the main cap for several days or weeks. Indentations in the rim may occur, with connecting cracks within the white area. Some regions free from the white deposit may suddenly become white again as if covered by a late hoarfrost. As a rule, the brightness of the deposit decreases in the course of each spring and summer, as if dust progressively contaminated the white surface. Each Martian year the same details are seen in the same locality during the same part of the Martian season; i.e., the developments show a very similar pattern, although some small delays or variations in detail may occur, as is characteristic of natural phenomena. The two planispheres shown in Plate 21 portray typical phases of the development of the north polar cap.

The main cause of the capricious breaking-up of the cap must be the Martian relief. From the estimated value of the vertical thermal gradient in the atmosphere, one computes that a plateau about 1000 meters high will suffice to account for the observed fragmentation of the rim. Often, right above these areas interpreted as elevated, some bright, small clouds appear in late spring, each day at the same locations. It is concluded that at this time of the year the tops of the hills, covered with the white deposit, are often overlain with cloud caps. In 1944 a veil of dust appeared to have covered part of the south polar cap. A thicker layer made the cap disappear completely in 1956 (Pl. 37); later a small cloud appeared very bright above the low-level dust. All polar deposits seemed reduced in brightness after the veil had cleared, as if polluted by the dust.

When during the spring a region emerges from the edge of the cap, the adjacent ground markings appear for a while abnormally dark, showing small, dark spots of great contrast. Through small telescopes or on photographs, this phenomenon appears as a dark belt around the cap, as is shown by Plates 11, 12, and 24. This behavior of the Martian surface is very surprising.

3.35. *Variations of the ground markings, general.*—The dark features on the planetary surface show changes in shape and intensity. These variations can be classed into several categories. Some transformations recur each Martian year, according to the same general evolution. Other regions appear at each opposition with a new behavior pattern. Still others show sudden changes followed by stable periods lasting from one to several Martian years.

As a rule, the seasonal variations seem to show a latitude effect. The most pronounced seasonal changes can be followed on the eight charts by de Mottoni reproduced in Plate 13. Photometric studies on the original plates were recently completed by Dr. Focas and are summarized in Section 3.39.

3.36. *Recurrent seasonal variations.*—At the beginning of spring for a given hemisphere the markings near the pole appear with high contrast. Later in spring the markings in the lower latitudes begin to darken, and the features at higher latitudes diminish. The evolution seems to follow a progression from the polar cap toward the equator. On Plate 22, taken at heliocentric longitude $\eta = 331°$ corresponding to the second part of spring, Mare Chronium is shown dark, just below the polar cap; while 34 days later, at the beginning of summer, when $\eta = 353°$, the western part has vanished. Visual observations showed the eastern part broken up into small dark spots. Plate 23 shows the evolution of the group of markings Depressio Hellespontica (350°–60°), Hellespontus (340°–50°), and Pandorae Fretum (340°–25°), a development recurring each Martian year. During the middle of southern spring (Pl. 23, *a*), Depression Hellespontica appears dark; Pandorae Fretum is still faint. At the beginning of September, 1956, a yellow dust cloud partly veiled this region, hampering the observations; but Plate 23, *b*, shows these markings faintly through the dust, while Pandorae Fretum is darker and broader than before. On October 8 (Pl. 23, *c*), at the beginning of the Martian southern summer, Pandorae Fretum is as dark as Sabaeus Sinus just below, with a width of about 6000 km. Hellespontus has completely disappeared, and Depressio Hellespontica is now faint. During mid-summer (Pl. 23, *d*) for $\eta = 30°$, Depressio Hellespontica cleared up more and more, Pandorae Fretum remained dark, and its eastern part, Mare Serpentis, strongly enhanced, appeared with great contrast. The maps of Plates 13 show that Mare Serpentis vanishes at the end of the southern autumn. Other annual changes occur during northern spring. Syrtis Major may be followed on the eight general maps of Plates 13 and also on Plates 24 and 32. After the summer solstice, $\eta = 177°$, Syrtis Major appears broader, the western part being dull, overlapping Moeris Lacus and Lybia. The edge reaches longitude 275°. The enhancement remains during the whole northern summer, slowly diminishing after $\eta = 260°$. During the northern winter the longitude of the eastern boundary of Syrtis is about 283°. The area affected by the periodic darkening is 500 km wide.

3.37. *Erratic seasonal variations.*—Some regions exhibit a fairly regular annual cycle of variations. In other regions the enhancement and fading of markings occurs differently each spring. As a consequence, new patterns are seen each year. Antoniadi has pointed out the frequent variations in shape of Solis Lacus. Plate 25 shows the changes that occurred in this region between the years 1941, 1943, and 1954. In 1954 the southern edge of Thaumasia looked like a bridge. Plate 26 shows the evolution of this region during the 1956 opposition; during mid-spring (Pl. 26, *a*), Thaumasia was sharply defined and light, Solis Lacus had the same contrast as Phrixi Regio; 62 days later (Pl. 26, *b*), at the southern summer solstice, Thaumasia had become darker, about as dark as Phrixi Regio. Its southern boundary had become enhanced by dark spots at (75°, −40°) and (90°, −45°). Other spots occurred at (115°, −60°) and (125°, −60°). Solis Lacus had become surprisingly dark, with a new boundary. Forty days later (Pl. 26, *c*), these few features remained, the dark spots being only slightly less prominent. The changes occurred as part of the seasonal wave of darkening in the southern hemisphere.

Plates 11, 12, 13, and 24 show the continuous changes in Nepenthes, Thoth, Casius, Utopia, and connected regions; each Martian year the region presents a new landscape, coincident with the northern seasonal wave.

Plate 27 shows the changes of Trivium Charontis and Cerberus, as seen with the high resolving power of visual observation. Plate 28 shows the writer's observations of Nilokeras, Plate 29 those of Propontis and Panchaia. Remarkable developments have taken place in Hesperia and Mare Tyrrhenum, as is seen on the photographic plates of Plates 11 and 12 and particularly the charts of Plate 13. In 1954 the pattern was greatly altered. It seems that all Martian features are, sooner or later, subject to such changes.

3.38. *Long-term variations.*—Large changes as observed in Mare Tyrrhenum in 1954 may remain for several decades, with only minor seasonal additional changes. The most prominent of the long-term modifications observed are those of the Boreosyrtis and Dioscuria regions. During each opposition since the beginning of the survey in 1941, this territory appeared dark and broad, as shown in Plate 24. However, according to the well-known earlier maps of Fournier, Antoniadi, and others, this aspect was new, while it has since remained at each opposition. Nevertheless, intensity variations do occur; for example, the large marking (280°, +55°), faint in 1946, appeared dark in 1948, vanished in 1950, and was again dull in 1952.

The northern edge of Mare Cimmerium, previously at latitude −10°, has shown since 1939 a prominent bay reaching latitude +5° at 235° (Pls. 13 and 15). The northeastern edge of Mare Cimmerium, above Zephyria, showed the structure of Plate 30, *a*, from 1939 to 1946 but returned to the previous structure in 1948 (cf. Pl. 30, *b*).

Plate 31 shows the unexpected appearance of a dark, round spot in the northern hemisphere, at about (140°, +35°) at the close of the 1959 southern spring.

A drastic change is shown in Plate 32. The large, bright circular region Hellas had completely vanished in 1954; it was observed again at the end of the 1956 southern summer, clearly defined, lighter than any other region of the Martian surface. Visual observations of Hellas showed distinctly brownish colors in a new pattern spread on a uniform background of light-red color.

3.39. *Photometric study of the seasonal variations.*—The following abbreviated report on the seasonal variations of Martian surface features based on a large number of Pic du Midi plates, was made by Dr. Focas for this chapter. A more detailed account was given elsewhere (Focas, 1959).

This study attempts to determine the progression of the wave that appears to activate the dark markings of Mars, from the pole toward the equator, during the spring of each hemisphere. For this purpose photometric measures were made of a selected set of surface markings.

It is noted that this wave progresses over a surface having the following basic properties: (i) a background of rather granular appearance and (ii) darker spots or groups of spots which appear to constitute the elements of the larger dark markings.

Figure 2 shows the intensity variations of the dark regions of Mars, plotted against heliocentric longitude η, based on measures from the Pic du Midi plate collection from 1943 to 1958. The regions are arranged vertically according to latitude starting at −65° on the edge of the south polar cap and terminating at +55°, the edge of the north polar cap. Each region is represented by a small-scale curve, showing also the individual measures as dots; the ordinates are the ratio B_s/B_c (B_s is the reflectivity of the area measured, B_c the reflectivity of a yellow marking at the center of the disk), and the abscissae the time expressed in terms of heliocentric longitude. Figure 3 shows the ratio B_s/B_c plotted against latitude. It is seen that for each hemisphere a darkening wave can be traced, comprised between the two arrows, traveling at a rate of 30 km/day from the circumpolar areas toward the equator and extending across the equator to latitudes 22° approximately in the opposite hemisphere. The dark markings of the equatorial zone appear to be affected by both darkening waves. These equatorial

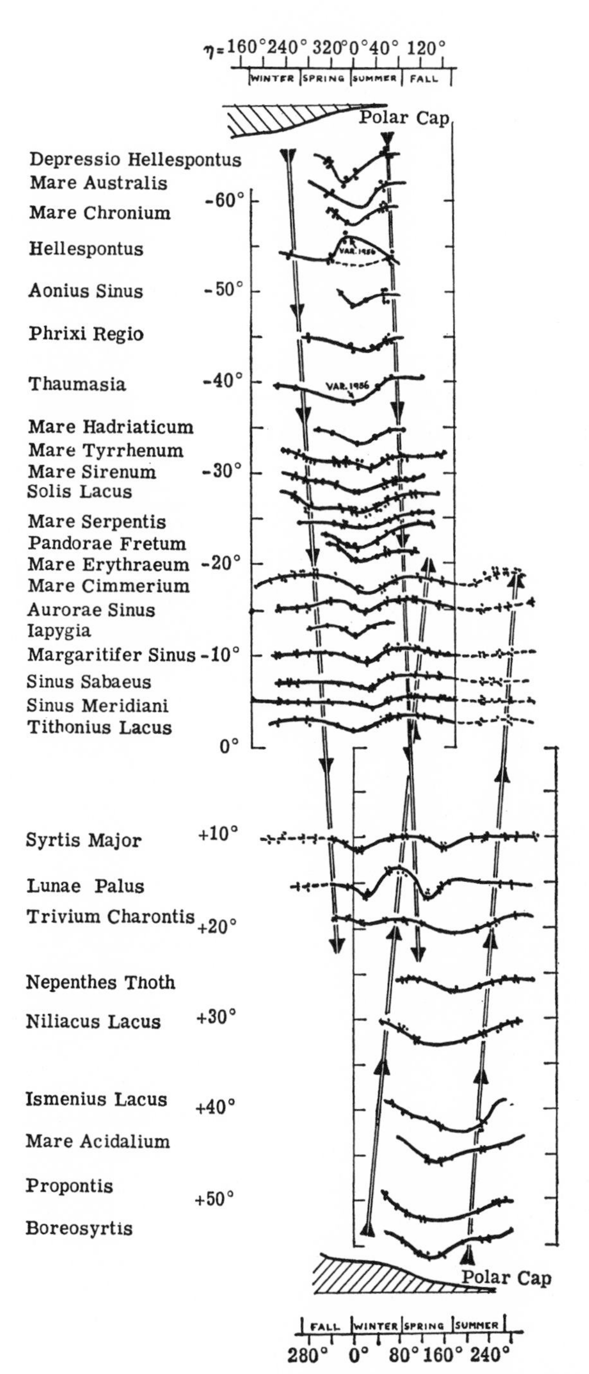

FIG. 2.—Intensity variations of dark regions of Mars plotted against heliocentric longitude η. Regions are arranged vertically according to latitude from edge of south polar cap to edge of north polar cap. Arrows encompass period of intensification of markings. (After Focas.)

areas are the darkest on the planet. Temperate and circumpolar areas may attain a maximum intensity equal to that of the equatorial areas, but they weaken or vanish as soon as the wave has passed.

If humidity is responsible for the darkening of the Martian areas, a permanent circulation of water vapor in each hemisphere must exist, the maximum quantity being diffused toward the equator after the complete evaporation of the polar clouds and the melting of the snow caps.

Mr. Focas' results are consistent with the circulation of the water vapor in the Martian atmosphere described in chapter 9, Section 4.49.

3.3.10. *Martian clouds, general.*—Visual and photographic observations undertaken in connection with polarimetric observations (chap. 9, Secs. 4.44 ff.) lead to a classification of the Martian cloud formations into sev-

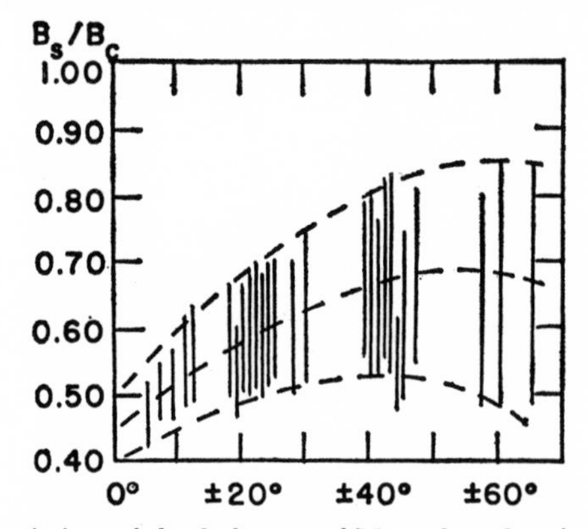

FIG. 3.—Intensity variations of the dark areas of Mars plotted against latitude. *Ordinates:* B_s/B_c, the intensity ratio of the dark areas in terms of bright regions near center of disk.

eral types, corresponding to different structures: (i) white clouds, fog, or haze due to ice crystals; (ii) dust storms; (iii) haze of droplets 2 or 3 μ in diameter diffusing blue light; and (iv) the faint, permanent ultraviolet haze, due to very small particles. One can compare original plates showing the same hemisphere and easily discover a great number of clouds or haze formations, often so faint that they are revealed only by a slight change in the brightness of surface markings. Visual observation with high resolution and polarimetric studies (chap. 9, Sec. 4.44) increase the power of the cloud survey and investigation.

3.3.11. *White clouds.*—Several categories of white clouds are observed.

a) White clouds in large formations, sometimes 2000 km long. They generally remain for several days, often for several weeks (Dollfus, 1956). They sometimes extend their edges on one side and dissolve on the other side, which causes apparent motion of the mass. Sometimes they move with the winds—we have observed winds as high as 35 km/hour. On the limb of the planet these formations may appear as bright prominences, as is shown in Plate 17, *c*. Under good seeing, masses of clouds are often resolved into small, white formations. Large systems of opaque clouds are often surrounded by a larger thin veil of haze, sometimes so faint that it may be discovered only with a polarimeter. The large, white clouds have a tendency to appear above particular regions, e.g., Ogygis and Argyre, Mare Acidalium (Pl. 35), and the central part of Amazonis (Pl. 17, *c*).

b) Bright, white clouds of small apparent size, isolated and generally remaining above the same locality. Some of these are only around 100 km in diameter. Plate 33 shows the evolution of a small complex formation above Edom in 1956, appearing each day with a new shape. Occasionally, shorter-lived clouds were also seen: on September 9 at (15°, −10°), on October 8 at (340°, −15°), and on October 13 at (355°, −15°). Plate 34 shows three permanent small formations—a bright elongated cloud between Aurorae Sinus and Mare Erythraeum (cf. also Pl. 26, *b*, near left edge); a larger but fainter uniform cloud above the southern part of Candor, masking the south component of Juventae Fons on September 30 and both components on October 4; a steady cloud above Pyrrhae Regio, surprisingly bright on October 8, wide and spread out above Margaritifer Sinus on October 4. These small, local, bright clouds seem to be of the same nature as the bright formations observed at the top of the mountains surrounding the polar cap, as described in Section 3.34, and are probably also due to surface elevations.

Small, bright, local clouds remaining fixed in position may be surrounded by larger and fainter formations, as is the case for Nix Olympica (130°, +20°) (Pl. 17, *c*). Thermal convection may play a role in these phenomena.

c) Other cloud formations, consisting of ice crystals formed by radiation cooling at low sun (i.e., near the limbs at opposition, when most of the observations are made) and at night. Prolonged observation discloses the rotational motion of the surface markings and their disappearance at the evening terminator or limb. By contrast, a faint haze or mist may persist there, not participating in the rotation. The haze may fluctuate in intensity and size, presumably depending on the humidity of the air above the region subject to the evening drop in temperature. Plate 35, *b*, shows a particularly bright afternoon fog above Mare Erythraeum in 1954. The phenomenon may also occur on the morning limb, where the haze usually disappears a few hours after sunrise. Some regions are found to be hazy on the limb for several successive days. Margaritifer Sinus showed this effect particularly well during the first 2 weeks of October, 1956. Plate 36, *a*, shows the disappearance of Margaritifer Sinus under morning haze in 1943. This case was more complex: a permanent cloud remained above the region, but this seemed to be reinforced at the rising limb by additional haze; cooling each night appeared to increase the condensation, with only partial evaporation during the daylight hours.

Small clouds may also appear locally on the limb each morning and disappear a few hours after sunrise. A pale filament, 300 km long and 100 km broad, remained during 7 consecutive mornings in 1952 between Mare Acidalium and Niliacus Lacus, with a slightly different contour each day. Occasionally, small formations also appeared during the afternoon, like the small patch formed every evening above the central part of Syrtis Major during 1 week in 1952. Perhaps this is caused by haze forming in depressions.

This morning and evening haze does not always become bright enough to be observed visually or photographically, but the far more sensitive polarimetric observations nearly always show, near the limb, variations in the amount of polarization due to haze. The large, yellow desert region Amazonis is an exception and often appears free from haze. The evening and morning haze appears most frequent in spring, presumably because of the greater water-vapor content of the atmosphere at that time. Usually in spring the polarimeter also shows thin haze around the polar cap.

In winter the polar regions are completely covered by a bright permanent cloud. This cloud cap is shown in Plate 36, *a* and *b*.

3.3.12. *Dust storms.*—Sometimes the surface features are hidden or appear softened by yellowish clouds or veils. This occurs rather rarely and almost exclusively near perihelion passage, during the southern spring. The yellow clouds are attributed to fine dust raised from the surface and suspended in the atmosphere.

During the period of observation at the Pic du Midi, such a veil was seen in January, 1944, toward the end of the Martian summer in the south polar region. Later, when the planet was observed far from perihelion, such veils were almost wholly absent, but in 1956, during the next perihelion opposition, the veils again made their appearance, on this occasion on a very large scale.

The 1956 development was as follows: From the start of observations at the Pic, in mid-July, toward mid-spring in the southern hemisphere, Mr. Focas noticed a strong perturbation of the planet's polarization, accompanied by low contrast of the surface features. Early in August these effects seemed to diminish progressively, and by August 15 the atmosphere had partly recovered its transparency. Two facts of importance remained:

(i) the sunset limb often showed a light-yellow border; (ii) the white cloud formations were unusually frequent and the morning haze very dense.

Suddenly, on August 30, the disturbance returned with much greater intensity. In Plate 37, *1*, a brilliant, opaque, yellowish cloud belt is seen to cover the planet between longitudes $-40°$ and $-55°$, above Chrysokeras, Ogygis, Argyre, and Noachis. Six hours later on the same night, the Mare Sirenum area was found to show a similar veil at the McDonald Observatory. A further 6 hours later, Japanese observers found the region south of Mare Cimmerium also completely covered, while a few days earlier Hellas had been found obscured. Hence this veil formed an almost continuous belt in the temperate latitudes. The south polar cap was also covered, but some features in intermediate latitudes, like Depressio Hellespontica and Hellespontus, showed up in marked contrast between this belt and the polar veil. From the timing and the latitude, the belt seems to have occurred in the region of greatest humidity, i.e., the zone where the annual seasonal variations were occurring at this time. Subsequently, the cloud formation became more extended: 4 days later (Pl. 37, *2* and *3*) the veil covered vast regions, softening the appearance of even remote markings. After another 4 days, Plate 37, *4*, the expansion still continued, the veil becoming more transparent as well as larger. Plate, 37, *5*, a few days later, shows Syrtis Major almost clear, the polar cap once more visible but greatly dimmed, and Mare Cimmerium still largely covered by diffuse bands, which are better shown in Plate 37, *6*.

According to Stokes's Law, under Martian conditions of gravity and atmospheric pressure, a spherical particle falls 3.8 times as fast as on the earth. The rate is 1 km per day for particles 20 μ in diameter and varies as the square of the diameter. Sifting therefore results, and only the finest particles remain suspended in the atmosphere for a long time and are blown by the winds to great distances. In blue light the planetary disk appears exceptionally bright, confirming the diffusion of light by minute particles. The dust did not reach great heights: in the polar and temperate regions, well-defined white clouds were observed above the veil. The spreading of the dust laterally and upward presumably occurred largely by turbulent mixing as a result of winds, though updraughts may also have been important and possibly electrostatic repulsion.

Toward the end of September and the beginning of October, 1956, the dust seemed to have largely settled, but a distinct yellow halo remained at the sunset limb, as for the earlier veils in August. This indicates that daylight heating raises some of the dust that had settled during the night.

Miyamoto (1957) attributed updrafts to increased abundance of atmospheric water vapor; infrared absorption then would give rise to heating of the air near the ground. The observations seem indeed to suggest a correlation between the appearance of such veils and the presence of water vapor. The veils were produced around the polar cap at the time of its maximum recession, when solar heating was pronounced, and they were often accompanied by haze and cloud formations.

The occasional occurrence of dust clouds during perihelion spring adds a further distinguishing feature between the seasonal cycles in the two hemispheres, the others being their lengths and difference in insolation resulting from the eccentric orbit.

After the veil had disappeared, no clear evidence was found of effects that might be attributed to dust deposits. The search was hindered by the appearance of new spots and topographic changes characteristic of the spring season. But the polar cap remained, in many parts, dimmed and dull.

The polarimetric properties of the dust storms are reviewed in chapter 9, Section 4.47.

3.3.13. *Blue veils.*—Visual observation of the blue veils and general atmospheric haze requires the use of a deep-blue filter and comparatively low magnification (around 200×) (cf. Pl. 38). The veils occur primarily near the limb and the terminator and may be regarded as tropical morning or evening "fog." Normally, they disappear 2 or 3 hours after sunrise and reappear a few hours before sunset. The limb haze changes continually in shape when observed during all-night runs (8–10) hours).

Sometimes the blue veils are connected with white clouds. In the great cloud formations, connected blue veils may be seen above the white clouds.

The polarimetric results summarized in chapter 9, Section 4.46, indicate that the veils are not ice crystals but small droplets, presumably water, having a diameter of 2–3 μ. They seem to be similar to the terrestrial noctilucent clouds but denser and more frequent in occurrence.

3.4 Jupiter

Jupiter viewed under perfect telescopic seeing is a beautiful sight. At opposition and with a magnification of 1000×, it subtends an angle of 14°; but many of the complex phenomena seen cannot be adequately followed visually. We have therefore studied Jupiter chiefly photographically.

About one hundred plates taken in yellow light were selected. Frames taken at short intervals suffice to show the rapid motions and transformation of the clouds and the periods of great activity, while those spaced at longer intervals record the slow changes in shape and the contrast of the belts. They also show the variations in intensity and extension of the dark polar areas.

Plate 39 shows two aspects of the planet in 1945, after the full recovery of the equatorial belts, which were faint in 1943. The equatorial zone shows delicate linear features and bright clouds. The Red Spot appears in Plate 39, *a*, and was observed visually as light red in color. The north equatorial belt structure is complex in both pictures. Plate 39, *b*, shows the shadow of Satellite II, Europa, sharply defined. Reference is made also to Lyot (1953).

3.41. *The Galilean satellites of Jupiter.*—Io, Europa, Ganymede, and Callisto were observed visually. Lyot examined them with the 38-cm refractor in 1941 and later with the 60-cm telescope. His observations of the markings on these bodies have largely been confirmed by Camichel and Gentili. Lyot selected more than 250 drawings for further study. The markings on the satellites are permanent, and the periods of rotation are found to be equal to the periods of revolution around Jupiter. The surface features were represented in four planispheres, and are shown in Plate 40. The centers of these maps correspond to the centers of the satellite disks when these are at inferior conjunction, projected on the disk of Jupiter (having the planet in the nadir).

Ganymede has shown whitening at the sunrise limb, covering permanent surface detail; this might be an indication of temporary light deposits on the ground or morning haze. Camichel (1953) has measured the diameter of the satellites, by comparing them with artificial disks projected in the field, of adjustable size, brightness, color, and limb darkening. Also, the writer (1954) has measured them with a double-image micrometer, correcting for limb darkening by increasing the aperture of the telescope by steps and extrapolating slightly to the value for infinite aperture. The results of these two sets of measurements agree around the following values: Io 3550 km; Europa 3100 km; Ganymede 5600 km; Callisto 5050 km. Ganymede is therefore nearly as large as Mars.

3.5 Saturn

The three rings surrounding Saturn are divided into sections by gaps and shades that often are of very low contrast. Their study requires fine seeing without loss of contrast. Lyot (1953) used a magnification of 900 under the best atmospheric conditions in order to see and draw them (Pl. 41).

According to Lyot, the outer Ring, *A*, is bright at its outer edge; next comes a narrow, well-marked minimum of light, a new maximum, and three wide minima which are fuzzy and of little contrast. They partly overlap, so as to reduce the brightness of the central region of Ring *A*. This distribution of light has previously been taken for a gap called the "Encke" division. The inner rim of Ring *A* is again bright.

The second Ring, designated as *B*, is separated from *A* by the wide "Cassini Division" and is divided into two parts by a narrow strip of low contrast; in the outer zone there appears a shallow minimum. In the interior zone Lyot found two narrow bands, only a quarter of a second apart, placed on a broad minimum which darkens the entire inner part of the Ring. The light was found to increase again at the inner edge of Ring *B*. The inner Ring, called *C* or "Crepe Ring," was reported to be separated from *B* by a gap a little more than half as wide as the Cassini Division.

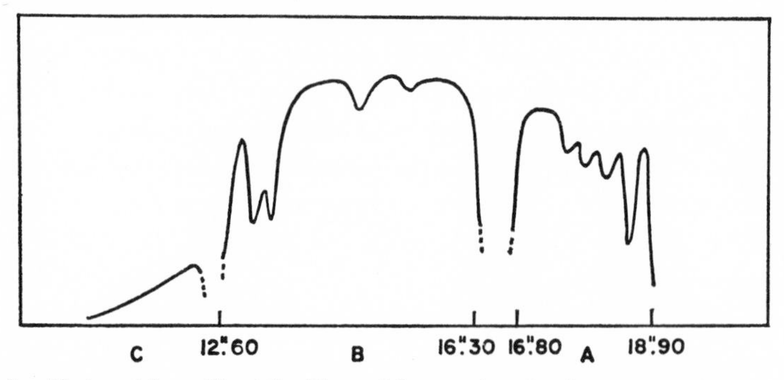

FIG. 4.—Photometric profile of the Rings of Saturn. Angular distances are given for Saturn placed at 10 a.u.

During the summer of 1957, the writer had the opportunity through the invitation of Dr. Kuiper to use the 82-inch telescope of the McDonald Observatory during 15 nights for observations of the Saturn Rings. With the aid of these observations and unpublished micrometer measurements of the dimensions of the Rings made at the Pic du Midi with a new double-image micrometer and photometric measurements by Dr. Camichel made on photographic plates, the writer has constructed a photometric profile of the Rings reproduced in Figure 4. Comparison with Lyot's drawing in Plate 41 shows good agreement.

Photographs of Saturn taken by Lyot and Camichel, when properly reproduced, bring out the main features of the Ring. In Plate 42, image *a*, a composite of 10 images, shows the outer Ring *A*. It is noted that the inner part of the Ring is bright and the central part darker and that the brightness decreases outward except along the outer edge, in agreement with Figure 4 if allowance is made for photographic blurring. Image *b*, made from two composites of 10 images each in which the inward intensity gradient was partly compensated for, shows the faint ripple near the center of the Ring and the wide dark band near the inner edge, with an increase in brightness along the edge itself. The resolving power is insufficient to show the two narrow dark components within this band. Image *c* has been copied to show the Crepe Ring and shows a strong intensity decrease inward.

The photographic records also show the evolution of the belts on the planet. Examples are shown in Plate 43. A white spot appeared in February, 1946, at latitude $-12\overset{\circ}{.}3$, visible on Plate 43, *c*. Camichel (1956*b*) succeeded in taking a series of good plates of this feature, giving a period of rotation of $10^{h}21\overset{m}{.}6$ for this latitude.

3.51. *Titan.*—Saturn's largest satellite, Titan, has a diameter of about $0\overset{''}{.}8$. The limb appears strongly darkened. Fuzzy spots, some bright and others dark, have been seen on the surface by Lyot, the writer, Dr. Camichel, and Mr. Bruch. They are shown in Plate 44; the drawings are in order of increasing planetocentric longitude, starting from opposition. These observations show variations from one drawing to the next. These variations could possibly be due to a period of rotation smaller than the period of revolution but, more probably, to an overlying atmosphere containing clouds. This last interpretation would be compatible with Kuiper's spectroscopic studies and the photometric results of Harris (chap. 8).

The diameter was found to be 5050 km by Camichel and 4950 km by the writer.

3.6. Uranus and Neptune

Uranus had its axis of rotation oriented toward the earth in 1946. Accordingly, during the last 15 years, one could observe only the polar area. Only a strong limb darkening was observed visually and recorded photographically.

On Neptune the faintness of the solar illumination makes it difficult to see details. The limb darkening is very pronounced. In addition, one finds very weak spots of irregular shape. No band structure appears. The three drawings of Plate 45, reproduced with increased contrast, are each the average of independent observations by Dr. Camichel, Mr. Bruch, and the writer (Lyot, 1953), which were, on the whole, accordant. Measurements of Neptune with the double-image micrometer gave 45,000 km for the diameter.

3.7. The Moon

Lunar photographs have been taken either at the direct focal length of 18.2 meters, at an enlarged focus of 30 meters, or with a wide-field camera constructed by Couder, which gives a focus of 50 meters. The plates have been obtained by Messrs. M. Gentili, J. Clastre, and H. Camichel. Selected reproductions are found in Volume **4** (in press) and Flammarion (1955).

The planetary researches described in this chapter were initiated by the late Dr. Lyot, whose influence has left its mark on the entire program. The work was facilitated in recent years by the active interest of Dr. Rösch, director of the Pic du Midi Observatory. I wish to thank Dr. Kuiper for arranging a stay at the Yerkes and McDonald Observatories which led to several improvements in this chapter and Dr. Van Biesbroeck and the Editors for making the translation of this paper from the original French text.

REFERENCES

Seeing and Techniques of Observation

Byram, G. 1944 *J. Opt. Soc. America*, **34**, 571, 718.
Couder, A. 1953 *C.R.*, **236**, 780.
Danjon, A. 1926 *C.R.*, **183**, 1032.
Danjon, A., and Couder, A. 1935 *Lunettes et télescopes* (Paris: Éditions de la Revue d'Optique).
Keller, G. 1955 *J. Opt. Soc. America*, **45**, 845.
Meinel, A. B. 1960 *Telescopes*, ed. G. P. Kuiper and B. M. Middlehurst (Chicago: University of Chicago Press), p. 154.
Protheroe, W. 1955 *J. Opt. Soc. America*, **45**, 851.
Rösch, J. 1955 *Ciel et Terre*, **71**, 205.
Stock, J., and Keller, G. 1960 *Telescopes*, ed. G. P. Kuiper and B. M. Middlehurst (Chicago: University of Chicago Press), p. 138.

Pic Du Midi Planetary Observations

Baume Pluvinel, A. de la, and Baldet, F. 1909 *C.R.*, **149**, 838–841.
Camichel, H. 1946 *Astronomie*, **60**, 161.
1953 *Ann. d'ap.*, **16**, 41.
1954 *Bull. Astr.*, **18**, 81, 191.
1956*a* *Ibid.*, **20**, 131.
1956*b* *Ibid.*, **20**, 141.

DOLLFUS, A. 1947 *Astronomie*, **61**, 258.
1948 *C.R.*, **226**, 996.
1953*a* *Astronomie*, **67**, 61.
1953*b* *Ibid.*, **67**, 85.
1954 *C.R.*, **238**, 1475.
1955 *Astronomie*, **69**, 415.
1956 *Météorologie*, **42**, 81.
FLAMMARION, C., *et al.* 1955 *L'Astronomie populaire* (Paris: Flammarion), Part V, "La Lune."
LYOT, B. 1943 *Astronomie*, **57**, 49, 63.
1953 *Ibid.*, **67**, 3.
MIYAMOTO, S. 1957 *Contr. Kwasan Obs., Kyoto*, No. 71.

GENERAL

ANTONIADI, E. M. 1930 *La Planète Mars* (Paris: Hermann & Cie).
FLAMMARION, C. 1892 *La Planète Mars* (Paris: Gauthier-Villars et fils).
FLAMMARION, C., *et al.* 1955 *L'Astronomie populaire* (Paris: Flammarion).
VAUCOULEURS, G. DE 1954 *Physics of the Planet Mars* (London: Faber & Faber).

CHAPTER 16

Photographs of Planets with the 200-Inch Telescope

By MILTON L. HUMASON
Mount Wilson and Palomar Observatories

THE photographs reproduced herewith, with one exception, were taken at the coudé focus of the 200-inch telescope. The scale was 1″.35 per millimeter. The exception is the photograph of Saturn, which was taken at the Cassegrain focus of the 100-inch telescope, with a scale of 5″.0 per millimeter. Fine-grain plates were used throughout, Eastman IV-E or IV-F for red light; Cramer Contrast and in one case Lantern Slide for blue light. The exposures in red light were taken through filters, while those in blue light were not.

Particulars are found in the legends of the figures. The longitudes were taken from the *Handbook of the British Astronomical Association.* The photographs reproduced are not composite; i.e., they were copied from single images. I am indebted to Dr. Gerard Kuiper, who has examined the photographs for interesting details. All the remarks concerning the surface features should be credited to him.

THE PLATES

PLATE 1.—Saturn in blue light, November 24, 1943, 9:38 U.T. Lantern Slide, no filter; 100-inch photograph; exposure 1.0 sec, seeing good.

Plate 2.—Venus in red light, September 22, 1951, 12:41 U.T. Eastman IV-E, no filter. Exposure 0.3 sec, seeing good.

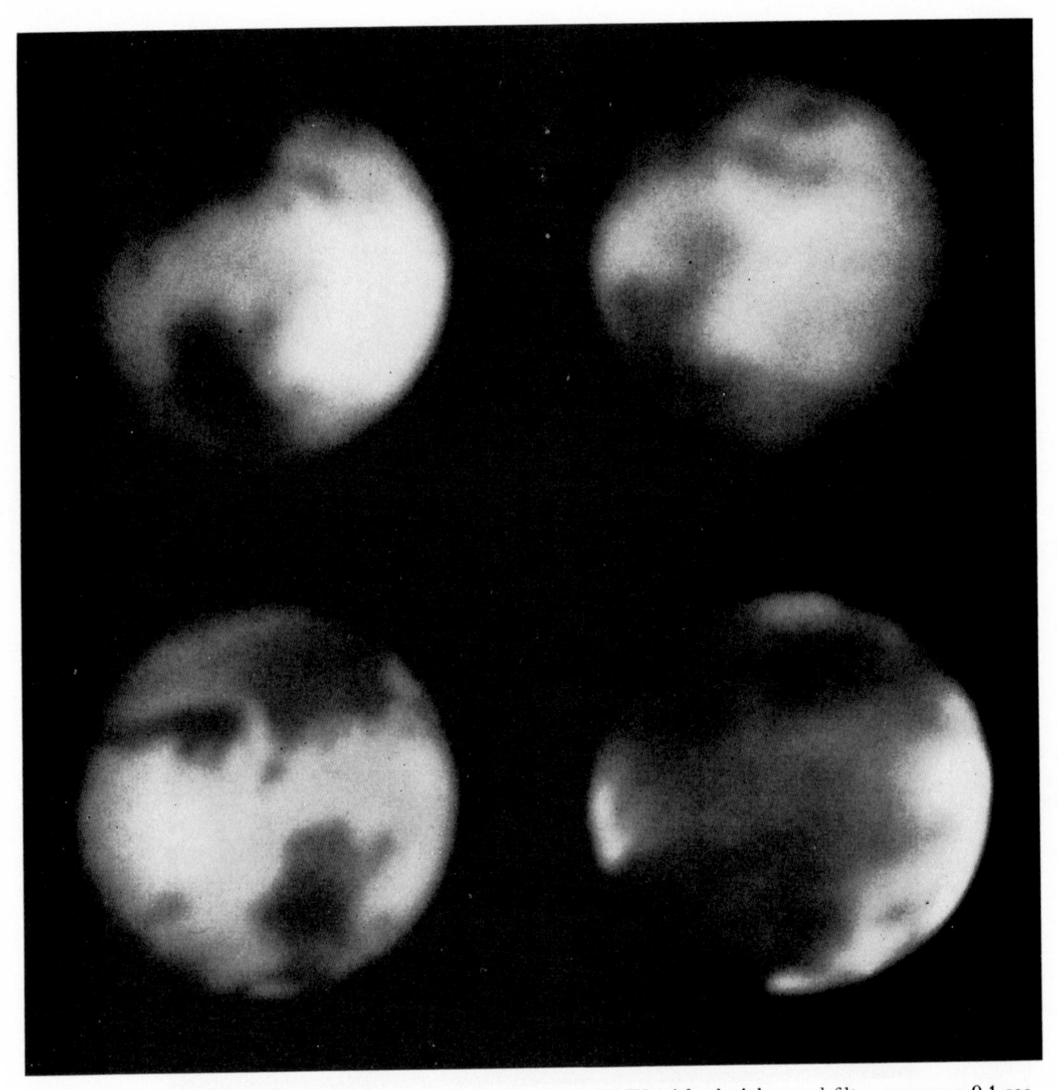

PLATE 3.—Mars. *Above:* photographs in red light, Eastman IV-F2 with plexiglass red filter, exposure 0.1 sec. Time interval about 1 month. Note bright yellow cloud on *left* photograph, taken on April 18, 1952, 8:30 U.T., $\omega = 58°$, seeing fair. *Right:* May 23, 1952, 7:06 U.T., $\omega = 89°$, seeing fair. *Below:* pair of red and blue photographs, April 22, 1952, 8:02 and 8:38 U.T., $\omega = 16°$ and $24°$, seeing good. Red photograph taken as above, blue on Cramer Contrast, no filter, exposure 1.0 sec. See legend of Pl. 5.

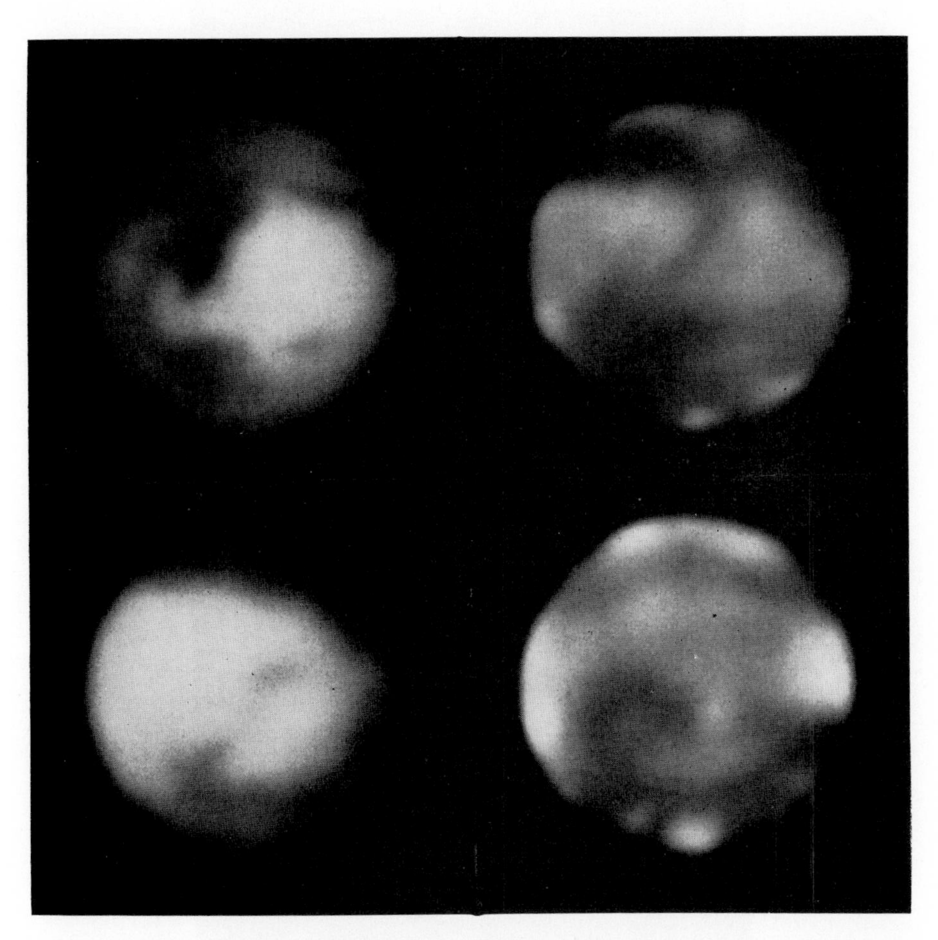

PLATE 4.—Mars, two pairs of red and blue photographs. Exposures, plates, and filters as in Pl. 3. *Above:* May 1, 1952, 8:42 and 7:41 U.T., ω = 306° and 292°, seeing fair. *Below:* May 11, 1952, 7:22 and 7:47 U.T., ω = 194° and 206°. Seeing poor and fair, respectively. See legend of Pl. 5.

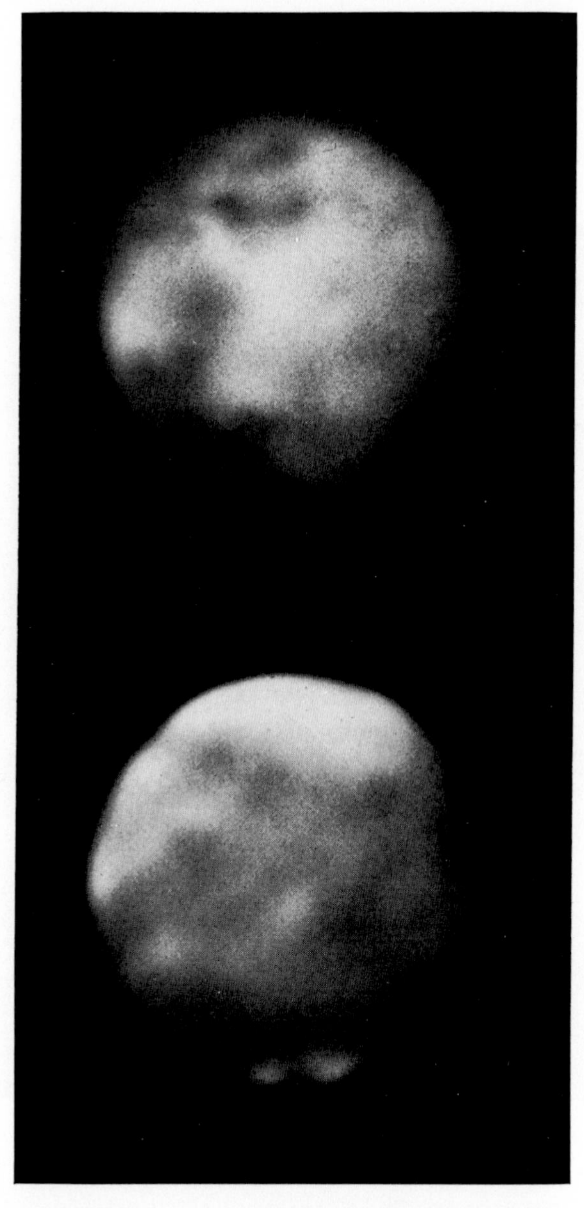

PLATE 5.—Mars, pair of red and blue photographs. May 20, 1952, 5:21 and 6:05 U.T., $\omega = 93°$ and 101°. Seeing fair and good, respectively. Red photograph shows some small circular defects due to spots on filter. Exposures, plates, and filters as in Pl. 3.

There is a partial correspondence between the dark surface features on the red photographs and the dark areas on the blue photographs on Pls. 3–5. This suggests local transparency of the atmosphere in blue light. However, several dark features on the blue photographs must be atmospheric, since they are absent in the red photographs. This throws some doubt on the reality of the atmospheric transparency in blue light where the dark features do agree in position. It is further noted that the blue haze on the right-hand photographs shows some tendency to occur in belts parallel to the equator.

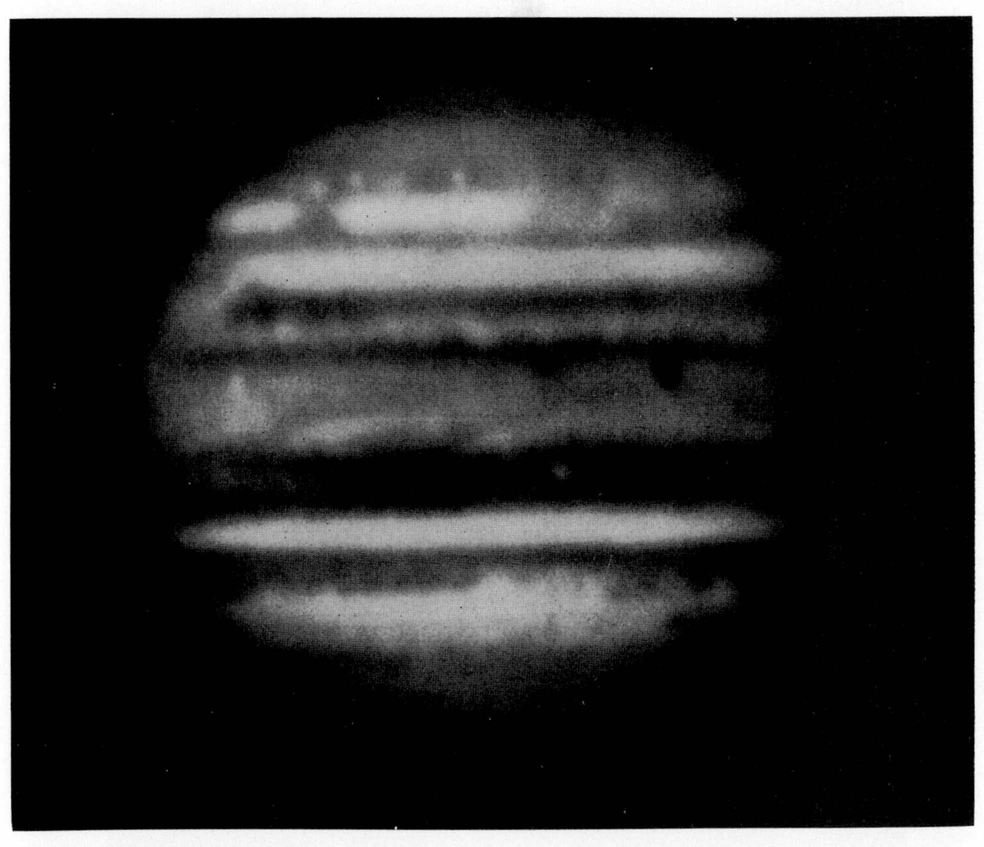

PLATE 6.—Jupiter, blue light, October 3, 1950, 5:03 U.T., exposure 2.0 sec, no filter, seeing good. Satellite II (bright), shown on disk 15 mm from left edge at same latitude as satellite shadow; ω_1 (System I, the equatorial system) = 253°; ω_2 = 313°. Red Spot visible on left limb. Cf. Pl. 8 (*left*), where Red Spot is bright also, and Pl. 10, where Red Spot is dark in blue light.

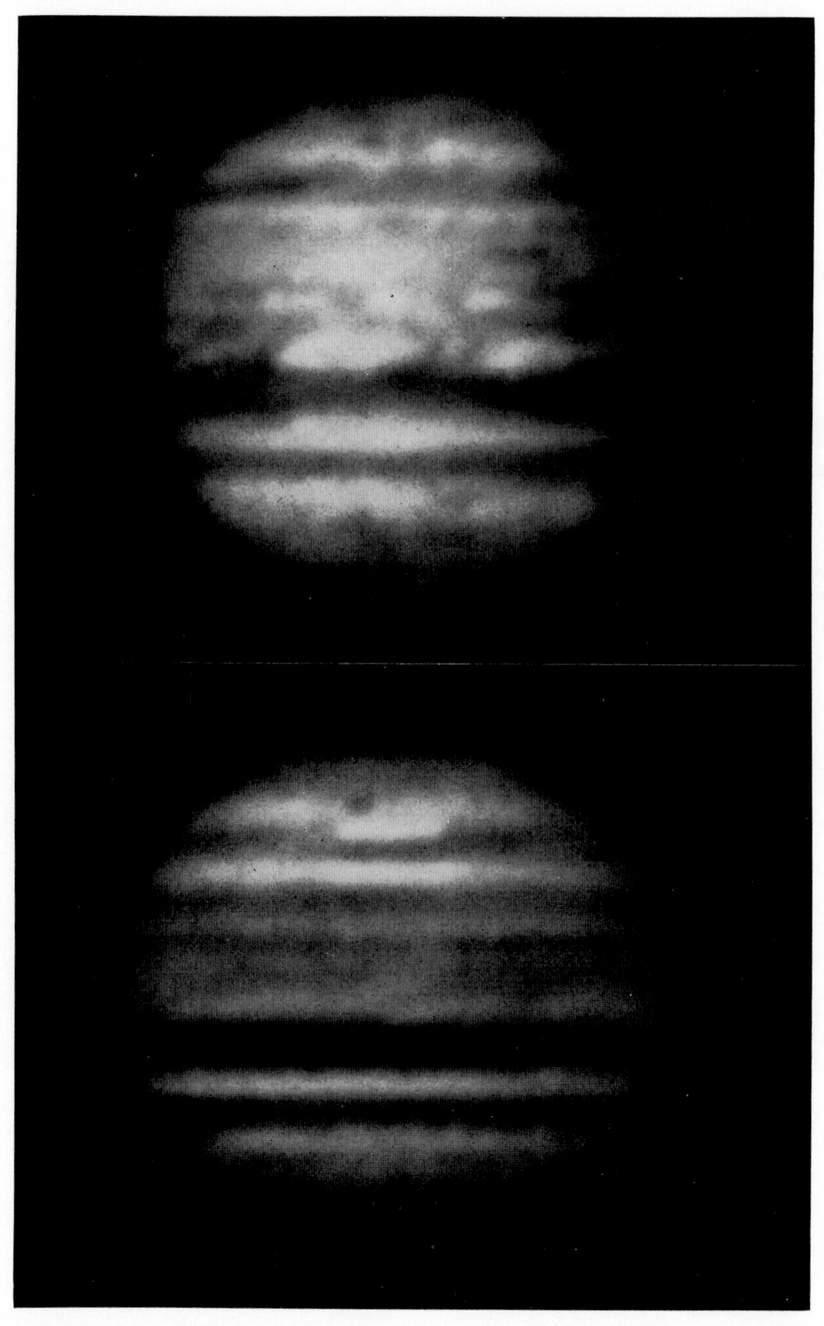

PLATE 7.—Jupiter, pair of red and blue photographs, October 4, 1950, differing by 52° in longitude; seeing fair. *Above:* Wratten No. 16 filter, 6:04 U.T., $\omega_1 = 88°$, $\omega_2 = 140°$, exposure 1.0 sec. *Below:* 4:39 U.T., $\omega_1 = 36°$, $\omega_2 = 88°$, exposure 1.0 sec. No satellites or shadows on disk. North-equatorial belt, being brown, is darker in blue light. Note enhanced contrast in red light of bright clouds in equatorial zone and in blue light of bright south-temperate cloud. Cramer Contrast plates, without filter, used on this and subsequent blue photographs of Jupiter; Eastman IV-F plus red filter, on red photographs.

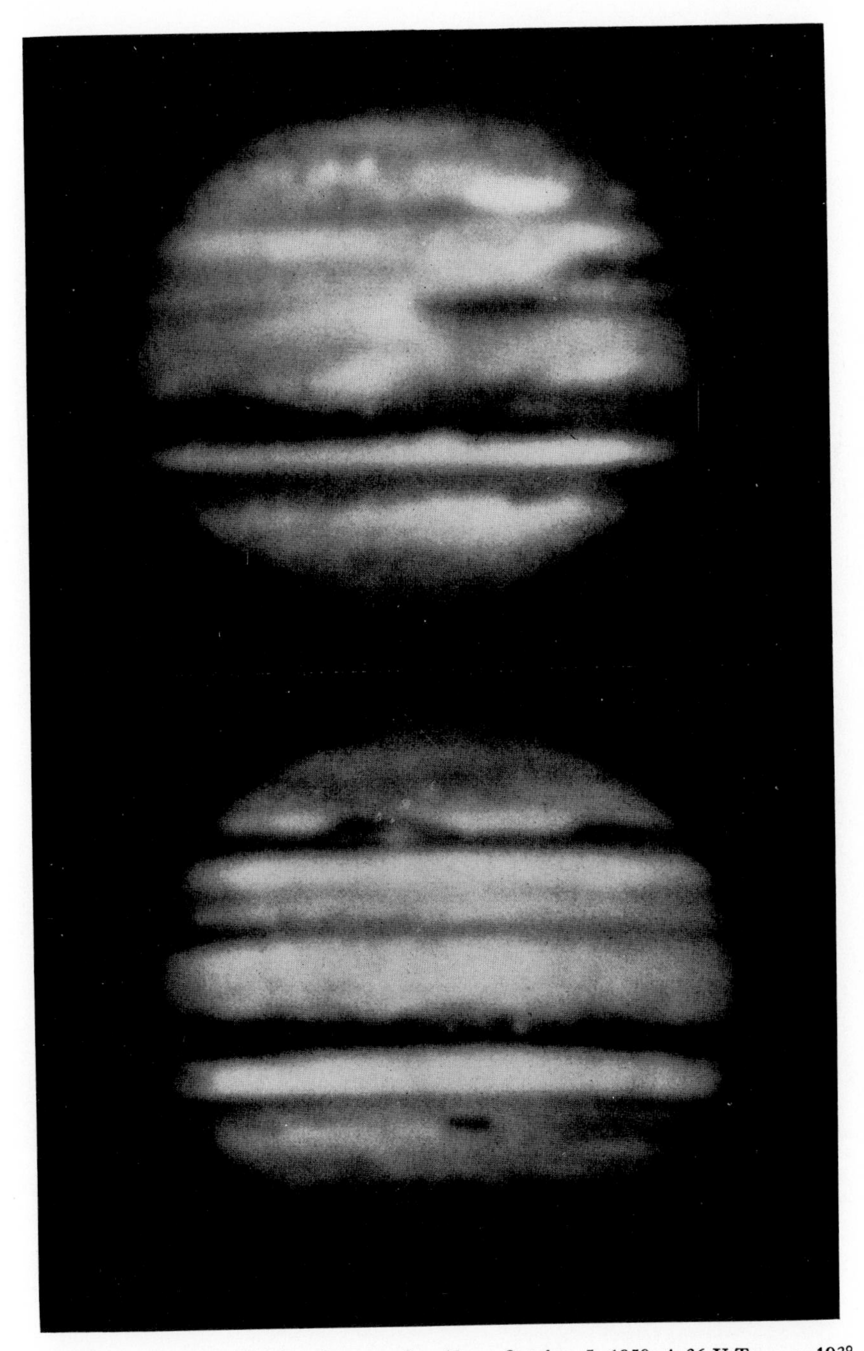

PLATE 8.—Jupiter, two blue photographs. *Above:* October 5, 1950, 4:36 U.T., $\omega_1 = 193°$, $\omega_2 = 237°$; exposure 1.5 sec, seeing good. Red Spot visible in veiled condition, appears associated with other veils crossing belts. *Below:* August 24, 1951, 12:22 U.T., $\omega_1 = 314°$, $\omega_2 = 51°$ exposure 2.0 sec.

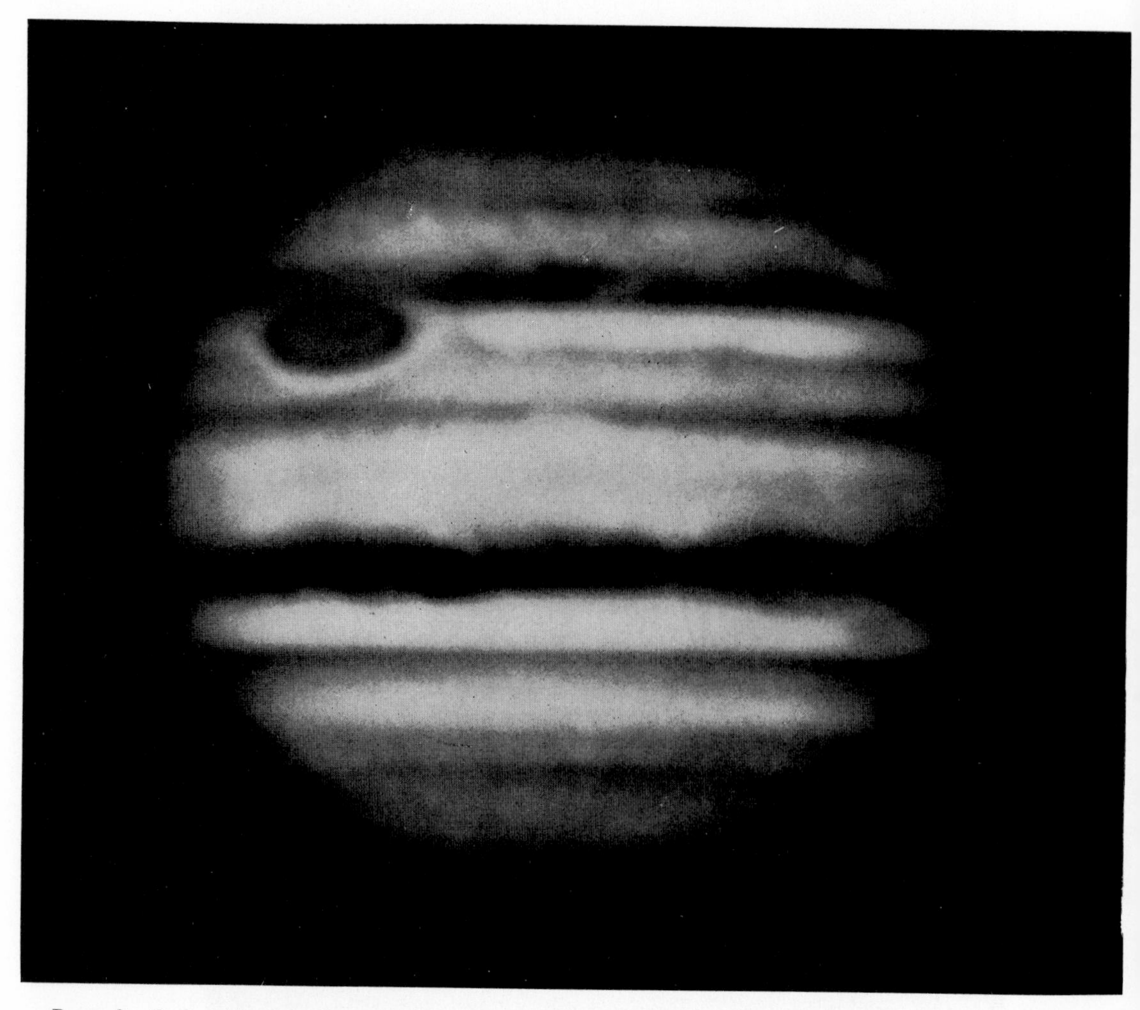

PLATE 9.—Jupiter, blue light, September 24, 1951, 9:34 U.T., $\omega_1 = 70°$, $\omega_2 = 292°$, exposure 1.5 sec, seeing good. Note structure of Red Spot; outer dark border and interior dark island are orange in color. Seeing good.

PLATE 10.—Jupiter, blue light, 21 hours or about 2 revolutions after Pl. 9. September 25, 1951, 6:30 U.T., $\omega_1 = 116°$, $\omega_2 = 331°$, exposure 1.5 sec, seeing good. Note remarkable transparency of Jupiter atmosphere above Red Spot on limb, fine structure of belts, and wealth of detail at high latitudes. (*Editor's note:* The resolution of this extraordinary photograph is about $\frac{1}{4}$ seconds of arc.)

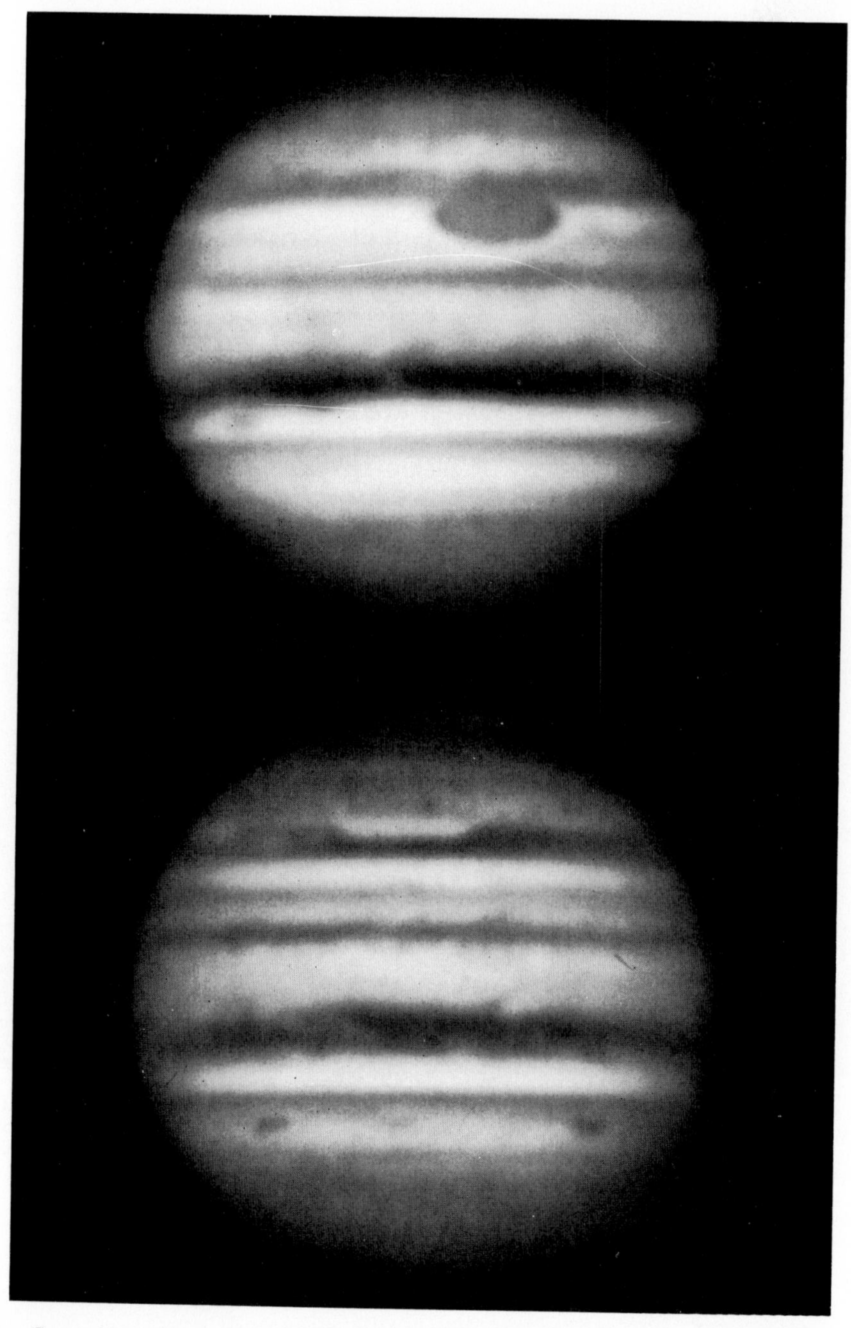

PLATE 11.—Jupiter, blue light, $2\frac{1}{2}$ revolutions apart, seeing fair. *Above:* October 9, 1951, 5:38 U.T., $\omega_1 = 137°$, $\omega_2 = 246°$, exposure 1.5 sec. *Below:* October 10, 1951, 5:44 U.T., $\omega_1 = 299°$, $\omega_2 = 40°$, exposure 1.0 sec. Note changed appearance of Red Spot; the continued presence of the two bar-shaped features shown in north-temperate zone of Pl. 10; and the changed structure and differential rotation of south-temperate belt and adjacent bright zone.

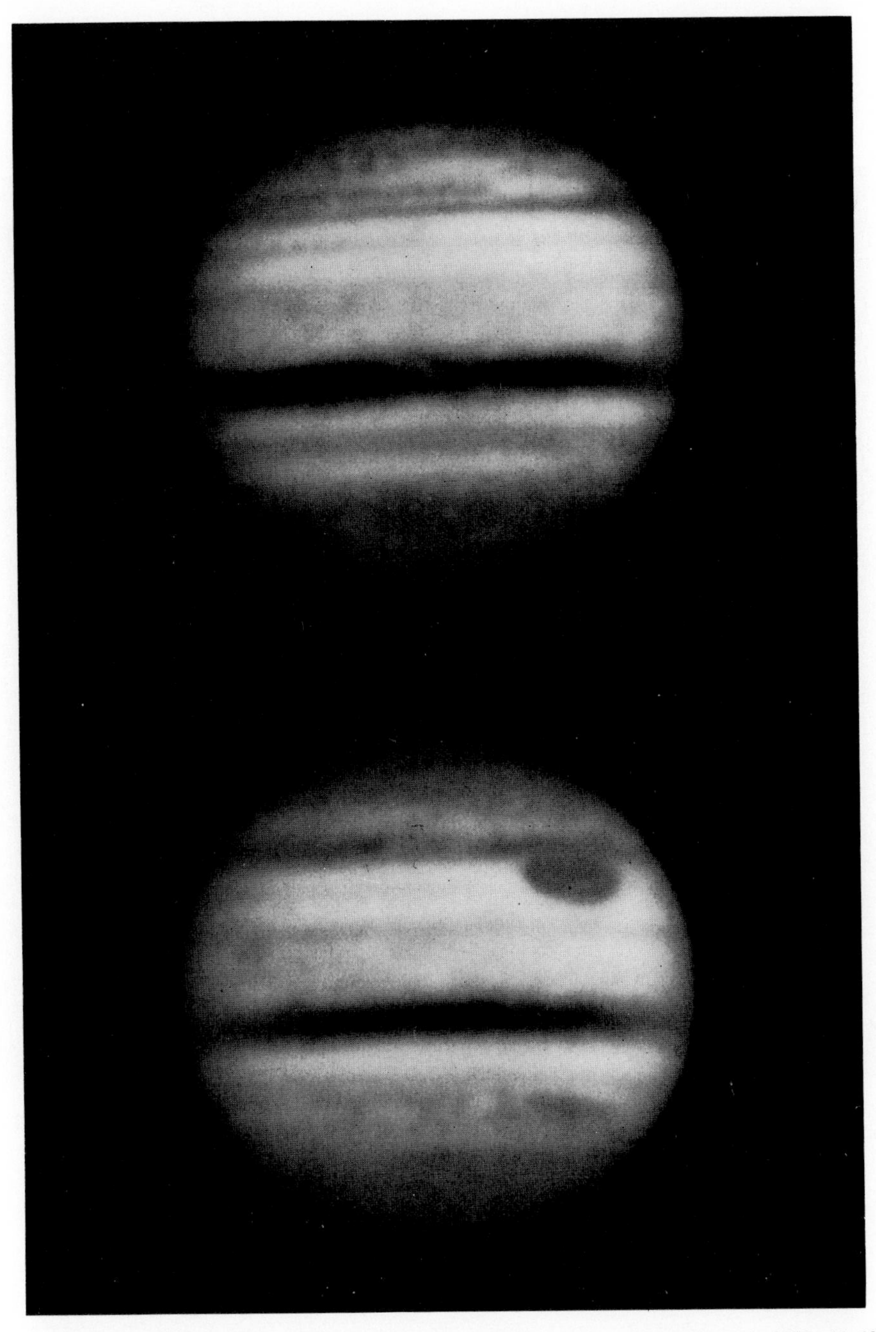

PLATE 12.—Jupiter, blue light, $2\frac{1}{2}$ revolutions apart. *Above:* September 16, 1952, 11:40 U.T., $\omega_1 = 113°$, $\omega_2 = 123°$, exposure 1.0 sec, seeing fair. *Below:* September 17, 1952, 10:31 U.T., $\omega_1 = 229°$, $\omega_2 = 231°$, exposure 1.0 sec, seeing fair to poor. Note high-latitude detail and division of south-temperate belt.

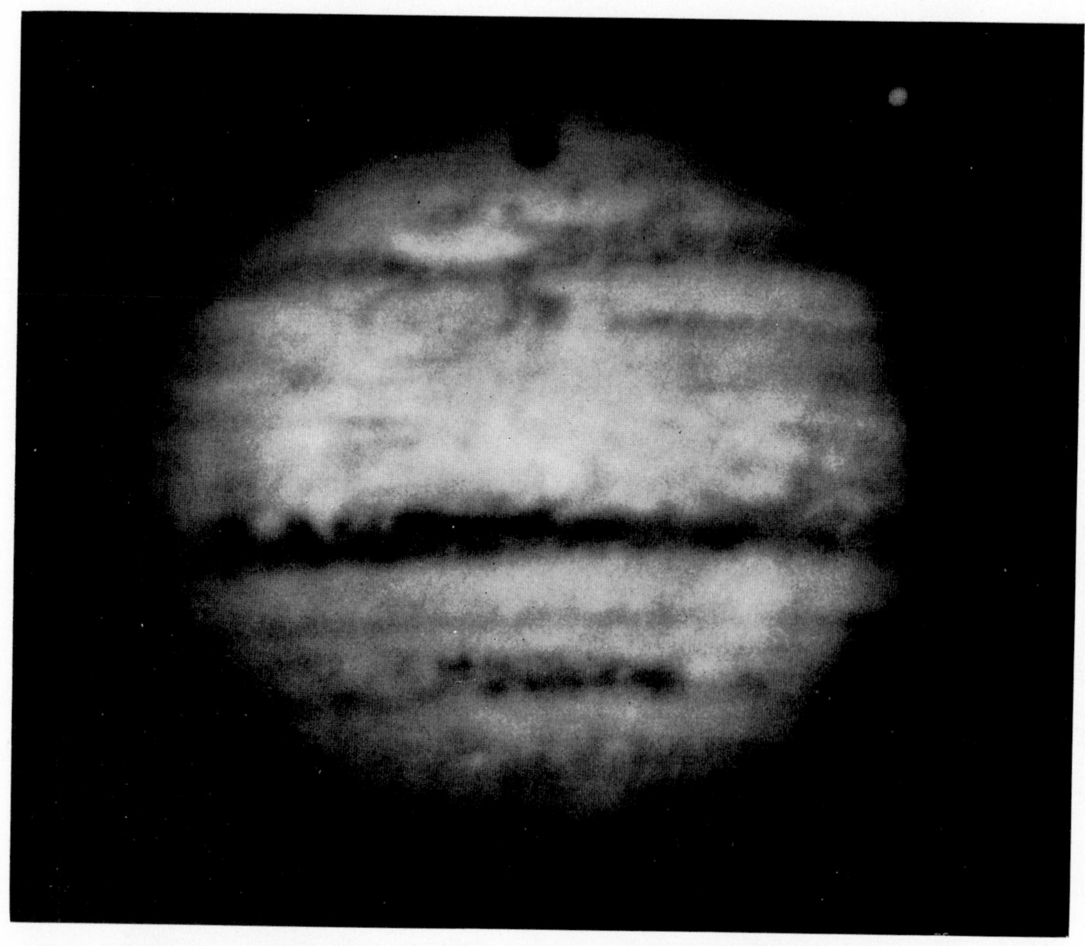

PLATE 13.—Jupiter, red light, plexiglass red filter, October 24, 1952, 7:22 U.T., $\omega_1 = 200°$, $\omega_2 = 282°$, exposure 1.0 sec, seeing good. Compare with Pl. 14.

PLATE 14.—Jupiter, blue light, October 24, 1952, 7:41 U.T., $\omega_1 = 211°$, $\omega_2 = 293°$, exposure 1.5 sec, seeing good. Compare with Pl. 13. Shadow of Satellite III on disk; satellite off upper-right limb. Red Spot somewhat irregular in red light, with central island distinctly redder than elliptical border. A "blue" feature follows Red Spot closely. Red photograph shows overlying cloud structure, linking belts. Many other colored features can be identified from intensity differences.

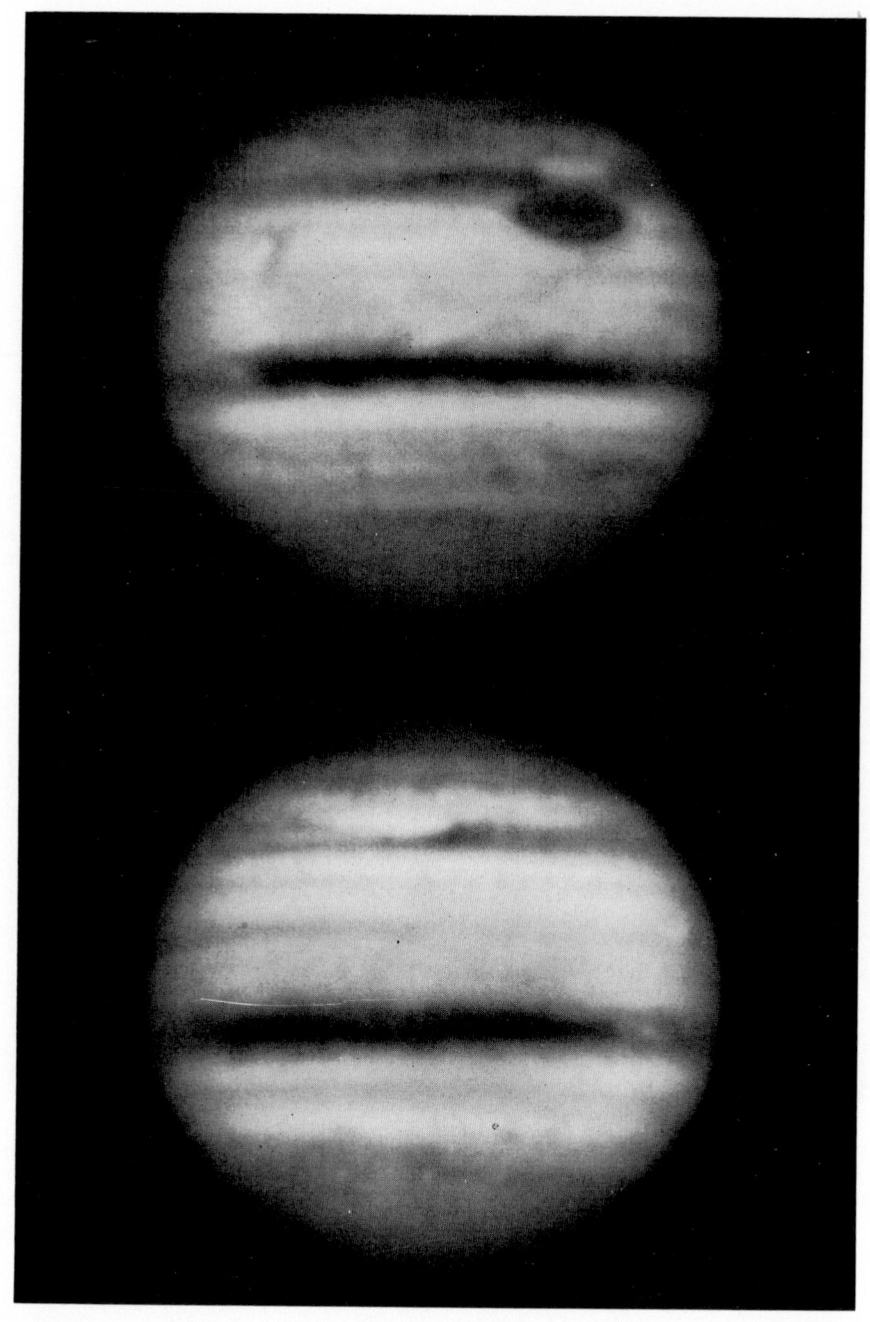

PLATE 15.—Jupiter, blue light, 4.7 revolutions apart, seeing fair. *Above:* October 21, 1952, 8:46 U.T., $\omega_1 = 138°$, $\omega_2 = 241°$, exposure 1.0 sec. *Below:* October 23, 1952, 7:34 U.T., $\omega_1 = 50°$, $\omega_2 = 138°$, exposure 1.5 sec. Note appearance of Red Spot and bright cloud in conjunction with it; feature crossing zone between components of south-equatorial belt, visible on both photographs, as is bright spot in north-equatorial belt; dotted appearance of south component of south-equatorial belt; the partial doubling of south-temperate belt; and small round spots at high latitude.

CHAPTER 17

Color Photographs of Mars

By W. S. FINSEN
Union Observatory, Johannesburg, South Africa

DURING the favorable 1939 opposition of Mars, the author made some tentative experiments in photographing the planet on 16-mm color film, with a view to exploring methods of composite enlargement and consequent reduction of grain. The results were sufficiently encouraging to warrant carrying the experiments further during the oppositions of 1954 and 1956.

The telescope used was the 26½-inch visual refractor, which was stopped down to 13¼ inches, as this had been found to give a general improvement in the seeing that more than compensated for the increase in exposure time. The image was magnified about 2.35 diameters by means of a Goerz Telenegative lens, for which the author is indebted to Mr. F. J. Hargreaves. The resulting scale value was approximately 1 mm = 8″.05. The camera, a Paillard Bolex, was equipped with a beam-splitter for visual monitoring of the image. At first, a pellicle was used, but this was replaced in 1956 by a prism retracted during the exposure by means of a cable release. Several types of color film were tried, but, on the whole, the best results were obtained with Kodachrome. Exposures in 1954 were 3 or 4 seconds, reduced in 1956 to as little as 1 second; this reduction was due mainly to various optical improvements, for example, the elimination of the pellicle, careful cleaning of all surfaces of the objective, and blooming of the Goerz lens. It was rather surprising and a little disappointing to note that the reduction in exposure time made no material improvement in image sharpness.

The exposures were made in rapid succession by hand, and it soon became evident that there was little to be gained by making more than 100 or so exposures at a time. Altogether, about 24,000 exposures were made in 1954 and 30,000 in 1956; many of the 1954 exposures were made without auxiliary magnification.

The best sequences in the processed films were examined frame by frame and the best 25 images selected for subsequent composite enlargement. Occasionally a larger number was used (in one case as many as 162), but the gain was not commensurate with the labor involved.

The enlarging camera underwent many improvements and modifications in the course of the work. In its final form it gives a fixed magnification of 7.7 diameters and consequently a scale value of the master negatives or transparencies of 1 mm = 1".05. The interchangeable graticule and film holder used in 1954 have been replaced by a prism-operated device whereby the image can be projected onto the graticule or the film at will. The use of cut film has been abandoned in favor of more convenient and economical 120 or Bantam-size roll film. Trouble had been experienced from the accumulation of dust on the emulsion surface; this has been effectively combated by mounting a wide, thin, soft brush, of the type used by gilders for handling gold leaf, on a swiveling arm operated externally, so that it can be passed over the film before each exposure.

The lens, a Taylor, Taylor, and Hobson *f*/1.9 of 1 inch in focal length, is mounted on a circular brass panel, sandwiched between spring-loaded annular plates, with two sets of three steel balls to provide kinematic restraint. Registration of successive images is accomplished by moving the lens panel in its own plane by means of two micrometer screws and an opposing spring spaced at 120°.

The 16-mm film is held in a gate of conventional type. Between the gate and the lamp house there is a sliding panel carrying a Compur shutter and filter holder. The light-source is a low-voltage head-lamp bulb, with diffusing screen, fed from a constant-voltage transformer and provided with a rheostat for control of color temperature.

The making of black-and-white composite negatives gave little trouble. Adox R 14 film with an Ilford Minus Blue filter, developed in D-11, gave a satisfactory rendering of Martian surface detail with suitable contrast. But satisfactory enlargements on color film proved much more difficult to obtain, and the author speedily made firsthand acquaintance with the many problems and difficulties that beset the worker in this field. Most of the available color processes, reversal and negative-positive, were tried, involving altogether approximately 1000 composite enlargements, with a very small percentage of successes.

The photographs reproduced here are of two kinds: (1) enlargements on Anscochrome Flash type roll film from Kodachrome originals, the blocks being made available by the courtesy of the Director of the South African State Information Service, and (2) black-and-white composites, as indicated by the legends.

(a) July 1, 1954, $21^{h}5^{m}$ U.T., ω 75°

(b) July 5, 1954, $20^{h}56^{m}$ U.T., ω 37°

(c) July 11, 1954, $20^{h}16^{m}$ U.T., ω 334°

(d) July 14, 1954, $18^{h}49^{m}$ U.T., ω 286°

PLATE 1.—*a–d:* Mars in 1954. Perhaps the most noteworthy feature of this opposition was the great change in the Thoth-Nepenthes region, well seen in ***d***. A blue haze-cap over the North Pole, discernible in all four figures, especially ***d***, is very striking on the composite transparencies, but hardly to be detected on the original Kodachrome images.

PLATE 2.—*a–f:* Mars in 1956. In *a* the appearance of the planet is quite normal. Frames *b–f* show the development of the remarkable yellow cloud masses characteristic of the opposition. The change from night to night is well shown in *b–e*. Note the cloud filament over Syrtis Major in *f*.

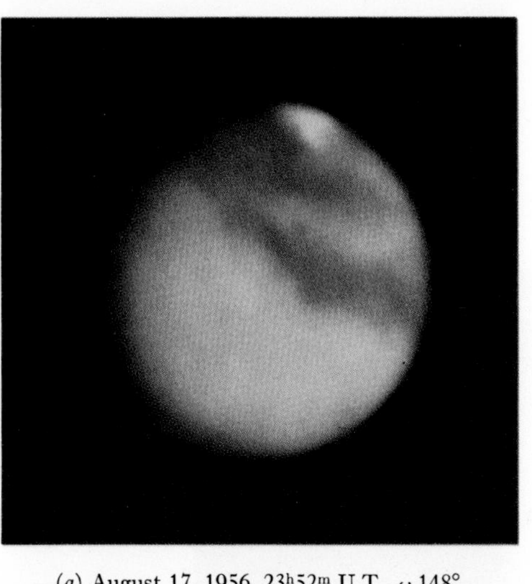

(*a*) August 17, 1956, 23^h52^m U.T., ω 148°

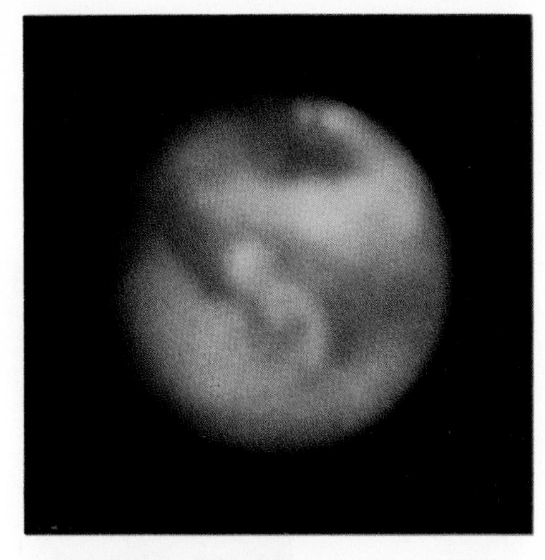

(*b*) August 29, 1956, 20^h39^m U.T., ω 354°

(*c*) August 29, 1956, 23^h24^m U.T., ω 34°

(*d*) August 30, 1956, 21^h56^m U.T., ω 4°

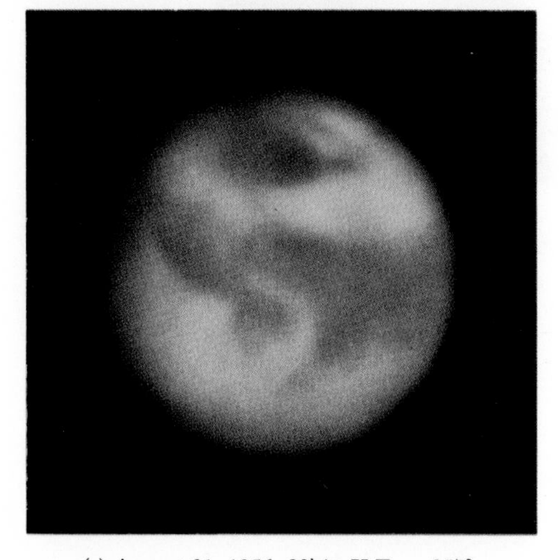

(*e*) August 31, 1956, 22^h4^m U.T., ω 357°

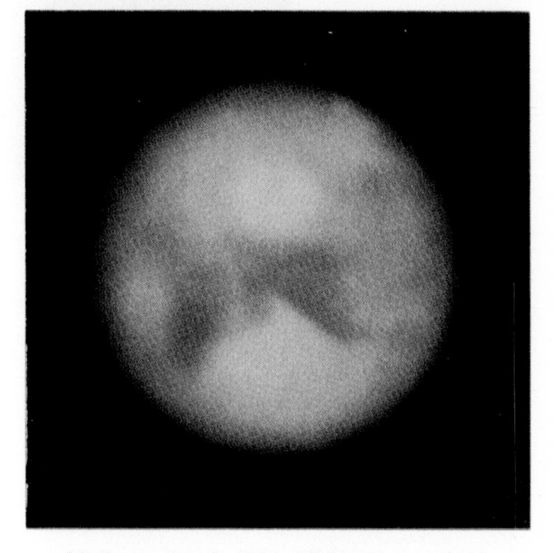

(*f*) September 7, 1956, 22^h59^m U.T., ω 308°

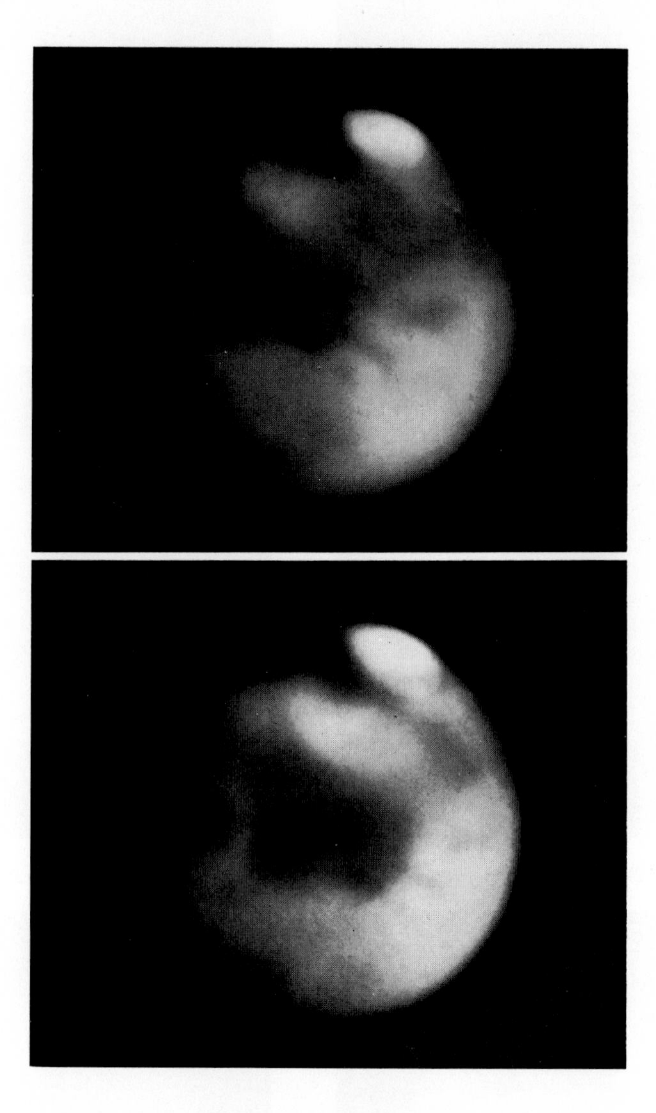

Plate 3.—Mars in 1956.

Upper: August 24, 1956, $21^{h}22^{m}$ U.T., $\omega = 49°$.
Lower: August 25, 1956, $20^{h}48^{m}$ U.T., $\omega = 31°$.

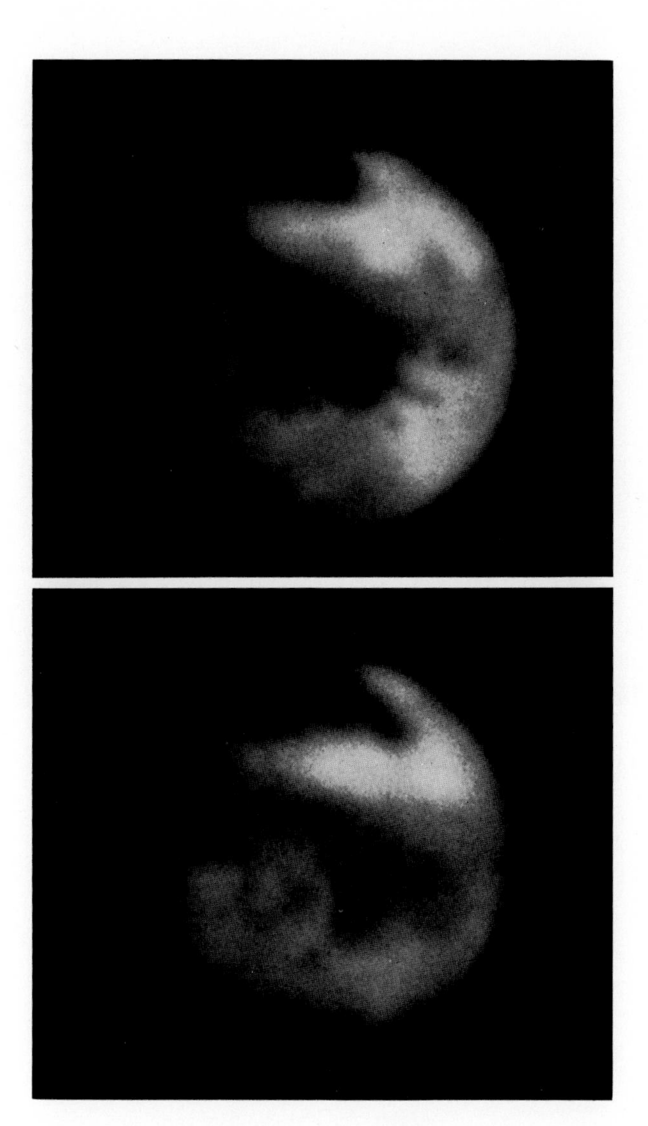

PLATE 4.—Mars in 1956.

Upper: August 29, 1956, 23^h24^m U.T., $\omega = 34°$.
Lower: August 30, 1956, 22^h02^m U.T., $\omega = 5°$.

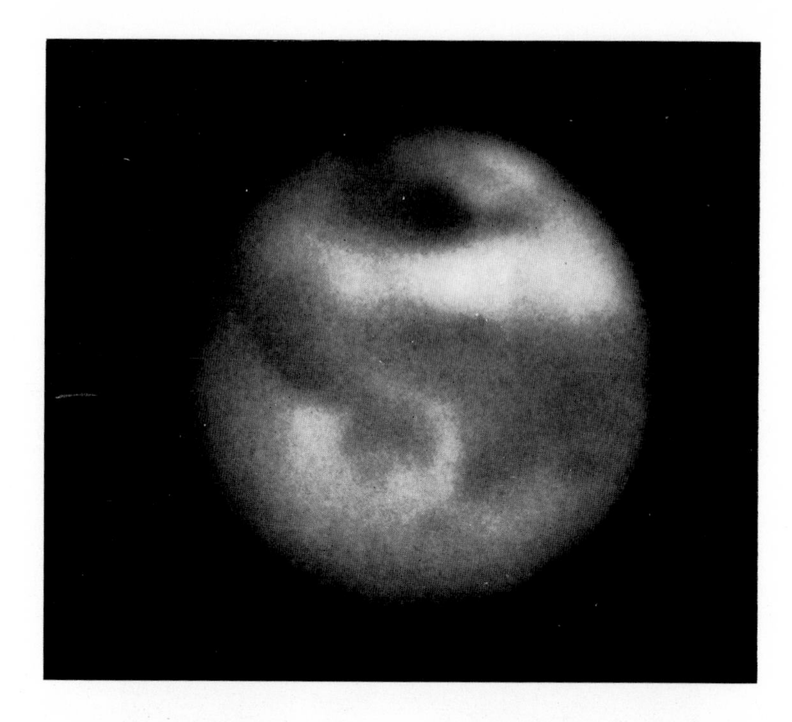

PLATE 5.—Mars in 1956. August 31, 1956, 22^h04^m U.T., $\omega = 357°$

CHAPTER 18

Limits of Completeness

By GERARD P. KUIPER
Lunar and Planetary Laboratory, University of Arizona

THE membership of the solar system is not completely known; the limits are set by such systematic surveys as have been made. In chapter 2 Mr. Tombaugh describes the extensive Lowell survey for transneptunian planets and his search for faint natural satellites of the earth. In this chapter the surveys for intramercurial planets are reviewed, as well as those for satellites around the other planets. The limits so far attained are of interest when the dynamics and the origin of the subsystems are considered. The fainter the limits can be pressed, the more adequate the resulting picture will be and the less uncertain its interpretation. The writer's surveys of the satellite systems are described here for the first time.

1. DEFINITIONS

There is no ambiguity as to what objects should be called *satellites*. These are bodies found to move around any one of the planets. Studies of their origin have led to the conclusion that not all objects that formed as satellites need still be captive. The Trojan "asteroids," as well as Hidalgo and Pluto, are probably such exceptions. The satellite searches yield, of course, only such objects as are satellites today. It is dynamically possible that the earth-moon system has captured some meteorites and retained them temporarily as tiny companions to the earth. Since such bodies have not been found (see chap. 2), the question as to whether such objects should be called "satellites" has not arisen.

Related to the satellites are the Rings of Saturn; and traces of rings may exist around the other Jovian planets. The search for *rings* of tiny satellites is included in this chapter; none have been found.

The term *planet* is less distinct. If the analogue of the term "satellite" were to be used, all bodies moving around the sun would be planets. However, the comets and their disintegration products—the meteor swarms and the zodiacal light—are so different from the planets proper that their classification as planets has no merit. The case of the asteroids or minor planets is less obvious. Where is the dividing line between minor and major planets—if the latter term is to denote the eight largest bodies of the solar system around the sun, Mercury to Neptune, and not merely the outer four, which may be called the *Jovian planets?* Is Pluto to be called a "minor planet" on the basis of its presumed origin? This would indeed be consistent with the use of that term for Hidalgo and the Trojans. A classification based on generic relationships is preferable to an arbitrary one; with improved understanding, it can be revised where necessary.

Accordingly, the following terminology is used: There are eight known planets, Mercury to Neptune, one of which is double, the earth-moon system. The question as to whether there are additional planets must be answered by intramercurial and transneptunian surveys, since gravitational tests (effects on orbits of other bodies) will normally be less sensitive. The existing surveys are discussed in this volume. The minor planets may be divided into two groups, the *asteroids*, formed largely or entirely between Mars and Jupiter and, with a few exceptions, still moving there, and the presumed *former satellites* of the Jovian planets—Pluto, Hidalgo, the Trojans, and probably others, yet unknown. The survey technique for former satellites of Uranus and Neptune is, of course, the same as that of transneptunian planets, since both are directed toward finding objects moving around the sun of smaller motion than the very numerous asteroids; therefore, they must be treated together. The only substantial transneptunian survey is the large Lowell Observatory survey made between 1929 and 1941. It is described by Mr. Tombaugh in chapter 2. The survey of asteroids presents a complex problem of its own; it is dealt with in Volume **4**. The small fragments called *meteorites* broken from asteroids in collisions are also considered in Volume **4**.

Satellite surveys are much less laborious than planet surveys, because only a small region in the sky needs to be examined. One can compute beforehand how large this region is, because the approximate limit for stable satellite orbits in the tidal field of the sun can be found. The author has examined with the 82-inch telescope at the McDonald Observatory the satellite regions around all the planets except (*a*) Mercury, for which the proximity to the sun poses special problems, and (*b*) the outer parts of the Jupiter field (beyond 25′ from the planet), which was thoroughly explored by Nicholson with the 100-inch telescope. Dr. Humason independently examined the Neptune and Pluto fields with the 200-inch telescope, and Dr. Baade had previously examined Pluto with the 100-inch telescope (verbal communications). The author is indebted to Drs. Humason, Nicholson, and Baade for making their records and plates available.

2. INTRAMERCURIAL PLANETS

Effective surveys for intramercurial planets have been made only during the brief and rare intervals of total solar eclipses. Perrine (1902, 1906, 1908, 1909) made extensive searches on plates taken at the 1901, 1905, and 1908 eclipses. Campbell and Trumpler (1923) used a 17 × 17-inch plate centered on the sun which showed somewhat fainter stars than had been photographed before. The focal length of the objective was 5 feet, its aperture 4 inches ($f/15$), the exposure time 1 minute, the plate Seed 23 (Eastman 33). Some 550 stars were visible on the plate that reached 10.2 mag. photographic but was probably complete only to 9.5 mag. The area examined was 15° × 15°, so that the maximum distance from the sun varied from 7°.5 to 10°.5. Perrine's earlier searches reached about 9.0 mag. and covered 8°.2 by 25°, with the long dimension parallel to the sun's equator. When Mercury is at perihelion, its maximum geocentric distance from the sun is 18°. A search for intramercurial planets up to 15° from the sun would therefore still be of interest.

A search for planets moving near the triangular points of Mercury has been made by Trumpler (1923). From twilight exposures he found no planets brighter than 11.0 mag. photographic near the Lagrangian point L_5; while from Perrine's eclipse plates he found none brighter than 8.2 photographic near L_4. With albedos equal to that of Mercury, the upper diameter limits for such bodies are 20 and 60 km, respectively. Bodies of these dimensions, when projected on the sun at inferior conjunction, would be only 0″.04 and 0″.13 in diameter and might easily be missed. Twilight searches from a station at very high altitude in the tropics (such as Chacaltaya, Bolivia) would therefore appear to be the best method of extending these surveys to still fainter limits.

3. THE SATELLITE SYSTEMS

A particle initially at rest in a co-ordinate system rotating with the planet around the sun will continue to belong to the planet if its distance from the planet is less than the mean *radius of action*, R_A, given by the interpolation formula,

$$\log \frac{R_A}{a} = +0.318 \log \mu - 0.327 , \tag{1}$$

if $\mu = M_p/(M_\odot + M_p)$ (Kuiper, 1951). Satellite orbits are not stable up to R_A because satellites are not at rest in the rotating frame. Since for a satellite with a near-circular orbit the surface of zero velocity is situated at roughly double its distance from the planet, 0.5 R_A may be taken as the average stability limit for satellites. Actually, for large orbits, retrograde motion is somewhat more stable than direct motion, and the outermost retrograde satellites of Jupiter extend slightly beyond 0.5 R_A. Furthermore, orbital eccentricities can cause a satellite to be projected on the sky at a distance appreciably greater than that which corresponds to its semimajor axis. Jupiter VIII, with its high eccentricity, is an example. For these reasons, satellite searches should be extended to about 0.75 R_A, if possible.

TABLE 1

RADIUS OF FIELD IN WHICH SATELLITES CAN OCCUR*

Planet	a (Astr. Units)	R_A/a	$\frac{1}{2}R_A$ (Astr. Units)	Satellite Boundary
Mercury......	0.387	0.0032	0.00063	2′.3
Venus........	0.723	.0077	.0028	14
Earth........	1.000	.0082	.0041	....
Mars.........	1.524	.0041	.0031	20
Jupiter......	5.203	.0515	.134	110
Saturn.......	9.539	.0350	.167	67
Uranus.......	19.19	.0193	.185	35
Neptune......	30.07	.0204	.307	36
Pluto........	39.52	<0.0082	<0.162	< 14

* Computed for a distance planet-earth of $a - 1$ for the outer planets (mean opposition distance) and $\sqrt{(1 - a^2)}$ for Venus and Mercury (quadrature). A factor of 1.5 may be applied to include retrograde orbits of high eccentricity (see text).

Table 1 lists the angular stability limits computed on the basis of the value 0.5 R_A. For the outer planets the distance planet-earth is assumed to be the mean opposition distance, $a - 1$, while for the inner planets, Venus and Mercury, one may use the value at mean maximum elongation from the sun, $\sqrt{(1 - a^2)}$. In that position the hypothetical satellites would be close to their maximum brightness. If a planet is examined at a distance different from that assumed in Table 1, the radius of the search field is readily found from the actual distance and the value of 0.5 R_A listed.

For increasingly distant planets ($a \gg 1$), the ratios R_A/a and $R_A(a - 1)$ will converge, with the result that the angular diameter of the satellite space as seen from the earth will become independent of a. Instead, it will depend on the mass of the planet only, roughly as the cube root (eq. [1]). This is verified by inspection of the last column in Table 1.

4. McDONALD OBSERVATORY SATELLITE SURVEY

With the exceptions referred to in Section 4.3, the surveys described here were made with the 82-inch telescope of the McDonald Observatory. Both the Cassegrain ($F/13.6$) and the prime focus ($F/4$) were used. The former is the better suited for recording close satellites, owing to its larger scale (1 mm = 7″.385), while the latter (scale 1 mm = 25″.4) is better suited elsewhere because of its greater speed. At the Cassegrain focus a series of graded exposures was made for each planet to insure that, for each distance from the bright and disturbing planet, an optimum exposure was available. The magnitude threshold thus varied from about 12 to 18 (exposure times about 5 seconds to 12 minutes). At the prime focus, $m = 20$ or 21 was easily reached in 15–30-minute exposures. However, the $F/4$ ratio limited the field in good definition to about 5 cm diameter. The field could be increased by reducing the free aperture of the mirror. Since the linear coma increases quadratically with the F-ratio, the area of the usable field increases with the fourth power, while the speed of the telescope decreases only with the square if the obscuration caused by the secondary mirror is neglected. Therefore, if areas larger than about 5 cm in diameter (or about 10 minutes of arc in radius) need to be covered, as is true for the outer planets (see Table 1, last column), it is useful to place a circular diaphragm over the mirror. Two diaphragms, of 66- and 54-inch aperture, were used. They make the telescope roughly $F/5$ and $F/6$. The 54-inch diaphragm increases the diameter of the good field $(82/54)^2 = 2.3$ fold, practically to the full width of a 5×7-inch plate (exposed area about 4.7×6.5 inches or about 50×70 minutes of arc.)

Another improvement was introduced at the Cassegrain focus, for the study of faint satellites close to bright planets, by using an eccentric diaphragm. Since the central obscuration by the mounting of the secondary cell is about 27 inches in diameter, the primary mirror as used is a ring about 27 inches wide; a 27-inch circular diaphragm can then be placed in any one of the four quadrants, which will give the theoretical diffraction pattern of a perfect circular aperture, without extra rays due to the supporting fins or irregularities in the central obscuration. The use of such a diaphragm was particularly useful in photographing Jupiter V and the Mars satellites.

4.1 Venus

Apart from a preliminary set taken on July 20, 1954 (when the planet was at low altitude), the following series were taken at the Cassegrain focus: plates 314–321, February 18; 326–333, February 20; 334–341, February 29; 355–356, March 1; and 357–361, March 2, all 1956. The emulsion used throughout was 103*a*-E, backed (maximum sensitivity around 0.65 μ), which reduced atmospheric extinction, the planet being about 4^h west at declination about $+4°$.

The plates were taken in pairs, with exposure times ranging from 3 seconds to 4 minutes. Occulting disks were used, in two sizes, 5 mm (37″) and 8 mm (59″), which intercepted the glare from the planetary image; but it is not certain that this precaution appreciably reduced the scattered light elsewhere. The best pairs were selected for blinking: 327 and 332 (30-second exposure); 328 and 331 ($1\frac{1}{2}$ minutes); and 329 with 330 (4 minutes). No satellites were found; the estimated limit is 15 mag. if allowance is made for the rapid motion of the planet in declination (5″ in 4 minutes), for which the plates were not guided. The field covered was 4.7×6.5 inches or 15×20 minutes of arc. Very close to the planet the best plates were 355 and 356, with exposures of 60 and 20 seconds (taken at about 3^h58^m west, declination $+9°$). From these plates the estimated limit for very close satellites is 14 mag.

At the prime focus, plates 3275–3277 were taken on March 6, 1956, using I N backed ($\lambda \simeq 0.8\ \mu$), and plates 3278–3281, using 103*a*-O ($\lambda \simeq 0.4\ \mu$). The exposure times were 4, $1\frac{1}{2}$, $\frac{1}{2}$, and 4, $1\frac{1}{2}$, $\frac{1}{2}$, $\frac{1}{2}$ minutes, respectively. The plates were blinked in various suitable combinations. No satellites were found. The limit is about 16 mag. except close to the planet, where the sky is strongly fogged by scattered light and the limit is less.

The distance of Venus from the earth was 1.10 astronomical units on February 20 and 1.00 units on March 6, 1956. Hence the stability radius of 0.0028 units (see Table 1) was seen at the angles 8′.8 and 9′.6, respectively, well covered by the plates.

On February 20 the fraction of the Venus disk illuminated was 0.736, and this must have been true for a hypothetical satellite. The upper bound on the size of such a satellite is found on the reasonable assumption that the albedo resembles that of Mercury, the moon, and the asteroids (about 7 per cent). Mercury had the same phase, 0.73, and the same distance as Venus from the earth on October 17, 1956, when its stellar magnitude was -0.7. Reduced to the Venus distance from the sun, Mercury's magnitude would have been $+1.0$, or 14–15 mag. above the plate limits found above. This corresponds to a ratio of nearly 1000 in the diameter or an upper bound for the diameter of the hypothetical Venus satellite of 5 km. For very close satellites the upper bound is about 12 km. With continued effort, these limits could be somewhat reduced, though the phase at which the present observations were made is close to optimum.[1]

4.2 Mars

The known satellites, Phobos and Deimos, were photographed on several nights each during the oppositions of 1954, 1956, and 1958. Plate 1 shows the improvement introduced by use of the eccentric diaphragm. The diffraction cross, having the diameter of the planet, is absent when the diaphragm is used. One of the several dozen 1956 photographs is shown in Plate 2. During that opposition, Phobos, about 1.2 mag. brighter than Deimos, could be followed, both visually and photographically, until very near contact with the planetary disk.

The search for fainter and more distant satellites was based on the following series, all obtained at the Cassegrain focus.

a) Cassegrain 226–229, June 28, 1954, 10-minute exposures guided on Mars, on 8 × 10-inch plates, 103*a*-O emulsion. The plates were taken in pairs, with the long dimension north-south, one pair east, and one west of the planet, with the planet itself just included in each case. The seeing was good in spite of the low declination, $-28°$. The estimated limiting magnitude was $17\frac{1}{2}$. Deimos was the only satellite found.

The radius of the computed satellite space was 1500″ or about 200 mm or 8 inches. Therefore, the plates covered the space in longitude, though not wholly in latitude.

b) Plates 238–241, July 19, 1954, 3-minute exposures centered and guided on Mars, taken on 5 × 7-inch plates, 103*a*-E emulsion, backed; with the long dimensions east-west. No new satellites were found; the limiting magnitude was about 17.

c) Plates 418–425, September 6, 1956, 4-minute exposures guided on Mars, on 8 × 10 inch plates, 103*a*-O emulsion, backed. The plates were taken in pairs, with Mars placed in each of the four corners, respectively; further, plates 426–427, centered on Mars, with 2-minute exposure times. The seeing was good. The plates were blinked in pairs. No new satellites were found. The limiting magnitude was found accurately by comparison with Selected Area 68; it was 17.1 mag. photographic.

The reduction of the third group is as follows: The distance from the earth was 0.378 astronomical unit, so the stability radius is $0.0031/0.378 = 0.0082$ radian or 28′ or 23 cm or 9 inches in the focal plane.

[1] In *Documentation des Observateurs*, Vol. 9, March, 1956, No. 3, the editor (R.R.) describes a doubtful Venus satellite of the 11th magnitude, observed on March 12, 1948, at a distance of 17′.4 from the planet. On that date the stability limit was about 10′, so that the object must have been a reflection in the optics used. Nor was it confirmed by the present survey.

The satellite space was therefore essentially covered by survey *c*. The plates were taken almost precisely at opposition, when the photographic magnitude of the planet was −1.4 mag., 18.5 mag., or 25,000,000 times, brighter than the survey limit. This corresponds to a diameter ratio of 5000 or an upper bound for the diameter of a hypothetical third satellite of 1.4 km (about 1 mile). The assumption made is that the photographic albedo is about 0.05, as is true for Mars and the moon; if it is greater, the diameter will be less.

4.3 Jupiter

The inner satellite region of Jupiter was photographed with the 82-inch telescope, in part to obtain good records of Jupiter V, the faint close satellite discovered visually by Barnard, that had not been photographed before. Plate 3 gives three typical exposures, with the overexposed images of Jupiter II and III nearby. These records were made with the eccentric diaphragm on the telescope. (In copying these exposures, some shading was done, using a straightedge vertically, to reduce the right-to-left gradient across the plates caused by scattered light of Jupiter.) The following plates were taken in 1955: Nos. 259–263 on January 25, 5^h41^m–5^h53^m U.T.; 267–269 on January 28, 4^h56^m–5^h17^m U.T.; and 280–287 on January 30, 5^h33^m–6^h04^m U.T. Earlier experiments made jointly with Professor G. van Biesbroeck had led to a series already published (van Biesbroeck, 1955).

Some plates were also taken of the bright satellites, in an effort to record them with exposure times (about $\frac{1}{25}$ second) that would at the same time show the surface detail of the planet. This is always possible when the seeing is good, since the satellites have disks of about 1″ and albedos similar to the planet. None of these exposures is reproduced, since Dr. Humason shows some examples in his chapter.

Jupiter VI was photographed on February 5, 1952, on II G with yellow filter (pl. 2680), and a comparison was made with Selected Area 68 (pl. 2679). The resulting visual magnitude was 14.0. The photographic magnitude was derived on March 6, 1956, from plates 3282 and 3283 (exposure times 1 and 4 minutes), calibrated with Selected Area 57 (pls. 3291, 3292). The result was 15.4. Reduced to mean opposition, the values are 13.9 visual and 15.3 photographic. Jupiter VII was recorded on PF plates 3289 and 3290 and found to be 17.1 photographic, or 17.0 reduced to mean opposition.

A region 20′ in radius centered on Jupiter was investigated at the prime focus on March 6, 1956. Three pairs were taken, all 103*a*-O backed, with exposure times 20 seconds, 1 minute, and 4 minutes. These plates were calibrated against Selected Area 57, which showed the limiting magnitudes to be 16.4, 17.4, and 18.4, respectively, except close to the planet. No new satellites were found; Satellite VI was present. Two plates centered on Jupiter VII, also taken on March 6, 1956 (7:18 and 7:30 U.T., each 2-minute exposure), failed to show any satellites other than VII.

At the Cassegrain a region about 7′ in radius centered on Jupiter was photographed on February 29, 1956, using 103*a*-E backed (pls. 350–354). The exposure times varied from 15 seconds to 6 minutes; the limiting magnitude was 17.

The principal survey for Jupiter satellites was made by S. B. Nicholson in 1938, using the Newtonian focus of the 100-inch Mount Wilson telescope (Nicholson, 1939). The photographs were on 8 × 10-inch plates, each covering 54′ × 68′, "and the settings were made so that the plates overlapped about one inch. The search covered an area of about ten square degrees, extending three degrees east and west and a degree and a quarter to the north and south of Jupiter." The arrangement of the plates and the satellites noted are shown in Figure 1, taken from a lantern slide kindly made available by Dr. Nicholson. Dr. Nicholson states:

> The telescope was equipped with a guiding eyepiece which could be moved in steps of 0″.5 (0.03 mm) parallel to the edge of the plate. The plateholder was adjusted with its edge parallel to the motion of Jupiter, and during the exposure the eyepiece was shifted at intervals corresponding to a displacement of this amount. Satellites not stationary with respect to Jupiter would of course make slightly elongated images. Satellites VI and VII may move as much as 5″ per hour relative to Jupiter, but their average motion is much less, and the more distant satellites move only about one-third that fast, even when in conjunction with Jupiter.
>
> Most of the exposures were made on Imperial Eclipse plates, although Eastman

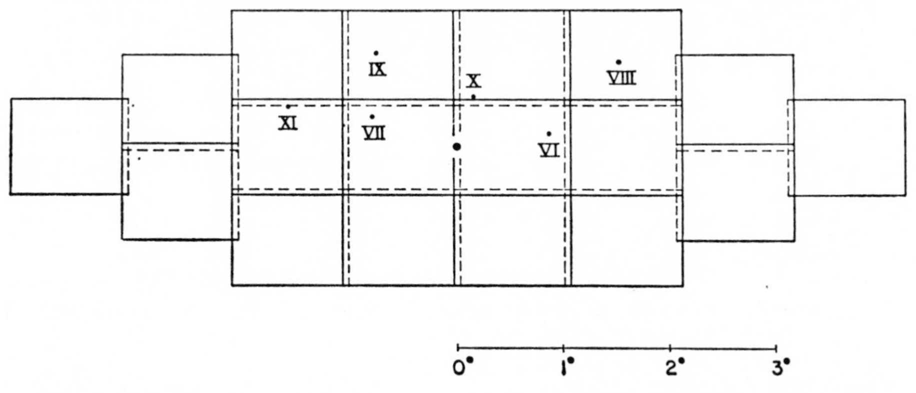

FIG. 1.—Coverage of S. B. Nicholson's survey of Jupiter satellites, 100-inch telescope, Mount Wilson Observatory, 1938.

Hα special plates hypersensitized with ammonia and Cramer Hi-speed plates were sometimes used. Satellites as faint as magnitude 20 should have registered, although on some nights this limit may not have been reached near the edges of the plates. Six nights were required for the complete survey, which was made on July 27 to August 1, inclusive.

Any satellite close to Jupiter on those nights might easily be missed because of the brightness of the field near the planet. Such satellites, however, would be either east or west of Jupiter at an earlier date, and to make the search more complete six fields, three on each side of Jupiter, were photographed on July 2–6. At that time Jupiter was moving slowly, and in an hour's exposure the star trails were too short to be easily distinguished from images of satellites. The plates were therefore duplicated and each pair of plates was examined in the blink comparator.

Satellites VI and VII, ephemerides of which are given in the *American Ephemeris and Nautical Almanac*, were easily located on the first night, and on the tenth plate of the search, taken on July 6, an image was found of another object with a motion resembling that of a satellite. On July 9, Mr. Joy, to whom the 100-inch had been assigned, interrupted his program to permit another observation of this object and thus avoid the danger of not finding it three weeks later when the main survey was made.

At first this object was thought to be J IX, which had not been observed for ten years and for which no ephemeris for 1938 had been computed, the intention being to make its rediscovery a test of the completeness of the survey. No ephemeris of J VIII was then available, but the new object was much fainter than that satellite. Its motion from July 6 to 9 was found to be much faster than that of J IX and it was therefore either an unknown satellite or an asteroid moving temporarily like a satellite. Trails of several asteroids were found on these plates with motions such that there was no question about their identification.

From July 27 to August 1, 1938, all the photographs for the search were obtained except two, which were to have been at the extreme east and west limits of the field, three degrees from Jupiter. About thirty-five moving objects were found on these plates, all except nine of which were easily identified as asteroids. Three of the nine were J VI, J VII, and the object discovered on July 6. An ephemeris of J VIII computed by Hertz of Yale University had then become available, and with its help one of the objects was identified as J VIII. Another was identified as J IX by means of an ephemeris computed from the mean elements of its orbit.

Careful measures identified three of the remaining four objects as asteroids and the last one as another new satellite. The first photograph of this satellite was made on July 30. Additional observations on August 24 and 25 confirmed the identification of the new satellites and their discovery was then announced.

Magnitudes. The photographic magnitudes of Jupiter X and XI were determined by comparison with Selected Area 68. Exposures of twenty minutes each were made with the 100-inch reflector on the satellites and on the Selected Area. Both satellites were so nearly stationary that no motion was set off and the images appear perfectly stellar. The resulting photographic magnitudes, corrected for atmospheric extinction, were 19.0 for Jupiter X and 18.4 for Jupiter XI. The exposures of Jupiter X were made on October 23 by Baade and those of Jupiter XI on October 18 by Nicholson. When reduced to mean opposition without applying a correction for phase, these magnitudes become 18.8 for Jupiter X and 18.4 for Jupiter XI. The photographic magnitude of Jupiter IX at mean opposition is 18.6.

Assuming a color index of one magnitude and a value of the albedo similar to that of the dark satellites and asteroids ($p = 0.1$), we find diameters of 17.5, 15.5, and 19.5 miles for Jupiter IX, X, and XI, respectively.

The history of the earlier discoveries may be briefly related (cf. Nicholson, 1952). Barnard (1892) found satellite V visually with the 36-inch refractor of the Lick Observatory, during prolonged systematic searches in 1892. C. D. Perrine (1906), in 1904 and 1905, found satellites VI and VII photographically with the 36-inch Crossley reflector of the Lick Observatory. P. J. Melotte (1908) at Greenwich, with the 30-inch reflector, found VIII in 1908, while photographing VI and VII, while S. B. Nicholson, with the Crossley reflector in 1914, found IX while photographing VIII. Following his 1938 survey, which yielded X and XI, he found XII while photographing X (Nicholson, 1951). Reference is made to a photograph of the very faint satellite XII published by Nicholson (1952).

The mean-opposition photographic magnitudes were listed by Nicholson (1952) as follows: VII, 17.0; VIII, 17.0; IX, 18.6; X, 18.8; XI, 18.4; and XII, 18.9 mag. These values were based on calibrations with stars in Selected Areas and are therefore subject to the small systematic corrections to the Selected Area system noted by W. A. Baum (unpublished) and others. The corrected values then are, respectively, 17.3, 17.3, 19.1, 19.3, 18.8, and 19.4 mag. The limiting magnitude of Nicholson's Jupiter satellite survey may then be taken as about 19.5 mag. If hypothetical faint satellites have the same photographic albedo as the moon, 0.05, the upper bound of the diameter of such bodies is found to be 18 km. If, instead, the albedo is 0.2, as is more likely for an object of that class at that distance from the sun, the upper bound is 9 km.

The question is unavoidable whether the slow convergence evident from these discoveries indicates that still further satellites might be present. An independent argument strengthens the belief that this may be the case. Satellites VI, VII, and X form one group of objects having very similar orbits, while satellites VIII, IX, XI, and XII form another. It is difficult to avoid the conclusion that these groups were formed as two, not seven, events, presumably by capture of two bodies that subsequently broke up (Kuiper, 1951, p. 719; 1956, p. 1661). Thus additional fainter members of these two groups are likely to be present.

4.4 Saturn

At the Cassegrain focus of the 82-inch telescope, seven short series of exposures were made of the inner satellite region, using II G plates and exposure times ranging from 2 seconds to 6 minutes. The dates are February 15 and 16, March 24 and 25, 1948, and February 24, 26, and 27, 1949. The plate numbers are, arranged by nights, 3–4; 7–8; 13–20 and 23–26; 37–42; 130–133; 142–144; and 148. In addition, seven plates were exposed on February 26 and 27, 1949, each bearing 6 exposures (F 192–198; II G backed). Two sample reproductions are shown in Plate 4. The plates showed no new satellites. For the longer exposures the limiting magnitude was 16–17 mag.

The more distant parts of the satellite space around Saturn (cf. Table 1) were examined with the prime-focus camera during three sessions independently: March 4, 1949; April 30—May 1, 1949; and January 24–25, 1950. Each examination was based on three pairs of plates, one centered on Saturn, one about 48′ west, and one 48′ east. All but the eastern pair of the third set were taken with the 66-inch diaphragm on the telescope; that pair was taken with the 54-inch diaphragm. The March 4, 1949, set consisted of 15-minute exposures on 103*a*-F plates (backed). The limiting magnitude was about 20 mag. near the plate centers and 18–19 mag. in the margins. The second set used II G and 30-minute exposures; the third set 103*a*-O and 15-minute exposures (central pair), 28 minutes (west pair) and 25 minutes (east pair, 54-inch diaphragm). The plates were all 5 × 7 inches, with the long dimensions east-west; the plate numbers are 2124–2129, 2131–2136, and 2198–2201 plus 2212–2213. The mean time interval between the plates within each pair was 22 minutes in the first set; 23 hours in the second set; and 19, 33, and 38 minutes for the three pairs of the third set.

No new satellites were found. Besides several of the closer satellites, Phoebe was recovered on each set, 1–3 mag. above the threshold, depending on its position on the plates. Each plate pair yielded about 2 asteroids, on the average.

Plate 5 shows the faintest of the known Saturn satellites, Phoebe, reproduced from PF plate 2683. (approx. mag. 17.3 pg).

The threshold of the present satellite searches at present is 19–20 mag. for the exterior parts to about 14 mag. very close to the rings and the planet. At mean opposition a nineteenth-magnitude satellite of albedo 0.2 would have a diameter of about 40 km. This may be regarded as an approximate upper bound to unknown satellites of Saturn.

4.5 Uranus

A plate at the Cassegrain focus of the 82-inch taken on February 16, 1948, for the relative magnitudes of the four known satellites showed a faint object that was confirmed as a new satellite on March 1, 1948

(Kuiper, 1949*a*). One of the discovery photographs (CC 10) is reproduced in Plate 6. A stronger photograph was obtained on CC 9 (10-minute exposure) which showed no additional satellites.

The series covering the five satellites was continued by the author and by D. L. Harris, who used the material in his doctoral thesis. For ready reference, the entire plate series is listed here (a few poor plates have been omitted): 1948: CC 9–10 (March 1), 11–12, 21–22 (March 24), 31–34 (March 25), 43–119 (October 19-November 11); 1949: 120–127 (February 24), 139, 145–147 (February 27); 1954: 236–239 (January 29); 1955: 274–279 (January 28), 288–295 (January 30); 1956: 322–323 (February 18), 343–349 (February 24), 362–364 (March 2); 1960: 464–473 (April 15), 474–488 (April 16); 1961: 539–544 (April 5), 546–549 (April 7).

Representative reproductions of some recent plates, taken at the time when the orbits were seen more highly inclined, are shown in Plate 7.

Fainter and more distant satellites were searched on four pairs of prime-focus plates, 5 × 7 inches in size, oriented east-west, and centered on the planet, just enough to cover the Uranus satellite space (cf. Table 1 and p. 579). These are, in 1948, plates 2165–2166 (October 28), 2170–2171 (October 29), and 2188–2189 (October 30); and, in 1950, 2193–2194 (January 24).

The diaphragms, emulsions, exposure times, and limiting magnitudes were as follows: *first pair*, 66 inches, 103*a*-F, 42 minutes, 20–20½ mag.; *second pair*, 54 inches, 103*a*-F, 60 minutes, 20½ mag.; *third pair*, 66 inches, 103*a*-O, 40 minutes, 20 mag.; *fourth pair*, 82 inches, 103*a*-F, 15 minutes, 19–20 mag. The second pair is probably the best. The central part of plate 2172, which is nearly identical with the pair PF 2170–2171, is reproduced in Plate 8. No new satellites were found on these four sets, but some asteroids were noted.

Reference is made to two earlier searches for Uranus satellites: (1) "Examination of Uranus for the Detection of New Satellites," by Holden and Schaeberle (1891), a search made visually with the 36-inch refractor of the Lick Observatory, and (2) "Report on a Search for Further Satellites of the Planets Uranus and Neptune," by Christie (1930), who used the 60-inch reflector at Mount Wilson photographically. Comparison of Christie's searches and those reported here show the usefulness of a series of graded exposures for close satellites and of prime-focus or Newtonian photography for faint and distant satellites. Further work is indicated, however, with infrared emulsions and appropriate filters, limiting photography to the strong methane absorption bands, which will reduce the scattered light of the planet.

4.6 Neptune

The following plates were taken at the Cassegrain focus to record Triton and any other close satellites in range: CC 134–138 (February 24, 1949), emulsions II G (yellow), 102*a*-E (red) and I N (near infrared), exposure times up to 15 minutes; and 489–495 (April 16, 1960), 103*a*-O, 30–45-second exposures.

Plate 9 shows reproductions of two plates, with 15- and 6-minute exposure times, and limiting magnitudes 18 and 17.5 approximately. No new satellites or any trace of a ring are seen.

Prime-focus search plates were taken, beginning on May 1, 1949, which led to the discovery of Nereid (Kuiper, 1949*b*). Plate 10 shows both satellites. The plate limit is about 21 mag., and the number of galaxies shown approximately equals the number of stars. No other distant satellites were found.

Dr. M. L. Humason in 1950 made an independent search for distant satellites with the 200-inch telescope and also found Nereid. Plate 11 is a reproduction of one of the search plates, kindly made available by Dr. Humason. Plates 10 and 11 have been reproduced on approximately the same scale.

The 82-inch series was continued by Professor G. van Biesbroeck, who in two papers derived the orbit of the satellite (van Biesbroeck, 1951, 1957) and, with it, the mass of Neptune. The orbit is shown graphically in Figure 2. The period is nearly a year (359.88 ± 0.02 days), the eccentricity exceptionally high for a satellite (0.7493 ± 0.0007), and the inclination with respect to the ecliptic, $4\overset{\circ}{.}97$. The unusual nature of this orbit is of great interest in connection with the interpretation of the irregular satellites (Kuiper, 1956, p. 1660).

While Plates 10 and 11 appear similar in limiting magnitude (partly because of the color sensitivity of the F plate used on the 82-inch), examination of the originals definitely favors the 200-inch plates. Dr. Humason permitted the author to re-examine his plates and also made available a calibration plate of Selected Area 57. Reblinking yielded no new satellites, but the following search limits were established with some care.

The limiting magnitude is 22.5 photographic except for a small region around the planet and at great distances. At 33″ from the planet the limit is estimated at 22.0 mag. because of the halation ring, recovering somewhat closer in, down to 16″ from the planet. At 11″ from the planet the limit has fallen to about 20.5. Within 10″ the images are too black to show satellites. The outer boundary of good definition has a radius of about 57 mm = 10ʹ.6. There is about a 2-mag. reduction at R = 80 mm = 14ʹ.8. The total exposed plate field is 120 × 171 mm.

The photographic magnitude of Nereid was well defined by the plates and Dr. W. A. Baum's (unpublished) magnitude sequence in Selected Area 57. The result is 19.5 mag.

In summary, the limiting magnitude of the searches is about $22\frac{1}{2}$ photographic for the distance range 16″–10ʹ from the planet; beyond 10ʹ it drops to about 20 mag. at about 16ʹ and 19 mag. at the stability boundary (about 35ʹ). A satellite at the distance of Neptune, having a photographic albedo of 0.1, will be of 22 mag. photographic if the diameter is 160 km. This may be regarded as the approximate upper bound of unknown satellites for the distance zones specified. The limit will be higher elsewhere, attaining 500 km very close to the planet.

4.7 Pluto

Pluto was photographed at the prime focus of the 82-inch telescope in 1950: January 24, PF 2195, 2196, 2197, all 103*a*-O, 30 minutes, full aperture;

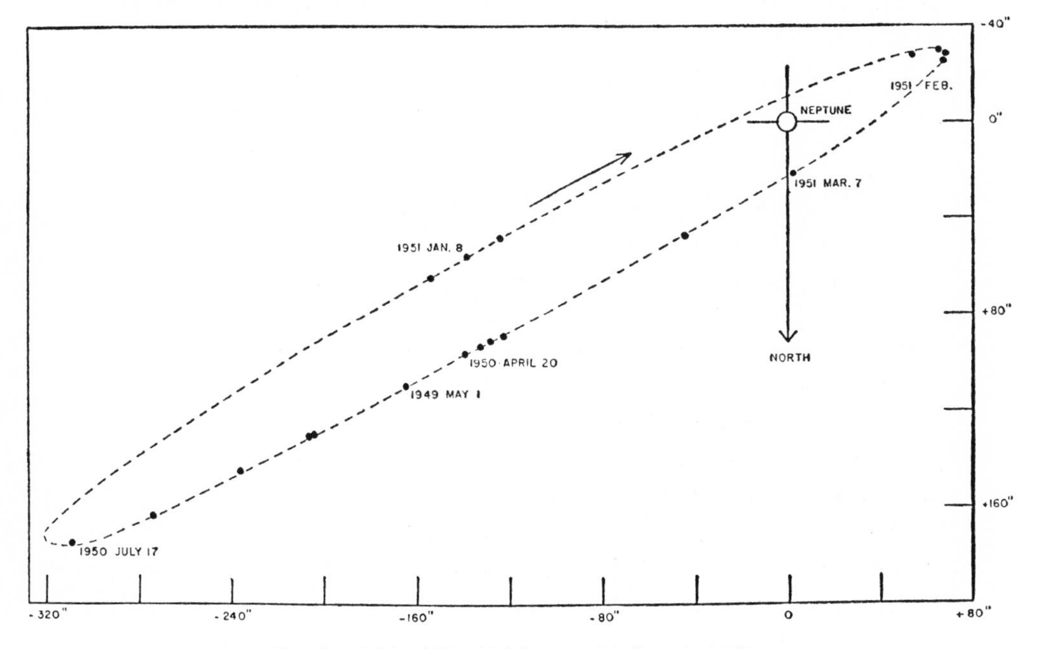

Fig. 2.—Orbit of Nereid (after van Biesbroeck, 1951)

January 25, PF 2210, 2211, both 103*a*-O, 60 minutes, 54-inch diaphragm. Limiting magnitude about 21 photographic in both cases. No satellites were found. Plate 12 shows two reproductions of the Pluto field.

Dr. Humason made a similar study during the same year, with the 200-inch telescope, equally with negative results. His plates were taken on April 15 and 16, 1950, 103*a*-O, 15-minute exposure each. He allowed the writer to re-examine the plates in order to establish their threshold. No satellites were found upon reblinking; the magnitude limit was 22.4 photographic as close as 2″ from the image center. The outer boundaries of the field were the same as those recorded above for Neptune. A third plate was available of 5-minute exposure time, on Eastman 33, taken with seeing 3 in a hazy sky. Its limiting magnitude was 18 or 19, and the inner boundary of the field covered by it was about 0″.5. The visual resolution during the writer's study of Pluto's diameter (Kuiper, 1950) was probably somewhat better, 19 mag. down to about 0″.3. Therefore, any satellite is believed to be fainter than 19 mag. for the interval 0″.3–2″ and fainter than 22.4 for the remainder of the stable satellite region.

If the hypothesis that Pluto originated as a satellite of Neptune is correct (e.g., Kuiper, 1956, pp. 1662–1664), this body should have no satellites of its own. It was regarded useful, however, to obtain here a purely empirical answer.

Part of the work recorded in this chapter was supported by a grant from the National Science Foundation.

The writer is indebted to Mr. E. A. Whitaker for assistance in the preparation of the plates for publication.

REFERENCES

Barnard, E. E. 1892 *A.J.*, Vol. **12**, No. 275.
Biesbroeck, G. van 1951 *Astr. J.*, **56**, 110–111.
1955 *Ibid.*, **60**, 275.
1957 *Ibid.*, **62**, 272–274.
Campbell, W. W., and Trumpler, R. J. 1923 *Pub. A.S.P.*, **35**, 214.
Christie, W. H. 1930 *Pub. A.S.P.*, **42**, 253.
Holden, E. S., and Schaeberle, J. M. 1891 *Pub. A.S.P.*, **3**, 285.
Kuiper, G. P. 1949*a* *Pub. A.S.P.*, **61**, 129.
1949*b* *Ibid.*, p. 175.
1950 *Ibid.*, **62**, 133.
1951 *Proc. Nat. Acad. Sci.*, **37**, 717.

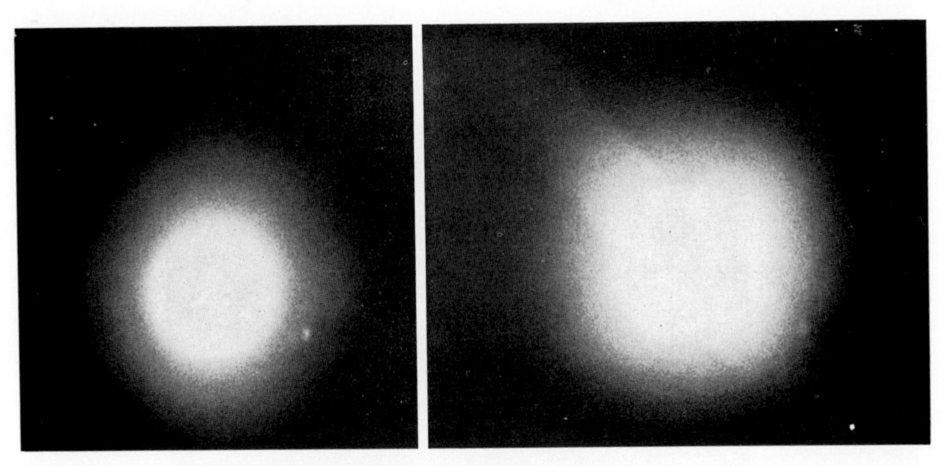

PLATE 1.—Recording Mars satellites with full aperture (*right*), showing diffraction cross, and with eccentric diaphragm (*left*). Plates CC 225, June 28, 1954, 7^h19^m U.T.; and CC 224, 6^h38^m U.T. Both 103*a*-O, exposures 7 and 60 seconds.

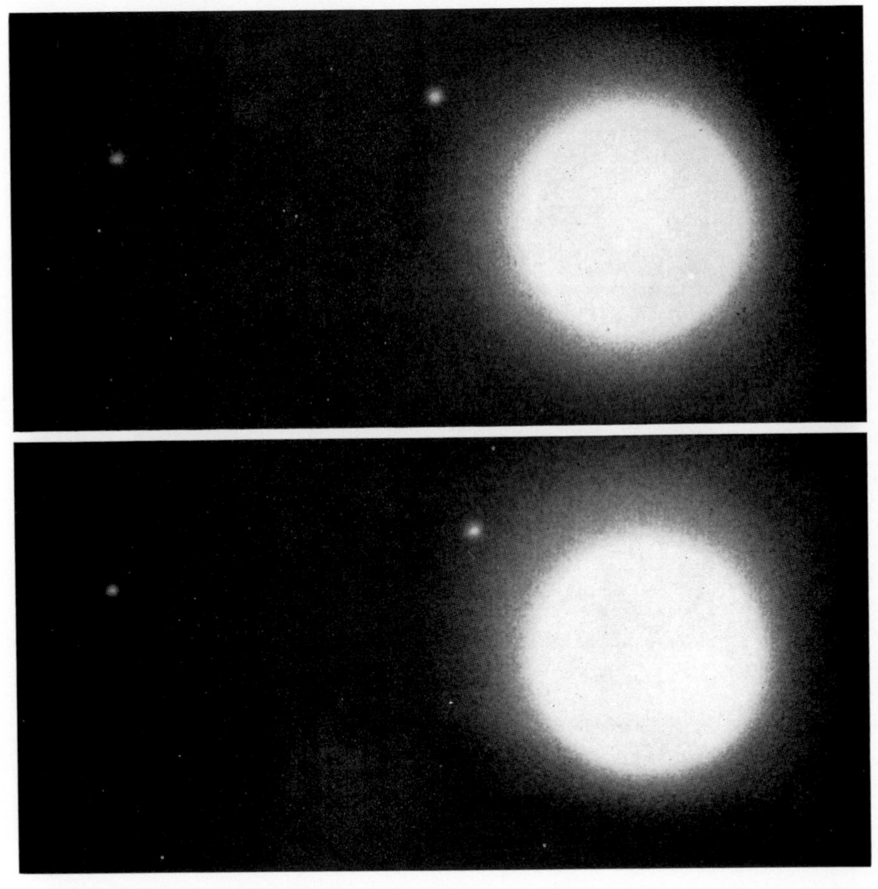

PLATE 2.—Satellites of Mars, September 14, 1956. *Above:* M299, 5^h59^m U.T.; *below:* M300, 6^h14^m U.T. Both 103*a*-O, 45-second exposure.

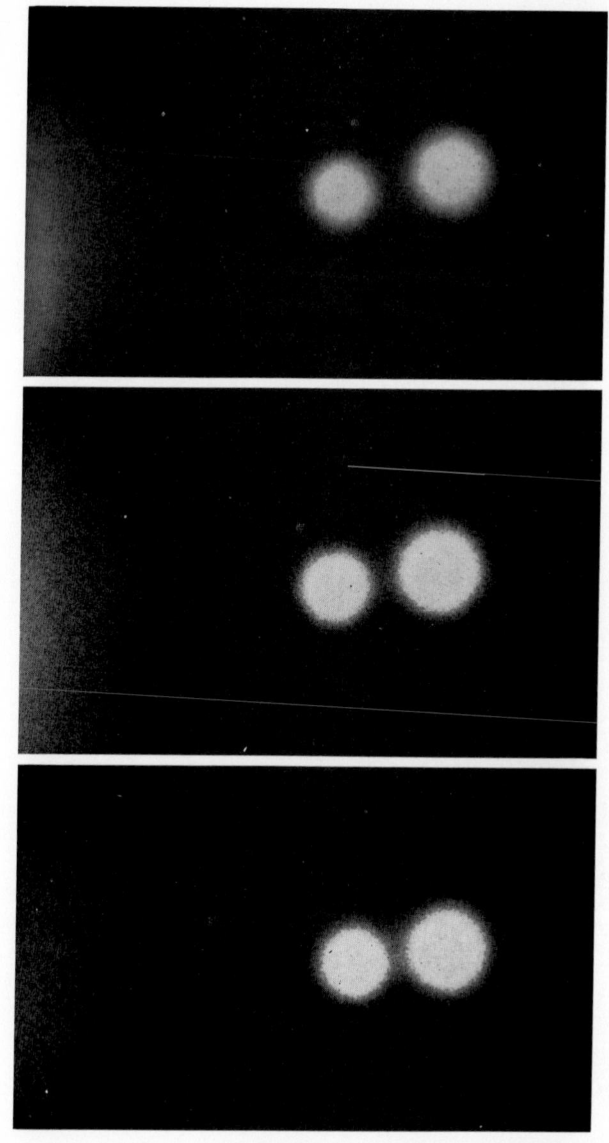

PLATE 3.—Jupiter V, also showing Jupiter II and III, January 30, 1955, on 103*a*-E, backed, 60-second exposures. *Above:* CC 280, 5^h33^m U.T.; *center:* CC 283, 5^h44^m U.T.; *below:* CC 287, 6^h04^m U.T.

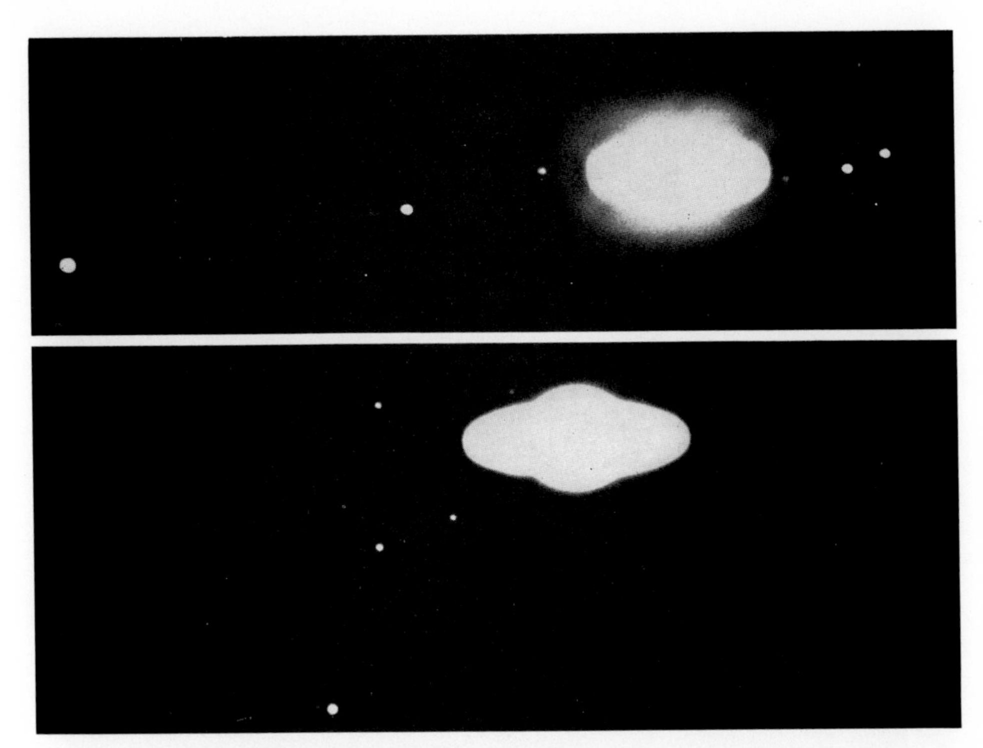

PLATE 4.—Saturn satellites, Cassegrain focus, 82-inch telescope. *Above:* inner 6 satellites; *left to right:* Titan, Rhea, Enceladus, Planet and Rings (showing diffraction cross), Mimas, Tethys, and Dione. Date February 15, 1948, $4^h30^m5^s$ U.T., exposure 30 seconds, II G emulsion, yellow filter. *Below:* same as above, except Mimas absent. The faint satellite above the ring is Enceladus; the others are, from top to bottom, Tethys, Dione, Rhea, and Titan. Date March 24, 1948, $3^h07^m14^s$ U.T., exposure 3 seconds through haze, II G.

PLATE 5.—The Saturn satellite, Phoebe, taken on February 5, 1952, 10^h46^m U.T., 5-minute exposure on 103*a*-O.

PLATE 6.—The satellites of Uranus, seen nearly normal to the common orbital plane. *In order of increasing distance:* Miranda (16.8 mag.), Ariel (14.8), Umbriel (15.4), Titania (13.9), and Oberon (14.3). Date March 1, 1948, 2^h46^m U.T., exposure $3\frac{1}{2}$ minutes on II G. Plate CC 10.

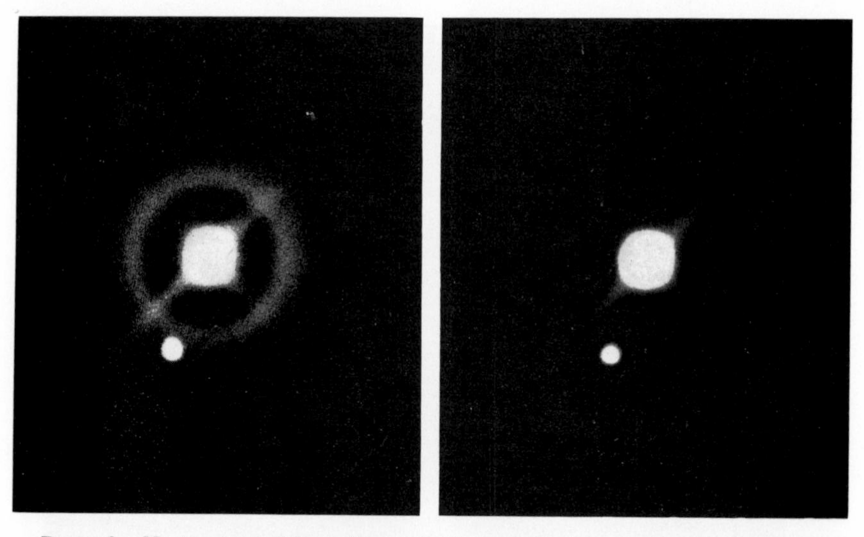

PLATE 9.—Neptune and Triton, Cassegrain focus, 82-inch. *Left:* CC 137, February 24, 1949, $9^h05^m40^s$ U.T. Exposure 15 minutes, 103*a*-E, no filter, showing absence of fainter and closer satellites to plate limit (about 18 mag.). *Right:* CC 134, same date, $7^h51^m0^s$ U.T., 6 minutes, II G, backed, no filter, plate limit $17\frac{1}{2}$.

PLATE 7.—The five satellites of Uranus, Cassegrain focus, 82-inch telescope. *Left:* April 15, 1960, 5^h02^m U.T. (CC 471); *center:* April 16, 1960, 3^h09^m U.T. (CC 480); *right:* April 5, 1961, 4^h46^m U.T. (CC 544).

PLATE 8.—Uranus field photographed at the prime focus, 82-inch telescope. Central portion of PF 2172, October 29, 1949, $10^h45^m30^s$ U.T., exposure 31 minutes, 103*a*-O. Oberon and Titania are shown in halation ring, below planet.

PLATE 10.—Neptune and its two satellites, Triton (below the planet) and Nereid (marked by second arrow); prime focus, 82-inch telescope. Date May 29, 1949, 5^h05^m U.T., exposure 30 minutes on 103*a*-F, backed. Fuzzy spots are galaxies.

PLATE 11.—Neptune and its two satellites, taken by M. L. Humason, prime focus, 200-inch telescope, date April 17, 1950, 6^h17^m U.T., exposure 20 minutes, on 103*a*-O.

PLATE 12.—Pluto and surroundings, 82-inch telescope on 103*a*-O plates. *Above:* PF 2197, January 24, 1950, 7^h01^m U.T., exposure 30 minutes, full aperture. *Below:* PF 2210, January 25, 1950, 6^h50^m U.T., exposure 60 minutes, 54-inch diaphragm. Note that Pluto has moved during the exposure. Many faint galaxies are shown.

	1956	*Vistas in Astronomy*, ed. A. BEER, Vol. **2** (London: Pergamon Press), "On the Origin of the Satellites and the Trojans," pp. 1631–1666.
MELOTTE, P. J.	1908	*M.N.*, **68**, 456.
NICHOLSON, S. B.	1939	*Astr. J.*, **48**, 129–132.
	1946	*Pub. A.S.P.*, **58**, 356.
	1951	*Ibid.*, **63**, 297–299.
	1952	*A.S.P. Leaflet*, No. 275.
PERRINE, C. D.	1902	*Lick Obs. Bull.*, **1**, 183.
	1906	*Ibid.*, **4**, 115.
	1908	*Ibid.*, **5**, 7.
	1909	*Ibid.*, **5**, 95.
TRUMPLER, R. J.	1923	*Pub. A.S.P.*, **35**, 313.

Index of Subjects and Definitions

[Page numbers for definitions are in italics.]